U0185040

谨以此书祝贺

张英伯教授 75 寿辰!

北京师范大学数学家文库

李仲来◎主编

张英伯文集

箭图和矩阵双模

张英伯◎著

JIANTU HE
JUZHEN SHUANGMO

北京师范大学出版集团
BEIJING NORMAL UNIVERSITY PUBLISHING GROUP
北京师范大学出版社

2021 · 北京

图书在版编目（CIP）数据

箭图和矩阵双模：张英伯文集 / 张英伯著，李仲来主编．– 北京：北京师范大学出版社，2021.7

（北京师范大学数学家文库）

ISBN 978-7-303-26824-5

Ⅰ．①箭… Ⅱ．①张… ②李… Ⅲ．①矩阵 – 文集
Ⅳ．① O151.21-53

中国版本图书馆 CIP 数据核字 (2021) 第 020160 号

箭 图 和 矩 阵 双 模： 张 英 伯 文 集
JIANTU HE JUZHEN SHUANGMO：ZHANG YINGBO WENJI

出版发行：北京师范大学出版社 www.bnup.com
北京市西城区新街口外大街 12-3 号
邮政编码：100088
印　　刷：鸿博昊天科技有限公司
经　　销：全国新华书店
开　　本：710 mm × 1 000 mm　1/16
印　　张：29
插　　页：4
字　　数：445 千字
版　　次：2021 年 7 月第 1 版
印　　次：2021 年 7 月第 1 次印刷
定　　价：99.00 元

策划编辑：岳昌庆　　责任编辑：岳昌庆
美术编辑：李向昕　　装帧设计：李向昕
责任校对：段立超　　责任印制：马　洁

版权所有　侵权必究

反盗版、侵权举报电话：010-58800697
北京读者服务部电话：010-58808104
外埠邮购电话：010-58808083
本书如有印装质量问题，请与印制管理部联系调换。
印制管理部电话：010-58805079

▲ 1993 年，在北京师范大学数学楼（左起，下同）：王世强、孟晓青、张英伯、刘绍学。

▲ 2004 年 4 月 27 日，在上海主持博士论文答辩之后。乐珏、章璞、张英伯、刘石平（加拿大）。

▲ 2005 年 3 月 22 日，Claus 和张英伯在绍兴，王羲之的兰亭序碑前。

▲ 2005 年 3 月 22 日，在绍兴参观鲁迅故居。

▲ 2010 年 8 月 13 日，参加日本国际代数表示论学术会议（International Conference on Representation Theory of Artin Algebras）。黄兆泳、Sato（日本）、韩阳、张英伯、Simson（波兰）、章璞。

▲ 2010 年 7 月 12 日，在北京师范大学数学楼。张英伯和王昆扬。

2010年7月12日，在《数学通报》编辑部。郑亚利、张英伯、李亚玲。

2012 年 6 月 4 日，在以色列 Mina 家中作客：王昆扬、Mr.Teicher、叶飞、张英伯、MinaTeicher。

2014年2月11日，在与乌克兰科学院院士 A.V.Drozd 学术合作期间，到他家中做客。Drozd 夫人、张英伯、Drozd。

2016 年 4 月 22 日，参加浙江大学召开的代数会议。王建磐、张英伯、王昆扬、李方、胡骏。

2016 年 6 月 14 日，在北京师范大学数学楼，为本科生讲高等代数课。

2017 年 5 月 20 日，在中国科技大学，学生为张英伯庆祝 70 周岁寿辰。
前排：王昆扬、张英伯。
中排：曾祥勇、叶彩娟、陈学庆、武月琴、赵德科、乐珏、魏丰、曹磊。
后排：董正林、刘根强、韩阳、韩德、徐运阁、徐帆。

自　序

作为“文化大革命”后的第一届硕士研究生，我1978年考入北京师范大学数学系.

在这之前，我读过小学和中学，在高中三年级时遭遇取消高考和史无前例的“文化大革命”，于1968年作为知识青年来到北国边陲北大荒. 1972年进入北京师范学院（现首都师范大学），成了极其稀有的、非“红五类”家庭出身的工农兵学员. 记得在极“左”思潮的巨大干扰下，北京师范学院的梅向明教授、林有浩教授断断续续地为我们讲过“解析几何”和一点“微积分”.

“文化大革命”终于结束，大学恢复招生. 现代科学知识几乎为零的我，壮着胆子与众多“老五届”大学生（即1966～1970届本科毕业生）、少数自学成才的年轻人一道报考了研究生. 那时我连研究生分成哪几个专业都搞不明白，因为曾在北京师范学院偶遇著名的环论专家、北京师范大学的刘绍学教授，就报了代数专业. 刘教授把我叫到工三楼他的家里，借给我两本书：范德瓦尔登的《代数学》和库洛什的《群论》. 对于我来说，这无异于“天书”.考研时笔试倒是达标了,但导师口试的问题基本没有听懂,就这样糊里糊涂地被录取了.

走进北京师范大学数学楼的第一天,就仿佛走进了科学的殿堂,心情特别激动,感觉特别神圣.在改革开放之初的1978年,大学教师和科学院的研究人员,都有着极其强烈的使命感和责任感.熬过十年文化浩劫的他们,在打开国门的那一刻,深切地体验到中国与世界在科学技术上的巨大差距.他们恨不能马上将国外的前沿领域学过来,把国内的年轻人培养上去.

因为师资力量不足,北京师范大学的刘绍学、中国科学院数学研究所的万哲先、北京大学的丁石孙三位导师为三个单位代数方向的研究生联合开课.我们骑自行车来往于北太平庄和中关村,分别学习了三位导师的环论、域论和交换代数.

1981年研究生毕业,我留在北京师范大学任教.1982年,数学系里的几位教授被国务院批准为首批博士生导师,与我同级的研究生纷纷继续读博士了.我没能跟上大家,直到1986年年底才和另外三位比我小十来岁的年轻人一起开始攻读博士学位,专业方向是代数表示论.

1988年10月,经过刘绍学教授的推荐,我来到当时的联邦德国,师从国际表示论界的知名专家Claus Michael Ringel教授,成为北京师范大学与比勒费尔德(Bielefeld)大学联合培养的博士生.那年我已经41岁,我的导师Ringel只比我大一岁.

直到这时,我才恍然大悟:自己真的是太差劲了,再不努力一辈子都赶不上去了.在德国短短的一年半时间,我拼命地学习.首先语言困难,别说德语根本不会,连英语的听力都很勉强,句子也说不完整.于是只能天天啃文献,在讨论班上竖着耳朵听各国专家和研究生的报告,快速地将公式和大意记录下来,回到自己的办公室反复推敲.

我永远忘不了撰写博士论文的那些日子.因为之前与德国教授讨论,发表过两篇有关Auslander-Reiten箭图(简称AR-箭图)结构的文章,Ringel给了我一个相当困难的题目:确定AR-箭图非循环稳定分支的结构.在这之前,Riedtmann给出了有限型代数模范畴AR-箭图的结构;Happel,Preiser,Ringel刻画了循环稳定分支的结构;而非循环稳定分支的结构问题从来没人碰过.

Ringel每两周跟我讨论一次,即便他出国开会,最多也只隔三周.有时一口气讨论三个小时，我都快站不住了，他仍然精神抖擞地在黑板上画着、讲着.夜里睡觉时，我的脑袋也不知不觉地围绕着这个问题转，似乎停不下来，突然猛地惊醒，觉得好像有点眉目了，等到清醒过来拿笔一算，又想错了.不知反复了多少次，总有上百次吧，问题终于解决了.可是因为英语表达太差，Ringel 不得不帮我从头到尾修改了一遍.我想署上他的名字，他坚决不肯.1990 年 3 月，我通过德国严格的论文答辩，被授予博士学位.

回国以后，我的工作主要围绕着有限维代数的表示型问题，特别是 AR-箭图与表示型之间的关系，所用工具基于矩阵双模问题的表示.AR-箭图的结构及其矩阵双模问题，构成本文集的主要部分.

20 年来，我始终担任本科生的线性代数和近世代数课教学，以及硕士研究生的基础课教学；同时指导博士生、硕士生、进修教师.令我感到十分欣慰的是，这些博士、硕士毕业之后，无论在科研单位或高等院校做研究工作，还是在中学、出版社投身教育事业，甚至去部队任教，到国防系统搞研发，都能够平实地做人，勤奋地做事，成为各自岗位上的佼佼者.

我与王恺顺教授合作，在北京师范大学出版社编写出版了教材《代数学基础》(上、下册)，在北京师范大学数学科学学院沿用至今.我还应高等教育出版社之邀，翻译了俄国莫斯科大学代数学家科斯特里金的教材《代数学引论》，出版后一直在中国科学院大学使用.

我曾于 1998～2012 年担任了 4 年《数学通报》副主编、10 年主编；于 2004～2012 年担任了两届中国数学会基础教育委员会主任.作为中国数学会刊物的《数学通报》，发表了大量文章，讨论如何学习西方先进的教育理念，同时保持中国数学教育严谨、求实的优良传统.基础教育委员会的成员参与了中学数学课程标准的修订.这些举措对阻止我国中学教育的“全面去数学化”起到了一定作用.

我曾写过若干文章谈“有教无类，因材施教”的教育理念：介绍欧美发达国家的数学英才教育、傅种孙教授的教育思想，并以此为题给全国各地的中学教师和师范院校作过近百场报告；还为热爱数学的中学生作过多次数学科普报告.2016 年，教育部委托中国教育学会进行“中国大学先修课程”的探索，我参加了数学专家组，一直在为高中生讲授高等代数.

张英伯

2019 年 1 月

目　录

Contents

中国科学：数学，

2018，48(11)：1 651～1 664.

AR-箭图的构造和矩阵双模问题①

The Structure of AR-Quivers and Matrix Bimodule Problems

献给刘绍学教授90华诞

摘要 本文简要介绍了 Artin 代数表示理论中的下述内容：Auslander-Reiten 箭图(简称 AR-箭图)稳定分支的结构，野型遗传代数稳定分支的性质，代数闭域上有限维代数 Λ 引出的矩阵双模问题、对偶的余双模问题及其相伴的具有余代数结构的双模(简称 box)，modΛ，$P_1(\Lambda)$及相伴 box 的表示范畴之间几乎可列序列的几乎一一对应，驯顺代数 Λ 上任意两个模之间态射集的维数和性质；最后介绍了代数的驯顺性与其模范畴的齐性.

关键词 AR-箭图的稳定分支；代数的表示型；矩阵双模问题；box 约化.

为了庆贺我的导师刘绍学先生90华诞，谨将我从硕士、博士研究生期间以及在北京师范大学任教期间的研究论文给出一个综述，作为向导师的汇报. 这些工作都是从考取先生的研究生开始的.

§1. 稳定分支的结构

在 Artin 代数的表示理论中，Auslander-Reiten 箭图(简称 AR-箭图)是一个重要的概念，它可以用图形和组合的方法刻画代数的模范畴.

① 国家自然科学基金(批准号：210100165)资助项目.

收稿日期：2017-10-20；接受日期：2018-06-20.

在 20 世纪七八十年代，对于有限表示型自入射代数的 AR-箭图，Reidtmann[1]给出了很好的结果，无限驯顺表示型代数 AR-箭图稳定分支的刻画则由 Happel 等[2]给出. 而野型 Artin 代数模范畴的 AR-箭图则鲜有结果.

20 世纪 80 年代末，读书期间，我的德国导师 Claus Michael Ringel 将这个问题作为我的论文题目. 事实上，这个问题是针对稳定赋值变换箭图进行讨论的. 下面对此问题做一个简要的介绍，参见文献[3].

定义 1.1　一个箭图 $Q=(V,A,s,t)$ 由两个集合 V 和 A 及两个映射 $s,t:A\to V$ 构成，其中 V 的元素称为箭图的顶点，A 的元素称为箭图的箭向. 如果 $\alpha\in A$，则 $s(\alpha)$ 称作箭向 α 的始点，而 $t(\alpha)$ 称作它的终点，并记

$$\alpha:s(\alpha)\to t(\alpha).$$

任取 $x\in V$：用 x^+ 表示所有始于 x 的箭向的终点集；用 x^- 表示所有终于 x 的箭向的始点集.

如果 $\forall x\in V$，两个集合 x^+ 和 x^- 都是有限集，那么称箭图 $Q=(V,A,s,t)$ 为局部有限的.

如果对于任意顶点对 $x,y\in V$，最多只有一个从 x 到 y 的箭向，那么称箭图 Q 没有重箭.

设 (V,A,s,t) 是一个没有重箭的箭图，并且存在两个映射：$d,d':A\to\mathbf{Z}^+$，则称 $\mathcal{Q}=(V,A,s,t,d,d')$ 是一个赋值箭图.

如果 $\forall\alpha\in A,d(\alpha)=1=d'(\alpha)$，那么称 $\mathcal{Q}$ 带有平凡赋值. 我们可以将任意的箭图 Q 看作一个带有平凡赋值的箭图. 如果 $\alpha:x\mapsto y$，我们有时记

$$d_{xy}=d(\alpha),\quad d'_{xy}=d'(\alpha).$$

设 $Q=(V,A,s,t)$ 是一个局部有限且没有重箭的箭图. 如果存在一一映射 $\tau:V\to V$，满足下述条件：$\forall z\in V,\tau(z)^+=z^-$，那么称 $\bar{Q}=(V,A,s,t,\tau)$ 是一个稳定平移箭图.

注意　在定义 1.1 中，箭向可以取圈，即始点和终点可以重合. 给定箭向 $\alpha:y\mapsto z$，将 τ 确定的唯一箭向记作 $\sigma(\alpha):\tau(z)\mapsto y$.

定义 1.2　$\Omega=(V,A,s,t,\tau,d,d')$ 称为一个稳定平移赋值箭图，如果 (V,A,s,t,d,d') 是赋值箭图，而 (A,V,s,t,τ) 是稳定平移箭图，并且 $\forall\alpha\in A,d(\alpha)=d'(\tau(\alpha)),d'(\alpha)=d(\tau(\alpha))$.

令 Ω 一个稳定平移赋值箭图，如果 $V\neq\varnothing$，而且 Ω 不能分解成两个稳

定平移赋值箭图的不交并，那么称Ω是连通的.

令Ω一个连通的稳定平移赋值箭图，$x\in V$. 如果存在正整数t，使得$\tau^t(x)=x$，那么称x为循环的. 注意到，如果Ω中有一个循环点，那么所有的点都是循环的，Ω也称作循环的，否则称作非循环的[2].

令Ω一个连通的稳定平移赋值箭图，如果它的赋值是平凡的，并且$\forall x\in V$，x^+恰有两个点，那么Ω称作光滑的.

连通稳定平移赋值箭图上的一个半加法函数l定义为

$$l:V\to \mathbf{Z}^+,\quad l(z)+l(\tau(z))\geqslant \sum_{y\in z^-} d'_{yz}l(y),\quad \forall z\in V.$$

如果定义中的等号永远成立，那么称其为Ω上的加法函数.

给定一个赋值箭图$S=(V,A,s,t,d,d')$，我们可以如下构造一个稳定平移赋值箭图$\mathbf{Z}S$：令$V_{\mathbf{Z}S}=Z\times V$，如果$S$中有一个箭向$\alpha:x\to y$，那么对于任意的$i\in\mathbf{Z}$，定义箭向$(i,\alpha):(i,x)\to(i,y)$并且$\sigma(i,\alpha):(i-1,y)\to(i,x)$；变换$\tau$定义为$\tau(i,x)=(i-1,x)$；赋值定义为$d_{\mathbf{Z}S}(i,\alpha)=d(\alpha)$，$d'_{\mathbf{Z}S}(i,\alpha)=d'(\alpha)$(参见文献[1]).

定理 1.1 令Ω是一个非循环连通的稳定平移赋值箭图，带有一个取值于非负整数集的非零半加法函数l，则Ω或者是光滑的，且l是有界加法函数；或者存在赋值箭图S，使得$\Omega=\mathbf{Z}S$.

在定理中的第一种情形下，稳定平移赋值箭图形如Π_{st}，其中$-s<t<0$. 这里叙述第二种情形$\Omega=\mathbf{Z}S$的一个重要性质.

设Ω是一个稳定平移赋值箭图，$n\in\mathbf{Z}^+$. Ω中箭向集$\{x\mid\alpha_1,\alpha_2,\cdots,\alpha_n\mid y\}$称为一条始于$x$、终于$y$的路，如果$s(\alpha_1)=x$，$s(\alpha_i)=t(\alpha_{i-1})$，$\forall 2\leqslant i\leqslant n$，$t(\alpha_n)=y$.

当$x=y$时，称之为循环路.

引理 1.1 令S是一个赋值图，则$\mathbf{Z}S$中的任意循环路都是一个截面. 如果S中没有循环路，那么$\mathbf{Z}S$中也没有循环路.

证 取$x,y\in V$，$i,j\in\mathbf{Z}$，使得$i>j$，那么在$\mathbf{Z}S$中，不存在从(i,x)到(j,y)的路. 所以，$\mathbf{Z}S$的循环路中仅包含形如(i,α)的箭向，其中$\alpha\in A$，$i\in\mathbf{Z}$，这只能是由S的循环路得到的截面. □

推论 1.1 设Ω是 Artin 代数Λ的稳定模范畴的 AR-箭图中一个非循环稳定分支，则$\Omega=\mathbf{Z}S$，其中S没有循环路，于是，Ω也没有循环路.

证 注意到Ω是非循环的连通稳定赋值箭图[2]，长度函数l是取值于正整数集的半加法函数. 这时 Ω 是代数 Λ 的 AR-箭图的一个完全分支,从而 l 是加法无界的. 根据 Auslander[4] 的一个定理可知,Ω 不是有限的. 这表明存在某个赋值箭图 S,使得 $\Omega=\mathbf{Z}S$.

因为 S 中的任意循环路都能够导出 Ω 的循环路,根据 Batista 和 Smalø[5] 的一个定理,这是不可能的. 再根据引理 1.1,得到最后一个论断. □

Artin 代数稳定 AR-箭图的结构已经被 Happle 等[2] 确定. 特别地,我们看到,Artin 代数 AR-箭图的正则分支或者是稳定管,或者形如 $\mathbf{Z}S$,其中 S 是一个赋值箭图.

§2.　野型遗传代数的 AR-箭图

关于野型遗传代数 AR-箭图稳定分支的性质,参见文献[6,7].

令 Λ 是一个连通的基(即 $\Lambda/\mathrm{rad}(\Lambda)$ 是互不同构的单模的直和)遗传 Artin 代数,则 Λ 的中心是一个域 k,Λ 是 k 上的有限维代数. 我们来考察有限维左模范畴 $\mathrm{mod}\Lambda$. 记 $P_1,P_2,\cdots,P_n$ 是 Λ 上不可分解投射模的完全集,$I_1,I_2,\cdots,I_n$ 是不可分解入射模的完全集. 设 $F_i=\mathrm{End}_\Lambda(P_i)=\mathrm{End}_\Lambda(I_i)$,$f_i=\dim_k(F_i)$,$\boldsymbol{F}=\mathrm{diag}(f_1,f_2,\cdots,f_n)$是一个对角矩阵.

令 $\boldsymbol{C}=(c_{ij})$是 Λ 的 Cartan 矩阵,则 $c_{ij}=\dim_{F_i}\mathrm{Hom}_\Lambda(P_i,P_j)$. 令 $\boldsymbol{B}=(b_{ij})$使得

$$b_{ij}=\dim_{F_j}\mathrm{Hom}_\Lambda(I_i,I_j),$$

因为 $\mathrm{Hom}_\Lambda(P_i,P_j)\simeq\mathrm{Hom}_\Lambda(I_i,I_j)$,所以有 $\boldsymbol{FC}=\boldsymbol{BF}$. 如果 $M,N\in\mathrm{mod}\Lambda$,维数向量为 $\underline{\boldsymbol{m}}$ 和 $\underline{\boldsymbol{n}}$,那么可以定义 $K_0(\Lambda)$上的双线性函数

$$\langle\underline{\boldsymbol{m}},\underline{\boldsymbol{n}}\rangle=\dim_k(M,N)-\dim_k\mathrm{Ext}^1_\Lambda(M,N).$$

设 $\boldsymbol{\phi}=-\boldsymbol{C}^{-1}\boldsymbol{B}$ 是 Coxeter 变换,则$\langle\underline{\boldsymbol{m}\boldsymbol{\phi}},\underline{\boldsymbol{n}\boldsymbol{\phi}}\rangle=\langle\underline{\boldsymbol{m}},\underline{\boldsymbol{n}}\rangle$.

令 $Q=(V,A,s,t,d,d')$是代数 Λ 的 Gabrial 箭图,见定义 1.1 赋值箭图. $\forall\, i,j\in A$,记 $d'_{ij}=d_{ji}$,我们得到 Cartan 矩阵

$$\boldsymbol{C}^{-1}+\boldsymbol{B}^{-1}=\begin{pmatrix}2 & -d_{12} & \cdots & -d_{1n}\\ -d_{21} & 2 & \cdots & -d_{2n}\\ \vdots & \vdots & & \vdots\\ -d_{n1} & -d_{n2} & \cdots & 2\end{pmatrix}$$

运用 Coxeter 函子和 Cartan 矩阵,不难证明下述定理.

定理 2.1 野型遗传 Artin 代数的任意 AR-分支的模被它们的合成因子唯一确定.

推论 2.1 在野型遗传 Artin 代数的任意一个 AR-分支中,维数不超过给定正整数 d 的模个数有限.

设 K 是 Artin 代数 Λ 的中心,任取 $M\in \mathrm{mod}\Lambda$,记 $l(M)$ 为 M 的 K 长度. 基于对 Coxeter 变换特征值的计算,我们给出下述定义.

定义 2.1 设 M 是一个 Λ 模,称上极限 $\overline{\lim\limits_{s\to+\infty}}\sqrt[s]{l(DTr^sM)}$ 为 M 的左生长数,记作 ρ_M^L; $\overline{\lim\limits_{s\to+\infty}}\sqrt[s]{l(TrD^sM)}$ 为 M 的右生长数,记作 ρ_M^R.

引理 2.1 令 $\mathfrak{C}$ 是 Artin 代数 Λ 的 AR-箭图中的一个稳定非循环分支,则 $\rho_M^L=\rho_N^L,\rho_M^R=\rho_N^R$ 对任意 $M,N\in\mathfrak{C}$ 成立.

定义 2.2 令 $\mathfrak{C}$ 是 Artin 代数 Λ 的 AR-箭图中的一个稳定非循环分支,$M\in\mathfrak{C}$,则 M 的左、右生长数,也叫作分支 $\mathfrak{C}$ 的左、右生长数,分别记作 $\rho_{\mathfrak{C}}^L$ 和 $\rho_{\mathfrak{C}}^R$.

定理 2.2 令 $\mathfrak{C}$ 是 Artin 代数 Λ 的 AR-箭图中的一个树无限的稳定非循环分支. 如果 $\mathfrak{C}$ 的生长数

$$\rho_{\mathfrak{C}}^L,\rho_{\mathfrak{C}}^R<\sqrt[3]{\frac{1}{2}+\sqrt{\frac{23}{108}}}+\sqrt[3]{\frac{1}{2}-\sqrt{\frac{23}{108}}},$$

那么 $\mathfrak{C}$ 形如 $\mathbf{Z}A_\infty,\mathbf{Z}A_\infty^\infty,\mathbf{Z}B_\infty,\mathbf{Z}C_\infty$ 或 $\mathbf{Z}D_\infty$.

§3. 矩阵双模问题及其具有余代数结构的双模

有限维代数的一些问题有时无法从代数本身找到答案,如表示型问题,这时需要转化成找到与之相关的其他代数结构. 本节的内容包含在文献[8]的前半部分中.

令 k 是一个代数闭域,Λ 是有限维基本(basic)k-代数,$J=\mathrm{rad}(\Lambda)$是 Λ 的幂零根,幂零指数为 m,记 $\mathrm{top}(\Lambda)=S=\Lambda/J$. 设$\{e_1,e_2,\cdots,e_h\}$是 Λ 的一个正交本原幂等元的完全集. 取向量空间 $e_i(J^l/J^{l+1})e_j$ 的一组 k-基,考虑基在典范投影 $J^l\to J^l/J^{l+1}$ 之下的原像,并规定它们的高度为 l. 当取 $l=m,m-1,\cdots,1$ 时,我们就得到了 J 的一组有序基(参见文献[9, 6.1]). 我们可以利用上述 Λ 的 k-基:

$$(a_n,a_{n-1},\cdots,a_2,a_1,e_1,e_2,\cdots,e_h)$$

构造代数 Λ 的一个左正则表示 $\bar{\Lambda}$. 如果规定幂等元的高度为 0,基按照高度从大到小的顺序排列.

记 $t=n+h$,即矩阵 $\bar{\boldsymbol{\Lambda}}$ 的阶数,或代数 Λ 的维数,集合 $T=\{1,2,\cdots,t\}$. 另一个集合 $\mathcal{T}=\{X_1,X_2,\cdots,X_h\}$ 是代数 Λ 顶点集的个数. 令

$$\mathcal{P}_1(\Lambda)=\{P\xrightarrow{\alpha}Q\mid P,Q\in \mathrm{proj}(\Lambda),\alpha(P)\subseteq \mathrm{rad}(Q)\},\qquad (3.1)$$

则 $\mathcal{P}_1(\Lambda)$ 构成一个范畴,像元为映射模同态 α,如果 $P'\xrightarrow{\beta}Q'$ 也是一个像元,射元为模同态对 (f_1,f_2),使得 $f_2\alpha=\beta f_1$,其中 $f_1:P\to P'$,$f_2:Q\to Q'$.

这是一个与代数 Λ 相关的范畴,是研究模范畴 $\mathrm{mod}(\Lambda)$ 的有力工具. 为此,我们来定义一般的概念:矩阵双模问题.

令 $\mathcal{T}=\mathcal{T}_0\uplus\mathcal{T}_1$ 是一个顶点集,其中子集 $\mathcal{T}_0$ 由平凡点构成,即 $\forall X\in\mathcal{T}_0$,存在一个 k-代数 $R_X\simeq k$ 带有单位元 1_X. $\mathcal{T}_1$ 由非平凡点组成,即 $\forall X\in\mathcal{T}_1$,存在一个多项式环 $k[x]$ 的局部代数 $R_X\simeq k[x,\phi_x(x)^{-1}]$,以 1_X 为单位元,其中 x 称作属于 $X\in\mathcal{T}_1$ 的参数.

称 k-代数 $R=\Pi_{X\in\mathcal{T}}R_X$ 为 $\mathcal{T}$ 上的极小代数,带有正交本原幂等元集 $\{1_X\mid X\in\mathcal{T}\}$,且

$$1_R=\sum_{X\in\mathcal{T}}1_X.$$

令 $\Delta_1=R\otimes_k R$ 是一个 R-R-双模,其左、右模作用分别为 $u\cdot(r_1\otimes_k r_2)=(ur_1)\otimes_k r_2$ 和 $(r_1\otimes_k r_2)\cdot u=r_1\otimes_k(r_2u)$,$\forall u\in R$,$r_1\otimes_k r_2=\in\Delta_1$.

定义 3.1　Δ_1 在 R 上生成一个分次代数 $\Delta=T_R(\Delta_1)$:

$$\Delta=\bigoplus_{p=0}^{\infty}\Delta_p,\text{其中 }\Delta_0=R,\Delta_p=R^{\otimes(p+1)}=\underbrace{R\otimes_k\cdots\otimes_k R}_{p+1},\bar{\Delta}=\bigoplus_{p=1}^{\infty}\Delta_p.$$

代数的乘法运算为 $\forall\alpha=r_1\otimes_k\cdots\otimes_k r_{p+1}\in\Delta_p$,$\beta=s_1\otimes_k\cdots\otimes_k s_{q+1}\in\Delta_q$,

$$\alpha\beta=\alpha\otimes_k\beta=r_1\otimes_k\cdots\otimes_k(r_{p+1}s_1)\otimes_k s_2\cdots\otimes_k s_{q+1}\in\Delta_{p+q}.$$

令正整数集 $T=\{1,2,\cdots,t\}$,顶点集 $\mathcal{T}$ 定义如上. 存在 T 的一个 $\mathcal{T}$-划分,及其划分映射:

$$b:T\to\mathcal{T},T=\biguplus_{X\in\mathcal{T}}T_X,T_X=b^{-1}(X),\forall X\in\mathcal{T}.\qquad (3.2)$$

考虑 $\mathbb{M}_t(\Delta)$ 的子集:

$$\begin{aligned}&\mathbb{M}_{(T,\mathcal{T})}=\bigoplus_{(X,Y)\in\mathcal{T}\times\mathcal{T}}\mathbb{M}_{XY},\\&\mathbb{M}_{XY}=\{(g_{ij})\mid g_{ij}\in 1_X\Delta 1_Y,\forall i\in T_X,j\in T_Y,\text{在其他情形下取 }0\}.\end{aligned}\qquad (3.3)$$

$\mathbb{M}_{(T,\mathcal{T})}$ 是 $\mathbb{M}_t(\Delta)$ 的子代数:使得 $\mathbb{M}_{XY}\ \mathbb{M}_{YZ}\subseteq\mathbb{M}_{XZ}$,$\mathbb{M}_{XY}\ \mathbb{M}_{Y'Z}=\{0\}$,$\forall Y\neq$

Y'. 记$\mathcal{E}_{ij} \in \mathbb{M}_t(\Delta)$为矩阵单位，它在$(i,j)$位置的元素为$1_R$，其他为0. 令

$$E_X = \sum_{i \in T_X} 1_X\, \mathcal{E}_{ii}, E_I = \sum_{X \in \mathcal{T}} E_X, \tag{3.4}$$

则$\{E_X \mid X \in \mathcal{T}\}$是代数$\mathbb{M}_{(T,\mathcal{T})}$中本原幂等元的完全集，$E_I$是恒等元.

定义 3.2 四元组$\mathfrak{A}=(R,\mathcal{K},\mathcal{M},H)$称为$(T,\mathcal{T})$上的一个矩阵双模问题，如果

(1) R是一个顶点集为$\mathcal{T}$的极小代数；

(2) $\mathcal{K}=\mathcal{K}_0 \oplus \bar{\Delta} \cdot \mathcal{K}_1$，其中$\mathcal{K}_0 \simeq R$是对角矩阵代数；$\mathcal{K}_1$是严格上三角、拟自由$\Delta_1$-双模，带有一组拟自由基$\mathcal{V}=\bigcup_{X,Y \in \mathcal{T}\times\mathcal{T}} \mathcal{V}_{XY}$，使得$\forall V \in \mathcal{V}_{XY}$，$1_X V 1_Y = V$，并且$\mathcal{K}$是$\Delta$上的一个$t \times t$上三角矩阵代数，乘法由映射$\mu_{11}:\mathcal{K}_1 \times \mathcal{K}_1 \to \mathcal{K}_2$诱导.

(3) $\mathcal{M}=\bar{\Delta} \cdot \mathcal{M}_1$，其中$\mathcal{M}_1$是拟自由$\Delta$-双模，带有一组平凡(即矩阵元素取自基础域的)正规化的拟自由基$\mathcal{A}=\bigcup_{X,Y \in \mathcal{T}\times\mathcal{T}} \mathcal{A}_{XY}$，使得$\forall A \in \mathcal{A}_{XY}$，$1_A 1_Y = A$. 并且$\mathcal{M}$是$\Delta$上的一个$t \times t$阶$\mathcal{K}$-$\mathcal{K}$-矩阵双模，左、右模运算分别由映射$\rho_{11}:\mathcal{K}_1 \times \mathcal{M}_1 \to \mathcal{M}_2$和$\tau_{11}:\mathcal{M}_1 \times \mathcal{K}_1 \to \mathcal{M}_2$诱导.

(4) $\boldsymbol{H}$是R上的$t \times t$矩阵，它确定一个导子$d:\mathcal{K} \to \mathcal{M}$，分别由映射$d_0=0:\mathcal{K}_0 \to \mathcal{M}_0$和$d_1:V_i \mapsto V_i H - H V_i (\forall V_i \in \mathcal{V})$所诱导.

现在来定义矩阵双模问题$\mathfrak{A}$的表示范畴$R(\mathfrak{A})$.

$(T,\mathcal{T})$上的非负整数向量$\underline{m}=(m_1,m_2,\cdots,m_t)$称作一个尺寸向量，如果当$i,j \in T_X$时，$m_i=m_j$. 这时$\underline{m}$确定了$\mathcal{T}$上的一个维数向量$\underline{d}=\{d_X=d_i \mid i \in T_X, \forall X \in \mathcal{T}\}$. 反之易见，任意维数向量也唯一确定一个尺寸向量.

定义 3.3 (1) 设域k上的Jordan矩阵$\boldsymbol{J}(\lambda)=\boldsymbol{J}_d(\lambda)^{e_d} \oplus \boldsymbol{J}_{r-1}(\lambda)^{e_{r-1}} \oplus \cdots \oplus \boldsymbol{J}_1(\lambda)^{e_1}$，其中$e_i$是非负整数. 记$m_j=e_r+e_{r-1}+\cdots+e_j$，$j=1$，2，$\cdots$，$r$，则$k$上相似于$\boldsymbol{J}(\lambda)$的分块矩阵$\mathbf{W}(\lambda)$称为以$\lambda$为特征值的域$k$上的Weyr矩阵：

$$\mathbf{W}(\lambda)=\begin{pmatrix} \lambda \boldsymbol{I}_{m1} & \mathbf{W}_{12} & \mathbf{0} & \cdots & \mathbf{0} & \mathbf{0} \\ & \lambda \boldsymbol{I}_{m2} & \mathbf{W}_{23} & \cdots & \mathbf{0} & \mathbf{0} \\ & & \lambda \boldsymbol{I}_{m3} & \cdots & \mathbf{0} & \mathbf{0} \\ & & & \ddots & \vdots & \vdots \\ & & & & \lambda \boldsymbol{I}_{m_{r-1}} & \mathbf{W}_{r-1,r} \\ & & & & & \lambda \boldsymbol{I}_{m_r} \end{pmatrix}_{r \times r}$$

其中 $\boldsymbol{W}_{j,j+1}=(\boldsymbol{I}_{m_{j+1}}\quad \boldsymbol{0})^{\mathrm{T}}$ 是一个 $m_j\times m_{j+1}$ 长方阵，T 表示转置.

(2) 具有不同特征值 λ_i 的 Weyr 矩阵的直和 $\boldsymbol{W}=\boldsymbol{W}(\lambda_1)\oplus\boldsymbol{W}(\lambda_2)\oplus\cdots\oplus\boldsymbol{W}(\lambda_s)$ 称作一个 k 上的 Weyr 矩阵.

(3) 如果对于 $X\in\mathcal{T}_1$，$\phi_X(W)$ 在 k 上可逆，那么 Weyr 矩阵 $\boldsymbol{W}$ 称为 R_X- 正则的. 对于 $X\in\mathcal{T}_0$，恒等矩阵 $\boldsymbol{I}=\mathrm{diag}(1,1,\cdots,1)$ 也称为 R_X- 正则 Weyr 矩阵.

取定 $X\in\mathcal{T}$，$m_X\in\mathbf{Z}^+$，以及 R_X- 正则 Weyr 矩阵 $\boldsymbol{W}_X\in\mathbb{M}_{m_X}(k)$：

$$\varphi_{W_X}:R_X=k[x]_{\phi x}\to\mathbb{M}_{m_X}(k),\quad h(x)\mapsto h(\boldsymbol{W}_X) \tag{3.5}$$

是一个代数同态. 以下记作“ • ”的模作用都是由 φ 诱导的.

定义 3.4 设 $\mathfrak{A}=(R,\mathcal{K},\mathcal{M},H)$ 是矩阵双模问题，$\underline{\boldsymbol{m}}$ 是 $(T,\mathcal{T})$ 上的尺寸向量. 如果存在 k 上的正则 Weyr 矩阵集 $\{\boldsymbol{W}_X\in\mathbb{M}_{m_X}(k)\mid\forall X\in\mathcal{T}\}$ 和矩阵集 $\{\boldsymbol{M}_i\in\mathbb{M}_{m_s(A_i)\times m_t(A_i)}(k)\mid i=1,2,\cdots,m\}$，那么称域 k 上的一个 $\underline{m}\times\underline{m}$-分块矩阵 $\overline{\boldsymbol{M}}$ 为 $\mathfrak{A}$ 在 k 上的一个表示：

$$\overline{\boldsymbol{M}}=\sum_{X\in\mathcal{T}}\boldsymbol{I}_{mX}\cdot\boldsymbol{H}_X+\sum_{i=1}^{m}\boldsymbol{M}_i\cdot\boldsymbol{A}_i. \tag{3.6}$$

如果 $\overline{\mathbf{N}}=\sum_{X\in\mathcal{T}}I_{n_X}\cdot\boldsymbol{H}_X+\sum_{i=1}^{m}\mathbf{N}_i\cdot\boldsymbol{A}_i$ 是尺寸向量 $\underline{\boldsymbol{n}}=(n_1,n_2,\cdots,n_t)$ 的一个表示，那么考察由 $f_X\in\mathbb{M}_{m_X\times n_X}(k)$ 和 $f_j\in\mathbb{M}_{m_s(V_j)\times n_t(V_j)}(k)$ 确定的 $\underline{m}\times\underline{n}$-矩阵

$$\bar{\boldsymbol{f}}=\sum_{X\in\mathcal{T}}\boldsymbol{f}_X\cdot\boldsymbol{E}_X+\sum_{j=1}^{m}\boldsymbol{f}_j\cdot\boldsymbol{V}_j. \tag{3.7}$$

不难定义两个同态的合成. 我们用符号 $R(\mathfrak{A})$ 记矩阵双模问题的表示范畴. 下述的余双模问题直到 box 的引出都是在文章构思和讨论的过程中由韩阳定义并证明的.

因为 $\mathcal{K}_1$ 和 $\mathcal{M}_1$ 都是拟自由 R-R-双模，我们可以分别构造它们的 $R\otimes_k R$-对偶结构 $\mathcal{C}_1$ 和 $\mathcal{N}_1$，带有 R-R-拟基 $\mathcal{V}^*$ 和 $\mathcal{A}^*$：

$$\begin{aligned}&\mathcal{C}_1=\mathrm{Hom}_{R^{\otimes 2}}(\mathcal{K}_1,R^{\otimes 2}),\mathcal{V}^*=\{v_1,v_2,\cdots,v_m\},\\&\mathcal{N}_1=\mathrm{Hom}_{R^{\otimes 2}}(\mathcal{M}_1,R^{\otimes 2}),\mathcal{A}^*=\{a_1,a_2,\cdots,a_n\}.\end{aligned} \tag{3.8}$$

记 $v:X\mapsto Y$（相应地，$a:X\mapsto Y$），如果 $V\in\mathcal{V}_{XY}$（相应地，$A\in\mathcal{A}_{XY}$）.

定义 3.5 设 $\mathfrak{A}=(R,\mathcal{K},\mathcal{M},H)$ 是一个矩阵双模问题，四元组 $\mathfrak{C}=(R,\mathcal{C},\mathcal{N},\partial)$ 称为一个相伴于 $\mathfrak{A}$ 的余双模问题，如果

(1) R 是一个以 $\mathcal{T}$ 为顶点集的极小代数；

(2) $\mathcal{C}=\mathcal{C}_0\oplus\Delta_1\cdot\mathcal{C}_1$，其中$\mathcal{C}_0=\mathrm{Hom}_R(\mathcal{K}_0,R)\simeq R$，带有与$\mathcal{K}$对偶的余代数结构，余乘由对偶映射 $\underline{\mu}_{11}:\mathcal{C}_1\to\mathcal{C}_1\otimes\mathcal{C}_1$ 确定；

(3) $\mathcal{N}=\mathcal{N}_1\cdot\overline{\Delta}$，带有与$\mathcal{K}$-$\mathcal{K}$-双模$\mathcal{M}$对偶的余双模结构，左、右余乘分别由 $\underline{\varrho}_{11}:\mathcal{N}_1\to\mathcal{C}_1\otimes\mathcal{N}_1$ 和$\underline{\tau}_{11}:\mathcal{N}_1\to\mathcal{N}_1\otimes\mathcal{C}_1$ 确定；

(4) $\underline{\mathrm{d}}:\mathcal{N}\to\mathcal{C}$是与导子 d 对偶的余导子，由 $\underline{\mathrm{d}}_0=0:\mathcal{N}_0\to\mathcal{C}_0$ 和 $\underline{\mathrm{d}}_1:\mathcal{N}_1\to\mathcal{C}_1$ 确定.

现在将伴随于双模问题 $\mathfrak{A}$ 的余双模问题$\mathfrak{C}=(R,\mathcal{C},\mathcal{N},\partial)$转化为更加易于表述和计算的另一种代数结构：box.

因为$\mathcal{N}$是一个 R-R-双模，所以，$\mathcal{N}$在 R 上生成一个分次代数

$\Gamma=T_R(\mathcal{N})$： $\Gamma=\bigoplus_{p=0}^{\infty}\Gamma_p$，其中 $\Gamma_0=R$，$\Gamma_p=\mathcal{N}^{\otimes p}=\mathcal{N}\otimes_R\cdots\otimes_R\mathcal{N}$，$\forall\, p\geqslant 1$.
定义一个 Γ-Γ-双模 $\Xi=\Gamma\otimes_R\mathcal{C}\otimes_R\Gamma$，带有一个由 $R\hookrightarrow\Gamma$ 诱导的余代数结构，记作$(\Xi,\mu_\Xi,\varepsilon_\Xi)$. 定义下述由左、右余模的余乘法及余导子给出的 R-R-双模映射：

$$\kappa_1:\mathcal{N}\xrightarrow{l}\mathcal{C}\otimes_R\mathcal{N}\xrightarrow{\cong}R\otimes_R\mathcal{C}\otimes_R\mathcal{N}\hookrightarrow\Gamma\otimes_R\mathcal{C}\otimes_R\Gamma,$$

$$\kappa_2:\mathcal{N}\xrightarrow{\tau}\mathcal{N}\otimes_R\mathcal{C}\xrightarrow{\cong}\mathcal{N}\otimes_R\mathcal{C}\otimes_R R\hookrightarrow\Gamma\otimes_R\mathcal{C}\otimes_R\Gamma,$$

$$\kappa_3:\mathcal{N}\xrightarrow{\partial}\mathcal{C}\xrightarrow{\cong}R\otimes_R\mathcal{C}\otimes_R\mathcal{N}\hookrightarrow\Gamma\otimes_R\mathcal{C}\otimes_R\Gamma.$$

引理 3.1 $\mathfrak{I}(\kappa_1-\kappa_2+\kappa_3)$是 Ξ 中的一个 Γ-余理想. 于是，$\Omega:=\Xi/\mathfrak{I}(\kappa_1-\kappa_2+\kappa_3)$是带有余代数结构的 Γ-Γ-双模.

回顾文献[9,10]，如上定义的 $\mathfrak{B}=(\Gamma,\Omega)$是一个 box，带有层

$$L=(R;\omega;a_1,a_2,\cdots,a_n;v_1,v_2,\cdots,v_m).$$

记 ε_Ω 和 μ_Ω 为诱导的余单位和余乘法，则 $\overline{\Omega}=\ker\varepsilon_\Omega$ 是由 $v_1,v_2,\cdots,v_m$ 自由生成的 Γ-Γ-双模，而双模的直和 $\Omega=\Gamma\oplus\overline{\Omega}$. 由此，我们得到嵌入：$\mathcal{C}_0\oplus\mathcal{C}_1\oplus\mathcal{N}\otimes\mathcal{C}_1\oplus\mathcal{C}_1\otimes\mathcal{N}\hookrightarrow\Gamma\otimes_R(\mathcal{C}_0\oplus\mathcal{C}_1\oplus\mathcal{N}\otimes_R\mathcal{C}_1\oplus\mathcal{C}_1\otimes_R\mathcal{N})\otimes_R\Gamma\subset\Omega$；以及同构 $\overline{\Omega}\otimes_R\overline{\Omega}\simeq\overline{\Omega}\otimes_\Gamma\overline{\Omega}$.

类群映射 $\omega:R\to\Omega$，$1_X\mapsto e_X$ 是 R-R-双模射. 回顾文献[9，定义 3.3]，注意到在 Ω 中，$(\iota_0(a_i)+\iota_1(a_i))-(\tau_0(a_i)+\tau_1(a_i))+\underline{d}_1(a_i)=(\kappa_1-\kappa_2+\kappa_3)(a_i)=0$，$\omega$ 确定的微分为 $\delta_1:\Gamma\to\overline{\Omega}$：

$\delta_1(1_X)=1_Xe_X-e_X1_X=0,\ x\in\mathcal{T}$,

$$\delta_1(a_i)=a_i\otimes_R e_t(a_i)-e_{s(a_i)}\otimes_R a_i=\underline{\varrho}_{11}(a_i)-\underline{\tau}_{11}(a_i)+\underline{d}(a_i),\ 1\leqslant i\leqslant n. \tag{3.9}$$

$\delta_2:\overline{\Omega}\mapsto\overline{\Omega}\otimes_\Gamma\overline{\Omega},\delta_2(v_j)=\mu_{11}(v_j),1\leqslant j\leqslant m$. 这时，box$\mathfrak{B}$ 称作伴随于矩阵双模问题 $\mathfrak{A}$ 的 box. 将矩阵双模问题及其伴随的 box 记作$(\mathfrak{A},\mathfrak{B})$，或者伴随对$(\mathfrak{A},\mathfrak{B})$.

特别地，将(3.1)中定义的矩阵双模问题$\mathcal{P}_1(\Lambda)$伴随的 box 称为 Drozd box.

分层 box $\mathfrak{B}$ 的表示是一个维数向量为 $\underline{d}$ 的左 Γ-模 P：

$$\begin{gathered}\{P_X=k^{d_X}\mid X\in\mathcal{T}\},\{P(x):P_X\to P_X\mid X\in\mathcal{T}\},\\ \{P(a_i):k^{d_{X_i}}\to k^{d_{Y_i}}\mid a_i:X_i\to Y_i,i=1,2,\cdots,n\}.\end{gathered}\tag{3.10}$$

从表示 P 到 Q 的同态是一个 Γ-映射 $f:\Omega\otimes_\Gamma P\to Q$. 显然，

$$\begin{aligned}\mathrm{Hom}_\Gamma(\overline{\Omega}\otimes_\Gamma P,Q)&\simeq\bigoplus_{j=1}^m\mathrm{Hom}_\Gamma(\Gamma 1_{s(v_j)}\otimes_k 1_{t(v_j)}P,Q)\\&\simeq\bigoplus_{j=1}^m\mathrm{Hom}_k(1_{t(v_j)}P,1_{s(v_j)}Q).\end{aligned}$$

将映射写成

$$f=\{f_X;f(v_j)\mid X\in\mathcal{T},1\leqslant j\leqslant m\},\tag{3.11}$$

则文献[11]证明了 f 是同态，当且仅当对于 $a_l\in\mathcal{A}^*,1\leqslant l\leqslant n$，有

$$\begin{aligned}P(a_l)f_{Y_l}-f_{Y_l}Q(a_l)&=\sum_{j<l,i}\eta_{ijl}\otimes_{R^{\otimes 3}}(f(v_i)\otimes_R Q(a_j))-\\&\sum_{i<l,j}\sigma_{ijl}\otimes_{R^{\otimes 3}}(P(a_i)\otimes_R f(v_j))+\sum_i\zeta_{il}\otimes_{R^{\otimes 2}}f(v_i).\end{aligned}\tag{3.12}$$

运用(3.12)，可以证明矩阵双模问题 $\mathfrak{A}$ 与伴随 box $\mathfrak{B}$ 的表示范畴是等价的，即 $R(\mathfrak{A})\simeq R(\mathfrak{B})$.

box 的最大优势在于存在几种称为约化的算法，根据这些算法，box $\mathfrak{B}$ 可以约化成一个诱导 box $\mathfrak{B}'$，并且存在一个表示范畴上的忠实满函子 $\vartheta:R(\mathfrak{B}')\to R(\mathfrak{B})$. 这使得我们能够从给定的 box 开始，按照需要一步步地得到一系列诱导 box. 例如，Drozd 在 Tame Wild 定理的讨论中，给出了野型 box 的两类条件，并证明了任意不满足这些条件的 box，对于每一个确定的维数 n，都可以用约化方法得到有限多个极小 box，于是可以断定原始的 box 是驯顺型的. 因而，有限维代数可以划分为驯顺型和野型两大类型.

还有一些问题用 box 不易解决，但回到产生 box 的矩阵双模问题反而可以保留原始代数更多的信息，见 §6. 为此，我们定义了矩阵双模问题的约化，它与伴随 box 的约化一一对应.

本节最后给出一个例子，说明如何从一个有限维代数得到对应的矩阵双模问题和 box.

例 3.1[12,13] 令 $Q = a$  b 是一个箭图，$I=\langle a^2, ba-ab, ab^2, b^3\rangle$ 是 kQ 的理想，代数 $\Lambda = kQ/I$. 将元素 e, a 和 b 在 Λ 中的剩余类仍然记作 e, a 和 b. 并记 $c=b^2, d=ab$. 我们得到了 Λ 的一组有序 k-基 $\{d, c, b, a, e\}$，对应的正则表示为 $\overline{\Lambda}$. 记 A, B, C 和 D 为 $\mathcal{M}_1$ 的 R-R-拟基，a, b, c 和 d 是 $\mathcal{N}_1$ 中的 R-R-对偶基.

我们有时将矩阵双模问题及其 box 的伴随对 $(\mathfrak{A}, \mathfrak{B})$ 写成一个形式方程

$$\begin{pmatrix} e & 0 & u_1 & u_2 & u_4 \\ & e & u_2 & 0 & u_3 \\ & & e & 0 & u_2 \\ & & & e & u_1 \\ & & & & e \end{pmatrix} = \begin{pmatrix} 0 & 0 & a & b & d \\ & 0 & b & 0 & c \\ & & 0 & 0 & b \\ & & & 0 & a \\ & & & & 0 \end{pmatrix} = \begin{pmatrix} 0 & 0 & a & b & d \\ & 0 & b & 0 & c \\ & & 0 & 0 & b \\ & & & 0 & a \\ & & & & 0 \end{pmatrix} \begin{pmatrix} f & 0 & v_1 & v_2 & v_4 \\ & f & v_2 & 0 & v_3 \\ & & f & 0 & v_2 \\ & & & f & v_1 \\ & & & & f \end{pmatrix}$$

其中 $e=e_X, f=e_Y$. $\mathfrak{A}$ 的伴随 box $\mathfrak{B}$ 有一个分层：$L=(R;\omega;a, b, c, d; u_1, u_2, u_3, u_4, v_1, v_2, v_3, v_4)$. $\mathfrak{B}$ 中实箭 $\{a, b, c, d\}$ 的微分可以从形式方程中读出，如图 3.1：

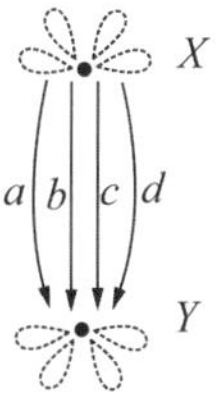

图 3.1

$$\begin{cases} \delta(a)=0, \\ \delta(b)=0, \\ \delta(c)=u_2 b - b v_2, \\ \delta(d)=u_1 b + u_2 a - b v_1 - a v_2. \end{cases}$$

§ 4. 几乎可列序列的对应

令 Λ 是代数闭域 k 上的有限维代数，我们在(3.1)中定义了范畴 $\mathcal{P}_1(\Lambda)$. 将 Λ 的单模、投射模和入射模分别记作 S_i, P_i 和 $I_i, i=1, 2, \cdots, h$. 进一步，如果 S_i 不是入射的，记 $N_i = DTr(S_i)$ 是 S_i 在 AR-变换下的像.

定义 4.1 设 Λ 是域 k 上的有限维基代数，使得 $\dim_k(\mathrm{top}(\Lambda))=h$.

(1)下述 $2h$ 个 $\mathcal{P}_1(\Lambda)$ 中的像元称为 Ext-投射的：

$$(0 \rightarrow P_i);(P_i \xrightarrow{\alpha_{N_i}} P_j^{m_{ij}}),i=1,2,\cdots,h,$$

其中第二式为模 N_i 的极小投射分解，这里 $N_i = DTr(S_i)$ 是非入射单模 S_i 在 AR-变换下的像；或者是入射单模 S_i 自身的极小投射分解，这时认为 $N_i = S_i$.

(2)下述 $2h$ 个$\mathcal{P}_1(\Lambda)$中的像元称为 Ext-入射的：

$$(P_i \rightarrow 0);(P_j^{m_{ij}} \xrightarrow{\alpha_{S_i}} P_i),i=1,2,\cdots,h,$$

其中第二式为模 S_i 的极小投射分解.

定义 4.2　$\mathcal{P}_1(\Lambda)$中的序列(如图 4.1)

$$\begin{array}{ccccc} P_1 & \xrightarrow{f_1} & W_1 & \xrightarrow{g_1} & Q_1 \\ \downarrow{\scriptstyle\alpha} & & \downarrow{\scriptstyle\gamma} & & \downarrow{\scriptstyle\beta} \\ P_0 & \xrightarrow[f_0]{} & W_0 & \xrightarrow[g_0]{} & Q_0 \end{array}$$

图 4.1

称为几乎可裂的，如果

(1) $P_1 \xrightarrow{\alpha} P_0$ 和 $Q_1 \xrightarrow{\beta} Q_0$ 都是不可分解的；

(2) $\begin{pmatrix} g_1 \\ g_0 \end{pmatrix}\begin{pmatrix} f_1 \\ f_0 \end{pmatrix}=\begin{pmatrix} 0 \\ 0 \end{pmatrix}$；

(3) $\begin{pmatrix} f_1 \\ f_0 \end{pmatrix}$是左极小几乎可裂映射，而$\begin{pmatrix} g_1 \\ g_0 \end{pmatrix}$是右极小几乎可裂映射.

引理 4.1　令 Λ 是完全域 k 上的有限维代数，则

(1) 任取$\mathcal{P}_1(\Lambda)$中不可分解的非 Ext-投射像元 $Q_1 \xrightarrow{\beta} Q_0$，存在$\mathcal{P}_1(\Lambda)$中的几乎可裂序列终于 $Q_1 \xrightarrow{\beta} Q_0$.

(2) 任取$\mathcal{P}_1(\Lambda)$中不可分解的非 Ext-入射像元 $P_1 \xrightarrow{\beta} P_0$，存在$\mathcal{P}_1(\Lambda)$中的几乎可裂序列始于 $P_1 \xrightarrow{\beta} P_0$.

(3) $\mathcal{P}_1(\Lambda)$中的任意几乎可裂序列都是正合的.

Zeng 和 Zhang[14]证明了下述定理.

定理 4.1　令 Λ 是完全域 k 上的有限维代数. 定义 4 个集合：

$\mathcal{B}$={mod-Λ 中始于非单模的几乎可裂序列}，

$\mathcal{D}$={$\mathcal{P}_1(\Lambda)$中终于$\mathcal{P}_2(\Lambda)$像元的几乎可裂序列}，

$\mathcal{B}_0=\{$中间项不包含入射直和项的$\mathcal{B}$的子集$\}$,

$\mathcal{D}_0=\{\mathcal{P}_2(\Lambda)$中的几乎可裂序列$\}$,

则

(1) $\mathcal{D}$与$\mathcal{B}$精确到序列的等价是一一对应的;

(2) 在函子 $Cok:\mathcal{D}_0\to\mathcal{B}_0$ 的作用下,$\mathcal{B}_0$ 与$\mathcal{D}_0$ 精确到序列的等价是一一对应的.

设 $\mathfrak{B}$ 是一个 box,带有分层 $L=(R;\omega;a_1,a_2,\cdots,a_n;v_1,v_2,\cdots,v_m)$. 回顾(3.11),$R(\mathfrak{B})$的态射 $f:M\to N$ 可以表示为 $f=\{f_X;f(v_j)\mid X\in\mathcal{T},1\leqslant j\leqslant m\}$. 记 $f_0=\{f_X\mid X\in\mathcal{T}\}:M\to N$ 是向量空间的映射.

定义 4.3 设 $\mathfrak{B}$ 是一个分层 box,则可以在 $R(\mathfrak{B})$中定义下述概念:

(1) 态射 $f:M\to N$ 称作真满(相应地,真单)射,如果 $f_0:M\to N$ 是满(相应地,单)射.

(2) N(相应地,M)$\in R(\mathfrak{B})$称为真投(相应地,真入)射的,如果任意真满(相应地,真单)射 $f:M\to N$ 是可裂的.

(3) 序列 $0\to M\xrightarrow{f}E\xrightarrow{g}N\to 0$ 称作真正合的,如果序列 $0\to M\xrightarrow{f_0}E\xrightarrow{g_0}N\to 0$ 作为极小代数 R 上的序列是正合的.

(4) 一个真正合序列 $0\to M\xrightarrow{f}E\xrightarrow{g}N\to 0$ 称为几乎可裂的,如果 $M\xrightarrow{f}E$ 是左极小几乎可裂的;$E\xrightarrow{g}N$ 是右极小几乎可裂的.

定理 4.2 设 $\mathfrak{B}$ 是引理 3.1 中由矩阵双模问题$\mathcal{P}_1(\Lambda)$的对偶基结构导出的 Drozd box,则存在等价函子 $\Sigma:\mathcal{P}_1(\Lambda)\to R(\mathfrak{B})$,使得

(1) $\mathcal{P}_1(\Lambda)$的像元 M 是 Ext-投(相应地,入)射的,当且仅当 $\Sigma(M)$在 $R(\mathfrak{B})$中是真投(相应地,入)射的.

(2) (e)是$\mathcal{P}_1(\Lambda)$中的几乎可裂序列,当且仅当 $\Sigma(e)$是 $R(\mathfrak{B})$中的几乎可裂序列.

根据定理 4.1 和 4.2 可以得知,对于代数闭域 k 上的有限维基代数 Λ,除掉有限多个等价类,mod(Λ),$\mathcal{P}_1(\Lambda)$和 $R(\mathfrak{B})$三个范畴中的几乎可裂序列精确到等价是一一对应的.

§5. 驯顺型代数的 Hom 空间

令 k 是代数闭域,Λ 是 k 上的有限维驯顺代数. 在 Λ 的有限维模范

畴 mod(Λ)中，对于每一个给定的正整数 d，我们可以刻画几乎所有维数不超过 d 的不可分解 Λ-模对的态射空间的结构，并确定空间的维数.

回顾驯顺代数的定义：代数 Λ 称为驯顺型的，如果对于任意正整数 d，都存在有限多个 Λ-$k[x]$-双模 $M_1, M_2, \cdots, M_n$，它们作为 $k[x]$-模是有限生成的，使得对于任意维数不超过 d 的不可分解 Λ-模 N，都存在某个 $1 \leqslant i \leqslant n$，以及元素 $\lambda \in k$，使得 $N \simeq M_i \otimes_{k[x]} k[x]/(x-\lambda)$.

根据文献[15]，对于每一个维数 d，维数不超过 d 的任意 Λ-模都可以被有限多个一般 Λ-模的同构类所“控制”，见下面的定理 5.1(1). 人们自然会问，那么这些模之间的态射集，即 Hom 集，是不是也能被这些一般模所控制呢？回答是肯定的.

定义 5.1　一个左 Λ-模 G 也可以看成是自同态环 $\mathrm{End}_\Lambda(G)$上的左模. 我们将 G 作为 $\mathrm{End}_\Lambda(G)$的长度，称为 G 的自同态(endo)长度. G 称作一般模，如果 G 是不可分解的，在 k 上的维数无限，但自同态长度有限.

设 G 是一般模，R 是 k 上有限生成的交换主理想整环，R 的分式域为 K. G 在 R 上的实现是一个有限生成的 Λ-R-双模 T，使得 $G \simeq T \otimes_R K$，并且 $\dim_K(T \otimes_R K)$等于 G 的自同态长度.

作为一个例子，我们来考虑由箭图 Q 确定的 Kronecker 代数 kQ，kQ 上的一般模 G 以及 G 在 $R=k[x]$上的实现 T：

$$Q: 1 \underset{b}{\overset{a}{\rightrightarrows}} 2, \quad G: k(x) \underset{id}{\overset{x}{\rightrightarrows}} k(x), \quad T: k[x] \underset{id}{\overset{x}{\rightrightarrows}} k[x].$$

我们用符号 Λ-Mod 表示 Λ-模的全体，Λ-mod 表示有限维模组成的 Λ-Mod 的满子范畴，Λ-ind 表示有限维不可约模组成的 Λ-mod 的满子范畴.

根据文献[15，定理 6.2]，如果 Λ 是驯顺表示型的，那么对于任意的一般 Λ-模 G，都存在相对于某个极小代数 R 的在下述意义之下的优良实现：

定义 5.2　一般 Λ-模 G 在某个极小代数 R 上的实现称为优良实现，如果

(1) T 是自由 R-模；

(2) 存在函子 $T \otimes - : R\text{-Mod} \to \Lambda\text{-Mod}$，保存同构类和不可分解性；

(3) 如果 $p \in R$ 是一个素元，$n \geqslant 1$，$(e_{p,n})$是正合序列：

$$0 \to R/(p^n) \xrightarrow{(p,\pi)} R/(p^{n+1}) \oplus R/(p^{n-1}) \xrightarrow{\binom{\pi}{-p}} R/(p^n) \to 0,$$

其中 p 是嵌入，π 是典范投射，那么 $T\otimes_R(e_{p,n})$ 是 mod-Λ 中的几乎可裂序列.

根据文献[15,定理 4.6]，如果 G 是一般 Λ-模，那么 $\mathrm{End}_\Lambda(G)=k(x)\oplus\mathrm{rad}(\mathrm{End}_\Lambda(G))$是可裂的. 这一分裂诱导出来 G 的左 $\Lambda^{k(x)}=\Lambda\otimes_k k(x)$-结构，称之为允许结构.

Bautista 等[16]的目的是证明下述定理.

定理 5.1 令 Λ 是代数闭域 k 上的有限维代数；d 是一个正整数，大于代数 Λ 在 k 上的维数，则存在一般 Λ-模 $G_1,G_2,\cdots,G_s$，带有左 $\Lambda^{k(x)}$-模允许结构，以及在某些 R_i 上的良好实现 T_i，其中 R_i 是 $k[x]$有限生成的局部化，$i=1,2,\cdots,s$；另一方面，存在维数不超过 t 的 Λ-模 $L_1,L_2,\cdots,L_t$，满足下述性质：

(1) 如果 M 是一个不可分解左 Λ-模，$\dim_k(M)\leqslant d$，那么存在整数 $j\in\{1,2,\cdots,t\}$，使得 $M\simeq L_j$；或者存在整数 $i\in\{1,2,\cdots,s\}$，使得对于某个素元 p 和正整数 m，有 $M\simeq T_i\otimes_{R_i}R_i/(p^m)$.

特别地. 如果 M 是不可分解的投射模，入射模或者单模，那么 $M\simeq L_j$ 属于第一种情形.

(2) 如果 $M=T_i\otimes_{R_i}R_i/(p^m)$，$N=T_{j\otimes R_j}R_j/(q^n)$，此处 $i,j\in\{1,2,\cdots,s\}$；m 和 n 是正整数；p 和 q 分别是 R_i 和 R_j 中的素元；$L_l^{k(x)}=L_{l\otimes k}k(x)$，那么，

$$\dim_k\mathrm{rad}^\infty(\mathrm{Hom}_\Lambda(M,N))=mn\dim_{k(x)}\mathrm{rad}(\mathrm{Hom}_{\Lambda k(x)}(G_i,G_j)),$$

$$\dim_k\mathrm{rad}^\infty(\mathrm{Hom}_\Lambda(L,M))=m\dim_{k(x)}\mathrm{rad}(\mathrm{Hom}_{\Lambda k(x)}(L^{k(x)},G_i)),$$

$$\dim_k\mathrm{rad}^\infty(\mathrm{Hom}_\Lambda(M,L))=m\dim_{k(x)}\mathrm{rad}(\mathrm{Hom}_{\Lambda k(x)}(G_i,L^{k(x)})).$$

(3) 如果 $i=j$ 并且 $p=q$，那么，

$$\mathrm{Hom}_\Lambda(M,N)\simeq\mathrm{Hom}_{R_i}(R_i/(p^m),R_i/(p^n))\oplus\mathrm{rad}^\infty(\mathrm{Hom}_\Lambda(M,N)).$$

如果 $i\neq j$ 或者 $p=q$，那么，

$$\mathrm{Hom}_\Lambda(M,N)\simeq\mathrm{rad}^\infty(\mathrm{Hom}_\Lambda(M,N)).$$

进一步，

$$\mathrm{Hom}_\Lambda(L_l,M)\simeq\mathrm{rad}^\infty(\mathrm{Hom}_\Lambda(L_l,M)),$$

$$\mathrm{Hom}_\Lambda(M,L_l)\simeq\mathrm{rad}^\infty(\mathrm{Hom}_\Lambda(M,L_l)).$$

定理的证明划分为几个步骤.

第 1 步 考虑 box 的约化. 给定一个 box $\mathfrak{B}$，存在下述几种约化算

法：正则化、点的删除、解圈、边约化和局部化. 通过其中任意一种，我们可以得到一个新的 box $\mathfrak{B}'$，以及约化函子 $\theta:\mathfrak{B}\to\mathfrak{B}'$，使得表示范畴之间的诱导函子 $\theta^*:R(\mathfrak{B}^*)\to R(\mathfrak{B})$ 是忠实满函子.

第 2 步　根据文献[9,12]的结果，给定一个维数 d，驯顺型 box $\mathfrak{B}$ 可以通过一系列约化得到有限多个极小 box $\mathfrak{B}_1,\mathfrak{B}_2,\cdots,\mathfrak{B}_s$ 以及诱导函子 $\theta_1^*,\theta_2^*,\cdots,\theta_s^*$，使得原 box $\mathfrak{B}$ 的任意一个维数不超过 d 的表示 $\overline{N}=\theta_i^*(M_i)$，其中 $i\in\{1,2,\cdots,s\}$. 极小 box 中的表示之间的态射可以用一般模之间的态射与两个表示之间的虚箭表达出来，其中后者在无穷根中.

第 3 步　从给定的有限维代数 Λ 构造范畴 $P_1(\Lambda)=\{P\xrightarrow{\alpha}Q\mid P,Q\in \mathrm{proj}(\Lambda),\alpha(P)\subseteq \mathrm{rad}(Q)\}$，见 §3 开头. 将第 2 步的结果转化到 mod-Λ 中.

§6.　模范畴的齐性与代数的驯顺性

在 20 世纪 80 年代后期，Crawley-Boevey[9] 证明了下述定理：设 Λ 是代数闭域 k 上的驯顺型有限维代数，则对于每一个维数 d，几乎所有(除掉有限多个同构类)维数不超过 d 的不可分解模 M 都是齐性的，即 M 的 AR-变换 $DTr(M)\simeq M$. 定理是通过 box 的语言给出证明的.

从此，很多作者试图证明 Crawley-Boevey 定理的逆. 希望仍然用 box 的手段证明：任何一个野型 box 都不是齐性的.

定义 6.1[17]　令 $\mathscr{E}$ 是一个正合范畴.

(1) $\mathscr{E}$ 中的一个不可分解像元 M 称为齐性的，如果存在 $\mathscr{E}$ 中的几乎可裂序列：

$$0\to M\to E\to M\to 0;$$

(2) 范畴 $\mathscr{E}$ 称为齐性的，如果对于每一个维数 d，几乎所有满足 $\dim_k(M)\leqslant d$ 的不可分解像元(除有限多个同构类之外)都是齐性的；

(3) 范畴 $\mathscr{E}$ 称为强齐性的，如果每一个维数有限的不可分解像元 M 都是齐性的.

出人预料的是，Zhang 等[18] 给出了一个野型 box 的例子，它的表示范畴是强齐性的.

例 6.1　令 $\mathfrak{B}$ 是一个局部 box，极小代数 $R=k[x,\phi(x)^{-1}]$，分层 $L=(R;\omega;a;v)$，微分 $\delta(a)=xv-vx$(如图 6.1)：

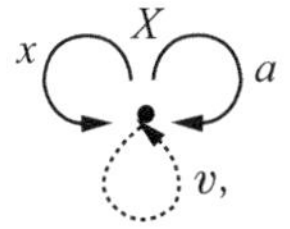

图 6.1

那么表示范畴 $R(\mathfrak{B})$是强齐性的.

接着,Bautista 等[19]构造了一大类强齐性野型范畴,特别地,这种范畴包含例 6.1 中的 box.

定义 6.2 设 k 是一个域,Λ 是有限维 k-代数.定义单变元多项式环 $\mathfrak{N}=\Lambda[z]$,其中变元 z 中心化 Λ.令 $\mathscr{C}$ 是左 Λ-模的一个子范畴,其中的模作为 Λ-模是有限维投射的.

引理 6.1 如果 Λ 是自入射的,那么范畴 $\mathscr{C}$ 有几乎可裂序列.

例如,取 $\Lambda=k$,$\mathfrak{N}=k[z]$是域上的多项式代数.如果 k 是代数闭域,那么 $\mathscr{C}=\text{mod-}\mathfrak{N}$ 是强齐性的,这是平凡的典型驯顺型代数.

一个最简单的非平凡例子是,取 $\Lambda=k[a]/(a^2)$,$\mathfrak{N}=\Lambda[z]$,这时 $\mathscr{C}$ 是强齐性的.

定理 6.1 令 $\mathfrak{N}=\Lambda[z]$,$\mathscr{C}$ 是定义 6.2 给出的范畴.如果 Λ 是自入射对称代数,那么 $\mathscr{C}$ 是强齐性的.

上述反例表明,Crawley-Boevey 定理的逆不能用 box 的方法去证明,必须另辟蹊径.尽管 box 的约化简单易行,但是在约化的过程中,有很多信息未能保存下来.

在这种情形下,Zhang 和 Xu[8]返回到原始的矩阵双模问题.

例 6.2 事实上,从例 3.1 给出的代数所对应的 box 出发,做一系列约化:.$a\mapsto(1)$,$b\mapsto\begin{pmatrix}0&1\\0&0\end{pmatrix}$,$c$ 和 d 分别变成了 2 阶方阵 $\begin{pmatrix}c_{11}&c_{12}\\c_{21}&c_{22}\end{pmatrix}$ 和 $\begin{pmatrix}d_{11}&d_{12}\\d_{21}&d_{22}\end{pmatrix}$.令 $c_{21}\mapsto(x)$,然后做三次正则化,得到 $c_{22}\mapsto\varnothing$,$u_{21}^2=xv$;$c_{11}\mapsto\varnothing$,$v_{21}^2=vx$;$c_{12}\mapsto\varnothing$,$u_{11}^2=v_{22}^2$.最后得到诱导 box $\mathfrak{B}'$,其中实箭的微分

$$\begin{cases}\delta(d_{21})=xv-vx,\\ \delta(d_{22})=u_{21}^1+u_{22}^2-v_{22}^2-d_{21}v,\\ \delta(d_{11})=u_{11}^2-v_{11}^2-v_{21}^1+vd_{21},\\ \delta(d_{12})=u_{11}^1+u_{12}^2-v_{12}^2-v_{22}^1-d_{11}v+vd_{22}\end{cases}$$

是一个野型强齐性 box.

为了探索原来的野型代数是否存在一个非齐性子范畴;我们只能利用产生这个 box 的原始代数的性质. 事实上是利用代数的矩阵双模问题能够保留原始代数更多信息的性质,在已知强齐性子范畴的基础上构造一个新的诱导矩阵双模问题,使得它的表示范畴,或等价地,相伴诱导 box 的表示范畴是非齐性的. 文献[8]的后半部分便是对这个目的的一次尝试.

参考文献

[1] Riedtmann C. Algebren, darstellungsköcher, ueberlagerungen und zurück. Comment Math Helv,1980,55:199-224.

[2] Happel D, Preiser U, Ringel C M. Vinberg's characterization of Dynkin diagrams using subadditive function with application to DTr-periodic modules. In: Lecture Notes in Mathematics, Vol. 832. Berlin: Springer-Verlag, 1980,280-294.

[3] Zhang Y B. The structure of stable components. Canada J Math, 1991, 43: 652-672.

[4] Auslander M. Applications of morphisms determined by objects. In: Proceedings of Conference on Representation Theory (Philadelphia, 1976). New York: Marcel Dekker, 1978, 245-327.

[5] Bautista R, Smalø S O. Nonexistent cycles. Comm Algebra, 1983, 11: 1 755-1 767.

[6] Zhang Y B. The modules in any component of the AR-quiver of a wild hereditary Artin algebra are uniquely determined by their composition factors. Arch Math, 1989, 53: 250-251.

[7] Zhang Y B. Eigenvalues of coxeter transformations and the structure of regular components of an auslander-reiten quiver. Comm Algebra, 1989, 17: 2 347-2 362.

[8] Zhang Y B, Xu Y G. Algebras with homogeneous module category are tame. ArXiv: 1407. 7576, 2014.

[9] Crawley-Boevey W W. On tame algebras and bocses. Proc Lond Math Soc, 1988, 56(3): 451-483.

[10] Rojter A V. Matrix problems and representations of BOCS's. In: Lectures

Notes in Mathematics, Vol. 831. Berlin: Springer Verlag, 1980: 3-38.

[11] Bautista R, Kleiner M. Almost split sequences for relatively projective modules. J Algebra, 1990, 135: 19-56.

[12] Drozd Yu A. On tame and wild matrix problems. Represent Quadratic Forms, 1979, 2: 104-114.

[13] Ringel C M. The representation type of local algebras. In: Lecture Notes in Mathematics, Vol. 488. Berlin: Springer-Verlag, 1975, 282-305.

[14] Zcng X, Zhang Y B. A correspondence of almost split sequences between some categories. Comm Algebra, 2001, 29: 557-582.

[15] Crawley-Boevey W W. Tame algebras generic modules. Proc Lond Math Soc, 1991, 63(3), 241-265.

[16] Bautista R, Drozd A, Zeng X Y, et al. On Hom-spaces of tame algebras. Cent Eur J Math, 2007, 5: 215-263.

[17] Dräxler P, Reiten I, Smalø S O, et al. Exact categories and vector space categories. Trans Amer Math Soc, 1999, 351: 647-682.

[18] Zhang Y B, Lei T G, Bautista R. A matrix description of a wild category. Sci China Ser A, 1998, 41: 461-475.

[19] Baulista R, Crawley-Boevey W W, Lei T, et al. On homogeneous exact categories. J Algebra, 2000, 230: 665-675.

Abstract We introduce briefly the following contents on representation of Artin algebras: the structure of stable components of AR-quivers; the property of stable components of wild hereditary algebras; the matrix bimodule problems introduced by finitely dimensional algebras over an algebraically closed field, the dual cobimodule problems, and associated boxes; almost one-by-one correspondence of almost split sequences between representation categories of modΛ, $P_1(\Lambda)$ and associated box; the dimension and property of Hom-spaccs of any two modules over an tame algebra Λ. We also give a brief introduction to the tameness and homogeneous property of Artin algebras.

Keywords the stable component of an AR-quiver; representation type of an algebra; matrix bimodule problem; box, reduction.

数学学报,1985,28(1):91-102.

一类二秩无扭 Abel 群的结构①

The Structure of a Class of Torsion Free Abelin Groups

摘要 本文以环论为工具,讨论了有最小型元素的二秩无扭群的结构和不变量,以及成同型与可分解的条件.

§1. 几种完备拓扑环

p 是素数,分母与 p 互素的全体有理数作成环 R_p. 在 R_p 中定义赋值 ρ: $\forall \frac{m}{n}\in R_p, m=m'p^k, (m',p)=1$,则 $\rho\left(\frac{m}{n}\right)=p^{-k}$. 由 ρ 确定的距离导出环 R_p 的拓扑结构,p-进整数环 $\mathscr{B}_p$ 是 R_p 的完备环. 仿此可引出另外几种完备拓扑环.

设 $p_1,p_2,\cdots,p_s,\cdots$ 是从小到大排列的某些素数,$n_1,n_2,\cdots,n_s,\cdots$ 是正整数,令 $R=\bigcap\limits_{s=1}^{\infty}R_s$,有环 R 的理想的降链:$R\supset p_1^{n_1}R\supset p_1^{n_1}p_2^{n_2}R\supset\cdots\supset p_1^{n_1}p_2^{n_2}\cdots p_s^{n_s}R\supset\cdots$. 易见 $\bigcap\limits_{s=1}^{\infty}p_1^{n_1}p_2^{n_2}\cdots p_s^{n_s}R=\{0\}$. 称 $p_1^{n_1}p_2^{n_2}\cdots p_s^{n_s}R$ 为 0 点的邻域,R 的邻域集

$$\Sigma=\{a+p_1^{n_1}p_2^{n_2}\cdots p_s^{n_s}R \mid \forall a\in R, s\in \mathbf{N}^*\}$$

作成环 R 的一族拓扑基.

称 R 的序列 $\xi=(a_1,a_2,\cdots,a_s,\cdots)$ 为基本序列,若 $\forall s$, $\exists N$,当 m, $m'>N$ 时,$a_m-a_{m'}\in p_1^{n_1}p_2^{n_2}\cdots p_s^{n_s}R$. 基本序列之间的加法和乘法按分量

① 收稿日期:1982-03-08;收修改稿日期:1983-07-20;收精简稿日期:1984-03-31.

进行，其全体作成环 $\widetilde{R}$. $(c_1,c_2,\cdots,c_s,\cdots)$称为 R 的 0 基本序列，若 $\forall s$，$\exists N$，当 $m>N$ 时，$c_m\in p_1^{n_1}p_2^{n_2}\cdots p_s^{n_s}R$. 0 基本序列的全体作成 $\widetilde{R}$ 的理想 $\widetilde{N}$，令

$$\bar{R}=\widetilde{R}/\widetilde{N}.$$

定理 1.1 环 $\bar{R}$ 是拓扑环 R 的完备环.

$\bar{R}$ 的元素$(a_1,a_2,\cdots,a_s,\cdots)+\widetilde{N}$ 可唯一地表作无穷级数

$$\lambda=t_0+t_1p_1^{n_1}+t_2p_1^{n_1}p_2^{n_2}+\cdots+t_{s-1}p_1^{n_1}p_2^{n_2}\cdots p_{s-1}^{n_{s-1}}+\cdots,$$

其中 $0\leqslant t_{s-1}<p_s^{n_s}$. λ 的前 s 项部分和记作 λ_s. 为便于区别，将此处定义的环 $\bar{R}$ 记作 $\mathscr{C}$.

命题 1.2 (1) $\lambda\in\mathscr{C}\lambda=0$，当且仅当 $\lambda_s\equiv0(\mathrm{mod}\ p_s^{n_s})$，$s\in\mathbf{N}^*$.

(2) $1=1+0p_1^{n_1}+\cdots+0\cdot p_1^{n_1}p_2^{n_2}\cdots p_{s-1}^{n_{s-1}}+\cdots$是 $\mathscr{C}$ 的单位元.

(3) $\lambda\in\mathscr{C}$，λ 可逆，当且仅当$(\lambda_s,p_s)=1$，$s\in\mathbf{N}^*$.

环 R 的定义同前，我们可以在 R 中定义另一种拓扑. 取 R 的子集

$$\Sigma'=\{a+p_1^sp_2^{s-1}\cdots p_sR:\forall a\in R,s\in\mathbf{N}^*\}.$$

Σ'也是 R 的一族拓扑基，R 的完备拓扑环记作 $\mathscr{D}$，其元 ϑ 可唯一地表作

$$\vartheta=t_0+t_1p_1+t_2p_1^2p_2+\cdots+t_{s-1}p_1^{t-1}p_2^{t-2}\cdots p_{s-1}+\cdots,$$

其中 $0\leqslant t_{s-1}<p_1p_2\cdots p_s$，$\vartheta$ 的前 s 项和记作 ϑ_s.

$\forall\vartheta\in\mathscr{D}$，$\left\{\dfrac{\vartheta_{i+s}-\vartheta_i}{p_1^ip_2^{i-1}\cdots p_i}\right\}_{s\in\mathbf{N}^*}$是 R 的基本序列，可得 $\mathscr{D}$ 的元素 $\bar{\vartheta}$，简记作$\dfrac{\vartheta-\vartheta_i}{p_1^ip_2^{i-1}\cdots p_i}$.

为讨论二秩无扭群的同构问题(§3)，我们分别在环 $\mathscr{B}_p$，$\mathscr{C}$，$\mathscr{D}$ 中引入元素之间的下述关系：Z 表整数加群，K 表任意一秩无扭群，看作有理加群 Q 的子群，$\chi^K(k)$记元素 k 的高.

定义 1.3 设 π，σ 是 p-进整数，若有 $a,b,c,d\in K$，满足

① 当 $\chi_{p_s}^K(1)<+\infty$，$p_s\neq p$ 时，矩阵$\begin{pmatrix}a & b\\ c & d\end{pmatrix}$及其逆阵的元素分母均不含因子 p_s；

② $a-b\pi$ 可逆；

③ $\sigma=-\dfrac{c-d\pi}{a-b\pi}$，则称 σ 与 π 为 K-分式线性相关. 当 $K=Z$，称之为

整分式线性相关.

定义 1.4 环 $\mathscr{C}$ 的元素 λ' 与 λ 称为 K-分式线性相关,若有 $a,b,c,d\in K$,满足

① 当 $\chi_q^K(1)<+\infty$,$\begin{pmatrix} a & b \\ c & d \end{pmatrix}$及其逆阵的元素分母不含因子 q;

② $a-b\lambda$ 可逆;

③ $\lambda'=-\dfrac{c-d\lambda}{a-b\lambda}$.

类似地有环 $\mathscr{D}$ 的元素 K-分式线性相关的定义. 为讨论二秩无扭群成齐次的条件(§4),仿照 $\mathscr{B}_p$ 中的 p-进无理整数,在环 $\mathscr{C}$ 与 $\mathscr{D}$ 中引入一类特殊元.

定义 1.5 设 $p_1,p_2,\cdots,p_s,\cdots$是从小到大排列的某些素数,给定非负整数序列$\{m_s\}$,任取正整数 n,设 $nm_s\equiv m_{n,s}\pmod{p_s}$,其中 $0\leqslant m_{n,s}<p$;$n\cdot m_s\equiv m'_{n,s}\pmod{p_s}$,其中$-p_s<m'_{n,s}\leqslant 0$,若对一切 n,序列$\{m_{n,s}\}$与$\{m'_{n,s}\}$中都没有无穷多个相等的数,则称$\{m_s\}$是关于$\{p_s\}$的零型序列.

引理 1.6 任给正整数 k,若 $\left[\dfrac{p_{i_s}}{i_s}\right]=k$,其中 i_s 是两两不同的正整数,p_{i_s} 是第 i_s 个素数,则 $s<+\infty$.

证 如若不然,存在无穷多个 i_s,使 $\left[\dfrac{p_{i_s}}{i_s}\right]=k$,设 $i_1<i_2<\cdots$,由数论中的不等式知[1]:有正常数 c_1,c_2,使 $c_1\ln j<\dfrac{p_j}{j}<c_2\ln j$ 对一切正整数 j 及第 j 个素数 p_j 成立. 由序列 i_s 之无限性,存在下标 t,使 $c_1\ln i_t>c_2\ln i_1+1$. 但$\dfrac{p_{i_t}}{i_t}>c_1\ln i_t$,$c_2\ln i_1>\dfrac{p_{i_1}}{i_1}$,故 $\left[\dfrac{p_{i_t}}{i_t}\right]>\left[\dfrac{p_{i_1}}{i_1}\right]$,与所设矛盾.

推论 1.7 取定正整数 n,l,若 $p_{i_s}-\left[\dfrac{p_{i_s}}{i_s}\right]=l$,则 $s<+\infty$.

定理 1.8 关于$\{p_s\}$的零型序列存在,其集合有连续统的势$\aleph_1$.

证 设 p_s 是第 i_s 个素数,令 $m_s=\left[\dfrac{p_s}{i_s}\right]$,则$\{m_s\}$是关于$\{p_s\}$的零型序列. 任取序列$\{\delta_s\}$,$\delta_s=0$ 或 1,如果$\{m_s\}$是零型序列,那么$\{m_s+\delta_s\}$也是.

解同余方程知,任取非负整数序列$\{m_s\}$,有环 $\mathscr{C}$ 的元素 λ,使

$$\lambda_s \equiv m_s \pmod{p_s}, s\in \mathbf{N}^*.$$

定义 1.9 若环 $\mathscr{C}$ 的元素 λ 满足 $\lambda_s \equiv m_s \pmod{p_s}$，而 $\{m_s\}$ 是关于 $\{p_s\}$ 的零型序列，称 λ 为 $\mathscr{C}$ 的零型元.

由定理 1.8 知，其集合有连续统的势.

环 $\mathscr{D}$ 的零型元要复杂些. $\forall \vartheta\in\mathscr{D}, p_i\in\{p_s\}_{s\in\mathbf{N}^*}$，$\forall k\in\mathbf{N}^*$，$p_i^k \mid (\vartheta_{i+k+n+m}-\vartheta_{i+k+n})$，其中 $\forall m,n\in\mathbf{N}^*$. 于是有

命题 1.10 $\forall \vartheta\in\mathscr{D}, p_i\in\{p_s\}_{s\in\mathbf{N}^*}$，序列 $\{\vartheta_s\}$ 作成某个 p_i-进整数 $\pi^{(i)}$ 的基本序列.

反之，取 p_s-进整数 $\pi^{(s)}$，$s\in\mathbf{N}^*$，记 $r_s=p_1^s p_2^{s-1}\cdots p_s$，则有 $a_1^{(s)}, a_2^{(s)}, \cdots, a_s^{(s)}\in\mathbf{Z}^*$，使

$$a_1^{(s)}\frac{r_s}{p_1^s}+a_2^{(s)}\frac{r_s}{p_2^{cs-1}}+\cdots+a_s^{(s)}\frac{r_s}{p_s}=1, \tag{1.1}$$

令

$$\vartheta_s'=a_1^{(s)}\pi_s^{(1)}\frac{r_s}{p_1^s}+a_2^{(s)}\pi_{t-1}^{(2)}\frac{r_s}{p_2^{s-1}}+\cdots+a_s^{(s)}\pi_1^{(s)}\frac{r_s}{p_s} \tag{1.2}$$

及

$$\vartheta_s\equiv\vartheta_s' \pmod{r_s}, 0\leqslant\theta_s<r_s \tag{1.3}$$

易知 $\vartheta_s\equiv\vartheta_s'\equiv a_i^{(s)}\pi_{t-i+1}^{(i)}\frac{r_s}{p_i^{s-i+1}}\equiv\pi_{s-i+1}^{(i)} \pmod{p_i^{s-i+1}}, i=1,2,\cdots,s.$

$$\tag{1.4}$$

$$\vartheta_s-\vartheta_{s-1}\equiv\vartheta_s'-\vartheta_{t-1}'\equiv\pi_{s-i+1}^{(i)}-\pi_{s-i}^{(i)}\equiv 0 \pmod{p_i^{r-i}}. \tag{1.5}$$

故 $r_{s-1}\mid\vartheta_s-\vartheta_{s-1}$，$\vartheta_s=\vartheta_{s-1}+t_{s-1}r_{s-1}$. 由(1.3)知 $0\leqslant t_{s-1}<p_1p_2\cdots p_s$，$s\in\mathbf{N}^*$. 令 $\vartheta=t_0+t_1p_1+\cdots+t_{s-1}p_1^{s-1}p_2^{s-2}\cdots p_{s-1}$，则 $\vartheta\in\mathscr{D}$，ϑ_s 恰为 ϑ 的前 s 项和，由(1.4)，$\{\vartheta_s\}$ 所成 p_i-进整数就是 $\pi^{(i)}$.

定义 1.11 环 $\mathscr{D}$ 的元素 ϑ 称为零型元，若①$\vartheta_s\equiv m_s \pmod{p_s}$，$\{m_s\}$ 是关于 $\{p_s\}$ 的零型序列，② $\forall p_i\in\{p_s\}_{s\in\mathbf{N}^*}$，$\{\vartheta_s\}$ 作成 p_i-进无理整数.

仍由解同余方程知，环 $\mathscr{D}$ 有零型元，其集合有连续统的势.

最后，为讨论二秩无扭群的分解问题，仿照 p-进有理整数，定义环 $\mathscr{C}$ 与 $\mathscr{D}$ 中的可分元.

定义 1.12 环 $\mathscr{C}(\mathscr{D})$ 的元素 $\lambda(\vartheta)$ 称为可分元，若集合 $\{p_s\}_{s\in\mathbf{N}^*}$ 可以分成互不相交的两组 $\{p_{i_r}\}_{r\in\mathbf{N}^*}$ 和 $\{q_{i_t}\}_{t\in\mathbf{N}^*}$，其中一组可以有限. 若存在

$a,b,c,d\in \mathbf{Z}$,满足

① $\begin{vmatrix} a & b \\ c & d \end{vmatrix}=\pm 1(\pm p_1^{K_1}\cdots p_m^{K_m})$;

② $a-b\lambda(a-b\vartheta)$可逆;

③ $(b+d)\lambda_{i_r}\equiv a+c(\bmod p_{i_r}^{n_{i_r}})$,$d\lambda_{j_t}\equiv c(\bmod p_{i_t}^{n_{i_t}})$($(b+d)\vartheta_{i_r+k-1}\equiv a+c(\bmod p_{i_r}^{k})$,$d\vartheta_{i_t+k-1}\equiv c(\bmod p_{i_t}^{k})$,其中$\forall k\in \mathbf{N}^*$).

§ 2. 有最小型元素的二秩无扭群的结构

将二秩无扭群 G 看作 $Q\oplus Q$ 的子群,其线性无关元 $x=(1,0)$,$y=(0,1)$,任意元

$$z=\frac{1}{r}(mx+ny).$$

命题 2.1 设 G 是二秩无扭群,则 G 可以用 $Q\oplus Q$ 的自由子群$\langle x\rangle\oplus\langle y\rangle$增添一系列形如$\frac{1}{p^k}(mx+ny)$的元素生成,此处 p 是素数,$m,n\in \mathbf{N}^*$. 若 $\chi^G(x)=(0,0,\cdots)$,则 G 可由$\langle x\rangle\oplus\langle y\rangle$增添形如$\frac{1}{p^k}(mx+y)$的元素生成.

以下假定二秩无扭群中含有高为 0 的元素 x,从 x 的系数入手讨论这类群的结构.

首先看一种最简单的情况:群 G 由 x,y 增添有限个元素生成. 设增添分母为 p 方幂的元 $x_1=\frac{1}{p}(l_1x+y)$,$x_2=\frac{1}{p}(l_2x+y)$,$\cdots$,$x_n=\frac{1}{p^n}(l_nx+y)$. 因 $x_j-px_{j+1}=-\frac{l_{j+1}-l_j}{p^j}x$,故 $l_{j+1}\equiv l_j(\bmod p^j)$,$x_j\in\langle x,x_{j+1}\rangle$,$j=0,1,\cdots,n-1$. 得 G 的子群的升链

$$\langle x,x_0\rangle\subset\langle x,x_1\rangle\subset\cdots\subset\langle x,x_n\rangle,$$

记 $y=x_0$. 这时可设 G 由 x,y 增添元素

$$z_1=\frac{1}{p_1^{n_1}}(m_1x+y),z_2=\frac{1}{p_2^{n_2}}(m_2x+y),\cdots,z_s=\frac{1}{p_s^{n_s}}(m_sx+y)$$

生成. 令 $r=p_s^{n_1}p_s^{n_2}\cdots p_s^{n_s}$,则有 $a_1,a_2,\cdots,a_s\in \mathbf{N}^*$,使

$$a_1\frac{r}{p_1^{n_1}}+a_2\frac{r}{p_2^{n_2}}+\cdots+a_s\frac{r}{p_s^{n_s}}=1. \tag{2.1}$$

令

$$l=a_1m_1\frac{r}{p_1^{n_1}}+a_2m_2\frac{r}{p_2^{n_2}}+\cdots+a_sm_s\frac{r}{p_s^{n_s}}, \tag{2.2}$$

$$l\equiv m_i(\bmod p_i^{n_i}). \tag{2.3}$$

若 $z=a_1z_1+a_2z_2+\cdots+a_sz_s=\frac{1}{r}(lx+y)$，则

$$\langle x,y,z_1,\cdots,z_s\rangle=\langle x,z\rangle=\langle x\rangle\oplus\langle z\rangle.$$

定理 2.2 $Q\oplus Q$ 的自由子群 $\langle x\rangle\oplus\langle y\rangle$ 增添有限个元素生成的群仍是二秩自由群.

其次看一种特殊的二秩无扭群[2]，设 G 由 x，y 以及

$$z_n=\frac{1}{p^n}(\pi_nx+y),0\leqslant\pi_n<p^n,n\in\mathbf{N}^*$$

生成，$\chi^G(x)=(0,0,\cdots)$，$\forall n\in\mathbf{N}^*$，$p^{n}z_{n+K}-z_n=\frac{1}{p^n}(\pi_{n+K}-\pi_n)x\in G$，所以 $p^n\mid\pi_{n+K}-\pi_n$. $\{\pi_n\}$是某 p-进整数 π 的基本序列. 记 $G=\langle x,y,\pi\rangle$. 反之，任取 p-进整数 π，我们有

命题 2.3 设 $G=\langle x,y,\pi\rangle$，

(1) $z_i=p^{K}z_{i+K}-\frac{\pi_{i+K}-\pi_i}{p^i}x$，对 $\forall i,k\in\mathbf{N}^*$ 成立；

(2)有 G 的子群的升链 $\langle x,z_0\rangle\subset\langle x,z_1\rangle\subset\cdots\subset\langle x,z_n\rangle\subset\cdots$，其中 $z_0=y$，$G=\bigcup\limits_{n=0}^{\infty}\langle x,z_n\rangle$；

(3) $\forall n$，x 与 z_n 线性无关；

(4) $\forall z\in G$，非 0 整数 $d\mid z$ 的充要条件是 $\exists n\in\mathbf{N}^*$，当 z 表成 x，z_n 的线性组合时，系数有公因子 d；

(5) $\chi^G(x)=(0,0,\cdots)$；

(6) $G=\langle x,z_i,\bar{\pi}\rangle$，其中 $\bar{\pi}=\frac{\pi-\pi_i}{p^i}\in\mathscr{B}_p$.

命题 2.4 设二秩无扭群 $G=\langle\langle x,y',\pi\rangle,\langle y\rangle\rangle$，$y=\frac{1}{r}(mx+y')$，$(r,p)=1$，则 $G=\langle x,y,\sigma\rangle$，$\sigma=\frac{\pi-m}{r}$ 也是 p-进整数.

第三,设 $p_1,p_2,\cdots,p_s,\cdots$ 是从小到大排列的某些素数,群 G 由 x,y 增添元素生成,$\chi^G(x)=(0,0,\cdots)$. $\forall p_s$,增添分母为 p_s 方幂的元有限个,可认为只添一个 $u_s=\frac{1}{p_s^{n_s}}(m_s x+y)$,$G=\langle x,y,u_1,\cdots,u_s,\cdots\rangle$. G 的子群 $\langle x,y,u_1,\cdots,u_s\rangle=\langle x,z_s\rangle$,$z_s=\frac{1}{p_1^{n_1}\cdots p_s^{n_s}}(\lambda_s x+y)$ 由(2.1)与(2.2)确定,并不妨假定 $0\leqslant\lambda_s<p_1^{n_1}\cdots p_s^{n_s}$,则 λ_s 是环 $\mathscr{C}$ 中某元素 λ 的前 s 项和. 记 $G=\langle x,y,\lambda\rangle$. 反之,$\forall\lambda\in\mathscr{C}$,$G=\langle x,y,\lambda\rangle$ 也有类似命题 2.3 的性质. 有 G 的子群的升链 $\langle x,z_0\rangle\subset\langle x,z_1\rangle\subset\cdots\subset\langle x,z_s\rangle\subset\cdots$,$G=\bigcup_{s=0}^{\infty}\langle x,z_s\rangle$;$\chi^G(x)=(0,0,\cdots)$;取定 $i\in\mathbf{N}$,若 $\bar{\mathscr{C}}$ 表 $\{p_{i+s}^{n_{i+s}}\}_{s\in\mathbf{N}^*}$ 确定的完备拓扑环,$\left\{\frac{\lambda_{i+s}-\lambda_i}{p_1^{n_1}p_2^{n_2}\cdots p_s^{n_s}}\right\}_{s\in\mathbf{N}^*}$ 是环 $\bigcap_{s=1}^{\infty}R_{p_{i+s}}$ 的基本序列,此序列确定 $\bar{\mathscr{C}}$ 的元素 $\bar{\lambda}$,

$$G=\langle x,z_i,\bar{\lambda}\rangle.$$

第四,设所添元素的分母涉及无穷多个素数,以每个素数方幂为分母的元素都有无穷多个,设为

$$x_1^{(1)}=\frac{1}{p_1}(\pi_1^{(1)}x+y),x_2^{(1)}=\frac{1}{p_1^2}(\pi_2^{(1)}x+y),\cdots,x_s^{(1)}=\frac{1}{p_1^s}(\pi_s^{(1)}x+y),\cdots$$

$$x_1^{(j)}=\frac{1}{p_j}(\pi_1^{(j)}x+y),x_2^{(j)}=\frac{1}{p_j^2}(\pi_2^{(j)}x+y),\cdots,x_s^{(j)}=\frac{1}{p_j^s}(\pi_s^{(j)}x+y),\cdots$$

这时 $\{\pi_s^{(j)}\}$ 是某 p_i-进整数 $\pi^{(j)}$ 的基本序列. 令 $G_j=\langle x,y,\pi^{(j)}\rangle$,则

$$G=\langle G_1,G_2,\cdots,G_j,\cdots\rangle.$$

令 $z_1=x_1^{(1)}$,$z_s=a_1^{(s)}x_s^{(1)}+a_2^{(s)}x_{s-1}^{(2)}+\cdots+a_s^{(s)}x_1^{(s)}$,$a_1^{(s)},a_2^{(s)},\cdots,a_s^{(s)}$ 的定义如(1.1). 由(1.2)知. $z_s=\frac{1}{r_s}(\vartheta'_s x+y)$,又由(1.3),不妨取 $z_s=\frac{1}{r_s}(\vartheta_s x+y)$. 根据定理 2.2 前的证明,$\langle x,y,x_s^{(1)},x_{s-1}^{(2)},\cdots,x_1^{(s)}\rangle=\langle x,y,z_s\rangle$,故 $G=\langle x,y,z_1,\cdots,z_s,\cdots\rangle$,$\{\vartheta_s\}$ 恰为环 $\mathscr{D}$ 的某元素 ϑ 的前 s 项和,记 $G=\langle x,y,\vartheta\rangle$. 反之,$\forall\vartheta\in\mathscr{D}$,令 $G=\langle x,y,\vartheta\rangle$,则有 G 的子群的升链 $\langle x,z_0\rangle\subset\langle x,z_1\rangle\subset\cdots\subset\langle x,z_s\rangle\subset\cdots$,且 $G=\bigcup_{s=0}^{\infty}\langle x,z_s\rangle$;

$$\chi^G(x)=(0,0,\cdots);G=\langle x,z_i,\bar{\vartheta}\rangle,\bar{\vartheta}=\frac{\vartheta-\vartheta_i}{r_i}\in\mathscr{D}.$$

命题 2.5 设 $G=\langle\langle x,y',\vartheta'\rangle,\langle t\rangle\rangle$，$y=\dfrac{1}{r}(mx+y')$，$(r,p_s)=1$，$s\in\mathbf{N}^*$，则 $G=\langle x,y,\vartheta\rangle$，此处 $\vartheta=\dfrac{\vartheta'-m}{r}$.

最后，上述群类可有如下组合：$G=\langle G_1,G_2,\cdots,G_k\rangle$，其中 $G_j=\langle x,y,\pi^{(j)}\rangle$；$G=\langle G_0,G_1,\cdots,G_k\rangle$，其中 $G_0=\langle x,y,\lambda\rangle,\lambda\in\mathscr{C},G_j=\langle x,y,\pi^{(j)}\rangle,\pi^{(j)}\in\mathscr{B}_{q_j},j=1,2,\cdots,K,\{p_s\}_{s\in\mathbf{N}^*}\cap\{q_j\}_{j\in\mathbf{N}^*},k=\varnothing$；$G=\langle G_0,G_1\rangle,G_0=\langle x,y,\lambda\rangle,\lambda\in\mathscr{C}$，由 $\{p_s^{n_s}\}$ 确定，$G_1=\langle x,y,\vartheta\rangle,\vartheta\in\mathscr{D},\mathscr{D}$ 由 $\{q_t\}$ 确定，$\{p_s\}_{s\in\mathbf{N}^*}\cap\{q_t\}_{t\in\mathbf{N}^*}=\varnothing$. 将全部有 0 高元的二秩无扭群列成表 2.1.

表 2.1

	群的类型	增添无穷多个元素的素数个数	增添有限个元素的素数个数
Ⅰ	$G=\langle x\rangle\oplus\langle y\rangle$	0	有限个
Ⅱ	$G=\langle G_1,G_2,\cdots,G_k\rangle,G_j=\langle x,y,\pi^{(j)}\rangle$，$\pi^{(j)}\in\mathscr{B}p_j$	有限个	有限个
Ⅲ	$G=\langle x,y,\vartheta\rangle,\vartheta\in\mathscr{D}$	$+\infty$	有限个
Ⅳ	$G=\langle x,y,\lambda\rangle,\lambda\in\mathscr{C}$	0	$+\infty$
Ⅴ	$G=\langle G_0,G_1,\cdots,G_k\rangle,G_0=\langle x,y,\lambda\rangle,G_j$ 同Ⅱ	有限个	$+\infty$
Ⅵ	$G=\langle G_0,G_1\rangle,G_0=\langle x,y,\lambda\rangle,G_1=\langle x,y,\vartheta\rangle$	$+\infty$	$+\infty$

定义 2.6 H 是任意二秩无扭群，$t^H(z)$ 表元素 z 在 H 中的型，$t(H)=\bigcap\limits_{z\in H}t^H(z)$ 称为群 H 的型.

任取 H 的线性无关元 x,y，$t(H)\leqslant t(x)\cap t(y)$. 反之，$\forall z\in H$，设 $rz=mx+ny$，$t(z)=t(rz)\geqslant t(x)\cap t(y)$，$t(H)=\bigcap\limits_{z\in H}t(z)\geqslant t(x)\cap t(y)$. 因此 $t(H)=t(x)\cap t(y)$. 全部二秩无扭群可分为两类，一是有 $x\in H$，使 $t^H(x)=t(H)$，称 H 为有最小型元素的群；二是群中任意元素的型均大于群的型. 现在我们讨论前一类群的结构.

法则 A 设 $r=p_{i_1}^{\beta_{i_1}}p_{i_2}^{\beta_{i_2}}\cdots p_{i_k}^{\beta_{i_k}}$，$p_{i_1},p_{i_2},\cdots,p_{i_k}$ 是第 $i_1,i_2,\cdots,i_k$ 个素数，不一定按序排列，$\beta_{i_1},\beta_{i_2},\cdots,\beta_{i_k}\in\mathbf{N}^*$. 若 $\alpha_{i_1},\alpha_{i_2},\cdots,\alpha_{i_k}$ 是一个固定数组，其中 $\alpha_{i_s}=$ 自然数或 $+\infty$，$s=1,2,\cdots,k$. 不妨假定 $\beta_{i_1}>\alpha_{i_1},\cdots,$

$\beta_{i_h}>\alpha_{i_h},\beta_{i_{h+1}}\leqslant\alpha_{i_{h+1}},\cdots,\beta_{i_k}\leqslant\alpha_{i_k}$. 令

$$l=p_{i_1}^{\alpha_{i_1}}\cdots p_{i_k}^{\alpha_{i_k}}p_{i_{k+1}}^{\beta_{i_{k+1}}}\cdots p_{i_k}^{\beta_{i_k}},r'=p_{i_1}^{\beta_{i_1}-\alpha_{i_1}}\cdots p_{i_h}^{\beta_{i_h}-\alpha_{i_h}},$$

则

$$\left(\frac{m}{r},\frac{n}{r}\right)=\frac{1}{l}\left(\frac{m}{r'},\frac{n}{r'}\right).$$

引理 2.7　H 是有最小型元素的二秩无扭群，$x\in H$，$\chi^H(x)=(\alpha_1,\alpha_2,\cdots,\alpha_s,\cdots)\in t(H)$. y 与 x 线性无关，$\chi^H(y)\geqslant\chi^H(x)$，$\forall u=\left(\frac{m}{r},\frac{n}{r}\right)\in H$，由法则 A 确定的元素 $\left(\frac{m}{r'},\frac{n}{r'}\right)$ 的集合记作 G，则 G 是 H 的子群，且 $\chi^G(x)=(0,0,\cdots)$. 又 $\frac{1}{l}$ 生成一秩无扭群 K，且 $\chi^K(1)=(\alpha_1,\alpha_2,\cdots,\alpha_s,\cdots)$.

证　$\forall z_1=\left(\frac{m_1}{r_1'},\frac{n_1}{r_1'}\right),z_2=\left(\frac{m_2}{r_2'},\frac{n_2}{r_2'}\right)\in G$. r_1',r_2' 的最大公因子

$$Q_0=p_{i_1}^{\varepsilon_{i_1}}\cdots p_{i_h}^{\varepsilon_{i_h}},\quad \frac{r_1'}{Q_0}=p_{j_1}^{\varepsilon_{j_1}}\cdots p_{j_l}^{\varepsilon_{j_l}}=Q_1,\quad \frac{r_2'}{Q_0}=p_{k_1}^{\eta_{k_1}}\cdots p_{k_g}^{\eta_{k_g}}=Q_2,$$

三组素数不交，指数为正. 再令

$$P_0=p_{i_1}^{\alpha_{i_1}}\cdots p_{i_k}^{\alpha_{i_k}},\quad P_1=p_{j_1}^{\alpha_{j_1}}\cdots p_{j_l}^{\alpha_{j_h}},\quad P_2=p_{k_1}^{\alpha_{k_1}}\cdots p_{k_g}^{\alpha_{k_g}}.$$

由法则 A 知 $\frac{1}{P_0P_1}z_1\in H$. 又因 $\chi^H(y)\geqslant\chi^H(x)$，故

$$\left(\frac{1}{P_2},0\right),\left(0,\frac{1}{P_2}\right)\in H.\ (Q_0P_0Q_1P_1,P_2)=1,\ \exists e,f\in\mathbf{Z},$$

使 $fQ_0P_0Q_1P_1+eP_2=1$，这时

$$e\left(\frac{m_1}{Q_0P_0Q_1P_1},\frac{n_1}{Q_0P_0Q_1P_1}\right)+f\left(\frac{m_1}{P_2},0\right)+f\left(0,\frac{n_1}{P_2}\right)=\frac{1}{P_0P_1P_2}z_1\in H.$$

同理 $\frac{1}{P_0P_1P_2}z_2\in H$. 故 $\frac{1}{P_0P_1P_2}(z_1-z_2)\in H$，自法则 A 知 $z_1-z_2\in G$，G 是子群. 又若 $\left(\frac{1}{p_s},0\right)\in G$，依法则 A，$\alpha_s<+\infty$，$\frac{1}{p_s^{\alpha_s}}\left(\frac{1}{p_s},0\right)\in H$，$p_s^{\alpha_s+1}\mid x$，与 $\chi_{p_s}^H(x)=\alpha_s$ 矛盾，故

$$\chi^G(x)=(0,0,\cdots).$$

最后，若 $\alpha_s<+\infty$，因 $\left(\frac{1}{p_s^{\alpha_s}},0\right)\in H$，依法则 A 知 $\frac{1}{p_s^{\alpha_s}}\in K$，而 $\frac{1}{p_s^{\alpha_s+1}}\notin K$. 若

$\alpha_s=+\infty, \forall t, \left(\frac{1}{p_s^t}, 0\right)\in H, \frac{1}{p_s^t}\in K$，故 $\chi_{p_s}^K(1)=\alpha_s$.

定理 2.8 条件同引理，由法则 A 得到群 K 和 G，则 $H=\{kz \mid \forall k\in K, z\in G\}$，记作 KG，且 $KG=K\otimes_2 G$.

§3. 有最小型元素的二秩无扭群的不变量

Курош 给出了可数无扭 Abel 群的不变量[2]，现局限于二秩群 G 简述如下：$\mathscr{K}_p$ 表 p-进数域，G_p^* 表 $\mathscr{B}_p$-模 $\mathscr{B}_pG$，x, y 与 x', y' 分别是 G 的两组线性无关元，

$$\begin{pmatrix}x'\\y'\end{pmatrix}=\mathfrak{B}\begin{pmatrix}x\\y\end{pmatrix},\quad \mathfrak{B}=\begin{pmatrix}a & b\\c & d\end{pmatrix}$$

是有理可逆阵.

(1) $k_p=2, l_p=0, G_p^*=\mathscr{K}_pv_1\oplus\mathscr{K}_pv_2, \begin{pmatrix}x\\y\end{pmatrix}=\mathfrak{M}_p\begin{pmatrix}v_1\\v_2\end{pmatrix}$. 如果

$$G_p^*=\mathscr{K}_pv_1'\oplus\mathscr{K}_pv_2',\quad \begin{pmatrix}v_1\\v_2\end{pmatrix}=\mathfrak{G}_p\begin{pmatrix}v_1'\\v_2'\end{pmatrix},\quad \mathfrak{G}_p=\begin{pmatrix}\gamma_{11} & \gamma_{12}\\\gamma_{21} & \gamma_{22}\end{pmatrix}$$

是 p-进可逆阵，则 $\mathfrak{M}_p'=\mathfrak{B}\mathfrak{M}_P\mathfrak{G}_p$.

(2) $k_p=1, l_p=1, G_p^*=\mathscr{K}_pv_1\oplus\mathscr{B}_pw, \begin{pmatrix}x\\y\end{pmatrix}=\mathfrak{M}_p\begin{pmatrix}v\\w\end{pmatrix}$. 若

$$G_p^*=\mathscr{K}_pv'\oplus\mathscr{B}_pw'\begin{pmatrix}v\\w\end{pmatrix}=\mathfrak{G}_p\begin{pmatrix}v'\\w'\end{pmatrix},\quad \mathfrak{G}_p=\begin{pmatrix}\gamma_1 & 0\\\gamma_2 & \delta\end{pmatrix},$$

γ_i 是 p-进数，δ 是 p-进可逆整数，$\mathfrak{G}_p^{-1}$ 有同样的形式；则

$$\mathfrak{M}_p'=\mathfrak{B}\mathfrak{M}_p\mathfrak{G}_p.$$

(3) $k_p=0, l_p=2, G_p^*=\mathscr{B}_pw_1\oplus\mathscr{B}_pw_2, \begin{pmatrix}x\\y\end{pmatrix}=\mathfrak{M}_p\begin{pmatrix}w_1\\w_2\end{pmatrix}$. 若

$$G_p^*=\mathscr{B}_pw_1'\oplus\mathscr{B}_pw_2',\quad \begin{pmatrix}w_1\\w_2\end{pmatrix}=\mathfrak{G}_p\begin{pmatrix}w_1'\\w_2'\end{pmatrix},\quad \mathfrak{G}_p=\begin{pmatrix}\delta_{11} & \delta_{12}\\\delta_{21} & \delta_{22}\end{pmatrix}$$

与 $\mathfrak{G}_p^{-1}$ 均为 p-进整数阵，则 $\mathfrak{M}_p'=\mathfrak{B}\mathfrak{M}_p\mathfrak{G}_p$.

$p_1, p_2, \cdots$ 是按序排列的全体素数，矩阵序列 $(\mathfrak{M}_{p_1}, \mathfrak{M}_{p_2}, \cdots)$ 与 $(\mathfrak{M}_{p_1}', \mathfrak{M}_{p_2}', \cdots)$ 称为等价，将等价序列类记作 $\mathfrak{M}_G$，则 $k_p, l_p, \mathfrak{M}_G$ 是二秩无扭群的完全不变系.

引理 3.1　设 G 由 x,y 增添元素生成，$\chi^G(x)=(0,0,\cdots)$. 若增添分母为 p 方幂的元素 $\frac{1}{p^n}(mx+y)$，则

$$k_p=0,l_p=2,G_p^*=\mathscr{B}_px\oplus\mathscr{B}_p\frac{1}{p^n}(mx+y),\quad \mathfrak{M}_p=\begin{pmatrix}1&0\\-m&p^n\end{pmatrix}.$$

若增添分母为 p 方幂的元无穷多，则

$$k_p=1,l_p=1,G_p^*=\mathscr{K}_p(\pi x+y)\oplus\mathscr{B}_px,\quad \pi\in\mathscr{B}_p,\quad \mathfrak{M}_p=\begin{pmatrix}0&1\\1&-\pi\end{pmatrix}.$$

证　① 设 $G'=\langle x,y,\frac{1}{p^n}(mx+y)\rangle=\langle x\rangle\oplus\langle\frac{1}{p^n}(mx+y)\rangle\subset G$. 令

$$w_1=x,\quad w_2=\frac{1}{p^n}(mx+y),$$

则 $G_p'^*=\mathscr{B}_pw_1\oplus\mathscr{B}_pw_2$. ②设 G 的子群 G' 由 x,y 及全体分母为 p 方幂的元素生成，则 $G'=\langle x,y,\pi\rangle,\pi\in\mathscr{B}_p$. $G_p'^*$ 中有 0 高元 x，及 $+\infty$ 高元 $\pi x+y$. 这时 $k_p=1,l_p=1$. 令 $v=\pi x+y,w=x,G_p'^*=\mathscr{K}_pv\oplus\mathscr{B}_pw$. ③若 G 与 G' 增添分母为 p 方幂的元素相同，则 $R_pG=R_pG'$，进而 $\mathscr{B}_pG=\mathscr{B}_pG'$.

引理 3.2　设 G 由 x,y 增添元素生成，$\chi^G(x)=(0,0,\cdots)$，当 G 看作由另一组线性无关元 $x',y',\chi^G(x')=(0,0,\cdots)$ 增添元素生成时，若 x,y 增添分母为 p 方幂的元素有限个(无穷多)，则后者亦然. 若前者增添 $\frac{1}{p_s^{n_s}}(m_sx+y),n_s>0$，后者增添 $\frac{1}{q_t^{k_t}}(m_t'x'+y'),k_s>0$.

则当 $\{p_s\}$ 取有限个(无穷多)素数时，$\{q_t\}$ 亦取有限个(无穷多)素数，在无限情况下，序列 $\{p_s^{n_s}\}$ 与 $\{q_t^{k_t}\}$ 几乎处处相等.

证　将 $\{p_s^{n_s}\}$ 与 $\{p_t^{k_t}\}$ 统一脚标，得 $\{p_j^{n_j}\},\{p_j^{k_j}\},0\leqslant n_j,k_j<+\infty$，有无穷多个 $n_j>0$. $\mathfrak{M}_{p_j}=\begin{pmatrix}1&0\\-m_j&p_j^{n_j}\end{pmatrix}$，$\mathfrak{M}'_{p_j}=\begin{pmatrix}1&0\\-m_j'&p_j^{k_j}\end{pmatrix}$. 由 Курош (3)，有 p_j-进整数可逆阵 $\mathfrak{G}_{p_j}$，使 $\mathfrak{M}'_{p_j}=\mathfrak{B}\mathfrak{M}_{p_j}\mathfrak{G}_{p_j}$. 取行列式有

$$p_j^{k_j}=(ad-bc)p_j^{n_j}\,|\mathfrak{G}_{p_j}|^r.$$

当既约分数 $ad-bc$ 的分子、分母均不含因子 p_j 时，$k_j=n_j$.

由引理易见，有零型元素的两个群 G 与 G' 若同构，必属表 2.1 中的同一类型.

命题 3.3 二秩无扭群 $G=\langle x,y,\pi\rangle$，$G'=\langle x',y',\sigma\rangle$，$\pi,\sigma\in\mathscr{B}_p$. 于是 $G\simeq G'$，当且仅当 σ 与 $\bar{\pi}$ 整分式线性相关，其中 $\bar{\pi}=\dfrac{\pi-\pi_i}{p^i}$.

证 必要性 不妨认为 $G'=G$，由命题 2.3 知，$\exists a,b,c,d\in\mathbf{N}^*$，使 $x'=ax+bz_i$，$y'=cx+dz_i$. 取 $\mathfrak{B}=\begin{pmatrix}a & b\\ c & d\end{pmatrix}$，$G=\langle x,z_i,\bar{\pi}\rangle=\langle x',y',\sigma\rangle$. $\forall q\neq p$，

$$\mathfrak{M}'_q=\mathfrak{M}_q=\begin{pmatrix}1 & 0\\ 0 & 1\end{pmatrix},$$

有 q-进整数可逆阵 $\mathfrak{G}_q$，使 $\mathfrak{M}'_q=\mathfrak{B}\mathfrak{M}_q\mathfrak{G}_q$，取行列式有

$$\begin{vmatrix}a & b\\ c & d\end{vmatrix}|\mathfrak{G}_q|=1,\quad q\nmid\begin{vmatrix}a & b\\ c & d\end{vmatrix},\quad \begin{pmatrix}a & b\\ c & d\end{pmatrix}=\pm p^m.$$

又 $\mathfrak{M}'_p=\begin{pmatrix}0 & 1\\ 1 & -\sigma\end{pmatrix}$，$\mathfrak{M}_p=\begin{pmatrix}0 & 1\\ 1 & -\bar{\pi}\end{pmatrix}$，$\mathfrak{M}'_p=\mathfrak{B}\mathfrak{M}_p\mathfrak{G}_p$. 令 $\mathfrak{G}_p^{-1}=\begin{pmatrix}\gamma_1 & 0\\ \gamma_2 & \delta\end{pmatrix}$，代入前式比较第二列有 $a-b\bar{\pi}=\delta$，$c-d\bar{\pi}=-\delta\sigma$. 依定义 1.4 立得.

充分性 取 $G=\langle x,z_i,\bar{\pi}\rangle$，$\mathfrak{B}=\begin{pmatrix}a & b\\ c & d\end{pmatrix}$. 令 $\begin{pmatrix}x'\\ y'\end{pmatrix}\mapsto\mathfrak{B}\begin{pmatrix}x\\ z_i\end{pmatrix}\in G$. $\forall q\neq p$，令 $\mathfrak{G}_q=\mathfrak{B}^{-1}$，则 $\mathfrak{G}_q$ 及其逆阵均为 q-进整数阵，满足 Курош(3). 令

$$\mathfrak{G}_p=(\mathfrak{B}\mathfrak{M}_p)^{-1}\mathfrak{M}'_p=\begin{pmatrix}\dfrac{a-b\bar{\pi}}{ad-bc} & 0\\ \dfrac{-b}{ad-bc} & \dfrac{1}{a-b\bar{\pi}}\end{pmatrix},$$

满足 Курош(2)，$\mathfrak{M}_G=\mathfrak{M}_{G'}$，$G\simeq G'$.

命题 3.4 设二秩无扭群 $G=\langle x,y,\lambda\rangle$，$G'=\langle x',y',\lambda'\rangle$，$\lambda,\lambda'\in\mathscr{C}$. 于是 $G\simeq C'$，$\begin{pmatrix}x\\ y\end{pmatrix}\mapsto\begin{pmatrix}a & b\\ c & d\end{pmatrix}\begin{pmatrix}x'\\ y'\end{pmatrix}$，$a,b,c,d\in\mathbf{Z}$，当且仅当 λ 与 λ' 整分式线性相关，相关系数是 a,b,c,d.

命题 3.5 设二秩无扭群 $G=\langle x,y,\vartheta\rangle$，$G'=\langle x',y',\vartheta'\rangle$，$\vartheta,\vartheta'\in\mathscr{D}$. 于是 $G\simeq G'$，当且仅当 $\bar{\vartheta}$ 与 ϑ' 整分式线性相关，$\bar{\vartheta}=\dfrac{\vartheta-\vartheta_i}{p_1^i p_2^{i-1}\cdots p_i}\in\mathscr{D}$.

对表 2.1 中的Ⅱ，Ⅴ，Ⅵ型群，既可借助于完备拓扑环 $\mathscr{B}_p$，$\mathscr{C}$，$\mathscr{D}$ 讨论其同构问题，也可另造新的完备拓扑环. 下面转入有最小型元素的群.

引理 3.6 H 是有最小型元素的二秩无扭群，x,y 与 x',y' 是 H 的两组线性无关元，$\boldsymbol{\chi}^H(x)=\boldsymbol{\chi}^H(x')=(\alpha_1,\alpha_2,\cdots,\alpha_s,\cdots)\in t(H)$. $\boldsymbol{\chi}^H(y)\geqslant\boldsymbol{\chi}^H(x)$，$\boldsymbol{\chi}^H(y')\geqslant\boldsymbol{\chi}^H(x')$. $H=KG$ 与 $H=KG'$ 分别依法则 A 确定 $G=\bigcup\limits_{s=0}^{\infty}\langle x,z_s\rangle$ 或 $\langle x,z_0\rangle$；$G'=\bigcup\limits_{s=0}^{\infty}\langle x',z'_s\rangle$ 或 $\langle x',z'_0\rangle$. 若 $\begin{pmatrix}x'\\y'\end{pmatrix}=\mathfrak{B}\begin{pmatrix}x\\z_i\end{pmatrix}$，$\mathfrak{B}=\begin{pmatrix}a&b\\c&d\end{pmatrix}$，$a,b,c,d\in K$. 则当 $\alpha_s<+\infty$ 时，$\mathfrak{M}'_{p_s}=\mathfrak{B}\mathfrak{M}_{p_s}\mathfrak{G}_{p_s}$，此处 $\mathfrak{M}_{p_s}$，$\mathfrak{M}'_{p_s}$ 分别是 $G^*_{p_s}$，$G'^*_{p_s}$ 的阵，$\mathfrak{G}_{p_s}$ 满足 Курош 条件.

证 视 G 由 x,z_i 增添元素生成. 若增添分母为 p_s 方幂的元素有限个，

$$G^*_{p_s}=\mathscr{B}_{p_s}w_1\oplus\mathscr{B}_{p_s}w_2,\quad \begin{pmatrix}x\\z_i\end{pmatrix}=\mathfrak{M}_{p_s}\begin{pmatrix}w_1\\w_2\end{pmatrix}.$$

又

$$H^*_{p_s}=\frac{1}{p_s^{\alpha_s}}G^*_{p_s}=\mathscr{B}_{p_s}\frac{1}{p_s^{\alpha_s}}w_1\oplus\mathscr{B}_{p_s}\frac{1}{p_s^{\alpha_s}}w_2,$$

$$\begin{pmatrix}x\\z_i\end{pmatrix}=\mathfrak{M}_{p_s}\begin{pmatrix}\dfrac{1}{p_s^{\alpha_s}}w_1\\\dfrac{1}{p_s^{\alpha_s}}w_2\end{pmatrix},$$

故 $\overline{\mathfrak{M}}=p_s^{\alpha_s}\mathfrak{M}_{p_s}$. 同理，$\overline{\mathfrak{M}}'_{p_s}=p_s^{\alpha_s}\mathfrak{M}'_{p_s}$. 在 $H^*_{p_s}$ 中，$\overline{\mathfrak{M}}'=\mathfrak{B}\overline{\mathfrak{M}}_{p_s}\mathfrak{G}_{p_s}$，$\mathfrak{G}_{p_s}$ 符合 Курош 条件(3)，两端同乘 $p_s^{-\alpha_s}$ 即可. 当增添分母为 p_s 方幂的元素无穷多时，同样得

$$\mathfrak{M}'_{p_s}=\mathfrak{B}\mathfrak{M}_{p_s}\mathfrak{G}_{p_s}.$$

对群 H，也有与 3.2 类似的引理，但只需讨论 $\alpha_s<+\infty$ 时，增添分母为 p_s 方幂的元.

定理 3.7 H 与 H' 是有最小型元素的二秩无扭群，元素 x,y 及 x'，y' 的条件与引理 3.6 同. $H=KG$，$H'=KG'$ 依法则 A 确定. 这时 $H\simeq H'$ 当且仅当

Ⅰ，$G=\langle x, z_0\rangle$，则 $G'=\langle x', z_0'\rangle$，$\begin{pmatrix}x'\\ z_0'\end{pmatrix}\mapsto\mathfrak{V}\begin{pmatrix}x\\ z_0\end{pmatrix}$，$\forall \alpha_s<+\infty$，$\mathfrak{V}$ 与 $\mathfrak{V}^{-1}$ 的元素分母不含因子 p_s.

Ⅱ，$G=\langle x, z_0, \pi\rangle$，则 $G'=\langle x', z_0', \sigma\rangle$，且 σ 与 $\bar{\pi}$ 为 K-分式线性相关，$\bar{\pi}=\dfrac{\pi-\pi_i}{p^i}$.

Ⅲ，$G=\langle x, z_0, \vartheta\rangle$，则 $G'=\langle x', z_0', \vartheta'\rangle$，且 ϑ' 与 $\bar{\vartheta}$ 为 K-分式线性相关，$\bar{\vartheta}=\dfrac{\vartheta-\vartheta_i}{p_1^i p_2^{i-1}\cdots p_i}$.

Ⅳ，$G=\langle x, z_0, \lambda\rangle$，$\begin{pmatrix}x\\ z_0\end{pmatrix}\mapsto\begin{pmatrix}a & b\\ c & d\end{pmatrix}\begin{pmatrix}x'\\ z_0'\end{pmatrix}$，则 $G'=\langle x', z', \lambda'\rangle$，且 λ' 与 λ 为 K-分式线性相关，相关系数是 a, b, c, d. 当 G 为Ⅱ，Ⅴ，Ⅵ类群时有类似的结果.

§ 4. 齐次群与有最小型元素的可分解群

定理 4.1 H 是有最小型元素的群，$H=KG$ 依法则 A 确定，于是 H 为齐次群，当且仅当 G 是零型群(即型为零的齐次群).

证 因 $H=K\otimes G$，故 $\forall u\in H$，$u=k\otimes g$，则 $\chi^H(u)=\chi^K(k)\chi^G(g)$，$t^H(u)=t^K(k).t^G(g)$[2](积表无穷序列对应项的和).

我们按表 2.1 中的分类讨论 G 成零型的条件.

定理 4.2 Ⅰ，$G=\langle x\rangle\oplus\langle y\rangle$是零型群.

Ⅱ，$G=\langle G_1, G_2, \cdots, G_k\rangle$是零型群，当且仅当 $\pi^{(j)}$ 是 p_j-进无理整数，$j=1,2,\cdots,k$.

Ⅲ，$G=\langle x, y, \vartheta\rangle$是零型群，当且仅当 ϑ 是环 $\mathscr{D}$ 的零型元.

Ⅳ，$G=\langle x, y, \lambda\rangle$是零型群，当且仅当 λ 是环 $\mathscr{C}$ 的零型元.

Ⅴ，$G=\langle G_0, G_1, \cdots, G_k\rangle$是零型群，当且仅当 λ 是环 $\mathscr{C}$ 的零型元，$\pi^{(j)}$ 是 p_j-进无理整数，$j=1,2,\cdots,k$.

Ⅵ，$G=\langle G_0, G_1\rangle$是零型群，当且仅当 λ, ϑ 分别是环 $\mathscr{C}$ 与 $\mathscr{D}$ 的零型元.

证 仅以 $G=\langle x, y, \pi\rangle$为例. 若 G 非零型，$\forall z\in G$ 由命题 2.3 知，

$$z=ax+bz_i=\left(a-b\frac{\pi_{i+k}-\pi_i}{p^i}\right)x+bp^k z_{i+k},$$

只可能 $\chi_p^G(z)=+\infty$. 这时, $\forall k_0$, $\exists k$, 使 $p^{k_0}\left|\left(a-b\dfrac{\pi_{i+k}-\pi_i}{p^i}\right)\right.$. 后者成 p-进整数 0 的基本序列, $a-b\dfrac{\pi-\pi_i}{p^i}=0$, π 是 p-进有理整数. 反之, 若 $\pi=\dfrac{a}{b}$ 是有理数, $(b,p)=1$. 令 $z=ax+by\in G$, $\forall k\in\mathbf{N}^*$, $z-bp^tz_k=ax+by-b(\pi_kx+y)=(a-b\pi_k)x$. 因 $a-b\pi=0$, 故 $p^k\mid a-b\pi_k$, 即 $p^k\mid z$, $\chi_p^G(z)=+\infty$, G 非零型.

命题 4.3　第Ⅰ类零型群在同构的意义下只有一个, Ⅱ～Ⅳ类的零型群在同构的意义下各有 $\aleph_1$ 个. 型为 $\{(\alpha_1,\alpha_2,\cdots,\alpha_s,\cdots)\}$, 至少有一个 $\alpha_s<+\infty$ 的齐次群在同构的意义下有 $\aleph_1$ 个, 而 $\{(+\infty,+\infty,\cdots)\}$ 型齐次群只有一个, 即 $Q\oplus Q$.

最后讨论有最小型元素的二秩无扭群的直和分解问题. 先看有零型元素的群. $\langle z\rangle_*$ 记元素 z 在 G 中生成的纯子群.

定理 4.4　$G=\langle x,y,\pi\rangle$ 可分解, 当且仅当 π 是 p-进有理整数. $G=\langle x,y,\lambda\rangle$ 可分解, 当且仅当 λ 是环 $\mathscr{C}$ 的可分元. $G=\langle x,y,\vartheta\rangle$ 可分解, 当且仅当 ϑ 是环 $\mathscr{D}$ 的可分元.

证　以 $G=\langle x,y,\vartheta\rangle$ 为例. 设 $G=A\oplus B$, $\exists u\in A$, $v\in B$, 使

$$\boldsymbol{\chi}^G(u)\cap\boldsymbol{\chi}^G(v)=(0,0,\cdots).\quad G=\langle u\rangle_*\oplus\langle v\rangle_*.$$

由引理 3.2 知, u,v 的分母可以取 $p_1,p_2,\cdots,p_s,\cdots$ 的任意次方幂, 对其他所有素数的积仅取有限值. 将 u,v 分别除以适当的整数得 u',v', 使 $\boldsymbol{\chi}^G(u')$ 与 $\boldsymbol{\chi}^G(v')$ 中只含 0 与 $+\infty$, 令 $x'=u'-v'$, $y'=v'$, $\boldsymbol{\chi}^G(x')=(0,0,\cdots)$, $G=\langle x',y',\vartheta'\rangle$. 将 $\{p_s\}$ 分为互不相交的两组 $\{p_{i_s}\}$ 与 $\{q_{j_t}\}$, 前者除尽 u', 后者除尽 v'. 因 $u'=x'+y'$, $v'=y'$, 由 (1.4) 知

$$\vartheta'_{i_s+k-1}\equiv1(\bmod p_{i_s}^k)_{k\in\mathbf{N}^*},\quad \theta'_{i_t+k-1}\equiv0(\bmod p_{j_t}^k)_{k\in\mathbf{N}^*}.$$

设 $\begin{pmatrix}x'\\y'\end{pmatrix}=\begin{pmatrix}a&b\\c&d\end{pmatrix}\begin{pmatrix}x\\y\end{pmatrix}$, 依命题 3.5, $\begin{vmatrix}a&b\\c&d\end{vmatrix}=\pm p_1^{k_1}p_2^{k_2}\cdots p_m^{k_m}$,

$$\vartheta'=-\frac{c-d\bar{\vartheta}}{a-b\bar{\vartheta}},\quad \bar{\vartheta}=\frac{\vartheta-\vartheta_i}{p_1^ip_2^{i-1}\cdots p_i}.$$

故 $(b+d)\bar{\vartheta}_{i_s+k-1}\equiv a+c(\bmod p_{i_s}^k)_{k\in\mathbf{N}^*}$, $d\bar{\vartheta}_{i_t+k-1}\equiv e(\bmod p_{j_t}^k)_{k\in\mathbf{N}^*}$. 由定义 1.12, $\bar{\vartheta}$ 是环 $\mathscr{D}$ 的可分元. 反之, 若 $\bar{\vartheta}$ 是可分元, 令 $\begin{pmatrix}x'\\y'\end{pmatrix}=\begin{pmatrix}a&b\\c&d\end{pmatrix}\begin{pmatrix}x\\z_i\end{pmatrix}$, $\vartheta'=-\dfrac{c-d\bar{\vartheta}}{a-b\bar{\vartheta}}$, 则 $G=\langle x',y',\vartheta'\rangle$. $\vartheta'_{i_s+k-1}-1\equiv$

$\left(-\frac{c-d\bar{\vartheta}}{a-b\bar{\vartheta}}-1\right)_{i_s+k-1}\equiv[(a+c)-(b+d)\bar{\vartheta}_{i_s+k-1}]\left(-\frac{1}{a-b\bar{\vartheta}}\right)_{i_s+k-1}\equiv 0$ $(\bmod\ p_{i_s}^k)_{k\in\mathbf{N}^*}$，同理 $\vartheta'_{i_t+k-1}\equiv 0(\bmod\ p_{j_t}^k)_{k\in\mathbf{N}^*}$. G 由 x',y' 增添形如 $\frac{1}{p_{i_s}^k}(x'+y')$，$\frac{1}{p_{j_t}^k}y'$ 的元素生成，故 $G=\langle x'+y'\rangle_*\oplus\langle y'\rangle_*$.

定理 4.5 设 H 是有最小型元素的二秩无扭群，$H=KG$ 依法则 A 确定. 若 G 可分解，则 H 可分解. 反之，若 $H=\bar{A}\oplus\bar{B}$，$\forall x'\in\bar{A}$，$y\in\bar{B}$，$x=x'-y$，则群 G 可分解.

证 若 $H=\bar{A}\oplus\bar{B}$，$\forall u_1\in\bar{A}$，$u_1=\frac{1}{l_1}z_1$ 依法则 A 得到，则全体形如 z_1 的元素作成 $\bar{A}$ 的子群 A. 同理有 $\bar{B}$ 的子群 B. $A\cap B=0$. $\forall z\in G$，$z=z_1+z_2$，$z_1\in\bar{A}$，$z_2\in\bar{B}$. 因 $H=\langle x'\rangle_*\oplus\langle y'\rangle_*$，可设

$$z_1=\frac{m}{r}(x+y),z_2=\frac{n}{r}y,z=\left(\frac{m}{r},\frac{m+n}{r}\right).$$

若 $r=p_{i_1}^{\beta_{i_1}-\alpha_{i_1}}p_{i_2}^{\beta_{i_2}-\alpha_{i_2}}\cdots p_{i_k}^{\beta_{i_k}-\alpha_{i_k}}$，令 $l=p_{i_1}^{\alpha_{i_1}}p_{i_2}^{\alpha_{i_2}}\cdots p_{i_k}^{\alpha_{i_k}}$，依法则 A，$\frac{1}{l}z\in H$. 设 $\frac{1}{l}z=u_1+u_2$，$lu_1=z_1$，$lu_2=z_2$. 故 $\frac{1}{l}z_1\in\bar{A}$，$\frac{1}{l}z_2\in\bar{B}$. 但 $\frac{1}{l}z_1=\frac{1}{l}\left(\frac{m}{r},\frac{n}{r}\right)$，依法则 A 知 $z_1\in A$，同理 $z_2\in B$，$G=A\oplus B$.

推论 4.6 设 H 是齐次二秩无扭群，$H=KG$，则 H 可分解，当且仅当 G 是自由群.

注 文献[3]等几篇文章用另外的方法讨论了二秩无扭群的不变量问题.

本文是在刘绍学教授的指导和罗里波老师的帮助下完成的，特此致谢.

参考文献

［1］ Apostol T M. Introduction to Analytic Number Theory. New York, Springer, 1976.

［2］ Fuch L. Abelian Groups. Publishing House of the Hungarian Academy of Sciences, Budapest, 1958.

［3］ Mutzbauer O. Klassifizierung torsionsfreier abelscher Gruppen des Ranges 2. Rend, Semin, Math Univ. Padova, 1976.

Arch. Math. ,1989,53:250-251.

野遗传 Artin 代数 AR-箭图的任意分支中的模由其合成因子唯一决定①

The Modules in Any Component of the AR-Quiver of a Wild Hereditary Artin Algebra are Uniquely Determined by Their Composition Factors

Let Λ be a connected basic hereditary artin algebra, then the center of Λ is a field k, and Λ is a finite dimensional k-algebra. We will consider the category of finite by generated left modules Λ-mod. Let P_1, $P_2, \cdots, P_n$; $Q_1, Q_2, \cdots, Q_n$ be the complete sets of indecomposable projective; injective modules respectively. Assume $\boldsymbol{F}_i = \mathrm{End}_\Lambda(P_i) = \mathrm{End}_\Lambda(Q_i)$, $\boldsymbol{f}_i = \dim_k \boldsymbol{F}_i$, and $\boldsymbol{F} = \begin{pmatrix} f_1 & & & \\ & f_2 & & \\ & & \ddots & \\ & & & f_n \end{pmatrix}$ to be a diagonal matrix. Let $\boldsymbol{C} = (c_{ij})_{n\times n}$ be the Cartan matrix of Λ, then $c_{ij} = \dim_{\boldsymbol{F}_i} \mathrm{Hom}_\Lambda(P_i, P_j)$. Let $\boldsymbol{B} = (b_{ij})_{n\times n}$, where $b_{ij} = \dim_{\boldsymbol{F}_i}(Q_i, Q_j)$. Because $\mathrm{Hom}_\Lambda(P_i, P_j) \cong \mathrm{Hom}_\Lambda(Q_i, Q_j)$, we have $\boldsymbol{FC} = \boldsymbol{BF}$. Let X, Y be any two Λ-modules and x, y be the dimensional types of X, Y. We can define a bilinear form in $K_0(\Lambda)$ as follows: $\langle x, y\rangle = \dim_k \mathrm{Hom}_\Lambda(X, Y) - \dim_k \mathrm{Ext}'_\Lambda(X, Y)$. Let $\phi = -\mathbf{C}^{-\mathrm{T}}\mathbf{B}$ be the Coxeter transformation, then $\langle x\phi, y\phi\rangle = \langle x, y\rangle$. Suppose $\Gamma = (\Gamma_0, \Gamma_1, d)$ to be the Gabriel quiver of Λ, where $\Gamma_0 = \{1, 2, \cdots, n\}$, $\Gamma_1 = \{i \circ \xleftarrow{(d_{ij}, d_{ji})} \circ j\}$. We know that Λ is of finite type, if and only if the

① Received: 1988-05-02.

underline graph $\bar{\Gamma}$ of Γ is Dynkin; Λ is of tame type, if and only if $\bar{\Gamma}$ is Euclidean [3, Thm. A. 1. 8]. We have also

$$\boldsymbol{C}^{-\mathrm{T}}+\boldsymbol{B}^{-1}=\begin{pmatrix} 2 & -d_{12} & \cdots & -d_{1n} \\ -d_{21} & 2 & \cdots & -d_{2n} \\ \vdots & \vdots & & \vdots \\ -d_{n1} & -d_{n2} & \cdots & 2 \end{pmatrix}$$

We assume that Λ is of wild type from now on. Because both the preprojective and preinjective components are directing, therefore the indecomposable modules in these two components are determined by their dimensional types.

Let $\mathscr{C}$ be an arbitrary regular component of the AR-guiver of Λ. Let M_0 be a quasisimple module in $\mathscr{C}$, such that the length $|M_0|$ is minimal. By [5], $\mathscr{C}$ is as fig. 1:

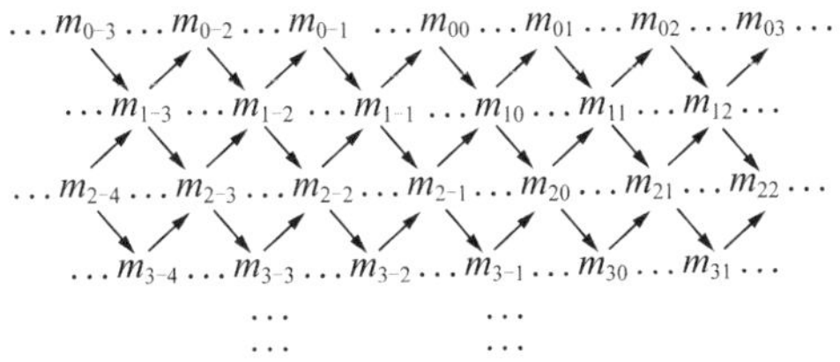

Fig. 1

where $m_{ij}=\dim M_{ij}$, $i=0,1,\cdots,j\in\mathbf{Z}$. The module M_{ij} is located at the i-th row and j-th column, $M_{00}=M_0$. It's not difficult, that

$$m_{ij}=m_{0j}+m_{0j+t}+\cdots+m_{0j+i},\quad \text{and}$$

$$m_{ij}\phi^{-1}=m_{jj+1}.$$

Assume $m_{ij}=m_{kl}$. If $i=k$, then $m_{ij}=m_{il}$. Let $j<l$, and

$$x=m_{ij}+m_{ij+1}+\cdots+m_{il-1},$$

then $x>0$ and $x\phi^{-1}=m_{ij+1}+\cdots+m_{il-1}+m_{il}=x$, or $x\phi=x$. In this case, $(x_1, x_2, \cdots, x_n)\begin{pmatrix} 2 & -d_{12} & \cdots & -d_{1n} \\ -d_{21} & 2 & \cdots & -d_{2n} \\ \vdots & \vdots & & \vdots \\ -d_{n1} & -d_{n2} & \cdots & 2 \end{pmatrix}=0$. By [2], $\bar{\Gamma}$ is a

Euclidean graph, we have a contradiction that Γ is of wild type. If $i \neq k$, we can assume $i < k$. Assume m_i to be of minimal length in the i-th row of $\mathscr{C}$, and $m_i = m_{ij}\phi^r$, we would have $|m_{ij}\phi^r| = |m_{kl}\phi^r| > |m_{il}\phi^r| \geqslant |m_i| = |m_{ij}\phi^r|$. Which is impossible.

Theorem The modules in any component of the AR-guiver of a wild hereditary artin algebra are uniquely determined by their composition factors.

Corollary Let d be a natural number, $\mathscr{C}$ be a component of the AR-guiver of a wild hereditary artin algebra. Then the number of isomorphism classes of indecomposable modules in $\mathscr{C}$ of length d is always finite.

I thank Prof I. Reiten for her kind help.

References

[1] Auslander M, Reiten I. Representation theory of artin algebras Ⅲ. Comm. Algebra, 1975, 3: 239-294.

[2] Berman S, Moody B. Cartan matrices with null roots and finite Cartan matrices. Indiana Univ. Math. J., 1972, 21(12): 1 091-1 099.

[3] Dlab V. An introduction to diagrammatical method in representation theory. Vorlesungen Fachbereich Math. Univ. Essen, 1981, 7.

[4] Platzeck M I, Auslander M. Representation theory of hereditary algebras. Lecture Notes Pure Appl. Math., 1978, 37: 389-424.

[5] Ringel C M. Finite dimensional hereditary algebras of wild representation type. Math. Z., 1978, 161: 235-255.

Communications in Algebra, 1989, 17(10): 2 347-2 362.

Coxeter 变换的特征值及其 AR-箭图正则分支的结构[①]

Eigenvalues of Coxeter Transformations and the Structure of Regular Components of an Auslander-Reiten Quiver

Let $c = \sqrt[3]{\frac{1}{2}+\sqrt{\frac{23}{108}}} + \sqrt[3]{\frac{1}{2}-\sqrt{\frac{23}{108}}}$. We show that any non-periodic regular component of the Auslander-Reiten quiver of an artin algebra, whose growth number is less than c, is of the form $\mathbf{Z}A_\infty$, $\mathbf{Z}A_\infty^\infty$, $\mathbf{Z}B_\infty$, $\mathbf{Z}C_\infty$ or $\mathbf{Z}D_\infty$. In this way, we give a partial solution to a problem presented by Prof. C. M. Ringel in his 1985 Durham lectures. The proof will be based on the calculation of the spectral radius of the Coxeter transformations of some labelled trees.

§ 1. The Coxeter transformation of a bipartite and symmetrizable valued quiver

1.1 Let $Q=(Q_0, Q_1, d, d')$ be a finite connected valued quiver, thus (Q_0, Q_1) is a connected finite quiver without multiple arrows, and $d, d': Q_1 \to \mathbf{N}_1$ are functions. We write $d_{xy}=d(\alpha), d'_{xy}=d'(\alpha)$ for $\alpha: x \circ\!\!\longrightarrow\!\!\circ\, y$. In case there is no arrow with starting point x and endpoint y, we write $d_{xy}=0=d'_{xy}$. Usually we assume $Q_0=\{1,2,\cdots,n\}$. We recall that Q is said to be symmetrizable, provided there are positive integers f_x, $x \in Q_0$, with $f_x \cdot d'_{xy}=d_{xy} \cdot f_y$, for all $x \circ\!\!\longrightarrow\!\!\circ\, y$. In this paper, we always assume, that Q is symmetrizable.

① Received: 1988-03; Revised: 1989-05.

Suppose there are no oriented cycles in Q, then the Coxeter transformation for Q is defined by $C=-P^{-1}I$; here, the ith-row of the matrix $\boldsymbol{P}$ is $\underline{\dim}\ P(i)$, where $(\underline{\dim}\ P(i))_i=1$, $(\underline{\dim}\ P(i))_j=\sum_s d_{il_1^{(s)}}d_{l_1^{(s)}l_2^{(s)}}\cdots d_{l m_s^{(s)}j}$, with $\{i\to l_1^{(s)}\to l_2^{(s)}\to\cdots\to l_{m_s}^{(s)}\to j\}_s$ the set of all paths from i to j, (if there is no path from i to j, then $(\underline{\dim}\ P(i))_j=0$) the ith-row of the matrix $\boldsymbol{I}$ is $\underline{\dim}\ I(i)$, where $(\underline{\dim}\ I(i))_i=1$, $(\underline{\dim}\ I(i))_j=\sum_s d'_{il_1^{(s)}}d'_{l_1^{(s)}l_2^{(s)}}\cdots d'_{l_{m_s}^{(s)}j}$, with $\{i\leftarrow l_1^{(s)}\leftarrow l_2^{(s)}\leftarrow\cdots\leftarrow l_{m_s}^{(s)}\leftarrow j\}_s$ the set of all paths from j to i (if there is no path from j to i, then $(\underline{\dim}\ I(i))_j=0$).

We call a connected quiver Q to be bipartite, provided $|Q_0|>1$, and any vertex of Q is not at the same time starting point of an arrow and endpoint of an arrow. If Q is bipartite, then there is no oriented cycle in Q. Let Q be a bipartite quiver, then Q_0 is the disjoint union of the set $\{a_1,a_2,\cdots,a_{n_1}\}$ of sources, and the set $\{b_1,b_2,\cdots,b_{n_2}\}$ of sinks, in particular $n_1+n_2=n$. Note that there is no arrow connecting a_i,a_j, for $1\leqslant i,j\leqslant n_1$, or b_i,b_j, for $1\leqslant i,j\leqslant n_2$.

1.2 We assume that Q is bipartite. Let $D=(d_{a_ib_j})_{n_1\times n_2}$, $\boldsymbol{E}^{\mathrm{T}}=(d'_{a_ib_j})_{n_1\times n_2}$, $F_a=\mathrm{diag}(f_{a_1},f_{a_2},\cdots,f_{a_{n_1}})$, $F_b=(f_{b_1},f_{b_2},\cdots,f_{b_{n_2}})$. We have $EF_a=(DF_b)^{\mathrm{T}}$, therefore $F_b^{-\frac{1}{2}}EF_a^{\frac{1}{2}}=(F_a^{-\frac{1}{2}}DF_b^{\frac{1}{2}})^{\mathrm{T}}$. If x is an eigenvector for DE with eigenvalue φ, then $xF_a^{\frac{1}{2}}$ is an eigenvector for the symmetric matrix $F_a^{-\frac{1}{2}}DEF_a^{\frac{1}{2}}$, with eigenvalue φ, in particular, φ is real. We can choose pairwise orthogonal eigenvectors $x_iF_a^{\frac{1}{2}}$ for $F_a^{-\frac{1}{2}}DEF_a^{\frac{1}{2}}$, such that $\{x_iF_a^{\frac{1}{2}}\}_{i=1,2,\cdots,n_1}$ is a basis of $\mathbf{R}^{n_1}$. We may call $\{X_i\}_{i=1,2,\cdots,n_1}$ to be an F_a-orthogonal basis of $\mathbf{R}^{n_1}$ of eigenvectors for DE. It follows that $x_iF_ax_j^{\mathrm{T}}=0$ for $i\neq j$.

Lemma rank $D=$ rank $E=$ rank $DE=$ rank ED.

Proof If $xD=0$, then $xDE=0$. Conversely, if $xDE=0$, then $xF_a^{\frac{1}{2}}(F_a^{-\frac{1}{2}}DF_b^{\frac{1}{2}})(F_b^{-\frac{1}{2}}EF_a^{\frac{1}{2}})(xF_a^{\frac{1}{2}})^{\mathrm{T}}=0$, so that $(xDF_b^{\frac{1}{2}})(xDF_b^{\frac{1}{2}})^{\mathrm{T}}=0$, thus $xDF_b^{\frac{1}{2}}=0$, therefore $xD=0$. On the other hand, rank $D=$

rank $F_a^{-\frac{1}{2}} D F_b^{\frac{1}{2}} =$ rank $F_b^{-\frac{1}{2}} E F_a^{\frac{1}{2}} =$ rank E. The proof is finished.

Let $x_1, x_2, \cdots, x_n$ be an Fa-orthogonal basis of eigenvectors for DE, with φ_i the eigenvalue for x_i. We assume that $\varphi_i \neq 0, 4$ for $1 \leqslant i \leqslant k$, $\varphi_i = 4$ for $k+1 \leqslant i \leqslant l$, and $\varphi_i = 0$ for $l+1 \leqslant i \leqslant n_1$. It is easily seen, that we obtain a basis of eigenvectors for ED by taking $x'_i = x_i D$ (with eigenvalue φ_i) for $1 \leqslant i \leqslant l$; and some x'_i with $x'_{iE} = 0, l+1 \leqslant i \leqslant n_2$ (its eigenvalue being zero). In this case, we compute the Coxeter transformation for Q as follows:

$$\boldsymbol{P} = \begin{pmatrix} \boldsymbol{I}_{n_1} & \boldsymbol{D} \\ \boldsymbol{0} & \boldsymbol{I}_{n_2} \end{pmatrix}, \quad \boldsymbol{I} = \begin{pmatrix} \boldsymbol{I}_{n_1} & \boldsymbol{0} \\ \boldsymbol{E} & \boldsymbol{I}_{n_2} \end{pmatrix}, \quad \boldsymbol{C} = \begin{pmatrix} -\boldsymbol{I}_{n_1} + \boldsymbol{DE} & \boldsymbol{D} \\ -\boldsymbol{E} & \boldsymbol{I}_{n_2} \end{pmatrix}_{n \times n}$$

where I_{n_1}, I_{n_2} are $n_1 \times n_1$ and $n_2 \times n_2$ unit matrices, respectively. The eigenvalues λ_i^j of C and a corresponding basis of $\mathbf{C}^n$ are as follows:

when $1 \leqslant i \leqslant k$,

$$\lambda_i^1 = \frac{1}{2}\varphi_i - 1 + \frac{1}{2}\sqrt{\varphi_i(\varphi_i - 4)}, \quad Y_i^1 = \left(x_i, \frac{1}{\lambda_i^1 + 1} x_i D\right),$$

$$\lambda_i^2 = \frac{1}{2}\varphi_i - 1 - \frac{1}{2}\sqrt{\varphi_i(\varphi_i - 4)}, \quad Y_i^2 = \left(x_i, \frac{1}{\lambda_i^2 + 1} x_i D\right),$$

$$\text{where} \quad Y_i^j C = \lambda_i^j Y_i^j, \quad \text{for} \quad j = 1, 2;$$

when $k+1 \leqslant i \leqslant l$,

$$\lambda_i^1 = 1, \quad Y_i^1 = \left(x_i, \frac{1}{2} x_i D\right),$$

$$\lambda_i^2 = 1, \quad Y_i^2 = \frac{1}{4}\left(x_i, -\frac{1}{2} x_i D\right),$$

$$\text{where} \quad Y_i^1 C = Y_i^1 \quad \text{and} \quad Y_i^2 C = Y_i^2 + Y_i^1;$$

when

$$l+1 \leqslant i \leqslant n_1, \lambda_i^1 = -1, Y_i^1 = (x_i, 0), \quad \text{where } x_i D = 0 \text{ and}$$

$$l+1 \leqslant i \leqslant n_2, \lambda_i^2 = -1, Y_i^2 = (0, x'_i), \quad \text{where } x'_i E = 0.$$

Note that the vectors Y_i^1 for $1 \leqslant i \leqslant n_1$, and Y_i^2, for $1 \leqslant i \leqslant k, l+1 \leqslant i \leqslant n_2$, are eigenvectors for C, and $Y_i^2 C = Y_i^2 + Y_i^1$, for $k+1 \leqslant i \leqslant l$, thus Y_i^1 and Y_i^2 give a 2×2 Jordan block.

This result is essentially due to [SS]. (In their paper, the authors assume that the underlying graph of Q is a tree without valuation.)

1.3 Since DE is non-negative, we may use some results about this kind of matrices, see [**S**], Theorem 1.1.

Definition We call a matrix (or vector) strictly positive, provided every entry (or component) is positive.

Lemma DE is a primitive matrix.

Proof Let $DE=X$, denote the (a_i, a_j) coefficient of X^m by $x_{a_ia_j}^{(m)}$. We have

$$x_{a_ia_j}^{(m)}=\sum_{l_1,l_2,\cdots,l_{2m-1}} d_{a_ib_{l_1}}\cdot d'_{a_{l_2}b_{l_1}}\cdot d_{a_{l_2}b_{l_3}}\cdot d'_{a_{l_1}b_{l_3}}\cdot\cdots\cdot d'_{a_jb_{l_{2m-1}}}.$$

Since Q is connected, there exists a positive integer m, such that $x_{a_ia_j}^{(m)}>0$. On the other hand, if $x_{a_ia_j}^{(m)}>0$, then

$$x_{a_ia_j}^{(m+1)}=\sum_{l_1,l_2} x_{a_ia_{l_1}}^{(m)} d_{a_{l_1}b_{l_2}}\cdot d'_{a_jb_{l_2}}\geqslant x_{a_ia_j}^{(m)} d_{a_jb_h}\cdot d'_{a_jb_h}>0,$$

where b_h is a neighbour of a_j. Thus for m large enough, for example $m=n_1-1$, X^m is strictly positive.

1.4 Definition Let $Q, \bar{Q}$ be valued quivers. We call Q smaller than $\bar{Q}$, provided $Q_0\subseteq\bar{Q}_0$, and $d_{xy}\leqslant\bar{d}_{xy}$, $d'_{xy}\leqslant\bar{d}'_{xy}$ for all $x, y\in Q_0$.

Proposition Assume that $Q, \bar{Q}$ are bipartite connected quivers, with Q smaller than $\bar{Q}$. Let ρ and $\bar{\rho}$ be the spectral radius for the Coxeter transformations for Q and $\bar{Q}$, respectively. Then $\rho\leqslant\bar{\rho}$. When $\bar{Q}$ is of wild type, then $\rho=\bar{\rho}$ only in case $Q=\bar{Q}$.

Proof When $x \circ\!\!\longrightarrow\!\!\circ\, y\in\bar{Q}_1\backslash Q_1$, let $d_{xy}=0=d'_{xy}$. Thus, we suppose that $Q_0=\bar{Q}_0$. Since $DE\leqslant\overline{DE}$, the maximal eigenvalue of DE is smaller than or equal to that of $\overline{DE}$, and they are equal, only in case $Q=\bar{Q}$. It has been computed, that the maximal eigenvalue of DE for Q of Euclidean type is 4. Therefore, the maximal eigenvalue of DE for Q of Dynkin type is smaller than 4 (but bigger than 0), and that for Q of wild type is bigger than 4.

1.5 Proposition Assume that Q is a bipartite connected valued quiver which is of wild type. Let C be the Coxeter transformation and ρ its spectral radius. Then there is an eigenvector y_+ of C with eigenvalue ρ, and an eigenvector y_- with eigenvalue ρ^{-1}. For any $z\in\mathbf{R}^n$, we have

$$z_+ := \lim_{s \to +\infty} \frac{zC^s}{\rho^s} = \alpha_+ y_+, \quad z_- := \lim_{s \to +\infty} \frac{zC^{-s}}{\rho^s} = \alpha_- y_-,$$

with $\alpha_+, \alpha_- \in \mathbf{R}$. If z is strictly positive, then $\alpha_+ + \alpha_- > 0$.

Proof The φ_1 be the maximal root of the characteristic equation of DE. Then φ_1 is positive, it is a simple root, and the eigenvector x_1 is strictly positive by the Perron-Frobenius theorem. Since Q is of wild type, the spectral radius ρ is equal to λ_1^1, and $\rho > 1$. We denote $y_+ = y_1^1$, and $y_- = y_1^2$, note that $\lambda_1^2 = \rho^{-1}$. For $z \in \mathbf{R}^n$, let

$$z = \alpha_+ y_+ + \alpha_- y_- + \sum_{i=2}^{k} (\alpha_i^1 y_i^1 + \alpha_i^2 y_i^2) + \sum_{i=k+1}^{l} (\alpha_i^1 y_i^1 + \alpha_i^2 y_i^2) + \sum_{i=l+1}^{n_1} \alpha_i^1 y_i^1 + \sum_{i=l+1}^{n_2} \alpha_i^2 y_i^2$$

be its expression in the basis

$$\{y_i^j, i=1,2,\cdots,l, j=1,2; y_i^1, i=l+1, l+2, \cdots, n_1; y_i^2, i=l+1, l+2, \cdots, n_2\} \text{ of } \mathbf{C}^n.$$

Observe that the coefficients α_+ and α_- are real. We have

$$zC^s = \alpha_+ \rho^s y_+ + \alpha_- \rho^{-s} y_- + \sum_{\substack{i=2,\cdots,k \\ j=1,2}} \alpha_i^j (\lambda_i^j)^s y_i^j + \sum_{i=k+1}^{l} [(\alpha_i^1 + s\alpha_i^2) y_i^1 + \alpha_i^2 y_i^2] + \sum_{i=l+1}^{n_1} (-1)^s \alpha_i^1 y_i^1 + \sum_{i=l+1}^{n_2} (-1)^s \alpha_i^2 y_i^2.$$

Since $\rho > |\lambda_i^j|$ for $i=2,3,\cdots,k, j=1,2, \rho>1$ and $\rho > \rho^{-1}$, we have

$$z_+ = \lim_{s \to +\infty} \frac{zC^s}{\rho^s} = \alpha_+ Y_+.$$

Similarly, we consider $\mathbf{C}^{-1} = \begin{pmatrix} -\boldsymbol{I} & -\boldsymbol{D} \\ \boldsymbol{E} & -\boldsymbol{I} + \boldsymbol{ED} \end{pmatrix}$. Its spectral radius is again ρ and the eigenvector of $\mathbf{C}^{-1}$ for ρ is $y_- = y_1^2$. It follows that

$$z_- = \lim_{s \to \infty} \frac{zC^{-s}}{\rho^s} = \alpha_- y_-.$$

Finally, assume z to be strictly positive. Let $z = \alpha_+ y_+ + \alpha_- y_- + z'$, let $x = (x^1, x^2, \cdots, x^{n_1})$ be the first n_1 components of z', and $x_1 = (x_1^1, x_1^2, \cdots, x_1^{n_1})$, then $(\alpha_+ + \alpha_-) x_1 + x$ is strictly positive. Since $x_1 F_a x^{\mathrm{T}} = 0$, thus $\sum_{i=1}^{n_1} x_1^i f_{a_i} x^i = 0$, and since all $x_1^i > 0$, either there exists some i_0, such that $x^{i_0}<0$, or $x^i=0$ for all i. In both cases, we have $\alpha_+ + \alpha_- > 0$.

1.6 Let Λ be a basic artin algebra, recall that the (ordinary) valued quiver $Q=(Q_0, Q_1, d, d')$ of Λ is defined as follows: $Q_0=\{1, 2, \cdots, n\}$, where $\{P(1), P(2), \cdots, P(n)\}$ is a complete set of indecomposable projective modules. There exists an arrow from i to j, if and only if $e_j(R/R^2)e_i \neq 0$, where R is the radical of Λ and $d_{ij}=\dim_{F_j}(e_j(R/R^2)e_i)$, $d'_{ij}=\dim(e_j(R/R^2)e_i)_{F_i}$, where $F_i=End(S(i))$, $S(i)$ the simple Λ-module, which is the top of $P(i)$. If $e_j(R/R^2)e_i=0$, we write $d_{ij}=0=d'_{ij}$.

Lemma Let Λ be a basic, connected, hereditary artin algebra, with valued quiver Q. Let $P(i)$ be an indecomposable projective Λ-module, then the dimension vector of $P(i)$ is $\underline{\dim}\, P(i)$ as defined for Q in 1.1.

Proof We prove it using induction on the Loewy length of $P(i)$. If $RP(i)=0$, then the vertex i is a sink in Q. The dimension vector of $P(i)$ is $(0, \cdots, 1, \cdots, 0)=\underline{\dim}\, P(i)$. Assume that the lemma is true for Loewy length smaller than k. If $R^{k-1}P(i)\neq 0$, but $R^kP(i)=0$. we have

$$RP(i)/R^2P(i) \simeq (R/R^2)e_i = \sum_j e_j(R/R^2)e_i \simeq \bigoplus_j d_{ij}S(j).$$

Since $RP_i \subset P_i$ is a projective module, $RP(i)=\bigoplus_j d_{ij}P(j)$, using the correspondence between projective modules and their tops. Now the Loewy length of $P(j)$ is smaller than k, so the lemma is proved.

Similarly, we prove that the dimension vector of an indecomposable injective Λ-module $I(i)$ is $\underline{\dim}\, I(i)$ as defined in 1.1. Therefore the Coxeter transformation of Λ see [PA, § 2] is that in 1.1.

Corollary Let Λ be an basic connected hereditary artin algebra, with ordinary valued quiver Q. Assume that Q is bipartite and of wild type. If z is the dimension vector of a preprojective Λ-module, then $\alpha_+<0$, $\alpha_+>0$ in the expression of proposition 1.5; if z is the dimension vector of a preinjective Λ-module, then $\alpha_+>0$, $\alpha_-<0$; if z is the dimension vector of a regular module, then $\alpha_+>0$, $\alpha_->0$.

Proof See [PA, Proposition 2.2.].

§ 2. Labelled trees

2.1 Lemma Let $Q, \bar{Q}$ be finite connected valued quivers with

$Q_0=\bar{Q}_0$ and $d_{xy}\cdot d'_{xy}=\bar{d}_{xy}\cdot\bar{d}'_{xy}$ for all x,y. Assume that the underlying graph of Q (and thus also of $\bar{Q}$) is a tree. Then the Coxeter transformations for Q and $\bar{Q}$ are similar.

Proof Take any arrow $\alpha:\circ\xrightarrow{(a,b)}\circ\in Q_1$. If we take a away (but keep the starting and endpoints of α), then two connected suquivers Q' and Q'' of Q remain. Assume that $Q'_0=\{i,i-1,\cdots,1\}$ is an admissable sequence for sinks, with i the starting point of α, and $Q''_0=\{i+1,i+2,\cdots,n\}$ is an admissable sequence for sources, with $i+1$ the endpoint of α. Let Q^0 and Q^1 be the valued quivers which have the same vertices, arrows and valuations as Q, only change $i\ \circ\xrightarrow{(a,b)}\circ\ i+1$ to $i\ \circ\xrightarrow{(1,1)}\circ\ i+1$ and $i\ \circ\xrightarrow{(1,ab)}\circ\ i+1$, fig. 2.1 respectively.

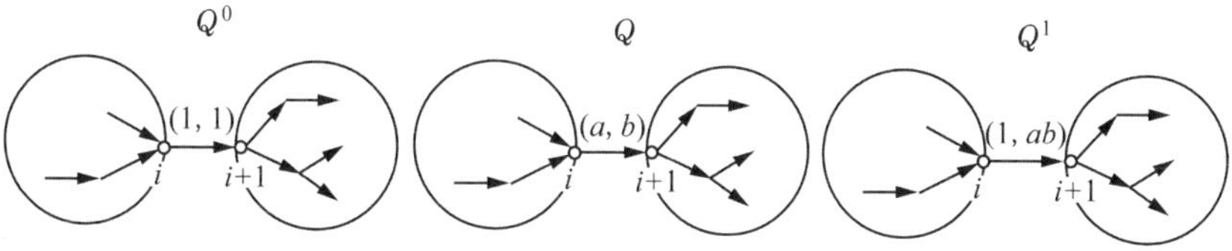

Fig. 2.1

Let $\boldsymbol{P}^0=\begin{pmatrix}\boldsymbol{X}&\boldsymbol{Z}\\ \boldsymbol{0}&\boldsymbol{Y}\end{pmatrix}$, $\boldsymbol{I}^0=\begin{pmatrix}\boldsymbol{A}&\boldsymbol{0}\\ \boldsymbol{C}&\boldsymbol{B}\end{pmatrix}$ for $\boldsymbol{Q}^0$, where the j-th row of $\boldsymbol{P}^0$ (or $\boldsymbol{I}^0$) is $\underline{\dim}\ \boldsymbol{P}^0(j)$ (or $\underline{\dim}\ \boldsymbol{I}^0(j)$), respectively. $\boldsymbol{X}$ and $\boldsymbol{A}$ are $i\times i$ matices, $\boldsymbol{Y}$ and $\boldsymbol{B}$ are $(n-i)\times(n-i)$ matrices. Then for Q, we have $\boldsymbol{P}=\begin{pmatrix}\boldsymbol{X}&a\boldsymbol{Z}\\ \boldsymbol{0}&\boldsymbol{Y}\end{pmatrix}$, $\boldsymbol{I}=\begin{pmatrix}\boldsymbol{A}&\boldsymbol{0}\\ b\boldsymbol{C}&\boldsymbol{B}\end{pmatrix}$. And for $\boldsymbol{Q}^1$, $\boldsymbol{P}^1=\begin{pmatrix}\boldsymbol{X}&\boldsymbol{Z}\\ \boldsymbol{0}&\boldsymbol{Y}\end{pmatrix}$, $\boldsymbol{I}^1=\begin{pmatrix}\boldsymbol{A}&\boldsymbol{0}\\ ab\boldsymbol{C}&\boldsymbol{B}\end{pmatrix}$.

Let $\boldsymbol{S}=\begin{pmatrix}a\boldsymbol{I}_i&\boldsymbol{0}\\ \boldsymbol{0}&\boldsymbol{I}_{n-1}\end{pmatrix}$, where $\boldsymbol{I}_i$ is $i\times i$ unit matrix. Then $\boldsymbol{PS}=\begin{pmatrix}a\boldsymbol{X}&a\boldsymbol{Z}\\ \boldsymbol{0}&\boldsymbol{Y}\end{pmatrix}=\boldsymbol{SP}^1$, and $\boldsymbol{IS}=\begin{pmatrix}a\boldsymbol{A}&\boldsymbol{0}\\ ab\boldsymbol{C}&\boldsymbol{B}\end{pmatrix}=\boldsymbol{SI}^1$. We have $\boldsymbol{S}^{-1}\boldsymbol{C}(\boldsymbol{Q})\boldsymbol{S}=-(\boldsymbol{PS})^{-1}\boldsymbol{IS}=-(\boldsymbol{SP}^1)^{-1}\boldsymbol{SI}^1=-(\boldsymbol{P}^1)^{-1}\boldsymbol{I}^1=\boldsymbol{C}(\boldsymbol{Q}^1)$. By induction, the lemma is proved.

2.2 Definition a A labelled tree $T=(T_0,T_1,t)$ is given by a set T_0 of vertices, and a function $t:T_0\times T_0\to\mathbf{N}_1\cup\{0\}$, satisfying $t(x,y)=t(y,x)$

and $t(x,x)=0$; if $t(x,y)\neq 0$, we connect the vertices x and y by an edge, and we assume that for $T_1=\{\underset{x}{\circ}\!\!-\!\!\underset{y}{\circ}: x,y\in T_0, t(x,y)\neq 0\}$, the graph (T_0, T_1) is a tree.

If $\underset{x}{\circ}\!\!-\!\!\underset{y}{\circ}\in T_1$, we write $t_{xy}=t(x,y)$ and use it as a label for the edge (when $t_{xy}=1$, we may delete the label).

Definition b A labelled tree $\overline{T}$ dominates the labelled tree T if $T_0\subseteq\overline{T}_0$, and $t_{xy}\leqslant\bar{t}_{xy}$ for all $x,y\in T_0$.

2.3 Proposition Let T be an infinite labelled tree, different from fig. 2.2.

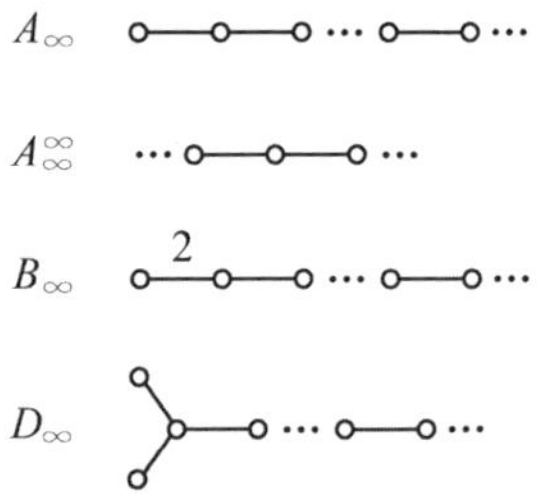

Fig. 2.2

Then T dominates one of the following labelled trees fig. 2.3.

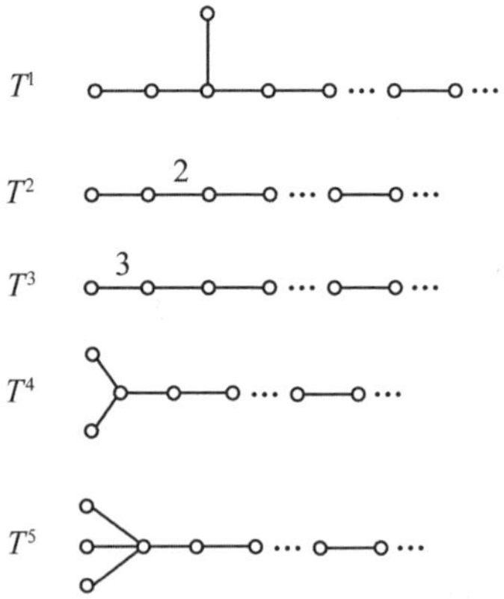

Fig. 2.3

Proof If T has a vertex with four branches, then T dominates T^5. Now we assume that the number of branches is at most 3 for any vertex of T. If the number of vertices with just 3 branches is at least 2, then T dominates T^1. Suppose that there exists just one vertex a with 3 branches. If at most one of the neighbours of a has only one neighbour, then T dominates T^1. Thus we may assume that the underlying graph of T is that of D_∞. Since $T \neq D_\infty$, we have an edge x ∘—∘ y with $t_{xy} \geqslant 2$. If this edge contains a vertex with only one neighbour then T dominates T^4. Otherwise, T dominates T^2. Now we assume that the number of branches is equal to 2 or 1, for any vertex of T. Because $T \neq A_\infty, A_\infty^\infty$, there exists an edge $\underset{x}{\circ}$—$\underset{y}{\circ}$ with $t_{xy} \geqslant 2$. If this edge does not contain a vertex with precisely one neighbour, we see that T dominates T^2. Otherwise T dominates T^3, because $T \neq B_\infty$.

2.4 Definition (1) Let Q be a valued quiver (finite or infinite), whose underlying graph is a tree. The underlying tree of Q is the labelled tree T, such that $T_0 = Q_0$, and $t_{xy} = d_{xy} \cdot d'_{xy}$ for all $\underset{x}{\circ}$—$\underset{y}{\circ} \in Q_1$.

Definition (2) Let $T = (T_0, T_1, t)$ be a labelled tree. If T is finite, let $\rho(T)$ be the spectral radius of the Coxeter transformation of any valued quiver Q with underlying tree T (according to Lemma 2.1, this is independent of Q). If T is infinite, let $\rho(T)$ be the supremum of all $\rho(T')$, where T' is a finite labelled tree dominated by T.

Remark If $\{T^{(m)}, m \in I\}$ is a set of labelled trees dominated by T, such that any finite labelled tree T' which is dominated by T is dominated by some $T^{(m)}$, then $\rho(T) = \sup\limits_{m \in I} \rho(T^{(m)})$.

Proof This is an immediate consequence of proposition 1.4.

2.5 Lemma $\rho(T^1) = \sqrt[3]{\dfrac{1}{2} + \sqrt{\dfrac{23}{108}}} + \sqrt[3]{\dfrac{1}{2} - \sqrt{\dfrac{23}{108}}}$.

Proof Let (fig. 2.4)

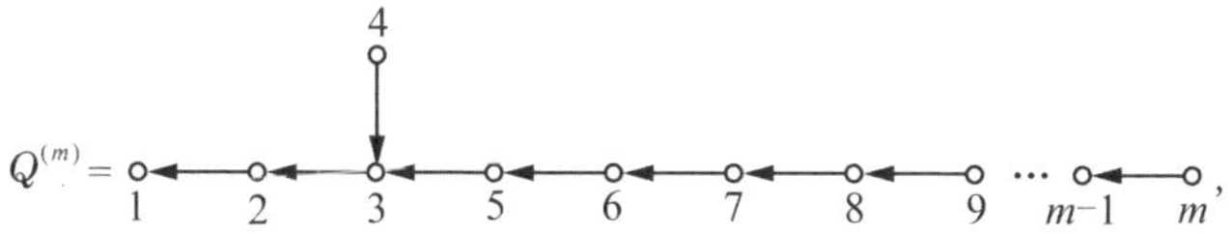

Fig. 2.4

and $T^{(m)}$ its underlying labelled tree.

$$P^{(m)}=\left(\begin{array}{ccccc|cccc}
1 & 0 & 0 & 0 & 0 & 0 & \cdots & 0 & 0 \\
1 & 1 & 0 & 0 & 0 & 0 & \cdots & 0 & 0 \\
1 & 1 & 1 & 0 & 0 & 0 & \cdots & 0 & 0 \\
1 & 1 & 1 & 1 & 0 & 0 & \cdots & 0 & 0 \\
1 & 1 & 1 & 0 & 1 & 0 & \cdots & 0 & 0 \\
\hline
1 & 1 & 1 & 0 & 1 & 1 & \cdots & 0 & 0 \\
\vdots & \vdots & \vdots & \vdots & \vdots & \vdots & & \vdots & \vdots \\
1 & 1 & 1 & 0 & 1 & 1 & \cdots & 1 & 1
\end{array}\right)$$

$$\boldsymbol{C}(\boldsymbol{Q}^{(m)})=-(\boldsymbol{P}^{(m)})^{-1}\boldsymbol{I}^{(m)}=$$

$$-\underbrace{\left(\begin{array}{ccccc|cccc}
1 & 0 & 0 & 0 & 0 & 0 & \cdots & 0 & 0 \\
-1 & 1 & 0 & 0 & 0 & 0 & \cdots & 0 & 0 \\
0 & -1 & 1 & 0 & 0 & 0 & \cdots & 0 & 0 \\
0 & 0 & -1 & 1 & 0 & 0 & \cdots & 0 & 0 \\
0 & 0 & -1 & 0 & 1 & 0 & \cdots & 0 & 0 \\
\hline
0 & 0 & 0 & 0 & -1 & 1 & \cdots & 0 & 0 \\
\vdots & \vdots & \vdots & \vdots & \vdots & \vdots & & \vdots & \vdots \\
0 & 0 & 0 & 0 & 0 & 0 & \cdots & -1 & 1
\end{array}\right)}_{\boldsymbol{A}}\cdot\underbrace{\left(\begin{array}{ccccc|cccc}
1 & 1 & 1 & 1 & 1 & 1 & \cdots & 1 & 1 \\
0 & 1 & 1 & 1 & 1 & 1 & \cdots & 1 & 1 \\
0 & 0 & 1 & 1 & 1 & 1 & \cdots & 1 & 1 \\
0 & 0 & 0 & 1 & 0 & 0 & \cdots & 0 & 0 \\
0 & 0 & 0 & 0 & 1 & 1 & \cdots & 1 & 1 \\
\hline
0 & 0 & 0 & 0 & 0 & 1 & \cdots & 1 & 1 \\
\vdots & \vdots & \vdots & \vdots & \vdots & \vdots & & \vdots & \vdots \\
0 & 0 & 0 & 0 & 0 & 0 & \cdots & 0 & 1
\end{array}\right)}_{\boldsymbol{B}}$$

$$=-\left(\begin{array}{ccccc|cccc}
1 & 1 & 1 & 1 & 1 & 1 & \cdots & 1 & 1 \\
-1 & 0 & 0 & 0 & 0 & 0 & \cdots & 0 & 0 \\
0 & -1 & 0 & 0 & 0 & 0 & \cdots & 0 & 0 \\
0 & 0 & -1 & 0 & -1 & -1 & \cdots & -1 & -1 \\
0 & 0 & -1 & -1 & 1 & 0 & \cdots & 0 & 0 \\
\hline
0 & 0 & 0 & 0 & -1 & 0 & \cdots & 0 & 0 \\
\vdots & \vdots & \vdots & \vdots & \vdots & \vdots & & \vdots & \vdots \\
0 & 0 & 0 & 0 & 0 & 0 & \cdots & -1 & 0
\end{array}\right)$$

$\boldsymbol{C}$　　　　$\boldsymbol{D}$

where $\boldsymbol{A}=-\boldsymbol{C}(\boldsymbol{Q})^{(5)}$, every column of $\boldsymbol{B}$ is equal to the last column of $\boldsymbol{A}$,

$$C=\begin{pmatrix}0 & -1\\ 0 & 0\end{pmatrix},\quad D=\begin{pmatrix}0 & & & & \\ -1 & 0 & & & \\ & -1 & & & \\ & & \ddots & & \\ & & & -1 & 0\end{pmatrix}.$$

We denote $\boldsymbol{E}^{(m)}(\lambda)=|\lambda\boldsymbol{I}-\boldsymbol{C}(\boldsymbol{Q}^{(m)})|$, where $\boldsymbol{I}$ is the unit matrix. Then $\boldsymbol{F}^{(m)}(\lambda)=\lambda\boldsymbol{E}^{(m-1)}(\lambda)-\lambda^3-\lambda^2+1$, for $m\geqslant 6$. And $\boldsymbol{E}^{(5)}(\lambda)=\lambda^5+\lambda^4+\lambda+1$. By induction, we have

$$E^{(m)}(\lambda)=(\lambda^m+\lambda^{m-1})-\left(\sum_{i=m-3}^{3}\lambda^i\right)+(\lambda+1).$$ We have also:

$$E^{(m)}(\lambda)=(\lambda^m+\lambda^{m-1})-\lambda^3\cdot\frac{\lambda^{m-5}-1}{\lambda-1}+(\lambda+1)$$
$$=\frac{\lambda^{m-2}}{\lambda-1}\left[(\lambda^3-\lambda-1)+\frac{\lambda^3+\lambda^2-1}{\lambda^{m-2}}\right].$$

We solve the cubic equation $\lambda^3-\lambda-1=0$, then

$$\rho(T^1)=\lim_{m\to+\infty}\rho(T^{(m)})=\sqrt[3]{\frac{1}{2}+\sqrt{\frac{23}{108}}}+\sqrt[3]{\frac{1}{2}-\sqrt{\frac{23}{108}}}.$$

2.6 Lemma $\rho(T^2)=\dfrac{1+\sqrt{5}}{2}$.

Proof Let fig. 2.5.

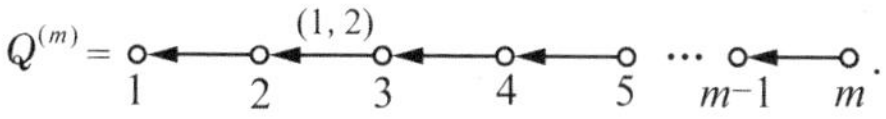

Fig. 2.5

$$\boldsymbol{P}^{(m)}=\left(\begin{array}{ccc:cccc}1 & 0 & 0 & 0 & \cdots & 0 & 0\\ 1 & 1 & 0 & 0 & \cdots & 0 & 0\\ 2 & 2 & 1 & 0 & \cdots & 0 & 0\\ \hdashline 2 & 2 & 1 & 1 & \cdots & 0 & 0\\ \vdots & \vdots & \vdots & \vdots & & \vdots & \vdots\\ 2 & 2 & 1 & 1 & \cdots & 1 & 1\end{array}\right),$$

$$C(Q^{(m)})=-(P^{(m)})^{-1}I^{(m)}=$$

$$-\left(\begin{array}{ccc|cccc} 1 & 0 & 0 & 0 & \cdots & 0 & 0 \\ -1 & 1 & 0 & 0 & \cdots & 0 & 0 \\ 0 & -2 & 1 & 0 & \cdots & 0 & 0 \\ \hline 0 & 0 & -1 & 1 & \cdots & 0 & 0 \\ \vdots & \vdots & \vdots & \vdots & & \vdots & \vdots \\ 0 & 0 & 0 & 0 & \cdots & -1 & 1 \end{array}\right)\left(\begin{array}{ccc|cccc} 1 & 1 & 1 & 1 & \cdots & 1 & 1 \\ 0 & 1 & 1 & 1 & \cdots & 1 & 1 \\ 0 & 0 & 1 & 1 & \cdots & 1 & 1 \\ \hline 0 & 0 & 0 & 1 & \cdots & 1 & 1 \\ \vdots & \vdots & \vdots & \vdots & & \vdots & \vdots \\ 0 & 0 & 0 & 0 & \cdots & 0 & 1 \end{array}\right)=$$

$$-\left(\begin{array}{ccc|cccc} 1 & 1 & 1 & 1 & \cdots & 1 & 1 \\ -1 & 0 & 0 & 0 & \cdots & 0 & 0 \\ 0 & -2 & -1 & -1 & \cdots & -1 & -1 \\ \hline 0 & 0 & -1 & 0 & \cdots & 0 & 0 \\ \vdots & \vdots & \vdots & \vdots & & \vdots & \vdots \\ 0 & 0 & 0 & 0 & \cdots & -1 & 0 \end{array}\right).$$

$E^{(m)}(\lambda)=\lambda E^{(m-1)}(\lambda)-\lambda^2-\lambda+1$, for $m\geqslant 4$, And $E^{(3)}(\lambda)=\lambda^3+1$, By induction, we have $E^{(m)}(\lambda)=\lambda^m-\left(\sum_{i=m-2}^{2}\lambda^i\right)+1$, or $E^{(m)}(\lambda)=\frac{1}{\lambda-1}[\lambda^{m-1}\cdot(\lambda^2-\lambda-1)+(\lambda^2+\lambda-1)]$. We solve the quadratic equation $\lambda^2-\lambda-1=0$, then $\rho(T^2)=\lim_{m\to+\infty}\rho(T^{(m)})=\frac{1+\sqrt{5}}{2}$.

2.7 Lemma $\rho(T^3)=2$.

Proof Let fig. 2.6.

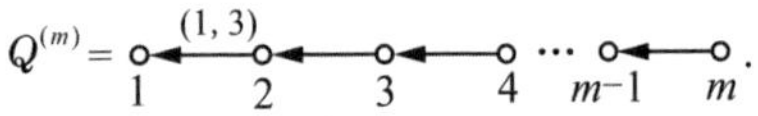

Fig. 2.6

$$P^{(m)}=\left(\begin{array}{cc|cccc} 1 & 0 & 0 & \cdots & 0 & 0 \\ 3 & 1 & 0 & \cdots & 0 & 0 \\ \hline 3 & 1 & 1 & \cdots & 0 & 0 \\ \vdots & \vdots & \vdots & & \vdots & \vdots \\ 3 & 1 & 1 & \cdots & 1 & 1 \end{array}\right),$$

$$C(Q^{(m)})=-\begin{pmatrix}1 & 0 & 0 & \cdots & 0 & 0\\ -3 & 1 & 0 & \cdots & 0 & 0\\ 0 & -1 & 1 & \cdots & 0 & 0\\ \vdots & \vdots & \vdots & & \vdots & \vdots\\ 0 & 0 & 0 & \cdots & -1 & 1\end{pmatrix}\cdot\begin{pmatrix}1 & 1 & 1 & \cdots & 1 & 1\\ 0 & 1 & 1 & \cdots & 1 & 1\\ 0 & 0 & 1 & \cdots & 1 & 1\\ \vdots & \vdots & \vdots & & \vdots & \vdots\\ 0 & 0 & 0 & \cdots & 0 & 1\end{pmatrix}$$

$$=-\begin{pmatrix}1 & 1 & 1 & \cdots & 1 & 1\\ -3 & -2 & -2 & \cdots & -2 & -2\\ 0 & -1 & 0 & \cdots & 0 & 0\\ \vdots & \vdots & \vdots & & \vdots & \vdots\\ 0 & 0 & 0 & \cdots & -1 & 0\end{pmatrix}$$

$E^{(m)}(\lambda)=\lambda E^{(m-1)}(\lambda)-2\lambda+1$, for $m\geqslant 3$. And $E^{(2)}(\lambda)=\lambda^2-\lambda+1$. Then $E^{(m)}=\lambda^m-\left(\sum_{i=m-1}^{1}\lambda^i\right)+1$, or $E(m)(\lambda)=\frac{1}{\lambda-1}[\lambda^m(\lambda-2)+(2\lambda-1)]$. We solve the linear equation $\lambda-2=0$, then $\rho(T^3)=\lim_{m\to+\infty}\rho(T^{(m)})=2$. The proof is finished.

In the same way, for T^4, we have $E^{(m)}(\lambda)=(\lambda+1)\cdot\left[\lambda^{m-1}-\left(\sum_{i=m-2}^{1}\lambda^i\right)+1\right]$, and $\rho(T^4)=\rho(T^3)=2$. For T^5, we have $E^{(m)}(\lambda)=(\lambda+1)^2\left[\lambda^{m-2}-\left(\sum_{i=m-3}^{1}\lambda^i\right)+1\right]$, $\rho(T^5)=\rho(T^3)=2$, in fig. 2. 7.

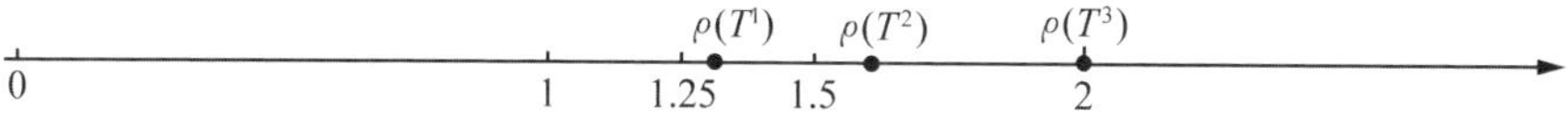

Fig. 2. 7

§ 3. The regular components with growth number smaller than $\sqrt[3]{\frac{1}{2}+\sqrt{\frac{23}{108}}}+\sqrt[3]{\frac{1}{2}-\sqrt{\frac{23}{108}}}$

3. 1 Let Λ be an artin algebra with center K. We denote the Λ-length of the module M by $l_\Lambda(M)$, and the K-length by $l(M)$. It has been proved, that $l_\Lambda(DTrM)\leqslant pql_\Lambda(M)$, where $p=\max l_\Lambda(P)$, with P running over the projective indecomposible modules, and $q=\max l_\Lambda(Q)$, with Q running over the injective indecomposible modules [R1]. We assume $a = \max l(S)$ and $b = \min l(S)$, where S runs over the simple modules. Then $l(DTrM) \leqslant a \cdot l_\Lambda(DTrM) \leqslant apq\, l_\Lambda(M) \leqslant \frac{apq}{b} \leqslant l(M)$.

We have $\sqrt[s]{l(DTr^sM)} \leqslant \frac{apq}{b} \sqrt[s]{l(M)}$, for any positive integers s.

Definition Let M be a Λ-module. We call $\overline{\lim\limits_{s\to+\infty}} \sqrt[s]{l(DTr^sM)}$ the left growth number of the module M, and denote it by ρ_M^L. We call $\lim\limits_{s\to+\infty} \sqrt[s]{l(TrD^sM)}$ the right growth number, and denote it by ρ_M^R.

3.2 Lemma Let C be a regular component of an Artin algebra Λ. then $\rho_M^L = \rho_N^L$, $\rho_M^R = \rho_N^R$ for all modules M, N in C.

Proof If M, N are located in the same τ-orbit, then $N = TrD^kM$ for a fixed integer k.

$$\begin{aligned}\rho_N^R &= \overline{\lim_{s\to+\infty}} \sqrt[s]{l(TrD^sN)} = \overline{\lim_{s\to+\infty}} \sqrt[s]{l(TrD^{s+k}M)} \\ &= \lim_{s+k\to+\infty} \left(\sqrt[s+k]{l(TrD^{s+k}M)}\right)^{\frac{s+k}{s}} = \rho_M^R.\end{aligned}$$

If there is an irreducible map $M \to N$, then $TrD^sM \to TrD^sN$ also. And $l_\Lambda(M) \leqslant (pq+1) l_\Lambda(N)$ [R1], thus $l(M) \leqslant \frac{a(pq+1)}{b} l(N)$, and

$$\rho_M^R \leqslant \overline{\lim_{s\to+\infty}} \sqrt[s]{\frac{a(pq+1)}{b} l(TrD^sN)} = \rho_N^R.$$

Similarly $\rho_N^R \leqslant \rho_M^R$. By induction, we obtain $\rho_M^R = \rho_N^R$ for all M, N in C.

Definition Let C be a regular component of the AR-quiver of an artin algebra, and M an indecomposible module in C. We call ρ_M^L the left growth number and ρ_M^R the right growth number of C, and we denote it by ρ_C^L or ρ_C^R respectively.

3.3 Theorem Let C be a non-periodic, tree-infinite regular component of the AR-quiver of an artin algebra. If the growth numbers ρ_C^L and ρ_C^R of C are smaller than $\sqrt[3]{\frac{1}{2} + \sqrt{\frac{23}{108}}} + \sqrt[3]{\frac{1}{2} - \sqrt{\frac{23}{108}}}$, then C is of one of the forms $\mathbf{Z}A_\infty$, $\mathbf{Z}A_\infty^\infty$, $\mathbf{Z}B_\infty$, $\mathbf{Z}C_\infty$ or $\mathbf{Z}D_\infty$.

Proof $C = \mathbf{Z}\Gamma/G$, where Γ is a valued oriented infinite tree, and G is an admissable automorphism group. The additive function l on $\mathcal{C}$ determined by the K-length of modules can be lifted to $\mathbf{Z}\Gamma$. Let $\rho_{\mathbf{Z}\Gamma}^{L}=\rho_{C}^{L}$, and $\rho_{\mathbf{Z}\Gamma}^{R}=\rho_{C}^{R}$. Assume T to be the underlying labelled tree of Γ. If T is not of the form A_{∞}, A_{∞}^{∞}, B_{∞}, or D_{∞}, then T dominates some T^{0}, where T^{0} is one of the five forms in proposition 2. 3. We can choose a bipartite subsection of $\mathbf{Z}\Gamma$, say Q^{0} such that T^{0} is the labelled tree of Q^{0}. For example, see fig. 2. 8.

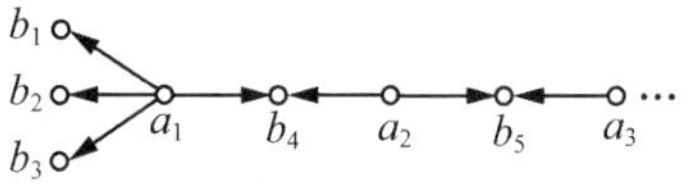

Fig. 2. 8

Let $Q^{(m)}$ be the subquiver of Q^{0} as in the proofs of Lemmas 2. 5. ~2. 7. , but having the corresponding orientations. Let $C_m=C(Q^{(m)})$, be the Coxeter transformation of $Q^{(m)}$ and $\rho_m=\rho(Q^{(m)})$ its spectral radius. Let $M^{(m)}=(l(M_1), l(M_2), \cdots, l(M_m))$, where M_i is the module located at the vertex i of Q^{0}. By Bautista's theorem [B], $(l(TrD^{s}(M_1)), \cdots, l(TrD^{s}(M_m)))=(M^{(m)})C_m^{-s}+P_s^{(m)}$ where $P_s^{(m)}$ is a positive vector. By proposition 1. 5, $\lim\limits_{s\to+\infty}\sqrt[s]{|(M^{(m)})C_m^{-s}|}=\rho_m$. Assume that $\alpha_{+}^{(m)}$ and $\alpha_{-}^{(m)}$ are the coefficients of the expression of $M^{(m)}$ in proposition 1. 5. Since $\alpha_{+}^{(m)}+\alpha_{-}^{(m)}>0$, we always have $\alpha_{+}^{(m)}>0$ or $\alpha_{-}^{(m)}>0$. There exists an infinite set of positive integers m, such that $\alpha_{+}^{(m)}>0$ or $\alpha_{-}^{(m)}>0$. In the second case,

$$\rho_{\mathbf{Z}\Gamma}^{R}=\overline{\lim_{s\to+\infty}}\sqrt[s]{\sum_{i=1}^{m}l(TrD^{s}M_i)}\geqslant\lim_{s\to+\infty}\sqrt[s]{|(M^{(m)})C_m^{-s}|}=\rho_m.$$

We have $\rho_{\mathbf{Z}\Gamma}^{R}\geqslant\lim\limits_{m\to+\infty}\rho_m=\rho(T^{0})$. This is a contradiction to the assumption that ρ_{C}^{R} is smaller than $\sqrt[3]{\frac{1}{2}+\sqrt{\frac{23}{108}}}+\sqrt[3]{\frac{1}{2}-\sqrt{\frac{23}{108}}}$. We can discuss the first case by $\rho_{\mathbf{Z}\Gamma}^{R}$ in the same way. Therefore $T=A_{\infty}$, A_{∞}^{∞}, B_{∞} or D_{∞}. In these cases, the admissable automorphism group is $G=\{1\}$ [BR]. The theorem is proved.

Remark After finishing this paper, I found out that Proposition 1. 5 is overlapping with Theorem 2. 5 of J. A. de la Peña and M. Takane's paper: Spectral properties of Coxeter transformations and applications (preprint).

Acknowledgements

The author would like to express her hearty thanks to Prof. O. Kerner. He's read the manuscript carefully, pointed out several errors, and discussed with the author helpfully. Her special thanks go to Prof. C. M. Ringel for many valuable suggestions. She acknowledges the financial support by the Deutsche Forschungs-gemeinschaft. She also would like to thank Mrs. M. Köllner for typewriting, and Prof. Shaoxue Liu for his many helps.

References

[B] Bautista R. Sectional in *AR*-quiver. LNM 832, 1979: 74-96.

[BLM] Berman S, Lee Y S, Moody R V. The spectrum of a Coxeter transformation, affine Coxeter transformation and the defect map. J. Algebra, 1989, 121: 339-357.

[BR] Butler M C R, Ringel C M. *AR*-sequences with few middle terms and applications to string algebras. Comm. in Algebra, 1987, 15(1—2): 145-179.

[HPR] Happel D, Preiser U, Ringel C M. Vinberg's characterization of Dynkin diagrams using subadditive functions with application to DT_1-periodic modules, LNM 832, 1979: 280-294.

[PA] Platzeck M I, Auslander M. Representation theory of hereditary artin algebra. Lecture Notes in Pure and Applied Math., 1978, 37: 389-424.

[R1] Ringel C M. Report on the Brauer-Thrall conjectures. LNM 831, 1979: 104-136.

[R2] Ringel C M, Tame algebra and integral quadratic forms. LNM 1099, 1984.

[R3] Ringel C M. Representation theory of finite dimensional algebras. London Math. Soc. Lect. Notes 116, 1986: 7-79.

[S] Seneta E. Non-negative matrices. London, 1973.

[SS] Subbotin V F, Stekol'shik R B. Jordan form of Coxeter transformations and applications to representations of finite graphs. Functional analysis and its aplications. 1978, 12 (1): 67-68.

Can. J. Math, 1991, 43(3): 652-672.

稳定分支的结构①

The Structure of Stable Components

Abstract Let A be an artin algebra. Let $\mathfrak{C}$ be a component of the stable Auslander-Reiten quiver of A. If $\mathfrak{C}$ is periodic, then the structure of $\mathfrak{C}$ is known. Here, we are going to consider the case when $\mathfrak{C}$ is non-periodic: we will show that $\mathfrak{C}$ is isomorphic to $\mathbf{Z}\mathfrak{S}$ with $\mathfrak{S}$ a valued quiver. In particular, there is no cyclic path in $\mathfrak{C}$.

§ 1. The main result

For basic concepts and notations in representation theory of algebras, we refer to [R]. Now we recall some of them. A quiver $Q=(V(Q), A(Q), s_Q, e_Q)$ is given by two sets $V(Q)$, $A(Q)$, and two maps $s_Q, e_Q: A(Q) \to V(Q)$. The elements of $V(Q)$ are called vertices or points, those of $A(Q)$ arrows. If $\alpha \in A(Q)$, then $s_Q(\alpha)$ is called its start vertex, $e_Q(\alpha)$ its end vertex, and we write $\alpha: s_Q(\alpha) \to e_Q(\alpha)$. A quiver Q is said to have no multiple arrows provided for any pair x, y of vertices there is at most one arrow $\alpha: x \to y$. Let Q be a quiver and $x \in V(Q)$, then x^+ is the set of end points with start point x, and x^- is the set of start points of arrows with end point x. We say that Q is locally finite provided both x^+ and x^- are finite sets, for any $x \in V(Q)$. A valued

① Received: 1990-05-03.

quiver is of the form $\mathfrak{S}=(V(\mathfrak{S}),A(\mathfrak{S}),s_{\mathfrak{S}},e_{\mathfrak{S}},d_{\mathfrak{S}},d'_{\mathfrak{S}})$, where $(V(\mathfrak{S}),A(\mathfrak{S}),s_{\mathfrak{S}},e_{\mathfrak{S}})$ is a quiver without multiple arrows, and $d_{\mathfrak{S}},d'_{\mathfrak{S}}:A(\mathfrak{S})\to\mathbb{N}_1$ are maps. In case $d_{\mathfrak{S}}(\alpha)=1=d'_{\mathfrak{S}}(\alpha)$ for all $\alpha\in A(\mathfrak{S})$, then we say that $\mathfrak{S}$ has trivial valuation, and we may consider any quiver as a valued quiver with trivial valuation. For $\alpha: x\to y$ in $A(\mathfrak{S})$, we write $d_{xy}=d_{\mathfrak{S}}(\alpha)$, $d'_{xy}=d'_{\mathfrak{S}}(\alpha)$.

A stable translation quiver $\Gamma=(V(\Gamma),A(\Gamma),s_\Gamma,e_\Gamma,\tau_\Gamma)$ is given by a quiver $(V(\Gamma),A(\Gamma),s_\Gamma,e_\Gamma)$ which is locally finite and has no multiple arrows, and a bijection $\tau_\Gamma:V(\Gamma)\to V(\Gamma)$, satisfying $z^-=(\tau_\Gamma z)^+$ for any $z\in V(\Gamma)$. (Note that we allow loops!) Given an arrow $\alpha: y\to z$, we denote by $\sigma\alpha$ the unique arrow: $\tau z\to y$. A valued stable translation quiver is of the form $\mathfrak{C}=(V(\mathfrak{C}),A(\mathfrak{C}),s_{\mathfrak{C}},e_{\mathfrak{C}},\tau_{\mathfrak{C}},d_{\mathfrak{C}},d'_{\mathfrak{C}})$, where, on the one hand, $(V(\mathfrak{C}),A(\mathfrak{C}),s_{\mathfrak{C}},e_{\mathfrak{C}},d_{\mathfrak{C}},d'_{\mathfrak{C}})$ is a valued quiver, whereas $(V(\mathfrak{C}),A(\mathfrak{C}),s_{\mathfrak{C}},e_{\mathfrak{C}},\tau_{\mathfrak{C}})$ is a stable translation quiver, and $d_{\mathfrak{C}}(\alpha)=d'_{\mathfrak{C}}(\sigma\alpha)$, $d'_{\mathfrak{C}}(\alpha)=d_{\mathfrak{C}}(\sigma\alpha)$ for any $\alpha\in A(\mathfrak{C})$. A valued stable translation quiver $\mathfrak{C}$ is said to be connected provided $V(\mathfrak{C})\neq\varnothing$, and $\mathfrak{C}$ cannot be written as a disjoint union of non-empty valued stable translation quivers. Let $\mathfrak{C}$ be a connected valued stable translation quiver. A vertex $x\in V(\mathfrak{C})$ is said to be periodic provided there is $t>0$ with $x=\tau^t x$. Note that the existence of any periodic vertex implies that all vertices are periodic and, in this case, $\mathfrak{C}$ is said to be periodic, otherwise non-periodic [HPR]. We say that $\mathfrak{C}$ is smooth provided its valuation is trivial and x^+ consists of precisely two vertices, for any $x\in V(\mathfrak{C})$. A subadditive function l on $\mathfrak{C}$ with values in $\mathbb{N}_0$ is a map $l:V(\mathfrak{C})\to\mathbb{N}_0$ satisfying

$$l(z)+l(\tau z)\geqslant\sum_{y\in z^-}d'_{yz}l(y),\text{for all } z\in V(\mathfrak{C});$$

and l is said to be additive provided we always have equality.

Given a valued quiver $\mathfrak{S}$, we construct a valued stable translation quiver $\mathbf{Z}\mathfrak{S}$ (following Riedtmann [Ri]) as follows: let $V(\mathbf{Z}\mathfrak{S})=\mathbf{Z}\times V(\mathfrak{S})$; given an arrow $\alpha:x\to y$ in $\mathfrak{S}$, there are arrows $(i,\alpha):(i,x)\to(i,y)$ and $\sigma(i,\alpha):(i-1,y)\to(i,x)$, for all $i\in\mathbf{Z}$; the translation τ is defined by $\tau(i,x)=(i-1,x)$, the valuations are given by

$$d_{\mathbf{Z}\mathfrak{S}}(i,\alpha)=d_{\mathfrak{S}}(\alpha),\quad d'_{\mathbf{Z}\mathfrak{S}}(i,\alpha)=d'(\alpha).$$

Theorem Let $\mathfrak{C}$ be a non-periodic connected valued stable translation quiver with a non-zero subadditive function l with values in $\mathbb{N}_0$. Then, either $\mathfrak{C}$ is smooth and l is both additive and bounded, or else $\mathfrak{C}$ is of the form $\mathbf{Z}\mathfrak{S}$ for some valued quiver $\mathfrak{S}$.

The precise structure of those stable translation quivers $\mathfrak{C}$ which are not of the form $\mathbf{Z}\mathfrak{S}$ for some valued quiver $\mathfrak{S}$, will be given below, see Section 4; they are of the form Π_{st} with $-s<t<0$.

Let Q be a quiver. A path in Q of length $t\geqslant 1$ is of the form $(x\,|\,\alpha_1,\alpha_2,\cdots,\alpha_t\,|\,y)$ with $s_Q(\alpha_1)=x$, $s_Q(\alpha_i)=e_Q(\alpha_{i-1})$ for $2\leqslant i\leqslant t$, and $e_Q(\alpha_t)=y$; in case $x=y$, this is called a cyclic path. If Γ is a translation quiver, a path $(x\,|\,\alpha_1,\alpha_2,\cdots,\alpha_t\,|\,y)$ in Γ is said to be sectional, provided $\sigma\alpha_i\neq\alpha_{i-1}$, for $2\leqslant i\leqslant t$. An important property of valued stable translation quiver of the form $\mathbf{Z}\mathfrak{S}$ is the following:

Lemma Let $\mathfrak{S}$ be a valued quiver. Then any cyclic path in $\mathbf{Z}\mathfrak{S}$ is sectional. If $\mathfrak{S}$ has no cyclic path, then $\mathbf{Z}\mathfrak{S}$ has no cyclic path.

Proof Given $x,y\in V(\mathfrak{S})$ and $i,j\in\mathbf{Z}$ with $i>j$, there is no path from (i,x) to (j,y) in $\mathbf{Z}\mathfrak{S}$. Thus any cyclic path in $\mathbf{Z}\mathfrak{S}$ involves only arrows of the form (i,α) with $\alpha\in A(\mathfrak{S})$ and some fixed $i\in\mathbf{Z}$, and thus it is a sectional path and is obtained from a cyclic path in $\mathfrak{S}$.

Corollary Let $\mathfrak{C}$ be a non-periodic component of the stable Auslander-Reiten quiver of an artin algebra A. Then $\mathfrak{C}$ is of the form $\mathbf{Z}\mathfrak{S}$ for some valued quiver $\mathfrak{S}$ without cyclic path. In particular, $\mathfrak{C}$ has no cyclic path.

Proof of the Corollary We note that $\mathfrak{C}$ is a non-periodic connected valued stable translation quiver (see [HPR]), and the length function l is a subadditive function on $\mathfrak{C}$ with values in $\mathbb{N}_1$. Either $\mathfrak{C}$ is a complete component of Auslander-Reiten quiver of A, then l is additive and unbounded (by a theorem of Auslander [A], since $\mathfrak{C}$ cannot be finite), or else l is not additive. It follows that $\mathfrak{C}$ is of the form $\mathbf{Z}\mathfrak{S}$ for some valued quiver $\mathfrak{S}$. Any cyclic path in $\mathfrak{S}$ would yield a sectional cyclic path in $\mathfrak{C}$, but this is impossible according to a theorem of Bautista-Smalø [BS]. The last assertion is a direct consequence of the Lemma.

We recall that the structure of the periodic components of the stable Auslander-Reiten quiver of an artin algebra has been determined by Happel-Preiser-Ringel [HPR]. In particular, we see that a regular component of the Auslander-Reiten quiver of an artin algebra is either a stable tube or else of the form $\mathbf{Z}\mathfrak{S}$, with $\mathfrak{S}$ a valued quiver.

For the proof of the main theorem, we may assume that we deal with a stable translation quiver with trivial valuation. For, assume the assertion of the main theorm has been shown for all $\mathfrak{C}$ with trivial valuation, and consider now a general $\mathfrak{C}$. Let $\Gamma=(V(\mathfrak{C}), A(\mathfrak{C}), s_{\mathfrak{C}}, e_{\mathfrak{C}}, \tau_{\mathfrak{C}})$ be the corresponding stable translation quiver (with trivial valuation). Since l is subadditive on $\mathfrak{C}$, it is subadditive on Γ. We can assume that the valuation of $\mathfrak{C}$ is non-trivial, thus l cannot be additive on Γ (assume $d'_{yz}>1$, for some arrow $y\rightarrow z$; according to Lemma 5.4 below, we find t such that $l(\tau^t y)\neq 0$, thus $\sum_{y\in z^-} d'_{yz} l(\tau^t y) > \sum_{y\in z^-} l(\tau^t y)$). This shows that $\Gamma=\mathbf{Z}Q$ for some quiver Q, according to the main theorem. Of course, we can transfer the valuation of $\mathfrak{C}$ to Q in order to obtain a valued quiver $\mathfrak{S}=(V(Q), A(Q), s_Q, e_Q, d_{\mathfrak{S}}, d'_{\mathfrak{S}})$, namely let $d_{\mathfrak{S}}(\alpha)=d_{\mathfrak{C}}(0,\alpha)$, $d'_{\mathfrak{S}}(\alpha)=d'_{\mathfrak{C}}(0,\alpha)$ for $\alpha\in A(Q)$ (and $(0,\alpha)\in A(\mathbf{Z}Q)=A(\Gamma)=A(\mathfrak{C})$), then $\mathfrak{C}=\mathbf{Z}\mathfrak{S}$.

§ 2. Preliminaries: quivers and graphs

Let Q be a quiver. An arrow α with $s_Q(\alpha)=e_Q(\alpha)$ is called a loop, and we denote by $L(Q)$ the set of all loops of Q. The paths of length ≥ 1 have been defined in Section 1; we should add that we also have to consider paths of length 0, they are of the form $(x|x)$ with $x\in V(Q)$. Note that any arrow $\alpha: x\rightarrow y$ may be considered as a path $\alpha=(x|\alpha|y)$ of length 1. Given two paths $w=(x|\alpha_1,\alpha_2,\cdots,\alpha_t|y)$ and $w'=(x'|\beta_1,\beta_2,\cdots,\beta_s|y')$ in Q, the product ww' is defined provided $y=x'$, and then $ww'=(x|\alpha_1,\alpha_2,\cdots,\alpha_t,\beta_1,\beta_2,\cdots,\beta_s|y')$. If $w=(x|\alpha_1,\alpha_2,\cdots,\alpha_t|y)$ is a path, we write $s_Q(w)=x$, and $e_Q(w)=y$, and say that w is a path from x to y in case $x=y$, and the lensth of w is at least 1, then w is called a cyclic path (starting at x). A cyclic path $(x|\alpha_1,\alpha_2,\cdots,\alpha_t|y)$ is said to be elementary provided $s_Q(\alpha_i)\neq s_Q(\alpha_j)$ for $i\neq j$. If $\mathfrak{U}$ is a subset of $A(Q)$, we denote by $\langle\mathfrak{U}\rangle$ the smallest subquiver of Q containing $\mathfrak{U}$, thus $V(\langle\mathfrak{U}\rangle)$ is the set of all start vertices and all end vertices of arrows in $\mathfrak{U}$, and $A(\langle\mathfrak{U}\rangle)=\mathfrak{U}$.

A graph $Y=(V(Y),A(Y),s_Y,e_Y,\iota_Y)$ is given by a quiver $(V(Y), A(Y),s_Y,e_Y)$ and a fixpoint free involution ι_Y of $A(Y)$. Thus $\iota_Y:A(Y)\to A(Y)$ is a map with $\iota_Y(\alpha)\neq\alpha$ and $\iota_Y^2(\alpha)=\alpha$ for all $\alpha\in A(Y)$. A path $(x|\alpha_1,\alpha_2,\cdots,\alpha_t|y)$ in Y is said to be reduced provided $\alpha_{i+1}\neq\iota\alpha_i$ for all $1\leqslant i\leqslant t-1$. A graph Y is called a tree provided for every pair $x,y\in V(Y)$ there is a unique reduced path from x to y, and, in this case, the unique reduced path from x to y is called the geodesic from x to y [D]. We will consider cyclic paths which are both reduced and elementary. Note that for a reduced, elementary, cyclic path $(x|\alpha_1,\alpha_2,\cdots,\alpha_t|x)$ in Y, we also have $\alpha_1\neq\iota\alpha_t$ (this is trivially true for $t=1$; if $t\geqslant 2$ and $\alpha_1=\iota\alpha_t$ then $s_Y(\alpha_t)=e_Y(\alpha_1)=s_Y(\alpha_2)$, thus $t=2$, but we assume $\alpha_2\neq\iota\alpha_1$).

(**Remark** the definition of a graph may look rather clumsy, the usual definition just identifies the arrows α and $\iota\alpha$ and calls this identified pair $\{\alpha,\iota\alpha\}$ an edge. There are several reasons for using the definition as given above: an edge joining two vertices x,y with $x\neq y$ can easily be oriented by specifying the order of x and y. However, we also will want to change the orientation of loops, replacing a loop α by $\iota\alpha$. Note that the orbit graph of a non-periodic stable translation quiver (as defined in Section 5 below) is always a graph in the sense defined above. The definition of a graph as specified above is due to Reidemeister [Re] who called this a "Streckenkomplex".)

Given a graph Y, an orientation Ω of Y is given by a subset Ω of $A(Y)$ such that Ω intersects any ι-orbit of $A(Y)$ in precisely one element. We denote by (Y,Ω) the subquiver $\langle\Omega\rangle$.

On the other hand, let Q be a quiver. The underlying graph $\bar{Q}$ of Q is the graph $\bar{Q}=(V(Q),A(\bar{Q}),s_{\bar{Q}},e_{\bar{Q}},\iota_{\bar{Q}})$, where $A(\bar{Q})$ consists of the disjoint union of two copies of $A(Q)$, one denoted by $A(Q)$, the other by $\{\alpha^* \mid \alpha\in A(Q)\}$, with $s_{\bar{Q}}(\alpha^*)=e_Q(\alpha)$, $e_{\bar{Q}}(\alpha^*)=s_Q(\alpha)$, and with $\iota_{\bar{Q}}(\alpha^*)=\alpha$, $\iota_{\bar{Q}}(\alpha)=\alpha^*$. By definition, a walk in Q is a path in $\bar{Q}$. (Note that we also will consider walks in $\bar{Q}$, thus paths in $\bar{\bar{Q}}$). Walks in Q will be denoted in the form $w=(x\mid\alpha_1^{\varepsilon_1},\alpha_2^{\varepsilon_2},\cdots,\alpha_n^{\varepsilon_n}\mid y)$, where $\alpha_1,\alpha_2,\cdots,\alpha_n$ belong to $A(Q)$, $\varepsilon_1,\varepsilon_2,\cdots,\varepsilon_n\in\{\pm1\}$,

$$\alpha_i^{\varepsilon_i}=\begin{cases}\alpha_i, & \text{for } \varepsilon_i=1,\\ \alpha_i^*, & \text{for } \varepsilon_i=-1\end{cases}$$

(and $s_Q(\alpha_1^{\varepsilon_1})=x$, $s_Q(\alpha_i^{\varepsilon_i})=e_Q(\alpha_{i-1}^{\varepsilon_{i-1}})$ for $2\leqslant i\leqslant n$, $e_Q(\alpha_n^{\varepsilon_n})=y$), and we write $x=s_Q(w)$, $y=e_Q(w)$. Such a walk $(x\mid\alpha_1^{\varepsilon_1},\alpha_2^{\varepsilon_2},\cdots,\alpha_n^{\varepsilon_n}\mid y)$ is reduced provided $\alpha_i=\alpha_{i+1}$ implies $\varepsilon_i=\varepsilon_{i+1}$, and it is elementary provided $s_Q(\alpha_i^{\varepsilon_i})\neq s_Q(\alpha_j^{\varepsilon_j})$ for $i\neq j$.

Let Y be a graph. We are going to define its first homology group $H_1(Y)$ (see [S]). Let $C_0(Y)$ be the free abelian group with basis $V(Y)$. Let $C_1(Y)$ be the factor group of the free abelian group with basis $A(Y)$ by the subgroup generated by all elements of the form $\alpha+\iota\alpha$, with $\alpha\in A(Y)$. Thus, in $C_1(Y)$, the element $\iota\alpha$, for $\alpha\in A(Y)$, is identified with $-\alpha$. If Ω is an orientation of Y, then $C_1(Y)$ may be identified with the free abelian group with basis $A(\langle\Omega\rangle)=\Omega$. We define $\delta: C_1(Y)\to C_0(Y)$ by $\delta(\alpha)=e_Y(\alpha)-s_Y(\alpha)$, for $\alpha\in A(Y)$, and $H_1(Y)$ is, by definition, the kernel of δ, The elements of $H_1(Y)$ are called cycles, thus a cycle is an element $c\in C_1(Y)$ such that $\delta(c)=0$. If $(x\mid\alpha_1,\alpha_2,\cdots,\alpha_t\mid x)$ is a cyclic path in Y, then $\sum_{i=1}^{t}\alpha_i$ belongs to $H_1(Y)$. An element of $H_1(Y)$ of the form $\sum_{i=1}^{t}\alpha_i$, where $(x\mid\alpha_1,\alpha_2,\cdots,\alpha_t\mid x)$ is an elementary reduced cyclic path in Y, is called an elementary cycle.

Given a quiver Q, let $H_1(Q)=H_1(\bar{Q})$, thus $H_1(Q)$ is the kernel of the map $\delta: C_1(Q) \to C_0(Q)$, where $C_1(Q)$ is the free abelian group with basis $A(Q)$, and $C_0(Q)$ is the free abelian group with basis $V(Q)$, with $\delta(\alpha) = e_Q(\alpha) - s_Q(\alpha)$ for $\alpha \in A(Q)$. The elements in $C_1(Q)$ will be written in the form $c = \sum_{\alpha \in A(Q)} c(\alpha)\alpha$ with $c(\alpha) \in \mathbf{Z}$ and almost all $c(\alpha) = 0$; the set $\operatorname{supp}(c) = \{\alpha \mid c(\alpha) \neq 0\}$ will be called its support and we write $|c| := \sum_{\alpha \in A(Q)} |c(\alpha)|$. Since we will have to deal with elementary cycles in $H_1(Q)$ rather frequently, let us repeat: let $(x \mid \alpha_1^{c(\alpha_1)}, \alpha_2^{c(\alpha_2)}, \cdots, \alpha_t^{c(\alpha_t)} \mid x)$ be an elementary reduced cyclic walk in Q (where $c(\alpha_i) \in \{\pm 1\}$), then $\sum_{i=1}^{t} c(\alpha_i)\alpha_i$ is an elementary cycle in $H_1(Q)$, and any elementary cycle is obtained in this way.

Let Y be a graph. On the set of paths in Y, we define the homotopy relation as the smallest equivalence relation $\sim$ with the following property: if w, w' are paths, with $e_Y(w) = s_Y(w') = s_Y(\alpha)$ for some arrow α, then

$$w \cdot \alpha \cdot \iota\alpha \cdot w' \sim w \cdot w'.$$

The equivalence class of the path w will be denoted by $\overline{w}$, and called its homotopy class. For $y \in V(Y)$, let $\pi_1(Y, y)$ be the set of homotopy classes of paths starting and ending at y, it is a group with respect to the composition $\overline{w} \cdot \overline{w'} = \overline{ww'}$, the fundamental group of Y at y. It is well-known that the fundamental group of a graph Y is a free group. Note that there is a canonical group homomorphism $\pi_1(Y, y) \to H_1(Y)$, sending the homotopy class of the path $(x \mid \alpha_1, \alpha_2, \cdots, \alpha_n \mid x)$ to $\sum_{i=1}^{n} \alpha_i$. The kernel of the homomorphism is the commutator subgroup, and this homomorphism is surjective provided Y is connected [S].

§ 3. Tempered maps

Let Y be a graph. A linear map $\vartheta: H_1(Y) \to \mathbf{Z}$ is called tempered provided it satisfies the following conditions: given any elementary cycle

c in $H_1(Y)$, then

$$\vartheta(c) \equiv |c| \pmod 2 \text{ and } |\vartheta(c)| \leqslant |c|.$$

Let Ω be any orientation of Y. We define $\vartheta_\Omega: H_1(Y) \to \mathbf{Z}$ by $\vartheta_\Omega(\sum_{\alpha\in\Omega} c(\alpha)\alpha) = \sum_{\alpha\in\Omega} c(\alpha)$. Then clearly ϑ_Ω is a tempered map on $H_1(Y)$. We will need the converse statement:

Proposition Let Y be a graph, and $\vartheta: H_1(Y) \to \mathbf{Z}$ a tempered map. Then there exists an orientation Ω on Y such that $\vartheta = \vartheta_\Omega$.

The proof of the proposition will be given in this section. We will start with a quiver Q such that $Y = \bar{Q}$, and we will construct a function $\eta: A(Q) \to \{\pm 1\}$ such that

$$\sum_{\gamma\in A(Q)} \eta(\gamma)c(\gamma) = \vartheta(c)$$

for any $c \in H_1(Q)$. Thus

$$\Omega = \{\alpha \mid \alpha \in A(Q), \eta(\alpha) = 1\} \cup \{\alpha^* \mid \alpha \in A(Q), \eta(\alpha) = -1\}$$

will be the required orientation of Y.

3.1 Shrinking of arrows.

Let Q be a quiver, $\mathfrak{U} \subseteq A(Q)$ some set of arrows. We define $Q/\mathfrak{U}$, the quiver obtained from Q by shrinking the arrow in $\mathfrak{U}$, as follows: given $x, y \in V(Q)$, write $x \sim y$ provided there exists a walk $(x \mid \alpha_1^{\varepsilon_1}, \alpha_2^{\varepsilon_2}, \cdots, \alpha_n^{\varepsilon_n} \mid y)$ from x to y with all arrows $\alpha_i \in \mathfrak{U}$. The equivalence class of x in $V(Q)$ with respect to $\sim$ is denoted by $[x]_{\mathfrak{U}}$, and $V(Q/\mathfrak{U})$ is the set of these equivalence classes. Let $A(Q/\mathfrak{U}) = A(Q)/\mathfrak{U}$ and given $\alpha \in A(Q/\mathfrak{U})$, let $s_{Q/\mathfrak{U}} = [s_Q(\alpha)]_{\mathfrak{U}}$, $e_{Q/\mathfrak{U}} = [e_Q(\alpha)]_{\mathfrak{U}}$. We stress that, by definition, the arrow set $A(Q/\mathfrak{U})$ of $Q/\mathfrak{U}$ is a subset of the arrow set $A(Q)$.

We define $\phi_i: C_i(Q) \to C_i(Q/\mathfrak{U})$ for $i = 0$ and I as follows: given $x \in V(Q)$, let $\phi_0(x) = [x]_{\mathfrak{U}}$; given $\alpha \in A(Q)$, let $\phi_1(\alpha) = \alpha$ provided $\alpha \notin \mathfrak{U}$ and $\phi_1(\alpha) = 0$ otherwise. Since the diagram (fig. 3.1)

$$\begin{array}{ccc} C_1(Q) & \xrightarrow{\phi_1} & C_1(Q/\mathfrak{U}) \\ \delta_Q \downarrow & & \downarrow \delta_{Q/\mathfrak{U}} \\ C_0(Q) & \xrightarrow{\phi_0} & C_0(Q/\mathfrak{U}) \end{array}$$

Fig. 3.1

obviously commutes, we obtain an induced map

$$H_1(Q) \xrightarrow{\phi} H_1(Q/\mathfrak{U})$$

between the kernels of the two δ-maps, we may call it the canonical map.

Note that the inclusion $A(Q/\mathfrak{U}) \subseteq A(Q)$ yields an embedding $C_1(Q/\mathfrak{U}) \subseteq C_1(Q)$, but $\delta_{Q/\mathfrak{U}}$ is usually not the restriction of δ_Q to $C_1(Q/\mathfrak{U})$.

We say that $T \subseteq A(Q)$ is cycle-free provided there does not exist a non-zero cycle c with $\operatorname{supp}(c) \subseteq T$, (or, equivalently, provided $H_1(\langle T \rangle) = 0$). Assume that T is cycle-free, then one easily shows that $H_1(Q/\mathfrak{U}) \simeq H_1(Q)$. We want to construct an explicit map

$$\xi_T : H_1(Q/T) \to H_1(Q)$$

(which will be shown to be an isomorphism). For any connected component $\langle T_t \rangle$ of $\langle T \rangle$ (where $T_t \subseteq T$), choose some vertex a_t of $\langle T_t \rangle$. If x is a vertex of $\langle T_t \rangle$, choose some walk $(a_t \mid \alpha_1^{\varepsilon_1}, \alpha_2^{\varepsilon_2}, \cdots, \alpha_n^{\varepsilon_n} \mid x)$ from a_t to x inside T_t, and let $c(x) = \sum_{i=1}^{n} \varepsilon_i \alpha_i$, whereas for $x \notin V(\langle T \rangle)$, let $c(x) = 0$. We define $\xi_0^a : C_0(Q/T) \to C_0(Q)$ by

$$\xi_0^a([x]_T) = \begin{cases} a_i, & \text{if } x \in V(\langle T_i \rangle) \\ x, & \text{if } x \in V(Q) \setminus V(\langle T \rangle), \end{cases}$$

for $x \in V(Q)$. And $\xi_1^a : C_1(Q/T) \to C_1(Q)$ by

$$\xi_1^a(\alpha) = \alpha + c(s(\alpha)) - c(e(\alpha)),$$

for $\alpha \in A(Q/T)$. We claim that the diagram (fig. 3. 2)

$$\begin{array}{ccc} C_1(Q/T) & \xrightarrow{\xi_1^a} & C_1(Q) \\ \Big\downarrow \delta_{Q/T} & & \Big\downarrow \delta_Q \\ C_0(Q/T) & \xrightarrow[\xi_0^a]{} & C_0(Q) \end{array}$$

Fig. 3. 2

commutes: given $\alpha \in A(Q/T)$, we have

$$(\delta_Q \xi_1^a)(\alpha) = \begin{cases} e(\alpha) - s(\alpha), & \text{if } s(\alpha) \notin V(\langle T \rangle), e(\alpha) \notin V(\langle T \rangle), \\ e(\alpha) - a_i, & \text{if } s(\alpha) \in V(\langle T_i \rangle), e(\alpha) \notin V(\langle T \rangle), \\ a_j - s(\alpha), & \text{if } s(\alpha) \notin V(\langle T \rangle), e(\alpha) \in V(\langle T_j \rangle), \\ a_j - a_i, & \text{if } s(\alpha) \in V(\langle T_i \rangle), e(\alpha) \in V(\langle T_j \rangle). \end{cases}$$

Consequently, ξ_1^a maps $H_1(Q/T)$ into $H_1(Q)$, thus we obtain an induced map

$$\xi_T : H_1(Q/T) \to H_1(Q),$$

which does not depend on the chosen vertices a_i (note that the map ξ_1^a already does not depend on the chosen walks, but, of course, it depends on the vertices a_i).

We stress that given $x \in C_1(Q/T)$, the element $\xi_1^a(c) - c$ has support in T, and thus the composition

$$C_1(Q/T) \xrightarrow{\xi_1^a} C_1(Q) \xrightarrow{\phi_1} C_1(Q/T)$$

is the identity; in particular, the composition

$$H_1(Q/T) \xrightarrow{\xi_T} H_1(Q) \xrightarrow{\phi} H_1(Q/T)$$

is the identity. Also the composition

$$H_1(Q) \xrightarrow{\phi} H_1(Q/T) \xrightarrow{\xi_T} H_1(Q)$$

is the identity, since for $c \in H_1(Q)$, the cycle $\xi_T\phi(c) - c$ has support in T, and thus vanishes.

Given a quiver Q, let $L(Q) \subseteq A(Q)$ be the set of loops. The quiver Q will be said to be reduced provided the support of any elementary cycle is a loop. If Q is a reduced quiver, then, clearly, $H_1(Q)$ is just the free abelian group generated by $L(Q)$. Thus, if Q is an arbitrary quiver, and T is a cycle-free subset of $A(Q)$ such that Q/T is reduced, then one obtains a free generating system for $H_1(Q)$ by

$$\{\xi_T(\alpha) \mid \alpha \in L(Q/T)\}.$$

In fact, since for $\alpha \neq \beta \in L(Q/T)$, we have $(\xi_T(\alpha))(\beta) = 0$, whereas $(\xi_T(\alpha))(\alpha) = 1$, we see that any $c \in H_1(Q)$ can be written in the form

$$c = \sum_{\alpha \in L(Q/T)} c(\alpha)\xi_T(\alpha).$$

3.2 Dealing with one elementary cycle.

Let Q be a quiver and $\vartheta : H_1(Q) \to \mathbf{Z}$ a tempered map. Note that for any loop α of Q, we have $|\vartheta(\alpha)| = 1$, since $|\vartheta(\alpha)| \leqslant 1$ and $\vartheta(\alpha) \equiv 1 \pmod 2$, thus

$$|\alpha| - |\vartheta(\alpha)| = 0.$$

We consider an elementary cycle c which is not a loop. By the definition of a tempered map, we know that $u := \frac{1}{2}(|c| - \vartheta(c))$ is a non-negative integer. Replacing, if necessary, c by $-c$, we assume $\vartheta(c) \geqslant 0$.

Lemma a Let c be an elementary cycle, not a loop, and assume $\vartheta(c) \geqslant 0$. Let $S = \mathrm{supp}(c)$, and $n = |c|$, and $u = \frac{1}{2}(n - \vartheta(c))$. Then there are C_n^u functions $\eta: S \to \{\pm 1\}$ such that

$$\sum_{\alpha \in S} \eta(\alpha) c(\alpha) = \vartheta(c). \tag{3.1}$$

Proof Given $\eta: S \to \{\pm 1\}$ satisfying (3.1), we have $\eta(\alpha)c(\alpha) \in \{\pm 1\}$ for all $\alpha \in S$. There are n summands on the left, all are in $\{\pm 1\}$, thus u of the summand $\eta(\alpha)c(\alpha)$ have to be -1, the remaining ones have to be 1, since $n - u = u + \vartheta(c)$. Thus $\mathcal{U} = \{\alpha \in S \mid \eta(\alpha)c(\alpha) = -1\}$ is a subset of S with u elements. Conversely, let $\mathcal{U}$ be a subset of S with u elements, then we define

$$\eta(\alpha) = \begin{cases} -c(\alpha), & \text{for } \alpha \in \mathcal{U}, \\ c(\alpha), & \text{for } \alpha \in S \setminus \mathcal{U}. \end{cases}$$

and we see that

$$\sum_{\alpha \in S} \eta(\alpha) c(\alpha) = \sum_{\alpha \in \mathcal{U}} (-1) + \sum_{\alpha \in S \setminus \mathcal{U}} 1 = -u + n - u = \vartheta(c),$$

thus (3.1) is satisfied.

Let $u_\vartheta = \min\{\frac{1}{2}(|c| - |\vartheta(c)|) \mid c$ an elementary cycle of Q, not a loop$\}$.

Lemma b Let c be an elementary cycle of Q with $u = u_\vartheta$. Let $\eta: S \to \{\pm 1\}$ be a function such that $\sum_{\alpha \in S} \eta(\alpha) c(\alpha) = \vartheta(c)$. Let $\alpha_0 \in S$ be a fixed arrow and $T := S \setminus \{\alpha_0\}$. Define $\eta_T: H_1(Q) \to \mathbf{Z}$ by

$$\eta_T(d) = \sum_{\beta \in T} \eta(\beta) d(\beta)$$

for $d \in H_1(Q)$. Then

$$\vartheta' = (\vartheta - \eta_T)\xi_T: H_1(Q/T) \to \mathbf{Z}$$

is a tempered map on $H_1(Q/T)$.

Proof　Let d be an elementary cycle of Q/T, we want to show that

$$\vartheta'(d) \equiv |d| \pmod 2 \text{ and } |\vartheta'(d)| \leqslant |d|.$$

Note that instead of d, we also may consider $-d$. We will consider the support of d as a subset of $A(Q/T) \subseteq A(Q)$, and we write ξ instead of ξ_T.

First, we assume $d = \alpha_0$. Since $\xi(d)$ is a cycle in $H_1(\langle S\rangle)$ with $\xi(d)(\alpha_0) = 1$, we have $\xi(d) = c(\alpha_0)c$. Now

$$\begin{aligned}\vartheta'(d) &= (\vartheta - \eta_T)(c(\alpha_0)c) = c(\alpha_0)\Big(\sum_{\alpha\in S}\eta(\alpha)c(\alpha) - \sum_{\alpha\in T}\eta(\alpha)c(\alpha)\Big)\\ &= c(\alpha_0)\eta(\alpha_0)c(\alpha_0) = \eta(\alpha_0),\end{aligned}$$

therefore, $|\vartheta'(d)| = 1$.

Thus, we can assume that $\mathrm{supp}(d) \neq \{\alpha_0\}$, and therefore $\alpha_0 \notin \mathrm{supp}(d)$, since d is an elementary cycle of Q/T. Let $([y]_T \mid \beta_1^{d(\beta_1)}, \beta_2^{d(\beta_2)}, \cdots, \beta_m^{d(\beta_m)} \mid [y]_T)$ be a cyclic walk in Q/T with $\mathrm{supp}(d) = \{\beta_1, \beta_2, \cdots, \beta_m\}$. Let $y_i = s_Q(\beta_i^{d(\beta_i)})$. There is at most one index i such that $y_i \in V(\langle T\rangle)$, and we can assume $y_i \notin V(\langle T\rangle)$ for all $2 \leqslant i \leqslant m$. Let $y = y_1$, and $z = e_Q(\beta_m^{d(\beta_m)})$. In case $z = y$, we see that d is an elementary cycle of Q itself and $\xi(d) = d$. Thus, we only have to consider the case $y \neq z$. In this case, both y, z belong to $\langle T\rangle$. Let $(x_1 \mid \alpha_1^{c(\alpha_1)}, \alpha_2^{c(\alpha_2)}, \cdots, \alpha_n^{c(\alpha_n)} \mid x_1)$ be a cyclic walk in Q with $\mathrm{supp}(c) = \{\alpha_1, \alpha_2, \cdots, \alpha_n\}$ and let $x_i = s_Q(\alpha_i^{c(\alpha_i)})$. Let $z = x_r, y = x_s$ with $1 \leqslant r, s \leqslant n$. We can assume $r < s$, otherwise we consider $-d$ instead of d. It follows that

$$\xi(d) = d + \sum_{i=r}^{s-1} c(\alpha_i)\alpha_i,$$

in particular, $|\xi(d)| = m + n'$, where $n' = s - r$. Let u' be the number of arrows $\alpha_i \in \mathfrak{U}$ with $r \leqslant i \leqslant s-1$, where $\mathfrak{U} = \{\alpha \in S \mid \eta(\alpha)c(\alpha) = -1\}$. Note that

$$\begin{aligned}\vartheta'(d) &= (\vartheta - \eta_T)(\xi(d)) = \vartheta(\xi(d)) - \sum_{i=r}^{s-1}\eta(\alpha_i)c(\alpha_i)\\ &= \vartheta(\xi(d)) + u' - (n' - u')\\ &= \vartheta(\xi(d)) + 2u' - n',\end{aligned}$$

since u' of the summands $\eta(\alpha_i)c(\alpha_i)$ are -1, the remaining ones are

1. Now, $\xi(d)$ is an elementary cycle of Q, and therefore $\vartheta(\xi(d)) \equiv |\xi(d)| = m + n'$ (mod 2), thus
$|\vartheta'(d)| = \vartheta(\xi(d)) + 2u' - n' \equiv m + n' + 2u' - n' \equiv m = |d|$ (mod 2).
Since $\xi(d)$ is an elementary cycle of Q and not a loop, we have

$$|\xi(d)| - |\vartheta(\xi(d))| \geqslant 2u_{\vartheta},$$

thus $\quad \vartheta(\xi(d)) \leqslant |\vartheta(\xi(d))| \leqslant |\xi(d)| - 2u_{\vartheta} = m + n' - 2u_{\vartheta},$

therefore

$$\begin{aligned} \vartheta'(d) &= \vartheta(\xi(d)) + 2u' - n' \leqslant m + n' - 2u_{\vartheta} + 2u' - n' \\ &= m - 2u_{\vartheta} + 2u' \leqslant m = |d| \end{aligned}$$

where we have used $u' \leqslant u = u_{\vartheta}$.

In order to show that we also have $-|d| \leqslant \vartheta'(d)$, let us consider $\xi(d) - c$. We have

$$\xi(d) - c = d - \sum_{i=1}^{r-1} c(\alpha_i)\alpha_i - \sum_{i=s}^{n} c(\alpha_i)\alpha_i,$$

in particular, also $\xi(d) - c$ is an elementary cycle of Q and not a loop. We have $|\xi(d) - c| = m + n''$, where $n'' = n - n'$. Since $\xi(d) - c$ is an elementary cycle of Q and not a loop,

$$|\xi(d) - c| - |\vartheta(\xi(d) - c)| \geqslant 2u_{\vartheta},$$

thus

$$\vartheta(\xi(d) - c) \geqslant -|\vartheta(\xi(d) - c)| \geqslant -|\xi(d) - c| + 2u_{\vartheta} = -m - n'' + 2u_{\vartheta}.$$

We also will use that

$$\vartheta(c) = |c| - 2u_{\vartheta} = n - 2u_{\vartheta}.$$

Altogether, we have

$$\begin{aligned} \vartheta'(d) &= \vartheta(\xi(d)) + 2u' - n' \\ &= \vartheta(\xi(d) - c) + \vartheta(c) + 2u' - n' \quad \text{(since } \vartheta \text{ is additive)} \\ &\geqslant -m - n'' + 2u_{\vartheta} + n - 2u_{\vartheta} + 2u' - n' \\ &= -m + 2u' \geqslant -m = -|d|. \end{aligned}$$

This shows that $|\vartheta'(d)| \leqslant |d|$, and therefore finishes the proof.

3.3 Transfinite induction.

Lemma Let $\vartheta: H_1(Q) \to \mathbf{Z}$ be a tempered map for a quiver Q. Then there exists a cycle-free subset T of $A(Q)$ and a function $\eta: T \to \{\pm 1\}$

such that

(1) X/T is reduced, and

(2) the map $\vartheta_T: H_1(Q/T) \to \mathbf{Z}$ defined by

$$\vartheta_T(c) = \vartheta(\bar{c}) - \sum_{\alpha \in T} \eta(\alpha)\bar{c}(\alpha)$$

for $c \in H_1(Q/T)$ and $\bar{c} = \xi_T(c)$ is tempered.

Proof Let $T_1 = \varnothing$, and $\vartheta_1 = \vartheta$. Assume there is some ordinal number λ such that for any ordinal number $\mu < \lambda$, we have constructed a subset $T_\mu \subseteq A(Q)$ and a map $\eta_\mu: T_\mu \to \{\pm 1\}$ with the following properties:

(a) If $\nu < \mu$ is an ordinal, then $T_\nu \subsetneqq T_\mu$, and $\eta_\nu = \eta_\mu | T_\nu$,

(b) T_μ is cycle-free,

(c) the map $\vartheta_\mu: H_1(Q/T_\mu) \to \mathbf{Z}$ defined by $\vartheta_\mu(c) = \vartheta(\bar{c}) - \sum_{\alpha \in T_\mu} \eta_\mu(\alpha)\bar{c}(\alpha)$ for $c \in H_1(Q/T_\mu)$ and $\bar{c} = \xi_{T_\mu}(c)$, is tempered.

First, assume λ is a limit ordinal. Let $T_\lambda = \bigcup_{\mu<\lambda} T_\mu$, and define η_λ by $\eta_\lambda | T_\mu = \eta_\mu$. By definition (a) is satisfied for $\mu = \lambda$. Since a filtered union of cycle-free subsets of $A(Q)$ is cycle-free, also (b) is satisfied for $\mu = \lambda$. In order to show (c) for $\mu = \lambda$, let c be an elementary cycle of $H_1(Q/T_\lambda)$, and $\bar{c} = \xi_{T_\lambda}(c)$. The support of $\bar{c} - c$ is a finite subset of T_λ, thus it lies in T_μ for some $\mu < \lambda$, and $\vartheta_\lambda(c) = \vartheta_\mu(c)$. Therefore $\vartheta_\lambda(c) \equiv |c| \pmod 2$ and $|\vartheta_\lambda(c)| \leqslant |c|$.

Now assume $\lambda > 1$ and that λ is not a limit ordinal, thus $\lambda - 1$ exists. Let $Q_{\lambda-1} := Q/T_{\lambda-1}$. In case $Q_{\lambda-1}$ is reduced, let $T = T_{\lambda-1}$, $\eta = \eta_{\lambda-1}$; clearly, all assertions of the lemma are satisfied in this way. So assume $Q_{\lambda-1}$ is not reduced. We write $\xi_{\lambda-1}$ instead of $\xi_{T_{\lambda-1}}$, and $u_{\lambda-1}$ instead of $u_{\vartheta_{\lambda-1}}$. We choose some elementary cycle of $Q_{\lambda-1}$, not a loop, with $\frac{1}{2}(|c| - |\vartheta_{\lambda-1}(c)|) = u_{\lambda-1}$, and we can assume $\vartheta_{\lambda-1}(c) \geqslant 0$. Let S be the support of c, fix some $\alpha_0 \in S$, and let $T := S \setminus \{\alpha_0\}$. According to Section 3.2, there exists a function $\eta: S \to \{\pm 1\}$ such that

$$\sum_{\alpha \in S} \eta(\alpha) c(\alpha) = \vartheta_{\lambda-1}(c),$$

for all $c \in H_1(X_\lambda)$, and the function $\vartheta'_{\lambda-1}: H_1(Q_{\lambda_1}/T) \to \mathbf{Z}$ defined by

$$\vartheta'_{\lambda-1}(d) = \vartheta_{\lambda-1}(\xi_T(d)) - \sum_{\alpha \in T} \eta(\alpha)(\xi_T(d))(\alpha),$$

for $d \in H(Q_{\lambda-1}/T)$, is a tempered map. Let $T_\lambda = T_{\lambda-1} \cup T \subseteq A(Q)$, let $\eta_\lambda: T_\lambda \to \{\pm 1\}$ be defined by $\eta_\lambda | T_{\lambda-1} = \eta_{\lambda-1}$, and $\eta_\lambda | T = \eta | T$, thus (a) is satisfied for $\mu = \lambda$. In order to see that T_λ is cycle-free, let c be a cycle with support in T_λ. Shrinking of $T_{\lambda-1}$ produces the cycle $c | T$ in $H_1(Q_{\lambda-1})$, however T is cycle-free in $Q_{\lambda-1}$, thus $c | T = 0$, or, equivalently, $\operatorname{supp}(c) \subseteq T_{\lambda-1}$. Since $T_{\lambda-1}$ is cycle-free, it follows that $c = 0$. For the proof of (c), we first note that clearly $Q/T_\lambda = Q_{\lambda-1}/T$ and we claim that $\vartheta_\lambda = \vartheta'_{\lambda-1}$, so that ϑ_λ is tempered. Let $\xi_\lambda = \xi_{T_\lambda}$. Let $d \in H_1(Q/T_\lambda)$ and note that $\xi_\lambda(d) = \xi_{\lambda-1}\xi_T(d)$, so that $(\xi_\lambda(d))(\alpha) = (\xi_T(d))(\alpha)$ for $\alpha \in T$. Thus

$$\begin{aligned}\vartheta_\lambda(d) &= \vartheta(\xi_\lambda(d)) - \sum_{\alpha \in T_\lambda} \eta_\lambda(\alpha)(\xi_\lambda(d))(\alpha) \\ &= \vartheta(\xi_{\lambda-1}\xi_T(d)) - \sum_{\alpha \in T_{\lambda-1}} \eta_{\lambda-1}(\alpha)(\xi_{\lambda-1}\xi_T(d))(\alpha) - \sum_{\alpha \in T} \eta(\alpha)(\xi_T(d))(\alpha) \\ &= \vartheta_{\lambda-1}(\xi_T(d)) - \sum_{\alpha \in T} \eta(\alpha)(\xi_T(d))(\alpha) \\ &= \vartheta'_{\lambda-1}(d).\end{aligned}$$

This finishes the proof of the induction step.

Since our algorithm produces a strictly increasing chain of subsets T_λ in $A(Q)$, it must stop. Thus, for some ordinal $\lambda > 1$, not a limit ordinal, $Q/T_{\lambda-1}$ has to be reduced. This yields the proof of the lemma.

3.4 Proof of the Proposition. Let X be a quiver, $\vartheta: H_1(Q) \to \mathbf{Z}$ a tempered map. Let T be a cycle-free subset of $A(Q)$ such that Q/T is reduced, and let $\eta: T \to \{\pm 1\}$ be a map such that ϑ_T (as defined in Section 3.3) is tempered. It remains to extend η to all of $A(Q)$.

If α is a loop of Q/T, let $\eta(\alpha) - \vartheta_T(\alpha)$, whereas for $\beta \in Q/T$, not a loop, we may choose $\eta(\beta)$ arbitrarily, say let $\eta(\beta) = 1$. Thus $\eta: A(Q) \to \{\pm 1\}$ is defined. Recall that $L(Q/T)$ denotes the set of loops of Q/T, and given $\alpha \in L(Q/T)$, let $\bar{\alpha} = \xi_T(\alpha) \in H_1(Q)$. We have

$$\sum_{\gamma \in A(Q)} \eta(\gamma)\bar{\alpha}(\gamma) = \eta(\alpha) + \sum_{\gamma \in T} \eta(\gamma)\bar{\alpha}(\gamma) \quad (\text{since } \bar{\alpha}(\alpha) = 1)$$

$$= \eta(\alpha) + \vartheta(\bar{\alpha}) - \vartheta_T(\alpha) \quad \text{(by the definition of } \vartheta_T\text{)}$$

$$= \vartheta(\bar{\alpha}).$$

Note that $\{\bar{\alpha} \mid \alpha \in L(Q/T)\}$ is a basis of $H_1(Q)$ as a free abelian group, in fact, given $c \in H_1(Q)$, we have $c = \sum_{\alpha \in L(Q/T)} c(\alpha)\bar{\alpha}$, and therefore

$$\sum_{\gamma \in A(Q)} \eta(\gamma)c(\gamma) = \sum_{\gamma}\sum_{\alpha} \eta(\gamma)c(\alpha)\bar{\alpha}(\gamma)$$

$$= \sum_{\alpha} c(\alpha)\Big(\sum_{\gamma} \eta(\gamma)\bar{\alpha}(\gamma)\Big)$$

$$= \sum_{\alpha} c(\alpha)\vartheta(\bar{\alpha}) = \vartheta(c). \quad \square$$

§ 4. Smooth non-periodic stable translation quivers

We are going to give a complete list of the non-periodic stable translation quivers which are smooth and to exhibit some of their properties.

4.1 The stable translation quiver Π_{st}. First, we note that an isomorphism $f: \Gamma \to \Gamma'$ of two stable translation quivers Γ, Γ' is given by two bijections $V(f): V(\Gamma) \to V(\Gamma')$, $A(f): A(\Gamma) \to A(\Gamma')$ which are compatible with s, e, τ, (thus,

$$s_{\Gamma'}A(f) = V(f)s_{\Gamma}, \quad e_{\Gamma'}A(f) = V(f)e_{\Gamma}, \quad \tau_{\Gamma'}V(f) = V(f)\tau_{\Gamma}).$$

Let Γ be a stable translation quiver. A group G of automorphisms of Γ is said to be admissible provided for $x \in V(\Gamma)$, any orbit of $V(\Gamma)$ under G intersects x^+ in at most one vertex. For an admissible group G of automorphisms of Γ we may form Γ/G, this is the translation quiver defined as follows: $V(\Gamma/G)$ is the set of G-orbits on $V(\Gamma)$, $A(\Gamma/G)$ is the set of G-orbits on $A(\Gamma)$, and $s_{\Gamma/G}, e_{\Gamma/G}\ \tau_{\Gamma/G}$ are induced by $s_{\Gamma}, e_{\Gamma}, \tau_{\Gamma}$, respectively. (Note that we follow Riedtmann [Ri], but with some slight changes in order to take care of the fact that our translation quivers are allowed to have loops).

Smooth stable translation quivers have been considered before. Following Butler-Ringel [BR], let us consider the stable translation quiver Π defined as follows: $V(\Pi) = \{(a, b) \in \mathbf{Z}^2 \mid a \equiv b \pmod 2\}$ with arrows $(a, b) \to (a+1, b+1)$, and $(a, b) \to (a+1, b-1)$ and with translation $\tau_{\Pi}(a,b)=(a-2,b)$, for all $(a,b)\in V(\Pi)$. Observe that Π is isomorphic to $\mathbf{Z}A_{\infty}^{\infty}$ as defined in [HPR]. Given $(s,t)\in\mathbf{Z}^2$, consider the automorphism g_{st} of Π defined by $g_{st}(a,b)=(a+s-t,b+s+t)$, let $\langle g_{st}\rangle$ be the group of automorphisms generated by g_{st}, and define $\Pi_{st}=\Pi/\langle g_{st}\rangle$. Since $g_{st}^{-1}=g_{-s,-t}$, we have $\Pi_{st}=\Pi_{-s,-t}$; also, the automorphism ι of Π defined by $(a,b)\to(a,-b)$ yields an isomorphism of Π_{st} and $\Pi_{-t,-s}$; thus any stable translation quiver of the form Π_{st} is isomorphic to one satisfying $0\leqslant s$ and $-s\leqslant t\leqslant s$. Of course, $\Pi_{00}=\Pi$.

Proposition The smooth non-period stable translation quivers are of the form Π_{st} with $0<s$ and $-s<t\leqslant s$, or of the form Π.

Proof Let Γ be a smooth stable translation quiver. Clearly, $\Gamma\simeq\Pi/G$ for some admissible group of automorphisms of Π (see Riedtmann [Ri]). Note that elements of the form $g_{ss}\iota$ with $s\in\mathbf{Z}$ cannot belong to G, since $g_{ss}\iota$ interchanges the two elements of $(s,s)^{+}$. On the other hand, for $s\neq t$ in $\mathbf{Z}$, we have $(g_{ss}\iota)^2(0,0)=(2s-2t,0)$, thus $g_{st}\iota$ cannot belong to G in case Π/G is non-periodic. Thus, assume Π/G is non-periodic. Let $g\in G$ and write $g(0,0)$ in the form $g(0,0)=(s-t,s+t)$ for some, $s,t\in\mathbf{Z}$, thus $g_{st}^{-1}g$ fixes $(0,0)$, therefore $g=g_{st}$ or $g=g_{st}\iota$, but as we have seen, the latter is impossible. Therefore all elements of G are of the form g_{st}. Also, if $(s,t),(s',t')\in\mathbf{Z}^2$, then $g_{st}^{-(s'+t')}\ g_{s't'}^{s+t}(0,0)=(2s't-2st',0)$. Both g_{st} and $g_{s't'}$ can belong to G only in case $s't=st'$, since otherwise Π/G would have a periodic vertex. It follows that G is a cyclic group, thus $\Gamma\simeq\Pi_{st}$ for some $0\leqslant s$ and $-s\leqslant t\leqslant s$. However, the cases $0<s=-t$ are impossible, since in these cases Π_{st} has periodic vertices.

Recall that $\widetilde{A}_{st}$ denotes the quiver(fig. 4. 1).

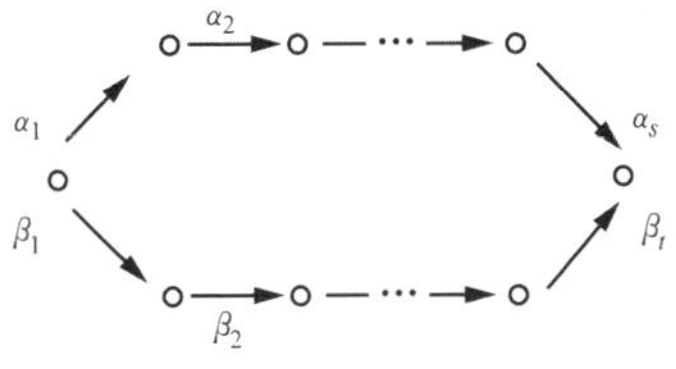

Fig. 4. 1

where $0\leqslant t\leqslant s$ and $0<s$ (note that we allow $t=0$; in this case we deal with an oriented cycle!). Also, A_{∞}^{∞} is the quiver with $V(A_{\infty}^{\infty})=\mathbf{Z}$, with arrows $z\rightarrow z+1$, for all $z\in\mathbf{Z}$. We observe that

$$\Pi_{st}\simeq\mathbf{Z}\bar{A}_{st}\text{ for }0<s\text{ and }0\leqslant t\leqslant s,$$

whereas, as we have noted above,

$$\Pi\simeq\mathbf{Z}A_{\infty}^{\infty}.$$

4.2 The cases $-s<t<0$. It remains to consider the stable translation quivers of the form Π_{st} with $-s<t<0$. They have a cyclic path which is not sectional, and thus they cannot be written in the form $\mathbf{Z}Q$, with Q a quiver.

Lemma Let $-s<t<0$. Let l be a subadditive function on Π_{st} with values in $\mathbb{N}_0$. Then l is additive and bounded.

Proof We write $\Pi_{st}=\Pi/\langle g_{st}\rangle$ and consider l as a g_{st}-invariant function $V(\Pi)\rightarrow\mathbb{N}_0$, thus l is a subadditive function on Π.

Given vertices x, y of Π, write $x\preceq y$ provided there is a path from x to y, and we denote by $[x, y]$ the set of all vertices z of Π satisfying $x\preceq z\preceq y$. The four neighbours of $x=(a, b)\in V(\Pi)$ will be denoted as follows: $x^{\cdot}=(a+1, b+1)$, $x_{.}=(a+1, b-1)$, ${}^{\cdot}x=(a-1, b+1)$, and ${}_{.}x=(a-1, b-1)$. If W is a finite subset of $V(\Pi)$, let $lW=\sum\limits_{w\in W}l(w)$.

Let $x\in V(\Pi)$. We claim that

$$l(x)+l(gx)\geqslant l(x+(s, s))+l(x+(-t, t)),$$

and that we have equality only in case

$$l(y)+l(\tau y)=l({}^{\cdot}y)+l({}_{.}y)\text{ for all }y\in[\tau^{-}x, gx].$$

For the proof, we add up the inequalities

$$l(y)+l(\tau y)\geqslant l({}^{\cdot}y)+l({}_{.}y),$$

with $y\in[\tau^{-}x, z]$, where $z=gx$. We obtain

$$l[\tau^{-}x, z]+l[x, \tau z]\geqslant l[x^{\cdot}, {}^{\cdot}z]+t\,[x_{.}, {}_{.}z].$$

The left hand side of this inequality is equal to

$$l(x)+l(z)+l[x^{\cdot}, {}_{.}z]+l[x_{.}, {}^{\cdot}z],$$

the right hand side is equal to

$$l(x+(s, s))+l(x+(-t, t))+l[x^{\cdot}, {}_{.}z]+l[x_{.}, {}^{\cdot}z].$$

We can subtract $l[x^{\cdot}, .z]+l[x., {}^{\cdot}z]$ from both sides in order to obtain the required inequality. Also, we see that we obtain an equality only in case we have started with equalities for all $y\in[\tau^{-}x, z]$.

Next, we claim $2l(x)\geqslant l(x+(s,s))+l(x-(s,s))$. For, we have

$$2l(x)=l(x)+l(gx)\geqslant l(x+(s,s))+l(x+(-t,t))$$
$$=l(x+(s,s))+l(x-(s,s)),$$

where we use that $g(x-(s,s))=x+(-t,t)$ and the fact that l is g-invariant.

It follows that for a fixed x, the map $f:\mathbf{Z}\to\mathbb{N}_0$, defined by $f(i)=l(x+(i\cdot s, i\cdot s))$ for $i\in\mathbf{Z}$ is constant: choose i_0 with $f(i_0)\leqslant f(i)$ for all $i\in\mathbf{Z}$ and conclude that $f(i_0+l)=f(i_0)=f(i_0-1)$, as in the proof of Lemma 3 of [HPR].

Thus, for all $x\in V(\Pi)$, we have $l(x+(s,s))=l(x)$. Similarly, we also have $l(x+(-t,t))=l(x)$. In particular, the only values taken by l are the numbers $l(y)$ with $y\in[\tau^{-}x, gx]$, and therefore l is bounded. Also, the equality $l(x)+l(gx)=l(x+(s,s))+l(x+(-t,t))$ implies $l(y)+l(\tau y)=l({}^{\cdot}y)+l(.y)$ for $y\in[\tau^{-}x, gx]$ thus l is additive.

Remark The results of this section establish the main theorem in case we deal with a smooth stable translation quiver. We summarize the discussion above as table 4. 1.

Table 4. 1

Conditions	Shape of Π_{st}	Is there a subadditive non-additive l?	Is additive l bounded?
$0<s, 0<t\leqslant s$	$\mathbf{Z}\widetilde{A}_{st}$	yes	yes
$0<s, t=0$	$\mathbf{Z}\widetilde{A}_{s0}$	no	can be unbounded
$0<s, -s<t<0$		no	yes
$0<s, -s=t$	$\mathbf{Z}A_{\infty}^{\infty}/(s)$	no	yes
$s=0, t=0$	$\mathbf{Z}A_{\infty}^{\infty}$	yes	can be unbounded

(In fact, when $s>0, t=0$, we consider $[x, gx.]$. Since $l(x)+l(gx.)=l(gx)+l(x.)$, we have $l(y)+l(\tau l)=l({}^{\cdot}y)+l(.y)$, for any $y\in[\tau^{-1}x, gx.]$. Therefore l is additive.)

§ 5. The non-periodic stable connected translation quivers

5.1 The orbit graph.

Let Γ be a non-periodic connected stable translation quiver. Give $x \in V(\Gamma)$, we denote by $x^{\mathcal{T}}$ its τ-orbit, given $\alpha \in A(\Gamma)$, we denote by $\alpha^{\mathcal{T}}$ its τ-orbit (by definition, $\tau\alpha = \sigma^2\alpha$). The orbit graph $\Gamma^{\mathcal{T}}$ of Γ is defined as follows: $V(\Gamma^{\mathcal{T}})$ is the set of τ-orbfis of vertices of Γ, and $A(\Gamma^{\mathcal{T}})$ is the set of τ-orbits of arrows of Γ, the maps $s_{\Gamma^{\mathcal{T}}}$ and $e_{\Gamma^{\mathcal{T}}}$ are induced by s_Γ and e_Γ, respectively, and $\iota_{\Gamma^{\mathcal{T}}}(\alpha^{\mathcal{T}}) = (\sigma\alpha)^{\mathcal{T}}$, for $\alpha \in A(\Gamma)$. In order to see that $\Gamma^{\mathcal{T}}$ is a graph, we only have to observe that $\alpha^{\mathcal{T}} \neq (\sigma\alpha)^{\mathcal{T}}$ for any $\alpha \in A(\Gamma)$. (For, assume $\sigma\alpha = \tau^t\alpha$ for $t \in \mathbf{Z}$, then $\tau\alpha = \sigma^2\alpha = \sigma\tau^t\alpha = \tau^t\sigma\alpha = \tau^{2t}\alpha$, therefore $\alpha = \tau^{2t-1}\alpha$. But this implies that $s_\Gamma(\alpha)$ and $e_\Gamma(\alpha)$ are periodic vertices).

Remark In case the group generated by the automorphism τ is admissible, we may consider instead of the orbit graph also the stable translation quiver $\Gamma_{\langle\tau\rangle}$. However, Π_{10} is an example of a non-periodic translation quiver where $\langle\tau\rangle$ is not admissible.

Lemma Let $w = (y \mid \beta_1^{\varepsilon_1}, \beta_2^{\varepsilon_2}, \cdots, \beta_m^{\varepsilon_m} \mid y')$ be a walk in $\Gamma^{\mathcal{T}}$. Let $x \in V(\Gamma)$ with $x^{\mathcal{T}} = y$. Then there exists a unique walk $w_x = (x \mid \alpha_2^{\varepsilon_1}, \alpha_2^{\varepsilon_2}, \cdots, \alpha_m^{\varepsilon_m} \mid x')$ in Γ with $\alpha_i^{\mathcal{T}} = \beta_i$, for all $1 \leqslant i \leqslant m$.

Proof First, we note the following: let $\alpha: x \to x'$ be an arrow of Γ. Then $\tau^t\alpha$ is the only arrow in $\alpha^{\mathcal{T}}$ starting at $\tau^t x$, and the only arrow in $\alpha^{\mathcal{T}}$ ending in $\tau^t x'$, since x and x' are non-periodic.

The proof of the lemma is by induction on m. The assertion is clear for $m = 0$. Consider now a walk $w = (y \mid \beta_1^{\varepsilon_1}, \beta_2^{\varepsilon_2}, \cdots, \beta_m^{\varepsilon_m} \mid y')$ in $\Gamma^{\mathcal{T}}$ with $m \geqslant 1$ and take some $x \in V(\Gamma)$ with $x^{\mathcal{T}} = y$. By induction there is a unique walk $(x \mid \alpha_1^{\varepsilon_1}, \alpha_2^{\varepsilon_2}, \cdots, \alpha_{m-1}^{\varepsilon_{m-1}} \mid x'')$ in Γ with $\alpha_i^{\mathcal{T}} = \beta_i$ for all $1 \leqslant i \leqslant m-1$. Now, $x'' = e_\Gamma(\alpha_{m-1}^{\varepsilon_{m-1}})$, therefore $(x'')^{\mathcal{T}} = e_{\Gamma^{\mathcal{T}}}(\beta_{m-1}^{\varepsilon_{m-1}}) = s_{\Gamma^{\mathcal{T}}}(\beta_m^{\varepsilon_m})$. Our first observation yields a unique arrow $\alpha_m \in \Gamma$ such that $\alpha_m^{\mathcal{T}} = \beta_m$ and

$s_\Gamma(\alpha_m^{\varepsilon_m}) = x''$. Let $e_\Gamma(\alpha_m^{\varepsilon_m}) = x'$; then $w_x = (x \mid \alpha_1^{\varepsilon_1}, \alpha_2^{\varepsilon_2}, \cdots, \alpha_m^{\varepsilon_m} \mid x')$ is the required walk.

5.2 The map ϑ_Γ.

Fix some vertex x of Γ. Therefore we have fixed some $x^{\mathcal{T}} \in \Gamma^{\mathcal{T}}$, and we may consider the fundamental group $\pi_1(\Gamma^{\mathcal{T}}, x^{\mathcal{T}})$.

Lemma Given a cyclic path $w = (x^{\mathcal{T}} \mid \beta_1, \beta_2, \cdots, \beta_m \mid x^{\mathcal{T}})$ in $\Gamma^{\mathcal{T}}$. Let $\vartheta_\Gamma(w) = m + 2t$, where $e_\Gamma(w_x) = \tau^t x$. Then $\vartheta_\Gamma : \pi_1(\Gamma^{\mathcal{T}}, x^{\mathcal{T}}) \to \mathbf{Z}$ is a group homomorphism.

Proof Consider $w = (x^{\mathcal{T}} \mid \beta_1, \beta_2, \cdots, \beta_m \mid x^{\mathcal{T}})$ and $\nu = (x^{\mathcal{T}} \mid \beta_{m+1}, \beta_{m+2}, \cdots, \beta_n \mid x^{\mathcal{T}})$. Let $w_x = (x \mid \alpha_1, \alpha_2, \cdots, \alpha_m \mid \tau^t x)$ and $v_x = (x \mid \alpha_{m+1}, \alpha_{m+2}, \cdots, \alpha_n \mid \tau^s x)$. Then

$$(wv)_x = (x \mid \alpha_1, \alpha_2, \cdots, \alpha_m, \tau^t \alpha_{m+1}, \tau^t \alpha_{m+2}, \cdots, \tau^t \alpha_n \mid \tau^{s+t} x).$$

Therefore

$$\vartheta_\Gamma(wv) = m + n + s + t = \vartheta_\Gamma(w) + \vartheta_\Gamma(v).$$

Proposition The map $\vartheta_\Gamma : \pi_1(\Gamma^{\mathcal{T}}, x^{\mathcal{T}}) \to \mathbf{Z}$ induces a group homomorphism $H_1(\Gamma^{\mathcal{T}}) \to \mathbf{Z}$ which is independent of x (and which will be denoted by ϑ_Γ, again).

Proof Since $H_1(\Gamma^{\mathcal{T}})$ is the commutator factor group of $\pi_1(\Gamma^{\mathcal{T}}, x^{\mathcal{T}})$, we see that $\vartheta_\Gamma : \pi_1(\Gamma^{\mathcal{T}}, x^{\mathcal{T}}) \to \mathbf{Z}$ factors through $H_1(\Gamma^{\mathcal{T}})$. In order to see that the induced map is independent of x, consider an arrow $\alpha : x' \to x$. Let $w = (x^{\mathcal{T}} \mid \beta_1, \beta_2, \cdots, \beta_m \mid x^{\mathcal{T}})$ be a cycle in $\Gamma^{\mathcal{T}}$, and let $w_x = (x \mid \alpha_1, \alpha_2, \cdots, \alpha_m \mid \tau^t x)$. Thus $\vartheta_\Gamma(w) = m + 2t$. Similarly, let $w' = ((x')^{\mathcal{T}} \mid \alpha^{\mathcal{T}}, \beta_1, \beta_2, \cdots, \beta_m, (\sigma\alpha)^{\mathcal{T}} \mid (x')^{\mathcal{T}})$. Then the canonical images of $\overline{w} \in \pi_1(\Gamma^{\mathcal{T}}, x^{\mathcal{T}})$ and $\overline{w}' \in \pi_1(\Gamma^{\mathcal{T}}, x'^{\mathcal{T}})$ in $H_1(\Gamma^{\mathcal{T}})$ coincide. On the other hand we have $w'_{x'} = (x' \mid \alpha, \alpha_1, \alpha_2, \cdots, \alpha_m, \tau^{t-1}\sigma\alpha \mid \tau^{t-1} x')$, since $\tau^{t-1}\sigma\alpha : \tau^t x \to \tau^{t-1} x'$ is the only arrow in $(\sigma\alpha)^{\mathcal{T}}$ starting at $\tau^t x$, therefore

$$\vartheta_\Gamma(w') = m + 2 + 2(t-1) = \vartheta_\Gamma(w).$$

5.3 A special subquiverof Γ.

Let $w = (x^{\mathcal{T}} \mid \beta_1^{\varepsilon_1}, \beta_2^{\varepsilon_2}, \cdots, \beta_m^{\varepsilon_m} \mid x^{\mathcal{T}})$ be a cyclic walk in $\Gamma^{\mathcal{T}}$. We define a corresponding cyclic path in $\Gamma^{\mathcal{T}}$ by

$$\widetilde{w}=(x^{\mathcal{T}}\mid \gamma_1,\gamma_2,\cdots,\gamma_m\mid x^{\mathcal{T}}),\text{where } \gamma_i=\begin{cases}\beta_i, & \text{for } \varepsilon_i=1,\\ \iota\beta_i, & \text{for } \varepsilon_i=-1,\end{cases}$$

and we define $\vartheta_\Gamma(w):=\vartheta_\Gamma(\widetilde{w})$. Given w and x, we have defined above the walk w_x in Γ, and we can use $\vartheta_\Gamma(w)$ in order to determine the endpoint of w_x.

Lemma a Let $x\in V(\Gamma)$, and $w=(x^{\mathcal{T}}\mid\beta_1^{\varepsilon_1},\beta_2^{\varepsilon_2},\cdots,\beta_m^{\varepsilon_m}\mid x^{\mathcal{T}})$ a cyclic walk in $\Gamma^{\mathcal{T}}$. Let $t=\frac{1}{2}\left(\vartheta_\Gamma(w)-\sum_{i=1}^m\varepsilon_i\right)$. Then $t\in\mathbf{Z}$ and $e(w_x)=\tau^t x$.

Proof We use induction on the number of indices i with $\varepsilon_i=-1$. If all $\varepsilon_i=1$, then $w=\bar{w}$, and by definition of $\vartheta_\Gamma(w)$ we have $\vartheta_\Gamma(w)=m+2t$ where $e(w_x)=\tau^t x$. Consider now some $w=(x^{\mathcal{T}}\mid\beta_1^{\varepsilon_1},\beta_2^{\varepsilon_2},\cdots,\beta_m^{\varepsilon_m}\mid x^{\mathcal{T}})$ with $\varepsilon_r=-1$ for some $1\leqslant r\leqslant m$. Let $\nu=(x^{\mathcal{T}}\mid\beta_1^{\varepsilon_1},\beta_2^{\varepsilon_2},\cdots,\beta_{r-1}^{\varepsilon_{r-1}},\iota\beta_r,\beta_{r+1}^{\varepsilon_{r+1}},\beta_{r+2}^{\varepsilon_{r+2}},\cdots,\beta_m^{\varepsilon_m}\mid x^{r})$; thus $\widetilde{w}=\tilde{\nu}$, and $\vartheta_\Gamma(w)=\vartheta_\Gamma(\nu)$. By induction, $e(\nu_x)=\tau^s x$, where $s=\frac{1}{2}\left(\vartheta_\Gamma(\nu)-\left(\sum_{i=1}^{r-1}\varepsilon_i+1+\sum_{i=r+1}^m\varepsilon_i\right)\right)=\frac{1}{2}\left(\vartheta_\Gamma(\mathrm{w})-\sum_{i=1}^m\varepsilon_i-2\right)$. Let $\nu_x=(x\mid\alpha_1^{\varepsilon_1},\alpha_2^{\varepsilon_2},\cdots,\alpha_{r-1}^{\varepsilon_{r-1}},\alpha_r,\alpha_{r+1}^{\varepsilon_{r+1}},\alpha_{r+2}^{\varepsilon_{r+2}}\cdots,\alpha_m^{\varepsilon_m}\mid\tau^s x)$. Then $w_x=(x\mid\alpha_1^{\varepsilon_1},\alpha_2^{\varepsilon_2},\cdots,\alpha_{r-1}^{\varepsilon_{r-1}},(\sigma\alpha_r)^{-1},(\tau\alpha_{r+1})^{\varepsilon_{r+1}},\cdots,(\tau\alpha_m)^{\varepsilon_m}\mid\tau^{s+1}x)$, therefore $e(w_x)=\tau^{s+1}x$, and

$$s+1=\frac{1}{2}\left(\vartheta_\Gamma(w)-\sum_{i=1}^m\varepsilon_i-2\right)+1=\frac{1}{2}\left(\vartheta_\Gamma(w)-\sum_{i=1}^m\varepsilon_i\right).$$

Let Ω be an orientation on $\Gamma^{\mathcal{T}}$. A cyclic walk $w=(x^{\mathcal{T}}\mid\beta_1^{\varepsilon_1},\beta_2^{\varepsilon_2},\cdots,\beta_m^{\varepsilon_m}\mid x^{\mathcal{T}})$ in $(\Gamma^{\mathcal{T}},\Omega)$ is a cyclic walk in $\Gamma^{\mathcal{T}}$ with all $\beta_i\in\Omega$; if we consider the corresponding cycle $\sum_{i=1}^m\varepsilon_i\beta_i$ in $H_1(\Gamma^{\mathcal{T}})$, then $\vartheta_\Omega\left(\sum_{i=1}^m\varepsilon_i\beta_i\right)=\sum_{i=1}^m\varepsilon_i$. Thus we may define $\vartheta_\Omega(w)=\sum_{i=1}^m\varepsilon_i$. The previous lemma can be reformulated in this case as follows:

Lemma b Let Ω be an orientation on $\Gamma^{\mathcal{T}}$. Let $x\in V(\Gamma)$, and $(x^{\mathcal{T}}\mid\beta_1^{\varepsilon_1},\beta_2^{\varepsilon_2},\cdots,\beta_m^{\varepsilon_m}\mid x^{\mathcal{T}})$ a cyclic walk in $(\Gamma^{\mathcal{T}},\Omega)$. Then $e(w_x)=\tau^t x$, where

$$t=\frac{1}{2}(\vartheta_\Gamma(w)-\vartheta_\Omega(w)).$$

Note that both maps $\vartheta_\Gamma, \vartheta_\Omega$ are defined on $H_1(\Gamma^T)$, and the case when $\vartheta_\Gamma = \vartheta_\Omega$ will be of great importance:

Proposition Let Ω be an orientation on Γ^T with $\vartheta_\Gamma = \vartheta_\Omega$. Let $x \in V(\Gamma)$. If w is a cyclic walk in (Γ^T, Ω) starting at x^T, then w_x is a cyclic walk in Γ starting at x.

Corollary Let Ω be an orientation on Γ^T with $\vartheta_\Gamma = \vartheta_\Omega$. Let $x \in V(\Gamma)$. Let Q be the following subquiver of Γ: its vertices are the end vertices of walks of the form w_x, its arrows are the arrows occurring in the walks of the form w_x, with w a walk in (Γ^T, Ω) starting at x^T. Then $V(Q)$ contains precisely one vertex of each τ-orbit of $V(\Gamma)$, and $A(Q)$ contains precisely one arrow of each σ-orbit of $A(\Gamma)$.

Proof Let w, v be walks in (Γ^T, Ω) starting at x^T and having the same end vertex. Let $e(v_x) = y, e(w_x) = \tau^t y$. We claim that $t = 0$. Now wv^{-1} is a cyclic walk in (Γ^T, Ω) (or of length zero), and $(wv^{-1})_x$ ends in $\tau^t x$. It follows from the previous proposition that $t = 0$.

5.4 If Γ is not smooth, then ϑ_Γ is tempered.

Let Γ be a non-periodic connected stable translation quiver, and let l be a non-zero subadditive function on Γ with values in $\mathbb{N}_0$.

Lemma For any vertex $x \in V(\Gamma)$, there is t with $l(\tau^t x) \neq 0$.

Proof Let $l(y) \neq 0$, for some $y \in V(\Gamma)$. If there is an arrow $y \to z$. Then $l(\tau z) + l(z) \geq l(y) > 0$. Thus $l(z) \neq 0$ or $l(\tau z) \neq 0$. It follows that the set of vertices y^T in Γ^T with $l(\tau^t y) \neq 0$ for at least one $t \in \mathbf{Z}$ is non-empty and closed under neighbours. Thus it is all of $V(\Gamma^T)$.

A stable translation subquiver Γ' of Γ is a stable translation quiver Γ' such that $V(\Gamma') \subseteq V(\Gamma)$, $A(\Gamma') \subseteq A(\Gamma)$ and $s_{\Gamma'}, e_{\Gamma'}, \tau_{\Gamma'}$ are the restrictions of $s_\Gamma, e_\Gamma, \tau_\Gamma$, respectively.

Corollary If Γ' is a stable translation subquiver of Γ, with $\Gamma' \neq \varnothing$ and $\Gamma' \neq \Gamma$, then $l \mid \Gamma'$ is not additive.

Proof Choose $y \in V(\Gamma) \setminus V(\Gamma')$, $z \in V(\Gamma')$ with an arrow $y \to z$. We can assume in addition, that $l(y) \neq 0$ (otherwise shift by some power of t). But then $l(\tau z)+l(z)\geqslant \sum_{y_i\in z^-} l(y_i)$, where z^- is the set of vertices y_i in $V(\Gamma)$ with an arrow $y_i\to z$. On Γ', we have to delete on the right side of the inequality at least $l(y)$. Thus we obtain a proper inequality.

Proposition Assume that Γ is not smooth. Then $\vartheta_\Gamma: H_1(\Gamma^{\mathcal{T}})\to \mathbf{Z}$ is tempered.

Proof Let $\mathcal{C}$ be an elementary cycle in $\Gamma^{\mathcal{T}}$, say $\sum_{i=1}^{m}\beta_i$, where $w=(y\,|\,\beta_1,\beta_2,\cdots,\beta_m\,|\,y)$ is a reduced, elementary, cyclic path in $\Gamma^{\mathcal{T}}$. Let $y_i=s_{\Gamma^{\mathcal{T}}}(\beta_i)$ for $1\leqslant i\leqslant m$. Let $\Gamma(\mathcal{C})$ be the stable translation subquiver of Γ with $V(\Gamma(\mathcal{C}))$ the set of all vertices $x\in y_i$ for some $1\leqslant i\leqslant m$, and with $A(\Gamma(\mathcal{C}))$ the set of all arrows α such that α or $\sigma\alpha$ belongs to β_i for some $1\leqslant i\leqslant m$. Clearly, $\Gamma(\mathcal{C})$ is smooth. Since we assume that Γ is not smooth, it follows that $\Gamma(\mathcal{C})\neq\Gamma$, thus $l\,|\,\Gamma(\mathcal{C})$ is not additive. According to the table of Section 4, we see that $\Gamma(\mathcal{C})$ is isomorphic to Π_{st} with $0<s$ and $0<t\leqslant s$. Since the definition of ϑ_Γ does not depend on the chosen base point, we may assume that our base point x satisfies $x^{\mathcal{T}}=y$. Without loss of generality, we can assume that w_x is either the image in Π_{st} of the path $(0,0)\to(1,1)\to\cdots\to(s+t,s+t)$ in Π, this is a path of length $m=s+t$, and $e(w_x)=\tau^{-t}x$, therefore $\vartheta_\Gamma(w)=s+t-2t=s-t$; or else that w_x is the image in Π_{st} of the path $(0,0)\to(1,-1)\to\cdots\to(s+t,-s-t)$ in Π, this is a path of length $m=s+t$, and $e(w_x)=\tau^{-s}x$, therefore $\vartheta_\Gamma(w)=s+t-2s=t-s$. Altogether, we see that $|\vartheta_\Gamma(w)|\equiv m \pmod 2$ and that $|\vartheta_\Gamma(w)|\leqslant m$.

5.5 Proof of the main theorem.

For the proof of the main theorem, we can assume that Γ is not smooth. By Proposition 5.4, the map $\vartheta_\Gamma: H_1(\Gamma^{\mathcal{T}})\to\mathbf{Z}$ is tempered. According to Section 3, there exists an orientation Ω on $\Gamma^{\mathcal{T}}$ such that $\vartheta_\Gamma=\vartheta_\Omega$. Corollary 5.3 yields a subquiver Q of Γ such that $V(Q)$ intersects every τ-orbit of $V(\Gamma)$ in precisely one vertex, and such that $A(Q)$ contains precisely one arrow of each σ-orbit of $A(\Gamma)$. Since Γ is non-periodic, it follows that $\Gamma \simeq \mathbf{Z}Q$. This completes the proof.

5.6 The existence of cyclic paths.

Corollary Let $\mathcal{C}$ be a non-periodic connected valued stable translation quiver, with a non-zero subadditive function l with values in $\mathbb{N}_0$. Assume that $\mathcal{C}$ has a cyclic path. Then $\mathcal{C}$ is smooth and l is additive.

Remark More precisely, we will show that under the given assumptions, either $\mathcal{C}=\Pi_{st}$ with $-s<t<0$, and the cyclic path is non-sectional, or else $\mathcal{C}=\mathbf{Z}\widetilde{A}_{s0}$ for some $0<s$, and the cyclic path is sectional.

Proof According to the main theorem mentioned in Section 1, and the table in Section 4, either $\mathcal{C}=\Pi_{st}$ for $0<s$, and $-s<t<0$, or $\mathcal{C}=\mathbf{Z}\mathfrak{S}$, for some valued quiver $\mathfrak{S}$. In the first case, there exits a cyclic path which is non-sectional. When $\mathcal{C}=\mathbf{Z}\mathfrak{S}$, the cyclic path has to be sectional and $\mathfrak{S}$ includes a cyclic path by the lemma of Section 1. Let $\Gamma=(V(\mathcal{C}), A(\mathcal{C}), s_{\mathcal{C}}, e_{\mathcal{C}}, \tau_{\mathcal{C}})$ be the corresponding stable translation quiver, with trivial valuation, and $\Gamma=\mathbf{Z}Q$, so that Q corresponds to $\mathfrak{S}$. Assume that $w_x=(x\mid\alpha_1,\alpha_2,\cdots,\alpha_n\mid x)$ is a cyclic path in Q, and let $s(\alpha_i), s(\alpha_{i+1}),\cdots, s(\alpha_j)$ be pairwise different, but $s(\alpha_i)=e(\alpha_j)$ for some $1\leqslant i\leqslant j\leqslant n$. Let $y=s(\alpha_i)$, and consider $w=(y^{\mathcal{T}}\mid\beta_i,\beta_{i+1},\cdots,\beta_j\mid y^{\mathcal{T}})$ in $\Gamma^{\mathcal{T}}$ with $\beta_k=\alpha_k^{\mathcal{T}}$ for $k=i, i+1, \cdots, j$. We want to show that w is reduced and elementary. First of all, $\beta_{k+1}\neq\iota\beta_k$, for $k=i, i+1,\cdots, j-1$, and $\beta_j\neq\iota\beta_i$, since $A(Q)$ meets each σ-orbit only once; and second, $s(\beta_k)\neq s(\beta_l)$, for $k\neq l$, since $V(Q)$ meets each τ-orbit only once (5.3, Corollary). Therefore $c=\sum\limits_{k=1}^{j}\beta_k$ is an elementary cycle. Consider the stable translation subquiver $\Gamma(\mathcal{C})$, with $V(\Gamma(\mathcal{C}))=\{x\mid x^{\mathcal{T}}=s(\beta_k)$, for some $i\leqslant k\leqslant j\}$, $A(\Gamma(\mathcal{C}))=\{\alpha\mid\alpha^{\mathcal{T}}$ or $(\sigma\alpha)^{\mathcal{T}}=\beta_k$ for some $i\leqslant k\leqslant j\}$. If $\mathcal{C}\neq\Gamma(\mathcal{C})$, $l|_{\Gamma(\mathcal{C})}$, is not additive. By the table of Section 4, we have $\Gamma(\mathcal{C})=\mathbf{Z}\widetilde{A}_{st}$, for $0<s$, $0<t\leqslant s$. But there is no cyclic path in $\widetilde{A}_{st}$, a contradiction. Thus $\mathcal{C}=\Gamma(\mathcal{C})$ is smooth. Since $\mathfrak{S}$ includes a cyclic path, the only possibility is $\mathfrak{S}=\widetilde{A}_{s0}$.

Appendix 1 The map $\vartheta_\Gamma: H_1(\Gamma^{\mathcal{T}})\to\mathbf{Z}$.

There is a more sophisticated way to construct ϑ_Γ. Following Bongartz-Gabriel [BG], we consider Γ as a simplicial set (the loops which we allow do not lead to difficulties). Under the assumption that Γ is connected and non-periodic, we easily see that the geometric realizations $|\Gamma|$ of Γ and of the orbit graph $\Gamma^{\mathcal{T}}$ are homotopic; just copy the proof in [BG]: first, we choose one fixed arrow in any σ-orbit of $A(\Gamma)$, and let X be the subspace of $|\Gamma|$ formed by these arrows and all "degree 2-arrows", so that X is a strong deformation retract; finally, we shrink all "degree 2-arrows" and obtain the geometric realization of orbit graph $\Gamma^{\mathcal{T}}$ (of course, for the geometric realization of a graph Y we have to identify α and $\sigma\alpha$, for any $\alpha \in A(Y)$.)

The first homology group $H_1(|\Gamma|)$ can be calculated as the homology group of the following complex

$$C_2(\Gamma) \xrightarrow{\delta_2} C_1(\Gamma) \xrightarrow{\delta_1} C_0(\Gamma),$$

where $C_2(\Gamma)$ is the free abelian group on the set of triangles, $C_1(\Gamma)$ the free abelian group on the set of all (degree 1 and degree 2) arrows, and $C_0(\Gamma)$ the free abelian group on the set of vertices, and δ_2, δ_1 are the corresponding boundary maps. To be more precise: any vertex z yields a "degree 2-arrow" $\tau z \dashrightarrow z$, and $\delta_1(\tau z \dashrightarrow z) = z - \tau z$, whereas $\delta_1(x \to y) = y - x$; any arrow $y \to z$ yields a triangle

$$\begin{array}{ccc} & y & \\ \nearrow & & \searrow \\ \tau z & \dashrightarrow & z \end{array}, \text{ and } \delta_2\left(\begin{array}{ccc} & y & \\ \nearrow & & \searrow \\ \tau z & \dashrightarrow & z \end{array}\right) = (\tau z \to y) + (y \to z) - (\tau z \dashrightarrow z).$$

The degree map $C_1(\Gamma) \xrightarrow{d} \mathbf{Z}$ with $d(x \to y) = 1$, $d(\tau z \dashrightarrow z) = 2$ vanishes on the image of δ_2, thus it yields a homomorphism $\bar{d}: H_1(\Gamma) \to \mathbf{Z}$. Combining $\bar{d}$ with the canonical homotopy equivalence $|\Gamma| \to |\Gamma^{\mathcal{T}}|$, we obtain $\vartheta_\Gamma: H_1(\Gamma^{\mathcal{T}}) \to \mathbf{Z}$.

Appendix 2　G-invariant complete sections

There is another way to construct the subquiver Q of Γ. A morphism of quiver $f: B \to Q$ is the disjoint union of two maps $V(f): V(B) \to V(Q)$, and $A(f): A(B) \to A(Q)$, with $f(s(\alpha)) = s(f(\alpha))$,

$f(e(\alpha))=e(f(\alpha))$. We call f a *covering* provided for any $x\in V(Q)$, the map $V(f)$ yields a bijection between x^+ and $f(x)^+$, and between x^- and $f(x)^-$. The quiver B is called an *oriented tree*, if $\overline{B}$ is a tree. A covering $f:B\to Q$ is said to be *universal*, if B is an oriented tree. If $f: B\to Q$ is a universal covering of quivers, then $\overline{f}:\overline{B}\to\overline{Q}$ is a universal covering of graphs. A *morphism of stable translation quivers* $f:\Gamma_1\to\Gamma_2$ is a morphism of quivers, with $f(\tau x)=\tau f(x)$ for any vertex x in Γ_1. If $f:\Gamma_1\to\Gamma_2$ is both a morphism of stable translation quivers and a covering of quivers, then it is called a covering of stable translation quivers, and f is a universal covering of stable translation quivers, if, in addition, $\Gamma_1^{\mathcal{T}}$ is a tree.

Let $\Gamma=\mathbf{Z}B/G$ be a non-periodic connected stable translation quiver with trivial valuation, where B is an oriented tree, and G is an admissible automorphism group. Here, $\mathbf{Z}B\xrightarrow{p}\Gamma$ is just the universal covering of Γ (see [BG]).

Let $q:\overline{B}\to\Gamma^{\mathcal{T}}$ be defined by sending $\alpha^{\mathcal{T}}$ to $p(a)^{\mathcal{T}}$, thus q is a universal covering of graphs, since $\overline{B}$ is a tree and q is bijective at any star. We have the following commutative (fig. 5.1).

$$\begin{array}{ccc} \mathbf{Z}B & \xrightarrow{\pi} & \overline{B} \\ {\scriptstyle p}\downarrow & & \downarrow{\scriptstyle q} \\ \Gamma & \xrightarrow[\pi]{} & \Gamma^{\mathrm{T}} \end{array}$$

Fig. 5. 1

where π is the orbit map, sending the vertex x to $x^{\mathcal{T}}$ and the arrow α to $\alpha^{\mathcal{T}}$.

According to algebraic topology, there is an isomorphism $\varphi:G\xrightarrow{\sim}\pi_1(\Gamma,x)$ of groups [S]. The isomorphism φ is defined as follows: take a fixed base point $a\in p^{-1}(x)\subseteq\mathbf{Z}B$, for any g in G, and consider $g(a)\in p^{-1}(x)$. Let $w(a,g(a))$ be any walk from a to $g(a)$ in $\mathbf{Z}B$. Thus $p(w(a,g(a)))$ is a cyclic walk at x in Γ or of length zero, and we take

$$\varphi(g)=\overline{p(w(a,g(a)))}.$$

We define a degree map $d:G\to\mathbf{Z}$, such that $d(g)=\sum_{i=1}^{n}\varepsilon_i d(\alpha_i)$, where $w(a, g(a)) = d(w(a \mid \alpha_1^{\varepsilon_1}, \alpha_2^{\varepsilon_2}, \cdots, \alpha_n^{\varepsilon_n} \mid g(a)))$ (see [BG] and Appendix 1). Since $\mathbf{Z}B$ is simply connected, d is independent of the choice of the walk. On the other hand, $d(gh) = d(w(a, gh(a))) = d(w(a, g(a)) \cdot w(g(a), gh(a))) = d(w(a, g(a))) + d(w(g(a), gh(a)))$. The walk $(g(a) \mid g(\alpha_1)^{\varepsilon_1}, g(\alpha_2)^{\varepsilon_2}, \cdots, g(\alpha_n)^{\varepsilon_n} \mid gh(a))$ is obtained from $(a \mid \alpha_1^{\varepsilon_1}, \alpha_2^{\varepsilon_2}, \cdots, \alpha_n^{\varepsilon_n} \mid h(a))$ by applying g. Thus we have $d(gh) = d(w(a, g(a))) + d(w(a, h(a))) = d(g) + d(h)$, and therefore d is a group homomorphism.

Since Γ and Γ^T are homotopy equivalent (see Appendix 1), we have an isomorphism $h: \pi_1(\Gamma, x) \simeq \pi_1(\Gamma^T, x^T)$ sending the homotopy class of $(x \mid \alpha_1^{\varepsilon_1}, \alpha_2^{\varepsilon_2}, \cdots, \alpha_n^{\varepsilon_n} \mid x)$ to the homotopy class of $(x^T \mid \beta_1^{\varepsilon_1}, \beta_2^{\varepsilon_2}, \cdots, \beta_n^{\varepsilon_n} \mid x^T)$, where $\beta_i = \alpha_i^T$, $i = 1, 2, \cdots, n$. We have also a map $f: \pi_1(\Gamma^T, x^T) \to H_1(\Gamma^T)$, sending the homotopy class of $(x^T \mid \beta_1^{\varepsilon_1}, \beta_2^{\varepsilon_2}, \cdots, \beta_n^{\varepsilon_n} \mid x^T)$ to the cycle $\sum_{i=1}^{n} \varepsilon_i \beta_i$, and we note that $\ker f = \pi_1(\Gamma^T, x^T)'$. Therefore the map ϑ_Γ is induced as fig. 5. 2.

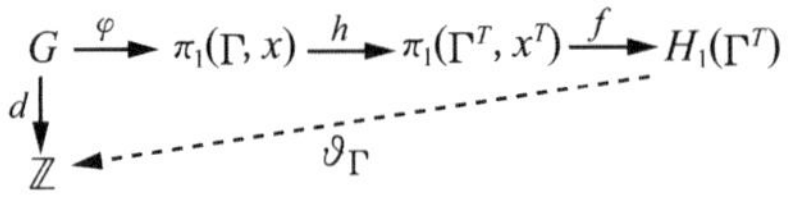

Fig. 5. 2

Assume that Γ is not smooth. Then ϑ_Γ is tempered (Proposition of Section 5. 4), and there exists an orientation Ω of Γ^T, with $\vartheta_\Gamma = \vartheta_\Omega$ (Proposition of Section 3).

Let B be an oriented tree. We call a subquiver U of $\mathbf{Z}B$ a section, if U is connected and intersects any τ-orbit of vertices at most once. If $\overline{U} = \overline{B}$, then U is said to be complete. If $GU = U$, then U is called G-invariant.

Proposition Let $\Gamma = \mathbf{Z}B/G$ be a non-periodic connected stable translation quiver, with a non-zero subadditive function l with values in $\mathbb{N}_0$ and assume that Γ is not smooth. Let Ω be an orientation of Γ^T, with $\vartheta_\Gamma = \vartheta_\Omega$, and let $(\overline{B}, \Omega)$ be the universal covering of (Γ^T, Ω). Then there exists a G-invariant complete section U of $\mathbf{Z}B$, which is homomorphic to $(\overline{B}, \Omega)$.

Proof Assume that we have a section U_0 of $\mathbf{Z}B$, which is isomorphic to a full oriented subtree $(\bar{B}_0, \Omega)$, of $(\bar{B}, \Omega)$, with $a \in V(U_0)$. Then $a^{\mathcal{T}} \in V(\bar{B}_0, \Omega)$, and $\alpha \in A(U_0)$. Then $\alpha^{\mathcal{T}} \in A(\bar{B}_0, \Omega)$. If $(\bar{B}_0, \Omega) \not\subseteq (\bar{B}_0, \Omega)$, then there exists a vertex $a \in V(U_0)$, such that $b^{\mathcal{T}} \cap V(U_0) = \varnothing$. But there is a σ-orbit between $a^{\mathcal{T}}$ and $b^{\mathcal{T}}$. Let $b = e(\alpha)$ in case we have $a^{\mathcal{T}} \to b^{\mathcal{T}}$ in $(\bar{B}, \Omega)$, and α is the unique arrow between $a^{\mathcal{T}}$ and $b^{\mathcal{T}}$ with $s(\alpha) = a$, and let $b = s(\alpha)$ in case we have $b^{\mathcal{T}} \to a^{\mathcal{T}}$ in $(\bar{B}, \Omega)$, and α is the unique arrow between $a^{\mathcal{T}}$ and $b^{\mathcal{T}}$ with $e(\alpha) = a$. Let U_1 be the subquiver with $V(U_1) = V(U_0) \cup \{b\}$, $A(U_1) = A(U_0) \cup \{\alpha\}$. Then U_1 is a section, and $U_1 \simeq (B_1, \Omega)$, where $V(B_1, \Omega) = V(B_0, \Omega) \cup \{b^{\mathcal{T}}\}$, $A(B_1, \Omega) = A(B_0, \Omega) \cup \{\alpha^{\mathcal{T}}\}$. By induction, we obtain a section $U \simeq (\bar{B}, \Omega)$. Since $\bar{U} = \bar{B}$, we see that U is a complete section of $\mathbf{Z}B$. It remains to prove that U is G-invariant.

For any $a \in V(U)$, and $g \in G$, assume that $g(a)^{\mathcal{T}} \cap V(U) = \{b\}$. Let $w(a, b) = (a \mid \alpha_1^{\varepsilon_1}, \alpha_2^{\varepsilon_2}, \cdots, \alpha_n^{\varepsilon_n} \mid b)$ be the unique geodesic from a to b inside U (see Section 2). We see that $\pi(w(a,b)) = (a^{\mathcal{T}} \mid \beta_1^{\varepsilon_1}, \beta_2^{\varepsilon_2}, \cdots, \beta_n^{\varepsilon_n} \mid b^{\mathcal{T}})$ is the unique geodesic from $a^{\mathcal{T}}$ to $b^{\mathcal{T}}$ in $(\bar{B}, \Omega)$, and $d(w(a,b)) = \sum_{i=1}^{n} \varepsilon_i$. On the other hand, let $w'(a, g(a))$ be any walk from a to $g(a)$ in $\mathbf{Z}B$. Since $b^{\mathcal{T}} = g(a)^{\mathcal{T}}$, it follows that $\pi\{w'(a, g(a))\}$ is equal to $w_1 \cdot w_1^{-1} \cdot \beta_1^{\varepsilon_1} \cdot w_2 \cdot w_2^{-1} \cdot \beta_2^{\varepsilon_2} \cdot \cdots \cdot \beta_n^{\varepsilon_n} \cdot w_{n+1} \cdot w_{n+1}^{-1}$, where w_i, $i = 1, 2, \cdots, n+1$, are walks in $(\bar{B}, \Omega)$, and where we also consider $\beta_i^{\varepsilon_i}$, $i = 1, 2, \cdots, n$, as walks. Therefore $\overline{q\pi(w')} = \overline{q\pi(w)}$, but $\pi p(w') = q\pi(w')$. Thus we have $\overline{\pi p(w')} = \overline{q\pi(w)}$. Now

$$\begin{aligned} d(g) &= \vartheta_\Gamma f \cdot h \cdot \varphi(g) = \vartheta_\Gamma f \cdot h \cdot \overline{p(w')} = \vartheta_\Gamma \cdot f\overline{(\pi p(w'))} \\ &= \vartheta_\Gamma \cdot f\overline{(q\pi(w))} = \vartheta_\Gamma\Big(\sum_{i=1}^{n} q(\beta_i)^{\varepsilon_i}\Big) = \sum_{i=1}^{n} \varepsilon_i, \text{ since } \vartheta_\Gamma = \vartheta_\Omega. \end{aligned}$$

Thus $d(w(a, g(a))) = d(w(a,b))$.

Since $g(a)^{\mathcal{T}} = b^{\mathcal{T}}$, we have $g(a) = b \in V(U)$ as required.

For the construction of Q, let $Q = U/(G|U)$. Then

$$\Gamma = \mathbf{Z}B/G = \mathbf{Z}U/G = \mathbf{Z}(U/(G|U)) = \mathbf{Z}Q.$$

Acknowledgements

This work is the author's thesis written under the direction of C. M. Ringel. He payed special attention to this paper and was most helpful with the mathematics and written language. The author wishes to express her gratitude to him, to D. Happel, J. A. de la Peña, and to her Chinese supervisor Liu Shao-Xue. She thanks Mrs. Köllner for typewriting. Finally she wishes to acknowledge the financial support by the Deutsche Forschungsgemeinschaft.

References

[A] Auslander M. Application's of morphisms determined by objects. Proc. Conf. on Representation Theory. Philadelphia, 1976, Marcel Dekker, 1978:245-327.

[B] Berge C. Graphs and hypergraphs. North-Holland Publishing Company, 1973.

[BG] Bongartz K, Gabriel P. Covering spaces in representation theory. Invent. Math. 1982, 65:331-378.

[BS] Bautista R, Smalø S O. Nonexistent cycles. Comm. Alg. 1983, 11:1 755-1 767.

[BR] Butler M C R, Ringel C M. AR-sequences with few middle terms and applications to string algebras, Comm. Alg. , 1987, 15:145-179.

[D] Dicks W. Groups, trees and projective modules. LNM 790, 1980.

[GZ] Gabriel P, Zisman M. Calculus of fractions and homology theory. Erg. Math. , 35, Berlin-Heildelberg-New York, Springer, 1967.

[HPR] Happel D, Preiser U, Ringel C M. Vinberg's characterization of Dynkin diagrams using subadditive function with application to DTr — periodic modules. LNM 832, 1979:280-294.

[Re] Reidemeister K. Einfuhrung in die kombinatorische Topologie. Reprint, Chelsea Publishing Company, New York, N. Y. , 1950.

[Ri] Riedtmann Chr. Algebren, Darstellungsköcher, Überlagerungen und zurück. Comm. Math. Helv. , 1980, 55:199-224.

[R] Ringel C M. Tame algebras and integral quadratic forms. Springer LNM 1099, 1984.

[S] Spanier E H. Algebraic Topology. McGraw Hill Book Company, 1966.

Science in China, 1996, 39A(5): 483-490.

对应于 Tame 遗传代数的 Bocses (Ⅰ)①

Bocses Corresponding to the Hereditary Algebras of Tame Type (Ⅰ)

Abstract The bocs corresponding to Kronecker algebra is shown and the corresponding minimal bocs with only irreducible maps for each dimension d is also given. The minimal bocs (after deleting two vertices) coincides with a full subquiver of the AR-quiver of Kronecker algebra, in which the indecomposable modules M have the property

$$\dim(\operatorname{top} M) < d.$$

Keywords bocs; reduction algorithm; Kronecker algebra; minimal bocs; AR-quiver.

§ 1. Preliminaries

A bocs $\mathscr{A}=(A, V)$ consists of a skeletally small category A (or equivalently an algebra A) and a coalgebra V over A. This structure was presented by V. A. Roiter at the end of the 1960s in studying matrix problems[1]. In the 1970s, Y. A. Drozd extended and completed this concept and gave an effective method to turn an algebra into the corresponding bocs of which the representation is equivalent to the

① Project supported by the National Natural Science Foundation of China and the State Education Commission.

Received: 1995-05-28.

本文与林亚南合作.

algebra. And he proved, using the technique of bocses successfully, that a finite-dimensional algebra over an algebraically closed field is either of representation tame type or wild type but not both. This is the well known theorem of Drozd[2,3]. In 1987, Crawley-Boevey restated the theory of bocses and described the AR-transformations of a tame algebra[4]. We used the notations and definitions from refs. [4,5] in this note.

The greatest advantage of bocses is that there exist five reduction algorithms. If a bocs is of representation tame type, then for each dimension d we may turn it into some minimal bocses using the reduction algorithms repeatedly. In terms of algebras, roughly speaking, for each dimension d we can turn a finite-dimensional algebra of representation tame type into some subcategories of its module category. This is just our expectation. Now we give the definition of a layered bocs and list their reduction algorithms which we need.

Let $\mathscr{A}=(A,V)$ be a bocs. Then V is an A-A-bimodule with a counite ε and comultiplication μ satisfying counite and co-associativity. Denote Ker ε by $\bar{V}$.

Definition 1[4] Let $\mathscr{A}=(A,V)$ be a bocs. We say that a collection $L=(A';\omega;a_1,a_2,\cdots,a_n;v_1,v_2,\cdots,v_m)$ is a layer for $\mathscr{A}$ (and that $\mathscr{A}$ is layered), provided that

(L1) A' is a minimal category;

(L2) A is freely generated over A' by indecomposable elements a_1, $a_2,\cdots,a_n$ (in the diagram of bocs $\mathscr{A}$, we draw them by solid arrows);

(L3) ω is a reflector for $\mathscr{A}$ relative to A';

(L4) $\bar{V}$ is freely generated as an A-A-bimodule by indecomposable elements $v_1,v_2,\cdots,v_m$ (dotted arrows);

(L5) writing A_1 for the subcategory of A generated by A' and a_1, $a_2,\cdots,a_i$, and δ for the differential given by ω; for any $0\leqslant i\leqslant n$, $\delta(a_{i+1})$ is contained in the A_i-A_i-sub-bimodules of $\bar{V}$ generated by $v_1,v_2,\cdots,v_m$.

Throughout the paper bocses are always layered. When $V=A$, bocs $\mathscr{A}=(A,A)$ is called a principal bocs.

Let $\mathscr{A}=(A,V)$ be a bocs, and denote by $R(\mathscr{A})$ the representation category of $\mathscr{A}$[4]. If $\theta: A \to B$ is a functor from A to a skeletally small category B, then we can construct an induced bocs $(B, {}^B V^B)$ denoted by $\mathscr{A}^B$, and an induced morphism of bocses $\theta_t = (\theta, \theta_1): \mathscr{A} \to \mathscr{A}^B$ with $\theta_1 = (\theta \otimes \mathrm{id} \otimes \theta)\tau$; $V \xrightarrow{\tilde{\tau}} A \otimes_A V \otimes_A A \to {}_A B \otimes_A V \otimes_A B_A$ being an A-A-bimodule homomorphism. Then from θ_l we obtain again an induced functor $\theta_l^*: R(\mathscr{A}^B) \to R(\mathscr{A})$, which is fully faithful (see 3.1 of ref. [4]).

The following definition describes a special case while $A = A'$.

Definition 2[4] A bocs $\mathscr{A} = (A, V)$ is called minimal, provided that A is a minimal category, $\bar{V}$ is a finitely generated projective A-A-bimodule, and there is an A-coalgebra map $\omega: A \to V$, such that the functor $(1, \omega)^*: R(\mathscr{A}) \to R(A)$ is a representation equivalence: it is full, dense and reflects isomorphisms.

Now we state some reduction algorithms.

(1) Regularization in 4.2 of ref. [4]. Let $\mathscr{A} = (A, V)$ be a bocs with layer $L = (A'; \omega; a_1, a_2, \cdots, a_n; v_1, v_2, \cdots, v_m)$ and suppose that $\delta(a_1) = v_1$. Let B be the subcategory of A generated by A' and $a_1, a_2, \cdots, a_n$, and $\theta: A \to B$ the functor acting as the identity on $A', a_2, a_3, \cdots, a_n$, and sending a_1 to zero. Then the induced bocs $\mathscr{A}^B$ has a layer $(A'; \omega \cdot \theta_1; a_2, a_3, \cdots, a_n; \theta_1(v_2), \theta_1(v_3), \cdots, \theta_1(v_m))$. θ_l^* is an equivalence.

(2) Edge reduction in ref. [4]. Let $\mathscr{A} = (A, V)$ be a bocs with layer $L = (A'; \omega; a_1, a_2, \cdots, a_n; v_1, v_2, \cdots, v_m)$. Suppose that $a_1 \in A(X, Y)$ has differential zero, that $X \neq Y$, and that $A'(X, X)$ and $A'(Y, Y)$ are both trivial. Then there is a category B and a functor $\theta; A \to B$ such that the induced bocs $\mathscr{A}^B$ is layered; θ_l^* is an equivalence.

(3) Unraveling in 4.7 of ref. [4]. Let $\mathscr{A} = (A, V)$ be a bocs with layer $L = (A'; \omega; a_1, a_2, \cdots, a_n; v_1, v_2, \cdots, v_m)$. Let X be an indecomposable object in A and suppose $A'(X, X) = k[x, f(x)^{-1}]$. Let g be a non-zero polynomial and r a positive integer. Then there is a category B and a functor $\theta: A \to B$ such that the induced bocs $\mathscr{A}^B$ is layered; every representation in $R(\mathscr{A})$ with norm at most r is isomorphic to $\theta_l^*(N)$ for some N in $R(\mathscr{A}^B)$.

The concrete structures of category B and functor θ for edge reduction and unraveling are given in reference [4].

Throughout the paper we assume that k is an algebraically closed field. Λ is a finitely dimensional algebra. If Λ is a hereditary algebra of representation tame type, then Λ belongs to one of the following three types according to the classification given by Dlab and Ringel[6]: (1) $\widetilde{A}_n$; (2) $\widetilde{D}_n$; (3) $\widetilde{E}_6$, $\widetilde{E}_7$, $\widetilde{E}_8$. We will show the corresponding bocses of hereditary algebras of types (1) (2) (3) respectively. And for each dimension d, we will give the corresponding minimal bocses with only dotted arrows of differentials zero. Except finite number of vertices the minimal bocs coincides completely with the AR-quiver of algebra Λ when d tends to infinite. The present paper is the first part of this work. In sec. 2, we give some results on bocses and reduction algorithms as a theoretical base. In sec. 3, we show the corresponding bocs and minimal bocs of Kronecker algebra, the simplest hereditary algebra of representation tame type. In the second part of the work, we will discuss the general cases. In the diagrams of bocses of the paper, we only draw the dotted arrows with differentials zero. They represent the irreducible maps between the indecomposables in a minimal bocs.

§ 2. Some lemmas

The category of finite-dimensional left Λ-modules is denoted by modΛ. Let $P_1(\Lambda)$ be the category whose objects are the triples (P, Q, α) with P and Q finite-dimensional projective left Λ-modules, and α a homomorphism $P \to \text{rad } Q$. $\dim(P, Q, \alpha) = (\dim(\text{top}P), \dim(\text{top}Q))$ by the definition.

Lemma 1 (cf. 6.4, 6.1 of ref. [4]) Let Λ be a finite-dimensional k-algebra, $P_2(\Lambda) = \{(P, Q, \alpha) \mid \text{Ker } \alpha \subset \text{rad } P\}$ be the full subcategory of $P_1(\Lambda)$. Then there is a representation equivalence Coker: $P_2(\Lambda) \to \text{mod}\Lambda$, taking (P, Q, α) to Cokerα. And $\dim \text{top}(\text{Coker}\alpha) = \dim \text{top}(Q) \leqslant \dim(P, Q, \alpha)$. On the other hand, there is a layered bocs $\mathscr{A}$ and a dimensional vector preserving equivalence $\Xi: R(\mathscr{A}) \to P_1(\Lambda)$.

Lemma 2 Let $\mathscr{C},\mathscr{D}$ be two categories, $F:\mathscr{C}\to\mathscr{D}$ be a representation equivalent functor (i. e. F is full, dense and reflects isomorphism). Assume that $v':M'\to N'$ is a map in $\mathscr{D}$. We have $M'=FM, N'=FN$, $v'=Fv$, If v' is an irreducible map in $\mathscr{D}$, then v is the same in $\mathscr{C}$. Therefore the set of the irreducible maps in $\mathscr{D}$ is embedding into the set $\{Fv \mid v$ is an irreducible map in $\mathscr{C}\}$.

Proof If $v: M\to N$ is not irreducible. Then there exist non-isomorphism $h:M\to L$ and $l:L\to N$ in $\mathscr{C}$, such that $v=l\cdot h$. Then $v'=F(v)=F(l)\cdot F(h)$, with $F(h), F(l)$ non-iso, since F reflects isomorphisms. Thus v' is not an irreducible map, a contradiction.

Let $\mathscr{A}=(A,V)$ be a bocs with layer $L=(A';\omega;a_1,a_2,\cdots,a_n;v_1,v_2,\cdots,v_m)$. For any indecomposable object X in A', let $S(X)$ be the simple representation of $R(\mathscr{A})$ with $S(X)(X)=k, S(X)(W)=0$. for $W\neq X$ and $S(X)(a_i)=0$ for $i=1,2,\cdots,n$.

Lemma 3 Let $\mathscr{A}=(A,V)$ be a bocs with layer $L=(A';\omega;a_1,a_2,\cdots,a_n;v_1,v_2,\cdots,v_m)$. Assume that $a_1:X\to Y$ with $X\neq Y, A'(X,X)=k=A'(Y,Y)$ and $\delta(a_1)=0$, $\mathscr{A}_1=(A_1,V_1)$ is the bocs obtained from $\mathscr{A}$ by edge reduction for a_1. $\theta:A\to A_1$ with $\theta'(X)=(Z_1\oplus Z_2), \theta'(Y)=Z_2\oplus Z_3$ (cf. 4. 9 of ref. [4]). Then

(1) $\theta_l^*(S(W))=S(W)$ for $W\neq Z_1,Z_2,Z_3$;

(2) $\theta_l^*(S(Z_1))=S(X), \theta_l^*(S(Z_3))=S(Y)$;

(3) $\theta_l^*(S(Z_2))=M$ where $M(X)=k=M(Y), M(a_1)=id, M(W)=0$ for $W\neq X,Y$ and $M(a_i)=0$ for $i\geqslant 2$.

Proof $\theta_l^*(S(W))=S(W)$ is obvious when $W\neq X$. $\theta_I^*(S(Z_1))(X)=S(Z_1)(\theta'(X))=S(Z_1)(Z_1\oplus Z_2)=k$. $\theta_I^*(S(Z_1))(W)=S(Z_1)(\theta'(W))=0$, for $W\neq X$. $\theta_I^*(S(Z_1))(a_1)=S(Z_1)(\theta(a_1))=0$. We have the similar argument for $\theta_I^*(S(Z_3))$. $\theta_I^*(S(Z_2))(X)=S(Z_2)(\theta'(X))=S(Z_2)(Z_1\oplus Z_2)=k$ and $\theta_I^*(S(Z_2))(Y)=k$. $\theta_I^*(S(Z_2))(a_1)=S(Z_2)(\theta(a_1))=id$. $\theta_I^*(S(Z_2))(W)=S(Z_2)(\theta(W))=0$. $\theta_I^*(S(Z_2))(a_i)=S(Z_2)(\theta(a_i))=0$, $i\geqslant 2$. This proves that $\theta_I^*(S(Z_2))=M$.

Lemma 4 Let A and A_1 be the same as in Lemma 3, $\mathscr{B}=(B,W)$ be a layered bocs and $\phi:B\to A$ be a functor. Then

(1) $\dim\varphi_I^* \circ \theta_I^* (S(W)) = \dim\varphi_I^* (S(W))$ for $W \neq Z_i, i=1,2,3$ in A_1;

(2) $\dim\varphi_I^* \circ \theta_I^* (S(Z_1)) = \dim\varphi_I^* (S(X))$ and $\dim\varphi_I^* \circ \theta_I^* (S(Z_3)) = \dim\varphi_I^* (S(Y))$;

(3) $\dim\varphi_I^* \circ \theta_I^* (S(Z_2)) = \dim\varphi_1^* (S(X)) + \dim\varphi_I^* (S(Y))$.

Proof (1) and (2) are obvious. For (3) denote $\theta_I^* (S(Z_2))$ by M. Let V be any indecomposable object in category B, and $\varphi(V) = \bigoplus_{i=1}^{l} W_i$, with W_i indecomposable in A'. Then

$$\varphi_I^* (M)(V) = M(\varphi'(V)) = M(W_1 \oplus \cdots \oplus W_1) = \bigoplus_{i=1}^{l} M(W_i).$$

But $M(W_i) = S(X)(W_i) \oplus S(Y)(W_i), i=1,2,\cdots,l$. Then (3) follows.

Lemma 5 (cf. 3.4, 6.6 of ref. [4]) Let $\mathscr{A}=(A,V)$ be a minimal bocs. Assume that $f: M \to N$ is an irreducible map in $R(\mathscr{A})$, and M, N are indecomposable. Then either $M = J(X, n, \lambda)$, $N = J(X, n \pm 1, \lambda)$ or $M = S(Y)$, $N = S(Z)$ with $A(Y,Y) = k = A(Z,Z)$. In the second case, we consider $M = Y$, $N = Z$, $f = v$ in $R(\mathscr{A})$ with $\delta(v) = 0$.

Lemma 6 Let $\mathscr{A}=(A,V)$ be a bocs, and be any positive integer. If there is a category B_d and functor $\theta_d: A \to B_d$ such that $\mathscr{B}_d = \mathscr{A}^{B_d}$ is minimal and every representation of $\mathscr{A}$ with dimension at most d is isomorphic to $\theta_d^* (N_d)$ for some representation N_d of $\mathscr{A}_d$. Assume that $v: M \to N$ is in $R(\mathscr{A})$ with M, N indecomposable, and $M = \theta_d^* (M_d)$, $N = \theta_d^* (N_d)$, $v = \theta_d^* (v_d)$ for $\max\{\dim M, \dim N\} < d$. If v_d is an irreducible map for each d, then v is the same in $R(\mathscr{A})$.

Proof If v is not an irreducible map, then there exist non-isomorphisms $h: M \to L$, $l: L \to N$, such that $v = lh$. Assume that $d_0 > \max\{\dim M, \dim N, \dim L\}$. Since $\theta_{d_0}^*: R(\mathscr{B}_{d_0}) \to R(\mathscr{A})$ is fully faithful, and reflecting isomorphism and $\dim X \leqslant \dim\theta_{d_0}^* (X)$ for any X in $R(\mathscr{A}_{d_0})$, there exist objects $M_{d_0}, N_{d_0}, L_{d_0}$ and morphisms $v_{d_0}, h_{d_0}, l_{d_0}$, such that $\theta_{d_0}^* (M_{d_0}) = M$, $\theta_{d_0}^* (N_{d_0}) = N$, $\theta_{d_0}^* (L_{d_0}) = L$ and $\theta_{d_0}^* (v_{d_0}) = v$, $\theta_{d_0}^* (h_{d_0}) = h$, $\theta_{d_0}^* (l_{d_0}) = l$. Then $v_{d_0} = l_{d_0} \cdot h_{d_0}$, and h_{d_0}, I_{d_0} are non-isomorphic. Thus v_{d_0} is not an irreducible map, a contradiction.

§ 3. Kronecker algebras

Let ordinary quiver (fig. 3.1)

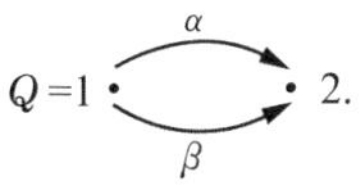

Fig. 3. 1

Then Kronecker algebra $\Lambda = kQ$, the path algebra of Q. Let $J = \langle \alpha, \beta \rangle$, the radical of Λ, $S = \Lambda / J \simeq k \times k$, $J^* = \mathrm{Hom}_S(J, S) = \langle \alpha^*, \beta^* \rangle$ be an S-S-bimodule, where $\alpha^*(\alpha) = e_2$, $\alpha^*(\beta) = 0$ and $\beta^*(\alpha) = 0$, $\beta^*(\beta) = e_2$. Let

$$\boldsymbol{A}' = \begin{pmatrix} S & 0 \\ 0 & S \end{pmatrix}, \quad \boldsymbol{A} = \begin{pmatrix} S & 0 \\ J^* & S \end{pmatrix}$$

which is freely generated by

$$\boldsymbol{a} = \begin{pmatrix} 1 & 0 \\ \alpha^* & 0 \end{pmatrix} \text{ and } \boldsymbol{b} = \begin{pmatrix} 1 & 0 \\ \beta^* & 0 \end{pmatrix}$$

over the trivial subcategory A'. And $\bar{V}$ is a freely generated A-A-bimodule with generators

$$\boldsymbol{v}_1 = \begin{pmatrix} 1 & 0 \\ 0 & 1 \end{pmatrix} \otimes_{A'} \begin{pmatrix} \alpha^* & 0 \\ 0 & 0 \end{pmatrix} \otimes_{A'} \begin{pmatrix} 1 & 0 \\ 0 & 1 \end{pmatrix},$$

$$\boldsymbol{v}_2 = \begin{pmatrix} 1 & 0 \\ 0 & 1 \end{pmatrix} \otimes_{A'} \begin{pmatrix} 0 & 0 \\ \alpha^* & 0 \end{pmatrix} \otimes_{A'} \begin{pmatrix} 1 & 0 \\ 0 & 1 \end{pmatrix},$$

$$\boldsymbol{v}_1' = \begin{pmatrix} 1 & 0 \\ 0 & 1 \end{pmatrix} \otimes_{A'} \begin{pmatrix} 0 & \beta^* \\ 0 & 0 \end{pmatrix} \otimes_{A'} \begin{pmatrix} 1 & 0 \\ 0 & 1 \end{pmatrix},$$

$$\boldsymbol{v}_2' = \begin{pmatrix} 1 & 0 \\ 0 & 1 \end{pmatrix} \otimes_{A'} \begin{pmatrix} 0 & 0 \\ 0 & \beta^* \end{pmatrix} \otimes_{A'} \begin{pmatrix} 1 & 0 \\ 0 & 1 \end{pmatrix}.$$

The corresponding bocs (fig. 3. 2) is

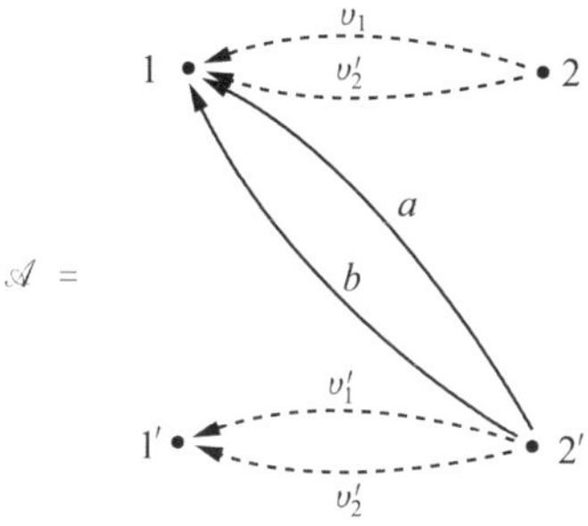

Fig. 3. 2

with differentials $\delta(a)=0, \delta(b)=0$, and $\delta(v_i)=0, \delta(v_i')=0$, for $i=1, 2$. Now we make reduction procedure for $\mathscr{A}$. First make an edge reduction for arrow b, obtain fig. 3. 3.

$\mathscr{A}_1 =$ 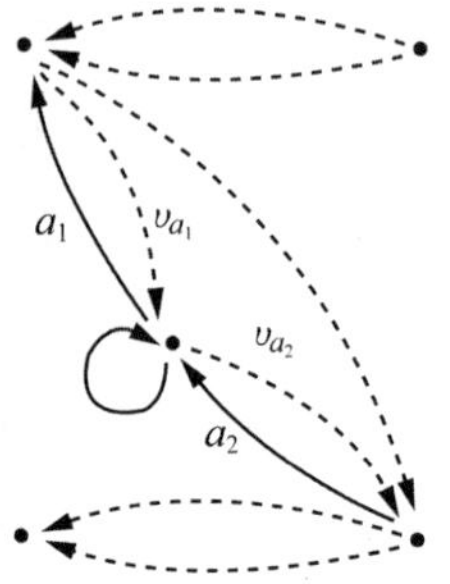

Fig. 3. 3

with differentials

$\delta(a)=0, \delta(a_1)=v_{a_2}\cdot a, \delta(a_2)=-a\cdot v_{a_1}, \delta(x)=-a_1\cdot v_{a_1}+v_{a_2}\cdot a_2.$

Then after edge reduction for a (fig. 3. 4),

$\mathscr{A}_2 =$ 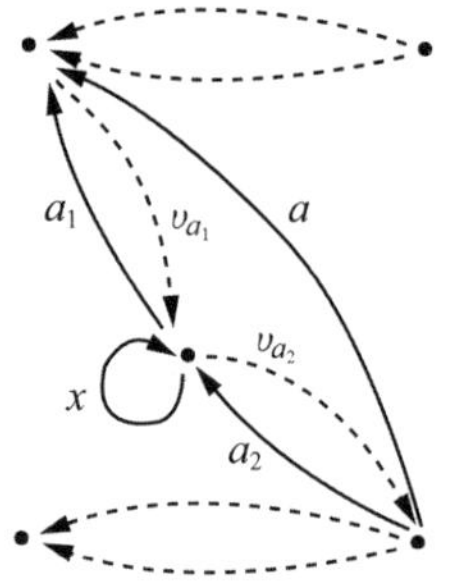

Fig. 3. 4

$$\delta(a_1)=0, \delta(a_2)=0, \delta(x)=-a_1\cdot v_{a_1}+v_{a_2}\cdot a_2.$$

After edge reduction for a_1 (fig. 3. 5),

$\mathscr{A}_3 =$ 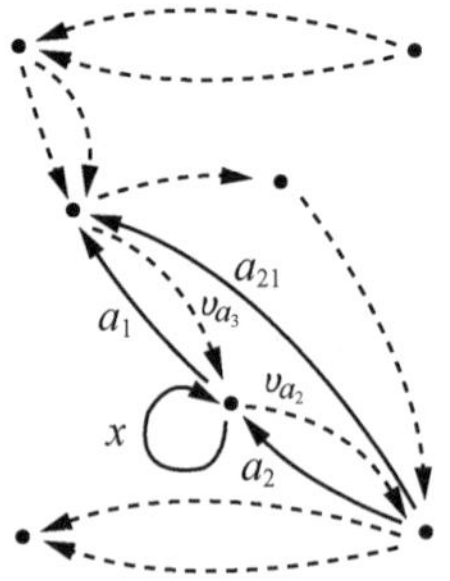

Fig. 3. 5

$\delta(a_{23})=0,\delta(a_2)=-a_{23}\cdot v_{a_3},\delta(a_3)=v_{a_2}\cdot a_{23},\delta(x)=v_{a_2}\cdot a_2-a_3\cdot v_{a_3}$.

After edge reduction for a_{23} and a_2 (fig. 3. 6).

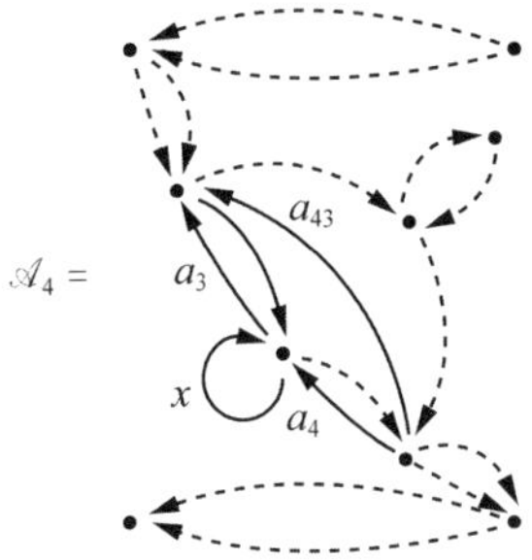

Fig. 3. 6

$\delta(a_{43})=0,\delta(a_3)=v_{a_4}\cdot a_{43},\delta(a_4)=-a_{43}\cdot v_{a_1},\delta(x)=-a_3\cdot v_{a_3}+v_{a_4}\cdot a_4$.

Repeat this procedure again and again. For any positive integer d, we finally obtain a minimal bocs (fig. 3. 7):

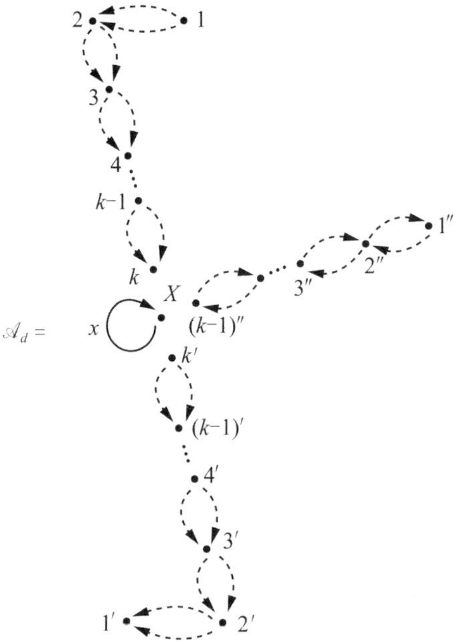

Fig. 3. 7

where

$$\Xi\circ\theta_d^*(X_1)=(0,P_2,0),\Xi\circ\theta_d^*(X_2)=(0,P_1,0),$$

$$\Xi\circ\theta_d^*(X_{1'})=(P_1,0,0),\Xi\circ\theta_d^*(X_{2'})=(P_2,0,0)$$

by Lemmas 1 and 3;

$$\dim(\theta_d^*(X_i))=1=\dim(\theta_d^*(X_{i'}))\text{ for } i=1,2;$$

$$\dim(\theta_d^*(X_i))=2i-3=\dim(\theta_d^*(X_{i'}))\text{ for } i\geqslant 3.$$

$$\dim(\theta_d^*(X_{i''}))=2i,\dim(\theta_d^*(X))=2$$

from Lemmas 3 and 4. Since $2(k-1)\leqslant d$ and $2k-3\leqslant d$, $k=\left[\frac{d+2}{2}\right]$. $\dim(\mathrm{top}(\mathrm{Cok}\circ\Xi\circ\theta_d^*(Y)))\leqslant\dim(\Xi\circ\theta_d^*(Y))\leqslant d$, for any indecomposable Y in $R(B_d)$. The dotted arrows correspond to the irreducible maps of $R(\mathscr{A})$ (Lemmas 5,6), therefore those of $P_1(\Lambda)$ (Lemma 1).

Delete two points on the bottom, and consider the case that d tends to infinite. Then by Lemmas 1 and 2, the component on the upper half plane just corresponds to the preprojective component of AR-quiver of Kronecker algebra, and that on the bottom half plane corresponds to the preinjective component, the component in the middle corresponds to "∞" homogeneous tube. And loop x represents $\mathbb{A}_1(k)$-family of homogeneous tubes. One-dimensional projective space

$$\mathbb{P}_1(k)=\mathbb{A}_1(k)\cup\{\infty\}.$$

Acknowledgements

The authors would like to express their thanks to supervisor Prof. Liu Shaoxue for his consistent encouragement.

References

[1] Roiter A V. Matrix problems and representations of bocses. Lecture Notes in Mathematics 831, Berlin: Springer, 1980, 288.

[2] Drozd Y A. Tame and Wild Matrix Problems, Matrix Problems. Kiev. Ukranian SSR., 1977: 104-114.

[3] Drozd Y A. Tame and Wild Matrix Problems, Representations and Quadratic Forms. Kiev. Ukranian SSR., 1979: 39-74.

[4] Crawley-Boevey W W. On the tame algebras and bocses. London Math. Soc., 1988, 56(3): 451.

[5] Ringel C M. Tame algebras and integral quadratic forms. Lecture Notes in Mathematics, 1 099. Berlin: Springer, 1984.

[6] Dlab V, Ringel C M. Indecomposable representations of graphs and algebras. Memories Amer. Math. Soc., 173, 1976.

Science in China, 1996, 39A(9): 909-918.

对应于 Tame 遗传代数的 Bocses (Ⅱ)①

Bocses Corresponding to the Hereditary Algebras of Tame Type (Ⅱ)

Abstract The bocs corresponding to each hereditary algebra of representation tame type is shown. And the minimal bocs obtained from it is given with only irreducible maps for each dimension d. The minimal bocs (after deleting finitely many vertices) coincides with a full subquiver of the AR-quiver of hereditary algebra, in which the indecomposable modules M have the property dim(top M)$<d$.

Keywords bocs; reduction algorithm; hereditary algebra of representation tame type; minimal bocs; AR-quiver.

Throughout the paper we assume that k is an algebraically closed field, Λ is a finitely dimensional k-algebra with identity element 1.

The algebraic structure of bocs was presented by Rojter at the end of the 1960s[1]. Later, Drozd, and Crawley-Boevey extended and completed this concept and solved successfully some important problems in representation theory of finitely dimensional algebras[2~4].

The major advantage of bocs is that it has five reduction algorithms. If a

---

① Project supported by the National Natural Science Foundation of China.

Received: 1995-06-26.

本文与林亚南合作.

bocs is of representation tame type, we may turn it into a minimal bocs using the reduction algorithms repeatedly for each dimension d. This minimal bocs is equivalent to a subcategory of the representation category of the original bocs consisting of the objects with dimension smaller than d. Concerning the method of turning an algebra to the associated bocs given by Drozd, we may turn an algebra of representation tame type to a subcategory of its module category under the sence of representation equivalence. Let Λ be a hereditary k-algebra of tame type. By the classification of Dlab and Ringel[5], we have three kinds of such algebras: (1) type $\widetilde{A}_n$; (2) type $\widetilde{D}_n$ ($n \geqslant 4$); (3) types $\widetilde{E}_6$, $\widetilde{E}_7$, $\widetilde{E}_8$. We will show the corresponding bocs of each algebra, and give the minimal bocs for any dimension d. When d tends to infinite, the minimal bocs with only the maps of differentials zero, i. e. the irreducible maps, coincides with the AR-quiver of the corresponding algebra except finite vertices. We have given the necessary basic theory on bocses and discussed the simplest hereditary algebra of tame type-Kronecker algebra in ref. [6]. In this paper, we will discuss the general hereditary algebras of tame type.

Bocs in the paper is always with a layer $L=(A';\omega;a_1,a_2,\cdots,a_n;v_1,v_2,\cdots,v_m)$. All concepts and terminologies come from ref. [4] or [6]. The real arrows represent $a_1, a_2, \cdots, a_n$; the dotted arrows represent $v_1, v_2, \cdots, v_m$ in the bigraph. For the sake of convenience, sometimes, we only show the dotted arrows with differentials zero. They represent the irreducible maps between indecomposable objects in a minimal bocs.

§ 1. Hereditary algebra of type $\widetilde{A}_n$

Let $\Lambda = kQ$ be a hereditary algebra of type $\widetilde{A}_n$, where Q is an oriented graph of type $\widetilde{A}_n$ (see fig. 1. 1(a)). When n is even, $p=n, q=n-1$; when n is odd, $p=n-1, q=n$. The corresponding bocs is in figl. 1(b), where we take $n=7$ as an example.

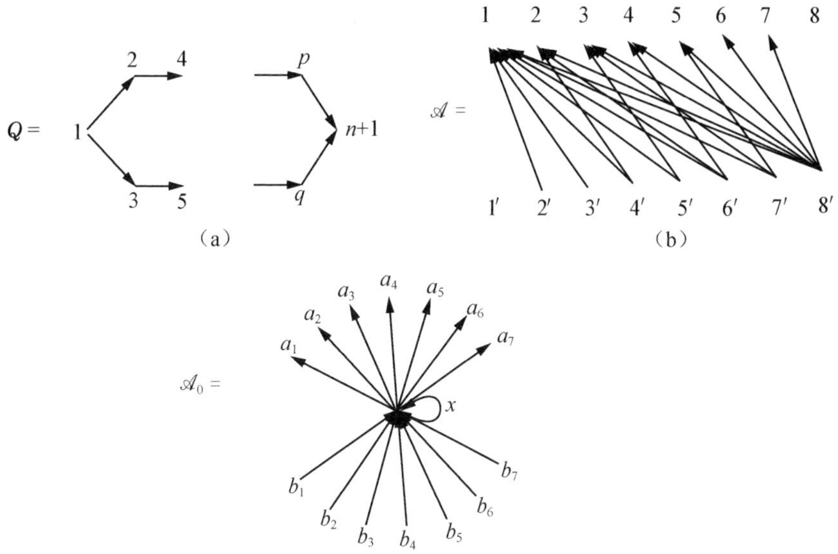

(a) (b) (c)

Fig. 1. 1

The number of arrows is $2+4+\cdots+(n+1)=\dfrac{(n+1)(n+3)}{4}$, when n is odd, and $1+3+\cdots+(n+1)=\left(\dfrac{n+2}{2}\right)^2$, when n is even.

$$\delta(a_{i+2,i})=0, i=1,2,\cdots,n_1, \delta(a_{21})=0, \delta(a_{n+1,n})=0;$$

$$\delta(a_{i+2k,i})=\sum_{1\leqslant k'\leqslant k-1}(v'_{i+2k,i+2k'}\cdot a_{i+2k',i}-a_{i+2k,i+2k'}\cdot v_{i+2k',i}),$$

$$i=1,2,\cdots,(n-1), 2\leqslant k\leqslant\left[\frac{n+1-i}{2}\right];$$

$$\delta(a_{2k,1})=\sum_{1\leqslant k'\leqslant k-1}(v'_{2k,2k'}\cdot a_{2k',1}-a_{2k,2k'}\cdot v_{2k',1}),\quad k\leqslant\left[\frac{n+1}{2}\right];$$

$$\delta(a_{n+1,n-2k+2})=\sum_{1\leqslant k'\leqslant k-1}(v'_{n+1,n-2k'+2}\cdot a_{n-2k'+2,1}-a_{n+1,n-2k'+2}\cdot v_{n-2k'+2,1}),$$

$$k=2,3,\cdots,\left[\frac{n}{2}\right];$$

$$\delta(a_{n+1,1})=\sum_{1\leqslant k'\leqslant\left[\frac{n}{2}\right]}(v'_{n+1,2k'}\cdot a_{2k',1}-a_{n+1,2k'}\cdot v_{2k',1});$$

$$\delta(a'_{n+1,1})=\sum_{1\leqslant k'\leqslant\left[\frac{n-1}{2}\right]}(v'_{n+1,2k'+1}\cdot a_{2k'+1,1}-a_{n+1,2k'+1}\cdot v_{2k'+1,1}),$$

where a_{ij} denotes the real arrow from i' to j, v_{ij} (resp. v'_{ij}) denotes the dotted arrow from i to j (resp. i' to j').

After a series of edge reductions and regularizations, we obtain a bocs $\mathscr{A}_0$, with $R(\mathscr{A}_0)$ equivalent to $R(\mathscr{A})$ (see fig. 1. 1(c)).

In the figure, we do not exhibit the dotted arrows, some vertices of which are not starting or ending points of any real arrows, and arrows c_{ij}, $i,j=1,2,\cdots,n$. If we denote the starting point of arrow a by $s(a)$ and the ending point by $e(a)$, then arrow c_{ij} is from $s(b_i)$ to $e(a_j)$, the dotted arrow $w_{ii'}$ from $s(b_i)$ to $s(b_{i'})$, $v_{jj'}$ from $e(a_j)$ to $e(a_{j'})$, v_{a_j} from $e(a_j)$ to $s(a_j)$, w_{b_i} from $e(b_i)$ to $s(b_i)$. The differentials are:

$$\delta(c_{ij})=\sum_{i'<i}w_{ii'}\cdot c_{i'j}-\sum_{j'<j}c_{ij'}v_{j'j},\ i,j=1,2,\cdots,\left[\frac{n}{2}\right]+1;$$

$$\delta(a_j)=\sum_{1\leqslant i\leqslant\left[\frac{n}{2}\right]+1}w_{b_i}\cdot c_{ij}-\sum_{j'<j}a_{j'}v_{j'j},\ i,j=1,2,\cdots,\left[\frac{n}{2}\right]+1;$$

$$\delta(b_j)=\sum_{1\leqslant j\leqslant\left[\frac{n}{2}\right]}c_{ij}\cdot v_{a_j}+\sum_{i'<i}w_{ii'}b_{i'},\ i,j=1,2,\cdots,\left[\frac{n}{2}\right]+1;$$

$$\delta(x)=-\sum_{1\leqslant j\leqslant\left[\frac{n}{2}\right]}a_j\cdot v_{a_j}+\sum_{1\leqslant i\leqslant\left[\frac{n}{2}\right]}w_{b_i}b_i;$$

$$\delta(c_{ij})=\sum_{i'<i}w_{ii'}\cdot c_{i'j}-\sum_{j'<j}c_{ij'}v_{j'j}-b_iv_{a_j},$$
$$i=1,2,\cdots,\left[\frac{n}{2}\right]+1,\ j=\left[\frac{n}{2}\right]+2,\left[\frac{n}{2}\right]+3,\cdots,n;$$

$$\delta(a_j)=\sum_{j'<j}a_{j'}\cdot v_{j'j}-x\cdot v_{a_j}+\sum_{1\leqslant i\leqslant\left[\frac{n}{2}\right]+1}w_{b_i}c_{ij},$$
$$j=\left[\frac{n}{2}\right]+2,\left[\frac{n}{2}\right]+3,\cdots,n;$$

$$\delta(c_{ij})=\sum_{i'<i}w_{ii'}\cdot c_{i'j}-\sum_{j'<j}c_{ij'}v_{j'j}+w_{b_i}\cdot a_j,$$
$$i=\left[\frac{n}{2}\right]+2,\left[\frac{n}{2}\right]+3,\cdots,n,\ j=1,2,\cdots,\left[\frac{n}{2}\right]+1;$$

$$\delta(b_i)=\sum_{i'<i}w_{ii'}b_{i'}+w_{b_i}\cdot x-\sum_{1\leqslant j\leqslant\left[\frac{n+1}{2}\right]}c_{ij}\cdot v_{a_j},$$
$$i=\left[\frac{n}{2}\right]+2,\left[\frac{n}{2}\right]+3,\cdots,n;$$

$$\delta(c_{ij})=\sum_{i'<j}w_{ii'}\cdot c_{i'j}-\sum_{j'<j}c_{ij'}\cdot v_{j'j}+w_{b_i}\cdot a_j-b_i\cdot v_{a_j},$$

$$i,j=\left[\frac{n}{2}\right]+2,\left[\frac{n}{2}\right]+3,\cdots,n.$$

For each dimension d, we have the following four components C_1, C_2, C_3, C_4 in minimal bocs $\mathscr{B}_d$. The first one comes from repeatedly making edge reductions for c_{ij}^1, $i,j=1,2,\cdots,\left[\frac{n}{2}\right]+1$, $l\in\mathbf{N}^*$ (see fig. 1.2).

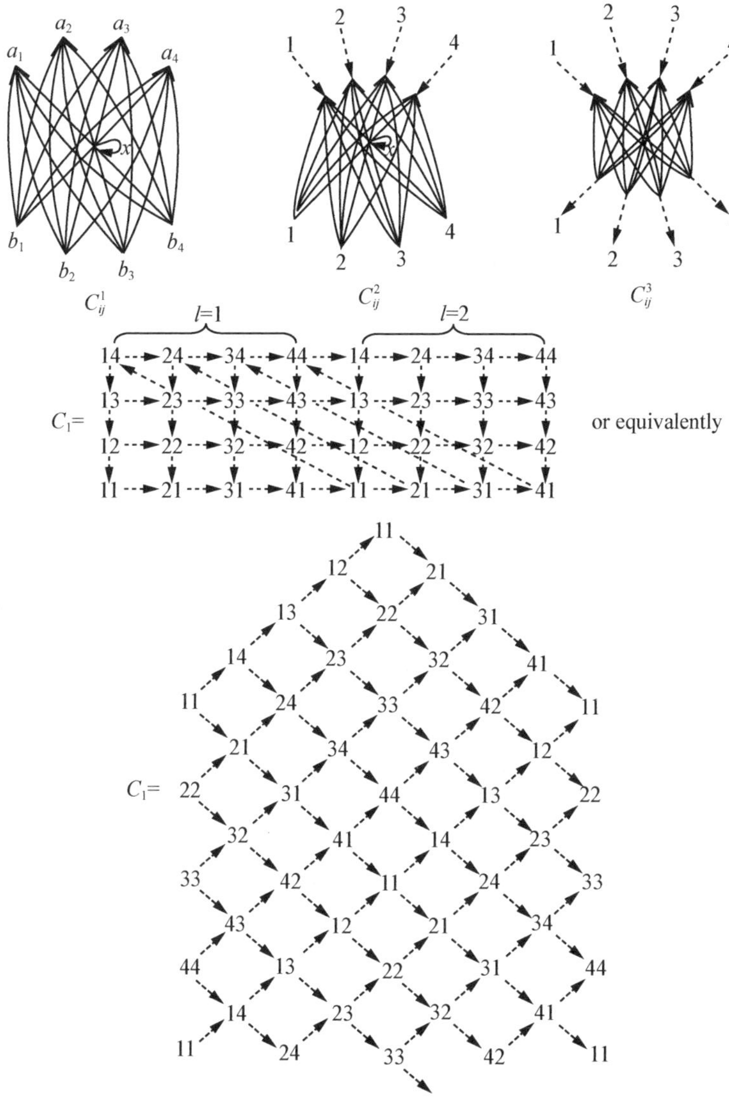

Fig. 1.2

The second one comes from edge reductions for a_j; then c_{ij}^1, $i=\left[\frac{n}{2}\right]+1, \left[\frac{n}{2}\right]+2, \cdots, n, j=1,2,\cdots,\left[\frac{n}{2}\right], l\in\mathbf{N}^*$. See fig. 1. 3, where in (b)

$$\delta(c_i^k) = -\sum_{k'<k} c_i^{k'}\cdot v_{k'k} + \sum_{i'<i} w_{ii'}\cdot c_{i'}^k,$$
$$i=\left[\frac{n}{2}\right]+2, \left[\frac{n}{2}\right]+3, \cdots, n, k\in\mathbf{N}^*.$$

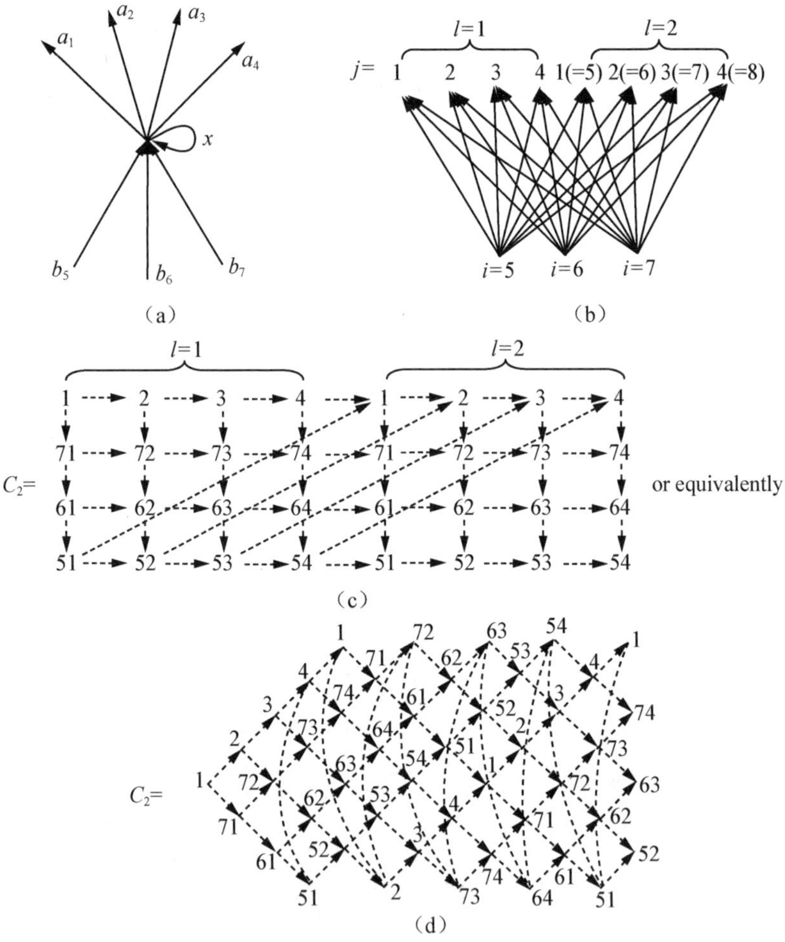

Fig. 1. 3

The third one comes from edge reductions for b_i; then c_{ij}^1, $i=1,2,\cdots,\left[\frac{n}{2}\right]+1, j=\left[\frac{n}{2}\right]+2, \left[\frac{n}{2}\right]+3, \cdots, n, l\in\mathbf{N}^*$. C_3 is dual to C_2.

The fourth one comes from unravelling for $g(x)=x$, and any positive integer r. See fig. 1. 4, where in (b)

$$\delta(a_j^k)=\sum_{k'<k}\xi_{kk'}\cdot a_j^{k'}-\sum_{j'<j}a_{j'}^k\cdot v_{j'j},\quad \delta(b_i^k)=\sum_{k'<k}b_i^{k'}\eta_{k'k}+\sum_{i'<i}w_{ii'}\cdot b_{i'}^k,$$

$\xi_{kk'}$ is a dotted arrow from k to k' and $\eta_{k'k}$ is that from k' to k, $k\in\mathbf{N}^*$.

Let d tend to $+\infty$. Then C_1 is just corresponding to the exceptional tube of width $\left[\frac{n+1}{2}\right]$. And C_4 is to that of width $\left[\frac{n}{2}\right]+1$. C_2, C_3 are corresponding to the preprojective component and preinjective component of the AR-quiver of Λ, respectively.

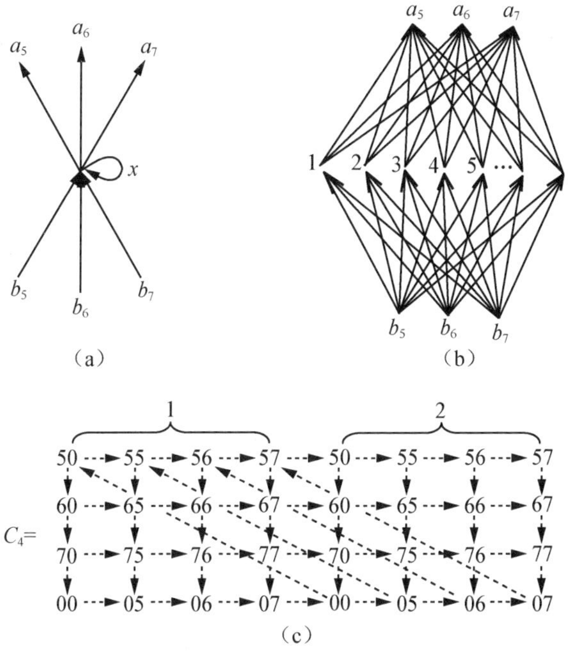

Fig. 1. 4

§ 2. Hereditary algebra of type $\widetilde{D}_n$ $(n\geqslant 4)$

Let Q be an oriented diagram of type $\widetilde{D}_n$ $(n\geqslant 4)$, and $\Lambda=kQ$, in this example $n=7$ (see fig. 2. 1(a)). The corresponding bocs is in fig. 2. 1(b), with $\delta(a_{i+1,i})=0$, $i=3,4,\cdots,n-2$, $\delta(a_{13})=0$, $\delta(a_{23})=0$, $\delta(a_{n,n-1})=0$, $\delta(a_{n+1,n-1})=0$, $\delta(a_{i+k,i})=\sum\limits_{i<i'<i+k}v'_{i+k,i'}\cdot a_{i',i}-\sum\limits_{i<i'<i+k}a_{i+k,i'}\cdot v_{i',i}$,

$i=3,4,\cdots,n-k-1, k=2,3,\cdots,n-4$, where a_{ij} is a real arrow from i' to j, v_{ij} (resp. v'_{ij}) is a dotted arrow from i to j (resp. i' to j').

After a series of edge reductions and regularizations, we obtain a bocs $\mathscr{A}_0$ with $R(\mathscr{A}_0)$ equivalent to $R(\mathrm{A})$ (see fig. 2.1(c)).

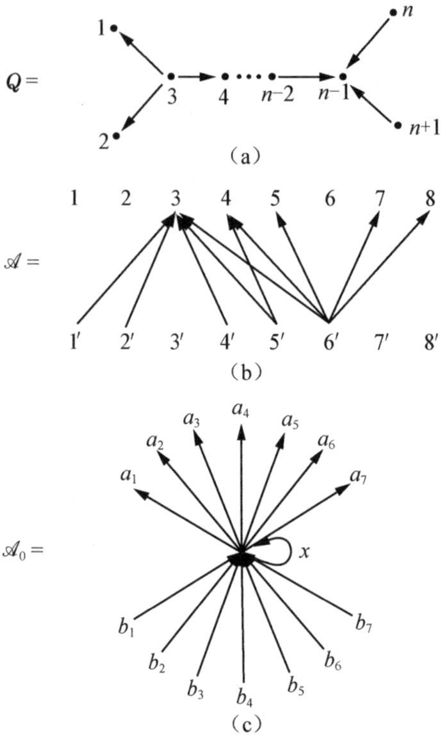

Fig. 2.1

In fig. 2.1(c), we do not exhibit the dotted arrows, some vertices of which are not starting or ending points of any real arrows, and arrows c_{ij} from $s(b_i)$ to $e(a_j)$, $i,j=1,2,\cdots,n$, but there are no $c_{n,n-1}, c_{n-1,n}$. Dotted arrow $w_{ii'}$ is from $s(b_i)$ to $s(b_{i'})$, $v_{jj'}$ from $e(a_j)$ to $e(a_{j'})$, v_{a_j} from $e(a_j)$ to $s(a_j)$, w_{b_i} from $e(b_i)$ to $s(b_i)$.

$$\delta(c_{ij})=\sum_{i'<i} w_{ii'}\cdot c_{i'j}-\sum_{j'<j} c_{ij}\cdot v_{j'j}, i,j=1,2,\cdots,(n-2);$$

$$\delta(a_j)=\sum_{1\leqslant i\leqslant n-2} w_{b_i}\cdot c_{ij}-\sum_{j'<j} a_{j'}\cdot v_{j'j}, j=1,2,\cdots,(n-2);$$

$$\delta(b_i)=-\sum_{1\leqslant j\leqslant n-2} c_{ij}\cdot v_{a_j}+\sum_{i'<i} w_{ii'}b_{i'}, i=1,2,\cdots,(n-2);$$

$$\delta(x)=-\sum_{1\leqslant j\leqslant n-2} a_j\cdot v_{a_j}+\sum_{1\leqslant i\leqslant n-2} w_{b_i}b_i;$$

$$\delta(c_{ij}) = \sum_{i'<i} w_{ii'} \cdot c_{i'j} - \sum_{1\leqslant j'\leqslant n-2} c_{ij'} v_{j'j} - b_i \cdot v_{a_j},$$
$$i = 1,2,\cdots,(n-2),\quad j = n-1, n;$$

$$\delta(a_{n-1}) = -\sum_{1\leqslant j\leqslant n-2} a_j \cdot v_{j,n-1} + \sum_{1\leqslant i\leqslant n-2} v_{b_i} \cdot c_{i,n-1} - x \cdot v_{a_{n-1}};$$

$$\delta(a_n) = -\sum_{1\leqslant j\leqslant n-2} a_j \cdot v_{j,n-1} + \sum_{1\leqslant j\leqslant n-1} v_{b_i} \cdot c_{in} - (x+1) \cdot v_{a_n};$$

$$\delta(c_{ij}) = \sum_{1\leqslant i'\leqslant n-2} w_{ii'} \cdot c_{i'j} - \sum_{j'<j} c_{ij'} v_{j'j} + w_{b_i} \cdot a_j,$$
$$i = n-1, n,\quad j = 1,2,\cdots,n-2;$$

$$\delta(b_{n-1}) = \sum_{1\leqslant j\leqslant n-2} w_{n-1,i} b_i - \sum_{1\leqslant j\leqslant n-2} c_{n-1,j} \cdot v_{a_j} + w_{b_{n-1}} \cdot x;$$

$$\delta(b_n) = \sum_{1\leqslant i\leqslant n-2} w_{n,i} \cdot b_i - \sum_{1\leqslant i\leqslant n-2} c_{n,j} \cdot a_j + w_{b_n}(x+1);$$

$$\delta(c_{ii}) = \sum_{1\leqslant i'\leqslant n-2} w_{ii'} \cdot c_{i'i} - \sum_{1\leqslant i'\leqslant n-2} c_{ii'} \cdot v_{i'i} + w_{b_i} \cdot a_i - b_i \cdot v_{a_i},$$
$$i = n-1, n.$$

For each dimension d, we have the following five components $C_1 \sim C_5$ in minimal bocs $\mathscr{B}_d$. C_1 comes from edge reductions for c_{ij}^l, $i, j = 1, 2, \cdots, n-2$, $l \in \mathbf{N}^*$. When d tends to infinite, it is corresponding to exceptional tube of width $n-2$. C_2 comes from edge reductions for a_j and c_{ij}^l, $i = n-1, n$, $j = 1, 2, \cdots, n-2$, $l \in \mathbf{N}^*$, which corresponds to the preprojective component of the AR-quiver of Λ. C_3 comes from edge reductions for b_i and c_{ij}^l, $i = 1, 2, \cdots, n-2$, $j = n-1, n$, $l \in \mathbf{N}^*$, which corresponds to the preinjective component. C_4 comes from unravelling for $g(x) = x(x+1)$ with eigenvalue 0, and subsequent edge reductions for a_{n-1}^l, b_{n-1}^l, $l \in \mathbf{N}^*$. C_5 comes from unravelling for $g(x) = x(x+1)$ with eigenvalue -1, and subsequent edge reductions for a_n^l, b_n^l. , $l \in \mathbf{N}^*$. C_4, C_5 correspond to the exceptional tubes of width 2 respectively.

§ 3. Hereditary algebras of types $\widetilde{E}_6, \widetilde{E}_7, \widetilde{E}_8$

Bocs $\mathscr{A}$(see fig. 3.1, an example of $\widetilde{E}_6$) is obtained from the bocs corresponding to

$\mathscr{A}=$ 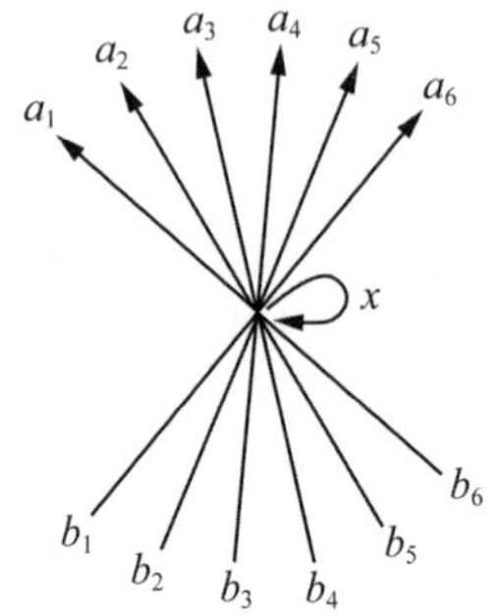

Fig. 3. 1

algebra kQ, with Q being an oriented diagram of type $\widetilde{E}_n (n=6,7,8)$. In the figure we do not exhibit dotted arrows, some vertices of which are not starting or ending points of any real arrows, and real arrows c_{ij}, $i=1,2,\cdots,n-3$, $j=1,2,\cdots,n-1$.

Let c_{ij} be a real arrow from $s(b_i)$ to $e(a_j)$, w_{ij} be a dotted arrow from $s(b_i)$ to $s(b_j)$, v_{ij} from $e(a_i)$ to $e(a_j)$, v_{a_j} from $e(a_j)$ to $s(a_j)$, w_{b_i} from $e(b_i)$ to $s(b_i)$. Then the bocs $\mathscr{A}$ has differentials:

$$\delta(c_{ij})=\sum_{i'<i} w_{ii'}\cdot c_{i'j}-\sum_{j'<j} c_{ij'}\cdot v_{j'j},\quad i,j=1,2,\cdots,n-3;$$

$$\delta(a_j)=\sum_{1\leqslant i\leqslant n-3} w_{b_i}\cdot c_{ij}-\sum_{j'<j} a_{j'}\cdot v_{j'j},\quad i=1,2,\cdots,n-3;$$

$$\delta(b_i)=-\sum_{1\leqslant j\leqslant n-3} c_{ij}\cdot v_{a_j}+\sum_{i'<i} w_{ii'}\cdot b_{i'}, j=1,2,\cdots,n-3;$$

$$\delta(x)=-\sum_{1\leqslant j\leqslant n-3} a_j\cdot v_{a_j}+\sum_{1\leqslant i\leqslant n-3} w_{b_i}\cdot b_i, j=1,2,\cdots,n-3;$$

$$\delta(c_{ij})=\sum_{i'<i} w_{ii'}c_{i'j}-\sum_{j'<j} c_{ij'}\cdot v_{j'j}-b_i\cdot v_{a_j},$$
$$i=1,2,\cdots,n-3,\quad j=n-2,n-1;$$

$$\delta(a_j)=\sum_{1\leqslant i\leqslant n-3} w_{b_i}\cdot c_{ij}-\sum_{j'<j} a_{j'}\cdot v_{j'j}-x\cdot v_{a_j}, j=n-2,n-1;$$

$$\delta(c_{ij})=\sum_{i'<i} w_{ii'}\cdot c_{i'j}-\sum_{j'<j} c_{ij'}\cdot v_{j'j}+w_{b_i}\cdot a_j,$$
$$i=n-2,n-1,\quad j=1,2,\cdots,n-3;$$

$$\delta(b_i)=-\sum_{1\leqslant j\leqslant n-3} c_{ij}\cdot v_{a_j}+\sum_{i'<i} w_{ii'}\cdot b_{i'}+w_{b_i}\cdot x, i=n-2,n-1;$$

$$\delta(c_{ij})=\sum_{i'<i} w_{ii'}\cdot c_{i'j}-\sum_{j'<j} c_{ij'}\cdot v_{j'j}+v_{b_i}\cdot a_j-b_i\cdot v_{a_j},$$
$$i,j=n-2,n-1;$$

$$\delta(c_{in}) = \sum_{i'<i} w_{ii'} \cdot c_{i'n} - \sum_{1\leqslant j\leqslant n-2} c_{ij} \cdot v_{jn} - b_i \cdot v_{a_n}, i=1,2,\cdots,n-3;$$

$$\delta(a_n) = \sum_{1\leqslant i\leqslant n-3} w_{b_i} \cdot c_{in} - \sum_{1\leqslant j\leqslant n-2} a_j \cdot v_{in} - (x-1) \cdot v_{a_n};$$

$$\delta(c_{nj}) = \sum_{1\leqslant i\leqslant n-2} w_{ni} \cdot c_{ij} - \sum_{j'<j} c_{nj'} \cdot v_{j'j} + w_{b_n} \cdot a_j, j=1,2,\cdots,n-3;$$

$$\delta(b_n) = -\sum_{1\leqslant i\leqslant n-3} c_{nj} \cdot v_{a_j} + \sum_{1\leqslant i\leqslant n-2} w_{ni} \cdot v_{j'j} + w_{b_n} \cdot (x-1);$$

$$\delta(c_{nn}) = \sum_{1\leqslant i\leqslant n-3} w_{ni} \cdot c_{in} - \sum_{1\leqslant j\leqslant n-3} c_{nj} \cdot v_{jn} + w_{b_n} \cdot a_n - b_n \cdot v_{a_n}.$$

Take $n = 6, 7, 8$. This bocs is representation equivalent to the hereditary algebra of types $\widetilde{E}_6, \widetilde{E}_7, \widetilde{E}_8$, respectively.

References

[1] Rojter A V. Matrix problems and representations of bocses. LNM. 1980,831: 288.

[2] Drozd Y A. Tame and wild matrix problems. Matrix Problems. Kiev,1977, 104-114.

[3] Drozd Y A. Tame and matrix problems. Representations and Quotriatic Forms,Kiev,1979,39-74.

[4] Crawley-Boevey W W. On the tame algebras and bocses. London Math. Soc. , 1988,56(3):451.

[5] Dlab V,Ringel C M. Indecomposable representations of graphs and algebras. Memoirs Amer. Math. Soc. ,1976,173.

[6] Lin Y N,Zhang Y B. Bocses corresponding to the hereditary algebras of tame type, Ⅰ. Science in China,1996,39A(5):483-490.

Chinese Science Bulletin,1997,42(2):108-112.

一个具有强齐性条件的野型 Bocs①

A Wild Bocs Having Strong Homogeneous Property

Keywords　bocs; wild representation type; Auslander-Reiten sequence; homogeneous property.

Bocs, which is the abbreviated form of bimodule over a categary with coalgebra structure, was introduced by Kleiner and Rojter in 1975[1] and developed by Drozd in 1979, then formulated by Crawley-Boevey in 1988. Let k be an algebraically closed field, Λ a finitely dimensional k-algebra. Then there exists a bocs $\mathscr{B}$ over k associated to Λ. From this relation Drozd proved one of the most important theorems in representation theory of algebra, namely, a finitely dimensional k-algebra is either of representation tame type or of representation wild type, but not both[2]. After 10 years, Crawley-Boevey proved the following theorem: if Λ is of representation tame type, then for each dimension d, almost all modules M with $\dim M \leqslant d$ have the property $D\mathrm{Tr}M \simeq$[3,4]. It is natural to ask the converse of the theorem: if for each dimension d, almost all modules M with $\dim M \leqslant d$ have the property $D\mathrm{Tr}M \simeq M$, is Λ of representation tame type? Or equivalently, if Λ is of

① 收稿日期:1996-10-28.
本文与雷天刚,Bautista R 合作.

representation wild type, does exist a dimension d and infinitely many modules M with $\dim M \leqslant d$ such that $D\mathrm{Tr}M \not\cong M$? Since every finitely dimensional k-algebra Λ is associated to a layered bocs $\mathscr{B}$ which has almost split sequences, we may consider this problem first in the case of bocses. In the consideration a special bocs, denoted by $\mathscr{A}$, appears.

§ 1. Bocs $\mathscr{A}$ and its representation category $R(\mathscr{A})$

Let $\mathscr{A}=(A,V)$ be a bocs defined over an algebraically closed field k, with layer $L=(A';\omega;x,a;v)$, where the set of indecomposables in A' consists of only one object X with $A'(X,X)=k$, and the differentials are $\delta(x)=0$, $\delta(a)=xv-vx$, $\delta(v)=0$. The bigraph of bocs $\mathscr{A}$ is[5] fig. 1. 1.

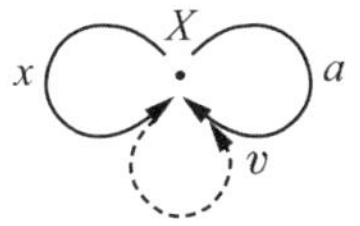

Fig. 1. 1

Denote by $R(\mathscr{A})$ the finitely dimensional representation category of bocs $\mathscr{A}$. Then for any M in $R(\mathscr{A})$, M consists of a finitely dimensional vector space M_X and two linear transformations $M(x), M(a): M_X \to M_X$. Denote by $M=(M_X; M(x), M(a))$.

If $N=(N_X; N(x), N(a))$ is also an object of $R(\mathscr{A})$, then a morphism φ in $R(\mathscr{A})$ from M to N consists of two linear transformations $\varphi_X, \varphi_v: M_X \to N_X$, satisfying the following conditions: $N(x)\varphi_X - \varphi_X M(x)=0$ from $\delta(x)=0$, and $N(a)\varphi_X-\varphi_X M(a)=N(x)\varphi_v-\varphi_v M(x)$ from $\delta(a)=xv-vx$. If $\psi: N \to L$ is another morphism in $R(\mathscr{A})$, then the composition $\psi\circ\varphi: M\to L$ consists of $(\psi\circ\varphi)_X=\psi_X\varphi_X$, and $(\psi\circ\varphi)_v=\psi_X\varphi_v+\psi_v\varphi_X$ from $\delta(v)=0$. Then identity morphism $\mathrm{id}: M\to M$ consists of $(\mathrm{id})_X=(\mathrm{id})_{M_X}$ and $(\mathrm{id})_v=0$ (fig. 1. 2).

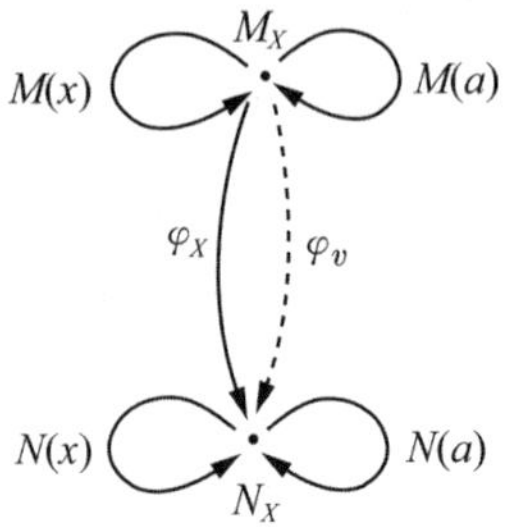

Fig. 1. 2

A morphism $\varphi: M \to N$ is called an isomorphism, if there exists a morphism $\psi: N \to M$ such that $\psi \circ \varphi = \mathrm{id}_M$ and $\varphi \circ \psi = \mathrm{id}_N$.

Since V is a left triangular tensor coalgebra over A, the idempotents split in $R(\mathscr{A})$[6]. Therefore $R(\mathscr{A})$ is a Krull-Schmidt category, i. e. every object has a direct decomposition which is unique up to isomorphisms of the components.

Let S and M be two objects of $R(\mathscr{A})$. S is called a direct summand of M, if there exist morphisms $\varphi: S \to M$ and $\psi: M \to S$ such that $\psi\varphi = \mathrm{id}_s$. An object M is called indecomposable, if M has no non-trivial direct summand, where the trivial direct summand means the direct summand isomorphic to M itself or O.

§ 2.　Objects and morphisms

Notations. (1) Denote by

$$\boldsymbol{J}_{d;p} = \begin{pmatrix} \mathbf{0} & \boldsymbol{I} & & & \\ & \mathbf{0} & \boldsymbol{I} & & \\ & & \ddots & \ddots & \\ & & & \mathbf{0} & \boldsymbol{I} \\ & & & & \mathbf{0} \end{pmatrix}$$

the $d \times d$ partitioned matrix with $\boldsymbol{I}$ $p \times p$ identity matrix and $\mathbf{0}$ $p \times p$ zero matrix.

(2) A $dp \times eq$ matrix $\mathbf{A}$ is called a $T(d \times e; p \times q)$-matrix, if $\mathbf{A}$ has the following partitioned expression:

$$\boldsymbol{A}=\begin{pmatrix} \boldsymbol{0} & \cdots & \boldsymbol{0} & \boldsymbol{A}^{(e-d+1)} & \boldsymbol{A}^{(e-d+2)} & \cdots & \boldsymbol{A}^{(e-1)} & \boldsymbol{A}^{(e)} \\ \boldsymbol{0} & \cdots & \boldsymbol{0} & \boldsymbol{0} & \boldsymbol{A}^{(e-d+1)} & \cdots & \boldsymbol{A}^{(e-2)} & \boldsymbol{A}^{(e-1)} \\ \vdots & \vdots & \vdots & \vdots & \vdots & & \vdots & \vdots \\ \boldsymbol{0} & \cdots & \boldsymbol{0} & \boldsymbol{0} & \boldsymbol{0} & \cdots & \boldsymbol{A}^{(e-d+1)} & \boldsymbol{A}^{(e-d+2)} \\ \boldsymbol{0} & \cdots & \boldsymbol{0} & \boldsymbol{0} & \boldsymbol{0} & \cdots & \boldsymbol{0} & \boldsymbol{A}^{(e-d+1)} \end{pmatrix} \text{ when } d\leqslant e,$$

and

$$\boldsymbol{A}=\begin{pmatrix} \boldsymbol{A}^{(1)} & \boldsymbol{A}^{(2)} & \cdots & \boldsymbol{A}^{(e-1)} & \boldsymbol{A}^{(e)} \\ \boldsymbol{0} & \boldsymbol{A}(1) & \cdots & \boldsymbol{A}^{(e-2)} & \boldsymbol{A}^{(e-1)} \\ \vdots & \vdots & \ddots & \vdots & \vdots \\ \boldsymbol{0} & \boldsymbol{0} & \cdots & \boldsymbol{A}^{(1)} & \boldsymbol{A}^{(2)} \\ \boldsymbol{0} & \boldsymbol{0} & \cdots & \boldsymbol{0} & \boldsymbol{A}^{(1)} \\ \boldsymbol{0} & \boldsymbol{0} & \cdots & \boldsymbol{0} & \boldsymbol{0} \\ \vdots & \vdots & & \vdots & \vdots \\ \boldsymbol{0} & \boldsymbol{0} & \cdots & \boldsymbol{0} & \boldsymbol{0} \end{pmatrix} \text{ when } d>e,$$

where $\boldsymbol{A}^{(i)}, i=e-\min\{d,e\}+1, e-\min\{d,e\}+2, \cdots, e$, and $\boldsymbol{0}$ are all $p\times q$ matrices.

(3) For $\underline{\boldsymbol{d}}=(d_1,d_2,\cdots,d_s)$, $\underline{\boldsymbol{p}}=(p_1,p_2,\cdots,p_s)$, denote

$$\boldsymbol{J}_{\underline{d};\underline{p}}=\boldsymbol{J}_{d_1;p_1}\oplus\cdots\oplus\boldsymbol{J}_{d_s;p_s}.$$

(4) Let $\underline{\boldsymbol{d}}=(d_1,d_2,\cdots,d_s)$, $\underline{\boldsymbol{p}}=(p_1,p_2,\cdots,p_s)$, $\underline{\boldsymbol{e}}=(e_1,e_2,\cdots,e_s)$, $\underline{\boldsymbol{q}}=(q_1,q_2,\cdots,q_t)$. A $\sum\limits_{i=1}^{s}d_ip_i\times\sum\limits_{j=1}^{t}e_jq_j$ matrix $\boldsymbol{F}$ is called a $\boldsymbol{T}(\underline{\boldsymbol{d}}\times\underline{\boldsymbol{e}};\underline{\boldsymbol{p}}\times\underline{\boldsymbol{q}})$-matrix, if $\boldsymbol{F}$ is partitioned to $\boldsymbol{F}=(\boldsymbol{F}_{ij})_{s\times t}$, such that $\boldsymbol{F}_{ij}$ is a $T(d_i\times e_j;p_i\times q_j)$-matrix.

(5) A $\boldsymbol{T}(\underline{\boldsymbol{d}}\times\underline{\boldsymbol{d}};\underline{\boldsymbol{p}}\times\underline{\boldsymbol{p}})$-matrix is simply denoted by $\boldsymbol{T}(\underline{\boldsymbol{d}};\underline{\boldsymbol{p}})$-matrix.

Lemma 2.1 (1) $\boldsymbol{J}_{d;p}$ is similar to $\mathrm{diag}(\boldsymbol{J}(0,d),\cdots,\boldsymbol{J}(0,d))_{p\times p}$ with $\boldsymbol{J}(0,d)$ the $d\times d$ Jordan form of eigenvalue 0.

(2) $\boldsymbol{A}$ is a $T(d\times e;p\times q)$-matrix, if and only if $\boldsymbol{A}\boldsymbol{J}_{e;q}=\boldsymbol{J}_{d;p}\boldsymbol{A}$.

(3) $\boldsymbol{F}$ is a $T(\underline{\boldsymbol{d}}\times\underline{\boldsymbol{e}};\underline{\boldsymbol{p}}\times\underline{\boldsymbol{q}})$-matrix, if and only if $\boldsymbol{F}\boldsymbol{J}_{\underline{e};\underline{q}}=\boldsymbol{J}_{\underline{d};\underline{p}}\boldsymbol{F}$.

Proposition 2.1[5] Let M be an object in $R(\mathscr{A})$. If $M(x)$ has at least two different eigenvalues, then M is decomposable.

Lemma 2.2[5] Let M, N be two objects in $R(\mathscr{A})$ with $M(x)=\lambda+\boldsymbol{J}_{\underline{e};\underline{q}}$, and $N(x)=\mu+\boldsymbol{J}_{\underline{d};\underline{p}}$. If $\lambda\neq\mu$, then there exists only a zero map from M to N.

Proposition 2.2[5] Let $\varphi: M\rightarrow N$ be a morphism on $R(\mathscr{A})$, with $M(x)=\lambda+\boldsymbol{J}_{\underline{e};\underline{q}}$ and $N(x)=\lambda+\boldsymbol{J}_{\underline{d};\underline{p}}$. Then $\varphi_X=\boldsymbol{F}$ is a $T(\underline{\boldsymbol{d}}\times\underline{\boldsymbol{e}};\underline{\boldsymbol{p}}\times\underline{\boldsymbol{q}})$-matrix.

Proposition 2.3[5] Let M be an indecomposable object in $R(\mathscr{A})$. Then M is isomorphic to an object N, with $N(x)=\lambda+\boldsymbol{J}_{\underline{d};\underline{p}}$ and $N(a)=(\boldsymbol{N}_{ij})_{s\times s}$, where $\underline{\boldsymbol{d}}=(d_1, d_2, \cdots, d_s)$, $\underline{\boldsymbol{p}}=(p_1, p_2, \cdots, p_s)$,

$$N_{ij}=\begin{pmatrix} N_{ij}^1 & 0 & \cdots & 0 \\ N_{ij}^2 & 0 & \cdots & 0 \\ \vdots & \vdots & & \vdots \\ N_{ij}^{d_i} & 0 & \cdots & 0 \end{pmatrix}_{d_i\times d_j}$$

with $N_{ij}^k, k=1,2,\cdots,d_i$, and $\boldsymbol{0}$ are $p_i\times p_j$ matrix. Furthermore $N_{ij}^1=N_{ij}^2=\cdots=N_{ij}^{d_i-d_j}=0$ when $d_i>d_j$.

§ 3. Endomorphisms of indecomposable objects

Lemma 3.1 Let F be a $T(\underline{\boldsymbol{d}};\underline{\boldsymbol{p}})$-matrix with $\underline{\boldsymbol{d}}=(d_1, d_2, \cdots, d_s)$ and $\underline{\boldsymbol{p}}=(p_1, p_2, \cdots, p_s)$. If $d_1, d_2, \cdots, d_s$ are pairwise different, then

$$\det \boldsymbol{F}=\prod_{i=1}^{s}\det \boldsymbol{F}_{ii}.$$

Key lemma Let F be a $T(\underline{\boldsymbol{d}};\underline{\boldsymbol{p}})$-matrix with $\underline{\boldsymbol{d}}=(d_1, d_2, \cdots, d_s)$ and $d_1<d_2<\cdots<d_s$. If $\boldsymbol{F}_{ii}^{(1)}=\widetilde{\boldsymbol{F}}_{ii}^{(1)}\oplus\boldsymbol{0}$ with $\widetilde{\boldsymbol{F}}_{ii}^{(1)}$ are $q_i\times q_i$ invertible matrices where $p_i\geqslant q_i\geqslant 0$ for all $i=1, 2, \cdots, s$. Then there exist a positive integer m and an invertible $T(\underline{\boldsymbol{d}};\underline{\boldsymbol{p}})$-matrix H such that $T(\underline{\boldsymbol{d}};\underline{\boldsymbol{p}})$-matrix $\boldsymbol{G}=\boldsymbol{H}\boldsymbol{F}^m\boldsymbol{H}^{-1}$ satisfies the following condition: $\boldsymbol{G}_{ij}^{(k)}=\widetilde{\boldsymbol{G}}_{ij}^{(k)}\oplus\boldsymbol{0}$, where $\widetilde{\boldsymbol{G}}_{ij}^{(k)}$ are $q_i\times q_j$ matrices, $k=d_j-\min\{d_i, d_j\}+1, \cdots, d_j, i, j=1, 2, \cdots, s$, and $\widetilde{\boldsymbol{G}}_{ii}^{(1)}$ is invertible.

Theorem 3.1 Let M be an indecomposable object in $R(\mathscr{A})$, with $M(x)=\lambda+J_{\underline{\boldsymbol{d}};\underline{\boldsymbol{p}}}$, $\underline{\boldsymbol{d}}=(d_1, d_2, \cdots, d_s)$, $d_1<d_2<\cdots<d_s$, $\underline{\boldsymbol{p}}=(p_1, p_2, \cdots,$

p_s). If $\varphi: M \to M$ is a morphism with $\varphi_X = \boldsymbol{F}$ a $T(\underline{\boldsymbol{d}}; \underline{\boldsymbol{p}})$-matrix, then $\boldsymbol{F}_{ii}^{(1)}$ has a unique eigenvalue and for $i = 1, 2, \cdots, s$, the eigenvalues are all equal. Moreover $\boldsymbol{F}$ has a unique eigenvalue.

§ 4. Bocs $\mathscr{A}$ has almost split sequences and strong homogeneous property

Definition 4.1 (1) Let $\mathscr{B}$ be a layered bocs. An object N in $R(\mathscr{B})$ is called proper projective if for any morphism $\varphi, M \to N$ in $R(\mathscr{B})$ with $(i, \omega)^* \varphi$ epimorphic, then φ is a split epimorphism. Dually an object M in $R(\mathscr{B})$ is called proper injective if for any morphism $\psi: M \to N$ in $R(\mathscr{B})$ with $(i, \omega)^* \psi$ monomorphism, then ψ is a split monomorphism.

(2) Let $\mathscr{B}$ be a layered bocs. $\mathscr{B}$ is said to have almost split sequences, if for any non properinjective M in $R(\mathscr{B})$, there exists a proper almost split sequence $(e): M \to E \to N$ in $R(\mathscr{B})$, and for any non proper-projective N in $R(\mathscr{B})$, there exists a proper almost split sequence $(e): M \to E \to N$ in $R(\mathscr{B})$ (proper means the sequence $(i, \omega)^* (e)$ exact).

(3) Let $\mathscr{B}$ be a layered bocs. We call an indecomposable object M in $R(\mathscr{B})$ homogeneous, if there exists a proper almost split sequence $M \xrightarrow{t} E \xrightarrow{\pi} M$ in $R(\mathscr{B})$. And we also call the sequence homogeneous.

(4) A layered bocs $\mathscr{B}$ is said to have the homogeneous property, if $\mathscr{B}$ has almost split sequences, and for any dimension d almost all (except finitely many) objects M with $\dim M \leqslant d$ are homogeneous.

(5) A layered bocs $\mathscr{B}$ is said to have the strong homogeneous property, if (1) $\mathscr{B}$ has almost split sequences; (2) there exists neither proper-projective, nor proper-injective in $R(\mathscr{B})$; (3) for any indecomposable M in $R(\mathscr{B})$, M is homogeneous.

Remark 1 Let Λ be a finitely dimensional k-algebra, $\mathscr{B} = (B, W)$ be the corresponding bocs. Then $\mathscr{B}$ has almost split sequences[6].

Remark 2 Let bocs $\mathscr{B} = (B, V)$ with layer $L = (B'; \omega; x, y)$, where B' is local and trivial, $\delta(x) = 0, \delta(y) = 0$, the bigraph of $\mathscr{B}$ is fig. 4.1.

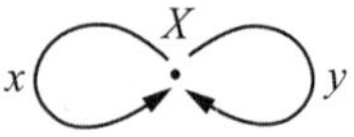

Fig. 4. 1

Denote by $\underline{r}$ the Jacobson radical of $R(\mathscr{B})$. Then $\underline{r}^2=\underline{r}$. Therefore there does not exist any irreducible map in $R(\mathscr{B})$, nor any almost split sequence in $R(\mathscr{B})$.

Main theorem Let $\mathscr{A}=(A,V)$ be the bocs defined in sec. § 1. If ch $k=0$, then $\mathscr{A}$ has almost split sequences and the strong homogeneous property.

Acknowledgements

Zhang Y B would like to express her gratitude to Prof. A. V. Rojter. The idea of using matrix method was suggested by him during the ICRA Ⅳ in Mexico. This work was supported by the National Natural Science Foundation of China (Grant No. 19331013).

References

[1] Kleiner M, Rojter A V. Representations of differential graded categories. Lecture Notes in Mathematics 488, Berlin: Springer, 1975, 316-340.

[2] Drozd Y A. Tame and wild matrix problems. Representations and Quadratic Forms (in Russian). Kiev: Institute of Mathematics, Academy of Sciences of Ukranian SSR, 1979, 39-74.

[3] Crawley-Boevey W W. On tame algebras and bocses. Proc. London Math. Soc., 1988, 56: 451.

[4] Auslander M, Reiten Ⅰ. Representation theory of Artin algebras Ⅲ. Comm. Algebra, 1977, 5: 443.

[5] Zhang Y B, Lei T G, Bautista R. The representation category of a bocs Ⅰ～Ⅳ. J. Beijing Normal Univ., 1995, 31: 313, 440.

[6] Bautista R, Kleiner M. Almost split sequences for relatively projective modules. J. Algebra, 1990, 135: 19.

Colloquium Mathematicum, 1998, 77(2): 271-292.

A_n 型路代数上例外序列的自同态代数①

Endomorphism Algebras of Exceptional Sequences over Path Algebras of Type $\mathbf{A}_n$②

Keywords exceptional sequences and their endomorphism algebras; gentle algebras; tilted algebras.

The notion of exceptional sequences originates from the study of vector bundles (see, for instance, [GR, B]) and was carried over to modules over hereditary artin algebras (see [CB, R2]). In this paper, we consider the following situation: let k be a commutative field, Q be a finite connected quiver without oriented cycles; then the path algebra $A=kQ$ is hereditary and we may study the exceptional sequences in the category mod A of finitely generated right A-modules. We recall that an indecomposable object E in mod A is called exceptional if $\operatorname{Ext}_A^1(E,E)=0$. A sequence $\mathcal{E}=(E_1,E_2,\cdots,E_t)$ of exceptional objects in mod$A$ is called an exceptional sequence if $\operatorname{Hom}_A(E_j,E_i)=0$ and $\operatorname{Ext}_A^1(E_j,E_i)=0$ for $j>i$. An exceptional sequence $\mathcal{E}=(E_1,E_2,\cdots,E_t)$ is called complete if t equals the number of isomorphism classes of simple A-modules, and

① Received: 1997-04-28; Revised: 1997-06-17, 1997-12-29.
本文与 Assem I 合作.
② Assem I gratefully acknowledges partial support form the NSERC of Canada. Zhang Yingbo would like to express her thanks to Professor Shiping Liu for his invitation to Canada, and the National Natural Science Foundation of China for its partial support.

connected if End($\bigoplus_{i=1}^{t} E_i$) (which we denote briefly by End $\mathcal{E}$) is a connected algebra. Ringel has asked whether, if $\mathcal{E}$ is a complete exceptional sequence in the module category over a representation-finite hereditary artin algebra, then End $\mathcal{E}$ is also representation-finite. This question was answered affirmatively in case $A=kQ$, where Q is of type $\mathbf{A}_n$, first by H. Yao [Y] in case Q has a linear orientation, then by H. Meltzer [M] in case Q has an arbitrary orientation. It is reasonable to generalise Ringel's question as follows: let $\mathcal{E}$ be a complete exceptional sequence in the module category over a tame path algebra; is it then true that End $\mathcal{E}$ is also tame? The objective of this paper is to answer this latter question affirmatively whenever $A=kQ$, where Q is of type $\widetilde{\mathbf{A}}_n$. More precisely, we prove the following theorem.

Theorem Let k be a commutative field, Q be a quiver with underlying graph $\mathbf{A}_n$, and $A=kQ$ be its path algebra. Let $\mathcal{E}$ be a complete exceptional sequence in mod A. Then End $\mathcal{E}$ is either a direct product of one tilted algebra of type $\mathbf{A}_m$ (with $m\leqslant n$) and tilted algebras of type $\mathbf{A}_l$ (with $l\leqslant n-m$), or a direct product of tilted algebras of type $\mathbf{A}_l$ (with $l\leqslant n+1$). Each connected subsequence of $\mathcal{E}$ is a partial tilting module.

We use essentially the description of the module category of a path algebra of type $\widetilde{\mathbf{A}}_n$, as in [DR, R1], and the structure of its indecomposable modules, as in [BR]. Notice that, if $(E_1, E_2, \cdots, E_t)$ is an exceptional sequence in mod A, where $A=kQ$, then, in particular, each E_i is exceptional, hence End $E_i=k$ (see, for instance, [K], (11.9)). If Q is an Euclidean quiver, this implies that E_i is postprojective, preinjective or regular lying in an exceptional tube of rank $m(>1)$, say, and, in this case, is of quasi-length at most $m-1$.

We use without further reference properties of the Auslander-Reiten translations $\tau=\mathrm{DTr}$ and $\tau^{-1}=\mathrm{TrD}$, and the Auslander-Reiten quiver $\Gamma(\mathrm{mod}\,A)$ of A as in [ARS, R1]. In particular, we frequently use the Auslander-Reiten formulae

$$\mathrm{Ext}_A^1(M,N)\cong D\,\mathrm{Hom}A(N,\tau M)\cong D\,\mathrm{Hom}_A(\tau^{-1}N,M).$$

For the classification results of tilted and iterated tilted algebras of type $\mathbf{A}_n$ and $\widetilde{\mathbf{A}}_n$, we refer to [A1, AH, AS, R, H].

§ 1. Regular exceptional modules

The aim of this section is to show that, if Γ is an exceptional tube of rank m, say, in the Auslander-Reiten quiver of the path algebra A of an Euclidean quiver, and $\mathcal{E}$ is a connected exceptional sequence all of whose terms lie in Γ, then End $\mathcal{E}$ is a tilted algebra of type $\mathbf{A}_t$.

In this situation, the tube Γ is standard, thus we may identify the points in Γ with the corresponding indecomposable A-modules. Each point in Γ will be given by two coordinates: the first is the quasi-length of the corresponding indecomposable A-module (thus is a positive integer), and the second represents its regular socle (and is chosen from $\mathbf{Z}_m$). The modules E_i being exceptional, they have quasi-length at most $m-1$. The figure on the next page shows the full translation subquiver Γ' of Γ consisting of all modules of quasi-length at most $m-1$. Associated to each point $M=(i,j)$ in Γ' are four sectional paths in Γ', these are (fig. 1. 1):

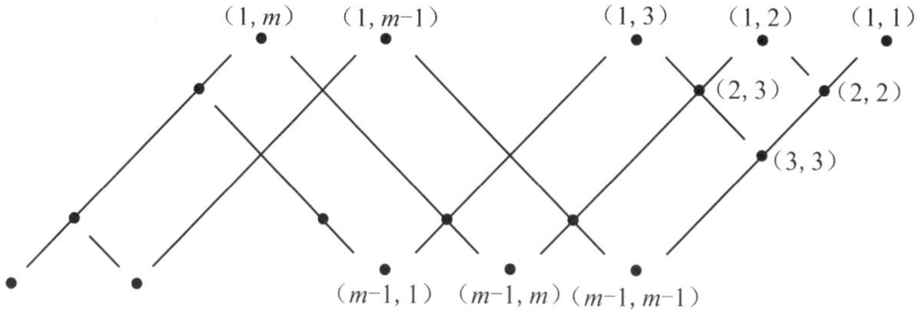

Fig. 1. 1

(1) ($M\nearrow$), the portion of coray from M to the mouth (that is, the sectional path from (i,j) to $(1,j-i+1)$),

(2) ($M\searrow$), the portion of ray from M to infinity in Γ' (that is, the sectional path from (i,j) to $(m-1,j)$),

(3) $(\searrow M)$, the portion of ray from the mouth to M (that is, the sectional path from $(1,j)$ to (i,j)), and

(4) $(\nearrow M)$, the portion of coray from infinity to M in Γ' (that is, the sectional path from $(m-1\ m-1+j-i)$ to (i,j)).

It also follows from the standardness of Γ that, if $M=(i,j)$ is in Γ', then the support Supp $\mathrm{Hom}_A(M,-)|_{\Gamma'}$ of the restriction to Γ' of the functor $\mathrm{Hom}_A(M,-)$ is a trapezoid with corners (i,j), $(1,j-i+1)$, $(m-1,j-i+1)$ and $(m-1,j)$, bounded by the sectional paths $(M\nearrow)$, $(M\searrow)$ and $((1,j-i+1)\searrow)$. Similarly, Supp $\mathrm{Hom}_A(-,M)|\Gamma'$ is a trapezoid with corners (i,j), $(m-1,m-1+j-i)$, $(1,j)$ and $(m-1,m-2+j)$, bounded by the sectional paths $(\searrow M)$, $(\nearrow M)$ and $(\nearrow(1,j))$.

Lemma 1.1 Let $M\in\mathcal{E}$, and M,N,L lie in Γ.

(1) Let $N\in\mathcal{E}$. Then $\mathrm{Hom}_A(M,N)\neq 0$ if and only if

$$N\in(M\nearrow)\cup(M\searrow).$$

(2) Let $L\in\mathcal{E}$. Then $\mathrm{Hom}_A(L,M)\neq 0$ if and only if

$$L\in(\searrow M)\cup(\nearrow M).$$

Proof We only show (1), since the proof of (2) is similar.

For $M,N\in\mathcal{E}$, $\mathrm{Hom}_A(M,N)\neq 0$ implies that (M,N) is a subsequence of $\mathcal{E}$ so that $\mathrm{Hom}_A(\tau^{-1}M,N)=0$, that is,

$$N\in \mathrm{Supp}\ \mathrm{Hom}_A(M,-)|\Gamma'$$

but

$$N\notin \mathrm{Supp}\ \mathrm{Hom}_A(\tau^{-1}M,-)|\Gamma'.$$

Therefore $N\in(M\nearrow)\cup(M\searrow)$. The converse is trivial. □

Lemma 1.2 There exists no path $M\to N\to L$ in Γ with $M=(i,j)$, $N=(i-l,j-l)$, $l\geqslant 1$, $L=(k,j-l)$, $k>i-l$, and $M,N,L\in\mathcal{E}$.

Proof Assume the contrary. Since $N\in(M\nearrow)\cup(M\searrow)$ by Lemma 1.1, we have in fact $N\in(M\nearrow)$. Similarly, $L\in(N\searrow)$. But then we obtain $L\in \mathrm{Supp}\ \mathrm{Hom}_A(\tau^{-1}M,-)|\Gamma'$, so that $\mathrm{Ext}_A^1(L,M)\neq 0$, a contradiction to the fact that (M,N,L) is a subsequence of $\mathcal{E}$. □

Lemma 1.3 Assume there exists a path $M\xrightarrow{f}N\xrightarrow{g}L$ in Γ with $M=(i,j)$, $N=(k,j)$, $k>i$, $L=(k-l,j-l)$, $1\leqslant l<k$ and $M,N,L\in$

$\mathcal{E}$. Then $gf=0$.

Proof By Lemma 1.1 and the hypothesis, we have $N\in(M\searrow)$. Also, since (M,N,L) is a subsequence of $\in\mathcal{E}$, we have

$L\notin \text{Supp Hom}_A(\tau^{-1}M,-)|_{\Gamma'}$, hence $L\notin \text{Supp Hom}_A(M,-)|_{\Gamma'}$. That is, $\text{Hom}_A(M,L)=0$. □

Lemma 1.4 Let $\mathcal{E}$ be a connected exceptional sequence lying in Γ. Then the quiver of End $\mathcal{E}$ is a tree.

Proof Assume the contrary; then the quiver of End $\mathcal{E}$ contains a cycle, which, by [Y], Proposition 3.2, is not an oriented cycle. Let thus $\mathcal{F}$ be a subsequence of $\mathcal{E}$ such that the quiver of End $\mathcal{F}$ is a cycle. We agree to say that $\mathcal{F}$ passes through two neighbouring corays $(\nearrow(1,j))$ and $(\nearrow(1,j-1))$ if there is an arrow α of the quiver of End $\mathcal{F}$ representing a sectional path $\alpha_1,\alpha_2,\cdots,\alpha_r$ where $\alpha_1,\alpha_2,\cdots,\alpha_r$ are arrows in Γ, and some $1\leqslant l\leqslant r$ such that α_l is the arrow in Γ from $(i,j+i-1)$ to $(i+1,j+i-1)$. We also denote by $\widetilde{\Gamma}'$ the universal covering of the full translation subquiver Γ' of Γ of all modules of quasi-length at most $m-1$ (thus $\widetilde{\Gamma}'\cong\mathbf{Z}\mathbf{A}_{m-1}$). We consider two cases:

(1) Assume that $\mathcal{F}$ passes through all pairs of neighbouring corays $(\nearrow(1,j))$ and $(\nearrow(1,j-1))$, where j ranges over $\mathbf{Z}_m$. Let $M\in\mathcal{F}$, then there exist two points M_0,M_1 in $\widetilde{\Gamma}'$ lifting M, and a path of length $m+1$ from M_0 to M_1. The corays passing through M_0 and M_1 determine a parallelogram $abcd$ in $\widetilde{\Gamma}'$ as shown (fig. 1.2):

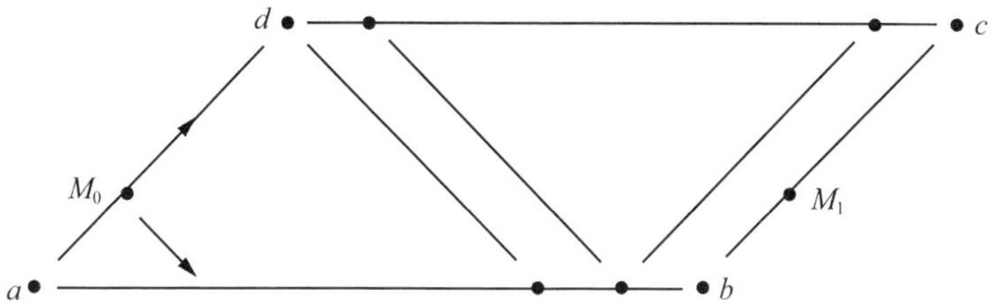

Fig. 1.2

There exists a walk $\widetilde{\mathcal{F}}$ inside $abcd$ lifting the non-oriented path $\mathcal{F}$. Since the horizontal size of $abcd$ is $m+1$, while its vertical size is $m-1$, the walk $\widetilde{\mathcal{F}}$ must necessarily contain a subpath as in Lemma 1.2. We thus obtain a contradiction.

(2) Assume that $\mathcal{F}$ does not pass through all pairs of neighbouring corays. Without loss of generality, we may suppose that $\mathcal{F}$ does not pass through the pair $(\nearrow(1,1)),(\nearrow(1,m))$ and that there exists a point M of $\mathcal{F}$ on the coray $(\nearrow(1,m))$. We may further assume that M is the point of $\mathcal{F}$ on $(\nearrow(1,m))$ having the largest first coordinate (that is, quasi-length). We construct as in (1) a point M_0 of $\widetilde{\Gamma}'$ lifting M, we consider the coray from $a=(m-1,m-2)$ to $d=(1,m)$ passing through M_0, then construct a parallelogram $abcd$, where $b=(m-1,m)$ and $c=(1,1)$. The hypothesis (2) says that there exists a lifting $\widetilde{\mathcal{F}}$ of $\mathcal{F}$ which is entirely contained inside $abcd$ (fig. 1.3).

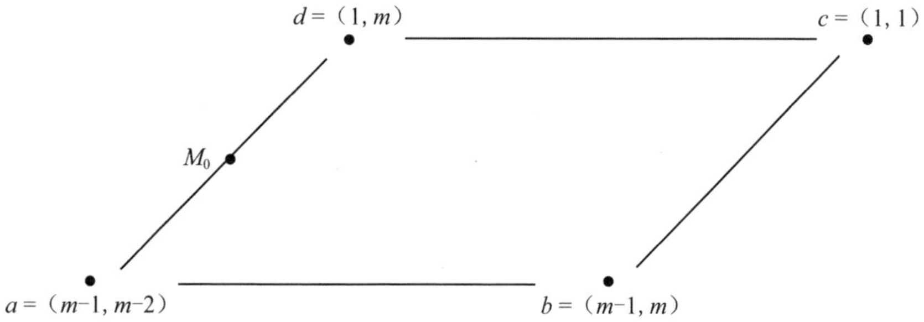

Fig. 1.3

We claim that $M_0\neq(1,m)$, $M_0\neq(m-1,1)$ and that M_0 is a source in $\widetilde{\mathcal{F}}$. Indeed, if $M_0=(1,m)$, then there is a single ray $(M_0\searrow)$ starting at M_0, no other paths in $abcd$ starting or ending at M_0, so that we cannot form a cycle. If $M_0=(m-1,1)$, then there is a single coray $(M_0\nearrow)$ starting at M_0, no other paths starting or ending at M_0, so that we cannot form a cycle. Finally, let $M_0\neq(1,m),(m-1,m-2)$. Then, by the choice of M, the only walks through M_0 which may lie in $\widetilde{\mathcal{F}}$ start with arrows from $(M_0\nearrow)\cup(M_0\searrow)$, that is, M_0 is a source in $\widetilde{\mathcal{F}}$. But then $\widetilde{\mathcal{F}}$ must contain a subpath as in Lemma 1.2, a contradiction. □

Theorem 1.5 Let Γ be an exceptional tube in the Auslander-Reiten quiver of the path algebra of an Euclidean quiver, and $\mathcal{E}=(E_1, E_2, \cdots, E_t)$ be a connected exceptional sequence whose terms lie in Γ. Then End $\mathcal{E}$ is a tilted algebra of type $\mathbf{A}_t$.

Proof By [A1, H], we must show that the bound quiver of End $\mathcal{E}$ is a gentle tree without double zeros. By Lemma 1.4, this quiver is a tree. It follows from Lemma 1.1 that the number of arrows entering or leaving a given point is at most two. By Lemma 1.2 and Lemma 1.3, the bound quiver of End $\mathcal{E}$ is gentle. Finally, Lemma 1.2 also implies that it has no double zeros. □

§ 2. Postprojective components

Let $A=kQ$ be the path algebra of a quiver Q of type $\widetilde{\mathbf{A}}_n$, with an arbitrary orientation. Assume that Q has p arrows in the counterclockwise sense, and q in the clockwise sense (thus $p+q=n+1$). We may clearly assume that $p \geqslant q$. Let Q' be the quiver of type $\widetilde{\mathbf{A}}_n$ having just one source 1, and one sink $n+1$, and having p arrows in the counterclockwise sense, and q in the clockwise sense, and let $B=kQ'$ (fig. 2. 1).

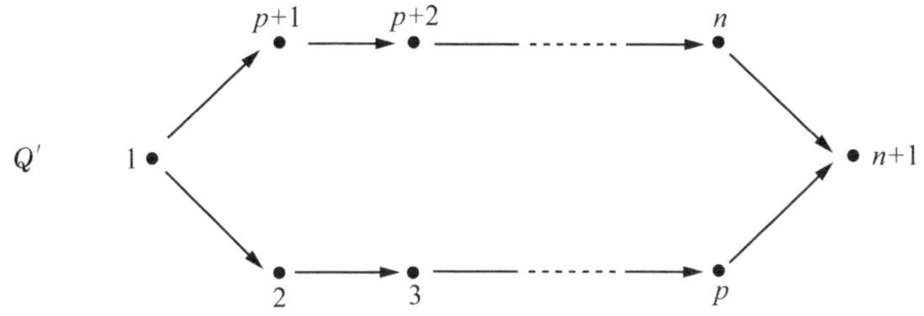

Fig. 2. 1

For a point i, we denote by P_i (or I_i) the corresponding indecomposable projective (or injective, respectively) module. There exists a tilting B-module T_B, which is the slice module of a complete slice in the post-projective component $\mathcal{P}$ of $\Gamma(\mathrm{mod}\ B)$, having as summand P_1, such that $A=\mathrm{End}\ T_B$. The tilting module T_B determines a torsion pair in each of mod B and mod A such that the full subcategory of mod A consisting of the postprojective A-modules is equivalent to the full subcategory of mod B consisting of the torsion postprojective B-modules [A2].

The postprojective component $\mathcal{P}$ contains two types of sectional paths, those parallel to the path from P_{n+1} to P_1 via P_p, which we call (q)-paths, and those parallel to the path from P_{n+1} to P_1 via P_n, which we call (p)-paths. We denote by Δ the full translation subquiver of $\mathcal{P}$ bounded by the two paths from P_{n+1} to P_1, and the (q)-path, and the (p)-path starting at P_1 (fig. 2.2).

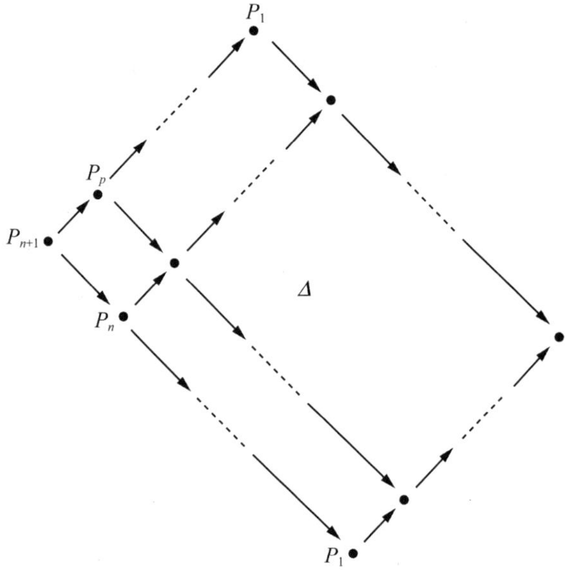

Fig. 2. 2

The indecomposable modules in $\mathcal{P}$ are described by lines in $\widetilde{Q}'$, the universal covering of Q' (see [BR]). Thus, for any $r \geqslant 0$, $\tau^{-r}P_{n+1}$ is given by the line (fig. 2.3).

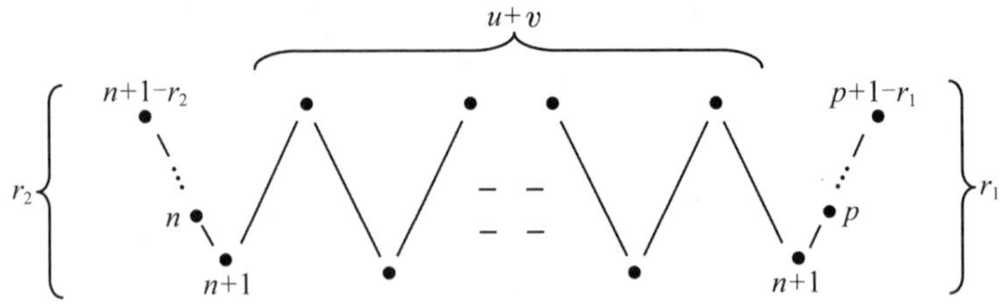

Fig. 2. 3

where the integers u, v, r_1, r_2 are defined by $r = pu + r_1 = qv + r_2, u, v \geqslant 0, 0 \leqslant r_1 < p, 0 \leqslant r_2 < q$. For $p < k \leqslant n+1$ and $r \geqslant 0, \tau^{-r} P_k$ is given by the line (fig. 2.4),

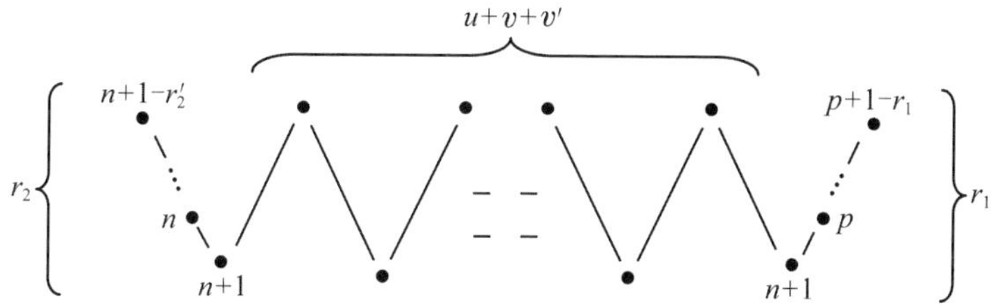

Fig. 2.4

where u, v, r_1, r_2 are as above, and v', r_2' are defined by $r_2 + (n+1-k) = v'q + r_2', v' \geqslant 0, 0 \leqslant r_2' < q$. Finally, for $1 \leqslant l \leqslant p$ and $r \geqslant 0, \tau^{-r} P_l$ is given by the line (fig. 2.5),

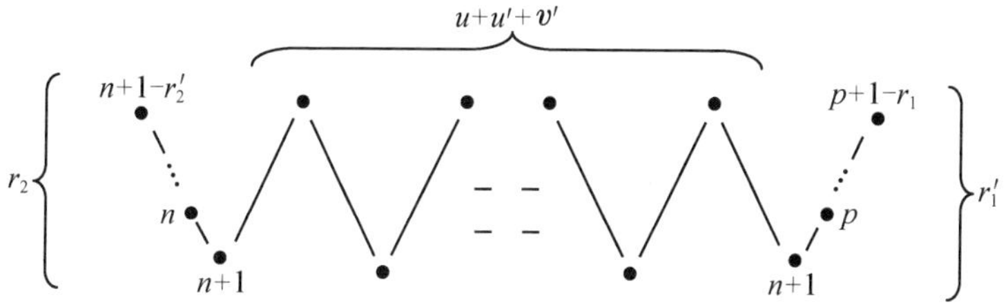

Fig. 2.5

where u, v, r_1, r_2 are as above and u', r_1' are defined by $r_1 + (p+1-l) = u'p + r_1', u' \geqslant 0, 0 \leqslant r_1' < p$.

We call $n+1-r_2$ (or $n+1-r_2'$) the left endpoint and $p+1-r_1$ (or $p+1-r_1'$) the right endpoint of the module.

Lemma 2.1 In $\mathcal{P}$, the modules lying on a(p)-path have the same right endpoint, and those lying on a (q)-path have the same left endpoint. Moreover, each path in the postprojective (or preinjective component) of Γ (mod A) is a monomorphism (or epimorphism, respectively).

Proof The first statement follows from the above description of the modules in $\mathcal{P}$, the second from this description and the tilting functor $\mathrm{Hom}_B(T,-):\mathrm{mod}\,B\to\mathrm{mod}\,A$. □

Let now $\mathcal{E}=(E_1,E_2,\cdots,E_t)$ be an exceptional sequence in the postprojective component $\mathcal{P}$ of $\Gamma(\mathrm{mod}\,B)$. Applying the functor $\tau=D\,\mathrm{Ext}_B^1(-,B)$, we may assume that one of the modules of $\mathcal{E}$ is projective. But now, if M,N are two modules in $\mathcal{P}$, we have $\mathrm{Hom}_B(M,N)\neq 0$ if and only if there exists a path from M to N in $\mathcal{P}$. Thus, (M,N) is a subsequence of $\mathcal{E}$ if and only if there exists a path from M to N in $\mathcal{P}$, but no path from M to τN. Since, for any indecomposable projective B-module P, and indecomposable module X which is not in Δ, there exists a path from P to τX, we deduce that $\mathcal{E}$ lies entirely in Δ.

Lemma 2.2 Let $\mathcal{E}=(E_1,E_2,\cdots,E_t)$ be an exceptional sequence in $\mathcal{P}$. Then there exists a complete slice $\mathcal{S}$ of $\mathcal{P}$ such that all terms of $\mathcal{E}$ lie on $\mathcal{S}$.

Proof Assume that E_i,E_j are two terms in $\mathcal{E}$. We claim that E_i, E_j belong to different τ-orbits in $\mathcal{P}$. Indeed, if this is not the case, then there exist an indecomposable projective module P_B and integers $r<s$ such that $E_i=\tau^{-r}P$, $E_j=\tau^{-s}P$. But then $\mathrm{Hom}_B(E_i,E_j)\neq 0$ implies that (E_i,E_j) is a subsequence of $\mathcal{E}$, and this contradicts

$$\mathrm{Ext}_B^1(E_j,E_i)\cong \mathrm{D}\,\mathrm{Hom}_B(\tau^{-1}E_i,E_j)\neq 0.$$

Let again E_i,E_j be two terms of $\mathcal{E}$. We may assume without loss of generality that (E_i,E_j) is a subsequence of $\mathcal{E}$ and such that the τ-orbits of E_i and E_j are neighbours among the orbits of the terms of $\mathcal{E}$ in the orbit graph of $\mathcal{P}$. Now $\mathrm{Hom}_B(\tau^{-1}E_i,E_j)=0$ implies that E_j is not a successor of $\tau^{-1}E_i$ in Δ and $\mathrm{Hom}_B(E_j,E_i)=0$ implies that E_j is not a predecessor of E_i. This shows that, if there exists a path from E_i to E_j, then this path is sectional. Consequently, E_i and E_j lie on a complete slice $\mathcal{S}$ of $\mathcal{P}$, and hence so do all terms in $\mathcal{E}$. □

Corollary 2.3 Let $\mathcal{E}=(E_1,E_2,\cdots,E_t)$ be an exceptional sequence in the postprojective component of $\Gamma(\mathrm{mod}A)$. Then End $\mathcal{E}$ is a direct product of path algebras of type $\mathbf{A}_m$ (with $m\leqslant t$), or is a connected path algebra of type $\mathbf{A}_{t-1}$. □

Lemma 2.4 If (M,N) is an exceptional sequence in mod A, with M postprojective and N preinjective, then $\mathrm{Hom}_A(M,N)=0$.

Proof Applying the functor τ^{-1}, we may assume that N is injective. By Lemma 2.1, there exists a monomorphism $f: M \to \tau^{-1}M$. Assume that $g: M \to N$ is non-zero. The injectivity of N implies the existence of $g': \tau^{-1}M \to N$ such that $g=g'f$. Thus $g' \neq 0$. Hence

$$\mathrm{Ext}_A^1(N,M) \cong \mathrm{D}\,\mathrm{Hom}_A(\tau^{-1}M,N) \neq 0,$$

a contradiction to the fact that (M,N) is an exceptional sequence. □

Lemma 2.5 Let $\mathcal{E}=(E_1,E_2,\cdots,E_t,F_1,F_2,\cdots,F_s)$ be an exceptional sequence in mod A, with the E_i postprojective and the F_j preinjective, and $t,s \geqslant 1$. Then $\mathrm{End}(\bigoplus_{i=1}^t E_i)$ is not the path algebra of a quiver of type $\widetilde{A}_{t-1}$.

Proof We assume that $\mathrm{End}(\bigoplus_{i=1}^t E_i)$ is the path algebra of a quiver of type $\widetilde{\mathrm{A}}_{t-1}$ and show that $\mathcal{E}$ cannot contain any preinjective term.

Let the quiver Q of A have sources $i_1,i_2,\cdots,i_r$ and sinks $j_1,j_2,\cdots,j_r$ such that we have paths from i_k to j_{k-1} and j_k, for each $1<k\leqslant r$, and paths from i_1 to j_r and j_1. Then, for each j lying on the reduced walk from j_{k-1} to j_k containing i_k, we have $\mathrm{Hom}_A(P_{i_k},I_j)\neq 0$. Let $m>0$ be an arbitrary integer. By Lemma 2.1, there exists an epimorphism $\tau^m I_j \to I_j$, hence an epimorphism $\mathrm{Hom}_A(P_{i_k},\tau^m I_j) \to \mathrm{Hom}_A(P_{i_k},I_j)$ so that $\mathrm{Hom}_A(P_{i_k},\tau^m I_j)\neq 0$. Furthermore, for any monomorphism $f: P_{i_k} \to X$ with X postprojective and morphism $g: P_{i_k} \to \tau^m I_j$, there exists a morphism $g': X \to \tau^m I_j$ such that $g'f=g$, because we may apply the functor τ^{-m} to these modules. Thus $\mathrm{Hom}_A(X,\tau^m I_j)\neq 0$.

It follows from the proof of Lemma 2.2 that $\mathrm{End}(\bigoplus_{i=1}^t E_i)$ is hereditary of type $\widetilde{\mathbf{A}}_{t-1}$ if and only if the terms E_i lie on a complete slice $\mathcal{S}$, of which all the sources and sinks are themselves terms of the sequence. If all of $P_{i_1},P_{i_2},\cdots,P_{i_r}$ are terms of $\mathcal{E}$, we are done. If P_{i_k} is not a term of $\mathcal{E}$, there exists a sink X of $\mathcal{S}$ that is a term of $\mathcal{E}$, and such that P_{i_k} is a submodule of X. Therefore $\mathcal{E}$ cannot contain any preinjective term. □

§ 3.　The arrows from postprojective to regular

In this section, we assume that A is a hereditary algebra of type $\widetilde{\mathbf{A}}_n$, and that $\mathcal{E}$ is an exceptional sequence in mod A such that some terms of $\mathcal{E}$ are postprojective, and some are regular. It follows from the considerations at the beginning of Section 2 that we may assume A to be given by the following quiver (fig. 3. 1):

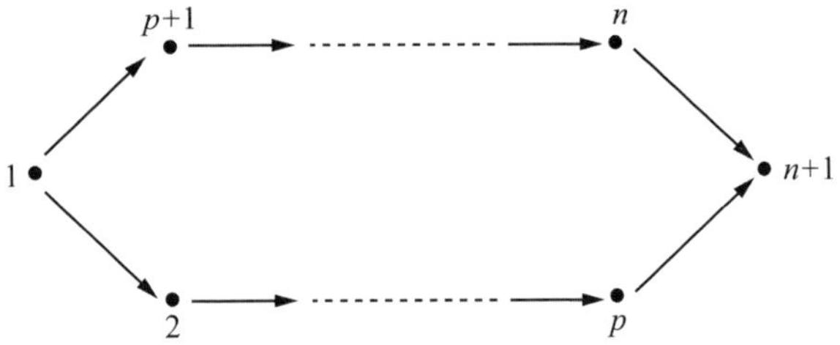

Fig. 3. 1

Then Γ(mod A) has two exceptional tubes Γ_p and Γ_q, of respective ranks p and q. We denote, as in Section 1, by Γ'_p and Γ'_q the full translation subquiver of Γ_p and Γ_q, respectively, consisting of the exceptional modules. We need one more notation: let M be a mouth module in an exceptional tube; the mitre $\widehat{M}$ of M is the full translation subquiver consisting of those exceptional modules N in the tube such that there exist sectional paths $X \to \cdots \to N$ for some X in $(\nearrow M)$ and $N \to \cdots \to Y$ for some Y in $(M \searrow)$.

Lemma 3. 1　Let (M, N) be an exceptional sequence with M postprojective and N regular. Assume the left endpoint of M is k (with $p+1 \leqslant k \leqslant n+1$) and its right endpoint is l (with $2 \leqslant l \leqslant p$ or $l = n+1$). Then $\mathrm{Hom}_A(M, N) \neq 0$ if and only if one of the following conditions is satisfied:

(1) $N \in (\nearrow(1, i))$ in Γ_q, where $i = 1$ whenever $k = n+1$ and $i = k - p+1$ whenever $p+1 \leqslant k \leqslant n$, or

(2) $N \in (\nearrow(1, i))$ in Γ_p, where $i = 1$ whenever $l = n+1$ and $i = l$

whenever $2 \leqslant l \leqslant p$.

Proof By the description [DR] of the indecomposable regular A-modules, $\mathrm{Hom}_A(M,N) \neq 0$ implies $N \in (\widehat{1,i})$ and, since (M,N) is an exceptional sequence, we have $\mathrm{Hom}_A(\tau^{-1}M, N) = 0$ so that $N \notin (\widehat{1,i-1})$. □

We shall need the dual of Lemma 3.1, which we state here for future reference.

Lemma 3.2 Let (M, N) be an exceptional sequence with M regular and N preinjective. Assume the right endpoint of N is k (with $k=1$ or $p+1 \leqslant k \leqslant n+1$) and the left endpoint of N is l (with $1 \leqslant l \leqslant p$). Then $\mathrm{Hom}_A(M,N) \neq 0$ if and only if one of the following conditions is satisfied:

(1) $M \in ((1,k)\searrow)$ in Γ_q, or

(2) $M \in ((1,l)\searrow)$ in Γ_q. □

Lemma 3.3 Let (M, N) be an exceptional sequence with M postprojective and N regular. If $\mathrm{Hom}_A(M,N) \neq 0$, there exists no $L \in (N\searrow)$ such that (M,N,L) is an exceptional sequence.

Proof Indeed, if this is the case, then $L \in (\widehat{1,i-1})$ so that we have $\mathrm{Hom}\ A(\tau^{-1}, M, L) \neq 0$, a contradiction. □

We shall again need the dual.

Lemma 3.4 Let (M, N) be an exceptional sequence with M regular and N preinjective. If $\mathrm{Hom}_A(M,N) \neq 0$, there exists no $L \in (\nearrow M)$ such that (L, M, N) is an exceptional sequence. □

Lemma 3.5 Let $\mathcal{E} = (E_1, E_2, \cdots, E_r, F_1, F_2, \cdots, F_s)$ be an exceptional sequence with the E_i postprojective, the F_j regular, and $(F_1, F_2, \cdots, F_s)$ connected. Then there exist a unique E_i and a unique F_j such that $\mathrm{Hom}_A(E_i, F_j) \neq 0$, and the non-zero morphisms from E_i to F_j factor through no other module in $\mathcal{E}$.

Proof Since $(F_1, F_2, \cdots, F_s)$ is connected, we may assume without loss of generality that the F_j lie in Γ_p. By Lemma 3.1, we must consider the right endpoint of any postprojective term of $\mathcal{E}$ which maps non-trivially to them.

Assume that $E_{i_1} \to E_{i_2} \to \cdots \to E_{i_u}$ in $\mathcal{E}$, where all these modules have the same right endpoint l; then, by Lemma 2.1, these modules are linearly ordered by inclusion. If these modules map non-trivially to some regular term in $\mathcal{E}$, then these regular terms $F_{j_1} \to F_{j_2} \to \cdots \to F_{j_v}$ belong to $(\nearrow(1,t))$, where $t=l$ if $2 \leqslant l \leqslant p$, or $t=1$ if $l=n+1$, and hence are linearly ordered by the quotient relation. Since

$$\operatorname{End}\mathcal{E} = [\bigoplus \operatorname{Hom}_A(E_{i_f}, E_{i_g})] \oplus [\bigoplus \operatorname{Hom}_A(E_{i_g}, E_{j_h})] \oplus [\bigoplus \operatorname{Hom}_A(F_{j_h}, F_{j_k})]$$

we choose $E_i = E_{i_u}$ and $F_j = F_{j_1}$. By construction, $\operatorname{Hom}_A(E_i, F_j) \neq 0$ and the non-zero morphisms from E_i to F_j factor through no other module in $\mathcal{E}$.

It remains to prove the uniqueness of the pair (E_i, F_j). Since, clearly, any pair satisfying the conditions of the statement is constructed in the above way, assume that there exist $E_{i'}$, with right endpoint $l' \neq l$, and $F_{j'}$, on the line $(\nearrow(1,t'))$ where $t'=l'$ if $2 \leqslant l' \leqslant p$ and $t'=1$ if $l'=n+1$.

Since $\operatorname{Hom}_A(\tau^{-1}E_i, F_{j'})=0$, we have $F_{j'} \notin (\widehat{1,t-1})$. Also, notice that $F_{j'} \notin (\nearrow(1,t))$ by construction of F_j. Since $\operatorname{Hom}_A(\tau^{-1}E_{i'}, F_j)=0$, we have similarly $F_j \in (\widehat{1,t'-1})$. Therefore, $F_{j'}$ belongs to the shaded area in the figure below (fig. 3.2).

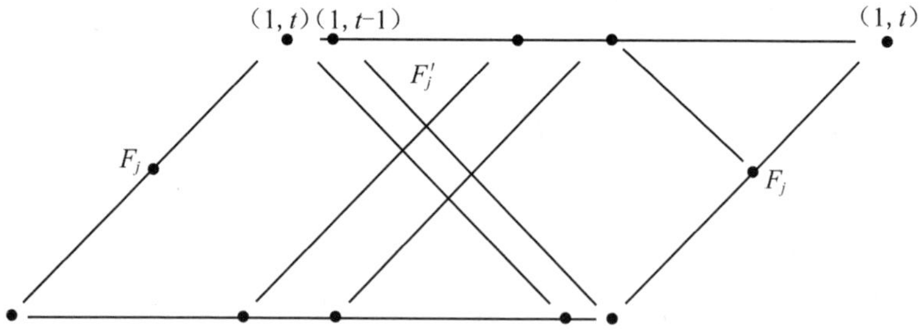

Fig. 3.2

By Lemma 3.3, there is no $L \in \mathcal{E}$ such that $L \in (\searrow F_j)$ or $L \in (\searrow F_{j'})$. By Lemma 1.2, there is no path $F_j \to L \to L'$ (or $F_{j'} \to L \to L'$) with $L \in$

$(F_i \nearrow)$ (or $L \in (F_{j'} \nearrow)$, respectively) and $L' \in (L \searrow)$. Therefore, F_j and $F_{j'}$ are disconnected in Γ_p, a contradiction. □

Lemma 3.6 With the assumptions and notation of Lemma 3.3, we have:

(1) If $\mathrm{End}(\bigoplus_{l=1}^r E_l)$ is representation-infinite, then E_i is a sink of $\mathcal{S}$.

(2) If we have two morphisms $f: E_l \to E_i$, $g: E_{l'} \to E_i$, where f is induced by a (q)-path, and g is induced by a (p)-path, and if $h: E_i \to F_j$ is a non-zero morphism, then $hf = 0$ whenever $F_j \in \Gamma_p$ and $hg = 0$ whenever $F_j \in \Gamma_q$.

Proof (1) This follows from the choice of E_i in Lemma 3.5, and the structure of the complete slice $\mathcal{S}$ (see Lemma 2.2).

(2) This follows from the description of the indecomposable A-modules. □

§ 4. Proof of the main result

Assume now that A is a tame hereditary algebra of type $\widetilde{\mathbf{A}}_n$ (with any orientation), and that $\mathcal{E} = (E_1, E_2, \cdots, E_{n+1})$ is a complete exceptional sequence in mod A. It follows easily from the considerations of Sections 2 and 3 that it suffices to consider the case where there exist t, s such that $(E_1, E_2, \cdots, E_t)$ are postprojective, $(E_{t+1}, E_{t+2}, \cdots, E_s)$ are regular and $(E_{s+1}, E_{s+2}, \cdots, E_{n+1})$ are preinjective.

We first recall the classification results from [AS, R, H] that will be needed. A triangular algebra is called gentle if it is isomorphic to a bound quiver algebra kQ/I, where (Q, I) satisfies:

(1) The number of arrows in Q with a given source or target is at most two.

(2) For any $\alpha \in Q_1$, there is at most one $\beta \in Q_1$ and one $\gamma \in Q_1$ such that $\alpha\beta, \gamma\alpha \notin I$.

(3) For any $\alpha \in Q_1$, there is at most one $\xi \in Q_1$ and one $\zeta \in Q_1$ such that $\alpha\xi, \zeta\alpha \in I$.

(4) I is generated by a set of paths of length two.

Then we have:

Theorem 4.1[AS] An algebra is iterated tilted of type $\widetilde{\mathbf{A}}_n$ if and only if it is gentle and its quiver contains a unique (non-oriented) cycle on which the number of clockwise oriented relations equals the number of counterclockwise oriented relations. □

Theorem 4.2[R][H] An iterated tilted algebra of type $\widetilde{\mathbf{A}}_n$ is tilted if and only if it contains no full subcategory of one of the following forms or their duals:

(1)

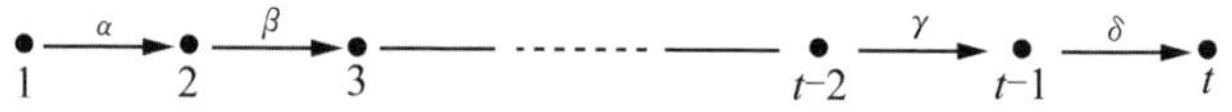

Fig. 4.1

with $t\geqslant 4$, $\alpha\beta=0$. $\gamma\delta=0$.

(2)

Fig. 4.2

with $t\geqslant 4$, $\beta\alpha=0$. $\gamma\delta=0$, 1 and 2 lie on the cycle while $t-1$ and t do not.

(3)

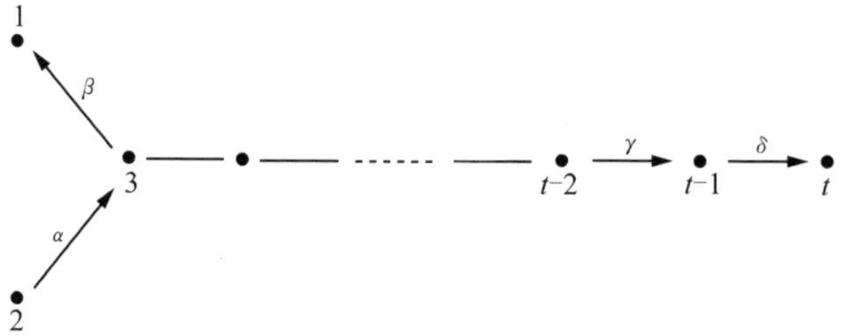

Fig. 4.3

with $t\geqslant 6$, $\alpha\beta=0$, $\gamma\delta=0$, 1, 2 and 3 lie on the cycle while $t-2$, $t-1$ and t do not.

(4)

(a)

Fig. 4.4

or

(b)

Fig. 4. 4

with $t \geqslant 5$, $\alpha\beta = 0$, $\gamma\delta = 0$, all points i with $2 \leqslant i \leqslant t-1$ lie on the cycle while 1 and t do not.

In each case, there are no other relations than the specified ones, and the arrows between 3 and $t-2$ are oriented arbitrarily. □

Lemma 4. 3 If (E, F) is an exceptional sequence in mod A with E post-projective and F preinjective, then:

(1) E belongs to the rectangle in the postprojective component $\mathcal{P}$ consisting of the (p)-paths starting at P_{n+1} and P_3, and the (q)-paths starting at P_{n+1} and P_{p+2}.

(2) F belongs to the rectangle in the preinjective component $\mathcal{Q}$ consisting of the (p)-paths ending at I_{p-1} and I_1, and the (q)-paths ending at I_{n-1} and I_1.

Proof This follows from the fact that, if E is a module of the form (fig 4. 5).

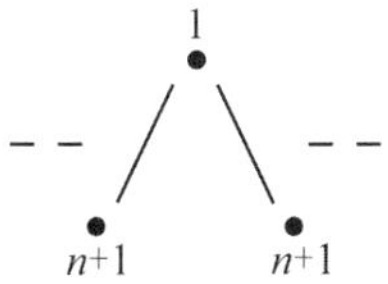

Fig. 4. 5

and F is any preinjective module, then $\mathrm{Hom}_A(E, F) \neq 0$. Dually, if E is any postprojective module, while F is a module of the form (fig. 4. 6).

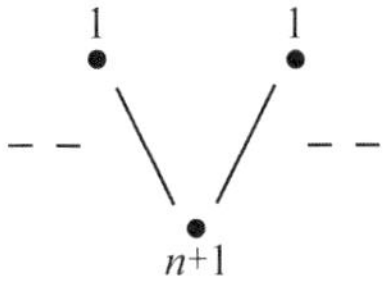

Fig. 4. 6

then $\mathrm{Hom}_A(E,F)\neq 0$. □

Lemma 4.4 Let $(E,M_1,M_2,\cdots,M_s,F)$ be a connected shortest walk in the exceptional sequence $\mathcal{E}$, with E postprojective, F preinjective and all the M_l regular lying in the same exceptional tube. Then $s=1$.

Proof We may assume that all the M_l belong to Γ_p. The connectedness of the given walk implies that $\mathrm{Hom}_A(E,M_i)\neq 0$ and $\mathrm{Hom}_A(M_j,F)\neq 0$ for some $1\leqslant i,j\leqslant s$. Let l be the right endpoint of E. Since (E,M_i) is an exceptional sequence with $\mathrm{Hom}_A(E,M_i)\neq 0$, we see, by Lemma 3.1, that $M_i\in(\nearrow(1,l))$ whenever $2\leqslant l\leqslant p$, and $M_i\in(\nearrow(1,1))$ whenever $l=n+1$. Dually, if k is the left endpoint of F, then, by Lemma 3.2, we have $M_j\in((1,k)\searrow)$. Denote by R_i the point $(1,l)$ if $2\leqslant l\leqslant p$, or $(1,1)$ if $l=n+1$, and by R_j the point $(1,k)$. By Lemma 3.5, we may assume that E is a sink (among the terms of $\mathcal{E}$) in a (p)-path, M_i is a source in $(\nearrow R_i)$, M_j is a sink in $(\searrow R_j)$, and F is a source in a (p)-path.

Since $\mathrm{Hom}_A(\tau^{-1}E,M_j)\cong D\,\mathrm{Ext}_A^1(M_j,E)=0$, it follows that $M_j\in(\widehat{1,l-1})$ when $2\leqslant l\leqslant p$, and $M_j\notin(\widehat{1,p})$ when $l=n+1$. Dually, since $\mathrm{Hom}_A(M_i,\tau F)\cong D\,\mathrm{Ext}_A^1(F,M_i)=0$, we have $M_i\notin(\widehat{1,k+1})$ when $k\neq p$, and $M_i\notin(\widehat{1,p})$ when $k=p$. Letting M be the module of least quasi-length in the intersection of $(\nearrow R_i)$ and $(\searrow R_j)$, we find that M_i, M_j lie on the sides of the triangle R_iMR_j. Similarly, if $1\leqslant l\leqslant s$, then M_l belongs neither to $(\widehat{1,l-1})$ when $2\leqslant l\leqslant p$, to $(\widehat{1,p})$ when $l=p+1$, nor to $(\widehat{1,k+1})$ when $k\neq p$, to $(\widehat{1,p})$ when $k=p$. The connectedness of the given walk then implies that M_l belongs to the triangle R_iMR_j (fig. 4.7).

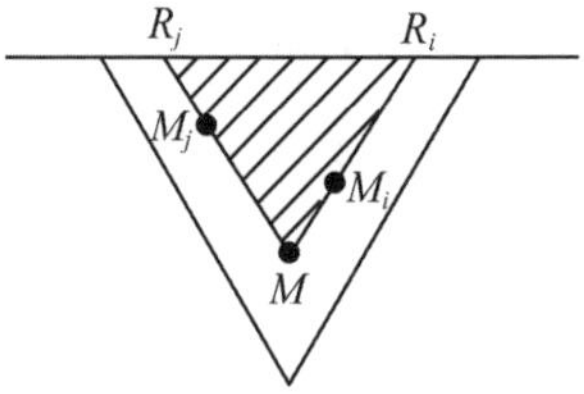

Fig. 4.7

We claim that $M_i = M_j$. Assume that $M_i \neq M_j$ and that $\mathrm{Hom}_A(M_j, M_i) \neq 0$. Then $M_i \in \mathrm{Supp}\ \mathrm{Hom}_A(M_j, -)$ but $M_i \notin \mathrm{Supp}\ \mathrm{Hom}_A(\tau^{-1}M_j, -)$. Hence $M_i = M$, and this contradicts the assumption that M_j is a sink in $(\searrow R_j)$. On the other hand, if $M_i \neq M_j$ and $\mathrm{Hom}_A(M_j, M_i) = 0$, then, by Lemma 1.1, there exists M_l inside the triangle R_iMR_j such that $\mathrm{Hom}_A(M_i, M_l) \neq 0$ or $\mathrm{Hom}_A(M_l, M_i) \neq 0$, that is, $M_l \in (M_i \nearrow)$ or $M_l \in (\searrow M_i)$, since M_i is a source in $(\nearrow R_i)$. By the connectedness of the given sequence, there exists M_h such that

$$\mathrm{Hom}_A(M_l, M_h) \neq 0 \text{ or } \mathrm{Hom}_A(M_h, M_l) \neq 0,$$

that is,

$$M_h \in (M_l \nearrow) \cup (M_l \searrow) \text{ or } M_h \in (\searrow M_l) \cup (\nearrow M_l).$$

By induction and Lemma 1.2, we obtain a walk of the form (fig. 4.8).

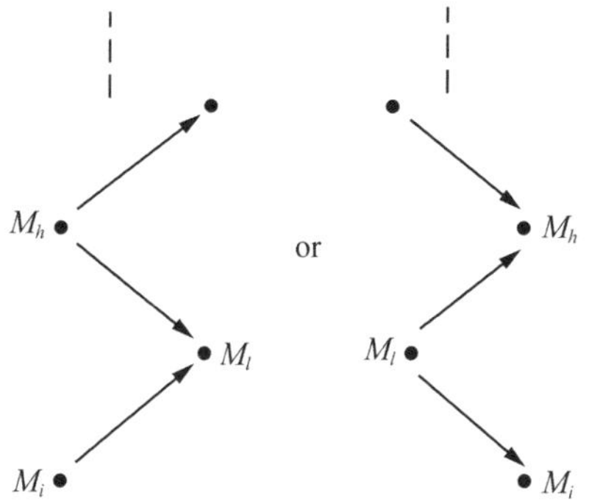

Fig. 4.8

Thus we cannot reach M_j, a contradiction. This shows that $M_i = M_j$. Hence $s = 1$. □

Lemma 4.5 Let (E, M, F) be a connected subsequence of $\mathcal{E}$, with E postprojective, M regular and F preinjective. Then the simple module S_{n+1} is a direct summand of the socle of M.

Proof We observe that S_{n+1} is a direct summand of soc M if and only if $M \notin \{i, i+1, \cdots, i+k\}$ (with $2 \leqslant i \leqslant p$, $i+k < p+1$ if $M \in \Gamma_p$ or $p+1 \leqslant i \leqslant n$, $i+k < n+1$ if $M \in \Gamma_q$), or, equivalently, if and only if $M \in \widehat{R}$ (where $R \in \{1, p+1, \cdots, n, n+1\}$ if $M \in \Gamma_p$, or $R \in \{1, 2, \cdots, p, n+1\}$, if $M \in \Gamma_q$). If S_{n+1} is not a direct summand of soc M, and $M \in \Gamma_p$, then the right endpoint of E is i, and the left endpoint of F is $i+k$. Therefore the left endpoint of τF is $i+k+1$. Hence $\mathrm{Ext}_A^1(F,E)=D\,\mathrm{Hom}_A(E,\tau F)\neq 0$, a contradiction. The proof is similar if $M\in\Gamma_q$. □

Lemma 4.6 Let (E_1,M_1,F_1) and (E_2,M_2,F_2) be two connected subsequences of $\mathcal{E}$, with E_1,E_2 postprojective, M_1,M_2 regular and F_1, F_2 preinjective. If $M_1\neq M_2$, then M_1 and M_2 lie in two different tubes.

Proof Assume that this is not the case, and that both M_1 and M_2 lie in Γ_p (say). Suppose the right endpoint of E_1, and therefore of M_1, is l_1 where $3\leqslant l_1\leqslant p$, or $l_1=n+1$, by Lemma 4.3, and similarly that the right endpoint of E_2, and therefore of M_2, is l_2, where $3\leqslant l_2\leqslant p$, or $l_2=n+1$.

(1) Assume $l_1=l_2=l$, say; then $M_1,M_2\in(\nearrow(1,l))$ when $l\leqslant p$, or $(\nearrow(1,1))$ whenever $l=n+1$. Without loss of generality, we may assume that $\mathrm{Hom}_A(M_1,M_2)\neq 0$. Now, $\mathrm{Hom}_A(M_2,F_2)\neq 0$, therefore $\mathrm{Hom}_A(F_2,M_1)=0$ (or, equivalently, (M_1,F_2) is a subsequence of $\mathcal{E}$). Letting k_2 denote the left endpoint of M_2 and F_2, where $k_2=1,2,\cdots,p-1$, we get $M_1\notin(\widehat{1,k_2})$ and this contradicts the fact that $M_1\in(\nearrow M_2)$.

(2) If $l_1<l_2$, then $M_1\in(\widehat{1,l_2})$ when $l_2\leqslant p$, or $M_1\in(\widehat{1,1})$ when $l_2=n+1$. Hence $\mathrm{Hom}_A(E_2,M_1)\neq 0$. On the other hand, $M_1\in(\widehat{1,l_2-1})$ when $l_2\neq n+1$, and $M_1\in(\widehat{1,p})$ when $l_2=n+1$, thus $\mathrm{Hom}_A(\tau^{-1}E_2,M_1)\neq 0$, that is, $\mathrm{Ext}_A^1(M_1,E_2)\neq 0$. This is impossible, since E_2,M_1 belong to the same exceptional sequence $\mathcal{E}$. □

Lemma 4.7 Let (E,M) be a connected subsequence of $\mathcal{E}$, with E post-projective and a sink ore a (p)-path (among the terms of $\mathcal{E}$), and $M\in\Gamma_p$. Then

(1) $\mathrm{Hom}_A(E',E)\neq 0$, with E' postprojective and in $\mathcal{E}$, implies that the path from E' to E is a (p)-path.

(2) $\mathrm{Hom}_A(E,E'')\neq 0$, with E'' postprojective and in $\mathcal{E}$, implies that the path from E to E'' is a (q)-path.

Furthermore, there cannot exist at the same time in $\mathcal{E}$ terms such as E' and E'' above.

Proof To show (1), assume that the path from E' to E is a (q)-path. The right endpoint of E' is larger than the right endpoint of E, and $M\in\widehat{R}$, where R is regular having the same right endpoint as that of $\tau^{-1}E'$, a contradiction. (2) is proven similarly. The last statement follows from the fact that, if E' and E'' both occur, then the points E', E, E'' cannot lie on a complete slice, a contradiction to Lemma 2.2. □

We shall also need the dual statement.

Lemma 4.8 Let (M, F) be a connected subsequence of $\mathcal{E}$, with F preinjective and a source on a (q)-path (among the terms of $\mathcal{E}$), and $M\in\Gamma_q$. Then

(1) $\mathrm{Hom}_A(F, F')\neq 0$, with F' preinjective and in $\mathcal{E}$, implies that the path from F to F' is a (q)-path.

(2) $\mathrm{Hom}_A(F'', F)\neq 0$, with F'' preinjective and in $\mathcal{E}$, implies that the path from F'' to F is a (p)-path.

Furthermore, there cannot exist at the same time in $\mathcal{E}$ terms such as F' and F'' above. □

Lemma 4.9 If (E, M, F) is a connected subsequence of $\mathcal{E}$, with E post-projective, M regular and F preinjective, then $\mathrm{Hom}_A(E, F)=0$. Further, if M_1, M_2 are regular and $\mathrm{Hom}_A(M_1, M)\neq 0$, $\mathrm{Hom}_A(M, M_2)\neq 0$, then $M_1\in(\searrow M)$, $M_2\in(M\nearrow)$ and $\mathrm{Hom}_A(M_1, M_2)=0$.

Proof The first statement is clear by Lemma 2.4. The second statement follows from Lemmas 1.1, 3.3, 3.4 and 1.3. □

Proposition 4.10 Let $A=kQ$ be a path algebra of type $\widetilde{\mathbf{A}}_n$, and $\mathcal{E}$ be an exceptional sequence in mod A. Assume that $\mathcal{E}$ contains a cycle $\mathcal{C}$ consisting of postprojective, regular and preinjective terms. Then the connected component of End $\mathcal{E}$ containing the cycle corresponding to $\mathcal{C}$ is a representation-finite tilted algebra of type $\widetilde{\mathbf{A}}_l$, with $l\leqslant n$.

Proof It follows from Lemmas 4.6, 4.4, 4.7, 4.8 and Theorem 1.5 that, if E belongs to $\mathcal{C}$, and E' belongs to $\mathcal{E}\setminus\mathcal{C}$, and both are postprojective, then

$$\mathrm{Hom}_A(E, E')=0 \text{ and } \mathrm{Hom}_A(E', E)=0,$$

and, dually, if F belongs to $\mathcal{C}$, and F' belongs to $\mathcal{E}\backslash\mathcal{C}$, and both are preinjective, then

$$\mathrm{Hom}_A(F, F') = 0 \text{ and } \mathrm{Hom}_A(F', F) = 0;$$

consequently, the quiver of the connected component of End $\mathcal{E}$ containing the points corresponding to the cycle $\mathcal{C}$ is as follows (fig. 4. 9).

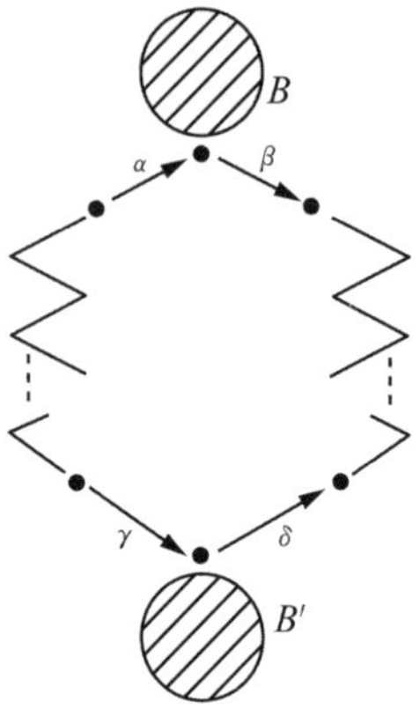

Fig. 4. 9

where $\alpha\beta = 0$, $\gamma\delta = 0$, all unoriented edges on the cycle may be oriented arbitrarily, and B, B' are tilted algebras of type $\mathbf{A}_m$. The statement then follows from Theorem 4. 2. □

We may thus assume that, if a cycle occurs in the bound quiver of End $\mathcal{E}$, then all the points of this cycle are postprojective (or, dually, preinjective). The main theorem follows from the next two lemma.

Lemma 4. 11 With the above notation, End $\mathcal{E}$ is either a direct product of one representation-infinite iterated tilted algebra of type $\mathbf{A}_n$ (with $m \leqslant n$) and iterated tilted algebras of tyre $\mathbf{A}_l$ (with $l \leqslant n-m$), or else a direct product of iterated tilted algebras of type A_l (with $l \leqslant n+1$).

Proof By Theorem 1. 5, Corollary 2. 3 and Lemmas 2. 5, 3. 5, the ordinary quiver of End $\mathcal{E}$ contains at most one cycle and, if it does, then this cycle is not bound by any relation. We thus only need to show that End $\mathcal{E}$ is a gentle algebra. Assume that $\mathcal{F} = (F_1, F_2, \cdots, F_t)$ is a connected subsequence of $\mathcal{E}$. If $\mathcal{F}$ lies entirely in the regular part, then, by Theorem 1.5, End $\mathcal{F}$ is gentle. If $\mathcal{F}$ lies in the postprojective (or the preinjective) component then, by Corollary 2.3, End $\mathcal{F}$ is also gentle. Assume that we have non-zero morphisms (fig. 4.10).

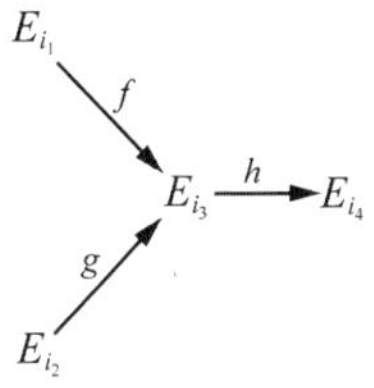

Fig. 4.10

where $E_{i_1}, E_{i_2}, E_{i_3}$ are postprojective, and E_{i_4} is regular (and E_{i_1} and E_{i_2} are not necessarily distinct). Then, by Lemma 3.6, we have either $hf=0$ or $hg=0$. Finally, assume that we have non-zero morphisms

$$E_{i_1} \xrightarrow{f} E_{i_2} \xrightarrow{g} E_{i_3}$$

which do not factor through other modules in $\mathcal{E}$, with E_{i_1} postprojective, and E_{i_2}, E_{i_3} regular. Then by Lemma 3.1, there exists no non-zero morphism $h: E_{i_2} \to E_{i_4}$ with $E_{i_4} \in \mathcal{E}$ regular and distinct from E_{i_3} and such that E_{i_2} does not factor through other modules in $\mathcal{E}$. Furthermore, if there exists a non-zero morphism $h: E_{i_4} \to E_{i_2}$ with $E_{i_4} \in \mathcal{E}$ regular and such that h does not factor through other modules in $\mathcal{E}$, then, by Lemma 3.1, 3.5, we have $gh=0$. Invoking the duality between postprojective and preinjective modules completes the proof. □

Lemma 4.12 With the notation above, each of the connected components of End $\mathcal{E}$ is in fact a tilted algebra.

Proof By Theorem 1.5, Corollary 2.3 and Lemma 3.5, if a cycle occurs in the bound quiver of End $\mathcal{E}$, then the corresponding terms of $\mathcal{E}$ are all postprojective (and then $\mathcal{E}$ has no preinjective terms, by Lemma 2.5) or all preinjective (and then, dually, $\mathcal{E}$ has no postprojective terms). Assume thus that a cycle occurs and that the corresponding terms of $\mathcal{E}$ all lie in $\mathcal{P}$. Then $\mathcal{E}=(E_1, E_2, \cdots, E_r, E_{r+1}, E_{r+2}, \cdots, E_{n+1})$ with $E_1, E_2, \cdots, E_r \in \mathcal{P}, E_{r+1}, E_{r+2}, \cdots, E_{n+1} \in \Gamma_p \vee \Gamma_q$ (here, $2 \leqslant r \leqslant n$) and End

$(\bigoplus_{i=1}^{r} E_i)$ is a path algebra of type $\widetilde{\mathbf{A}}_{r-1}$. In order to show our claim, we need to prove that the bound quiver of End $\mathcal{E}$ contains no full bound subquiver of one of the forms (1)~(4) listed in Theorem 4.2.

We first notice that the arrows between $\mathcal{P}$ and $\Gamma_p \vee \Gamma_q$ are all from $\mathcal{P}$ to $\Gamma_p \vee \Gamma_q$, therefore case (3) cannot occur. Assume that (1) occurs, that is, there exists a walk of the form (fig. 4.11).

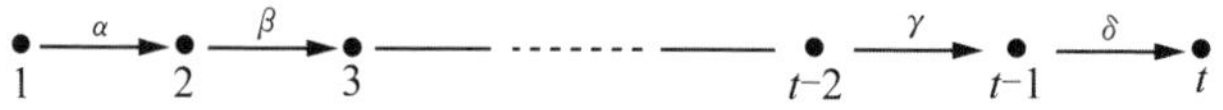

Fig. 4.11

with $\alpha\beta = 0$, $\gamma\delta = 0$ and $t \geqslant 4$, in the bound quiver of End $\mathcal{E}$. Then, by Theorem 1.5 and Corollary 2.3, not all the terms of $\mathcal{E}$ corresponding to the points of this walk lie in the same component. Since End$(\bigoplus_{i=1}^{r} E_i)$ is hereditary, this means that the terms corresponding to $1, 2, \cdots, t-2$ are all regular. But this is impossible by Lemmas 1.2, 3.3. Thus (1) does not occur. Finally, for (2) and (4), we notice that the only possibility of occurrence of two zero-relations in the same walk pointing in different directions is of the form (fig. 4.12).

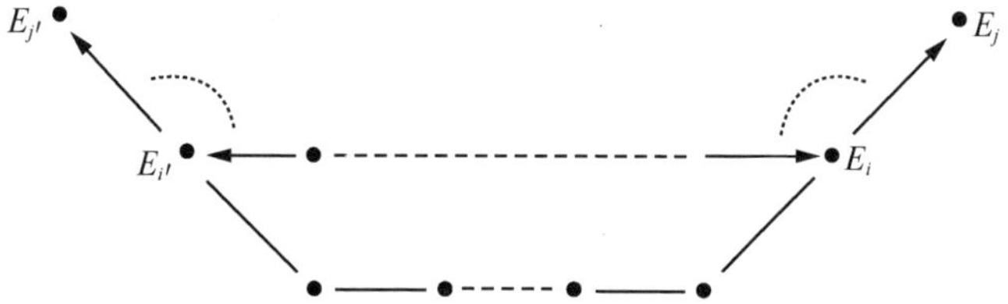

Fig. 4.12

where the dotted lines indicate zero-relations. Therefore (2) and (4) do not occur. This completes the proof in case the bound quiver of End $\mathcal{E}$ contains a cycle.

If the bound quiver of End $\mathcal{E}$ contains no cycle we need to show that it contains no walk of the form (fig. 4.13).

Fig. 4.13

with $\alpha\beta=0$, $\gamma\delta=0$ and $t\geqslant 4$. If there is no path from $\mathcal{P}$ to the preinjective component $\mathcal{I}$ in $\mathcal{E}$, we are done by the argument above. If there exists such a path, then we have a subsequence (E,F,G) of $\mathcal{E}$ with $E\in\mathcal{P}$, $F\in\Gamma_p\vee\Gamma_q$, $G\in\mathcal{I}$ and $\mathrm{Hom}_A(E,F)\neq 0$, $\mathrm{Hom}_A(F,G)\neq 0$ by Lemmas 3.3, 3.4. Assume that $F\in\Gamma_p$ (the other case is similar) with right endpoint l, and $2\leqslant l\leqslant p$, then E has right endpoint l, by Lemma 3.1, and G has left endpoint l, by Lemma 3.2. Then, if $l\neq 2$, $\tau^{-1}E$ has right endpoint $l-1$ or $\tau^{-1}E$ is given by a line of the form (fig. 4.14).

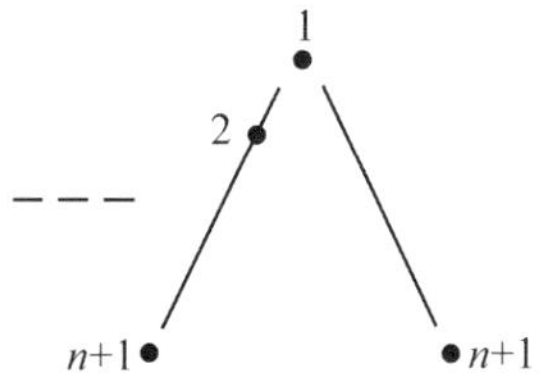

Fig. 4.14

(see [BR]). Since G has l as left endpoint, we have $\mathrm{Hom}_A(\tau^{-1}E,G)\neq 0$, a contradiction. □

Remark 4.13 (1) With the above notation $\bigoplus_{i=1}^{n+1}E_i$ is generally not a tilting module; for instance, if A is given by the quiver (fig. 4.15).

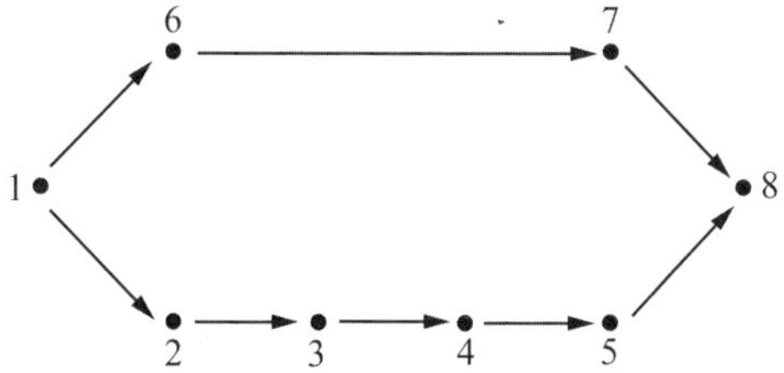

Fig. 4.15

then the sequence (S_2,S_3) consisting of the simple modules corresponding to the points 2, 3 is clearly exceptional, but $\mathrm{Ext}_A^1(S_2,S_3)\neq 0$ shows that $S_2\oplus S_3$ is not a partial tilting module.

(2) The methods of Section 1 can be used with only slight modifications

to prove the following theorem:

Theorem Let k be a commutative field, Q be a quiver with underlying graph $\mathbf{A}_n$, and $A = kQ$ be its path algebra. Let $\mathcal{E}$ be an exceptional sequence in mod A. Then End $\mathcal{E}$ is a direct product of tilted algebras of type $\mathbf{A}_l$ (with $l \leqslant n$). Each connected subsequence of $\mathcal{E}$ is a partial tilting module. □

This strengthens the result of [Y]. We omit the proof, since we learned later that it was proved independently by Meltzer [M], using the derived category. In the same paper, Meltzer gives an example showing that a similar statement does not hold for other Dynkin diagrams.

References

[A1] Assem I. Tilted algebras of type $\mathbf{A}_n$. Comm. Algebra, 1982, 10: 2 121-2 139.

[A2] Assem I. Tilting theory—an introduction. in: Topics in Algebra, Banach Center Publ. 26, Part 1, PWN, Warszawa, 1990: 127-180.

[AH] Assem I, Happel D. Generalized tilted algebras of type $\mathbf{A}_n$. Comm. Algebra, 1981, 9: 2 101-2 125.

[AS] Assem I, Skowroński A. Iterated tilted algebras of type $\widetilde{\mathbf{A}}_n$. Math. Z., 1987, 195: 269-290.

[ARS] Auslander M, Reiten I, Smalø S O. Representation Theory of Artin Algebras. Cambridge Stud. Adv. Math., 36, Cambridge Univ. Press, 1995.

[B] Bondal A I. Representations of associative algebras and coherent sheaves. Izv. Akad. Nauk SSSR Ser. Mat., 1989, 53: 25-44 (in Russian); English transl.: Math. USSR-Izv. 1990, 34: 23-45.

[BR] Butler M, Ringel C M. Auslander-Reiten sequences with few middle terms and applications to string algebras. Comm. Algebra, 1987, 15: 145-179.

[CB] Crawley-Boevey W W. Exceptional sequences of representations of quivers. in: Representations of Algebras, Sixth Internat. Conf., Ottawa, 1992, CMS Conf. Proc. 14, Amer. Math. Soc., 1993: 117-124.

[DR] Dlab V, Ringel C M. Indecomposable representations of graphs and algebras. Mem. Amer. Math. Soc., 173, 1976.

[GR] Gorodentsev A L, Rudakov A N. Exceptional vector bundles on projective spaces. Duke Math. J., 1987, 54: 115-130.

[H] Huard F. Tilted gentle algebras. Comm. Algebra, 1998, 26: 63-72.

[K] Kerner O. Representations of wild quivers. in: Representation Theory of Algebras and Related Topics, Seventh Internat. Conf., Mexico, 1994, CMS Conf. Proc. 19, Amer. Math. Soc., 1996: 65-107.

[M] Meltzer H. Exceptional sequences and tilting complexes for hereditary algebras of type $\mathbf{A}_n$. Algebras and Modules Ⅱ, 1996.

[R1] Ringel C M. Tame Algebras and Integral Quadratic Forms. Lecture Notes in Math. 1099, Springer, 1984.

[R2] Ringel C M. The braid group action on the set of exceptional sequences of a hereditary artin algebra. in: Abelian Group Theory and Related Topics (Oberwolfach, 1993), Contemp. Math. 171, Amer. Math. Soc., 1994, 339-352.

[R] Roldán O. Tilted algebras of type $\widetilde{\mathbf{A}}_n$, $\widetilde{\mathbf{B}}_n$, $\widetilde{\mathbf{C}}_n$ and $\widetilde{\mathbf{BC}}_n$. Ph. D. thesis, Carleton Univ., 1983.

[Y] Yao H. Endomorphism algebras of exceptional sequences of type $\mathbf{A}_n$. Algebra Colloq., 1996, 3: 25-32.

Journal of Algebra,2000,230:665-675.

关于齐性范畴①

On Homogeneous Exact Categories

Abstract In the present paper we prove that a certain subcategory $\mathscr{C}$ of the module category over some infinite-dimensional algebra R has almost split sequences and strongly homogeneous property; i. e., for each indecomposable module M in $\mathscr{C}$, there is an almost split sequence starting and also ending at M. It is also proved that except for a trivial case, $\mathscr{C}$ is of wild representation type.

Let $\mathscr{C}$ be a k-additive category and k be a field. $\mathscr{C}$ is called a Krull-Schmidt category, provided the endomorphism ring of each indecomposable object is local. Then the decomposition of any object into indecomposables is unique up to isomorphism [R, p. 52]. The category $\mathscr{C}$ is said to be exact if $\mathscr{C}$ is a subcategory of an abelian category $\mathscr{A}$, and $\mathscr{C}$ is closed under extensions in $\mathscr{A}$[R, p. 59]. Then we have in $\mathscr{C}$ the notions of projective and injective objects, as well as almost split sequences.

Denote by Λ an Artin algebra and by mod Λ the category of finitely

① Project supported by National Natural Science Foundation of China, 19831070, and the research fund for the Doctoral Program of Higher Education of China.

Received: 1999-02-14.

本文与 Bautista R, Carwley-Boevey, 雷天刚合作.

generated left Λ-modules. The existence of almost split sequences in mod Λ has been proved in [AR1]. Later several authors have proved the existence of almost split sequences in some subcategories of mod Λ, for example, the representation category of a partially ordered set [R, Chap. 2], the category of relative projective modules [BK], etc. Of special interest are those modules M; the almost split sequences starting at M also end at M. We may call M homogeneous, following the term of homogeneous tubes.

In the late 1980s Crawley-Boevey proved the following theorem. If k is an algebraically closed field, Λ a tame, finite dimensional k-algebra, then for each dimension d, almost all (except finitely many iso-classes) modules M with $\dim M \leqslant d$ are homogeneous using the method of bocses [CB1, Theorem D]. Since that time several experts have tried to prove the converse of the theorem. Under the hypothesis that the same method would work for the converse, the expectation was that none of the minimal wild bocses appearing in the consideration of Drozd and Crawley-Boevey is homogeneous. Contrary to this Zhang et al. [ZLB] showed that one of the minimal wild bocs is in fact homogeneous. The converse of the Crawley-Boevey theorem is still open. In the present paper we describe a large class of wild homogeneous categories that contains in particular the category of representations of the above minimal bocs. In Section 1 certain subcategories $\mathscr{C}$ of the module categories over a class of infinite dimensional algebras R are defined, where the R are polynomial rings in one indeterminate over some finite dimensional k-algebras Λ. Then $\mathscr{C}$ has almost split sequences if Λ is self-injective. In Section 2 we assume moreover that Λ is symmetric; then $\mathscr{C}$ is strongly homogeneous, i. e., for each indecomposable module M in $\mathscr{C}$, there is an almost split sequence starting and also ending at M. In Section 3 we prove that $\mathscr{C}$ is tame if Λ is semi-simple and $\mathscr{C}$ is wild if Λ is not semi-simple, under the assumption that k is an algebraically closed field. For this we introduce some categories equivalent to $\mathscr{C}$. Vossieck has found a similar method to construct a larger class of strongly homogeneous categories (so-called symmetric bimodule problems) independently.

Throughout the paper we denote by k a field, by Λ a finite dimensional k-algebra, by R an infinite dimensional k-algebra, and mod R the category of finite dimensional R-modules, Mod R, the category of all R-modules. We write composition of paths and morphisms from right to left.

§ 1. A subcategory having almost split sequences

1.1 Let k be a field, Λ be a finite dimensional k-algebra, and $R = \Lambda[t]$ be a ring of polynomials in one indeterminate t centralizing Λ, which is an infinite dimensional k-algebra. Equivalently, R is the tensor ring over Λ of the bimodule Λ. Then there exists a "universal" exact sequence of R-modules,

$$0 \to R \otimes_\Lambda R \xrightarrow{d} R \otimes_\Lambda R \to R \to 0,$$

in which $d(1 \otimes 1) = t \otimes 1 - 1 \otimes t$, and the last map is multiplication [C, Vol. 3, Chap. 2. 2]. It splits as a sequence of right (or left) R-modules. Therefore it remains exact after tensoring with an arbitrary left (or right) R-module.

Now let $\mathscr{C}$ be the category of left R-modules, which are finite dimensional projectives as left Λ-modules. Then any module M in $\mathscr{C}$ has a canonical R-projective resolution

$$0 \to R \otimes_\Lambda R \otimes_R M \xrightarrow{d \otimes 1} R \otimes_\Lambda R \otimes_R M \to R \otimes_R M \to 0,$$

or simply

$$0 \to R \otimes_\Lambda M \xrightarrow{d \otimes 1} R \otimes_\Lambda M \to M \to 0,$$

Taking $\mathrm{Hom}_R(-, R)$, we obtain a sequence

$$0 \to \mathrm{Hom}_R(M, R) \to \mathrm{Hom}_R(R \otimes_\Lambda M, R) \xrightarrow{\mathrm{Hom}(d \otimes 1, R)} \mathrm{Hom}(R \otimes_\Lambda M, R).$$

We define the transpose of M to be the cokernel of $\mathrm{Hom}_R(d \otimes 1, R)$, denoted by $\mathrm{Tr}_R(M)$.

Let $M^* = \mathrm{Hom}_\Lambda(M,\Lambda)$. Then M^* has a natural right R-module structure. It is clear that M^* is a finite dimensional projective right Λ-module.

1.2 Lemma $\mathrm{Tr}_R(M) \simeq M^*$ as R-module.

Proof There is a fig. 1.1 with exact rows,

$$\begin{array}{ccccccc}
M^*\otimes_\Lambda R & \xrightarrow{1\otimes d} & M^*\otimes_\Lambda R & \longrightarrow & M^* & \longrightarrow & 0 \\
\downarrow{\scriptstyle\vartheta} & & \downarrow{\scriptstyle\vartheta} & & \downarrow & & \\
\mathrm{Hom}_R(R\otimes_\Lambda M,R) & \xrightarrow{(d\otimes 1)^*} & \mathrm{Hom}_R(R\otimes_\Lambda M,R) & \longrightarrow & \mathrm{Tr}_R(M) & \longrightarrow & 0
\end{array}$$

Fig. 1.1

where the top line is the projective resolution of the right R-module M^*, $\vartheta: \mathrm{Hom}_\Lambda(M,\Lambda)\otimes_\Lambda R \to \mathrm{Hom}_R(R\otimes_\Lambda M,R)$ is given by $\vartheta(f\otimes r)(s\otimes m) = sf(m)r$ for any $r, s \in R$, $m \in M$, $f \in \mathrm{Hom}_\Lambda(M,\Lambda)$. ϑ is an isomorphism, since M is a finite dimensional projective left module. Moreover

$$\begin{aligned}
&[\vartheta(1\otimes d)(f\otimes_R r_1\otimes_\Lambda r_2)](s_1\otimes_\Lambda s_2\otimes_R m)\\
&\quad=[\vartheta(f\otimes_R(r_1t\otimes_\Lambda r_2 - r_1\otimes_\Lambda tr_2))](s_1\otimes_\Lambda s_2 m)\\
&\quad=s_1(f(r_1t))(s_2m)r_2 - s_1(fr_1)(s_2m)tr_2,
\end{aligned}$$

and

$$\begin{aligned}
&[(d\otimes 1)^*\vartheta(f\otimes_R r_1\otimes_\Lambda r_2)](s_1\otimes_\Lambda s_2\otimes_R m)\\
&\quad=[\vartheta((fr_1)\otimes_\Lambda r_2)](s_1t\otimes_\Lambda s_2 - s_1\otimes_\Lambda ts_2)\otimes_R m\\
&\quad=s_1t(fr_1)(s_2m)r_2 - s_1(fr_1)(ts_2m)r_2,
\end{aligned}$$

It follows that $\vartheta(1\otimes d) = -(d\otimes 1)^*\vartheta$, therefore $\mathrm{Tr}_R(M)\simeq M^*$.

1.3 Theorem If Λ is self-injective, then the category $\mathscr{C}$ has almost split sequences.

Proof Take any indecomposable module $M\in\mathscr{C}$. Then M is finitely presented by its projective resolution given in Subsection 1.1. We claim that

$$\mathrm{Hom}_R(M,R)=0. \tag{1.1}$$

Since t^i, $i\in\mathbf{N}^*$, are not zero divisors, any submodule of R is infinite dimensional. But $\dim_k M < +\infty$, Thus we obtain (1.1). In particular M

is not a projective R-module. Since mod R is a Krull-Schmidt category, $\mathrm{End}_R(M)$ is local. Denote by $\mathscr{P}_R(M,M)$ the ideal of $\mathrm{End}_R(M)$ consisting of the morphisms decomposed through R-projectives. Then $\mathscr{P}_R(M,M)=0$ still by (1.1).

From [A, II, 5.1], there exists a unique (up to isomorphism) almost split sequence ending at M in Mod R,

$$0\to \mathrm{Hom}_\Gamma(\mathrm{Tr}_R(M),I)\to E\to M\to 0,$$

where $\Gamma=\mathrm{End}(\mathrm{Tr}_R(M))$, and I is the injective envelope of a simple Γ-madule. So I is a direct summand of $\mathrm{D}\Gamma$, where $\mathrm{D}=\mathrm{Hom}_k(-,k)$. But Γ is a local ring, therefore is indecomposable as a projective Γ-module, so is $\mathrm{D}\Gamma$ as an injective Γ-module. Thus $I=\mathrm{D}\Gamma$. On the other hand $\mathrm{Hom}_\Gamma(\mathrm{Tr}_R(M),\mathrm{D}\Gamma)\simeq \mathrm{D}\,\mathrm{Tr}_R(M)$. Moreover $\mathrm{Tr}_R(M)\simeq M^*$ by Lemma 1.2, so it is a finite dimensional projective right Λ-module. Now Λ being self-injective implies that $\mathrm{DTr}_R(M)\simeq \mathrm{D}M^*$ is a projective left Λ-module, so is in $\mathscr{C}$. E is also in $\mathscr{C}$, since the sequence splits over Λ. Thus the sequence is an almost split sequence in $\mathscr{C}$. It can be written as

$$0\to \mathrm{DTr}_R(M)\to E\to M\to 0.$$

§ 2.　A strongly homogeneous subcategory

2.1　Definitions　Let $\mathscr{C}$ be an exact category.

(1) An indecomposable object M in $\mathscr{C}$ is called homogeneous if there is an almost split sequence

$$0\to M\xrightarrow{\iota} E\xrightarrow{\pi} M\to 0$$

in $\mathscr{C}$.

(2) The category $\mathscr{C}$ is said to be homogeneous if $\mathscr{C}$ has almost split sequences, and for each dimension d, almost all (except finitely many iso-classes) indecomposables M with dim $M\leqslant d$ are homogeneous.

(3) $\mathscr{C}$ is said to be strongly homogeneous if (a) $\mathscr{C}$ has almost split sequences; (b) there exist neither projectives nor injectives in $\mathscr{C}$; (c) any indecomposable in $\mathscr{C}$ is homogeneous.

Remark　Take any finite dimensional k-algebra Λ. If k is algebraically closed and Λ is tame, then mod Λ is homogeneous [CB1, Theorem D].

Let $\Lambda=k$, $R=k[t]$ be the polynomial algebra in one determinate t over any field k. Then $\mathscr{C}=\text{mod } R$ is strongly homogeneous. When k is algebraically closed, this is a trivial and typical tame algebra.

A simplest non-trivial example is given by $R=\Lambda[t]$, $\Lambda=k[\varepsilon]/(\varepsilon^2)$, a 2-dimension symmetric algebra. The subcategory $\mathscr{C}$ defined in Subsection 1.1 is strongly homogeneous. The above results are consequences of the following theorem.

2.2 Theorem Let $R=\Lambda[t]$ and $\mathscr{C}$ be defined in Subsection 1.1. If Λ is symmetric, then the category $\mathscr{C}$ is strongly homogeneous.

Proof The category $\mathscr{C}$ has almost split sequences (Theorem 1.3). Any indecomposable M in $\mathscr{C}$ is neither projective nor injective by (1.1) in the proof of Theorem 1.3 $\text{Tr}_R(M)\simeq M^*$ (Lemma 1.2) and $M^*=\text{Hom}_\Lambda(M,\Lambda)\simeq\text{Hom}_k(M,k)=\text{D}M$ as R-modules, since Λ is symmetric. Thus $\text{DTr}_R(M)\simeq M$.

2.3 If a category $\mathscr{C}$ has almost split sequences, then $\mathscr{C}$ has an Auslander-Reiten quiver as in the case of Artin algebras. The vertices are iso-classes of indecomposable objects $[M]$, and we put an arrow $[M]\to[N]$, if there is an irreducible map $\varphi: M\to N$ [AR2].

Lemma Let $\mathscr{C}$ be a strongly homogeneous category. Then for any indecomposable M in $\mathscr{C}$ and any positive integer n, there exists a chain of indecomposables and irreducible maps,

$$M_n\xrightarrow{\varphi_n}M_{n-1}\to\cdots\xrightarrow{\varphi_2}M_1\xrightarrow{\varphi_1}M_0=M$$

in $\mathscr{C}$, such that $\varphi_1\varphi_2\cdots\varphi_n\neq 0$.

Proof Take any indecomposable N in $\mathscr{C}$. There is a minimal right almost split epimorphism [AR2, p. 454], $L_1\oplus L_2\oplus\cdots\oplus L_t\xrightarrow{\pi_1,\pi_2,\cdots,\pi_t}N$, with each π_j, $j=1,2,\cdots,t$, irreducible. By induction, we have a sequence

$$E_n\xrightarrow{f_n}E_{n-1}\to\cdots\xrightarrow{f_2}E_1\xrightarrow{f_1}E_0=M,$$

where f_i is the direct sum of minimal right almost split maps to the indecomposable summands of E_{i-1}. Thus for each $1\leqslant i\leqslant n$, the matrix component of f_i related to the indecomposable summands of E_i and E_{i-1} are either zero or irreducible. The composition $f_1 f_2\cdots f_n$ is an $1\times n$ matrix related to the indecomposable summands of E_n and E_0, such that each component is either zero or the composition of irreducible maps of length n. Since $f_1 f_2\cdots f_n$ is an epimorphism, there exists at least one component, say $\varphi_1\varphi_2\cdots\varphi_n\neq 0$, with $\varphi_i: M_i\rightarrow M_{i-1}$ an irreducible map between indecomposables.

2.4 Proposition(a corollary of Theorem 2.2) Let $\mathscr{C}$ be the category defined in Theorem 2.2. Then each connected component of the Auslander-Reiten quiver of $\mathscr{C}$ is a homogeneous tube.

Proof Since there is neither projectives nor injectives in $\mathscr{C}$, the Auslander-Reiten quiver of $\mathscr{C}$ is a stable valued translation quiver. Let Γ be any connected component of the quiver. Then there is an additive function $\dim_k: \Gamma\rightarrow \mathbf{N}$. Γ is periodic. The Cartan class of Γ is either a Dynkin diagram, an Euclidean diagram, or one of A_∞, A_∞^∞, B_∞, C_∞, D_∞ by [HPR, Theorem p. 289]. But the additive function $\dim_k$ on Γ is unbounded, otherwise Subsection 2.3 would contradict the Harada-Sai lemma. Since the Dynkin diagram and Euclidean diagram are finite, and any additive function on A_∞^∞, B_∞, C_∞, D_∞ is bounded, the only possible Cartan class of Γ is A_∞. Therefore $\Gamma=\mathbf{Z}A_\infty/\langle\tau\rangle$ is a homogeneous tube.

§ 3. Equivalent categories and representation type of $\mathscr{C}$

3.1 Recall from [CB2] that a bimodule problem consists of a k-category K, a K-K-bimodule M, and a derivation $i: K\rightarrow M$. The representation category $\mathrm{Mat}(K, M, i)$ has the objects Xm, with $X\in \mathrm{ob}\, K$ and $m\in M(X, X)$, the morphisms $f: Xm\rightarrow X'm'$ with $f\in K(X, X')$, and $mf-fm'=i(f)$.

Now let K be the category of finitely generated projective Λ-modules, M be the K-K-bimodule with $M(X, Y)=K(X, Y)=\mathrm{Hom}_\Lambda(X, Y)$ for any projective Λ-modules $X, Y, i=0: K \to M$.

Proposition The category $\mathscr{C}$ defined in Subsection 1.1 is equivalent to $\mathrm{Mat}(K, M, i)$.

Proof For any $Xm \in \mathrm{Mat}(K, M)$, X is a projective Λ-module with an action $X(t)=m: X \to X$, which commutes with the action of Λ on X, since $m \in M(X, X)=\mathrm{Hom}_{\Lambda}(X, X)$. Therefore X is an $\Lambda[t]$-module. Any morphism $f: Xm \to X'm'$ is a morphism in $\mathrm{Hom}_{\Lambda}(X, X')$, satisfying $mf-fm'=i(f)=0$, i. e., $X(t)f=fX(t)$, f is an $\Lambda[t]$-map. Then $F: \mathrm{Mat}(K, M, i) \to \mathscr{C}$ defined above is a functor. Conversely, take any $X \in \mathscr{C}$, X is an $\Lambda[t]$-module, which is projective as an Λ-module. Let $m=X(t)$; then $Xm \in \mathrm{Mat}(K, M, i)$. $G: \mathscr{C} \to \mathrm{Mat}(K, M, i)$ by sending X to Xm is a functor. It is obvious that $GF=I_{\mathrm{Mat}(K, M, i)}$ and $FG=l_{\mathscr{C}}$, where I stands for the identity functor of a category.

3.2 Since (K, M, i) is a Krull-Schmidt bimodule problem with K having only finitely many indecomposable objects, we may construct a differential biquiver (Q, d) related to (K, M, i) according to [CB2, 3.1].

From now on, we assume that Λ is a split basic algebra with ordinary quiver Q. Then $\Lambda=kQ/I$, where I is an admissible ideal [D2, 3.6]. Denote by J the Jacobson radical of Λ. Then $S=\Lambda/J \simeq k \times k \times \cdots \times k$, n copies of k. Let $e_1, e_2, \cdots, e_n$ be a complete set of the orthogonal primitive idempotents of Λ. Then $\Lambda=J \oplus S$, $S=ke_1 \times ke_2 \times \cdots \times ke_n$ is a subalgebra of Λ.

Since kQ/I is generated over k by paths, we may take a k-basis of Λ consisting of paths. Take first $e_1, e_2, \cdots, e_n$, i. e., the paths of length 0. Second take a k-basis $\bar{\alpha}$ of $e_q(J/J^2)e_p$ with arrows $\alpha: p \to q$ in Q for $1 \leqslant p, q \leqslant n$. Inductively if we have already a k-basis $\bar{\beta}$ of Λ/J^i, with paths β of length $<i$, then take a k-basis $\bar{\gamma}$ of $e_q(J^i/J^{i+1})e_p$, with path $\gamma: p \to q$ of length i for all $1 \leqslant p, q \leqslant n$. Since J is nilpotent, we obtain finally a k-basis of Λ. Denote by $e_1, e_2, \cdots, e_n, \zeta_1, \zeta_2, \cdots, \zeta_s$, or $\rho_1, \rho_2, \cdots, \rho_n, \rho_{n+1}, \rho_{n+2}, \cdots, \rho_{s'}$, $s'=s+n$, this basis according to the above

ordering. Suppose we have the multiplication table

$$\rho_i\rho_j = \sum_l k_{ij}^l \rho_l.$$

It is obvious that $k_{ij}^l = 0$ if $l > i$ or $l > j$.

Draw a biquiver with n vertices and solid arrows $\rho_l^0 : q \to p$ if we have a base element $\rho_l : p \to q$, $1 \leqslant l \leqslant n+s$. On the other hand, if there is a base element $\zeta_l : p \to q$, then put a dotted arrow $\zeta_l^* : q \to p$. Define

$$d(\rho_l^0) = \sum_{i,j} k_{ij}^l (\zeta_{j-n}^* \otimes \rho_i^0 - \rho_j^0 \otimes \zeta_{i-n}^*),$$

$$d(\zeta_{l-n}^*) = \sum_{i,j} k_{ij}^l (\zeta_{j-n}^* \otimes \zeta_{i-n}^*).$$

Finally we obtain a differential biquiver (Q, d).

3.3 The representation category $R(Q,d)$ of (Q,d) is equivalent to the representation category $R(\mathscr{A})$ according to [CB2, 4.3], where $\mathscr{A} = (A, V)$, A is freely generated by the solid arrows of Q over $A' = S$, $V = A \oplus \bar{V}$ is an $A-A$-bimodule, and $\bar{V}$ is freely generated by dotted arrows of Q. $\mathscr{A}$ has a layer

$$L = (A'; \omega; e_1^0, e_2^0, \cdots, e_n^0, \zeta_1^0, \zeta_2^0, \cdots, \zeta_s^0; \zeta_1^*, \zeta_2^*, \cdots, \zeta_s^*),$$

or a layer

$$L_1 = (A_1'; \omega_1; \zeta_1^0, \zeta_2^0, \cdots, \zeta_s^0; \zeta_1^*, \zeta_2^*, \cdots, \zeta_s^*)$$

with $A_1'(p,p) = k[e_p^0] \simeq k[x]$, $1 \leqslant p \leqslant n$.

Now we give explicitly the equivalent functor $E: R(\mathscr{A}) \to \mathscr{C}$. Take any base element $\rho_l : p \to q$ in Λ and denote by $\tilde{\rho}_l : \Lambda e_q \to \Lambda e_p$ the right multiplication by ρ_l. Let $X = \bigoplus_{p=1}^n X_p \in R(\mathscr{A})$. Define $E(X) = \bigoplus_{p=1}^n \Lambda e_p \bigotimes_k X_p$. The action of Λ on $E(X)$ is the natural left multiplication. The action of t on $E(X)$ is given by

$$E(X)(t) = \sum_{l=1}^{n+s} \tilde{\rho}_l \bigotimes_k X(\rho_l^0) : E(X) \to E(X).$$

Then the action of Λ and the action of t on $E(X)$ are commutative; thus $E(X) \in \mathscr{C}$. Moreover if $\phi : X \to Y$ is a morphism in $R(\mathscr{A})$, $E(\phi)$ is given by

$$\sum_{p=1}^n \tilde{e}_p \bigotimes_k \phi_p^0 + \sum_{l=1}^s \tilde{\zeta}_l \bigotimes_k \phi(\zeta_1^*)$$

$$E(\phi)E(X)(t)$$
$$=\Big[\sum_{p=1}^{n}\tilde{e}_p\otimes\phi_p^0+\sum_{j=n+1}^{n+s}\tilde{\zeta}_{j-n}\otimes\phi(\zeta_{j-n}^*)\Big]\Big[\sum_{i=1}^{n+s}\tilde{\rho}_i\otimes X(\rho_i^0)\Big]$$
$$=\sum_{p,l}\tilde{e}_p\tilde{\rho}_l\otimes\phi_p^0X(\rho_l^0)+\sum_{j,i}\tilde{\zeta}_{j-n}\tilde{\rho}_i\otimes\phi(\zeta_{j-n}^*)X(\rho_i^0)$$
$$=\sum_{l=1}^{n+s}\tilde{\rho}_l\otimes\Big[\phi_p^0X(\rho_l^0)+\sum_{i,j}k_{ij}^l\phi(\zeta_{j-n}^*)X(\rho_i^0)\Big];$$
$$E(X)(t)E(\phi)$$
$$=\Big[\sum_{j=1}^{n+s}\tilde{\rho}_j\otimes Y(\rho_j^0)\Big]\Big[\sum_{q=1}^{n}\tilde{e}_q\otimes\phi_q^0+\sum_{i=n+1}^{n+s}\zeta_{i-n}\otimes\phi(\zeta_{i-n}^*)\Big]$$
$$=\sum_{l,q}\tilde{\rho}_l\tilde{e}_q\otimes Y(\rho_l^0)\phi_q^0+\sum_{j,i}\tilde{\rho}_i\tilde{\zeta}_{i-n}\otimes Y(\rho_j^0)\phi(\zeta_{i-n}^*)$$
$$=\sum_{l=1}^{n+s}\tilde{\rho}_l\otimes\Big[Y(\rho_l^0)\phi_q^0+\sum_{i,j}k_{ij}^lY(\rho_j^0)\phi(\zeta_{i-n}^*)\Big].$$

They are equal by the formula of $d(\rho_l^0)$ given in Subsection 3. 2. Thus

$$E(\phi)\in\mathrm{Hom}_R(E(X),E(Y)).$$

In a similar way, we can prove that if $\phi: X\to Y$ and $\phi': Y\to Z$ are morphisms in $R(\mathscr{A})$, then $E(\phi\phi')=E(\phi)E(\phi')$ in $\mathscr{C}$.

E is dense. In fact, take any $M=\bigoplus_{p=1}^{n}(\Lambda e_p)^{m_p}$ in $\mathscr{C}$, let $X=\bigoplus_{p=1}^{n}X_p$, where X_p is a k-vector space of dimension m_p. Then $M=\bigoplus_{\rho_l:p\to q}\rho_lX_q$. Let $X(\rho_l^0)=M(t)|_{e_qX_q\to\rho_lX_q}:X_q\to X_p$. Then $E(X)=M$.

Finally we prove that E is fully faithful. If ϕ is a morphism in $R(\mathscr{A})$, with $E(\phi)=0$, then by the definition of $E(\phi)$, $\phi_p=0$, $p=1,2,\cdots,n$, $\phi_l^*=0$, $l=1,2,\cdots,s$, i. e., $\phi=0$. If $X,Y\in R(\mathscr{A})$, with $E(X)=M$, $E(Y)=N$, $\psi: M\to N$ is a morphism in $\mathscr{C}$. Let $\phi: X\to Y$ defined by $\phi_p^0=\psi|_{e_pX_p-e_qY_q}:X_p\to Y_p$, $p=1,2,\cdots,n$, $\phi(\zeta_l^*)=\psi|_{e_qX_q\to\zeta_lY_p}:X_q\to Y_p$, $l=1,2,\cdots,s$. Then

$$E(\phi)=\psi.$$

3. 4 We still assume that Λ is a basic algebra. Moreover k is an algebraically closed field.

Definition[CB1, 3. 9] Let Σ be the category of finitely generated free $k\langle x,y\rangle$-modules. We say that a bocs $\mathscr{A}=(A,V)$ is wild if there is a functor $F: A\to\Sigma$, such that the induced functor $(F,\varepsilon\circ F)^*: R(\Sigma)\to$

$R(\mathscr{A})$ preserves isomorphism classes and indecomposability.

Definition Let T be a $R-k\langle x,y\rangle$-bimodule, which as Λ-$k\langle x,y\rangle$-bimodule has the form $T=\bigoplus_{p=1}^{n}\Lambda e_p\bigotimes_k T_p$, where T_p is a free $k\langle x,y\rangle$-module. The action of t on T is given by $\sum_{l=1}^{n+s}\bar{\rho}_l\bigotimes_k T(\rho_l^0)$ with $T(\rho_l^0)\in \mathrm{Hom}_{k\langle x,y\rangle}(T_q, T_p)$ for $\rho_l: p\to q$. The category $\mathscr{C}$ is called of wild representation type if there exists such a bimodule T, such that the functor $T\bigotimes_{k\langle x,y\rangle}-$: mod $k\langle x,y\rangle\to\mathscr{C}$ preserves indecomposability and isomorphism classes.

Proposition The category $\mathscr{C}$ is wild if and only if $R(\mathscr{A})$ is wild.

Proof Suppose $R(\mathscr{A})$ is wild and $F:A\to\Sigma$ is a functor defined in the first definition. Construct an $\Lambda-k\langle x,y\rangle$-bimodule

$$T=\bigoplus_{p=1}^{n}\Lambda e_p\bigotimes_k F(p).$$

The action of t is defined by

$$T(t)=\sum_{l=1}^{n+s}\tilde{\rho}\bigotimes_k F(\rho_l^0).$$

Thus T is a $R-k\langle x,y\rangle$-bimodule. Moreover we have the fig. 3.1.

$$\begin{array}{ccc} R(\Sigma) & \xrightarrow{(F,F\varepsilon)^*} & R(\mathscr{A}) \\ \Big\downarrow{\scriptstyle \mathrm{Id}} & & \Big\downarrow{\scriptstyle E} \\ \mathrm{mod}\ k\langle x,y\rangle & \xrightarrow{T\bigotimes_{k\langle x,y\rangle}-} & \mathscr{C} \end{array}$$

Fig. 3.1

where Id stands for the identity functor, $R(\Sigma)=\mathrm{mod}\ k\langle x,y\rangle$. Take any $N\in\mathrm{mod}\ k\langle x,y\rangle$ and suppose $(F,F\varepsilon)^* N=X$. Then $X_p=E(p)N$ and $X(\rho_l^0)=F(\rho_l^0): X_q=F(q)N\to F(p)N=X_p$ is given by the left multiplication. Then

$$T\bigotimes_{k\langle x,y\rangle}N=\bigoplus_{p=1}^{n}\Lambda e_p\bigotimes_k F(p)\bigotimes_{k\langle x,y\rangle}N=\bigoplus_{p=1}^{n}\Lambda e_p\bigotimes X_p=E(X).$$

Thus the above diagram commutes.

Conversely if $\mathscr{C}$ is wild, $T=\bigoplus_{p=1}^{n}\Lambda e_p\bigotimes T_p$ as in the second definition. Let $F:A\to\Sigma$ given by $F(p)=T_p$, $F(\rho_l^0)=T(\rho_l^0)$. F is an algebra homomorphism, since $T(t)$ is a module action,

$$F(\rho_i^0\rho_j^0)=T(\rho_i^0\rho_j^0)=T(\rho_i^0)T(\rho_j^0)=F(\rho_i^0)F(\rho_j^0).$$

Again we have the above commutative diagram, and then $R(\mathscr{A})$ is wild.

3.5 Theorem Let k be an algebraically closed field, Λ be a basic finite dimensional k-algebra, and $\mathscr{C}$ be the category defined in Subsection 1.1. Then $\mathscr{C}\simeq \text{mod } k[t]\times\cdots\times \text{ mod } k[t]$ when Λ is semi-simple; and $\mathscr{C}$ is wild when Λ is not semi-simple.

Proof If $J\neq 0$, then $\delta(\zeta_1^0)=\zeta_1^*\otimes e_q^0-e_p^0\otimes\zeta_1^*$, where $\zeta_1: p\to q$ satisfies the condition (2) of [CB1, 3.10]. Then $R(\mathscr{A})$ is wild; consequently $\mathscr{C}$ is wild.

Acknowledgements

This is a team work. The original idea and proof were given by Zhang, Lei, and Bautista by a concrete description of the category $R(\mathscr{A})$. The present proof was given by Crawley-Boevey and completed by Zhang. Bautista and Zhang thank the Beijing Normal University and the Universidad Nacional Autónoma de Mexico for supporting visits to Beijing and Mexico, respectively.

References

[A] Auslander M. Functors and morphisms determined by modules. in: Representation Theory of Algebras. Proceedings, Philadelphia Conference, 1976:1-244.

[AR1] Auslander M, Reiten I. Representation theory of Artin algebras, Ⅲ. Comm. Algebra, 1975, 3:239-294.

[AR2] Auslander M, Reiten I. Representation theory of Artin algebras, Ⅳ. Comm, Algebra, 1977, 5:443-518.

[BK] Bautista R, Kleiner M. Almost split sequences for relatively projective modules. J. Algebra, 1990, 135:19-56.

[C] Cohn P M. Algebra. Wiley, New York, 1977.

[CB1] Crawley-Boevey W W. On tame algebras and bocses. Proc. London Math. Soc., 1988, 56:451-483.

[CB2] Crawley-Boevey W W. Matrix problems and Drozd's theorem. in: Topics in Algebra. Banach Center Publications. PWN, Warsaw, 1990, 26(part 1):199-222.

[D1] Drozd Y A. Tame and wild matrix problems. in: Representations and Quadratic Forms. Institute of Mathematics, Academy of Sciences, Ukranian SSR, Kiev, 1979:39-74. [In Russian]

[D2] Drozd Y A, Kirichendo V. Finite Dimensional Algebras. Springer-Verlag, New York/Berlin, 1994.

[H] Harada M, Sai Y. On categories of indecomposable modules, Ⅰ. Osaka J. Math., 1970, 7:323-344.

[HPR] Happel D, Preiser U, Ringel C M. Vinberg's characterization of Dynkin diagrams using subadditive functions with application to DTr-periodic modules. in: Lecture Notes in Math., Vol. 831; 280-294, Springer-Verlag, Berlin, 1979.

[K] Kleiner M. Induced modules and comodules and representations of Bocs's and DGC's representations of algebras. in: Lecture Notes in Math., Vol. 903:168-185, Springer-Verlag, Berlin, 1981.

[R] Ringel C M. Tame algebras and integral quadratic forms. in: Lecture Notes in Math., Vol. 1 099, Springer-Verlag, New York/Berlin, 1984.

[Ro] Roiter A V. Matrix problems and representations of BOCS's. in: Lecture Notes in Math., Vol. 831:288-324, Springer-Verlag, Berlin, 1980.

[RK] Roiter A V, Kleiner M. Representations of differential graded categories. in: Lecture Notes in Math., Vol. 488: 316-340, Springer-Verlag, Berlin, 1975.

[ZLB] Zhang Y B, Lei T G, Bautista R. The representation category of a bocs, Ⅰ～Ⅳ. J. Beijing Normal Univ, 1995, 31:313-316, 440-445; 1996, 32: 143-149, 289-295.

Science in China, 2001, 44A(12): 1 515-1 522.

不可分解性与链环数①

Indecomposability and the Number of Links

Abstract Let $(\mathscr{K}, \mathscr{M})$ be a linear matrix problem induced from a finite dimensional algebra Λ. Then an $\underline{n} \times \underline{n}$ matrix $\boldsymbol{M}$ in $R(\mathscr{K}, \mathscr{M})$ is indecomposable if and only if the number of links in the canonical form $\boldsymbol{M}^{(\infty)}$ of $\boldsymbol{M}$ is equal to $\mathscr{M}$-dim $\underline{n}-1$. On the other hand, the dimension of the endomorphism ring of $\boldsymbol{M}$ is equal to $\mathscr{K}$-dim $\underline{n}$-$\sigma(\boldsymbol{M})$.

Keywords linear matrix problem; canonical form; indecomposability; link; endomorphism ring.

Let k be an algebraically closed field, and let Λ be a finite dimensional basic k-algebra. Then there exists an induced linear matrix problem $(\mathscr{K}, \mathscr{M})$ (see sec. 1) such that the problem of classifying Λ-modules can be reduced to the problem of classifying canonical forms of the representations of $(\mathscr{K}, \mathscr{M})$ under some admissible transformations. Such a canonical form can be considered as a generalization of Jordan normal form. Denote by $R(\mathscr{K}, \mathscr{M})$ the representation category of $(\mathscr{K}, \mathscr{M})$ consisting of finite dimensional objects. The aim of the present paper is to give some numerical description of indecomposability and calculate the dimension of the endomorphism ring for an object of $R(\mathscr{K}, \mathscr{M})$ based

① Received: 2001-03-03;
本文与徐运阁合作.

on Belitskii's reduction algorithms (see sec. 2). The proof of Main Theorems A and B is given in sec. 3.

Theorem A Let Λ be a finite dimensional k-algebra, and $(\mathscr{K}, \mathscr{M})$ the linear matrix problem induced from Λ. Assume that M is any object of dimension $\underline{n}$ in $\mathrm{R}(\mathscr{K}, \mathscr{M})$ with the canonical form $(\boldsymbol{M}^{(\infty)}, \boldsymbol{E}^{(\infty)})$ of dimension vector $\underline{n}^{(\infty)}$. Then the following statements are equivalent:

(1) $\boldsymbol{M}$ is indecomposable;

(2) $\boldsymbol{E}^{(\infty)}$ is a local algebra;

(3) $\mathscr{M}\text{-}\dim\underline{n}^{(\infty)} = 1$, i. e. $\underline{n}^{(\infty)} = (1, 1, \cdots, 1)$ and all indices are equivalent;

(4) the number of links in $\boldsymbol{M}^{(\infty)}$ equals $\mathscr{M}\text{-}\dim\underline{n} - 1$;

(5) $\dim(\widetilde{\boldsymbol{M}}^{\infty}) = 1$ when $\theta_{\infty}^{*}(\widetilde{\boldsymbol{M}}^{(\infty)}) = \sum(\boldsymbol{M})$.

Theorem B Let Λ be a finite dimensional k-algebra, and let $(\mathscr{K}, \mathscr{M})$ be the linear matrix problem induced from Λ. Assume that $\boldsymbol{M}$ is any object of dimension $\underline{n}$ in $\mathrm{R}(\mathscr{K}, \mathscr{M})$ with the canonical form $\boldsymbol{M}^{(\infty)}$. Then

$$\dim_k \mathrm{End}_{(\mathscr{K}, \mathscr{M})}(\boldsymbol{M}) = \mathscr{M}\text{-}\dim\underline{n} - \sigma(\boldsymbol{M}).$$

§ 1. Category R(K, M)

Let k be an algebraically closed field, and let Λ be a finite dimensional basic k-algebra. Then there exists a bimodule problem (K, M) induced from Λ[1]. Let $K = (\Lambda\text{-proj}) \times (\Lambda\text{-proj})$, and $M = \mathrm{rad}(\Lambda\text{-proj})$. M is a K-K-bimodule by restriction via projections of K onto the first and second factors respectively. The objects of $\mathrm{R}(K, M)$ consist of $(P_1 \xrightarrow{\alpha} P_0)$ for any $(P_1, P_0) \in K$ and $\alpha \in \mathrm{rad}_{\Lambda}(P_1, P_0)$. The morphisms are given by the commutative fig. 1. 1:

$$\begin{array}{ccc} P_1 & \xrightarrow{\alpha} & P_0 \\ {\scriptstyle f_1}\downarrow & & \downarrow{\scriptstyle f_0} \\ P_1' & \xrightarrow{\alpha'} & P_0' \end{array}$$

Fig. 1. 1

with $(f_1, f_0) \in \mathrm{Hom}_K((P_1, P_1'), (P_0, P_0'))$. It is easy to see that $R(K, M)$ is in fact the category $P_1(\Lambda) = \{(P_1 \xrightarrow{\alpha} P_0) \mid \mathrm{im}(\alpha) \subset \mathrm{rad}(P_0)\}$.

In order to calculate the objects of $R(K, M)$, a linear matrix problem $(\mathscr{K}, \mathscr{M})$ based on the bimodule problem (K, M) has been constructed in ref. [2] as follows. Denote by J the Jacobson radical of Λ, and $\{e_1, e_2, \cdots, e_s\}$ a complete set of primitive idempotents of Λ. Let Q be the ordinary quiver of Λ with s vertices and the arrows determined by J/J^2. The multiplica tion of paths of Q is written from left to right. Take a k-basis of J: If $J^m \neq 0, J^{m+1}=0$, fix a k-basis of $e_p J^m e_q$, $1 \leqslant p$, $q \leqslant s$. Without loss of generality, we may take them as paths in Q. Suppose we already have a basis of J^i, then extend it to a k-basis of J^{t-1} such that the images of the added elements form a k-basis of $e_p(J^{i-1}/J^i)e_q$, $1 \leqslant p, q \leqslant s$. The obtained k-basis of J is denoted by $B=\{\xi\}$ with a fixed ordering as above. Finally adding paths $e_1, e_2, \cdots, e_s$ of length 0, we obtain a k-basis of Λ. Constructing the left regular representation of Λ using this basis, we obtain a $t \times t$ upper—triangular matrix algebra $\widetilde{\boldsymbol{\Lambda}}$ with $t=\dim\Lambda$, which is isomorphic to Λ. Let

$$\mathscr{K}=\begin{pmatrix} \widetilde{\boldsymbol{\Lambda}} & \mathbf{0} \\ \mathbf{0} & \widetilde{\boldsymbol{\Lambda}} \end{pmatrix}, \quad \mathscr{M}=\begin{pmatrix} \mathbf{0} & \operatorname{rad} \widetilde{\boldsymbol{\Lambda}} \\ \mathbf{0} & \mathbf{0} \end{pmatrix}.$$

Then $(\mathscr{K}, \mathscr{M})$ is a linear matrix problem and $\mathrm{R}(\mathscr{K}, \mathscr{M})$ stands for its representation category.

Let $(\mathscr{K}, \mathscr{M})$ be a linear matrix problem constructed from a finite dimensional algebra Λ. Following ref. [2], we let $T_0=\{1,2,\cdots,t\}$ which is partitioned into s parts, $\mathscr{T}_p=\{l_1, l_2, \cdots, l_{t(p)}\}$, provided idempotent e_p corresponds to $E_{l_1 l_1}+E_{l_2 l_2}+\cdots+E_{l_{l(p)} l_{l(p)}}\}$, where E_{ij} stands for the $t \times t$ matrix with (i, j)-th element 1, others zero, and $p=1,2,\cdots,s$, $t=\sum_{p=1}^{s} t(p)$. The partition determines an equivalent relation $\sim_0$. Let $T_1=\{i' \mid i \in T_0\}$, $T=T_1 \cup T_0=\{1', 2', \cdots, t', 1, 2, \cdots, t\}$. For any $1 \leqslant i$, $j \leqslant t$, $i \sim j$ and $i' \sim j'$ if and only if $i \sim_0 j$ as given above.

Definition 1 A vector of nonnegative integers $\underline{\boldsymbol{n}}=(m_{1'}, m_{2'}, \cdots, m_{t'}; n_1, n_2, \cdots, n_t)$ is called a $(\mathscr{K}, \mathscr{M})$-dimension vector if $m_{i'}=m_{j'}$, and $n_i=n_j$ whenever $i \sim_0 j$. For simplicity, we still call $\underline{\boldsymbol{n}}$ a dimension vector.

Definition 2 Let $\underline{n}$ be a dimension vector, a corresponding (K, M)-dimension vector is defined by $(m_{\mathcal{T}_1'}, m_{\mathcal{T}_2'}, \cdots, m_{\mathcal{T}_s'}, n_{\mathcal{T}_1}, n_{\mathcal{T}_2}, \cdots, n_{\mathcal{T}_t})$, where $m_{\mathcal{T}'} = m_{i'}$ for $i' \in \mathcal{T}'$ and $n_{\mathcal{T}} = n_i$, for $i \in \mathcal{T}$.

Definition 3 Let $\underline{n}$ be a dimension vector. The corresponding $\mathscr{M}$-dimension is defined by

$$\mathscr{M}\text{-dim}\ \underline{n} = \sum_{\mathcal{T}, \mathcal{T} \in T_0/\sim} (m_{\mathcal{T}} + n_{\mathcal{T}}),$$

and $\mathscr{K}$-dimension by

$$\mathscr{K}\text{-dim}\ \underline{n} = \sum_{\mathcal{T}, \mathcal{T} \in T/\sim} (m_{\mathcal{T}'}^2 + n_{\mathcal{T}}^2) + \sum_{\zeta \in B} (m_{s(\zeta)'} m_{e(\zeta)'} + n_{s(\zeta)} n_{e(\zeta)}),$$

where $s(\zeta)$ and $e(\zeta)$ stand for the starting and ending vertices of a path ζ respectively, and B is a basis of rad Λ constructed above.

Next we give the definition of $(\mathscr{K}, \mathscr{M})$ and also the objects and morphisms of $\mathrm{R}(\mathscr{K}, \mathscr{M})$ in terms of matrix.

Denote by

$$\boldsymbol{X} = \begin{pmatrix} x_{11} & x_{12} & \cdots & x_{1t} \\ & x_{22} & \cdots & x_{2t} \\ & & \ddots & \vdots \\ & & & x_{tt} \end{pmatrix}$$

a $t \times t$ upper-triangular matrix. $\widetilde{\Lambda}$ consists of all $t \times t$ matrices $\boldsymbol{X}$ such that $x_{ii} = x_{jj}$ if $i \sim_0 j$, and the off-diagonal elements satisfy a collection of k-linear equations:

$$\sum_{\mathcal{T} \ni i > j \in \mathcal{J}} c_{ij}^{(l)} x_{ij} = 0, \quad \forall\, \mathcal{T}, \mathcal{J} \in T/\sim$$

for each pair $\mathcal{T}, \mathcal{J} \in T/\sim$, which are obtained from the multiplication table of the basis elements of J. Thus

$$\mathscr{K} = \left\{ \begin{pmatrix} \boldsymbol{X} & \boldsymbol{0} \\ \boldsymbol{0} & \boldsymbol{Y} \end{pmatrix} \middle| \boldsymbol{X}, \boldsymbol{Y} \in \widetilde{\boldsymbol{\Lambda}} \right\}, \quad \mathscr{M} = \left\{ \begin{pmatrix} \boldsymbol{0} & \boldsymbol{M} \\ \boldsymbol{0} & \boldsymbol{0} \end{pmatrix} \middle| \boldsymbol{M} \in \mathrm{rad}\, \widetilde{\boldsymbol{\Lambda}} \right\},$$

and the pair $(\mathscr{K}, \mathscr{M})$ satisfies the following conditions:

(1) The index set and its partition $(T, \sim)$ are defined in the third paragraph of this section;

the elements of matrices $\boldsymbol{X}, \boldsymbol{Y}$ satisfy

(2) $x_{i'i'} = x_{j'j'}$ if $i' \sim j'$ and $y_{ii} = y_{jj}$ if $i \sim j$;

(3) $\sum_{\substack{\mathcal{T}\ni i'<j'\in\mathcal{J}' \\ i<j}} c_{i'j'}^{(l)} x_{i'j'}=0 \quad \sum_{\mathcal{T}\ni i<j\in\mathcal{J}'} c_{ij}^{(l)} y_{ij}=0$;

the elements of M satisfy

(4) $\sum_{\substack{i'\in\mathcal{T}, j\in\mathcal{J} \\ i<j}} c_{i'j}^{(l)} a_{i'j}=0$.

Given any dimension vector $\underline{\boldsymbol{n}}$, an object of $\mathrm{R}(\mathcal{K},\mathcal{M})$ is a partitioned matrix $\begin{pmatrix}\mathbf{0} & \boldsymbol{M}\\ \mathbf{0} & \mathbf{0}\end{pmatrix}$ with $\boldsymbol{M}=(\boldsymbol{M}_{i'j}), 1\leqslant i<j\leqslant t$. $\boldsymbol{M}_{i'j}$ is an $m_{i'}\times n_j$ matrix satisfying equation system (4). We sometimes denote it just by $\boldsymbol{M}$ for simplicity.

If $\begin{pmatrix}\mathbf{0} & \boldsymbol{N}\\ \mathbf{0} & \mathbf{0}\end{pmatrix}$ is also an object of dimension $n'=(m'_{1'}, m'_{2'}, \cdots, m'_{t'}; n'_1, n'_2, \cdots, n'_t)$, then a morphism between them is a partitioned matrix

$$\begin{pmatrix}\boldsymbol{X} & \mathbf{0}\\ \mathbf{0} & \boldsymbol{Y}\end{pmatrix}$$

such that

$$\begin{pmatrix}\boldsymbol{X} & \mathbf{0}\\ \mathbf{0} & \boldsymbol{Y}\end{pmatrix}\begin{pmatrix}\mathbf{0} & \boldsymbol{N}\\ \mathbf{0} & \mathbf{0}\end{pmatrix}=\begin{pmatrix}\mathbf{0} & \boldsymbol{M}\\ \mathbf{0} & \mathbf{0}\end{pmatrix}\begin{pmatrix}\boldsymbol{X} & \mathbf{0}\\ \mathbf{0} & \boldsymbol{Y}\end{pmatrix},$$

where $\boldsymbol{X}=(\boldsymbol{X}_{i'j'})$ is a $t\times t$ partitioned matrix with each block $\boldsymbol{X}_{i'j'}$, an $m_{i'}\times m'_{j'}$ matrix, for $1\leqslant i\leqslant j\leqslant t$, and $\boldsymbol{Y}=(\boldsymbol{Y}_{ij})$ is also a $t\times t$ partitioned matrix, $\boldsymbol{Y}_{ij}$ being $n_{i'}\times n'_j$ matrices, and both $\boldsymbol{X}$ and $\boldsymbol{Y}$ satisfying (2) and (3).

Example 1 Let $\Lambda=kQ$ be a path algebra of Dynkin type, $Q=1\xrightarrow{a}2\xrightarrow{b}3$ with $c=ab$. Take a basis $\{c, a, b, e_1, e_2, e_3\}$. Then $T=\{1', 2', \cdots, 6', 1, 2, \cdots, 6\}$, $\mathcal{J}'_1=\{1', 2', 4'\}$, $\mathcal{J}'_2=\{3', 5'\}$, $\mathcal{J}'_3=\{6'\}$, $\mathcal{J}_1=\{1, 2, 4\}$, $\mathcal{J}_2=\{3, 5\}$, $\mathcal{J}_3=\{6\}$.

$$\boldsymbol{X}=\begin{pmatrix} x_{11} & 0 & x_{13} & 0 & 0 & x_{16}\\ & x_{22} & 0 & 0 & x_{25} & 0\\ & & x_{33} & 0 & 0 & x_{36}\\ & & & x_{44} & 0 & 0\\ & & & & x_{55} & 0\\ & & & & & x_{66}\end{pmatrix},$$

where $x_{11}=x_{22}=x_{44}$; $x_{33}=x_{55}$ by the equivalent relation. The equation system is $x_{13}=x_{25}$ and $x_{ij}=0$ when $(ij)\neq(13)(16)(25)$ or (36).

There exists a differential biquiver (Q^b,δ) induced from (K,M), or equivalently from $(\mathscr{K},\mathscr{M})$ by 3. 1 of ref. [1].

The following structure is given in ref. [3]: Put on the top the veritices $1',2',\cdots,s'$ corresponding to the equivalent classes $\mathscr{J}'_1,\mathscr{J}'_2,\cdots,\mathscr{J}'_s$, and on the bottom the vertices $1,2,\cdots,s$ corresponding to the equivalent classes $\mathscr{J}_1,\mathscr{J}_2,\cdots,\mathscr{J}_s$.

Denote by $DB=\{\zeta^*\}$ the dual basis of $DJ=\mathrm{Hom}_k(J,k)$. Let $M(i,j)=e_iJe_j$, and $A_{i'j}\subseteq DB$, the k-basis of $DM(i,j)$, and then draw a set of solid arrows $A_{i'j}$ from i' to j.

Let $J(i,j)=e_iJe_j$, $\Phi_{i'j'}\subseteq DB$, and $\Phi_{ij}\subseteq DB$ the k-basis of $DJ(i,j)$, and then draw two sets of dotted arrows $\Phi_{i'j'}$ from i' to j' and Φ_{ij} from i to j.

Thus we obtain a biquiver Q^b, and the differentials δ of solid and dotted arrows in Q^b can also be defined, which is exactly the Drozd bocs $\mathscr{A}$ induced from algebra Λ. Denote by $R(Q^b,\delta)$ the representation category of (Q^b,δ), which is the same as the representation category $\mathrm{R}(\mathscr{A})$ of bocs $\mathscr{A}$.

The differential biquiver of the example given in Example 1 is fig. 1. 2.

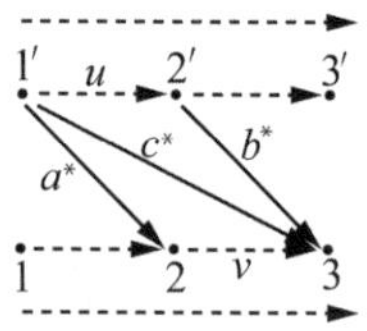

Fig. 1. 2

with differentials $\delta(a^*)=0,\delta(b^*)=0,\delta(c^*)=ub^*-a^*v$.

Lemma 1 There exists an equivalent functor $\Sigma: R(\mathscr{K},\mathscr{M})\simeq P_1(\Lambda)=R(K,M)\to R(Q^b,\delta)\simeq R(\mathscr{A})$ (cf. refs. [1,3]).

§ 2. Reductions

Let $\underline{n}$ be a dimension vector, and $\boldsymbol{E}=\mathscr{K}_{\underline{n}}$, be the set of $\underline{n}\times\underline{n}$ upper triangular partitioned matrices whose blocks satisfy (2) and (3) of sec. 1. $\boldsymbol{M}=\overline{\mathscr{M}}_{\underline{n}}$ is the set of $\underline{n}\times\underline{n}$ partitioned matrices whose blocks satisfy (4). There are three reduction algorithms[2]: regularization, edge reduction and loop reduction, reducing a pair $(\boldsymbol{M},\boldsymbol{E})$ to a canonical form $(\boldsymbol{M}^{(\infty)},\boldsymbol{E}^{(\infty)})$. We obtain a reduction sequence step by step:

$$(\boldsymbol{M},\boldsymbol{E})=(\boldsymbol{M}^{(0)},\boldsymbol{E}^{(0)}),(\boldsymbol{M}^{(1)},\boldsymbol{E}^{(1)}),\cdots,(\boldsymbol{M}^{(p)},\boldsymbol{E}^{(p)})=(\boldsymbol{M}^{(\infty)},\boldsymbol{E}^{(\infty)}).$$

Let $\boldsymbol{M}^{(\infty)}$ be the structured E-canonical form[2], partitioned into rectangular parts $\boldsymbol{M}_1,\boldsymbol{M}_2,\cdots,\boldsymbol{M}_v$. Among them $\boldsymbol{M}_{q_1}<\boldsymbol{M}_{q_2}<\cdots<\boldsymbol{M}_{q_p}$ are called free blocks provided they are actually reduced in the reduction sequence, i. e. $\boldsymbol{M}_{q_k}$ is obtained by reducing the first non-stable block $\boldsymbol{M}_{lr}^{(k-1)}$ in $\boldsymbol{M}^{(k-1)}$. Then each $\boldsymbol{M}_{q_k}$; falls into three cases: $\boldsymbol{\phi}$, $\begin{pmatrix}\boldsymbol{0} & \boldsymbol{I}\\ \boldsymbol{0} & \boldsymbol{0}\end{pmatrix}$ or a Weyr matrix[2]. And we also call $\boldsymbol{M}_{lr}^{(k-1)}$ a free block in $\boldsymbol{M}^{(k-1)}$.

Remark 1 It is possible that there are some blocks which are linear combinations of free blocks (called dependent blocks). Considering that the k-th step of reduction from $(\boldsymbol{M}^{(k-1)},\boldsymbol{E}^{(k-1)})$ to $(\boldsymbol{M}^{(k)},\boldsymbol{E}^{(k)})$, $\boldsymbol{M}_{lr}^{(k-1)}$ may appear in some dependent blocks as summands, these summands are reduced into one of the above three forms with $\boldsymbol{M}_{lr}^{(k-1)}$ simultaneously.

Remark 2 Let $\mathscr{A}$ be the Drozd bocs induced from Λ. By ref. [1], the reduction sequence corresponding to the reductions of bocses is

$$\mathscr{A}=\mathscr{A}^{(0)},\mathscr{A}^{(1)},\cdots,\mathscr{A}^{(p)}=\mathscr{A}^{(\infty)},$$

with $\mathscr{A}^{(\infty)}$ a minimal bocs. Denote by $\theta_k:\mathscr{A}\to\mathscr{A}^{(k)}$ the composition functor of reductions from the first step to the k-step, $\theta_{kl}^{*}:\mathrm{R}(\mathscr{A}^{(k)})\to\mathrm{R}(\mathscr{A})$ the induced functor. When $k=p$, θ_k is denoted by θ_∞.

Lemma 2 Denote by $\boldsymbol{J}_m(\lambda)$ a Jordan block of size m and eigenvalue λ. Let $U_i=\bigoplus_{j=1}^{u}\boldsymbol{J}_{m_{ij}}(\lambda)^{p_{ij}}$ be a direct sum of Jordan blocks, with positive integers $d_i=m_{i1}>m_{i2}>\cdots>m_{iu}>0$ and positive integers $p_{i1},p_{i2},\cdots,$

p_{iu}, $U=\bigoplus_{i=1}^{l}U_i$. Let $V=\bigoplus_{i=1}^{l}V_i$, $V_i=\mathrm{diag}\{J_{m_{ij},p_{ij}}(\lambda_i),\cdots,J_{m_{in},p_{in}}(\lambda_i)\}$, where

$$J_{m_{ij},p_{ij}}(\lambda_i)=\begin{pmatrix}\lambda_i I_{p_{ij}} & I_{p_{ij}} & & & \\ & \lambda_i I_{p_{ij}} & I_{p_{ij}} & & \\ & & \ddots & \ddots & \\ & & & \lambda_i I_{p_{ij}} & I_{p_{ij}} \\ & & & & \lambda_i I_{p_{ij}}\end{pmatrix}_{m_{ij}\times m_{ij}}$$

Then V is similar to U.

Let $W=\bigoplus_{i=1}^{l}W_i$ with

$$W_i=\begin{pmatrix}\lambda_i I_{m'_t 1} & W_{i2} & & & \\ & \lambda_i I_{m'_t 2} & W_{i3} & & \\ & & \ddots & \ddots & \\ & & & \lambda_i I_{m'_t d_t-1} & W_{id_t} \\ & & & & \lambda_i I_{m'_t d_t}\end{pmatrix}$$

where $m'_{ih}=\sum_{h\leqslant m_{ij}\leqslant d_i}p_{ij}$, $h=1,2,\cdots,d_i$, $W_{ih}=\begin{pmatrix}I_{m'_t h}\\ 0\end{pmatrix}$, or $\begin{pmatrix}I\\ 0\end{pmatrix}$ for simplicity, and $h=2,3,\cdots,d_i$. Then W is similar to V. And W is called a Weyr matrix[2].

Definition 4 Let (M,E) be defined in the first paragraph of this section, $M^{(\infty)}$ be the structured Λ-canonical form, and M_{q_k} be any free rectangular block of $M^{(\infty)}$, which is not ϕ. Then $M_{q_k}=\begin{pmatrix}0 & I\\ 0 & 0\end{pmatrix}$ or $M_{q_k}=$ W is a Weyr matrix The elements I's in $\begin{pmatrix}0 & I\\ 0 & 0\end{pmatrix}$ or $W_{ih}=\begin{pmatrix}I\\ 0\end{pmatrix}$ are called links. Denote by $\tau(M_{q_k})$ the number of links in M_{q_k}.

Definition 5 Let $\mathscr{F}=\{M_{q_1},M_{q_2},\cdots,M_{q_p}\}$, the set of free blocks in a structured canonical form $M^{(\infty)}$ given in the first paragraph of this section. Define a map σ from $\mathscr{F}$ to $\mathbf{Z}$, the set of integers, as follows:

(1) $\sigma(M_{q_k})=mn$ if M_{q_k} is a regularization block with size $m\times n$;

(2) $\sigma(\boldsymbol{M}_{q_k}) = d(m+n-d)$ if $\boldsymbol{M}_{q_k}$ is an edge reduction block $\begin{pmatrix} \mathbf{0} & \boldsymbol{I}_d \\ \mathbf{0} & \mathbf{0} \end{pmatrix}$ with size $m \times n$;

(3) $\sigma(\boldsymbol{M}_{q_k}) = m^2 - \sum_{i=1}^{l} \sum_{h,j=1}^{u} p_{ih} p_{ij} \cdot \min\{m_{ih}, m_{ij}\}$ if $\boldsymbol{M}_{q_k}$ is a Wyer matrix $\boldsymbol{W}$ of size m defined in Lemma 2. It is easy to see that

$$m = \sum_{i=1}^{l} \sum_{j=1}^{u} m_{ij} p_{ij}.$$

Furthermore, we denote

$$\sigma(\boldsymbol{M}) = \sum_{k=1}^{p} \sigma(\boldsymbol{M}_{q_k}).$$

Let Λ be a finite dimensional k-algebra, $(\mathscr{K}, \mathscr{M})$ the linear matrix problem induced from Λ, and $\mathscr{A}$ the Drozd hocs of Λ. Suppose $\Sigma: \mathrm{R}(\mathscr{K}, \mathscr{M}) \to \mathrm{R}(\mathscr{A})$ to be the equivalent functor defined in Lemma 1. Then $\forall \boldsymbol{M} \in \mathrm{R}(\mathscr{K}, \mathscr{M})$ with dimension vector $\underline{\boldsymbol{n}}$ and $\Sigma(\boldsymbol{M}) = \overline{\boldsymbol{M}}$, $\dim(\overline{\boldsymbol{M}}) = \mathscr{M}$-dim $\underline{\boldsymbol{n}}$ since Σ preserves dimensions. Assume that $(\boldsymbol{M}, \boldsymbol{E})$ goes to $(\boldsymbol{M}', \boldsymbol{E}')$ under some reduction, and $\boldsymbol{M}'$ has dimension vector $\underline{\boldsymbol{n}}'$. Suppose $\theta: \mathscr{A} \to \mathscr{A}'$ to be the corresponding functor and $\theta_I^*(\overline{\boldsymbol{M}}' = \overline{\boldsymbol{M}})$. Then

$$\dim \underline{\boldsymbol{n}}' = \dim(\overline{\boldsymbol{M}}') \leqslant \dim(\overline{\boldsymbol{M}}) = \dim \underline{\boldsymbol{n}}.$$

§ 3. Calculation of dimensions

Lemma 3 Assume that a regularization reduces $(\boldsymbol{M}^{(k-1)}, \boldsymbol{E}^{(k-1)})$ to $(\boldsymbol{M}^{(k)}, \boldsymbol{E}^{(k)})$ and the obtained free block is $\boldsymbol{M}_{q_k} = \boldsymbol{M}_{lr}^{(k-1)'} = \varnothing$. Then

$$\mathscr{M}\text{-dim } \underline{\boldsymbol{n}}^{(k)} = \mathscr{M}\text{-dim } \underline{\boldsymbol{n}}^{(k-1)} \text{ and } \mathscr{K}\text{-dim } \underline{\boldsymbol{n}}^{(k)} = \mathscr{K}\text{-dim } \underline{\boldsymbol{n}}^{(k-1)} - \sigma(\boldsymbol{M}_{q_k}).$$

Furthermore, $\dim_k(\overline{\boldsymbol{M}}^{(k)}) = \mathscr{M}$-dim $\underline{\boldsymbol{n}}^{(k)}$, where $\theta_{kl}^*(\overline{\boldsymbol{M}}^{(k)}) = \overline{\boldsymbol{M}}$ defined in the last paragraph of sec. 2.

Proof It is clear that $\mathscr{M}$-dim $\underline{\boldsymbol{n}}^{(k)} = \mathscr{M}$-dim $\underline{\boldsymbol{n}}^{(k-1)}$.

Since the transformation algebra $\boldsymbol{E}^{(k-1)}$ can be regarded as the solution space of the matrix equations:

$$\sum_{\mathscr{T} \ni i < j \in \mathscr{J}} c_{ij}^{(l)} \boldsymbol{S}_{ij}^{(k-1)} = \mathbf{0}, 1 \leqslant l \leqslant q_{\mathscr{T}\mathscr{J}}, \mathscr{T}, \mathscr{J} \in T^{(k-1)} / \sim^{(k-1)}.$$

The only change from $\boldsymbol{E}^{(k-1)}$ to $\boldsymbol{E}^{(k)}$ by regularization is adding the following matrix equation to the above equation system:

$$a_{l1}\boldsymbol{S}_{1r}^{(k-1)}+\cdots+a_{l,r-1}\boldsymbol{S}_{r-1,r}^{(k-1)}-\boldsymbol{S}_{l,l+1}^{(k-1)}a_{l+1,r}-\cdots-\boldsymbol{S}_{lt}^{(k-1)}a_{tr}=\boldsymbol{0},$$

which is equivalent to the system of mn equations that are linearly independent. Thus

$$\mathscr{K}\text{-dim } \underline{\boldsymbol{n}}^{(k)}=\mathscr{K}\text{-dim } \underline{\boldsymbol{n}}^{(k-1)}-mn=\mathscr{K}\text{-dim } \underline{\boldsymbol{n}}^{(k-1)}-\sigma(\boldsymbol{M}_{q_k}) \text{ (see sec. 2)}.$$

Lemma 4 Assume that an edge reduction reduces $(\boldsymbol{M}^{(k-1)},\boldsymbol{E}^{(k-1)})$ to $(\boldsymbol{M}^{(k)},\boldsymbol{E}^{(k)})$ and the obtained free block is $\boldsymbol{M}_{q_k}=\boldsymbol{M}_{lr}^{(k-1)'}=\begin{pmatrix}\boldsymbol{0} & \boldsymbol{I}_d\\ \boldsymbol{0} & \boldsymbol{0}\end{pmatrix}$. Then

$\mathscr{M}\text{-dim } \underline{\boldsymbol{n}}^{(k)}=\mathscr{M}\text{-dim } \underline{\boldsymbol{n}}^{(k-1)}-\tau(\boldsymbol{M}_{q_k})$ and $\mathscr{K}\text{-dim } \underline{\boldsymbol{n}}^{(k)}=\mathscr{K}\text{-dim } \underline{\boldsymbol{n}}^{(k-1)}-\sigma(\boldsymbol{M}_{q_k})$. Furthermore, $\dim_k(\overline{\boldsymbol{M}}^{(k)})=\mathscr{M}\text{-dim } \underline{\boldsymbol{n}}^{(k)}$, where $\theta_{kI}^{*}(\overline{\boldsymbol{M}}^{(k)})=\overline{\boldsymbol{M}}$, as defined in the sec. 2.

Proof Assume that $l\in\mathscr{T}^{(k-1)}$, $r\in\mathscr{J}^{(k-1)}$, $\mathscr{T}^{(k-1)}$, $\mathscr{J}^{(k-1)}$ in $T^{(k-1)}/\sim^{(k-1)}$.

The only change in the diagonal blocks from $\boldsymbol{E}^{(k-1)}$ to $\boldsymbol{E}^{(k)}$ is

$$\boldsymbol{S}_{ll}=\begin{pmatrix}\boldsymbol{P}_1 & \boldsymbol{P}_2\\ \boldsymbol{P}_3 & \boldsymbol{P}_4\end{pmatrix},\quad \boldsymbol{S}_{rr}=\begin{pmatrix}\boldsymbol{Q}_1 & \boldsymbol{Q}_2\\ \boldsymbol{Q}_3 & \boldsymbol{Q}_4\end{pmatrix},$$

with $\boldsymbol{P}_1=\boldsymbol{Q}_4$ and $\boldsymbol{P}_3=\boldsymbol{0}$, $\boldsymbol{Q}_3=\boldsymbol{0}$. Therefore (m), a component of vector $\underline{\boldsymbol{n}}^{(k-1)}$, is changed into $(d,m-d)$, and n, another component of $\underline{\boldsymbol{n}}^{(k-1)}$, into $(n-d,d)$. Thus $m+n$ decreases to

$$d+(m-d)+(n-d)=m+n-d=\tau(\boldsymbol{M}_{q_k})$$

and

$$\mathscr{M}\text{-dim } \underline{\boldsymbol{n}}^{(k)}=\mathscr{M}\text{ dim } \underline{\boldsymbol{n}}^{(k-1)}-\tau(\boldsymbol{M}_{q_k}).$$

Moreover, the added matrix equations $\boldsymbol{P}_1=\boldsymbol{Q}_4$ and $\boldsymbol{P}_3=\boldsymbol{0}$, $\boldsymbol{Q}_3=\boldsymbol{0}$ are equivalent to $d^2+(m-d)d+(n-d)d=d(m+n-d)$ many of linearly independent equations; thus $\mathscr{K}\text{ dim } \underline{\boldsymbol{n}}^{(k)}=\mathscr{K}\text{-dim } \underline{\boldsymbol{n}}^{(k-1)}-\sigma(\boldsymbol{M}_{q_k})$.

Lemma 5 Assume that a loop reduction reduces $(\boldsymbol{M}^{(k-1)},\boldsymbol{E}^{(k-1)})$ to $(\boldsymbol{M}^{(k)},\boldsymbol{E}^{(k)})$ and the obtained free block is $\boldsymbol{M}_{q_k}=\boldsymbol{M}_{lr}^{(k-1)'}=\boldsymbol{W}$, a Weyr matrix defined in Lemma 2. Then $\mathscr{M}\text{-dim } \underline{\boldsymbol{n}}^{(k)}=\mathscr{M}\text{-dim } \underline{\boldsymbol{n}}^{(k-1)}-\tau(\boldsymbol{M}_{q_k})$ and $\mathscr{K}\text{-dim } \underline{\boldsymbol{n}}^{(k)}=\mathscr{K}\text{-dim } \underline{\boldsymbol{n}}^{(k-1)}-\sigma(\boldsymbol{M}_{q_k})$. Furthermore, $\dim_k(\overline{\boldsymbol{M}}^{(k)})=\mathscr{M}\text{-dim } \underline{\boldsymbol{n}}^{(k)}$, where $\theta_{kI}^{*}(\overline{\boldsymbol{M}}^{(k)})=\overline{\boldsymbol{M}}$ defined in sec. 2.

Proof The only change in the diagonal blocks from $\boldsymbol{E}^{(k-1)}$ to $\boldsymbol{E}^{(k)}$

is S_{ll} with $l\sim^{(k-1)}r\in\mathscr{T}^{(k-1)}$, which is changed into $\mathbf{S}'_{ll}$. For convenience, we consider $\mathbf{V}$ (see Lemma 2) instead of Weyr matrix $\boldsymbol{W}$. It is easy to see that

$$\tau(\boldsymbol{M}_{q_k})=\sum_{i=1}^{l}\sum_{j=1}^{u_i}p_{ij}(m_{ij}-1).$$

On the other hand, since m is the size of $\mathbf{V}$, then (m), a component of vector $\underline{\boldsymbol{n}}^{(k-1)}$, changes into

$$(\cdots,\underbrace{p_{i1},\cdots,p_{i1}}_{m_{i1}},\cdots,\underbrace{p_{iu_i},\cdots,p_{iu_i}}_{m_{iu_i}},\cdots),$$

where the same footnotes are equivalent in $T^{(k)}$. Then m decreases to $\sum_{i=1}^{l}\sum_{j=1}^{u_i}p_{ij}$, But

$$m=\sum_{i=1}^{l}\sum_{j=1}^{u_i}p_{ij}m_{ij}.$$

Thus $\mathscr{M}$-dim $\underline{\boldsymbol{n}}^{(k)}=\mathscr{M}$-dim $\underline{\boldsymbol{n}}^{(k-1)}-\tau(\boldsymbol{M}_{q_k})$.

Moreover, the diagonal block in $\boldsymbol{E}^{(k)}$ which commutes with $\mathbf{V}$ equals

$$\boldsymbol{Z}=\bigoplus_{i=1}^{l}\boldsymbol{Z}_i,\boldsymbol{Z}_i=(\boldsymbol{Z}_{hj}),h,j=1,2,\cdots,u_i,$$

$$\boldsymbol{Z}_{hj}=\begin{pmatrix}\boldsymbol{Z}_{hj}^1 & \boldsymbol{Z}_{hj}^2 & \cdots & \boldsymbol{Z}_{hj}^{m_{ij}}\\ \mathbf{0} & \boldsymbol{Z}_{hj}^1 & \cdots & \boldsymbol{Z}_{hj}^{m_{ij}-1}\\ \vdots & \vdots & & \vdots\\ \mathbf{0} & \mathbf{0} & \cdots & \boldsymbol{Z}_{hj}^1\\ \mathbf{0} & \mathbf{0} & \cdots & \mathbf{0}\\ \vdots & \vdots & & \vdots\\ \mathbf{0} & \mathbf{0} & \cdots & \mathbf{0}\end{pmatrix}_{m_{ih}\times m_{ij}};$$

when $h\leqslant j$, with $\boldsymbol{Z}_{hj}^g$ being $p_{ih}\times p_{ij}$ matrices, $g=1,2,\cdots,m_{ij}$,

$$\boldsymbol{Z}_{hj}=\begin{pmatrix}\mathbf{0} & \cdots & \mathbf{0} & \boldsymbol{Z}_{hj}^1 & \boldsymbol{Z}_{hj}^2 & \cdots & \boldsymbol{Z}_{hj}^{m_{ij}}\\ \mathbf{0} & \cdots & \mathbf{0} & \mathbf{0} & \boldsymbol{Z}_{hj}^1 & \cdots & \boldsymbol{Z}_{hj}^{m_{ij}-1}\\ \vdots & & \vdots & \vdots & \vdots & & \vdots\\ \mathbf{0} & \cdots & \mathbf{0} & \mathbf{0} & \mathbf{0} & \cdots & \boldsymbol{Z}_{hj}^1\end{pmatrix}_{m_{ih}\times m_{ij}};$$

when $h>j$, with $\boldsymbol{Z}_{hj}^g$ being $p_{ih}\times p_{ij}$ matrices, $g=1,2,\cdots,m_{ij}$. Then the number of new added equations from $\boldsymbol{E}^{(k-1)}$ to $\boldsymbol{E}^{(k)}$ is

$$m^2 - \sum_{i=1}^{q} \sum_{h,j=1}^{a_i} p_{ih} p_{ij} \cdot \min\{m_{ih}, m_{ij}\}.$$

Thus $\mathscr{K}$-dim $\underline{\boldsymbol{n}}^{(k)} = \mathscr{K}$-dim $\underline{\boldsymbol{n}}^{(k-1)} - \sigma(\boldsymbol{M}_{q_k})$.

The proof of Theorem A The equivalence of (2) and (3) is obvious. Since $M \simeq_{\mathrm{E}} \boldsymbol{M}^{(\infty)}$ and $\boldsymbol{E}^{(\infty)}$ is the endomorphism ring of $\boldsymbol{M}^{(\infty)}$, the equivalence of (1) and (2) holds[2]. Now assume that we have the reduction sequence from $(\boldsymbol{M}, \boldsymbol{E})$ to $(\boldsymbol{M}^{(\infty)}, \boldsymbol{E}^{(\infty)})$. Using induction and Lemmas 3～5, the equivalence of (3) and (4) follows. The equivalence of (4) and (5) follows from Remark 2 and the last paragraph of sec. 2.

The proof of Theorem B Since $\dim_k \mathrm{End}_{R(\mathscr{K},\mathscr{M})}(\boldsymbol{M}) = \dim_k \boldsymbol{E}^{(\infty)}$, the theorem follows from Lemmas 3～5.

Acknowledgements

This work was supported by the National Natural Science Foundation of China (Grant No. 19831070) and the Doctoral Foundation of Institution of Higher Education.

References

[1] Crawley-Boevey W W. Matrix problems and Drozd's theorem, Topics in Algebra, 1990, 26: 199-222.

[2] Sergeichuk V V. Canonical matrices for linear matrix problems. Linear Algebra and Its Applications, 2000, 317(1-3): 53-102.

[3] Zeng X Y, Zhang Y B. A correspondence of almost split sequences between some categories. Comm. Alg., 2001, 29(2): 557-582.

Communications in Algebra, 2001, 29(2): 557-582.

几乎可列序列在若干相关范畴中的对应①

A Correspondence of Almost Split Sequences Between Some Categories

Abstract In the present paper we describe Ext-projectives, Ext-injectives and almost split sequences of category $P_1(\Lambda)$. We show that all almost split sequences of category $P_1(\Lambda)$ are exact. And we also give a correspondence of projectives, injectives, and almost split sequences between categories mod Λ, $P_1(\Lambda)$ and some other categories.

Keywords Almost split sequence; Ext-projective; Ext-injective.

§ 1. The category $P_1(\Lambda)$ and some homological lemmas

Throughout the paper we write the composition of morphisms from left to right.

1.1 Let k be any perfect field and Λ be a basic finite dimensional k-algebra. We use Λ-mod to denote the category of finite dimensional left Λ-modules and Λ-proj to denote the full subcategory of Λ-mod consisting of projective modules. Let $P(\Lambda)=\{(P_1 \xrightarrow{\alpha} P_0) \mid P_1, P_0 \in \Lambda\text{-proj}, \alpha \in \mathrm{Hom}_\Lambda(P_1, P_0)\}$ and the morphisms (f_1, f_0) form $(P_1 \xrightarrow{\alpha} P_0)$

① Received: 1998-12; Revised: 2000-01, 2000-08.
本文与曾祥勇合作.

to $(Q_1 \xrightarrow{\beta} Q_0)$ with $f_1 \in \mathrm{Hom}_\Lambda(P_1, Q_1)$, $f_0 \in \mathrm{Hom}_\Lambda(P_0, Q_0)$ are defined by the commutative fig. 1. 1.

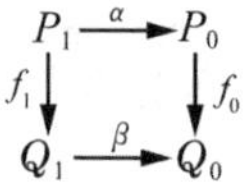

Fig. 1. 1

There are a full subcategory

$$P_1(\Lambda)=\{(P_1 \xrightarrow{\alpha} P_0) \mid \mathrm{im}\alpha \subseteq \mathrm{rad}P_0\} \subseteq P(\Lambda),$$

and a full subcategory

$$P_2(\Lambda)=\{(P_1 \xrightarrow{\alpha} P_0) \mid \ker\alpha \subseteq \mathrm{rad}P_1\} \subseteq P_1(\Lambda).$$

It is well known that the functor Cok: $P_2(\Lambda) \to \Lambda$-mod is a representation equivalence, that is dense, full, and reflects isomorphisms (not necessarily faithful).

Lemma Let (fig. 1. 2)

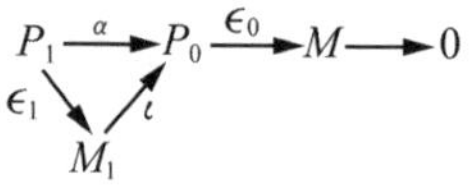

Fig. 1. 2

be a projective presentation, where $M_1 = \ker \epsilon_0$ and $\alpha = \epsilon_1 \cdot \iota$.

(1) $(P_1 \xrightarrow{\alpha} P_0) \in P_1(\Lambda)$ if and only if ϵ_0 is a projective cover of M; if and only if $(P_1 \xrightarrow{\alpha} P_0)$ has no nonzero direct summand $(P \xrightarrow{\mathrm{iso}} P)$.

(2) Assume that $(P_1 \xrightarrow{\alpha} P_0) \in P_1(\Lambda)$, then $(P_1 \xrightarrow{\alpha} P_0) \in P_2(\Lambda)$ if and only if ϵ_1 is a projective cover of M_1; if and only if $(P_1 \xrightarrow{\alpha} P_0)$ has no nonzero direct summand $(P \xrightarrow{0} 0)$.

(3) $(P_1 \xrightarrow{\alpha} P_0) \in P_2(\Lambda)$ if and only if $P_1 \xrightarrow{\alpha} P_0 \xrightarrow{\epsilon_0} M \longrightarrow 0$ is a minimal projective presentation.

Proof We only prove (1), (2) is similar, (3) is an obvious consequence. If $(P_1 \xrightarrow{\alpha} P_0) \in P_1(\Lambda)$, then $\mathrm{im}\alpha \subseteq \mathrm{rad}P_0$. But $\ker \epsilon_0 =$

$\mathrm{im}\alpha$, so

$$\mathrm{top}M = M/\mathrm{rad}M \simeq (P_0/\ker\epsilon_0)/(\mathrm{rad}P_0/\ker\epsilon_0) \simeq P_0/\mathrm{rad}P_0 = \mathrm{top}P_0,$$

thus $\epsilon_0: P_0 \to M$ is a projective cover of M.

Conversely, if ϵ_0 is a projective cover of M, then $\ker\epsilon_0 \subseteq \mathrm{rad}P_0$ since the preimage of $\mathrm{rad}M_0$ under ϵ_0 is $\mathrm{rad}P_0$. Therefore $\mathrm{im}\alpha = \ker\epsilon_0 \subseteq \mathrm{rad}P_0$. □

1.2 Denote by mod-Λ the category of finite dimensional right Λ-modules, which is equivalent to Λ^{op}-mod, where Λ^{op} is the opposite algebra of Λ. Denote by proj-Λ the full subcategory of mod-Λ consisting of projectve modules. Then proj-$\Lambda \simeq \Lambda^{op}$-proj. Denote by $P'(\Lambda)$ the category $P(\Lambda^{op})$.

Lemma (1) The functor $* = \mathrm{Hom}_\Lambda(-,\Lambda): \Lambda\text{-proj} \to \text{proj-}A$ is a duality.

(2) $* = \mathrm{Hom}_\Lambda(-,\Lambda): P(\Lambda) \to P'(\Lambda)$ by sending $(P_1 \xrightarrow{\alpha} P_0)$ to $(P_0^* \xrightarrow{\alpha^*} P_1^*)$ is a duality.

(3) $*|_{P_1(\Lambda)}: P_1(\Lambda) \to P_1'(\Lambda)$ is a duality.

Proof (1)[Chapter 2.4.3, Ref. (1)].

(2) If $P \in \Lambda$-proj, then $P = \Lambda e$ for some idempotent $e \in \Lambda$. Take any $f \in \mathrm{Hom}_\Lambda(P,\Lambda)$, suppose $(e)f = a \in \Lambda$, then $(be)f = b\cdot(e)f = ba$, $\forall b \in \Lambda$, is a right multiplication map by a, furthermore $a = (e)f = (ee)f = e\cdot(e)f = ea \in e\Lambda$. We may identify $\mathrm{Hom}_\Lambda(P,\Lambda)$ with $e\Lambda$ via identifying f with $\cdot(ea)$ the right multiplication by ea. On the other hand $\forall P_1, P_0 \in \Lambda$-proj, $P_1 = \Lambda e_1$, $P_0 = \Lambda e_0$, $\forall \alpha \in \mathrm{Hom}_\Lambda(\Lambda e_1, \Lambda e_0)$, suppose $(e_1)\alpha = a \in \Lambda e_0$, then $(be_1)\alpha = b(e_1)\alpha = ba$, $\forall b \in \Lambda$, is a right multiplication map by a, furthermore $a = e_1 a = ae_0$, thus $a \in e_1\Lambda e_0$. It is easy to see that $(P_1 \xrightarrow{\alpha} P_0) = (\Lambda e_1 \xrightarrow{\cdot(e_1ae_0)} \Lambda e_0) \in P_1(\Lambda)$, then $(P_0^* \xrightarrow{\alpha^*} P_1^*) = (e_0\Lambda \xrightarrow{(e_1ae_0)\cdot} e_1\Lambda) \in P_1'(\Lambda)$ with $(e_1ae_0)\cdot$ is a left multiplication map by (e_1ae_0).

(3) $(P \xrightarrow{\cdot(eae)} P) \in P(\Lambda)$ with $P = \Lambda e$ indecomposable and $\cdot(eae)$ invertible if and only if $(P^* \xrightarrow{(eae)\cdot} P^*) \in P'(\Lambda)$ with $P^* = e\Lambda$ indecomposable and (eae). invertible. Thus an object belongs to

$P_1(\Lambda)$ if and only if whose image under $*$ belongs to $P_1'(\Lambda)$. □

1.3 It is known that $P_1(\Lambda)$ is a k-additive category with the idempotents split but not an abelian category.

Definition The sequence $\begin{pmatrix} e_1 \\ e_0 \end{pmatrix}$ (fig. 1.3) in $P_1(\Lambda)$

$$\begin{array}{lccccccccc} (e_1): & 0 & \to & P_1 & \to & W_1 & \to & Q_1 & \to & 0 \\ & & & \downarrow & & \downarrow & & \downarrow & & \\ (e_0): & 0 & \to & P_0 & \to & W_0 & \to & Q_0 & \to & 0 \end{array}$$

Fig. 1.3

is called exact provided both (e_1) and (e_0) are split in Λ-mod.

1.4 Lemma The following diagram (fig. 1.4) is exact and commutative in Λ-mod.

$$\begin{array}{lccccccccc} (e_1): & 0 & \to & P_1 & \xrightarrow{(1,0)} & P_1\oplus Q_1 & \xrightarrow{(1,0)^{\mathrm{T}}} & Q_1 & \to & 0 \\ & & & \downarrow{\scriptstyle \epsilon_{M_1}} & & \downarrow{\scriptstyle \epsilon_{E_1}} & & \downarrow{\scriptstyle \epsilon_{N_1}} & & \\ & 0 & \to & M_1 & \longrightarrow & E_1 & \longrightarrow & N_1 & \to & 0 \\ & & & \downarrow{\scriptstyle \iota_{M_1}} & & \downarrow{\scriptstyle \iota_{E_1}} & & \downarrow{\scriptstyle \iota_{N_1}} & & \\ (e_0): & 0 & \to & P_0 & \xrightarrow{(1,0)} & P_0\oplus Q_0 & \xrightarrow{(1,0)^{\mathrm{T}}} & Q_0 & \to & 0 \\ & & & \downarrow{\scriptstyle \epsilon_{M}} & & \downarrow{\scriptstyle \epsilon_{E}} & & \downarrow{\scriptstyle \epsilon_{N}} & & \\ (e): & 0 & \to & M & \xrightarrow{f} & E & \xrightarrow{g} & N & \to & 0 \\ & & & \downarrow & & \downarrow & & \downarrow & & \\ & & & 0 & & 0 & & 0 & & \end{array}$$

Fig. 1.4

Suppose that ϵ_M and ϵ_N are projective covers of M and N respectively, the sequence (e_0) is constructed by Horseshoe Lemma. $M_1=\ker(\epsilon_M)$, $N_1=\ker(\epsilon_N)$, $E_1=\ker(\epsilon_E)$, ι_{M_1}, ι_{E_1}, ι_{N_1} are embeddings. ϵ_{M_1} and ϵ_{N_1} are projective covers of M_1 and N_1 respectively, the sequence (e_1) is constructed also by Horseshoe Lemma.

Then the sequence $\begin{pmatrix} e_1 \\ e_0 \end{pmatrix}$ (fig. 1.5)

$$\begin{array}{lccccccccc} (e_1): & 0 & \to & P_1 & \xrightarrow{(1,0)} & P_1\oplus Q_1 & \xrightarrow{(1,0)^{\mathrm{T}}} & Q_1 & \to & 0 \\ & & & \downarrow{\scriptstyle \alpha_{M}} & & \downarrow{\scriptstyle \alpha_{E}} & & \downarrow{\scriptstyle \alpha_{N}} & & \\ (e_0): & 0 & \to & P_0 & \xrightarrow{(1,0)} & P_0\oplus Q_0 & \xrightarrow{(1,0)^{\mathrm{T}}} & Q_0 & \to & 0 \end{array}$$

Fig. 1.5

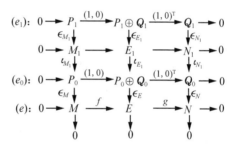

with $\alpha_M=\epsilon_M\cdot\iota_M$, $\alpha_E=\epsilon_E\cdot\iota_E$, $\alpha_N=\epsilon_N\cdot\iota_N$ is in $P(\Lambda)$.

(1) $\begin{pmatrix} e_1 \\ e_0 \end{pmatrix}\in P_1(\Lambda)$ if and only if ϵ_E is a projective cover of E; if and only if $\mathrm{top}E\simeq\mathrm{top}M\oplus\mathrm{top}N$.

(2) If $\begin{pmatrix} e_1 \\ e_0 \end{pmatrix}\in P_1(\Lambda)$, then $\begin{pmatrix} e_1 \\ e_0 \end{pmatrix}\in P_2(\Lambda)$ if and only if ϵ_{E_1} is a projective cover of E_1; if and only if $\mathrm{top}E_1\simeq\mathrm{top}M_1\oplus\mathrm{top}N_1$.

Proof $\epsilon_E: P_0\oplus Q_0\to E$ is a projective cover if and only if

$$\mathrm{top}E\simeq\mathrm{top}(P_0\oplus Q_0)\simeq\mathrm{top}P_0\oplus\mathrm{top}Q_0\simeq\mathrm{top}M\oplus\mathrm{top}N.$$

So we obtain (1), (2) can be obtained similarly. □

1.5 Lemma Consider the diagrams in 1.4, suppose that (e) is an almost split sequence irt Λ-mod, then $\begin{pmatrix} e_1 \\ e_0 \end{pmatrix}\in P_1(\Lambda)$ if and only if M is nonsimple.

Proof If M is not simple, then

$$0\to\mathrm{top}M\to\mathrm{top}E\to\mathrm{top}N\to 0$$

is exact by [Proposition 4.5. (g), Ref. (2)]. If M is simple, then $f(M)\subseteq\mathrm{rad}E$, and $\mathrm{top}E\simeq\mathrm{top}N$. We obtain the desired conclusion by 1.4 Lemma (1). □

1.6 Lemma Consider the diagrams in 1.4, suppose that (e) is an almost split sequence in Λ-mod with M nonsimple.

(1) If M is not projective, and that the middle term of the almost split sequence starting at $(M)\mathrm{TrD}$ does not contain any nonzero injective direct summands. Then $\begin{pmatrix} e_1 \\ e_0 \end{pmatrix}\in P_2(\Lambda)$.

(2) If M is projective. Then $\begin{pmatrix} e_1 \\ e_0 \end{pmatrix}\in P_2(\Lambda)$.

Proof $\begin{pmatrix} e_1 \\ e_0 \end{pmatrix}\in P_1(\Lambda)$ by 1.5.

(1) From (e) we obtain an almost split sequence

$$0\to(M)\mathrm{TrD}\to(E)\mathrm{TrD}\to(N)\mathrm{TrD}\simeq M\to 0$$

by assumption and a remark after [Proposition 2. 2. (c), Ref. (3)]. And by [Proposition 4. 5. (c) Ref. (4)], we know that

$$0 \to \mathrm{soc}((M)\mathrm{TrD}) \to \mathrm{soc}((E)\mathrm{TrD}) \to \mathrm{soc}((N)\mathrm{TrD}) \to 0$$

is exact. But

$$\mathrm{soc}((M)\mathrm{TrD}) \simeq \mathrm{top}M_1, \mathrm{soc}((E)\mathrm{TrD}) \simeq \mathrm{top}E_1, \mathrm{soc}((N)\mathrm{TrD}) \simeq \mathrm{top}N_1$$

by [Proposition 5. 3. (a), Ref. (4)].

Therefore $\mathrm{top}E_1 = \mathrm{top}M_1 \oplus \mathrm{top}N_1$, $\begin{pmatrix} e_1 \\ e_0 \end{pmatrix} \in P_2(\Lambda)$.

(2) If M is projective, we only need to show that an almost split sequence starting at a nonsimple, noninjective, indecomposable projective module P induces an exact sequence in $P_2(\Lambda)$ (fig. 1. 6).

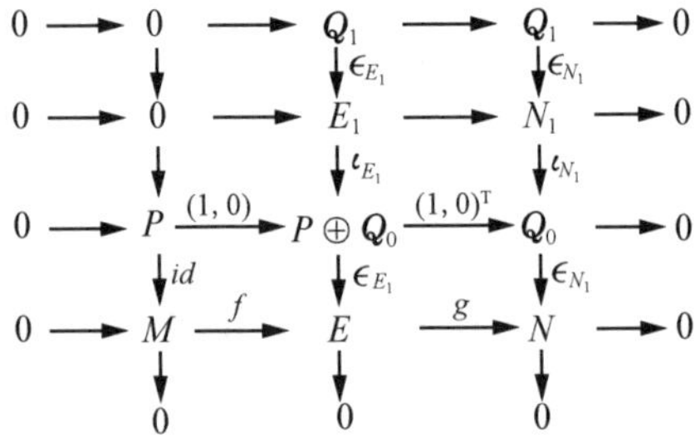

Fig. 1. 6

Since $\epsilon_{N_1}: Q_1 \to N_1$ is a projective cover, and $E_1 \simeq N_1$ by Snake Lemma. $\epsilon_{E_1}: Q_1 \to E_1$ is also a projective cover, thus $(Q_1 \xrightarrow{\alpha_E} P \oplus Q_0) \in P_2(\Lambda)$ by 1. 4 Lemma. □

1. 7 Assume that Λ is basic and connected. Denoted by J the Jacobson radical of Λ, $S = \Lambda / J$, $e(1), e(2), \cdots, e(t)$ the complete set of orthogonal primitive idempotents of Λ. Denote by $S(i), P(i), I(i), i = 1, 2, \cdots, t$, the representatives of iso-classes of simple, projective, injective Λ-modules respectively. Let Q be the ordinary quiver of Λ, then $\Lambda = kQ^* / I$, with Q^* the opposite quiver of Q, and I an admissible ideal [Ref. (4)]. We write the composition of paths of Q from left to right. Suppose that $a_{ij}^{(w)}: i \to j$ is an arrow in the ordinary quiver, define a map $(a_{ij}^{(w)}) \cdot : P(j) \to P(i)$ given by the left multiplication by $a_{ij}^{(w)}$. Let

$m_{ij}=\dim_k \mathrm{Ext}_{\Lambda}^1(S(j),S(i))=$ the number of arrows from i to j in Q.

Lemma Consider the diagrams in 1.4, suppose that $M=S(i)$ is a noninjective simple module, $N=N(i)\simeq(S(i))DTr$, (e) is an almost split sequence.

(1) $S(i)$ has a minimal projective presentation:

$$\bigoplus_{j=1}^{t}P(j)^{m_{ij}}\xrightarrow{\alpha_{S(i)}}P(i)\xrightarrow{\epsilon_{S(i)}}S(i)\to 0,$$

with $\sigma_{S(i)}=(A_{i1},A_{i2},\cdots,A_{it})$ and

$$A_{ij}=(a_{ij}^{(1)},a_{ij}^{(2)},\cdots,a_{ij}^{(m_{ij})}):P(j)^{m_{ij}}\to P(i).$$

(2) $N(i)$ has a minimal projective presentation:

$$P(i)\xrightarrow{\alpha_{N(i)}}\bigoplus_{j=1}^{t}P(j)^{m_{ji}}\xrightarrow{\epsilon_{N(i)}}N(i)\to 0,$$

with $\alpha_{N(i)}=(B_{1i},B_{2i},\cdots,B_{ti})^{\mathrm{T}}$ and

$$B_{ji}=(a_{ji}^{(1)},a_{ji}^{(2)},\cdots,a_{ji}^{(m_{ji})})^{\mathrm{T}}:P(i)\to P(j)^{m_{ji}}.$$

Proof (1) $P(i)\xrightarrow{\epsilon_{S(i)}}S(i)\to 0$ is a projective cover of $S(i)$, and

$$\mathrm{top}(\ker(\epsilon_{S(i)}))=\bigoplus_{j=1}^{t}S(j)^{m_{ij}}$$

the minimality of the presentation follows from 1.1.

(2) Assume that $Q_1\xrightarrow{\alpha_N}Q_0\to N\to 0$ is a minimal projective presentation of N, then

$$0\to\mathrm{Hom}(N,\Lambda)\to\mathrm{Hom}(Q_0,\Lambda)\xrightarrow{\alpha_N^*}\mathrm{Hom}(Q_1,\Lambda)\to Coka_N^*\to 0$$

is an exact sequence in mod-Λ and the latter 3 terms form a minimal projective presentation of $Coka_N^*$ [p. 248, Ref. (2)]. On the other hand $S(i)=(N)\mathrm{TrD}=(Coka_N^*)\mathrm{D}$, therefore $Coka_N^*=(S(i))D$ is a simple right Λ-module at i. By the similar argument to that in (1), $\mathrm{Hom}(Q_1,\Lambda)\simeq e_i\Lambda$ and $\mathrm{Hom}(Q_0,\Lambda)\simeq\bigoplus_{j=1}^{t}(e_j\Lambda)^{m_{ji}}$. Finally $Q_1=P(i)$ and $Q_0=\bigoplus_{j=1}^{t}P(j)^{m_{ji}}$, and $\alpha_{N(i)}$ has the given expression above. □

1.8 Lemma If $S(i)$ is an injective simple module, then

(1) $S(i)$ has a minimal projective presentation in $P_1(\Lambda)$ with the same expression as in 1.7.(1);

(2) i is a source in Q and $m_{ji}=0$ for all $1\leqslant j\leqslant t$. □

§ 2. Projectives, injectives and almost split sequences in $P_1(\Lambda)$

2. 1 Definition Let Λ be any basic connected finite dimensional k-algebra. The following $2t$ objects in $P_1(\Lambda)$ are called Ext-projectives:

$$(0 \xrightarrow{0} P(i)), i=1,2,\cdots,t \quad : \left(P(i) \xrightarrow{\alpha_{N(i)}} \bigoplus_{j=1}^{t} P(j)^{m_{ji}}\right), i=1,2,\cdots,t.$$

defined in 1. 7. Lemma (2) and 1. 8. Lemma (2).

The following $2t$ objects in $P_1(\Lambda)$ are called Ext-injectives:

$$(P(i) \xrightarrow{0} 0), i=1,2,\cdots,t \quad : \left(\bigoplus_{j=1}^{t} P(j)^{m_{ji}} \xrightarrow{\alpha_{S(i)}} P(i)\right), i=1,2,\cdots,t.$$

defined in 1. 7. Lemma (1) and 1. 8. Lemma (1).

An explination of Ext-projective (resp. Ext-injective) will be given in 3. 8.

2. 2 Definition A sequence (fig. 2. 1)

$$\begin{array}{ccccc} P_1 & \xrightarrow{f_1} & W_1 & \xrightarrow{g_1} & Q_1 \\ {\scriptstyle\alpha}\downarrow & & \downarrow{\scriptstyle\gamma} & & \downarrow{\scriptstyle\beta} \\ P_0 & \xrightarrow{f_0} & W_0 & \xrightarrow{g_0} & Q_0 \end{array}$$

Fig. 2. 1

in $P_1(\Lambda)$ is called an almost split sequence provided

(a) both $(P_1 \xrightarrow{\alpha} P_0)$ and $(Q_1 \xrightarrow{\beta} Q_0)$ are indecomposable;

(b) $\begin{pmatrix} f_1 \\ f_0 \end{pmatrix} \cdot \begin{pmatrix} g_1 \\ g_0 \end{pmatrix} = \begin{pmatrix} 0 \\ 0 \end{pmatrix}$;

(c) $\begin{pmatrix} f_1 \\ f_0 \end{pmatrix}$ is a left minimal almost split map and $\begin{pmatrix} g_1 \\ g_0 \end{pmatrix}$ is a right minimal almost split map [p. 454, Ref. (5)].

2. 3 Lemma Consider the diagrams in 1. 4, suppose that (e) is an almost split sequence with M nonsimple, then $\begin{pmatrix} e_1 \\ e_0 \end{pmatrix}$ is an almost split sequence in $P_1(\Lambda)$.

Proof (1) Since $(P_1 \xrightarrow{\alpha_M} P_0)$ and $(Q_1 \xrightarrow{\alpha_N} Q_0)$ belong to $P_2(\Lambda)$, whose images M and N under functor Cok are indecomposable in Λ-mod, therefore both $(P_1 \xrightarrow{\alpha_M} P_0)$ and $(Q_1 \xrightarrow{\alpha_N} Q_0)$ are indecomposable.

(2) $(1,0) \cdot (0,1)^{\mathrm{T}} = \mathbf{0}$.

(3) We prove that $\begin{pmatrix} f_1 \\ f_0 \end{pmatrix}$ is a left minimal almost split map, $\begin{pmatrix} g_1 \\ g_0 \end{pmatrix}$ is a right minimal almost split map can be proved dually. Where $f_1 = (1, 0)$, $f_0 = (1,0)$ over f, and $g_1 = (0,1)^{\mathrm{T}}$, $g_0 = (0,1)^{\mathrm{T}}$ over g given in 1.4.

First (f_1, f_0) is not a split monomorphism. Otherwise $(P_1 \xrightarrow{\alpha} P_0)$ is a direct summand of $(P_1 \oplus Q_1 \xrightarrow{\alpha_E} P_0 \oplus Q_0) \simeq (U_1 \xrightarrow{\alpha'} U_0) \oplus (P \xrightarrow{0} 0)$ with $(U_1 \xrightarrow{\alpha'} U_0)$ a minimal projective presentation of E. Therefore $(P_1 \xrightarrow{\alpha} P_0)$ is a direct summand of $(U_1 \xrightarrow{\alpha'} U_0)$ and M is a direct summand of E under functor Cok, a contradiction.

Secondly (f_1, f_0) is a left almost split map. Take any indecomposable object $(W_1 \xrightarrow{\beta} W_0) \in P_2(\Lambda)$ denote by $L = Cok\beta$. And take any morphism (h_1, h_0) from $(P_1 \xrightarrow{\alpha} P_0)$ to $(W_1 \xrightarrow{\beta} W_0)$ in $P_1(\Lambda)$ which is not a split monomorphism, (see the picture below). Then (h_1, h_0) induces an Λ-map $h : M \to L$ which is not a split monomorphism, since both $(P_1 \xrightarrow{\alpha} P_0)$ and $(W_1 \xrightarrow{\beta} W_0)$ are in $P_2(\Lambda)$, and functor Cok reflects isomorphisms (fig. 2.2).

$$
\begin{array}{ccccccc}
P_1 & \xrightarrow{\alpha} & P_0 & \xrightarrow{\epsilon_M} & M & \longrightarrow & 0 \\
{\scriptstyle h_1}\downarrow & & {\scriptstyle h_0}\downarrow & & {\scriptstyle h}\downarrow & & \\
W_1 & \xrightarrow{\beta} & W_1 & \xrightarrow{\epsilon_L} & L & \longrightarrow & 0
\end{array}
$$

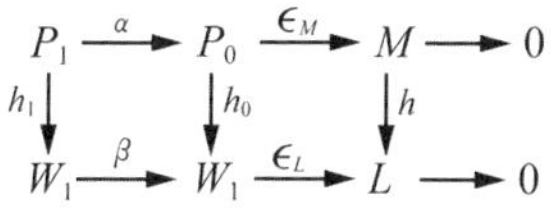

Fig. 2. 2

We obtain a commutative diagram with exact rows in Λ-mod since (e) is an almost split sequence (fig. 2.3).

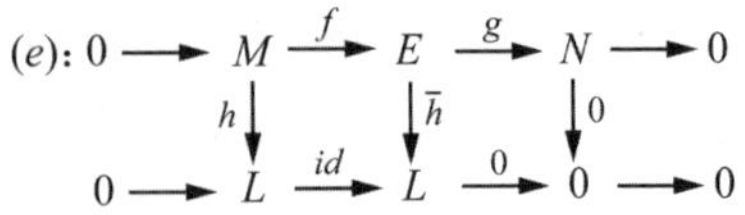

Fig. 2. 3

By a theorem in homological algebra [6. 24. Lamma, Ref. (6)], there exists some chain maps between projective resolutions with exact rows in the category of complexes of mod-Λ (fig. 2. 4).

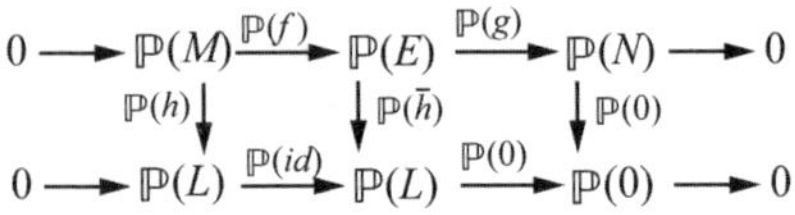

Fig. 2. 4

Where the first two maps of ℙ(h) being (h_1, h_0), and those of ℙ(0) being (0,0). Denote by $(\bar{h}_1, \bar{h}_0)$ the first two maps of ℙ$(\bar{h})$, then

$$(h_1, h_0) = (f_1, f_0) \cdot (\bar{h}_1, \bar{h}_0).$$

On the other hand take any indecomposable $(P \xrightarrow{0} 0) \in P_1(\Lambda) \backslash P_2(\Lambda)$, and (h_1, h_0) is a morphism in $P_1(\Lambda)$ which is not a split monomorphism, (see the fig. 2. 5). Let $\bar{h}_1 = (h_1, 0)$ and $\bar{h}_0 = (0,0)$, then $f_1 \cdot \bar{h}_1 = (1,0) \cdot (h_1, 0)^{\mathrm{T}} = h_1$ and $f_0 \cdot \bar{h}_0 = 0$.

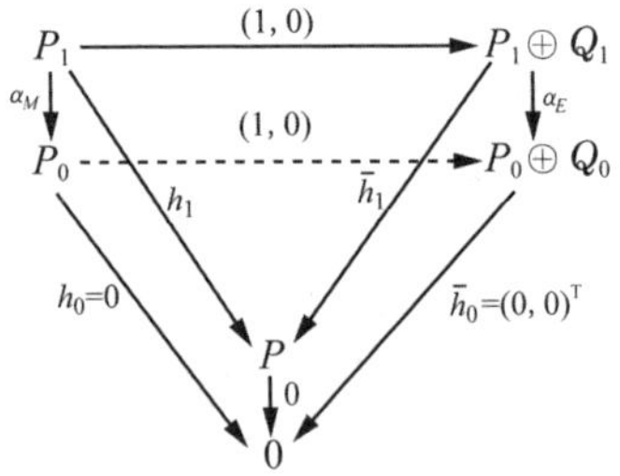

Fig. 2. 5

Summaring up the above two cases, (f_1, f_0) is a left almost split map in $P_1(\Lambda)$.

Thirdly (f_1, f_0) is a left minimal map. If we are given a morphism

(h_1, h_0) from$(P_1 \oplus Q_1 \xrightarrow{\alpha_E} P_0 \oplus Q_0)$ to $(P_1 \oplus Q_1 \xrightarrow{\alpha_E} P_0 \oplus Q_0)$ in $P_1(\Lambda)$ with $(f_1, f_0) \cdot (h_1, h_0) = (f_1, f_0)$. Let $h: E \to E$ be the induced map of (h_1, h_0). Then $f \cdot h = f$ and h is an isomorphism, since (e) is an almost split sequence, let $l: N \to N$ be induced by h, then l is also an isomorphism by Five Lemma. Thus we have the following commutative diagram with the exact rows in Λ-mod (fig. 2. 6):

$$
\begin{array}{ccccccccc}
0 & \longrightarrow & P_0 & \xrightarrow{f_0} & P_0 \oplus Q_0 & \xrightarrow{g_0} & Q_0 & \longrightarrow & 0 \\
 & & \downarrow{\scriptstyle \epsilon_M} & & \downarrow{\scriptstyle \epsilon_E} & & \downarrow{\scriptstyle \epsilon_N} & & \\
0 & \longrightarrow & M & \xrightarrow{f} & E & \xrightarrow{g} & N & \longrightarrow & 0 \\
 & & \downarrow{\scriptstyle id} & & \downarrow{\scriptstyle h} & & \downarrow{\scriptstyle l} & & \\
0 & \longrightarrow & M & \xrightarrow{f} & E & \xrightarrow{g} & N & \longrightarrow & 0 \\
 & & \uparrow{\scriptstyle \epsilon_M} & & \uparrow{\scriptstyle \epsilon_E} & & \uparrow{\scriptstyle \epsilon_N} & & \\
0 & \longrightarrow & P_0 & \xrightarrow{f_0} & P_0 \oplus Q_0 & \xrightarrow{g_0} & Q_0 & \longrightarrow & 0
\end{array}
$$

Fig. 2. 6

There are two maps $id: P_0 \to P_0$ with $\epsilon_M \cdot id_M = id_{P_0} \cdot \epsilon_M$, and $h_0: P_0 \oplus Q_0 \to P_0 \oplus Q_0$ with $\epsilon_E \cdot h = h_0 \cdot \epsilon_E$. Since h is an induced map. Define $l_0: Q_0 \to Q_0$ given by sending $(x)g_0$ to $(x)(h_0 \cdot g_0)$, $\forall x \in P_0 \oplus Q_0$, that is $g_0 \cdot l_0 = h_0 \cdot g_0$. Then $l_0 \cdot \epsilon_N = \epsilon_N \cdot l$. In fact

$g_0 \cdot l_0 \cdot \epsilon_N = h_0 \cdot g_0 \cdot \epsilon_N = h_0 \cdot \epsilon_E \cdot g = \epsilon_E \cdot h \cdot g = \epsilon_E \cdot g \cdot l = g_0 \cdot \epsilon_N \cdot l$.

We obtain $l_0 \cdot \epsilon_N = \epsilon_N \cdot l$, since g_0 is surjcctive. Furthermore l_0 is an isomorphism, since $\epsilon_N: Q_0 \to N$ is a projective cover, and l is an isomorphism.

An isomorphism $l_1: Q_1 \to Q_1$ induced from h_1, can be obtained similarly based on the commutative diagram (fig. 2. 7):

$$
\begin{array}{ccccccccc}
0 & \longrightarrow & M_1 & \xrightarrow{f_1} & E_1 & \xrightarrow{g_1} & N_1 & \longrightarrow & 0 \\
 & & \downarrow{\scriptstyle id} & & \downarrow{\scriptstyle h_{0|E_1}} & & \downarrow{\scriptstyle l_{0|N_1}} & & \\
0 & \longrightarrow & M_1 & \xrightarrow{f_1} & E_1 & \xrightarrow{g_1} & N_1 & \longrightarrow & 0
\end{array}
$$

Fig. 2. 7

Thus by Five Lemma h_0 is an isomorphism, since id and l_0 are isomorphisms, h_1 is an isomorphism, since id and h_1 are isomorphisms, i. e. (h_1, h_0) is an isomorphism in $P_1(\Lambda)$. Therefore (f_1, f_0) is a left minimal map. □

2.4 Lemma Let $I(i)$ be an indecomposable injective module.

(1) If $I(i)$ is nonsimple, then there exists an almost split sequence in $P_1(\Lambda)$ (fig. 2.8):

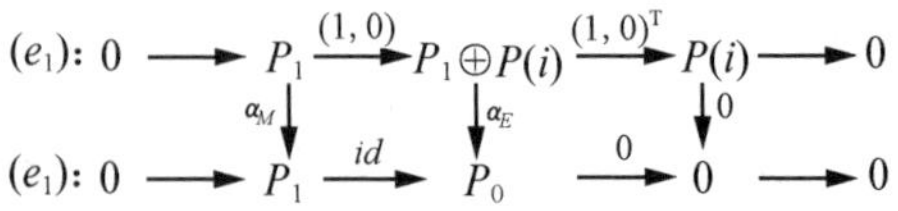

Fig. 2.8

with $M=I(i)$ and $E=I(i)/S(i)$, $P_1 \xrightarrow{\alpha_M} P_0 \xrightarrow{\epsilon_M} M \to 0$ a minimal projective presentation of M.

(2) If $I(i)$ is simple, then $(P(i)\to 0)$ is Ext-projective in $P_1(\Lambda)$.

Proof (1) Consider the following exact and commutative diagram (fig. 2.9):

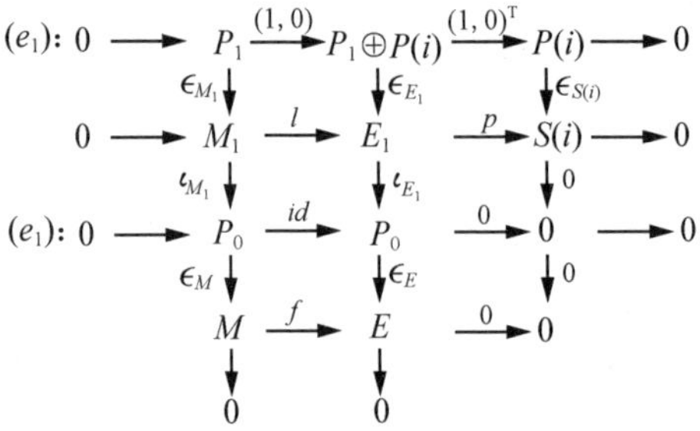

Fig. 2.9

where $M_1=\ker(\epsilon_M)$, ι_{M_1} is the natural embedding, ϵ_{M_1} is the projective cover of M_1 and $\epsilon_{M_1}\cdot\iota_{M_1}=\alpha_M$. Since $\mathrm{top}M=\mathrm{top}I(i)\simeq\mathrm{top}I(i)/S(i)=\mathrm{top}E$, $\epsilon_E:P_0\to E$ is a projective cover of E. Let $E_1=\ker(\epsilon_E)$, ι_{E_1} the natural embedding. Because

$P_0/E_1=E=M/S(i)=(P_0/M_1)/((S(i))\epsilon_M^{-1})/M_1=P_0/(S(i))\epsilon_M^{-1}$,

we have $(S(i))\epsilon_M^{-1}=E_1$. Thus $E_1/M_1=(S(i))\epsilon_M^{-1}/(0)\epsilon_M^{-1}\simeq S(i)$. Denote by $l:M_1\to E_1$ the embedding, $p:E_1\to S(i)$ the projection. Let $\epsilon_{S(i)}:P(i)\to S(i)$ be the projective cover of $S(i)$, ϵ_{E_1} be constructed by Horseshoe Lemma, and $\alpha_E=\epsilon_{E_1}\cdot\iota_{E_1}$.

Denote by $f_1=(1,0)$, $f_0=id$.

We first prove that $\begin{pmatrix} f_1 \\ f_0 \end{pmatrix}$ in 2. 4. (1) is a left almost split map. We claim that (f_1,f_0) is not a split monomorphism, otherwise $(P_1 \xrightarrow{\alpha_M} P_0)$ is a direct summand of $(P_1 \oplus P(i) \xrightarrow{\alpha_E} P_0) \simeq (U_1 \xrightarrow{\alpha'} U_0) \oplus (P \to 0)$ with $(U_1 \xrightarrow{\alpha'} U_0) \in P_2(\Lambda)$ a minimal projective presentation of E. Thus $(P_1 \xrightarrow{\alpha_M} P_0)$ is a direct summand of $(U_1 \xrightarrow{\alpha'} U_0)$, and M is a direct summand of E under functor Cok, a contradiction.

$\forall (W_1 \xrightarrow{\beta} W_0) \in P_2(\Lambda)$, let $L=Cok\beta$, $L_1=\ker\beta$, ι_{L_1} the natural embedding and ϵ_{L_1} the projection (fig. 2. 10):

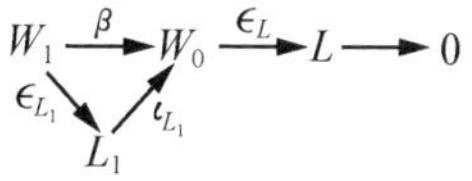

Fig. 2. 10

Suppose that $(h_1,h_0):(P_1 \xrightarrow{\alpha_M} P_0) \to (W_1 \xrightarrow{\beta} W_0)$ is not a split monomorphism, $h:M \to L$ is induced by (h_1,h_0) , then h is not a split monomorphism, since functor *Cok* reflects isomorphisms, (see fig. 2. 11). There exists a map $\bar{h}:E \to L$ with $f \cdot \bar{h}=h$, since f is a left almost split map. Let $\bar{h}_0=h_0$ which is a lifting of $\bar{h}$, in fact

$$\bar{h}_0 \cdot \epsilon_L = h_0 \cdot \epsilon_L = \epsilon_M \cdot h = \epsilon_M \cdot f \cdot \bar{h} = \epsilon_E \cdot \bar{h}.$$

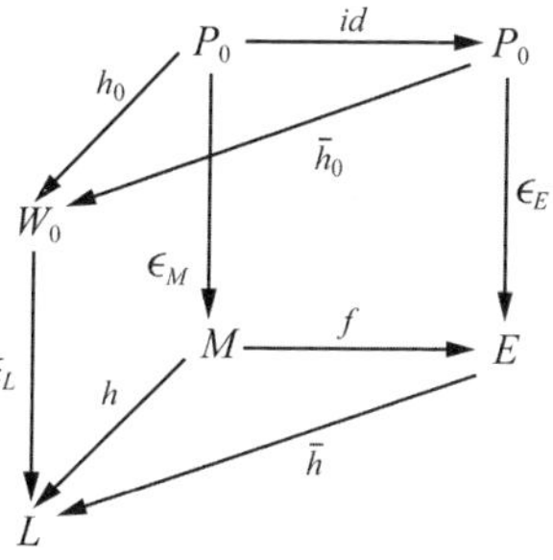

Fig. 2. 11

Let $h_0':M_1\to L_1$ be the restriction of h_0 on M_1, and $\bar{h}_0':E_1\to L_1$ the restriction of $\bar{h}_0$ on E_1, then [6.23, Ref. (6)] gives the following commutative diagram with exact rows (fig. 2.12):

$$\begin{array}{ccccccccc} 0 & \longrightarrow & M_1 & \xrightarrow{l} & E_1 & \xrightarrow{p} & S(i) & \longrightarrow & 0 \\ & & \downarrow{\scriptstyle h_0'} & & \downarrow{\scriptstyle \bar{h}_0'} & & \downarrow{\scriptstyle 0} & & \\ 0 & \longrightarrow & L_1 & \xrightarrow{id} & L_1 & \xrightarrow{o} & 0 & \longrightarrow & 0 \end{array}$$

Fig. 2. 12

Using [6.24. Lemma, Ref. (6)] again we obtain a morphism $\bar{h}_1:P_1\oplus P(i)\to W_1$ such that $f_1\cdot\bar{h}_1=h_1$ in Λ-mod. And then a morphism $(\bar{h}_1,\bar{h}_0):(P_1\oplus P(i)\xrightarrow{\alpha_E}P_0)\to(W_1\xrightarrow{\beta}W_0)$ with $\begin{pmatrix}f_1\\f_0\end{pmatrix}\cdot\begin{pmatrix}\bar{h}_1\\\bar{h}_0\end{pmatrix}=\begin{pmatrix}h_1\\h_0\end{pmatrix}$.

$\forall(P\xrightarrow{0}0)\in P_1(\Lambda)\backslash P_2(\Lambda)$ and a morphism $(h_1,h_0):(P_1\xrightarrow{\alpha_M}P_0)\to(P\xrightarrow{0}0)$, there is a morphism $(\bar{h}_1,\bar{h}_0):(P_1\oplus P(i)\xrightarrow{\alpha_E}P_0)\to(P\xrightarrow{0}0)$, such that $\begin{pmatrix}f_1\\f_0\end{pmatrix}\cdot\begin{pmatrix}\bar{h}_1\\\bar{h}_0\end{pmatrix}=\begin{pmatrix}h_1\\h_0\end{pmatrix}$. In fact, $h_0=0,\bar{h}_0=0,\bar{h}_1=(h_1,0)$, Summarsing up the above two cases $\begin{pmatrix}f_1\\f_0\end{pmatrix}$ is a left almost split map.

We prove secondly that there is an exact almost split sequence in $P_1'(\Lambda)$, over an almost split sequence in mod-Λ ending at $Cok\alpha^*$.

Since $P_1\xrightarrow{\alpha}P_0\to I(i)\to 0$ is a minimal projective presentation with $\alpha=\alpha_M$, then $P_0^*\xrightarrow{\alpha^*}P_1^*\to Cok\alpha^*\to 0$ is a minimal projective presentation of $Cok\alpha^*$ by [p. 248, Ref. (2)]. Now we caculate $(Cok\alpha^*)$ TrD$\in$ mod-Λ. Since $(P_0^{**}\xrightarrow{\alpha^{**}}P_1^{**})=(P_0\xrightarrow{\alpha}P_1)$ by 1.2, $(Cok\alpha^*)\mathrm{Tr}=I(i)$ and $(Cok\alpha^*)\mathrm{TrD}=(I(i))\mathrm{D}=e(i)\Lambda\in$ proj-Λ, since $I(i)=(e(i)\Lambda)\mathrm{D}$ with the primitive idempotent $e(i)$. Suppose that $0\to e(i)\Lambda\to E'\to Cok\alpha^*\to 0$ is an almost split sequence ending at $Cok\alpha^*$ in mod-Λ. By 2.3 there exists an almost split sequence in $P_1'(\Lambda)$ constructed by Horseshoe

Lemma, since $e(i)\Lambda$ is not a simple module in mod-Λ (fig. 2. 13):

$$\begin{array}{lccccccccc}
(e_1'): & 0 & \longrightarrow & 0 & \xrightarrow{0} & P_0^* & \xrightarrow{id} & P_0^* & \longrightarrow & 0 \\
 & & & {\scriptstyle 0}\downarrow & & \downarrow{\scriptstyle \alpha_{E'}} & & \downarrow{\scriptstyle \alpha^*} & & \\
(e_0'): & 0 & \longrightarrow & e(i)\Lambda & \xrightarrow{(0,1)} & P_1^*\oplus e(i)\Lambda & \xrightarrow{(1,0)^{\mathrm{T}}} & P_1^* & \longrightarrow & 0 \\
 & & & {\scriptstyle id}\downarrow & & \downarrow & & \downarrow & & \\
(e'): & 0 & \longrightarrow & e(i)\Lambda & \longrightarrow & E' & \longrightarrow & \mathrm{Cok}\alpha^* & \longrightarrow & 0 \\
 & & & \downarrow & & \downarrow & & \downarrow & & \\
 & & & 0 & & 0 & & 0 & &
\end{array}$$

Fig. 2. 13

Thirdly we prove that $\begin{pmatrix} e_1' \\ e_0' \end{pmatrix}^*$ is an almost split sequence in $P_1(\Lambda)$ and isomorphic to $\begin{pmatrix} e_1 \\ e_0 \end{pmatrix}$ in 2. 4. (1). $\begin{pmatrix} e_1' \\ e_0' \end{pmatrix}^*$ (fig. 2. 14):

$$\begin{array}{ccccccccc}
0 & \longrightarrow & P_1 & \xrightarrow{(1,0)} & P_1\oplus P(i) & \xrightarrow{(0,1)^{\mathrm{T}}} & P(i) & \longrightarrow & 0 \\
 & & \downarrow{\scriptstyle \alpha} & & \downarrow{\scriptstyle \alpha_{E'}^*} & & \downarrow{\scriptstyle 0} & & \\
0 & \longrightarrow & P_0 & \xrightarrow{id} & P_0 & \xrightarrow{0} & 0 & \longrightarrow & 0
\end{array}$$

Fig. 2. 14

(1) $(P_1 \xrightarrow{\alpha} P_0)$ and $(P(i) \xrightarrow{0} 0)$ are indecomposable in $P_1(\Lambda)$;

(2) $(1,0)\cdot(0,1)^{\mathrm{T}}=0$ and $id\cdot 0=0$;

(3) Denote by $f'_1=(1,0)$, $f'_0=id$. Since $\begin{pmatrix} id \\ (1,0)^{\mathrm{T}} \end{pmatrix}$ is a right minimal almost split map in $P'_1(\Lambda)$ and a morphism $\begin{pmatrix} h_1 \\ h_0 \end{pmatrix}$ is not a split monomorphism in $P_1(\Lambda)$ if and only if $\begin{pmatrix} h_0^* \\ h_1^* \end{pmatrix}$ is not a split epimorphism in $P'_1(\Lambda)$, $\begin{pmatrix} f'_1 \\ f'_0 \end{pmatrix}$ is a left minimal almost split map follows from the duality given in 1. 2. Similarly $\begin{pmatrix} (0,1)^{\mathrm{T}} \\ 0 \end{pmatrix}$ is a right minimal almost split map. Thus $\begin{pmatrix} e_1' \\ e_0' \end{pmatrix}^*$ is an almost split sequence.

Now we consider the following commutative diagram (fig. 2. 15).

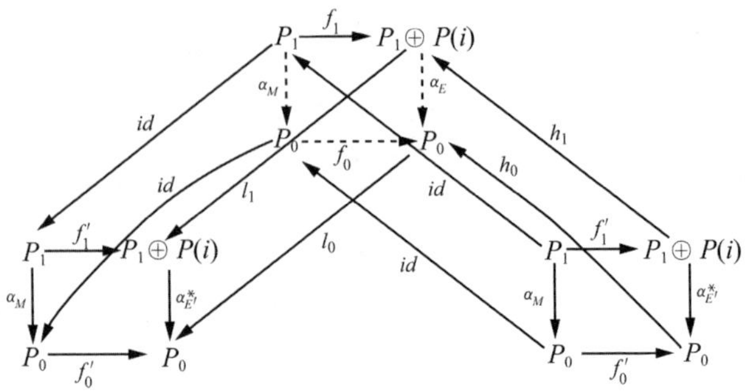

Fig. 2. 15

Since $\begin{pmatrix} f_1 \\ f_0 \end{pmatrix}$ is not a split monomorphism there exists a morphism $\begin{pmatrix} h_1 \\ h_0 \end{pmatrix}$ such that $\begin{pmatrix} f_1' \\ f_0' \end{pmatrix} \cdot \begin{pmatrix} h_1 \\ h_0 \end{pmatrix} = \begin{pmatrix} f_1 \\ f_0 \end{pmatrix}$, since $\begin{pmatrix} e_1' \\ e_0' \end{pmatrix}^*$ is an almost split sequence. On the other hand there exists a morphism $\begin{pmatrix} l_1 \\ l_0 \end{pmatrix}$ such that $\begin{pmatrix} f_1 \\ f_0 \end{pmatrix} \cdot \begin{pmatrix} l_1 \\ l_0 \end{pmatrix} = \begin{pmatrix} f_1' \\ f_0' \end{pmatrix}$, since $\begin{pmatrix} f_1 \\ f_0 \end{pmatrix}$ in 2. 4. (1) is a left almost split map. Thus $\begin{pmatrix} f_1' \\ f_0' \end{pmatrix} \cdot \begin{pmatrix} h_1 \\ h_0 \end{pmatrix} \cdot \begin{pmatrix} l_1 \\ l_0 \end{pmatrix} = \begin{pmatrix} f_1' \\ f_0' \end{pmatrix}$ and $\begin{pmatrix} h_1 \\ h_0 \end{pmatrix} \cdot \begin{pmatrix} l_1 \\ l_0 \end{pmatrix}$ is an isomorphism in $P_1(\Lambda)$, since $\begin{pmatrix} e_1' \\ e_0' \end{pmatrix}^*$ is an almost split sequence. Therefore both $\begin{pmatrix} h_1 \\ h_0 \end{pmatrix}$ and $\begin{pmatrix} l_1 \\ l_0 \end{pmatrix}$ are isomorphisms. Let $(\bar{h}_1, \bar{h}_0)$ from $(P(i) \xrightarrow{0} 0)$ to $(P(i) \xrightarrow{0} 0)$ be induced by (h_1, h_0), then $(\bar{h}_1, \bar{h}_0)$ is also an isomorphism by Five Lemma. Finally we obtain an isomorphism from $\begin{pmatrix} e_1' \\ e_0' \end{pmatrix}^*$ to $\begin{pmatrix} e_1 \\ e_0 \end{pmatrix}$.

Thus $\begin{pmatrix} e_1 \\ e_0 \end{pmatrix}$ in 2. 4. (1) is an almost split sequence in $P_1(\Lambda)$.

(2) By 1. 8 and 2. 1 directly. □

2.5 Theorem Let Λ be a finite dimensional algebra over a perfect field k. Then

(1) $\forall(Q_1 \xrightarrow{\beta} Q_0) \in P_1(\Lambda)$ which is indecomposable non-Ext-projective, then there is an almost split sequence in $P_1(\Lambda)$ ending at $(Q_1 \to Q_0)$;

(2) $\forall(P_1 \xrightarrow{\alpha} P_0) \in P_1(\Lambda)$ which is indecomposable non-Ext-injective, then there is an almost split sequence in $P_1(\Lambda)$ starting at $(P_1 \xrightarrow{\alpha} P_0)$;

(3) All almost split sequences in $P_1(\Lambda)$ are exact.

Proof (1) By 2.3 and 2.4.

(2) By (1) and the duality given in 1.2.(3).

2.6 Theorem Let Λ be a finite dimensional algebra over a perfect field k. Let

$\mathcal{B}$ = {almost split sequences in Λ-mod starting at nonsimple modules},

$\mathcal{C}$ = {almost split sequences in $P_1(\Lambda)$ ending at objects in $P_2(\Lambda)$},

$\mathcal{B}_0$ = {almost split sequences in $\mathcal{B}$ satisfy the assumptions in 1.6},

$\mathcal{C}_0$ = {almost split sequences in $P_2(\Lambda)$}.

Then (1) the sets $\mathcal{B}$ and $\mathcal{C}$ are 1-1 correspondent up to isomorphisms given in 2.3. (2) the sets $\mathcal{B}_0$ and $\mathcal{C}_0$ are 1-1 correspondent up to isomorphisms under functor Cok.

Proof By 2.3 and 1.6. □

2.7 An example Let $\Lambda = kQ$, Q is fig. 2.16,

Fig. 2.16

be a quiver of Dynkin type. The AR-quiver of Λ is the fig. 2.17:

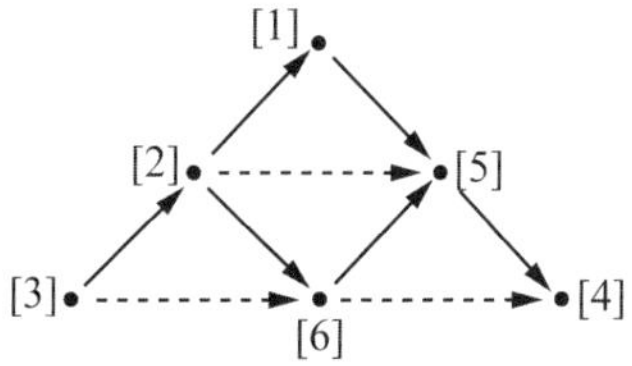

Fig. 2.17

where [1] $P_1=I_3$, with the dimension vector $(1,1,1)$, [2] P_2, $(0,1,1)$, [3] $P_3=S_3$, $(0,0,1)$, [4] $I_1=S_1$, $(1,0,0)$, [5] I_2, $(1,1,0)$, [6] S_2, $(0,1,0)$, with P, I, S being projective, injective, simple modules respectively. The correspondent differential biquiver (i. e. the bocs, see 3. 5 below) is fig. 2. 18.

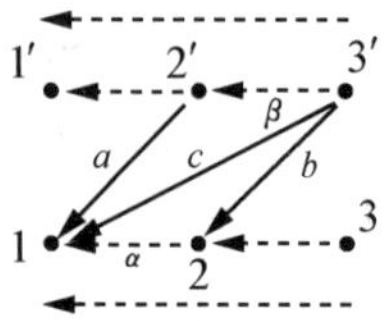

Fig. 2. 18

(1) $\delta=0$,　(2) $\delta=0$,　(3) $\delta=\beta a-b\alpha$.

The AR-quiver of $P_1(\Lambda)$ is fig. 2. 19.

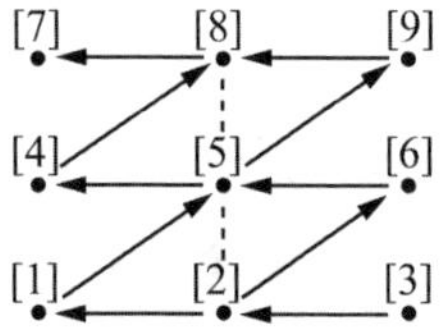

Fig. 2. 19

Ext-projective objects: [1] $(0\to P_1)$, [2] $(0\to P_2)$;

Ext-projective and Ext-injective objects:

[3] $(0\to P_3)$, [4] $(P_2\to P_1)$, [6] $(P_3\to P_2)$, [7] $(P_1\to 0)$;

Ext-injective objects: [8] $(P_2\to 0)$, [9] $(P_3\to 0)$

AR-sequences in $P_1(\Lambda)$ are:

$$0\to[2]\to[1]\oplus[6]\to[5]\to 0,$$

$$0\to[5]\to[4]\oplus[9]\to[8]\to 0.$$

§ 3.　Almost split sequences in some categories

Throughout this section we assume that k is an algebraically closed field.

3. 1　Some categories

Let Λ be a basic, connected, finite dimensional k-algebra. We will give the corresponding relation of projectives, injectives, and almost split sequences in the following categories.

(1) Λ-mod.

(2) $P_1(\Lambda)$. This is in fact the representation category $R(K,M)$ of a bimodule problem $(K,M,0)$, [Ref. (2), $R(K,M)$ is denoted by M at $(K,M,0)$ there]. Let $K=(\Lambda\text{-proj})\times(\Lambda\text{-proj})$, $M=\mathrm{rad}(\Lambda\text{-proj})$, then M is an K-K-bimodule by restriction via projections of K onto the first and second factors respectively. The objects of $R(K,M)$ consisting of $(P_1 \xrightarrow{\alpha} P_0)$ for any $(P_1, P_0) \in K$ and $\alpha \in \mathrm{rad}_\Lambda(P_1, P_0)$. The morphisms are given by the commutative diagram (fig. 3.1):

$$\begin{array}{ccc} P_1 & \xrightarrow{\alpha} & P_0 \\ {\scriptstyle f_1}\downarrow & & \downarrow{\scriptstyle f_0} \\ Q_1 & \xrightarrow{\beta} & Q_0 \end{array}$$

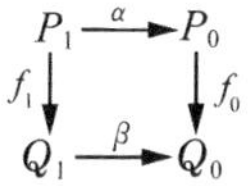

Fig. 3.1

with $(f_1, f_0) \in \mathrm{Hom}_K((P_1,Q_1),(P_0,Q_0))$.

(3) $R(Q^b,\delta)$, the representation category of a differential biquiver (Q^b,δ) constructed from $P_1(\Lambda)$, see 3.5 below.

(4) $R(\mathcal{A})$, the representation category of the Drozd bocs $\mathcal{A}=(A, V)$ with layer $L=(A';\omega;a_1,a_2,\cdots,a_n;v_1,v_2,\cdots,v_m)$ which is induced from Λ [6.1, Ref. (8)]. $R(\mathcal{A})$ is equivalent to $R(Q^b,\delta)$ by [4.3, Ref. (7)]. On the other hand there is a dimension vector preserving equivalence functor $\Xi:R(\mathcal{A})\to P_1(\Lambda)$ by [6.1, Ref. (8)].

(5) $p(^*V,A)$, the category of relatively projective modules defined in [Ref. (9), Ref. (12)], where $^*V=\mathrm{Hom}_A(V,A)$ is an algebra. Ext-projective and Ext injective modules have been defined in $p(^*V,A)$ and it has been proved that if M is a non-Ext-injective module in $p(^*V,A)$ then there is an almost split sequence $0\to M\to E\to N\to 0$ starting at M; if N is a non-Ext-projective module in $p(^*V,A)$ then there is an almost split sequence $0\to M\to E\to N\to 0$ ending at N.

3.2 Definitions (1) A morphism $\varphi:G\to H$ in $R(\mathcal{A})$ is called a proper epimorphism (resp. proper monomorphism), if

$$(\varphi)(i,\omega)^*:(G)(i,\omega)^*\to(H)(i,\omega)^*$$

is surjective (resp. injective) in A'-mod.

(2) An object H of $R(\mathcal{A})$ is called properly projective if any proper epimorphism $\varphi:G\to H$ splits. Dually an object F of $R(\mathcal{A})$ is called properly injective if any proper monomorphism $\phi:F\to G$ splits.

(3) A sequence, $F\xrightarrow{\iota}G\xrightarrow{\pi}H$, is called a properly exact sequence, if $0\to(F)(i,\omega)^*\xrightarrow{(\iota)(i,\omega)^*}(G)(i,\omega)^*\xrightarrow{(\pi)(i,\omega)^*}(H)(i,\omega)^*\to 0$ is exact in A'-mod.

(4) A sequence $F\xrightarrow{\iota}G\xrightarrow{\pi}H$ is called an almost split sequence, provided (a) F and H are indecomposable; (b) $\iota\cdot\pi=0$; (c) $\iota:F\to G$ is a left minimal almost split map and $\pi:G\to H$ is a right minimal almost split map [p. 454, Ref. (5)]. In particular an almost split sequence is called a proper almost split sequence if it is properly exact.

3.3 Let Λ, $\mathcal{A}=(A,V)$, $R(\mathcal{A})$, $p({}^*V,A)$ be defined in 3.1. Let $\epsilon:V\to A$, and $\mu:V\to V\otimes V$ be the counite and comultiplication of A-coalgebra V respectively. Then $\ker\epsilon=\overline{V}\simeq\bigoplus_{j=1}^{m}Q_{j\otimes k}P_j$, where Q_j, P_j are projective A-modules.

$R(\mathcal{A})$ is exactly the Kleisli category $(\text{mod }A)_{V\otimes -}$ induced by the comonad $(V\otimes -,\mu,\epsilon)$, which is equivalent to the Kleisli category $(\text{mod})_{{}^*V\otimes -}$ induced by a monad $({}^*V\otimes -,m,e)$. In fact

$$\text{Hom}_A(V\otimes_A G,H)\xrightarrow[\sim]{(\alpha_1)_{G,H}}\text{Hom}_A(G,\text{Hom}_A(V,H))$$

by adjoint functors $V_{\otimes A-}$ and $\text{Hom}_A(V,-)$. On the other hand, since V is a projective left A-module, $\text{Hom}(V,-)\xleftarrow[\sim]{\beta}{}^*V\otimes -$ by [Formula (3.3), Ref. (10)]. Therefore

$$\text{Hom}_A(G,\text{Hom}(V,H))\xrightarrow[\sim]{(\alpha_2)_{G,H}}\text{Hom}_A(G,{}^*V\otimes H)$$

induced by β.

Again by [p. 34, Ref. (9)] there is an equivalence $U:(\text{mod }A)_{{}^*V\otimes -}\to$

$(\mathrm{mod}A)_0^{*V\otimes -}$, from a Kleisli category to a subcategory of Eilenberg-Moore category $(\mathrm{mod}A)^{*V\otimes -}$. Take any G in mod$-A$, $(G)U=({}^*V\otimes G, m_G)$ is a free $*V\otimes_A-$algebra. If $\eta: G\to H$ is a morphism in $(\mathrm{mod}A)_{*V\otimes -}$, then $\eta: G\to {}^*V\otimes A\ H$ is an A-map and

$$(\eta)U: {}^*V\otimes_A G \xrightarrow{id\otimes\eta} {}^*V\otimes_A({}^*V\otimes_A H)\simeq({}^*V\otimes_A {}^*V)\otimes_A H \xrightarrow{m\otimes id} {}^*V\otimes_A H.$$

The full subcategory of $(\mathrm{mod}-A)^{*V\otimes -}$ consisting of the direct summands of objects in $(\mathrm{mod}A)_0^{*V\otimes -}$ is exactly $p({}^*V,A)$, the category of relatively projective modules. We finally obtain a functor

$$\Delta=\alpha_1\circ\alpha_2\circ U: R(\mathcal{A})\to p({}^*V,A),$$

which is equivalent. The functor $\mathrm{Hom}_A(-,A)$ acts on the exact sequence:

$$0\to\bigoplus_{j=1}^{m}Q_j\otimes_k P_j\to V\xrightarrow{\epsilon}A\to 0,$$

obtain an exact sequence

$$0\to A\to{}^*V\to\bigoplus_{j=1}^{m}I_j\otimes_k W_j\to 0$$

by [Lemma 3.3, Ref. (9)], with each I_j left injective A-modules and each W_j right projective modules, which implies that $p({}^*V,A)$ has almost split sequences by [Ref. (9)].

Proposition There is an equivalent functor $\Delta: R(\mathcal{A})\to p({}^*V,A)$. □

3.4 Theorem An indecomposable object H is properly projective (respectively properly injective) in $R(\mathcal{A})$ if and only if $(H)\Delta$ is Ext-projective (respectively Ext-injective) in $p({}^*V,A)$.

Proof Sufficiency. Assume that $(H)\Delta$ is Ext-projective in $p({}^*V,A)$ for an indecomposable object H in $R(\mathcal{A})$. Let $\varphi: V\otimes_A G\to H$ be any proper epimorphism. We show that φ splits.

There exists an object G' in $R(\mathcal{A})$ and an isomorphism $\rho: V\otimes G'\to G$ such that the composition $\phi=\rho\cdot\varphi$ is essentially an A-map, i. e. $\varphi|_{\overline{V}\otimes_A G'}=0$ by [Lemma 3.5, Ref. (11)]. And $\varphi^0=\varphi|_{A\otimes_A G'}$ is surjective as an A-map. Without loss of generality we denote G' still by G.

Consider $(\varphi)\Delta: (G)\Delta\to(H)\Delta$, we claim that $(\varphi)\Delta=id\otimes\varphi^0$:

$${}^*V\otimes G\to{}^*V\otimes H$$

is essentially determined by an A-map. If it is so, we then have an exact sequence $0 \to \ker \varphi^0 \xrightarrow{\iota^0} G \xrightarrow{\varphi^0} H \to 0$ in mod$A$, therefore an exact sequence in $p(^*V, A)$:

$$0 \to *V \otimes_A \ker \varphi^0 \xrightarrow{id \otimes \iota^0} *V \otimes_A G \xrightarrow{id \otimes \varphi^0} *V \otimes_A H \to 0.$$

Since $(H)\Delta = *V \otimes_A H$ is Ext-projective, $id \otimes \varphi^0 = (\varphi)\Delta$ is a split epimorphism in $p(^*V, A)$, so is φ in $R(\mathcal{A})$ since Δ is a category equivalence.

Take any indecomposable object X in A', and any g_X in G_X, then $f \otimes g_X$ is in $*V \otimes_A G$. Assume that $(g_X)\varphi_0 = h_X$, which is in H_X. We show in the following that $(f \otimes g_X)((\varphi)\Delta) = f \otimes h_X$.

$(g_X)((\varphi)\alpha_1): V \to H$ (see 3.3) via

$$(v_j)[(g_X)(\varphi)\alpha_1] = (v_j \otimes g_X)\varphi = 0, \quad j = 1, 2, \cdots, m,$$
$$((1_X)\omega)[(g_X)((\varphi)\alpha_1)] = ((1_X)\omega \otimes g_X)\varphi = (g_X)\varphi^0 = h_X,$$
$$((1_Y)\omega)[(g_X)(\varphi)\alpha_1] = ((1_Y)\omega \otimes g_X)\varphi = 0$$

for any indecomposable Y in A', $Y \neq X$.

Recall counite $\epsilon: V \to A$, we have $\epsilon|_{\bar{V}} = 0$, $((1_X)\omega)\epsilon = 1_X$. For any $\epsilon \otimes h_X \in {}^*V \otimes H$, $(\epsilon \otimes h_X)\beta: V \to H$ via $(v)[(\epsilon \otimes h_X)\beta] = (v)\epsilon \cdot h_X$. But $(v_j)\epsilon \cdot h_X = 0$, $j = 1, 2, \cdots, m$, $((1_X)\omega)\epsilon \cdot h_X = 1_X \cdot h_X = h_X$, $((1_Y)\omega)\epsilon \cdot h_X = 1_Y \cdot h_X = 0$, when $Y \neq X$. Thus $(g_X)(\varphi)\alpha_1 = (\epsilon \otimes h_X)\beta$. On the other hand, $(\varphi)(\alpha_1 \cdot \alpha_2) = ((\varphi)\alpha_1)\beta^{-1}$, we have $(g_X)[\varphi(\alpha_1 \cdot \alpha_2)] = \epsilon \otimes h_X$.

Finally $(f \otimes g_X)(\varphi\Delta) = (f \otimes g_X)\{[(\varphi)(\alpha_1 \cdot \alpha_2) \cdot U]\} = (f \otimes g_X)\{[(\varphi)(\alpha_1 \circ \alpha_2) \cdot U]\} = f \otimes (g_X)[\varphi(\alpha_1 \circ \alpha_2)] = f \otimes (\epsilon \otimes h_X) = (f \otimes \epsilon) \otimes h_X = (f \cdot \epsilon) \otimes h_X = (\mu(1 \otimes \epsilon) f) \otimes h_X = f \otimes h_X$, and $\varphi\Delta = \mathrm{id} \otimes \varphi^0$, we are done.

Necessity. We assume that H is properly projective in $R(\mathcal{A})$. Let $\varphi: P \to H$ be the projective cover in mod-A. Define $\hat{\varphi}: V \otimes_A P \to H$ by $\hat{\varphi}|_{\bar{V} \otimes_A P} = 0$, $\hat{\varphi}|_{A \otimes_A P} = \varphi$, then $\hat{\varphi}$ is a morphism in $R(\mathcal{A})$, and is properly epimorphic. Then $\hat{\varphi}$ splits in $R(\mathcal{A})$, $(\hat{\varphi})\Delta: (P)\Delta \to (H)\Delta$ splits in $p(^*V, A)$. But $(P)\Delta = {}^*V \otimes_A P$ is ${}^*V$-projective, so is $(H)\Delta$ as a direct summand of $(P)\Delta$. Therefore $(H)\Delta$ is Ext-projective in $p(^*V, A)$ by [Proposition 2.3.(d), Ref.(9)].

Similarly we show that H is properly injective in $R(\mathcal{A})$ if $(H)\Delta$ is Extinjective in $p(^*V,A)$. And if H is properly injective in $R(\mathcal{A})$, then $(H)\Delta$ is Extinjective in $p(^*V,A)$ by [proposition 4.1.(c) Ref. (9)]. □

3.5 Now we give the differential biquiver induced from $P_1(\Lambda)$ [3.1, Ref. (7), Ref. (13), Ref. (14), Ref. (15), Ref. (16), Ref. (17)]. Before doing that, we construct first a k-basis of J, the Jacobson radical of Λ. Since J is nilpotent, we take inductively a k-basis of $e(i)(J^p/J^{p+1})e(j)$ for p from large to small represented by some paths of length p of Q^*, the opposite quiver of the ordinary quiver Q. Collect such paths such that if a path ζ of length $(p+1)$ has been chosen, then any subpaths with lenth p of ζ must be chosen as the representatives of a basis of J^p/J^{p+1}. Denote by B a collection of the above paths which is a k-basis of J.

From now on we identify $\mathrm{Hom}_\Lambda(P(i),P(j))$ with $e(i)\Lambda e(j)$ by sending $f:P(i)\to P(j)$ to $(e(i))f=a\in e(i)\Lambda e(j)$ (see 1.2).

Let $P(i)\xrightarrow{f}P(l)\xrightarrow{g}P(j)$ be a composition of the morphisms f and g, then $f\cdot g$ goes to ab where $b=(e(l))g\in e(l)\Lambda e(j)$. This yields an isomorphism from proj-Λ to addΛ with the usual multiplication in Λ.

Remember we write the composition of the paths of Q^* from left to right. Suppose that the multiplication table is the following:

$$\zeta_{il}^{u}\cdot\zeta_{lj}^{v}=\sum_{w}\gamma_{\binom{u}{il}\binom{v}{lj}}^{\binom{w}{il}}\zeta_{ij}^{w}$$

with $\zeta_{il}^{u},\zeta_{lj}^{v}$ be any two basis elements from i to l and l to j respectively, ζ_{ij}^{w} runs over the all basis elements from i to j.

Consider any two objects in $P_1(\Lambda)$ and any morphism between them (fig. 3.2):

$$\begin{array}{ccc} P_1 & \xrightarrow{\alpha} & P_0 \\ {\scriptstyle f}\downarrow & & \downarrow{\scriptstyle g} \\ Q_1 & \xrightarrow{\beta} & Q_0 \end{array}$$

Fig. 3.2

where

$$P_1=\bigoplus_{i=1}^{t}P(i)^{m_i},\quad P_0=\bigoplus_{l=1}^{t}P(l)^{n_l},\quad \alpha=\sum_{r,p,u,i,l}\alpha_{il}^{u}(p,r)\zeta_{il}^{u}(p,r) \qquad (3.1)$$

with $p=1,2,\cdots,m_i$; $r=1,2,\cdots,n_l$, u runs over the basis elements from i to l, $1\leqslant i,j\leqslant t$, $\alpha_{il}^{u}(p,r)\in k$, $\zeta_{il}^{u}(p,r)=\zeta_{il}^{u}$. Let

$$Q_1=\bigoplus_{l=1}^{t}P(l)^{\overline{m}_l},Q_0=\bigoplus_{j=1}^{t}P(j)^{\overline{n}_j},\beta=\sum_{q,r,v,l,j}\beta_{lj}^{v}(r,q)\zeta_{lj}^{v}(r,q),$$

$$f=\sum_{r,p,i}f_{ii}^{0}(p,r)e(i)+\sum_{r,p,u,i,l}f_{il}^{u}(p,r)\zeta_{il}^{u}(p,r),$$

$$g=\sum_{q,r,j}g_{jj}^{0}(r,q)e(j)+\sum_{q,r,v,l,j}g_{lj}^{v}(r,q)\zeta_{lj}^{v}(r,q). \tag{3.2}$$

Then $\alpha\cdot g=f\cdot\beta$, if and only if

$$\sum_{r,w}\alpha_{ij}^{w}(p,r)\zeta_{ij}^{w}(p,r)\,g_{jj}^{0}(q,r)+\sum_{r,u,v,l}\alpha_{il}^{u}(p,r)\zeta_{il}^{u}(p,r)\,g_{lj}^{v}(r,q)\zeta_{lj}^{v}(r,q)$$
$$=\sum_{r,w}f_{ii}^{0}(p,r)\beta_{ij}^{w}(r,q)\zeta_{ij}^{w}(r,q)+\sum_{r,u,v,l}f_{il}^{u}(p,r)\zeta_{il}^{u}(p,r)\beta_{lj}^{v}(r,q)\zeta_{lj}^{v}(r,q) \tag{3.3}$$

Using the multiplication table, it yields

$$\sum_{r}\alpha_{ij}^{w}(p,r)\,g_{jj}^{0}(r,q)-\sum_{r}f_{ii}^{0}(p,r)\beta_{ij}^{w}(r,q)$$
$$=\sum_{r,l}\gamma_{\binom{u}{il}\binom{v}{lj}}^{\binom{w}{ij}}\left(f_{il}^{u}(p,r)\beta_{lj}^{v}(r,q)-\alpha_{il}^{u}(p,r)\,g_{lj}^{v}(r,q)\right) \tag{3.4}$$

for all possible basis elements ζ_{ij}^{w}.

The differential biquiver is constructed as follows. Put on the bottom the vertices $1,2,\cdots,t$ corresponding to the indecomposables objects $(0,P(1)),(0,P(2)),\cdots,(0,P(t))$ in K, and on the top the vertices $1',2',\cdots,t'$ corresponding to the indecomposables objects $(P(1),0),(P(2),0),\cdots,(P(t),0)$ in K.

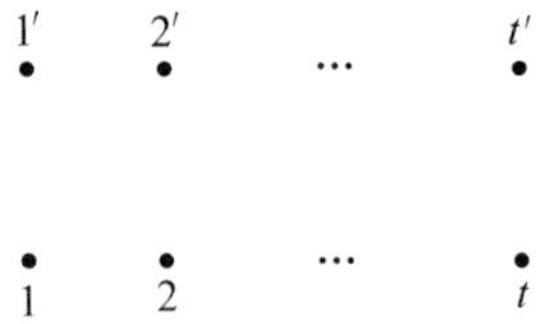

Denote by $(B)D=\{(\zeta_{ij}^{w})^{*}\}$ the dual basis of $(J)D=\mathrm{Hom}_k(J,k)$. Let $M(i,j)=\mathrm{rad}_{\Lambda}(P(i),P(j))$, and $A_{i'j}\subseteq(B)\mathrm{D}$, the k-basis of $(M(i,j))\mathrm{D}$, then draw a set of solid arrows $A_{i'j}$ from i' to j.

Let $J(i,j)=\mathrm{rad}_{\Lambda}(P(i),P(j))$, and $\Phi_{i'j'}\subseteq(B)\mathrm{D}$, $\Phi_{ij}\subseteq(B)\mathrm{D}$ the k-basis of $(J(i,j))\mathrm{D}$, then draw two sets of dotted arrows $\Phi_{i'j'}$ from i' to j' and Φ_{ij} from i to j (fig. 3.3).

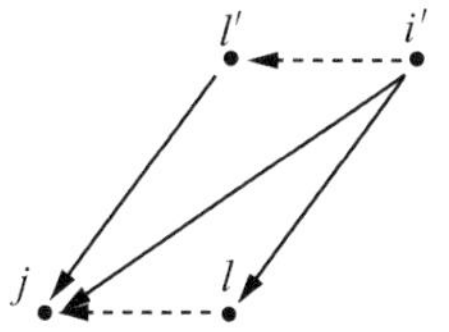

Fig. 3. 3

Define the differentials:

$$(\zeta^{w}_{i'j})^{*}\delta=\sum_{u,v,l}\gamma^{\binom{w}{ij}}_{\binom{u}{il}\binom{v}{lj}}((\zeta^{u'}_{i'l'})^{*}\otimes(\zeta^{v}_{l'j})^{*}-(\zeta^{u}_{i'l})\otimes(\zeta^{v}_{lj})^{*}).$$

Define the comultiplications:

$$(\zeta^{w'}_{i'j'})^{*}\mu=\sum_{u,v,l}\gamma^{\binom{w}{ij}}_{\binom{u}{il}\binom{v}{lj}}(\zeta^{u'}_{i'l'})^{*}\otimes(\zeta^{v'}_{l'j'})^{*},$$

$$(\zeta^{w}_{ij})^{*}\mu=\sum_{u,v,l}\gamma^{\binom{w}{ij}}_{\binom{u}{il}\binom{v}{lj}}(\zeta^{u}_{il})^{*}\otimes(\zeta^{v}_{lj})^{*}.$$

The objects X in $R(Q^{b},\delta)$ are the A-modules, that is

$$X=\{(X_{1'},X_{2'},\cdots,X_{t'};X_{1},X_{2},\cdots,X_{t});X(\zeta^{u}_{i'l})^{*}:X_{i'}\to X_{l},\ \forall\,\zeta^{u}_{il}\in B\}\tag{3.5}$$

with $X_{i'}=k^{m_i}$, $X_l=k^{n_l}$, $1\leqslant i,l\leqslant t$, the k-vector spaces, and $X(\zeta^{u}_{i'l})^{*}$ the k linear maps.

If $Y\in R(Q^{b},\delta)$, with

$$Y=\{(Y_{1'},Y_{2'},\cdots,Y_{t'};Y_{1},Y_{2},\cdots,Y_{t});Y(\zeta^{v}_{l'j})^{*}:Y_{l'}\to Y_{j},\ \forall\,\zeta^{v}_{lj}\in B\},$$

a morphism $\varphi:X\to Y$ is given by linear maps:

$$\{\varphi^{0}_{i'}:X_{i'}\to Y_{i'},\varphi^{0}_{j}:X_{j}\to Y_{j},1\leqslant i,j\leqslant t;$$

$$\varphi(\zeta^{w'}_{i'j'})^{*}:X'_{i}\to Y'_{j},\varphi(\zeta^{w}_{ij})^{*}:X_{i}\to Y_{j},\ \forall\,\zeta^{w}_{ij}\in B\}$$

such that

$$X(\zeta^{w}_{i'j})^{*}\varphi^{0}_{j}-\varphi^{0}_{i'}Y(\zeta^{w}_{i'j})^{*}$$

$$=\sum_{u,v,l}\gamma^{\binom{w}{ij}}_{\binom{u}{il}\binom{v}{lj}}(\varphi(\zeta^{u'}_{i'l'})^{*}Y(\zeta^{v}_{l'j})^{*}-X(\zeta^{u}_{i'l})\varphi(\zeta^{v}_{lj})^{*})\tag{3.6}$$

If $X\xrightarrow{\varphi}Y\xrightarrow{\psi}Z$, are two morphisms in $R(\mathcal{A})$ the composition $\varphi\cdot\psi=\chi$, then $\chi^{0}_{i'}=\varphi^{0}_{i'}\cdot\psi^{0}_{i'}$, $\chi^{0}_{j}=\varphi^{0}_{j}\cdot\psi^{0}_{j}$ and

$$(\zeta^{w'}_{i'j'})^{*}\chi=\varphi^{0}_{i'}\cdot(\zeta^{w'}_{j'i'})^{*}\psi+(\zeta^{w'}_{j'i'})^{*}\psi\cdot\varphi^{0}_{j'}+\sum_{u,v,l}\gamma^{\binom{w}{ij}}_{\binom{u}{il}\binom{v}{lj}}(\zeta^{u'}_{i'l'})^{*}\varphi\cdot(\zeta^{v'}_{l'j'})^{*}\psi,$$

$$(\zeta_{ij}^{w})^{*}\chi=\varphi_{i}^{0}(\zeta_{ij}^{w})^{*}\psi+(\zeta_{ij}^{w})^{*}\psi\varphi_{j}^{0}+\sum_{u,v,l}\gamma_{\binom{u}{il}\binom{v}{lj}}^{\binom{w}{ij}}(\zeta_{il}^{u})^{*}\varphi(\zeta_{jl}^{v})^{*}\psi.$$

$$(3.7)$$

There exists a functor $\Sigma: P_1(\Lambda)\to R(Q^b,\delta)\simeq R(\mathcal{A})$ by sending $(P_1\xrightarrow{\alpha}P_0)$ in formula (3.1) to X in formula (3.5) with

$$X(\zeta_{i'l}^{u})^{*}=(\alpha_{il}^{u}(p,r))_{m_i\times n_l}.$$

$$(3.8)$$

If $(Q_1\xrightarrow{\beta}Q_0)$ is also an object in $P_1(\Lambda)$ which goes to Y under Σ, and (f,g) is a morphism from $(P_1\xrightarrow{\alpha}P_0)$ to $(Q_1\xrightarrow{\beta}Q_0)$, then $(f,g)\Sigma=\varphi$ is given by

$$\varphi_{i'}^{0}=(f_{i'i'}'(p',r'))_{m_i\times\tilde{m}_i};\quad \varphi_{j}^{0}=(g_{jj'}^{0}(r,q))_{n_j\times\tilde{n}_j};$$
$$(\zeta_{i'l}^{u'})^{*}\varphi=(f_{i'l'}^{u'}(p',r'))_{m_i\times\tilde{m}_l};\quad (\zeta_{jl}^{v})^{*}\varphi=(g_{jl}^{v}(r,q))_{n_l\times\tilde{n}_j}. \qquad (3.9)$$

φ is a morphism, since Equations (3.8), (3.9) and (3.4) implies Equation (3.6). $(\varphi\cdot\psi)\Sigma=(\varphi)\Sigma\cdot(\psi)\Sigma$ that is Equation (3.7) holds following from Equation (3.9) and the composition of morphisms in $P_1(\Lambda)$. If $P_1=Q_1$, $P_0=Q_0$ and $f=id$, $g=id$, then $(id)\Sigma=id$ by Equation (3.9). Thus Σ is a functor.

Σ is an inverse functor of $\Xi: R(\mathcal{A})\to P_1(\Lambda)$. The definition of the latter one is the following. If $X\in R(\mathcal{A})$ is given by Equation (3.5) with $X(\zeta_{i'l}^{u})^{*}$ given by Equation (8). Then $(X)\Xi=(P_1\xrightarrow{\alpha}P_0)$ is defined by Equation (1).

If $Y\in R(\mathcal{A})$ is also an object with $(Y)\Xi=(Q_1\xrightarrow{\beta}Q_0)$, and $\varphi:X\to Y$ is a morphism defined by Equation (3.9). Let $f:P_1\to Q_1$ and $g:P_0\to Q_0$ given by Equation (3.2). Then Equation (3.6) implies Equation (3.4) and Equation (3.4) implies Equation (3.3). $(\varphi)\Xi=(f,g):$ $(P_1\xrightarrow{\alpha}P_0)\to(P_0\xrightarrow{\beta}Q_0)$ is a morphism in $P_1(\Lambda)$.

Lemma　If fig. 3.4

$$\begin{CD} 0 @>>> P_1 @>f_1>> P_1\oplus Q_1 @>g_1>> Q_1 @>>> 0 \\ @. @V\alpha_1VV @V\alpha_2VV @V\alpha_3VV @. \\ 0 @>>> P_0 @>f_0>> P_0\oplus Q_0 @>g_0>> Q_0 @>>> 0 \end{CD}$$

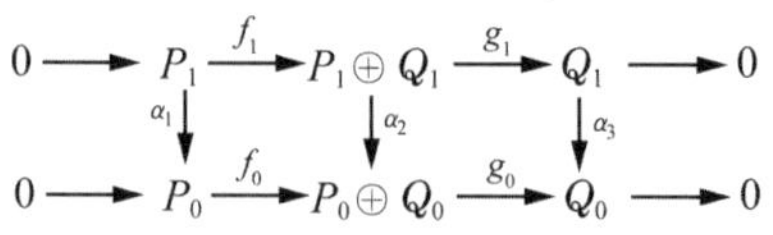

Fig. 3.4

is an exact sequence in $P_1(\Lambda)$. Then the image sequence under $\Sigma: 0 \to F \xrightarrow{\iota} G \xrightarrow{\pi} H \to 0$ in $R(\mathcal{A})$ is properly exact.

Proof Let $\iota = (f_1, f_0)\Xi$ and $\pi = (g_1, g_0)\Xi$. Then

$$\mathrm{rk}(\mathrm{im}(\iota^0_{i'})) = m_i, \quad \mathrm{rk}(\mathrm{im}(\iota^0_l)) = n_l,$$

and

$$\mathrm{rk}(\ker(\pi^0_{l'})) = \widetilde{m}_l, \quad \mathrm{rk}(\ker(\pi^0_j)) = \bar{n}_j \quad \text{for all } 1 \leqslant i, l, j \leqslant t.$$

ι^0 is injective and π^0 is surjective as linear maps. And $\iota^0 \cdot \pi^0 = 0$, that is $\mathrm{im}(\iota^0) \subseteq \ker(\pi^0)$. But $\dim G = \dim F + \dim H$ by the definition of functor Ξ.

$$\dim(\ker(\pi^0)) = \dim G - \dim H = \dim F = \dim(\mathrm{im}(\iota^0))$$

thus $\ker(\pi^0) = \mathrm{im}(\iota^0)$.

Therefore $0 \to F \xrightarrow{\iota} G \xrightarrow{\pi} H \to 0$ in $R(\mathcal{A})$ is properly exact. □

3.6 Proposition (1) The iso-classes of indecomposable properly projective objects in $R(\mathscr{A})$ are:

$\bar{P}(i)$, the simple projective A-module at i, $i = 1, 2, \cdots, t$;

$\bar{N}(i')$, a direct summand of indecomposable projective A-module $\bar{P}(i')$ with $\bar{N}(i')(\zeta^u_{i'l})^* = 1$ if $(\zeta^u_{li'})^*\delta = 0$, $\bar{N}(i')(\zeta^u_{i'l})^* = 0$ if $(\zeta^u_{li'})^*\delta \neq 0$, for all possible u, l. Where $i' = 1', 2', \cdots, t'$.

Furthermore

$$(\bar{P}(i))\Xi = (0 \xrightarrow{0} P(i)), \text{and } (\bar{N}(i'))\Xi = (P(i) \xrightarrow{\alpha_{N(i)}} \bigoplus_{j=1}^{t} P(j)^{m_{ji}}),$$

(see 2.1).

(2) The iso-classes of indecomposable properly injective objects in $R(\mathcal{A})$ are:

$\bar{I}(i')$, the simple injective A-module at i', $i = 1, 2, \cdots, t$;

$\bar{M}(i)$, a direct summand of indecomposable injective A-module $\bar{I}(i)$ with $\bar{M}(i)(\zeta^l_{ij'})^* = 1$ if $(\zeta^l_{ij'}) * \delta = 0$, $M(i)(\zeta^l_{ij'})^* = 0$ if $(\zeta^l_{ij'})^*\delta \neq 0$, for all possible u, l. Where $i = 1, 2, \cdots, t$.

Furthermore

$$(\bar{I}(i'))\Xi = (P(i) \xrightarrow{0} 0), \text{and } (\bar{M}(i))\Xi = (\bigoplus_{j=1}^{t} P(j)^{m_{ij}} \xrightarrow{\alpha_{S(i)}} P(i))$$

(see 2.1).

Proof (1) Let H be any indecomposable properly projective object in $R(\mathcal{A})$, $\varphi: P \to H$ be the projective cover of H in A-mod. Defined $\hat{\varphi}: V \otimes_A P \to H$ by $\hat{\varphi}^0 = \varphi$ and $\hat{\varphi} \mid \overline{V}_{\otimes_A P} = 0$, then $\hat{\varphi}$ is a proper epimorphism in $R(\mathcal{A})$, which is split, that is H is a direct summand of P. By 3.3, there are altogether $2t$ indecomposable properly projective objects in $R(\mathcal{A})$, since there are $2t$ Ext-projective modules in $p({}^*V, A)$. It is sufficient to determine them if we find all indecomposable direct summands of $2t$ indecomposable projective A-modules.

Since i is a sink of the ordinary quiver of A, $\overline{P}(i)$ is a simple A-module which is already indecomposable (fig. 3.5).

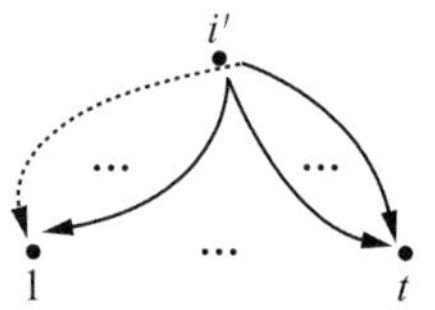

Fig. 3. 5

$\forall\, 1 \leqslant i \leqslant t$, $\overline{P}(i')$ has the following structure: $\overline{P}(i')(\zeta^w_{i'j})^*: x \mapsto x^w_{i'j}$, where x forms a k-basis of $\overline{P}(i')_{i'}$, and $\{x^w_{i'j} \mid \forall\, \zeta^w_{ij} \in B\}$ form a k-basis of $\overline{P}(i')_j$. If $(\zeta^w_{ji'})^*\delta \neq 0$, there is some l and u, v, such that $\gamma^{\binom{w}{ij}}_{\binom{u}{il}\binom{v}{lj}} \neq 0$ and $(\zeta^u_{li'})^*\delta = 0$ by the choice of the basis B in 3.5. Define a map $\varphi: \overline{P}(i') \to \overline{P}(l)$ (see fig. 3.6) given by $(x^w_{i'j})\varphi^0_j = x_j$, x_j forms a k-basis of simple module $\overline{P}(j)$, and other basis elements of $\overline{P}(i')_j$ go to 0 under φ^0_j; $(x^u_{i'l})(\zeta^v_{lj})^*\varphi = -(\gamma^{\binom{w}{ij}}_{\binom{u}{il}\binom{v}{lj}})^{-1} x_j$, and other basis elements of $\overline{P}(i')_j$ go to 0 under $(\zeta^v_{lj})^*\varphi$; $\varphi^0_m = 0$ for all $m \neq j$, $\varphi^0_{m'} = 0$ for all $1 \leqslant m \leqslant t$, and $(\xi)^*\varphi = 0$ for all $\zeta \neq \zeta^v_{lj}$. Then φ is a morphism in $R(\mathcal{A})$. In fact

$$\overline{P}(i')(\zeta^w_{i'j})^* \cdot \varphi^0_j - \varphi'^0_i \cdot \overline{P}(j)(\zeta^w_{i'j})^* = 1 = -\gamma^{\binom{w}{ij}}_{\binom{u}{il}\binom{v}{lj}} \overline{P}(i')(\zeta^u_{li'})^* \cdot (\zeta^v_{lj})^*\varphi$$
$$= ((\zeta^w_{ji'})^*\delta)\varphi.$$

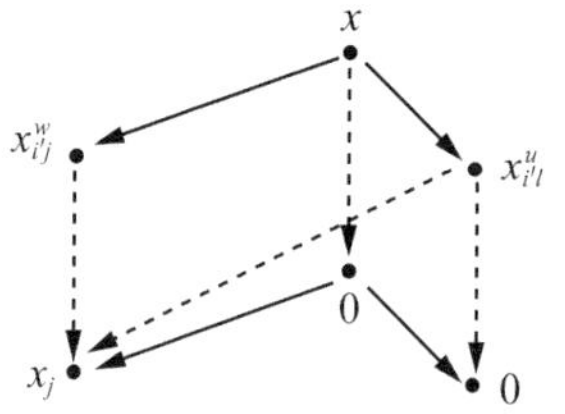

Fig. 3. 6

Define again a morphism in $R(\mathcal{A})$ say $\psi: \bar{P}(j) \to \bar{P}(i')$ by $(x_j)\psi_j^0 = x_{i'j}^w$, $\psi_m^0 = 0$, $m \neq j$, $\psi_{m'}^0 = 0$, $m = 1, 2, \cdots, t$, and $(\zeta)^* \psi = 0$, $\forall \zeta \in B$. Then $\psi \cdot \varphi = id_{\bar{P}(j)}$ and $kx_{i'j}^w \simeq kx_j = \bar{P}(j)$ is a direct summand of $\bar{P}(i')$. $\bar{P}(i')$ is a direct sum of $\bar{N}(i')$ and some $\bar{P}(j)$'s. $\bar{N}(i')$ is indecomposable in $R(\mathcal{A})$ since the decomposibility of $\bar{N}(i')$ in $R(\mathcal{A})$ is the same as in A-mod.

Finally $(\bar{N}(i'))\Xi = (P(i) \xrightarrow{\alpha_{N(i)}} \bigoplus_{j=1}^{t} P(j)^{m_{ji}})$ by the structure of the functor Ξ given in 3. 5.

(2) Dually to (1). □

3. 7 Proposition Let $\mathcal{A} = (A, V)$ be the Drozed bocs of a finite dimensional k-algebra Λ.

(1) For any indecomposable module H in $R(\mathcal{A})$ which is not properly projective, there is a proper almost split sequence ending at H.

(2) For any indecomposable module F in $R(\mathcal{A})$ which is not properly injective, there is a proper almost split sequence starting at F.

Proof If $H \in R(\mathcal{A})$ is indecomposable and not properly projective, then 3. 6. Proposition shows that $(H)\Xi = (Q_1 \xrightarrow{\alpha_3} Q_0)$ is not Ext-projective. There exists an exact almost split sequence by 2. 5. Theorem (fig. 3. 7):

$$
\begin{array}{ccccccccc}
e_1: & 0 & \longrightarrow & P_1 & \xrightarrow{f_1} & P_1 \oplus Q_1 & \xrightarrow{g_1} & Q_1 & \longrightarrow & 0 \\
 & & & {\scriptstyle \alpha_1}\downarrow & & \downarrow{\scriptstyle \alpha_2} & & \downarrow{\scriptstyle \alpha_3} & & \\
e_0: & 0 & \longrightarrow & P_0 & \xrightarrow{f_0} & P_0 \oplus Q_0 & \xrightarrow{g_0} & Q_0 & \longrightarrow & 0
\end{array}
$$

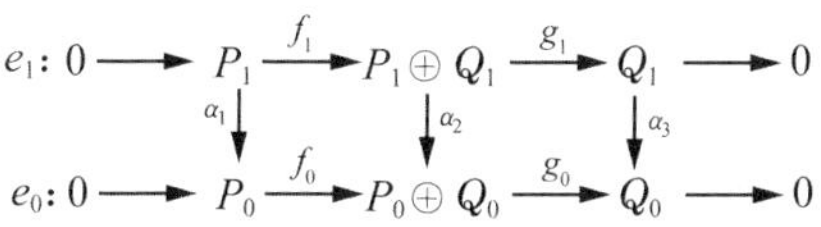

Fig. 3. 7

with $f_1=(1,0), g_1=(0,1)^T, f_0=(1,0), g_0=(0,1)^T$. Denote by

$$(P_1 \xrightarrow{\alpha_1} P_0)\Sigma=F, \quad (P_1 \bigoplus Q_1 \xrightarrow{\alpha_2} P_0 \bigoplus Q_0)\Sigma=G,$$

$$(Q_1 \xrightarrow{\alpha_3} Q_0)\Sigma=H, \quad (f_1,f_0)\Sigma=\iota, \quad (g_1,g_0)\Sigma=\pi,$$

$F \xrightarrow{\iota} G \xrightarrow{\pi} H$ is an almost split sequence in $R(\mathcal{A})$, since $\Sigma: P_1(\Lambda) \to R(\mathcal{A})$ is an equivalence. On the other hand, $0 \to F \xrightarrow{\iota} G \xrightarrow{\pi} H \to 0$ is properly exact by 3.5. Lemma.

The assertion for indecomposable noninjective object F follows dually. $\square$

3.8 Proposition (1) An object $(Q_1 \xrightarrow{\beta} Q_0)$ in $P_1(\Lambda)$ is Ext-projective if and only if any exact sequence ending at it splits in $P_1(\Lambda)$.

(2) An object $(P_1 \xrightarrow{\alpha} P_0)$ in $P_1(\Lambda)$ is Ext-injective if and only if any exact sequence starting at it splits in $P_1(\Lambda)$.

Proof (1) Suppose that $(Q_1 \xrightarrow{\beta} Q_0)$ is Ext-projective in $P_1(\Lambda)$ and fig. 3.8

$$\begin{array}{ccccccccc}
0 & \longrightarrow & P_1 & \xrightarrow{f_1} & P_1 \oplus Q_1 & \xrightarrow{g_1} & Q_1 & \longrightarrow & 0 \\
& & \downarrow{\alpha_1} & & \downarrow{\alpha_2} & & \downarrow{\alpha_3} & & \\
0 & \longrightarrow & P_0 & \xrightarrow{f_0} & P_0 \oplus Q_0 & \xrightarrow{g_0} & Q_0 & \longrightarrow & 0
\end{array}$$

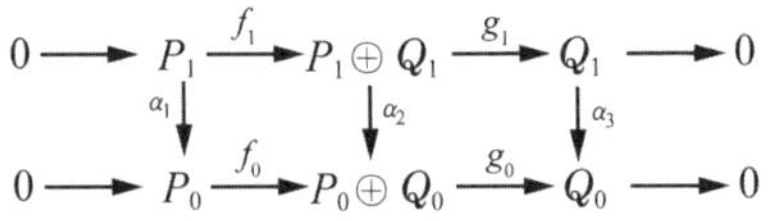

Fig. 3.8

is an exact sequence in $P_1(\Lambda)$. Acting the functor Σ, we obtain a proper exact sequence in $R(\mathcal{A})$ by 3.5. Lemma:

$$0 \to F \xrightarrow{\iota} G \xrightarrow{\pi} H \to 0.$$

And H is properly projective in $R(\mathcal{A})$ by 3.6. Lemma. Therefore π is a split epimorphism, so is (g_1,g_0). Thus the original sequence in $P_1(\Lambda)$ splits.

Conversely if $(Q_1 \xrightarrow{\beta} Q_0)$ is not Ext-projective in $P_1(A)$, then there exists an almost split sequence in $P_1(\Lambda)$ ending at it by 2.5. Theorem, which is not split in $P_1(\Lambda)$.

(2) Dually. $\square$

Remark (1) It is easy to see that ($Q_1 \xrightarrow{\beta} Q_0$) is Ext-projective if and only if any surjective map (g_1, g_0) splits(fig. 3. 9).

(2) But the dual statement for Ext-injective is not true. Consider the example in 2. 7 (fig. 3. 10).

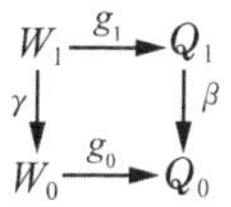

Fig. 3. 9

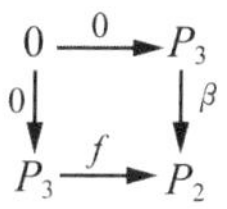

Fig. 3. 10

where β and f are natural embeddings. Then $(0, f)$ is an injective morphism and ($0 \xrightarrow{0} P_3$) is Ext-injective, but $(0, f)$ is not a split monomorphism.

3. 9 Theorem Let Λ be a basic connected finite dimensional algebra over an algebraical closed field k with t iso-classes of simples.

There exists $2t$ Ext-projective (respectively, Ext-injective) objects in $P_1(\Lambda)$; $2t$ properly projective (respectively, properly injective) objects in $R(\mathcal{A})$, where $\mathcal{A}=(A, V)$ is the Drozd bocs induced from Λ; $2t$ Ext-projective (respectively, Ext-injective) modules in $p({}^{*}V, A)$. They are 1-1 correspondent under the functors Σ and Δ respectively.

The exact almost split sequences in $P_1(\Lambda)$; proper almost split sequences in $R(\mathcal{A})$ and almost split sequences in $p({}^{*}V, A)$ are 1-1 correspondent under the functors Σ and Δ respectively. □

Acknowledgements

This is a Master thesis of X Y Zeng. Y B Zhang would like to thank Professor C . M. Ringel for his attention to this paper and his hospitality during her stay in Bielefeld, the paper was improved during that time. Project supported by the National Science Foundation of China (19831070), the Research Fund of the Doctoral Program of Higher Education and by the Volkswagen Foundatin of Germany under the joint Project "Darstellungstheorie von Algebren."

References

[1] Auslander M, Reiten. I, Smalo, S. Representation Theory of Artin Algebras. Cambridge University Press, Cambridge, 1995.

[2] Auslander M, Reiten I. Representation theory of Artin algebras Ⅲ. Comm. Alg, 1975, 3(3), 239-294.

[3] Auslander M, Reiten, I. Representation theory of Artin algebras V. Comm. Alg. 1975, 5 (5), 519-554.

[4] Ringel C. Tame Algebras and Integral Quadratic Forms; Springer Verlag, Berlin, 1984. LNM Vol. 1 099.

[5] Auslander M, Reiten I. Representation theory of Artin algebras Ⅳ. Comm. Alg. , 1975, 5 (5), 433-518.

[6] Rotman J. An Introduction to Homological Algebra. Academic Press, New York, 1979.

[7] Crawley-Boevey W W. Matrix Problems and Drozd's Theorem. In: Topics in Algebra, Vol. 26. Banach Center Publications, 1990; 199-222.

[8] Crawley-Boevey W W. On tame algebras and bocses. Proc. London Math. Soc. , 1988, 56(3), 451-483.

[9] Bautista R, Kleiner M. Almost split sequences for relatively projective modules. J. Alg. , 1990, 135(1), 19-56.

[10] Kleiner M. Induced Modules and Comodules and Representations of Bocs's and DGC's, 1980: 168-185. LMN Vol. 903.

[11] Ovsienko S A. Generic Representations of Rree Bocses. Univ. Bielefeld. (preprint 93-010).

[12] Burt W L, Butler M C R. In: Almost Split Sequence for Bocses. Canadian Math. Soc. Conference Proceedings, 1991, Vol. Ⅱ: 89-121.

[13] Drozd Y A. Matrix Problem. Ukranian SSK: Kiev, 1977: 104-114.

[14] Lin Y N, Zhang Y B. Bocses corresponding to the hereditary algebras of tame type (Ⅰ). Science in China, Ser A, 1996, 39(5), 483-490.

[15] Rojter A V. Matrix Problem and Representation of Bocses. 1980: 288-321. LNM Vol. 831.

[16] Zhang Y B, Lin Y N. Bocses corresponding to the hereditary algebras of tame type (Ⅱ). Science in China, 1996, 39(9), 909-918.

[17] Zhang Y B, Lei T G, Bautista R. The representation category of a bocs (Ⅱ). J. BNU, 1995, 31(4), 438-445.

Journal of Algebra, 2001, 239: 675-704.

仿射箭图上 Ringel-Hall 代数的极小生成系[①]

Minimal Generators of Ringel-Hall Algebras of Affine Quivers

Let $\mathscr{H}(kQ)$ be the Ringel-Hall algebra of affine quiver Q. All indecomposable representations of Q, which can be generated inside $\mathscr{H}(kQ)$ by "smaller" representations of Q, are classified; systems of minimal homogenous generators of $\mathscr{H}(kQ)$ are explicitly written out.

§ 1. Introduction

Let k be a finite field and let A be a finite-dimensional hereditary k-algebra, with all simple A-modules $S(1), S(2), \cdots, S(n)$, up to isomorphism, Let A-*mod* be the category of finite-dimensional left A-modules, which is exactly the category of A-modules with finitely many elements. The Grothendieck group $K_0(A)$ of the finite A-modules modulo short exact sequences can be identified with $\mathbf{Z}^n$, such that the image of $S(i)$ is the ith coordinate vector $\mathbf{e}_i$. For $M \in A$-*mod*, denote its isoclass by $[M]$, and its image in $K_0(A)$ by dim M, which is called the dimension vector of M.

Let $\mathbf{R}$ be the field of real numbers. By definition [R2] (see also

① This work was done in part at Universität Bielefeld and was supported by the Volkswagen-Stiftung, Germany and the National Science Foundation of China.

Received: 2000-07-12.

Dedicated to Idun Reiten on her sixtieth birthday

本文与章璞、郭晋云合作.

[Mac], but only for discrete valuation rings), the Ringel-Hall algebra $\mathscr{H}(A)$ of A is an $\mathbf{R}$-space with basis the set of isoclasses $[M]$ of all finite modules, with multiplication given by

$$[M]\cdot[N]:=\sum_{[L]} g_{M,N}^{L}[L],$$

where the structure constant $g_{M,N}^{L}$ is the number of submodules V of L such that $V\cong N$ and $L/V\cong M$. Then $\mathscr{H}(A)$ is an $\mathbf{N}_0^n$-graded associative $\mathbf{R}$-algebra with identity $[0]$, where $\mathbf{N}_0$ denotes the set of non-negative integers, and for $\mathbf{d}\in\mathbf{N}_0^n$, the homogeneous component $\mathscr{H}(A)_{\mathbf{d}}$ is the $\mathbf{R}$-space with basis $\{[M]\mid \dim M=\mathbf{d}\}$. In particular $\mathscr{H}(A)_0=\mathbf{R}$. Here we use the untwisted multiplication on $\mathscr{H}(A)$; however, all considerations hold for the twisted one introduced in [R5].

By definition [R3] the composition algebra $\mathscr{C}(A)$ of A is the subalgebra of $\mathscr{H}(A)$ generated by all isoclasses of simple A-modules $[S(1)],[S(2)],\cdots,[S(n)]$. For $\mathbf{d}=(d_1,d_2,\cdots,d_n)\in\mathbf{N}_0^n$ with $l=d_1+d_2+\cdots+d_n$, let $\mathscr{C}(A)_{\mathbf{d}}$ be the $\mathbf{R}$-space spanned by all monomials $[S(i_1)],[S(i_2)],\cdots,[S(i_l)]$, such that the number of occurrences of i in the sequence $i_1,i_2,\cdots,i_l$ is exactly d_i for $1\leqslant i\leqslant n$. Then $\mathscr{C}(A)=\bigoplus_{\mathbf{d}}\mathscr{C}(A)_{\mathbf{d}}$ is an $\mathbf{N}_0^n$-graded, $\mathbf{R}$-subalgebra of $\mathscr{H}(A)$.

Note that there is a one-to-one correspondence between the valuation diagrams of A and the symmetrizable generalized Cartan matrices Δ. Under this correspondence, by the work of Ringel and Green (see [R3, R6, G, R7]), composition algebra $\mathscr{C}(A)$ is isomorphic to the positive part of Drinfeld-Jimbo's quantum group of type Δ (see [L]).

In this paper, we mainly consider the case where the valuation diagram of A is simply laced, or, equivalently, the corresponding Δ is symmetric. This is exactly the case when A is the path algebra of a finite quiver Q without oriented cycles. (But in Section 4 we deal with the general case.) By the work of Sevenhant and Van den Bergh, one can construct the corresponding Borcherds-Cartan matrix; via Ringel-Hall algebra $\mathscr{H}(kQ)$, the corresponding Borcherds form B is exactly the symmetrization of the Ringel's form as defined in Section 2.2 below.

From this Borcherds-Cartan matrix B one has the corresponding generalized Kac-Moody algebra (see [B]) and its quantized enveloping algebra (see [Kang]). Then by [SV]$\mathscr{H}(kQ)$ is exactly the positive part of thc corresponding quantized generalized Kac-Moody algebra. It is proved in [SV] that the degrees of the real simple roots of B are exactly the coordinate vectors $\mathbf{e}_i$; and the degrees of the imaginary simple roots arc exactly the dimension vectors of minimal homogeneous generators of $\mathscr{H}(A)$; moreover, if Q is affine, then the degrees of the imaginary simple roots are of form $\lambda\mathbf{n}$, where λ are positive integers, and $\mathbf{n}$ is the minimal positive imaginary root of Q.

We are interested in the following two questions:

(1) How does one classify all indecomposable representations M of Q which can be generated inside $\mathscr{H}(kQ)$ by some representations of Q with strictly smaller dimensions; and

(2) How does one write out explicit systems of minimal homogeneous generators of $\mathscr{H}(A)$?

If Q is of Dynkin type, then $\mathscr{H}(kQ)$ coincides with its subalgebra $\mathscr{C}(kQ)$, and hence the answers to the questions above are clear: For

(1) it is all indecomposable non-simple representations;

(2) it is exactly all isoclasses of simple representations.

If Q is not of Dynkin type, then $\mathscr{H}(kQ) \neq \mathscr{C}(kQ)$. The aim of this paper is to answer the two questions for the case where Q is an arbitrary affine quiver. The answer for (1) is given in Theorem 5.3, and the answer for (2) is given in Theorems 6.3 and 6.4.

As a corollary, we also get the formula for the number of the imaginary simple roots at degrce $\lambda\mathbf{n}$, in the corresponding Borcherds-Cartan matrix associated with affine quiver Q, see Corollary 6.5. This formula has been given by Hua and Xiao in [HX]: they obtained this from a character formula. This formula in [HX] also stimulate us to consider the question (2) above.

The methods we used are the classifications of representations of affine quivers Q over finite fields (see Section 2), reduction technique to tubes by using the triangular decomposition for composition algebra $\mathscr{C}(kQ)$ (see Section 3), and the structure of the Ringel-Hall algebras of tubes, particularly the Gelfand-Kirillov dimensions of some subalgebras of $\mathscr{H}(kQ)$ (see the proof of Theorem 4.5).

Since $\mathscr{H}(kQ)$ is isomorphic to the positive part of the corresponding quantized generalized Kac-Moody algebra [SV], our results also give new insights into the corresponding quantized generalized Kac-Moody algebra via the representations and combinatorics of quivers.

Throughout this paper, let k be a finite field with q elements, and let Q be an affine quiver, i.e., Q is of type $\widetilde{A}_n$ $(n\geqslant 1)$, or $\widetilde{D}_n$ $(n\geqslant 4)$, or $\widetilde{E}_n$ $(n=6,7,8)$, with arbitrary orientation, except in the case of type $\widetilde{A}_n$ $(n\geqslant 1)$, where we exclude an oriented cycle. Denote by $\mathbf{n}$ the minimal positive imaginary root of Q.

§ 2. Representations of affine quivers over finite fields

Let us recall some basics in the representation theory of Q over finite fields, which is needed later, from [DR, ARS, R1]. Mainly, we need the formula for the number of homogeneous quasi-simples with fixed dimension vector (Proposition 2.7). This formula is more or less well known: in [K] Kac has known the number of indecomposable representations of Q over a finite field, with fixed dimension vector; combining this with the table in [DR], one gets the formula. However, here we provide a proof by using a perpendicular category and 2×2 matrices.

2.1 Let $A=kQ$ be the path k-algebra of Q. For A-modules X, Y, define

$$\langle X, Y\rangle=\dim_k \operatorname{Hom}_A(X, Y)-\dim_k \operatorname{Ext}_A^1(X, Y). \tag{2.1}$$

By a homological argument it is clear that $\langle X, Y\rangle$ depends only on $\dim X$ and $\dim Y$, so it can be bilinearly extended to $\mathbf{Z}^n$, where n is the number of vertices of Q, Denote by $(-,-)$ the symmetric, bilinear form on $\mathbf{Z}^n$ given by

$$(X,Y)=\langle X,Y\rangle+\langle Y,X\rangle, \tag{2.2}$$

and by q_A the quadratic form on $\mathbf{Z}^n$ given by $q_A(x)=\langle x,x\rangle$. Then $q_A(x)$ is positive semi-definite but not positive definite, and $\{z\in\mathbf{Z}^n \mid q_A(z)=0\}=\mathbf{Zn}$(see [DR, R1]).

Let $\tau=\mathrm{DTr}=D\ \mathrm{Ext}_A^1(-,A)$ and $\tau^{-1}=\mathrm{TrD}=\mathrm{Ext}_A^1(D(A),-)$, where $D=\mathrm{Hom}_k(-,k)$, be the Auslander-Reiten translates (see, e. g., [ARS]). An indecomposable A-module M is said to be preprojective (resp. preinjective) provided that there exists a positive integer m such that $\tau^m(M)=0$ (resp. $\tau^{-m}(M)=0$), and to be regular otherwise. An arbitrary A-module X is said to be preprojective (resp. regular, preinjective) provided that every indecomposable direct summand of X is so.

If P, R, and I are respectively preprojective, regular, and preinjective modules, then there holds the nice property

$$\mathrm{Hom}_A(R,P)=\mathrm{Hom}_A(I,P)=\mathrm{Hom}_A(I,R)=0 \tag{2.3}$$

and

$$\mathrm{Ext}_A^1(P,R)=\mathrm{Ext}_A^1(P,I)=\mathrm{Ext}_A^1(R,I)=0, \tag{2.4}$$

which is frequently used for calculation in $\mathscr{H}(A)$.

Define the defect $\partial(M)$ of a module M to be the integer $\langle\mathbf{n},\dim M\rangle$. Then an indecomposable module M is preprojective (resp. regular, preinjective) if and only if $\partial(M)<0$ (resp. $\partial(M)=0$; $\partial(M)>0$).

With indecomposables as vertices, and using irreducible maps between indecomposables to attach arrows, we obtain the Auslander-Reiten quiver of A (see [ARS]). By [DR], the Auslander-Reiten quiver of A has one preprojective component, which consists of all indecomposable preprojective modules and one preinjective component, which consists of all indecomposable preinjective modules; all other components turn out to he "tubes," which are of the form $T=\mathbf{ZA}_\infty/m$, where m is called the rank of T. If $m=1$, then T is called a homogeneous tube, and if otherwise, it is a non-homogeneous tube. The ranks of non-homogeneous tubes of A is completely determined by the type of Q, except in the case of type $\widetilde{A}_n(n\geqslant1)$. That is, type $\widetilde{D}_n(n\geqslant4)$ has three non-homogeneous tubes of ranks $n-2,2,2$; type $\widetilde{E}_n(n=6,7,8)$ has three non-homogeneous tubes of ranks $2,3,n-3$. For type $\widetilde{A}_n(n\geqslant1)$, by iteratedly using reflections of quivers, we can assume that Q has n_1 arrows going clockwise and n_2 arrows going anticlockwise. Then the ranks of non-homogeneous tubes of A are completely determined by the pair (n_1,n_2), $n_1,n_2\geqslant1$. That is, if $n_1=n_2=1$, then it is the Kronecker algebra and it has no non-homogeneous tubes; if $n_1>n_2=1$, then A has a unique non-homogeneous tube of rank n_1; if $n_1\geqslant n_2>1$, then A has two non-homogeneous tubes of ranks n_1 and n_2.

Note that indecomposable modules in different tubes have no non-zero homomorphisms and no non-trivial extensions; all regular modules form an extension-closed abelian subcategory of A-mod. The simple objects in this subcategory will be called quasi-simple modules; any indecomposable regular module M is regular uniserial, and hence M is uniquely determined by its quasi-top and quasi-length.

The dimension vectors of all the quasi-simples in non-homogeneous tubes have been listed in [DR, Tables]. Thus. we need the number of homogeneous quasi-simples with fixed dimension vector $\lambda\mathbf{n}$, where λ is an arbitrary, positive integer.

2.2 An indecomposable module M is called a stone provided $\mathrm{Ext}_A^1(M,M)=0$. Any indecomposable non-regular module is a stone; there are no stones in a homogeneous tube; and an indecomposable M in a non-homogeneous tube of rank m is a stone if and only if the quasi-length of M is less than m. Note that the endomorphism algebra of a stone is always the base field k, and that the existence of a stone with a fixed dimension vector does not depend on the base field (see [HHKU]).

Let X be a stone. Recall the perpendicular category $X^{\perp}$, introduced by Geigle-Lenzing in [GL], and Schofield in [S], is the full subcategory of A-mod given by

$$X^{\perp}=\{M\in A\text{-mod}\mid \mathrm{Hom}_A(X,M)=0=\mathrm{Ext}_A^1(X,M)\}. \qquad (2.5)$$

Then $X^{\perp}$ is equivalent to B-mod, where B is again a path algebra with $n-1$ simple modules, and n is the number of simple A-modules. The embedding functor B-mod→A-mod is exact and induces the isomorphisms on both Hom and Ext.

If algebra A is specified, then replace $\mathbf{n}$ with $\mathbf{n}_A$, and dim $\mathbf{M}$ with $\dim_A M$.

Lemma Let $S_1, S_2, \cdots, S_m$ be the all pairwise non-isomorphic simple B-modules.

(1) If $M\in X^{\perp}$ with $\dim_B M=(d_1,d_2,\cdots,d_m)$, then

$$\dim_A M=d_1\dim_A S_1+d_2\dim_A S_2+\cdots+d_m\dim_A S_m. \qquad (2.6)$$

In particular, if $M, N\in X^{\perp}$ with $\dim_B M=\dim_B N$, then $\dim_A M=\dim_A N$.

(2) If both A and B are tame, then

$$\mathbf{n}_A=n_1\dim_A S_1+n_2\dim_A S_2+\cdots+n_m\dim_A S_m, \qquad (2.7)$$

where $\mathbf{n}_B=(n_1,n_2,\cdots,n_m)$.

(3) If both A and B are tame, and $M\in X^{\perp}$, then $\dim_B M=\lambda\mathbf{n}_B$ if and only if $\dim_A M=\lambda\mathbf{n}_A$.

Proof (1) This follows from the definition of dimension vectors.

(2) Since both A and B are tame, it follows that X is regular. Let $X\in T$, where T is a non-homogeneous tube of A. Choose an indecomomposable regular B-module N with $\dim_B N=\mathbf{n}_B$. Then N is also indecomposable regular as an A-module, and

$$\begin{aligned} q_A(\dim_A N)&=\dim_k \mathrm{Hom}_A(N,N)-\dim_k \mathrm{Ext}_A^1(N,N)\\ &=\dim_k \mathrm{Hom}_B(N,N)-\dim_k \mathrm{Ext}_B^1(N,N)\\ &=q_B(\dim_B N)=q_B(\mathbf{n}_B)=0, \end{aligned}$$

it follows that $\dim_A N=\lambda\mathbf{n}_A$ for some positive integer λ.

Now choose an indecomposable regular A-module M with $\dim_A M=\mathbf{n}_A$ and $M\in X^{\perp}$. It is easy to see that such an M exists. Again we have $\dim_B M=t\mathbf{n}_B$ for some positive integer t. Then by (1) we have

$$\lambda \mathbf{n}_A = \dim_A N = n_1 \dim_A S_1 + n_2 \dim_A S_2 + \cdots + n_m \dim_A S_m,$$

and

$$\mathbf{n}_A = \dim_A M = tn_1 \dim_A S_1 + tn_2 \dim_A S_2 + \cdots + tn_m \dim_A S_m = t\lambda \mathbf{n}_A,$$

and therefore $t = \lambda = 1$.

(3) If $\dim_B M = \lambda \mathbf{n}_B$, then $\dim_A M = \lambda \mathbf{n}_A$ by (1) and (2). Conversely, if $\dim_A M = \lambda \mathbf{n}_A$, then $\dim_B M = t\mathbf{n}_B$ for some positive integer t. Again by (1) and (2) we get $\lambda = t$. □

2.3 Denote by $t_\lambda(A)$ the number of homogeneous quasi-simple A-modules X with $\dim_A X = \lambda \mathbf{n}_A$.

Let T be a non-homogeneous tube of A, and let E be a quasi-simple stone in T with $E^\perp$ equivalent to B-mod.

Lemma (1) If $\mathrm{rank}(T) > 2$, then $t_\lambda(A) = t_\lambda(B)$ for any positive integer λ.

(2) If $\mathrm{rank}(T) = 2$, then

$$t_\lambda(A) = \begin{cases} t_\lambda(B), & \lambda \neq 1; \\ t_1(B) - 1, & \lambda = 1. \end{cases}$$

Proof Notice that any homogeneous quasi-simple A-module X with $\dim_A X = \lambda \mathbf{n}_A$ is a homogeneous quasi-simple B-module, with $\dim_B X = \lambda \mathbf{n}_B$ by Lemma 2.2(3). Other possible homogeneous quasi-simple B-modules must belong to T as A-modules; however, if $\mathrm{rank}(T) = m > 2$, then those indecomposable modules in T which belong to $E^\perp$ form a non-homogeneous tube T' of B with $\mathrm{rank}(T') = m - 1$, and hence $t_\lambda(A) = t_\lambda(B)$.

Let X be a homogeneous quasi-simple B-module with $\dim_B X = \lambda \mathbf{n}_B$. Then $\dim_A X = \lambda \mathbf{n}_A$ by Lemma 2.2(3). Such an X is not a homogeneous quasi-simple A-module if and only if $\mathrm{rank}(T) = 2, \lambda = 1$, and as an indecomposable A-module X is in T with quasi-top E and quasi-length 2. In particular, such an X is unique. This proves (2). □

Denote by K the Kronecker k-algebra; i.e., K is the path k-algebra of the quiver $1 \cdot \overset{\rightarrow}{\rightarrow} \cdot 2$.

Corollary 2.4 We have $t_\lambda(A)=t_\lambda(K)$ for $\lambda\neq1$; and

$$t_1(A)=\begin{cases} q, & A \text{ of type } \widetilde{A}_{(n_1,1)}, \quad n_1>1, \\ q-1, & A \text{ of type } \widetilde{A}_{(n_1,n_2)}, \quad n_1,n_2>1, \\ q-2, & A \text{ of type } \widetilde{D}_n \text{ or } \widetilde{E}_n. \end{cases}$$

Proof The assertion follows from Lemma 2.3, by iteratedly using perpendicular reductions, and the fact that $t_1(K)=q+1$. □

2.5 In this subsection we will determine $t_\lambda(K)$, the number of homogeneous quasi-simple K-modules with dimension vector (λ,λ), where K is the Kronecker algebra over k. It is dear that $t_1(K)=q+1$, so we assume $\lambda\geqslant2$ in the following.

Denote by $N(q,\lambda)$ the number of monic irreducible polynomials of degree λ over the field of q elements. Then we have the well-known formula due to Gauss:

$$N(q,\lambda)=\frac{1}{\lambda}\sum_{d\mid\lambda}\mu\left(\frac{\lambda}{d}\right)q^d, \tag{2.8}$$

where μ is the Möbius function, i.e. $\mu(1)=1$, $\mu(n)=0$ if n has a square factor, and $\mu(n)=(-1)^s$ if $n=p_1,p_2,\cdots,p_s$, p_i distinct primes.

Lemma If $\lambda\geqslant2$, then $t_\lambda(K)=N(q,\lambda)$.

Proof Recall that K-mod can be identified with the category $\mathcal{K}$, whose objects are quartets $M=(k^m,\alpha,\beta,k^n)$, where $\alpha,\beta:k^m\to k^n$ are k-linear maps, For two objects $M_1=(k^{m_1},\alpha_1,\beta_1,k^{n_1})$ and $M_2=(k^{m_2},\alpha_2,\beta_2,k^{n_2})$, the morphism set $\mathrm{Hom}_{\mathcal{K}}(M_1,M_2)$ is defined to he the set of the pairs $f=(f_1,f_2)$, where $f_1:k^{m_1}\to k^{m_2}$ and $f_2=k^{n_1}\to k^{n_2}$ are k-linear maps, such that (we write the composition of morphisms from left to right)

$$f_1\alpha_2=\alpha_1f_2, \quad f_1\beta_2=\beta_1f_2.$$

Such a morphism $f=(f_1,f_2)$ is an isomorphism of K-modules if and only if both f_1 and f_2 are invertible.

Note that $M_1\oplus M_2=(k^{m_1}\oplus k^{m_2},\alpha_1\oplus\alpha_2,\beta_1\oplus\beta_2,k^{n_1}\oplus k^{n_2})$.

Now, let $M=(k^\lambda,\alpha,\beta,k^\lambda)$ be an indecomposable K-module of dimension vector (λ,λ), with $\text{rank}(\alpha)=r$. Then we have invertible $\lambda\times\lambda$ matrices $\boldsymbol{g}_1$ and $\boldsymbol{g}_2$ such that

$$\boldsymbol{g}_1\alpha\boldsymbol{g}_2=\begin{pmatrix}\boldsymbol{I}_r & \boldsymbol{0}\\ \boldsymbol{0} & \boldsymbol{0}\end{pmatrix},$$

where $\boldsymbol{I}_r$ is the $r\times r$ identity matrix. It is easy to see that M is isomorphic to the K-module

$$\left(k^\lambda,\begin{pmatrix}\boldsymbol{I}_r & \boldsymbol{0}\\ \boldsymbol{0} & \boldsymbol{0}\end{pmatrix},\boldsymbol{g}_1\beta\boldsymbol{g}_2,k^\lambda\right).$$

It follows that we can assume that M is of the form

$$M=\left(k^\lambda,\begin{pmatrix}\boldsymbol{I}_r & \boldsymbol{0}\\ \boldsymbol{0} & \boldsymbol{0}\end{pmatrix},\beta,k^\lambda\right).$$

Case (1): $r=\lambda$. In this case, let g be the rational canonical form of β, and let h be a $\lambda\times\lambda$ invertible matrix such that $h\beta h^{-1}=g$. Then it is easy to see that M is isomorphic to the K-module

$$(k^\lambda,I_\lambda,g,k^\lambda).$$

Since M is indecomposable, it follows that g has to be an indecomposable matrix, and that g has to be invertible. Thus, g is the companion matrix of a polynomial $(\varphi(x))^d$ of degree λ, where $\phi(x)$ is a monic irreducible polynomial in $k[x]$, and d is a positive factor of λ, and if $\deg(\phi(x))=1$, then $\phi(x)\neq x$ (see, e. g., [FIS, Theorems 6.11 and 6.13]). Therefore, we can identify M with the K-module given by the pair

$$(k[x]/((\phi(x))^d),id,m,k[x]/((\phi(x))^d)),\qquad (2.9)$$

where $m:k[x]/(\phi(x)^d)\to k[x]/(\phi(x)^d)$ is the k-map given by $m(a)=a\bar{x}$, where $\bar{x}$ is the coset $x+(\phi(x)^d)$. Since g is the companion matrix of a polynomial $(\phi(x))^d$ of degree λ, it is easy to see m coincides with g.

Using this presentation of M we can easily see that

$$\text{End}_K M\cong k[x]/((\phi(x))^d).$$

In fact, any endomorphism in $\text{End}_K M$ is of the form (f,f), where f is completely determined by $f(1)$; i. e., we have

$$f(\bar{x}^i)=f(1)\bar{x}^i,\quad i=1,2,\cdots,\lambda-1.$$

It follows that $\mathrm{End}_K M$ is a field if and only if $d=1$. In this case, $\mathrm{End}_K M \cong k[x]/(\phi(x))$ and hence $[\mathrm{End}_K M : k] = \deg(\phi(x)) = \lambda$, where $\dim_K M = \lambda(\mathrm{I}, 1)$.

Notice that if $\phi_1(x)$ and $\phi_2(x)$ are different monic irreducible polynomials of degree λ in $k[x]$, then the corresponding modules $M_1 = (k[x]/(\phi_1(x)), id, m, k[x]/(\phi_1(x)))$ and $M_2 = (k[x]/(\phi_2(x)), id, m, k[x]/(\phi_2(x)))$ are not isomorphic. In this way we have already obtained $N(q,\lambda)$ homogeneous quasi-simple K-modules of dimension vector (λ,λ). In the next case, we shall see that this is the complete list of homogeneous quasi-simples of dimension vector (λ,λ).

Case(2): $r<\lambda$. Since $\lambda \geqslant 2$ and M is indecomposable, it follows that $r \geqslant 1$. Let M be given by the pair

$$M=\left(k^{\lambda}, \begin{pmatrix} \boldsymbol{I}_r & \boldsymbol{0} \\ \boldsymbol{0} & \boldsymbol{0} \end{pmatrix}, g, k^{\lambda}\right),$$

where

$$\boldsymbol{g}=\begin{pmatrix} \boldsymbol{g}_1 & \boldsymbol{g}_2 \\ \boldsymbol{g}_3 & \boldsymbol{g}_4 \end{pmatrix}$$

is a $\lambda \times \lambda$ matrix and $\boldsymbol{g}_1$ is a $r \times r$ matrix.

We want to prove that $\mathrm{End}_K M$ is not a field, and hence M is not a homogeneous quasi-simple K-module. For this purpose, consider the pair $\boldsymbol{f}=(\boldsymbol{f}_1, \boldsymbol{f}_2)$ of $\lambda \times \lambda$ matrices, where

$$\boldsymbol{f}_1=\begin{pmatrix} \boldsymbol{0} & \boldsymbol{g}_2 \\ \boldsymbol{0} & \boldsymbol{g}_4 \end{pmatrix}, \quad \boldsymbol{f}_2=\begin{pmatrix} \boldsymbol{0} & \boldsymbol{0} \\ \boldsymbol{g}_3 & \boldsymbol{g}_4 \end{pmatrix}.$$

Then it is easy to see $\boldsymbol{f}=(\boldsymbol{f}_1, \boldsymbol{f}_2) \in \mathrm{End}_K M$, namely, there hold the following equalities:

$$\begin{pmatrix} \boldsymbol{0} & \boldsymbol{g}_2 \\ \boldsymbol{0} & \boldsymbol{g}_4 \end{pmatrix} \cdot \begin{pmatrix} \boldsymbol{I}_r & \boldsymbol{0} \\ \boldsymbol{0} & \boldsymbol{0} \end{pmatrix} = \boldsymbol{0} = \begin{pmatrix} \boldsymbol{I}_r & \boldsymbol{0} \\ \boldsymbol{0} & \boldsymbol{0} \end{pmatrix} \cdot \begin{pmatrix} \boldsymbol{0} & \boldsymbol{0} \\ \boldsymbol{g}_3 & \boldsymbol{g}_4 \end{pmatrix};$$

$$\begin{pmatrix} \boldsymbol{0} & \boldsymbol{g}_2 \\ \boldsymbol{0} & \boldsymbol{g}_4 \end{pmatrix} \cdot \begin{pmatrix} \boldsymbol{g}_1 & \boldsymbol{g}_2 \\ \boldsymbol{g}_3 & \boldsymbol{g}_4 \end{pmatrix} = \begin{pmatrix} \boldsymbol{g}_2\boldsymbol{g}_3 & \boldsymbol{g}_2\boldsymbol{g}_4 \\ \boldsymbol{g}_4\boldsymbol{g}_3 & \boldsymbol{g}_4\boldsymbol{g}_4 \end{pmatrix} - \begin{pmatrix} \boldsymbol{g}_1 & \boldsymbol{g}_2 \\ \boldsymbol{g}_3 & \boldsymbol{g}_4 \end{pmatrix} \cdot \begin{pmatrix} \boldsymbol{0} & \boldsymbol{0} \\ \boldsymbol{g}_3 & \boldsymbol{g}_4 \end{pmatrix}.$$

It is clear that $\boldsymbol{f}$ is not an automorphism of K-module M; and since

M is indecomposable, it follows that g_2 and g_3 cannot be zero simultaneously, and hence $f \neq 0$.

This completes the proof. □

From the proof of Lemma 2.5, Case (1), and using Lemma 2.2(3), we see

2.6 Corollary Let E be a homogeneous quasi-simple A-module of dimension vector $\lambda \mathbf{n}_A$. Then $\mathrm{End}_A E$ is the field with $[\mathrm{End}_K E : k] = \lambda$.

By Corollary 2.4 and Lemma 2.5, we get

2.7 Proposition Let $t_\lambda(A)$ be the number of homogeneous quasi-simple A-modules of dimension vector $\lambda \mathbf{n}$. Then

$$t_\lambda(A) = N(q, \lambda), \quad \lambda > 1;$$

and

$$t_1(A) = \begin{cases} q+1, & A \text{ of type } \widetilde{A}_{(1,1)}, \\ q, & A \text{ of type } \widetilde{A}_{(n_1,1)}, \quad n_1 > 1, \\ q-1, & A \text{ of type } \widetilde{A}_{(n_1,n_2)}, \quad n_1, n_2 > 1, \\ q-2, & A \text{ of type } \widetilde{D}_n \text{ or } \widetilde{E}_n. \end{cases}$$

Since $N(q,\lambda) \geqslant 1$ for all q and λ, and since $N(q,\lambda) = 1$ if and only if $q = \lambda = 2$, we have

2.8 Corollary (1) $t_\lambda(A) = 0$ if and only if $\lambda = 1, q = 2$, and A is of type $\overline{D}_n$ or $\overline{E}_n$.

(2) $t_\lambda(A) = 1$ if and only if one of the following cases occurs:

① $\lambda = 1, q = 3$, A of type $\overline{D}_n$ or $\overline{E}_n$;

② $\lambda = 1, q = 2$, A of type $\overline{A}_{(n_1,n_2)}$, $n_1, n_2 \geqslant 2$;

③ $\lambda = q = 2$.

(3) In the remaining cases we have $t_\lambda(A) \geqslant 2$.

Remark As we mentioned in the beginning of this section, Proposition 2.7 can be also obtained by combining a result in Kac's paper[K] and the table in Dlab-Ringel's paper [DR].

Using the Euler φ-function (see, e.g., [J]) $\phi(\lambda) = \sum_{s|\lambda} s\mu^{(\frac{\lambda}{s})}$, where $\varphi(\lambda)$ is the number of positive integers $<\lambda$ which are relatively prime to λ, one can rewrite

$$\sum_{s|\lambda} N(q,s)=\frac{1}{\lambda}\sum_{s|\lambda}\varphi\left(\frac{\lambda}{s}\right)q^{s}. \tag{2.10}$$

§ 3. The space $B_{\mathbf{d}}(A)$

3.1 Let $A=kQ$ with $K_0(A)=\mathbf{Z}^n$, and let $\mathscr{H}(A)$ and $\mathscr{C}(A)$ be the Ringel-Hall algebra and Ringel's composition algebra of A, respectively. For $0\neq\mathbf{d}\in\mathbf{N}_0^n$, define

$$B_{\mathbf{d}}(A):=\sum_{x+y=\mathbf{d};x,y\neq 0}\mathscr{H}(A)_x\mathscr{H}(A)_y\subseteq\mathscr{H}(A)_{\mathbf{d}} \tag{3.1}$$

and $B_0:=\mathbf{R}[0]$. By definition $B_{e_i}=0, i=1,2,\cdots,n$, where $e_i, i=1,2,\cdots,n$, are the coordinate vectors.

The spaces $B_{\mathbf{d}}(A)$ reflect the structure of $\mathscr{H}(A)$. We are interested in the following two questions.

(1) How do we classify all indecomposable modules M with the property $[M]\in B_{\dim M}(A)$? This will be answered in Theorem 5.3.

(2) How do we write out explicit systems of minimal homogeneous generators of $\mathscr{H}(A)$? This will be done in Section 6.

For both purposes, first we need to give sufficient and necessary conditions for an element in $B_{\mathbf{d}}(A)$; this is the aim of this section (see Theorem 3.7).

Lemma For $\mathbf{d}\neq\mathbf{e}_i, i=1,2,\cdots,n$, there holds

$$\mathscr{C}(A)_{\mathbf{d}}\subseteq B_{\mathbf{d}}(A)\subseteq\mathscr{H}(A)_{\mathbf{d}}, \tag{3.2}$$

and $B_{\mathbf{d}}(A)=\mathscr{H}(A)_{\mathbf{d}}$ for all $\mathbf{d}\neq\mathbf{e}_i, i=1,2,\cdots,n$, if and only if

$$\mathscr{C}(A)=\mathscr{H}(A).$$

3.2 For $\mathbf{d}\in K_0(A)=\mathbf{Z}^n$, define the following element in $\mathscr{H}(A)$:

$$r_{\mathbf{d}}:=\sum_{[M]}[M],$$

where M runs over all regular modules with $\dim M=\mathbf{d}$.

Note that this is a finite sum since k is a finite field. If there are no regular modules M with $\dim M=\mathbf{d}$, then set $r_{\mathbf{d}}:=0$. Set $r_0:=[0]$. In

[Z1, Theorem 1], we have proved that $r_{\mathbf{d}} \in \mathscr{C}(A)$ for all $\mathbf{d}$. Let $\mathscr{T}$ denote the subalgebra of $\mathscr{H}(A)$ generated by all elements $r_{\mathbf{d}}$ with $\mathbf{d} \in \mathbf{N}_0^n$. Let $\mathscr{P}$ and $\mathscr{I}$ denote the subalgebra of $\mathscr{H}(A)$ generated by preprojectives and by preinjectives, respectively. Let $\mathscr{P} \cdot \mathscr{T} \cdot \mathscr{I}$ be thc $\mathbf{R}$-subspace of $\mathscr{H}(A)$ spanned by all products $[P] \cdot r_{\mathbf{d}_1}, r_{\mathbf{d}_2}, \cdots, r_{\mathbf{d}_m} \cdot [I]$, where P (resp. I) runs over the preprojectives (resp. preinjectives), $\mathbf{d}_1, \mathbf{d}_2, \cdots, \mathbf{d}_m \in \mathbf{N}_0^n$, and $m \in \mathbf{N}_0$.

We need the following triangular decomposition theorem for $\mathscr{C}(A)$ proved in [Z3].

Theorem We have $\mathscr{C}(A) = \mathscr{P} \cdot \mathscr{T} \cdot \mathscr{I} = \mathscr{P} \otimes_M \mathscr{T} \otimes_M \mathscr{I}$.

3.3 Terminologies Let T be a tube of A, and let M be an A-module. By $M \in T$ we mean that every indecomposable direct summand of M belongs to T.

Let $x = \sum_{[M]} c_M [M] \in \mathscr{H}(A)$. If $c_M \neq 0$, then $[M]$ is said to be a term of x with coefficient c_M. If $[M]$ is a term of x, and M is indecomposable, then $[M]$ is said to be an indecomposable term of x. If $[M]$ is a term of x, and $M \in T$, then $[M]$ is said to be a T-term of x. Define the regular part of x to be

$$r(x) := \sum_{M \text{ is regular}} c_M [M]. \tag{3.3}$$

Define the T-part of x to be

$$r_T(x) := \sum_{M \in T} c_M [M]. \tag{3.4}$$

For $\mathbf{d} \in \mathbf{N}_0^n$, denote by $r_{\mathbf{d}}(T)$ the T-part of $r_{\mathbf{d}}$, i. e.,

$$r_{\mathbf{d}}(T) := \sum_{[M]} [M] \tag{3.5}$$

where M runs over all modules in T with $\dim M = \mathbf{d}$.

Lemma 3.4 Let $x \in \mathscr{H}(A)_{\mathbf{d}}$, where $\mathbf{d} \neq \mathbf{e}_i$, $i = 1, 2, \cdots, n$. Then $x - r(x) \in B_{\mathbf{d}}(A)$. In particular, $x \in B_{\mathbf{d}}(A)$ if and only if $r(x) \in B_{\mathbf{d}}(A)$.

Proof Note that any term of $x - r(x)$ is of the form $[P \oplus R \oplus I]$, such that at least one of P and I is not zero, where P, R, and I are, respectively, preprojective, regular, and preinjective. Since $[P \oplus R \oplus I] =$

$[P]\cdot[R]\cdot[I]$, it follows that if any two of P, R, and I are not zero module, then $[P\oplus R\oplus I]\in \mathrm{B}_{\mathbf{d}}(A)$. If $R=0=I$, then $[P]\in \mathscr{C}(A)_{\mathbf{d}}\subseteq B_{\mathbf{d}}(A)$ since $\mathbf{d}\neq \mathbf{e}_i$, $i=1,2,\cdots,n$. Similarly for the case $P=0=R$, Thus,

$$x-r(x)\in B_{\mathbf{d}}(A). \quad \square$$

Lemma 3.5 Lt $x\in \mathscr{K}(A)_{\mathbf{d}}$, where $\mathbf{d}\neq \mathbf{e}_i$, $i=1,2,\cdots,n$. Then there holds

(1) $r(x)-\sum\limits_{\text{tube } T} r_T(x)\in B_{\mathbf{d}}(A)$.

In particular, $x\in B_{\mathbf{d}}(A)$ if and only if $\sum\limits_{\text{tube } T} r_T(x)\in B_{\mathbf{d}}(A)$.

(2) $\sum\limits_{\text{tube } T} r_{\mathbf{d}}(T)\in B_{\mathbf{d}}(A)$.

Proof Since any term of $r(x)-\sum\limits_{\text{tube } T} r_T(X)$ is of the form $[R_1\oplus R_2]$, where both R_1 and R_2 are non-zero regular modules, such that $R_1\in T$ for some tube T, but R_2 has no direct summands in T, it follows that $[R_1\oplus R_2]=[R_1]\cdot[R_2]$, and hence (1) follows.

If $\mathbf{d}\neq \mathbf{e}_i$, $i=1,2,\cdots,n$, then $r_{\mathbf{d}}\subseteq \mathscr{C}(A)_{\mathbf{d}}\subseteq B_{\mathbf{d}}(A)$, and hence (2) follows from (1). $\square$

3.6 Let X, Y be non-zero A-modules. Consider the regular part $r([X]\cdot[Y])$. Let $X=P\oplus R\oplus I$, $Y=P'\oplus R'\oplus I'$, with P, P' preprojective; R, R' regular, and I, I' preinjective. Thus

$$[X]\cdot[Y]=[P]\cdot[R]\cdot[I]\cdot[P']\cdot[R']\cdot[I'],$$

and it follows that if $P\neq 0$, or $I'\neq 0$, then $r([X]\cdot[Y])=0$. Now assume $P=0=I'$. Then by comparing defects we see

$$r([X]\cdot[Y])=r([R]\cdot[I]\cdot[P']\cdot[R'])=[R]\cdot r([I]\cdot[P'])\cdot[R'].$$

While by the Theorem in 3.2 we have $[I]\cdot[P']\in \mathscr{C}(A)=\mathscr{P}\cdot\mathscr{T}\cdot\mathscr{I}$ and $r([I]\cdot[P'])\in\mathscr{T}$, it follows that if $R=0=R'$, then $r([X]\cdot[Y])=R([I]\cdot[P'])$ is of the form

$$r([X]\cdot[Y])=cr_{\dim X+\dim Y}+\sum_{t\geqslant 2} c_{\mathbf{d}_1,\mathbf{d}_2,\cdots,\mathbf{d}_t} r_{\mathbf{d}_1} r_{\mathbf{d}_2}\cdots r_{\mathbf{d}_t},$$

with $\mathbf{d}_1+\mathbf{d}_2+\cdots+\mathbf{d}_t=\dim X+\dim Y$, and $c, c_{\mathbf{d}_1,\mathbf{d}_2,\cdots,\mathbf{d}_t}\in\mathbf{R}$; and that if $R\neq 0$; or $R'\neq 0$, then $r([X]\cdot[Y])$ is of the form $\sum c_{M,N}[M]\cdot[N]$, where M, N are non-zero regular modules and $c_{M,N} \in \mathbf{R}$. This proves the following.

Lemma Let X, Y be non-zero A-modules. Then the regular part $r([X]\cdot[Y])$ is of the form

$$r([X]\cdot[Y]) = cr_{\dim X+\dim Y} + \sum c_{M,N}[M]\cdot[N], \tag{3.6}$$

where M, N are non-zero regular modules with

$$\dim M + \dim N = \dim X + \dim Y, \text{and } c, c_{M,N} \in \mathbf{R}.$$

Since $B_{\mathbf{e}_i} = 0$ for $i = 1, 2, \cdots, n$, the following result gives a description of all spaces $B_{\mathbf{d}}(A)$, $\mathbf{d} \in \mathbf{N}_0^n$. The expressions (3.7) and (3.8) below play an important role in Sections 5 and 6.

Theorem 3.7 Let $x \in \mathscr{H}(A)_{\mathbf{d}}$, where $\mathbf{d} \neq \mathbf{e}_i$, $i = 1, 2, \cdots, n$. Then the followings are equivalent:

(1) $x \in B_{\mathbf{d}}(A)$;

(2) $r(x) \in B_{\mathbf{d}}(A)$;

(3) $r(x)$ is of the form

$$r(x) = cr_{\mathbf{d}} + \sum c_{M,N}[M]\cdot[N], \tag{3.7}$$

where M, N are non-zero regular modules with $\dim M + \dim N = \mathbf{d}$, and $c, c_{M,N} \in \mathbf{R}$.

(4) There exists *a* $c \in \mathbf{R}$ such that for every tube T of A, $r_T(x)$ is of the form

$$r_T(x) = cr_{\mathbf{d}}(T) + \sum c_{M,N}[M]\cdot[N], \tag{3.8}$$

where M, N are non-zero modules in T with $\dim M + \dim N = \mathbf{d}$, and $c_{M,N} \in \mathbf{R}$.

Proof The implication (1)⇒(2) follows from Lemma 3.4.

(2)⇒(3): If $r(x) \in B_{\mathbf{d}}(A)$, then we can write

$$r(x) = \sum c_{X,Y}[X]\cdot[Y],$$

where X, Y are non-zero modules with $\dim X + \dim Y = \mathbf{d}$, and $c_{X,Y} \in \mathbf{R}$. Now, taking the regular parts from both sides of the preceeding equality, and then using Lemma in 3.6, we get

$$r(x) = r(r(x)) = \sum c_{X,Y} r([X]\cdot[Y]) = cr_{\mathbf{d}} + \sum c_{M,N}[M]\cdot[N],$$

where M, N are non-zero regular modules with $\dim M + \dim N = \mathbf{d}$, and $c, c_{M,N} \in \mathbf{R}$.

(3)⇒(4): This follows from by taking the T-parts from the both sides of (3.7).

(4)⇒(1): If there exists **a** $c \in \mathbf{R}$ such that for every tube of A, $r_T(x)$ is of the form (3.8), then

$$\sum_{\text{tube } T} r_T(x) = c \sum_{\text{tube } T} r_{\mathbf{d}}(T) + y,$$

with $y \in B_{\mathbf{d}}(A)$. While $\sum_{\text{tube } T} r_{\mathbf{d}}(T) \in B_{\mathbf{d}}(A)$ by Lemma 3.5(2), and hence $\sum_{\text{tube } T} r_T(X) \in B_{\mathbf{d}}(A)$, therefore, $x \in B_{\mathbf{d}}(A)$ by Lemma 3.5(1). □

Remark (1) As pointed out by Sevenhant and Van den Bergh in [SV], the imaginary simple roots of $\mathscr{H}(A)$ are of the form $\lambda\mathbf{n}$, where λ is a positive integer, i.e., if $\mathbf{d} \neq \mathbf{e}_i$, $i = 1, 2, \cdots, n$, and if $B_{\mathbf{d}}(A) \neq \mathscr{H}(A)_{\mathbf{d}}$, then $\mathbf{d} = \lambda\mathbf{n}$. Therefore, Theorem 3.7 is only used for $\mathbf{d} = \lambda\mathbf{n}$.

(2) Let $x \in \mathscr{H}(A)_{\lambda\mathbf{n}}$. As pointed out in Lemma 3.5, if $r_T(x) \in B_{\lambda\mathbf{n}}(A)$ for any tube T, then $x \in \mathrm{B}_{\lambda\mathbf{n}}(A)$. But the converse is not true.

For example, let K be the Kronecker algebra, and let $N_1, N_2, \cdots, N_{q+1}$ be all the indecomposable modules of dimension vector (1,1). Then $r_{(1,1)} = \sum_{1 \leqslant i \leqslant q+1} [N_i] \in \mathscr{C}(A)_{(1,1)} \subseteq B_{(1,1)}(A)$, but every $[N_i] \notin B_{(1,1)}(A)$.

§ 4.   Ringel-Hall algebras of tubes

The aim of this section is to study the Ringel-Hall algebras of tubes, for application in the next section: The main results are Theorems 4.5 and 4.10, Corollary 4.8, and Lemma 4.9.

4.1 In this section, A is an arbitrary tame hereditary algebra over a finite field k, with minimal positive imaginary root $\mathbf{n}$. Let T be a tube of A. Denote by $\mathscr{H}(T)$ the subspace of $\mathscr{H}(A)$ with basis $\{[M] \mid M \in T\}$. Then $\mathscr{H}(T)$ is also an $\mathbf{N}_0^n$-graded algebra with homogeneous component $\mathscr{H}(T)_{\mathbf{d}}$ being the space with basis $\{[M] \mid M \in T, \dim M = \mathbf{d}\}$. Set

$$B_{\mathbf{d}}(T) = \sum_{x+y=\mathbf{d};\, x, y \neq 0} \mathscr{H}(T)_x \cdot \mathscr{H}(T)_y. \tag{4.1}$$

Then $B_{\mathbf{d}}(T) \subseteq B_{\mathbf{d}}(A) \cap \mathscr{H}(T)_{\mathbf{d}}$.

The motivation of introducing $B_{\mathbf{d}}(T)$ is as follows. For $x \in \mathscr{H}(T)_{\lambda\mathbf{n}}$, we want to reduce the criterion of $x \in B_{\lambda\mathbf{n}}(A)$ to the one of $x \in B_{\lambda\mathbf{n}}(T)$. The advantage of this reduction is that, inside a tube more combinatorial techniques could be used, and results in [DR] can be used more efficiently; particularly in [R4] the structure of $\mathscr{H}(T)$ has been extensively studied for non-homogeneous tube T. As we will see in Theorem 5. 2. this idea works.

4. 2 We fix the following notations. Let T be a tube with rank$(T) = m$; let $E_1, E_2, \cdots, E_m$ be the quasi-simples in T with $\tau(E_i) = E_{i+1}, 1 \leqslant i \leqslant m-1$; and let $\tau(E_m) = E_1$. Let $E_i(j)$ denote the indecomposable in T with quasi-length j and quasi-top E_i. If $m \geqslant 2$, then

$$\dim E_1(m) + \dim E_2(m) + \cdots + \dim E_m(m) = g\mathbf{n}, \qquad (4.2)$$

where g is a positive integer with $1 \leqslant g \leqslant 3$ (see [DR] or [M]). In particular, if A is the path algebra of an affine quiver, then $g = 1$ (see [DR, Tables]).

Let $x, x_1, x_2, \cdots, x_t \in \mathscr{H}(A)$. We say that x is generated by $x_1, x_2, \cdots, x_t$, provided that x is a **R**-combination of some products with all divisors being in $\{x_1, x_2, \cdots, x_t\}$. A subset of $\mathscr{H}(A)$ is said to be generated by $x_1, x_2, \cdots, x_t$, provided that every element in it can be generated by $x_1, x_2, \cdots, x_t$.

Lemma 4. 3 [**GP**] (1) $\mathscr{H}(A)$ is generated by all isoclasses of indecomposable A-modules.

(2) $\mathscr{H}(T)$ is generated by all isoclasses of indecomposables in T.

Lemma 4. 4 If $m \geqslant 2$, M is an indecomposable in T with quasi-length λm, and N is an arbitrary indecomposable in T with quasi-length λm, say,

$$N = \tau^i(M), \text{then } [N] \in i[M] + B_{\lambda g\mathbf{n}}(T). \qquad (4.3)$$

Proof Denote by L the unique maximal regular submodule of M and by E the quasi-top of M. Since $\dim_k \operatorname{Ext}_A^1(E, L) = 1 = \dim_k \operatorname{Ext}_A^1(L, E)$ and $\operatorname{Hom}_A(L, E) = 0 = \operatorname{Hom}_A(E, L)$, it follows that

$$[E]\cdot[L]=[M]+[E\oplus L]$$

and

$$[L]\cdot[E]=[\tau(M)]+[E\oplus L].$$

It follows that

$$[\tau(M)]=[M]+[L]\cdot[E]-[E]\cdot[L]\in[M]+B_{\lambda g\mathbf{n}}(T),$$

and hence the assertion follows by repeating this process. □

Theorem 4.5 Let T be a non-homogeneous tube with rank m. Then for any positive integer λ there holds

$$\mathscr{H}(T)_{\lambda g\mathbf{n}}=\mathbf{R}[M_\lambda]\oplus B_{\lambda g\mathbf{n}}(T), \tag{4.4}$$

where M_λ is an arbitrary indecomposable in T with quasi-length λm.

Proof By Lemma 4.3, $\mathscr{H}(T)_{\lambda g\mathbf{n}}$ is generated by the isoclasses of indecomposable modules in T. It follows from Lemma 4.4 that

$$\mathscr{H}(T)_{\lambda g\mathbf{n}}=\mathbf{R}[M_\lambda]+B_{\lambda g\mathbf{n}}(T).$$

It remains to prove that $B_{\lambda g\mathbf{n}}(T)\neq\mathscr{H}(T)_{g\lambda\mathbf{n}}$ for all $\lambda\geqslant 1$.

For an A-module M, let $\zeta(M)$ be the number of the indecomposable direct summands of M.

For $\lambda\geqslant 1$ and $1\leqslant i\leqslant m$, let

$$T_\lambda(i)=E_i((\lambda-1)m+1)\oplus\bigoplus_{t\neq i,1\leqslant t\leqslant m}E_t.$$

Note that $T_1(1)=T_1(2)=\cdots=T_1(m)=E_1\oplus E_2\oplus\cdots\oplus E_m$, $\dim T_\lambda(i)=\lambda g\mathbf{n}$, and $\zeta(T_\lambda(i))=m$.

Define

$\Omega_\lambda=\{M\in T\mid \text{there exists } i, 1\leqslant i\leqslant m, \text{such that } T_\lambda(i)\preceq M\preceq E_i(\lambda m)\}$,

where the order $\preceq$ is defined as in [R4, p. 520, 4.7], and as in [Guo, Sect. 2] (there he used the symbol $<_0$).

If $M\in\Omega_\lambda$, then by construction we have

(1) $\dim M=\lambda g\mathbf{n}$.

(2) $1\leqslant\zeta(M)\leqslant m$.

(3) $\zeta(M)=1$ if and only if $M=E_i(\lambda m)$ for some i; and $\zeta(M)=m$ if and only if $M=T_\lambda(i)$ for some i.

(4) M is multiplicity free.

Define

$$c_\lambda = \sum_{M \in \Omega_\lambda} (1-q)^{\zeta(M)-1}[M].$$

Notice that $c_\lambda \in \mathcal{H}(T)_{\lambda g\mathbf{n}}$.

For example,

$$c_1 = \sum_{M \in T: \dim M = g\mathbf{n}} (1-q)^{\zeta(M)-1}[M].$$

(We point out a fact which is not needed in this paper, i. e. , c_1 is in the center of $\mathcal{H}(T)$, but c_λ is not in thc center of $\mathcal{H}(T)$ for $\lambda \geqslant 2$ (see [Guo, Proposition 3. 2]).)

It is proved in [Guo] that c_λ is in the centralizer of $\mathcal{C}(T)$ for $\lambda \geqslant 1$, where $\mathcal{C}(T)$ is the subalgebra of $\mathcal{H}(T)$ generated by all isoclasses of quasi-simples in T.

Denote by $\mathcal{H}(\lambda)$ the subalgebra of $\mathcal{H}(T)$ generated by $\{E_i(j) \mid 1 \leqslant i \leqslant m, 1 \leqslant j \leqslant \lambda\}$. (Note that $\mathcal{H}(\lambda m)$ is different from $\mathcal{H}(T)_{\lambda g\mathbf{n}}$.) By definition and Lemma 4. 3 we have

$$B_{\lambda g\mathbf{n}}(T) \subseteq \mathcal{H}(\lambda m - 1). \tag{4. 5}$$

It is well known that $\mathcal{H}(l) = \mathcal{C}(T)$ for $1 \leqslant l \leqslant m-1$. By Proposition 3. 4 and Theorem 4. 7 of [Guo], we also know that for positive integer λ and $0 \leqslant l \leqslant m-1$,

$$\begin{aligned} \mathcal{H}(\lambda m + l) &= \langle \mathcal{C}(T), c_1, c_2, \cdots, c_\lambda \rangle \\ &= \mathcal{C}(T)[c_1, 1, 0][c_2, 1, \delta_2] \cdots [c_\lambda, 1, \delta_\lambda] \end{aligned} \tag{4. 6}$$

is an iterated Ore extension of $\mathcal{C}(T)$, where $\delta_i(c_j) = c_i c_j - c_j c_i$ for $1 \leqslant j < i \leqslant \lambda$; it follows that the Gelfand-Kirillov dimension of $\mathcal{H}(\lambda m + l)$ over $\mathcal{C}(T)$ is λ. (We pointed out that one of the points of the proof in Theorem 4. 7 of [Guo] is to use the basis of $\mathcal{C}(T)$ at degree $\lambda\mathbf{n}$ in [R4].)

We claim that

$$c_\lambda \notin B_{\lambda g\mathbf{n}}(T)$$

for any positive integer λ.

In fact, otherwise, by (4. 5) and (4. 6) we have

$$\begin{aligned} c_\lambda &\in B_{\lambda g\mathbf{n}}(T) \\ &\subseteq \mathcal{H}(\lambda m - 1) \\ &= \mathcal{H}((\lambda - 1)m + m - 1) \\ &= \langle \mathcal{C}(T), c_1, c_2, \cdots, c_{\lambda-1} \rangle \end{aligned}$$

(where if $\lambda=1$, then the equality is understood as $c_1\in\mathscr{C}(T)$). This means

$$\mathscr{K}(\lambda m)=\langle\mathscr{C}(T),c_1,c_2,\cdots,c_{\lambda-1},c_\lambda\rangle=\langle\mathscr{C}(T),c_1,c_2,\cdots,c_{\lambda-1}\rangle,$$

and then the Gelfand-Kirillov dimension of $\mathscr{K}(\lambda m)$ over $\mathscr{C}(T)$ is $\lambda-1$, a contradiction. This completes the proof. □

By Lemma 4.4 and Theorem 4.5 we have

Corollary 4.6 Let T be a non-homogeneous tube of rank m. Denote by $s_\lambda(T)=\Sigma[N]$ where N runs over indecomposable modules in T with quasilength λm. Then for all $\lambda\geqslant 1$ there holds

$$s_\lambda(T)\notin B_{\lambda g\mathbf{n}}(T).$$

4.7 Denote by Σ_m the symmetric group of degree m. Let $\sigma=(1,2,\cdots,m)\in\Sigma_m$. For $[E_{i_1}(j_1)][E_{i_2}(j_2)]\cdots[E_{i_t}(j_t)]\in\mathscr{K}(T)$, where $1\leqslant i_1,i_2,\cdots,i_t\leqslant m$; $t\geqslant 1$; $j_1,j_2,\cdots,j_m\geqslant 1$, define

$$\sigma([E_{i_1}(j_1)][E_{i_2}(j_2)]\cdots[E_{i_t}(j_t)])=[E_{\sigma(i_1)}(j_1)][E_{\sigma(i_2)}(j_2)]\cdots[E_{\sigma(i_t)}(j_t)].) \tag{4.7}$$

We introduce the following element in $\mathscr{K}(T)_{g\mathbf{n}}$:

$$c_{g\mathbf{n}}:=\sum_{1\leqslant i,j\leqslant m}\sigma^i([E_1][E_2]\cdots[E_{m-j+1}(j)]). \tag{4.8}$$

Thus

$$c_{g\mathbf{n}}=\sum_{1\leqslant i\leqslant m}\sigma^i([E_1][E_2]\cdots[E_m])+\sigma^i([E_1][E_2]\cdots[E_{m-1}(2)])+\cdots+\sigma^i([E_1(m)]).$$

Notice that by definition we have

$$c_{g\mathbf{n}}=s_1(T)+x,\quad \text{with } x\in B_{g\mathbf{n}}(T). \tag{4.9}$$

Recall that by $r_{g\mathbf{n}}(T)$ we have denoted the sum of the isoclasses of the all modules in T with dimension vector $g\mathbf{n}$(see 3.3).

Lemma Let T be a non-homogeneous tube of rank m. Then the following holds:

$$c_{g\mathbf{n}}(T)=mr_{g\mathbf{n}}(T). \tag{4.10}$$

Proof Let M be an arbitrary module in T with din $M=g\mathbf{n}$. Since dim E_1, dim E_2, $\cdots$, dim E_m are $\mathbf{Z}$-linear independent (see [DR, Tables], or [R1, p. 146]), it follows that M can he uniquely written as

$$M=E_i(v_1)\oplus E_{i+v_1}(v_2)\oplus\cdots\oplus E_{i+v_1+\cdots+v_{j-1}}(v_j),\qquad(4.11)$$

with $v_1,v_2,\cdots,v_j\geqslant 1, j\geqslant 1$, and $v_1+v_2+\cdots+v_j=m$. We take the low indices modulo m. Thus $E_{i+v_1+\cdots+v_{j-1}}(v_j)=E_{i-v_j}(v_j)$.

We claim that $[M]$ is a term of $c_{\mathbf{gn}}$ with coefficients m, and hence the assertion follows.

In fact, by the presentation (4.11) of M, we can easily analyze the types of filtrations of M from the Auslander-Reiten quiver. Note that $c_{\mathbf{gn}}$ is a sum of m^2 monomials. Those monomials in which $[M]$ is a term are exactly in the following list (this can be seen geometrically from the structure of a tube):

$[E_i]\cdots[E_{i+v_1-1}]\cdots([E_{i-v_j}]\cdots[E_{i-1}])=\sigma^{i-1}([E_1]\cdots[E_m])$,

$[E_i]\cdots[E_{i+v_1-1}]\cdots([E_{i-v_j}]\cdots[E_{i-3}]\cdot[E_{i-2}(2)])=$
$\sigma^{i-1}([E_1]\cdots[E_{m-2}]\cdot[E_{m-1}(2)])$,

…

$[E_i]\cdots[E_{i+v_1-1}]\cdots([E_{i-v_j}(v_j)])=$
$\sigma^{i-1}([E_1]\cdots[E_{m-v_j}]\cdot[E_{m-v_j+1}(v_j)])$;

$[E_{i-v_j}]\cdots[E_{i-1}]\cdots([E_{i-v_j-v_{j-1}}]\cdots[E_{i-v_j-1}])=$
$\sigma^{i-v_j-1}([E_1]\cdots[E_m])$,

$[E_{i-v_j}]\cdots[E_{i-1}]\cdots([E_{i-v_i-v_j-1}]\cdots[E_{i-v_j-3}]\cdot[E_{i-v_j-2}(2)])=$
$\sigma^{i-v_j-1}([E_1]\cdots[E_{m-2}]\cdot[E_{m-1}(2)])$,

…

$[E_{i-v_j}]\cdots[E_{i-1}]\cdots([E_{i-v_j-v_{j-1}}(v_{j-1})])=$
$\sigma^{i-v_j-1}([E_1]\cdots[E_{m-v_j}]\cdot[E_{m-v_{j-1}+1}(v_{j-1})])$;

…

$[E_{i+v_1}]\cdots[E_{i+v_1+v_2-1}]\cdots([E_i]\cdots[E_{i+v_1-1}])=\sigma^{i+v_1-1}([E_1]\cdots[E_m])$,

$[E_{i+v_1}]\cdots[E_{i+v_1+v_2-1}]\cdots([E_i]\cdots[E_{i+v_1-3}]\cdot[E_{i+v_1-2}(2)])=$
$\sigma^{i+v_1-1}([E_1]\cdots[E_{m-2}]\cdot[E_{m-1}(2)])$,

…

$[E_{i+v_1}]\cdots[E_{i+v_1+v_2-1}]\cdots([E_i(v_1)])=$
$\sigma^{i+v_1-1}([E_1]\cdots[E_{m-v_1}]\cdot[E_{m-v_1+1}(v_1)])$.

Altogether we have $v_j + v_{j-1} + \cdots + v_1 = m$ such monomials. Note that the coefficients of $[M]$ in every monomial in the preceeding list are 1, since $\mathrm{Hom}_A(E_i, E_j) = 0$ for $i \neq j$. This proves that $[M]$ is a term of $c_{g\mathbf{n}}$ with coefficients m. □

By (4.10), (4.9), and Corollary 4.6 we have

Corollary 4.8 We have

$$r_{g\mathbf{n}}(T) \notin B_{g\mathbf{n}}(T).$$

Remark We do not know how to prove $r_{\lambda g\mathbf{n}}(T) \notin B_{\lambda g\mathbf{n}}(T)$ for $\lambda > 1$. But for $\lambda = 2$ and $\mathrm{rank}(T) = 2$, we have the following fact by direct calculations, which is needed in the next section.

Lemma 4.9 Let T be a tube with $\mathrm{rank}(T) = 2$. Then $r_{2g\mathbf{n}}(T) \notin B_{2g\mathbf{n}}(T)$, where $r_{2g\mathbf{n}}(T)$ is the sum of the isoclasses of all modules in T with dimension vector $2g\mathbf{n}$.

Proof Let E_1, E_2 be the quasi-simples in T, and let N_1, M_1, L_1 be the indecomposables in T with quasi-top E_1, and with quasi-length 2, 3, 4, respectively. Set $N_2 = \tau N_1$, $M_2 = \tau M_1$, and $L_2 = \tau L_1$.

By Lemma 4.4 we have

$$[L_2] \in [L_1] + B_{2g\mathbf{n}}(T). \tag{4.12}$$

Since

$$[M_2 \oplus E_1] = [E_1] \cdot [M_2] - [L_1],$$
$$[M_1 \oplus E_2] = [E_2] \cdot [M_1] - [L_2],$$

it follows that

$$[M_2 \oplus E_1], [M_1 \oplus E_2] \in -[L_1] + B_{2g\mathbf{n}}(T). \tag{4.13}$$

Since

$$(q+1)[N_1^2] = [N_1] \cdot [N_1] - [L_1],$$
$$(q+1)[N_2^2] = [N_2] \cdot [N_2] - [L_2],$$

it follows that

$$[N_1^2], [N_2^2] \in -\frac{1}{q+1}[L_1] + B_{2g\mathbf{n}}(T). \tag{4.14}$$

Since

$$[N_1] \cdot [N_2] = (q-1)[M_1 \oplus E_2] + q[N_1 \oplus N_2],$$

it follows from (4.13) that

$$[N_1 \oplus N_2] \in \frac{q-1}{q}[L_1] + B_{2g\mathbf{n}}(T). \tag{4.15}$$

Since

$$[N_1] \cdot [E_1 \oplus E_2] = [M_1 \oplus E_2] + q[N_1 \oplus E_1 \oplus E_2]$$

and

$$[N_2] \cdot [E_1 \oplus E_2] = [M_2 \oplus E_1] + q[N_2 \oplus E_1 \oplus E_2],$$

it follows from (4.13) that

$$[N_1 \oplus E_1 \oplus E_2], [N_2 \oplus E_1 \oplus E_2] \in \frac{1}{q}[L_1] + B_{2g\mathbf{n}}(T). \tag{4.16}$$

Since

$$[E_1] \cdot [E_1 \oplus E_2^2] = [N_1 \oplus E_1 \oplus E_2] + (q+1)[E_1^2 \oplus E_2^2],$$

it follows from (4.16) that

$$[E_1^2 \oplus E_2^2] \in -\frac{1}{q(q+1)}[L_1] + B_{2g\mathbf{n}}(T). \tag{4.17}$$

By (4.12)~(4.17) we have

$$\begin{aligned} r_{2g\mathbf{n}}(T) = & [L_1] + [L_2] + [M_1 \oplus E_2] + [M_2 \oplus E_1] + [N_1^2] + [N_2^2] + \\ & [N_1 \oplus N_2] + [N_1 \oplus E_1 \oplus E_2] + [N_2 \oplus E_1 \oplus E_2] + [E_1^2 \oplus E_2^2] \\ \in & \frac{q}{q+1}[L_1] + B_{2g\mathbf{n}}(T), \end{aligned}$$

and hence $r_{2g\mathbf{n}}(T) \notin B_{2g\mathbf{n}}(T)$, since $[L_1] \notin B_{2g\mathbf{n}}(T)$ by Theorem 4.5. □

Now, we consider homogeneous tubes.

Theorem 4.10 Let T be a homogeneous tube with quasi-simple E and $\dim E = s\mathbf{n}$. Then for any positive integer λ there holds.

$$\mathscr{H}(T)_{\lambda s\mathbf{n}} = \mathbf{R}[M_\lambda] \oplus B_{\lambda s\mathbf{n}}(T), \tag{4.18}$$

where M_λ is the indecomposable in T with quasi-length λ.

Proof Denote by M_i the indecomposable in T with quasi-length i. Let $\mathbf{P}(\lambda)$ be the set of partitions of λ. A partition p of λ is denoted by

$$p = (\lambda_1^{n_1}, \lambda_2^{n_2}, \cdots, \lambda_t^{n_t}), \text{i. e.},$$

$$n_1\lambda_1 + n_2\lambda_2 + \cdots + n_t\lambda_t = \lambda; 0 < \lambda_1 < \lambda_2 < \cdots < \lambda_t; n_1, n_2, \cdots, n_t > 0; t > 0.$$

For every partition $p = (\lambda_1^{n_1}\ \lambda_2^{n_2} \cdots \lambda_t^{n_t}) \in \mathbf{P}(\lambda)$, set

$$[M(p)] = [M_{\lambda_1}^{n_1} \oplus M_{\lambda_2}^{n_2} \oplus \cdots \oplus M_{\lambda_t}^{n_t}] \in \mathscr{H}(T)_{\lambda s\mathbf{n}},$$

and

$$m_p = [M_{\lambda_1}]^{n_1}[M_{\lambda_2}]^{n_2}\cdots[M_{\lambda_t}]^{n_t} \in \mathcal{H}(T)_{\lambda s\mathbf{n}}.$$

Then

$$\{[M(p)] \mid p \in \mathbf{P}(\lambda)\}$$

is a basis of $\mathcal{H}(T)_{\lambda s\mathbf{n}}$. Since $\mathcal{H}(T)$ is a commutative algebra (see [M, p. 183] or [Z2]). and since every isoclass of a module in T is generated by isoclasses of indecomposables in T. it follows that $B_{\lambda s\mathbf{n}}(T)$ = the space spanned by m_p, $p \in \mathbf{P}(\lambda)$, $p \neq (\lambda)$, and that $\{m_p \mid p \in \mathbf{P}(\lambda)\}$ is a generating system of $\mathcal{H}(T)_{\lambda s\mathbf{n}}$, and hence is also a basis of $\mathcal{H}(T)_{\lambda s\mathbf{n}}$. In particular we have $m_{(\lambda)} = [M(\lambda)] \notin B_{\lambda s\mathbf{n}}(T)$, i. e.,

$$\mathcal{H}(T)_{\lambda s\mathbf{n}} = \mathbf{R}[M_\lambda] \oplus B_{\lambda s\mathbf{n}}(T). \quad \square$$

§ 5. Indecomposable modules which can be generated

Let $A = kQ$ with Q an affine quiver. The aim of this section is to classify all indecomposables M which can be generated inside $\mathcal{H}(A)$ by "smaller" modules, i. e., $[M] \in B_{\dim M}(A)$ (see Theorem 5. 3). We do this by reducing the criterion of $x \in B_{\mathbf{d}}(A)$. where $x \in \mathcal{H}(T)_{\mathbf{d}}$, to the one of $x \in B_{\mathbf{d}}(T)$ (see Theorem 5. 2).

Lemma 5. 1 Let E be a homogeneous quasi-simple with dim $E = \lambda\mathbf{n}$. Then $[E] \notin B_{\lambda\mathbf{n}}(A)$.

Proof Otherwise, by Theorem 3. 7 (3) there exists non-zero regular modules M, N, with dim M + dim $N = \lambda\mathbf{n}$, and $c, c_{M,N} \in \mathbf{R}$, such that

$$[E] = cr_{\lambda\mathbf{n}} + \sum_{c_{M,N}} [M] \cdot [N]. \qquad (5.\,1)$$

Since E is quasi-simple, it follows that $[E]$ is not a term of $\sum_{c_{M,N}}[M][N]$; but $[E]$ is a term of $r_{\lambda\mathbf{n}}$, and hence by comparing the coefficients of $[E]$ on both sides of (5. 1) we get $c = 1$. This forces $t_\lambda(A) = 1$ in the sense of 2. 3, i. e., E has to be the unique homogeneous quasi-simple A-module with dim $E = \lambda\mathbf{n}$(otherwise, let E' be a homogeneous quasi-simple with $\dim E' = \lambda\mathbf{n}$. $E' \not\cong E$. Then by comparing the coefficients of $[E']$ on the two sides of (5.1) we get the contradiction $c=0$). Then $\lambda=2$ or $\lambda=1$, according to Corollary 2.8(2).

If $\lambda=2$ and $t_1(A)\neq 0$, then let E' be a homogeneous quasi-simple module with $\dim E'=\mathbf{n}$. Let L he the (homogeneous) indecomposable with quasi-socle E' and $\dim L=2\mathbf{n}$. Then both $[L]$ and $[E'\oplus E']$ are terms of $r_{2\mathbf{n}}$, with coefficient 1. But, since E' is homogeneous quasi-simple, it follows that product $[M]\cdot[N]$ on the right side of (5.1), such that $[L]$ or $[E'\oplus E']$ is a term of $[M]\cdot[N]$, is unique and has to be $[E']\cdot[E']$. Note that

$$[E']\cdot[E']=[L]+(q+1)[E'\oplus E'].$$

Then by comparing the coefficients of $[L]$ and $[E'\oplus L']$ on both sides of (5.1) we get a contradiction,

$$0=1+c_{E',E'} \quad \text{and} \quad 0=1+(q+1)c_{E',E'}.$$

If $\lambda=2$ and $t_1(A)=0$, then A is of type $\widetilde{D}_n$ or $\widetilde{E}_n$ by Corollary 2.8(1), and hence A has a tube T of rank 2. Taking the T-part in both sides of (5.1). and noticing that if M, N are regular, then $[M]\cdot[N]$ has a T-term if and only if $M, N\in T$, we then get

$$r_{2\mathbf{n}}(T)=-\sum_{c_{M,N}}[M]\cdot[N]\in B_{2\mathbf{n}}(T),$$

which contradicts Lemma 4.9.

If $\lambda=1$, then A is not the Kronecker algebra since $t_1(A)=1$, and then there exists a non-homogeneous tube T'. Taking the T'-part on both sides of (5.1), we then get

$$-r_{\mathbf{n}}(T')=\sum_{c_{M,N}}[M]\cdot[N]\in B_{\mathbf{n}}(T'),$$

which contradicts Corollary 4.8. □

5.2 Theorem For any tube T of A, and any positive integer λ, there holds

$$B_{\lambda\mathbf{n}}(T)=B_{\lambda\mathbf{n}}(A)\cap\mathscr{H}(T)_{\lambda\mathbf{n}}. \tag{5.2}$$

Thus, given an $x\in\mathscr{H}(T)_{\lambda\mathbf{n}}$, then $x\in B_{\lambda\mathbf{n}}(T)$ if and only if $x\in B_{\lambda\mathbf{n}}(A)$.

Proof Let $0\neq x\in B_{\lambda\mathbf{n}}(A)\cap\mathscr{H}(T)_{\lambda\mathbf{n}}$. Then by Theorem 3.7(3) we

have
$$x=r(x)=cr_{\lambda\mathbf{n}}+\sum c_{M,N}[M]\cdot[N], \tag{5.3}$$
where M,N are non-zero regular modules with $\dim M+\dim N=\lambda\mathbf{n}$, and $c, c_{M,N}\in\mathbf{R}$.

First, assume that $t_\lambda(A)\neq 0$; i. e., there exists a homogeneous quasi-simple E with $\dim E=\lambda\mathbf{n}$. It is easy to see that $[E]$ is not a term of x; otherwise, $E\in T$ and then T is a homogeneous tube. Since E is the quasi-simple in T with $\dim E=\lambda\mathbf{n}$, it follows that $x=a[E]$ with $a\neq 0$. But by Lemma 5.1 we have $[E]\notin B_{\lambda\mathbf{n}}(A)$.

Thus, by comparing the coefficients of $[E]$ on both sides of (5.3) we get $c=0$. Now, taking the T-part on both sides of (5.3), we then get
$$x=\sum c_{M,N}[M]\cdot[N]\in B_{\lambda\mathbf{n}}(T).$$

Second, if $t_\lambda(A)=0$, then $\lambda=1$ and A is of type $\widetilde{D}_n$ or $\widetilde{E}_n$ by Corollary 2.8(1), and hence A has a non-homogeneous tube T' such that $T'\neq T$. Taking the T'-part on both sides of (5.3), we then get
$$cr_{\mathbf{n}}(T')\in B_{\mathbf{n}}(T');$$
by Corollary 4.8 this forces $c=0$, and then $x\in B_{\mathbf{n}}(T)$. □

We say that an A-module M can be generated, provided that
$$[M]\in B_{\dim M}(A).$$

Theorem 5.3 Let M be an indecomposable A-module. Then M can be generated if and only if $\dim M\neq e$, $i=1,2,\cdots,n$, and $\dim M\neq\lambda\mathbf{n}$ for any non-negative integer λ.

In particular, $B_{\mathbf{d}}(A)=H(A)_{\mathbf{d}}$ if and only if $\mathbf{d}\neq \boldsymbol{e}_i$, $i=1,2,\cdots,n$, and $\mathbf{d}\neq\lambda\mathbf{n}$ for any non-negative integer λ.

Proof If M can be generated, then by Theorems 4.5, 4.10, and 5.2 we know that $\dim M\neq\lambda\mathbf{n}$ for any non-negative integer λ, and $\dim M\neq e_i$, $i=1,2,\cdots,n$.

The converse is already known from [SV]; here we give a more direct proof. If $\dim M\neq e_i$, $i=1,2,\cdots,n$, and $\dim M\neq\lambda\mathbf{n}$ for any non-negative integer λ, then by Lemma 3.4 we may assume that $M\in T$, where T is a non-homogeneous tube with rank m, and the quasi-length of M is not a multiple of m. If the quasi-length of M is smaller than m, then $[M]$ is generated by the quasi-simples in T (see [Z3, Theorem 1.1]). Let the quasi-length of M be $\lambda m+l$ with $1\leqslant l\leqslant m-1$. Let $L\in T$ be the indecomposable submodule of M, with quasi-length λm. Then it is easy to see that

$$[M/L]\cdot[L]=[M]+[L\oplus M/L];[L]\cdot[M/L]=[L\oplus M/L].$$

It follows that

$$[M]=[M/L]\cdot[L]-[L]\cdot[M/L]\in B_{\dim M}(A).$$

§ 6. Minimal homogeneous generators of $\mathscr{K}(\mathrm{A})$

The aim of this section is to explicitly write out systems of minimal homogeneous generators of $\mathscr{K}(kQ)$, especially all systems of minimal generators of $\mathscr{K}(kQ)$ consisting of isoclasses of indecomposable modules (see Theorems 6.3 and 6.4). in particular, we get the formula of the number of the imaginary, simple roots at degree $\lambda\mathbf{n}$, in the corresponding Borcherds-Cartan matrix associated with affine quiver Q (see Corollary 6.5).

6.1 For a positive integer λ, let $\mathbf{T}(\lambda)$ be the set consisting of all non-homogeneous tubes, and those homogeneous tubes T with quasi-simple in T having dimension vector $s\mathbf{n}, s\mid\lambda$. For $T\in\mathbf{T}(\lambda)$, take $M_\lambda(T)$ to be an arbitrary indecomposable in T with dimension vector $\lambda\mathbf{n}$. Then by Theorems 4.5 and 4.10, there exists a unique $c_\lambda(T)\in\mathbf{R}$ and a unique $x\in B_{\lambda\mathbf{n}}(T)$, such that

$$r_{\lambda\mathbf{n}}(T)=c_\lambda(T)[M_\lambda(T)]+x. \tag{6.1}$$

Define $b_{\lambda\mathbf{n}}$ to be the element

$$b_{\lambda\mathbf{n}}:=\sum_{T\in\mathbf{T}(\lambda)}c_\lambda(T)[M_\lambda(T)]\in\mathscr{K}(A)_{\lambda\mathbf{n}}. \tag{6.2}$$

Lemma For any positive integer λ, we have

(1) $b_{\lambda\mathbf{n}}\neq 0$.

(2) $b_{\lambda\mathbf{n}}\in B_{\lambda\mathbf{n}}(A)$.

Proof (1) If there exists a homogeneous tube T with quasi-simple

E such that $\dim E=\lambda\mathbf{n}$, then $r_{\lambda\mathbf{n}}(T)=[E]$, and hence by definition we have $c_\lambda(T)=1$, it follows that $b_{\lambda\mathbf{n}}\neq 0$.

Otherwise, by Corollary 2.8(i) we have $\lambda=1$; then by Lemma 4.8 $r_{\mathbf{n}}(T)\notin B_{\mathbf{n}}(T)$ for any non-homogeneous tube T. Thus $c_1(T)\neq 0$ by (6.1) and Theorems 4.5.

(2) By definition we have

$$b_{\lambda\mathbf{n}}-\sum_{T\in\mathbf{T}(\lambda)} r_{\lambda\mathbf{n}}(T)=b_{\lambda\mathbf{n}}-\sum_{\text{tube } T} r_{\lambda\mathbf{n}}(T)\in B_{\lambda\mathbf{n}}(A),$$

and then the assertion follows from Lemma 3.5(2). □

Remark We conjecture that $c_\lambda(T)\neq 0$ for all $T\in\mathbf{T}(\lambda)$ and all positive integers λ; or equivalently, $r_{\lambda\mathbf{n}}(T)\notin B_{\lambda\mathbf{n}}(T)$ for all $T\in\mathbf{T}(\lambda)$ and all positive integers λ (cf. the Remark preceding Lemma 4.9).

Theorem 6.2 Let V_λ be the $\mathbf{R}$-space with basis $\{[M_\lambda(T)]\mid T\in\mathbf{T}(\lambda)\}$. Then we have

(1) $\mathscr{H}(A)_{\lambda\mathbf{n}}=V_\lambda+B_{\lambda\mathbf{n}}(A)$.

(2) $V_\lambda\cap B_{\lambda\mathbf{n}}(A)=\mathbf{R}b_{\lambda\mathbf{n}}$.

(3) $\dim_{\mathbf{R}}V_\lambda=1+\sum_{s\mid\lambda}N(q,s)$.

Proof The assertion (1) follows from Lemmas 3.4 and 3.5 and Theorems 4.5 and 4.10.

(2) By Lemma 6.1 we only need to prove that if $x=\sum_{T\in\mathbf{T}(\lambda)} a_T[M_\lambda(T)]\in V_\lambda\cap B_{\lambda\mathbf{n}}(A)$, then $x=cb_{\lambda\mathbf{n}}$ for some $c\in\mathbf{R}$.

In fact, by Theorem 3.7(4), $x\in B_{\lambda\mathbf{n}}(A)$ means that there exists a $c\in\mathbf{R}$ such that for every tube T of A, $r_T(x)$ is of the form

$$r_T(x)=cr_{\lambda\mathbf{n}}(T)+\sum c_{M,N}[M]\cdot[N],$$

where M,N are non-zero modules in T with $\dim M+\dim N=\lambda\mathbf{n}$, and $c_{M,N}\in\mathbf{R}$, and hence by (6.1) we have

$$r_T(x)=cc_\lambda(T)[M_\lambda(T)]+y$$

for some $y\in B_{\lambda\mathbf{n}}(T)$. Since $r_T(x)=a_T[M_\lambda(T)]$; and $[M_\lambda(T)]\notin B_{\lambda\mathbf{n}}(T)$ by Theorems 4.5 and 4.10, it follows that $a_T=cc_\lambda(T)$ for $T\in\mathbf{T}(\lambda)$, i.e., $x=cb_{\lambda\mathbf{n}}$.

(3) Let m be the number of non-homogeneous tubes of A. and let $t_s(A)$ be the number of homogeneous quasi-simples with dimension vector $s\mathbf{n}$. Then by Proposition 2.7 we have

$$\begin{aligned}\dim_{\mathbf{R}} V_\lambda &= |\mathbf{T}(\lambda)| \\ &= \sum_{s\mid\lambda} t_s(A) + m \\ &= m + t_1(A) + \sum_{s\mid\lambda, s>1} N(q,s) \\ &= 1 + q + \sum_{s\mid\lambda, s>1} N(q,s) \\ &= 1 + \sum_{s\mid\lambda} N(q,s). \quad \square\end{aligned}$$

6.3 For any positive integer λ, let W_λ be an arbitrary complement of $\mathbf{R}b_{\lambda\mathbf{n}}$ in V_λ, i.e.,

$$V_\lambda = W_\lambda \oplus \mathbf{R}b_{\lambda\mathbf{n}}. \tag{6.3}$$

Then by Theorem 6.2 we have

$$\dim_{\mathbf{R}} W_\lambda = \sum_{s\mid\lambda} N(q,s) = \frac{1}{\lambda}\sum_{s\mid\lambda}\varphi\left(\frac{\lambda}{s}\right)q^s, \tag{6.4}$$

and

$$\mathscr{K}(A)_{\lambda\mathbf{n}} = W_\lambda \oplus B_{\lambda\mathbf{n}}(A). \tag{6.5}$$

This proves the following.

Theorem Taking a basis G_λ of W_λ, the set

$$G = \{[S(i)] \mid 1 \leqslant i \leqslant n\} \cup \bigcup_\lambda G_\lambda \tag{6.6}$$

is a system of minimal homogeneous generators of $\mathscr{K}(A)$, where $S(1)$, $S(2), \cdots, S(n)$ are all simple A-modules.

6.4 Choose an arbitrary tube $T' \in \mathbf{T}(\lambda)$, such that $c_\lambda(T') \neq 0$. Then the set

$$\{[M_\lambda(T)] \mid T \in \mathbf{T}(\lambda), T \neq T'\}$$

spans a complement of $\mathbf{R}b_{\lambda\mathbf{n}}$ in V_λ, and hence by the theorem above we have

Theorem The set

$$G = \{[S(i)] \mid 1 \leqslant i \leqslant n\} \cup \bigcup_\lambda \{[M_\lambda(T)] \mid T \in \mathbf{T}(\lambda), T \neq T'\} \tag{6.7}$$

is a system of minimal generators of $\mathscr{K}(A)$; moreover, any system of minimal generators of $\mathscr{K}(A)$, which consists of isoclasses of indecomposable modules, is of this form.

6.5 As a consequence, we get the number of minimal homogeneous generators of $\mathscr{K}(A)$ at degree $\lambda\mathbf{n}$. In [HX], Section 5, Hua and Xiao has obtained the following formula in different terminology, using character formula.

Corollary Let $\mathbf{d}\in\mathbf{N}_0^n$. Then

$$\operatorname{codim}_{\mathbf{R}} B_{\mathbf{d}}(A)=\begin{cases}1, & \mathbf{d}=\mathbf{e}_i, i=1,2,\cdots,n;\\ \sum\limits_{s\mid\lambda} N(q,s), & \mathbf{d}=\lambda\mathbf{n}, \lambda \text{ a positive integer},\\ 0, & \text{otherwise}.\end{cases}$$

Proof If $\mathbf{d}=\mathbf{e}_i$, $i=1,2,\cdots,n$, then the assertion follows from $B_{\mathbf{d}}(A)=0$ and $\dim_{\mathbf{R}}\mathscr{K}(A)_{\mathbf{d}}=1$. If $\mathbf{d}=\lambda\mathbf{n}$ for a positive integer λ, then the assertion follows from (6.4) and (6.5).

In the remaining case, $B_{\mathbf{d}}(A)=\mathscr{K}(A)_{\mathbf{d}}$ by Theorem 5.3. □

Acknowledgements

This work was done in part while the authors visited Universität Bielefeld, supported by a grant from Volkswagen-Stiftung. They thank Professor C. M. Ringel for his hospitality and helpful conversations, and the Fakultät für Mathematik for the working facilities.

References

[ARS] Auslander M, Reiten I. and Smaløs. Representation Theory of Artin Algebras, Cambridge Studies in Advanced Mathematics. Vol. 36, Cambridge Univ. Press, Cambridge, UK, 1994.

[B] Borcherds R. Generalized Kac-Moody algebras. J. Algebra, 1988, 115: 501-512.

[DR] Dlab V, Ringel C M. Indecomposable representations of graphs and algebras. Mem. Amer. Math, Soc, 173, 1976.

[FIS] Friedberg S H, Insel A J, Spence L E, Lincar Algebra. Prentice-Hall, Englewood Cliffs. NJ, 1979.

[G] Green J A. Hall algebras, hereditary algebras and quantum groups. Invent. Math. , 1995, 120: 361-377.

[GL] Geigle W, Lenzing H. Perpendicular categories with applications to representations and sheaves. J. Algebra, 1991, 144: 273-343.

[GP] Guo J Y, Peng I G. Universal PBW-basis of Hall-Ringel algebras and Hall polynomials. J. Algebra, 1997, 198: 339-351.

[Guo] Guo J Y. PBW-basis for the composition algebras of affine type $\widetilde{A}_n$, Preprint.

[HHKU] Happel D, Hartlicb S, Kerner Q Unger L. On perpendicular categories of stones over quiver algebras. Comment. Math. Helv. 1996, 71: 463-474.

[HX] Hua J, Xiao J. On Ringel-Hall algebras of tame hereditary algebras. SFB(1999－124), Universität Bielefeld.

[J] Jacobson N, Basic Algebra 1, 2nd ed. Freeman, New York, 1985.

[K] Kac V G, Infinite root systems, representations of graphs and invariant theory. Invent, Math, 1980, 56: 57-92.

[Kang] Kang S J. Quantum deformations of generalized Kac-Moody algebras and then modules. J. Algebra, 1995, 175: 1 041-1 066.

[L] Lusztig G. Introduction to Quantum Groups. Birkhäuser. Boston. 1993.

[Mac] MacDonald I G. Symmetric Functions and Hall Polynomials, 2nd ed. Oxford Univ. Press. Oxford/Basel/New York, 1995.

[M] Moody R V. Euclidean Lie algebras. Canad. J. Math. 1969, 21: 1 432-1 454.

[R1] Ringel C M. Tame Algebras and Integral Quadratic Forms. Lecture Notes in Mathematics, Vol. 1 099, Springer-Verlag, Berlin/New York, 1984.

[R2] Ringel C M. Hall algebras. in: Topics in Algebra, Banach Center Publ, 1990, 26: 433-447.

[R3] Ringel C M. Hall algebras and quantum groups. Invent. Math. , 1990, 101: 583-592.

[R4] Ringel C M. The composition algebra of a cyclic quiver. Proc. London Math. Soc, 1993, 66(3): 507-537.

[R5] Ringel C M. Hall algebras revisited. Israel Math. Conf. Proc. , 1993, 7: 171-176.

[R6] Ringel C M. PBW-bases of quantum groups. J. Reine Angew. Math., 1996, 470:51-88.

[R7] Ringel C M. Green's Theorem on Hall algebras. in: Canadian mathematical Society Conference Proceedings. Vol. 19 (R. Bautista, R. Martinez-Villa and J. A. de la Peña, Eds.), 185-245, Am. Math. Soc., Providence, 1996.

[S] Schofield A. Semi-invariants of quivers. J. London Math. Soc., 1991, 43:385-395.

[SV] Sevenhant B, van den Bergh M. A relation between a conjecture of Kac and the structure of the Hall Algebra. J. Pure Appl. Algebra, 2001, 160:319-332.

[Z1] Zhang P. Triangular decomposition of the composition algebra of the Kronecker algebra. J. Algebra, 1996, 184:159-174.

[Z2] Zhang P. Ringel-Hall algebras of standard homogeneous tubes. Algebra Colloq, 1997, 4(1):89-94.

[Z3] Zhang P. Composition algebras of affine types. J. Algebra, 1998, 206:505-540.

[Z4] Zhang P. Representations as elements in affine composition algebras. Trans. Amer. Math. Soc., 2001, 353:1 221-1 249.

Journal of Algebra,2003,267:342-358.

域 k 的有理函数域上 k-代数的表示①

Representations of a k-algebra over the Rational Functions over k

Abstract For Λ a finite-dimensional k-algebra, k a field, we study the relations between the category of all left Λ-modules, Mod Λ, and the category of finitely generated $\Lambda \otimes_k k(x) = \Lambda^{k(x)}$-modules, mod $\Lambda^{k(x)}$. In particular, we consider those $G \in$ mol $\Lambda^{k(x)}$ such that ${}_\Lambda G$ is indecomposable. We prove that such modules are in the mouth of components in mod $\Lambda^{k(x)}$ which are tubes or have the shape $\mathbf{Z}A_\infty$, $\mathbf{Z}D_\infty$, or $\mathbf{Z}B_\infty$.

§ 1. Introduction

Let Λ be a finite-dimensional k-algebra, k a field. By Mod Λ we denote the category of left Λ-modules, mod Λ denote the full subcategory of Mod Λ whose objects are the left Λ-modules which are finite-dimensional over k.

It is known from [5] that if k is algebraically closed and Λ of tame representation type, for any generic Λ-module there is a splitting $\mathrm{End}_\Lambda(G) = k(x) \oplus \mathrm{radEnd}_\Lambda(G)$, where $k(x)$ is the field of rational functions in x with coefficients in k. Chosen such a splitting G has a

① Received: 2002-10-22.
本文与 Bautista R 合作.

structure of $\Lambda^{k(x)}=\Lambda\otimes_k k(x)$-module. In fact, G is indecomposable in mod $\Lambda^{k(x)}$. This fact suggests a close relation between Mod Λ and mod $\Lambda^{k(x)}$. We will see that for finite representation type there is a simple relation between mod Λ and mod $\Lambda^{k(x)}$. For infinite representation type we will study those $G\in$ mod $\Lambda^{k(x)}$ such that ${}_\Lambda G$ is indecomposable (in this case ${}_\Lambda G$ is a generic Λ-module). We will prove that these G are sitting in the mouth of components of the form $\mathbf{Z}A_\infty$, $\mathbf{Z}D_\infty$, $\mathbf{Z}B_\infty$, or in tubes. In particular, if k is algebraically closed, Λ is of tame representation type and $\mathrm{End}_\Lambda(G)/\mathrm{radEnd}_\Lambda(G)\cong k(x)$ then G lies in the mouth of a homogeneous tube.

§ 2. Field extensions

Let Λ be a finite-dimensional k-algebra and $k\subseteq K$ a field extension, take the K-algebra $\Lambda^K=\Lambda\otimes_k K$.

Consider the functor

$$\otimes_k K:\mathrm{Mod}\ \Lambda\to\mathrm{Mod}\ \Lambda^K.$$

If $M\in$ Mod Λ we put $M^K=M\otimes_k K$. Here Λ is artinian, therefore any projective (injective) Λ-module is coproduct of finitely generated Λ-modules, then from [10], Lemmas 2.3 and 2.4, we have:

Lemma 2.1 (1) If P is projective in Mod Λ then P^K is projective in Mod Λ^K. Moreover, each indecomposable projective in Mod Λ^K is a direct summand of P^K for some projective $P\in$ Mod Λ.

(2) If E is injective in Mod Λ, E^K is injective in Mod Λ^K. Moreover, any indecomposable injective in Mod Λ^K is a direct summand of E^K for some injective in Mod Λ.

(3) For $M,N\in$ mod Λ and $i\geqslant 0$ the canonical homomorphism

$$\mathrm{Ext}^i_\Lambda(M,N)\otimes_k K\to\mathrm{Ext}^i_{\Lambda^K}(M^K,N^K)$$

is an isomorphism.

(4) If $M,N\in$ mod Λ then $M\cong N$ iff $M^K\cong N^K$.

We are now going to consider the extension $k\subseteq k(x)$, where $k(x)$ is the field of rational functions on x. In the following, a rational algebra is an algebra $\Gamma=k[x]_{h(x)}=\{f(x)/h(x)^m \mid m\in\mathbf{N}, f(x)\in k[x]\}$ for $h(x)\in k[x]$. We put $S(\Gamma)=\{\lambda\in k \mid h(\lambda)\neq 0\}$. For Γ a rational algebra we also consider the algebra $\Lambda\otimes_k\Gamma=\Lambda^\Gamma$. Now if $\lambda\in S(\Gamma)$, $S_\lambda=k[x]/(x-\lambda)$ is a Γ-module. If $L\in \text{Mod } \Lambda^\Gamma$, $L\otimes_\Gamma S_\lambda\in \text{Mod } \Lambda$. If L is free finitely generated over Γ then $L\otimes_\Gamma S_\lambda$ is finite-dimensional over k.

Definition 2.1 A field extension $k\subseteq K$ is Mac Lane separable if char $k=0$ or char $k=p$ and K is linearly disjoint from kp^{-1} over k, that is the natural homomorphism

$$K\otimes_k k^{p^{-1}}\to Kk^{p^{-1}} \text{ is bijective,}$$

where $k^{p^{-1}}$ is the subfield of the algebraic closure of k consisting of the elements x with $x^p\in k$. Here K and $k^{p^{-1}}$ are considered as subfields of $\overline{K}$ the algebraic closure of K.

In [8, Theorem 8.37(3)] is proved that $k\subset K$ is Mac Lane separable iff K has a trascendency base B over k, such that K is separable algebraic over $k(B)$. The extension $k\subset k(x)$ is a particular case of separable extension in the sense of Mac Lane. In [9] is proved that global dimension of a k-algebra is preserved under this last class of extension. In [10] the behaviour of almost split sequences under Mac Lane separable extensions is described. For the convenience of the reader we include Lemmas 2.2, 2.3 and Corollaries 2.1~2.3 which have been obtained in [10]. In case $k\subset k(x)$ one can obtain sharper results as those in Lemma 2.4 and in Section 3 of this paper.

Lemma 2.2 If $k\subseteq K$ is a Mac Lane separable field extension and A is a semisimple finite-dimensional k-algebra then A^K is a semisimple K-algebra.

A consequence of the above is

Lemma 2.3 If $k\subseteq K$ is a Mac Lane separable field extension, then

$$(\text{rad } \Lambda)^K=\text{rad}(\Lambda^K).$$

Corollary 2.1 Let $X\in \text{mod } \Lambda$.

(1) If

$$P_1\xrightarrow{\phi}P_2\xrightarrow{\eta}X\to 0$$

is a minimal projective presentation of X, then

$$P_1^K \xrightarrow{\phi \otimes 1} P_2^K \xrightarrow{\eta \otimes 1} X^K \to 0$$

is a minimal projective presentation of X^K.

(2) If

$$0 \to X \xrightarrow{\sigma} E_1 \xrightarrow{\rho} E_2$$

is a minimal injective copresentation of X then

$$0 \to X^K \xrightarrow{\sigma \otimes 1} E_1^K \xrightarrow{\rho \otimes 1} E_2^K$$

is a minimal injective copresentation of X^K.

Corollary 2.2 For $X \in \text{mod}\Lambda$ there are isomorphisms

$$(D\ \text{tr}_\Lambda X)^K \cong D\ \text{tr}_{\Lambda^K} X^K, \quad (\text{tr}\ D_\Lambda X)^K \cong (\text{tr}\ D_{\Lambda^K}) X^K.$$

In the rest of the paper we put $K = k(x)$.

Lemma 2.4 Let D be a division ring containing the field k in its center and of finite dimension over k. Then $D \otimes_k K = D^K$ is a division ring.

Proof First we will prove that in D^K the product of nonzero elements is nonzero. We consider $D \otimes_k k[x]$, the elements of this ring are polynomials $\sum a_i X^i$ with $a_i \in D$ and $X = 1 \otimes x$. Clearly, the product of two nonzero elements in this ring is nonzero. We have an inclusion $D \otimes_k k[x] \xrightarrow{i} D^K$, we will identify $a \in D \otimes_k k[x]$ with its image $i(a)$. With this identification if $u, v \in D^K$,

$$u = a(1 \otimes 1/h(x)), \quad v = b(1 \otimes 1/g(x))$$

with $h(x), g(x) \in k[x]$, $a, b \in D \otimes_k k[x]$. Then, if $0 = uv = ab(1 \otimes 1/h(x)g(x))$ we have $ab = ab(1 \otimes 1) = 0$. Therefore, either $a = 0$ or $b = 0$, thus either $u = 0$ or $v = 0$. On the other hand, by Lemma 2.2, D^K is a semisimple finite-dimensional K-algebra where the product of nonzero elements is nonzero, this implies by the Wedderburn theorem that D^K is a division ring. □

Lemma 2.5 If $M \in \text{mod}\ \Lambda$ is indecomposable, then M^K is indecomposable in mod Λ^K.

Proof Since M is indecomposable $F(M) = \text{End}_\Lambda(M)/\text{radEnd}_\Lambda(M)$

is a division ring.

Then by Lemma 2.1(3):

$$\begin{aligned}&\mathrm{End}_{\Lambda^K}(M^K)/\mathrm{radEnd}_{\Lambda^K}(M^K)\cong\mathrm{End}_{\Lambda}(M^K)/\mathrm{rad}(\mathrm{End}_{\Lambda}(M))^K\\&\cong\mathrm{End}_{\Lambda}(M)^K/(\mathrm{radEnd}_{\Lambda}(M))^K\cong(\mathrm{End}_{\Lambda}(M)/\mathrm{radEnd}_{\Lambda}(M))^K\\&\cong F(M)\otimes_k K=F(M)^K,\end{aligned}$$

by Lemma 2.4 this is a division ring and consequently $\mathrm{End}_{\Lambda^K}(M^K)$ is local ring which implies that M^K is indecomposable. □

Corollary 2.3 (1) P indecomposable in mod Λ^K is projective iff $P\cong P_1^K$ for some indecomposable projective P_1 in mod Λ.

(2) E indecomposable in mod Λ^K is injective iff $E\cong E_1^K$ for some indecomposable injective E_1 in mod Λ.

§ 3. Almost split sequences

Here we will study the relation between almost split sequences in mod Λ and mod Λ^K. We will need the following useful result.

Proposition 3.1 Let R be a finite-dimensional k-algebra $M\in$ mod R indecomposable then

$$0\to D\ \mathrm{tr}_R\, M\xrightarrow{f}E\xrightarrow{g}M\to 0$$

is an almost split sequence if for any $u\in\mathrm{rad\ End}_R(M)$ there is $h:M\to E$ with $gh=u$.

Proof It follows from the fact that almost split sequences correspond to elements in the socle of $\mathrm{Ext}_R(M,D\ \mathrm{tr}_R\, M)$ as right $\mathrm{End}_R(M)$-module (see [1, Proposition 2.1, Chapter V]). □

In the rest of the paper we will assume that if Λ is a finite-dimensional k-algebra, Λ is connected, this is, 1 is the only nonzero central idempotent of Λ.

Proposition 3.2 If $0\to N\xrightarrow{f}E\xrightarrow{g}M\to 0$ is an almost split sequence in mod Λ, then

$$0\to N^K\xrightarrow{f\otimes 1_K}E^K\xrightarrow{g\otimes 1_K}M^K\to 0$$

is an almost split sequence in mod Λ^K.

Proof Here $N \cong D\ \mathrm{tr}_\Lambda M$ and $N^K \cong D\ \mathrm{tr}_\Lambda{}^K(M^K)$ by Corollary 1.3. Now by Lemmas 2.1 and 2.3 $\mathrm{rad}(\mathrm{End}_{\Lambda^K}(M^K)) = \mathrm{rad}(\mathrm{End}_\Lambda(M))^K$. Therefore, any morphism $u \in \mathrm{End}_{\Lambda^K}(M)$ is a sum of the form $u = \sum_j u_j \otimes c_j$ with $u_j \in \mathrm{radEnd}_\Lambda(M)$, $c_j \in K$. Since the first sequence is almost split then there are $h_j : M \to E$ with $u_j = gh_j$, thus $u = (g \otimes 1)(\sum_j h_j \otimes c_j)$ with $\sum_j h_j \otimes c_j : M^K \to E^K$. Then by our previous proposition we obtain our claim. □

We recall that the valued Auslander-Reiten quiver of Λ has by vertices the isomorphism classes of indecomposable objects in mod Λ, if $[X]$, $[Y]$ are isoclasses of indecomposables we put an arrow $\alpha : [X] \to [Y]$ if there is a minimal right almost split morphism $E \to Y$, with X direct summand of E, in this case also there is a minimal left almost split morphism $X \to F$, with Y direct summand of F. The valuation of the arrow is $(a(\alpha), a'(\alpha))$ where $a(\alpha)$ is the multiplicity of Y as direct summand of F and $a'(\alpha)$ is the multiplicity of X as direct summand of E. We recall that in the case that Y is indecomposable nonprojective there is an arrow $\alpha : [X] \to [Y]$ iff there is an arrow $\beta : [D\ \mathrm{tr}\ Y] \to [X]$, moreover, $a(\alpha) = a'(\beta)$, $a'(\alpha) = a(\beta)$.

Theorem 3.1 Suppose k is any field. If $\mathcal{C}$ is a connected component of the Auslande-Reiten quiver of Λ and $[X]$ is in $\mathcal{C}$, then the map $[Y] \to [Y^K]$ induces an isomorphism of valued quivers between $\mathcal{C}$ and $\mathcal{C}^K$, where the latter is the component containing the isoclass $[X^K]$ in mod Λ^K.

Proof It follows from (1), (2) of Corollary 2.3, (4) of Lemma 2.1, Lemma 2.5 and Proposition 3.2. □

As a consequence we have

Theorem 3.2 (Compare [9]) Suppose k is anyfield. Then Λ is coffinite representation type iff Λ^K is of finite representation type. In this case the corresponding valued Auslander-Reiten quivers are isomorphic.

Proof Observe that Λ^K is connected. If Λ is of finite representation type, its valued Auslander-Reiten quiver has only a finite connected component but this implies that the valued Auslander-Reiten quiver of Λ^K has a finite connected component, which implies, since Λ^K is connected, that Λ^K has only one finite connected component, so Λ^K is of finite representation type. Conversely, if Λ^K has finite representation type the valued Auslander-Reiten quiver has only one connected component and this component is finite. By our previous result Λ has also only one component and this is finite. □

Definition 3.1 We will say that $X \in \text{mod}\ \Lambda^K$ is defined over k provided $X \cong Y^K$ for some $Y \in \text{mod}\ \Lambda$.

Theorem 3.3 Let Λ be a k-algebra, k algebraically closed $K = k(x)$. Then the following statements are equivalent:

(1) Λ is of finite representation type;

(2) Λ^K is of finite representation type;

(3) any $M \in \text{mod}\ \Lambda^K$ is defined over k.

Proof By Theorem 3.2 (1) and (2) are equivalent. Now if Λ^K is of finite representation type by the second part of Theorem 3.2 all indecomposables Λ^K-modules are defined over k. Thus any $M \in \text{mod}\ \Lambda^K$ is defined over k. We only need to prove that (3) implies (1). So suppose any $M \in \text{mod}\ \Lambda^K$ is defined over k. Assume Λ is of infinite representation type. Since k is algebraically closed by [2] and [3] there are infinitely many nonisomorphic indecomposable modules with the same k-dimension. Then by the equivalence (2)~(4) in Theorem 9.6 in [6] there is a $G \in \text{mod}\ \Lambda^K$ such that ${}_\Lambda G$ is indecomposable, but then G is not defined over k, therefore Λ is of finite representation type. □

§ 4. Minimal projective presentations

We recall that if $M \in \text{Mod}\ \Lambda$, a minimal projective presentation of M is given by an exact sequence

$$P \xrightarrow{\phi} Q \xrightarrow{\eta} M \to 0$$

such that $\text{Im}\ \phi \subset (\text{rad}\ \Lambda)Q$ and $\text{Ker}\ \phi \subset (\text{rad}\ \Lambda)P$.

Lemma 4.1 Suppose $P^K \xrightarrow{\phi} Q^K \xrightarrow{\eta} M \to 0$ is a minimal projective presentation of M as Λ^K-modules. Then ${}_\Lambda P^K \xrightarrow{\phi} {}_\Lambda Q^K \xrightarrow{\eta} {}_\Lambda M \to 0$ is a minimal projective presentation for ${}_\Lambda M$ in Mod Λ.

Proof we have $\mathrm{lm}\phi \subset (\mathrm{rad}\ \Lambda^K)Q^K = (\mathrm{rad}\ \Lambda)_\Lambda Q^K$. In the same way we obtain $\ker\phi \subset (\mathrm{rad}\ \Lambda)_\Lambda P^K$, this proves our claim. □

In the following by Proj Λ we denote the full subcategory of Mod Λ whose objects are projectives, similarly Inj Λ is the full subcategory of Mod Λ whose objects are injectives. By proj Λ we denote the full subcategory of Proj Λ whose objects are in mod Λ, similarly for inj Λ.

There is an equivalence of categories (see [11]): F: Proj $\Lambda \to$ Inj Λ given by $F(-) = D(\Lambda) \otimes_\Lambda -$, where for $M \in$ mod Λ, $D(M) = \mathrm{Hom}_k(M, k)$. Now if $M \in$ Mod Λ, there is a minimal projective presentation $P \xrightarrow{\phi} Q \xrightarrow{\eta} M \to 0$. Following [11] we define $A(M) = \ker(F(\phi))$; $F(\phi): F(P) \to F(Q)$. Now if $M \xrightarrow{i} I \xrightarrow{u} J$ is a minimal injective copresentation of M, as in [11] we define $A^{-1}(M) = \mathrm{Coker}(\mathrm{Hom}_\Lambda(D(A), u))$.

Clearly, if $M \in$ mod Λ, $A\ (M) \cong D\ \mathrm{tr}_\Lambda M$, $A^{-1}(M) \cong \mathrm{tr}\ D_\Lambda M$.

On the other hand, we have the restriction functor res: mod $\Lambda^K \to$ Mod Λ. This restricts to functors,

$$\mathrm{res}: \mathrm{proj}\ \Lambda^K \to \mathrm{Proj}\ \Lambda, \mathrm{res}: \mathrm{inj}\ \Lambda^K \to \mathrm{Inj}\ \Lambda.$$

For $M \in$ mod Λ^K, we have the dual with respect to the field K, $D_x(M) = \mathrm{Hom}_{\Lambda^K}(M, K)$. As above there is an equivalence:

$$F^x: \mathrm{proj}\ \Lambda^K \to \mathrm{inj}\ \Lambda^K, \text{given by } F^x(-) = D_x(\Lambda^K) \otimes_{\Lambda^K} -.$$

If we consider $P^K \in$ mod Λ^K, $F(\mathrm{res}\ P^K) = D(\Lambda) \otimes_\Lambda P^K$ is a Λ-K-bimodule.

Remark $\mathrm{res}(F^x(P^K)) \cong F(\mathrm{res}(P^K))$ as Λ-K-bimodules.

Proof For P^K consider the following composition of morphisms:

$$F^x(P^K) = \mathrm{Hom}_K(\Lambda^K, K) \otimes_{\Lambda^K} P^K \cong \mathrm{Hom}_\Lambda(\Lambda, k) \otimes_k K \otimes_{\Lambda^K} P \otimes_k K$$

$$\xrightarrow{\sigma} \mathrm{Hom}_k(\Lambda, k) \otimes_\Lambda (P \otimes_k K),$$

with $\sigma(a \otimes b \otimes p \otimes c) = a \otimes (p \otimes bc)$. The above composition is natural in

P^K, if we put $P=\Lambda$ we obtain an isomorphism, then since any P is a direct sum of summands of Λ we obtain our result. □

Now if $P^K \xrightarrow{\phi} Q^K \xrightarrow{\eta} M$ is a minimal projective presentation, res(ϕ) is a minimal projective presentation of ${}_\Lambda M$, therefore, $A({}_\Lambda M)=$ ker $F(\mathrm{res}\phi)$ has a structure of Λ-K-bimodule.

Proposition 4.1 If $M\in \mathrm{mod}\ \Lambda^K$, ${}_\Lambda D\ \mathrm{tr}_{\Lambda^K} M\cong A({}_\Lambda M)$, as Λ-K-bimodules.

Proof Take $P^K \xrightarrow{\varphi} Q^K \xrightarrow{\eta} M\to 0$ a minimal projective presentation of M. Then by our previous remark, $D\ \mathrm{tr}_{\Lambda^K} M=\ker F^x(\phi)\cong$ ker $F(\mathrm{res}\phi)=A({}_\Lambda M)$ isomorphism of Λ-K-bimodules. □

Proposition 4.2 If $M\in \mathrm{mod}\ \Lambda^K$, then $A^{-1}({}_\Lambda M)\cong {}_\Lambda \mathrm{tr}D_{\Lambda^K} M$.

Proof In the following if $f:X\to Y$ and $f':X'\to Y'$ are morphisms in a category we will say that they are isomorphic if there are isomorphisms

$$u:X\to X' \text{ and } v:Y\to Y' \text{ such that } f'u=vf.$$

Now for $X\in \mathrm{mod}\ \Lambda^K$, we have:

$\mathrm{Hom}_{\Lambda^K}(\Lambda^K/\mathrm{rad}\ \Lambda^K, X)\cong \mathrm{Hom}_{\Lambda^K}((\Lambda/\mathrm{rad}\ \Lambda)^K, X)\cong \mathrm{Hom}_\Lambda(\Lambda/\mathrm{rad}\ \Lambda, X)$.

The composition of the above isomorphisms is natural in X. Therefore, if $f:X\to Y$ is a morphism in mod Λ^K, the morphisms $\mathrm{Hom}_{\Lambda^K}(\Lambda^K/\mathrm{rad}\ \Lambda^K, f)$ and $\mathrm{Hom}_\Lambda(\Lambda/\mathrm{rad}\ \Lambda, f)$ are isomorphic in the category of k-vector spaces.

Now recall that a morphism $X\xrightarrow{i} I$ in Mod Λ with I injective is an injective envelope iff $\mathrm{Hom}_\Lambda(\Lambda/\mathrm{rad}\ \Lambda, i)$ is isomorphism. Therefore, if we have an exact sequence $0\to X\xrightarrow{i} I\xrightarrow{\phi} J\xrightarrow{\eta} C\to 0$ with I, J injectives, $X\xrightarrow{i} I\xrightarrow{\phi} J$ is a minimal injective copresentation iff

$$\mathrm{Hom}_\Lambda(\Lambda/\mathrm{rad}\ \Lambda, \phi)=0 \text{ and } \mathrm{Hom}_\Lambda(\Lambda/\mathrm{rad}\ \Lambda, \eta)=0.$$

Consider now $M\xrightarrow{i} E_1^K\xrightarrow{\varphi} E_2^K$ a minimal injective copresentation of $M\in \mathrm{Mod}\ \Lambda$. Take $\eta: E_2^K\to C$ a cokernel of ϕ. Then

$$\mathrm{Hom}_\Lambda(\Lambda/\mathrm{rad}\ \Lambda, \phi)\cong \mathrm{Hom}_{\Lambda^K}(\Lambda^K/\mathrm{rad}\ \Lambda^K, \phi)=0,$$
$$\mathrm{Hom}_\Lambda(\Lambda/\mathrm{rad}\ \Lambda, \eta)\cong \mathrm{Hom}_{\Lambda^K}(\Lambda^K/\mathrm{rad}\ \Lambda^K, \eta)=0,$$

consequently ${}_{\Lambda}M \xrightarrow{i} {}_{\Lambda}E_1^K \xrightarrow{\phi} {}_{\Lambda}E_2^K$ is a minimal injective copresentation in mod Λ.

We have

$$A^{-1}({}_{\Lambda}M) \cong \mathrm{Coker}(\mathrm{Hom}_{\Lambda}(D(\Lambda), \phi)),$$

$$\mathrm{tr}\, D_{\Lambda^K} M \cong \mathrm{Coker}(\mathrm{Hom}_{\Lambda^K}(D_x \Lambda^K, \phi)).$$

On the other hand, $D_x \Lambda^K = \mathrm{Hom}_K(\Lambda, K) \cong \mathrm{Hom}_k(\Lambda, k) \otimes_k K = D(\Lambda)^K$.

Therefore,

$$\mathrm{Hom}_{\Lambda^K}(D_x \Lambda^K, Y) \cong \mathrm{Hom}_{\Lambda^K}(D(\Lambda)^K, Y) \cong \mathrm{Hom}_{\Lambda}(D(\Lambda), {}_{\Lambda}Y)$$

as Λ-modules. The composition of the above isomorphisms is natural in Y. Consequently, $\mathrm{Hom}_{\Lambda}(D(\Lambda), \phi) \cong \mathrm{Hom}_{\Lambda^K}(D_x \Lambda^K, \phi)$ as morphisms of Λ-modules. This implies our proposition. □

We recall from [5] that G an indecomposable Λ-module is called generic if G has finite endolength, this is, G has finite length as left $\mathrm{End}_{\Lambda}(G)$-module, and G has infinite dimension over k. In this case $\mathrm{End}_{\Lambda}(G)$ is a local ring and its radical is nilpotent [6, Proposition 4.4, Lemma 4.2].

From [11, Propositions 5.10, 5.11] we have:

Proposition 4.3 If G is a generic Λ-module, then $A(G)$ and $A^{-1}(G)$ are also generic Λ-modules.

Lemma 4.2 If H is in mod Λ^K, then ${}_{\Lambda}H$ is a finite endolength module. If, moreover, ${}_{\Lambda}H$ is indecomposable then ${}_{\Lambda}H$ is a generic Λ-module.

Proof We have $K \subset \mathrm{End}_{\Lambda^K}(H) \subset \mathrm{End}_{\Lambda}({}_{\Lambda}H)$ and H has finite length over K, then also has finite length over $\mathrm{End}_{\Lambda}({}_{\Lambda}H)$, thus ${}_{\Lambda}H$ has finite endolength. If, moreover, ${}_{\Lambda}H$ is indecomposable, ${}_{\Lambda}H$ is generic by definition. □

Proposition 4.4 If $H \in$ mod Λ^K is such that ${}_{\Lambda}H$ is indecomposable, then ${}_{\Lambda}D\ \mathrm{tr}_{\Lambda^K} H$ and ${}_{\Lambda}\mathrm{tr}\ D_{\Lambda^K} H$ are indecomposables.

Proof Here $A({}_{\Lambda}H) \cong {}_{\Lambda}D\ \mathrm{tr}_{\Lambda^K} H$ and $A^{-1}({}_{\Lambda^K}H) \cong {}_{\Lambda}\mathrm{tr}D_{\Lambda K}H$. Thus, if ${}_{\Lambda}H$ is indecomposable, is generic, consequently $A({}_{\Lambda}H)$ and $A^{-1}({}_{\Lambda}H)$ are generic, then indecomposables. This proves our claim. □

§ 5. A-periodic modules

Let G be in mod Λ^K with ${}_\Lambda G$ indecomposable. We are interested in $\mathcal{C}$ which is the component of the Auslander-Reiten quiver of Λ^K containing [G]. For this we first will consider the case in which G is A-periodic, this is, there is a positive integer s such that $A^s({}_\Lambda G)\cong{}_\Lambda G$. Observe that if G is A-periodic G is not necessarily $D\ \mathrm{tr}_{\Lambda^K}$-periodic, since there are Λ-isomorphisms which are not K-linear. Nevertheless, we will prove that if G is A-periodic then for some s, $\dim_K(D\ \mathrm{tr}_{\Lambda^K})^s G=\dim_K G$.

First, we consider some general facts. Let B be a R-algebra, R a field. Consider $\mathcal{A}$ the category Mod B or mod B. If $\mathcal{D}$ is an additive subcategory of $\mathcal{A}$, the category $\mathcal{A}_{\mathcal{D}}$ is the category with objects those of $\mathcal{A}$ and the space of morphisms is given by $\mathrm{Hom}_{\mathcal{A}_{\mathcal{D}}}(X,Y)=\mathrm{Hom}_{\mathcal{A}}(X,Y)/\mathcal{F}_{\mathcal{D}}(X,Y)$, where $\mathcal{F}_{\mathcal{D}}(X,Y)$ is the subspace of all morphisms which factorizes through some object in $\mathcal{D}$.

We define

$$\underline{\mathrm{Mod}}\,B=\mathrm{Mod}\,B_{\mathrm{Proj}\,B},\quad \overline{\mathrm{Mod}}\,B=\mathrm{Mod}\,B_{\mathrm{Inj}\,B},$$

$$\underline{\mathrm{mod}}\,B=\mathrm{mod}\,B_{\mathrm{proj}\,B},\quad \overline{\mathrm{mod}}\,B=\mathrm{mod}\,B_{\mathrm{inj}\,B}.$$

We recall from [11] that the assignment $M\mapsto AM$ induces an equivalence $A:\underline{\mathrm{Mod}}\,\Lambda\to\overline{\mathrm{Mod}}\,\Lambda$ and $X\mapsto D\ \mathrm{tr}_{\Lambda^K}X$ induces an equivalence $D\ \mathrm{tr}_{\Lambda^K}:\underline{\mathrm{mod}}\,\Lambda^K\to\overline{\mathrm{mod}}\,\Lambda^K$. Now observe that for $M\in\mathrm{mod}\,\Lambda^K$, the K-structure of M induces a morphism of rings $l:K\to\mathrm{End}_\Lambda({}_\Lambda M)$. On the other hand, A induces an isomorphism $\eta:\underline{\mathrm{End}}_\Lambda({}_\Lambda M)\to\overline{\mathrm{End}}_\Lambda(A({}_\Lambda M))$. The K-structure of $A({}_\Lambda M)$ induces a morphism of rings $l':K\to\mathrm{End}_\Lambda(A({}_\Lambda M))$, we have the commutative diagram (fig. 5. 1):

$$\begin{array}{ccccc}
K & \xrightarrow{\ l\ } & \mathrm{End}_\Lambda({}_\Lambda M) & \longrightarrow & \underline{\mathrm{End}}_\Lambda({}_\Lambda M)\\
\Big\| & & & & \big\downarrow{\scriptstyle\eta}\\
K & \xrightarrow{\ l'\ } & \mathrm{End}_\Lambda(A({}_\Lambda M)) & \longrightarrow & \overline{\mathrm{End}}_\Lambda(A({}_\Lambda M)).
\end{array}$$

Fig. 5. 1

Now $A({}_\Lambda M)$ and $D\ \mathrm{tr}_{\Lambda^K} M$ are isomorphic as Λ-K-bimodules, thus $\overline{\mathrm{End}}_\Lambda({}_\Lambda D\ \mathrm{tr}_{\Lambda^K} M) \cong \overline{\mathrm{End}}_\Lambda(A({}_\Lambda M))$ as K-modules, therefore we obtain an isomorphism of K-modules $\phi_M: \underline{\mathrm{End}}_\Lambda({}_\Lambda M) \to \overline{\mathrm{End}}_\Lambda({}_\Lambda D\ \mathrm{tr}_{\Lambda^K} M)$.

For H a generic Λ-module we put $F(H) = \mathrm{End}_\Lambda(H)/\mathrm{radEnd}_\Lambda(H)$. We denote by $\mathrm{el}(H)$ the endolength of H. Observe that for $G \in \mathrm{mod}\ \Lambda^K$, $\mathrm{End}_\Lambda({}_\Lambda G)$ is a left K-module. Moreover, if ${}_\Lambda G$ is indecomposable, $\mathrm{rad}(\mathrm{End}_\Lambda({}_\Lambda G))$ is a left K-module, thus $F({}_\Lambda G)$ is a left K-module.

Lemma 5.1 Suppose G is an object in mod Λ^K with ${}_\Lambda G$ indecomposable, then $\dim_K F({}_\Lambda G)$ is finite and $\dim_K G = \mathrm{el}({}_\Lambda G)\dim_K F({}_\Lambda G)$.

Proof Here G has finite endolength, there is a chain of $\mathrm{End}_\Lambda(G)$-submodules of G:

$$0 \subset G_s \subset G_{s-1} \subset \cdots \subset G_1 \subset G$$

such that each $G_i/G_{i+1} \cong F({}_\Lambda G)$ as left $\mathrm{End}_\Lambda(G)$-modules, consequently as K-modules. From here we obtain our result. □

Lemma 5.2 If $G \in \mathrm{mod}\ \Lambda^K$ and ${}_\Lambda G$ is indecomposable,

$$\dim_K F({}_\Lambda G) = \dim_K F({}_\Lambda D\mathrm{tr}_{\Lambda^K} G).$$

Proof Since ${}_\Lambda G$ is a non projective indecomposable any morphism which factorizes through some projective is in the radical. Therefore, ϕ_G induces an isomorphism of K-modules $F({}_\Lambda G) \to F({}_\Lambda D\ \mathrm{tr}_{\Lambda^K} G)$. This proves our lemma. □

Lemma 5.3 With the above hypothesis. If ${}_\Lambda(D\ \mathrm{tr}_{\Lambda^K})^S G \cong {}_\Lambda G$ for some s, then

$$\dim_K (D\ \mathrm{tr}_{\Lambda^K})^s G = \dim_K G.$$

Proof Putting $G^s = (D\ \mathrm{tr}_{\Lambda^K})^s G$, one obtains

$$\dim_K G^s = \mathrm{el}({}_\Lambda G^s)\dim_K F({}_\Lambda G^s) = \mathrm{el}({}_\Lambda G)\ \dim_K F({}_\Lambda G) = \dim_K G. \quad \square$$

Proposition 5.1 Let G be in mod Λ^K, with ${}_\Lambda G$ indecomposable. Then if G is A-periodic, $\mathcal{C}$, the Auslander-Reiten component of Λ^K containing $[G]$ is a tube or it has shape $\mathbf{Z}A_\infty$.

Proof By Theorem 3.1 if X is defined over k, $[X] \in \mathcal{C}$ then for all $[Z] \in \mathcal{C}$, Z is defined over k. Now, by Corollary 2.3 indecomposable injectives and indecomposable projectives are defined over k. Observe that any Λ^K-module of the form M^K with M a Λ-module when restricted to Λ is isomorphic to an infinite sum of copies of M. Here ${}_\Lambda G$ is indecomposable, G is not defined over k. Therefore, $\mathcal{C}$ cannot contain isoclasses of projectives or injectives, consequently our component is regular. In case $\mathcal{C}$ is periodic, $\mathcal{C}$ is a tube, by [7] or [14]. In case $\mathcal{C}$ is not periodic, by our previous lemma, in the $D\,\mathrm{tr}_{\Lambda^K}$-orbit of G, there are infinitely many nonisomorphic modules with the same K-dimension. Thus, by [13, Theorem 5.3], $\mathcal{C}$ has shape $\mathbf{Z}A_\infty$. □

§ 6. Components in mod Λ^K

Let G be an object in mod Λ^K with ${}_\Lambda G$ is indecomposable and $\mathcal{C}$ the component of the Auslander-Reiten quiver of Λ^K containing $[G]$. Before we have described the shape of $\mathcal{C}$ when G is A-periodic. Now we will consider the general case.

Proposition 6.1 Let H be indecomposable in mod Λ^K and

$$0\to H'\xrightarrow{u}E\xrightarrow{v}H\to 0 \tag{6.1}$$

an almost split sequence in mod Λ^K, then if H is not defined over k, the exact sequence

$$0\to {}_\Lambda H'\xrightarrow{u}{}_\Lambda E\xrightarrow{v}{}_\Lambda H\to 0 \tag{6.2}$$

splits in Mod Λ.

Proof Since ${}_\Lambda H'$ has finite endolength, thus by [12], H is pure injective. For proving our result it is enough to prove that (6.2) is a pure exact sequence. For this, by [12, Lemma 3.3], we need to prove that for any $X\in$ mod Λ, the induced sequence

$$0\to \mathrm{Hom}_\Lambda(X, {}_\Lambda H^1)\xrightarrow{u_*}\mathrm{Hom}_\Lambda(X, {}_\Lambda E)\xrightarrow{v_*}\mathrm{Hom}_\Lambda(X, {}_\Lambda H)\to 0$$

is exact. For this we only need to prove that v_* is an epimorphism.

Take $h: X\to {}_\Lambda H$ a morphism of Λ-modules. We have the morphism $\tilde{h}: X^K\xrightarrow{h\otimes 1}{}_\Lambda H\otimes_k K\xrightarrow{m}H$ in mod $\Lambda^{k(x)}$, where $m: {}_\Lambda H\otimes_k K\to H$ is given by the multiplication by elements of K using the K-structure of H. The morphism $\tilde{h}$ is not split epimorphism, otherwise H would be a direct summand of X^K, which is not possible, since H is not defined over k. Now the sequence (6.2) is an almost split sequence, thus there is a $g: X^K \to E$ such that $vg = \tilde{h}$. On the other hand, we have the morphism of Λ-modules $\sigma: X \to X^K$ given by $\sigma(x) = x \otimes 1$. Then $vg\sigma(x) = \tilde{h}(x \otimes 1) = h(x)$. Therefore, v_* is epimorphism, which proves our claim. □

Remark If $M \in \mathrm{mod}\ \Lambda^K$, ${}_\Lambda M$ has finite endolength, therefore ${}_\Lambda M \cong \bigoplus_{i \in I} M_i$ with M_i indecomposable of finite endolength. Now, since endomorphism rings of indecomposable finite endolength modules are local, if I is finite and ${}_\Lambda M \cong \bigoplus N_{j \in J}$, N_j indecomposable, then J is finite and there is a bijective function $\sigma: I \to J$ such that for all $i \in I$, $N_{\sigma(i)} \cong M_i$ (see [6, Propositions 4.4, 4.5]).

Proposition 6.2 Let G be an indecomposable in mod Λ^K, and $\mathcal{C}$ the component of $[G]$ in the Auslander-Reiten quiver of Λ^K. Suppose ${}_\Lambda G$ is indecomposable, then:

(1) $\mathcal{C}$ is a regular component;

(2) if $[Z] \in \mathcal{C}$, ${}_\Lambda Z$ is a finite direct sum of generic Λ-modules Y_i with $Y_i \cong {}_\Lambda H_i$, H_i in the $D\ \mathrm{tr}_{\Lambda K}$-orbit of G;

(3) the function $s: \mathcal{C} \to \mathbf{N}$, defined by $s([Z])$ = number of indecomposable summands of ${}_\Lambda Z$ is a $D\ \mathrm{tr}_{\Lambda^K}$-invariant, additive function in the sense of [7].

Proof The part (1) is proved as in Proposition 5.1.

For proving (2), denote by T the family of generic Λ-modules Z which are isomorphic to some ${}_\Lambda Y$ with Y in the $D\ \mathrm{tr}_{\Lambda^K}$-orbit of G. Thus, for such modules $A(Z) \cong A({}_\Lambda Y) \cong {}_\Lambda D\ \mathrm{tr}_{\Lambda^K} Y$ and $A^{-1}(Z) \cong \mathrm{tr}\ D_{\Lambda^K} Y$ by Propositions 4.1 and 4.2. Therefore, if Z is in T, $A(Z)$ and $A^{-1}(Z)$ are also in T.

For $[X] \in \mathcal{C}$ consider the almost split sequences:

$$0 \to X \to \bigoplus_{i=1}^{n} F_i \to \mathrm{tr}\ D_{\Lambda^K} X \to 0, \quad 0 \to D\ \mathrm{tr}_{\Lambda^K} X \to \bigoplus_{j=1}^{n} E_j \to X \to 0$$

with the F_i and E_j indecomposables. If ${}_\Lambda X$ is a direct sum of objects in T then ${}_\Lambda D\ \mathrm{tr}_{\Lambda^K} X \cong A({}_\Lambda X)$ and ${}_\Lambda\ \mathrm{tr}\ D_{\Lambda^K} X \cong A^{-1}({}_\Lambda X)$ are also finite direct sums of objects in T. By Proposition 6.1 $\bigoplus_{i=1}^{n}({}_{\Lambda}F_i)\cong{}_{\Lambda}X\oplus A^{-1}({}_{\Lambda}X)$. Thus, by the remark before Proposition 6.2 each ΛF_j is a finite direct sum of objects in T. In similar way each ${}_{\Lambda}E_j$ is a finite direct sum of objects in T. Now by induction on the minimal number of arrows connecting $[Z]$ with $[G]$ we can prove that for $[Z]\in\mathcal{C}$, ${}_{\Lambda}Z$ is a finite direct sum of objects in T. This proves (2).

Now observe that by Proposition 6.1 $s([X])+s([D\ \mathrm{tr}_{\Lambda^K}X])=\sum_i s([E_j])$, therefore s is an additive function. Moreover, if ${}_{\Lambda}X\cong\bigoplus_{i=1}^{n}H_i$ with $H_i\in T$, ${}_{\Lambda}D\ \mathrm{tr}_{\Lambda^K}X\cong A({}_{\Lambda}X)\cong\bigoplus_{i=1}^{n}A(H_i)$ thus $s([D\ \mathrm{tr}_{\Lambda^K}X])=s([X])$, therefore s is $D\mathrm{tr}_{\Lambda}{}^{K}$-invariant, proving (3). □

Lemma 6.1 Suppose $G\in\mathrm{mod}\ \Lambda^K$ is such that ${}_{\Lambda}G$ is indecomposable, then if $0\to G^1\xrightarrow{u}E\xrightarrow{v}G\to 0$ is an almost split sequence in mod Λ^K, E is an indecomposable Λ^K-module.

Proof By Proposition 6.1 and the Remark before Proposition 6.2, $s(E)=2$, therefore, if E is decomposable, $E\cong Z_1\oplus Z_2$, with $s(Z_1)=s(Z_2)=1$, thus Z_1 and Z_2 are both indecomposables Λ^K-modules. Our almost split sequence restricted to Λ splits, therefore, we have ${}_{\Lambda}E\cong{}_{\Lambda}G\oplus{}_{\Lambda}G^1$. But then we may assume ${}_{\Lambda}Z_1\cong{}_{\Lambda}G$. Take $Z_1\xrightarrow{v_1}G$ the restriction of v, v_1 is an irreducible morphism in mod Λ^K. Therefore, v_1 is not isomorphism. We have an isomorphism in Mod Λ, $G\xrightarrow{\sigma}Z_1$, then the morphism $v_1\sigma:{}_{\Lambda}G\to{}_{\Lambda}G$ is not isomorphism, since $\mathrm{End}_{\Lambda}(G)$ is local, then $v_1\sigma$ is in the radical which implies that $v_1\sigma$ is nilpotent [6, Lemma 4.2]. But v_1 is either monomorphism or epimorphism, the same is true for $v_1\sigma$, but a nilpotent endomorphism cannot be neither epimorphism nor monomorphism. Thus, E is indecomposable. □

In the following $G\in\mathrm{mod}\ \Lambda^K$ with ${}_{\Lambda}G$ indecomposable, denote by E_2 the middle term of the almost split sequence ending in G and $\mathcal{C}$ the corresponding Auslander-Reiten component of G. Also, if $Z\in\mathcal{C}$, we put

$$Z^s=D\ \mathrm{tr}^s_{\Lambda^{k(x)}}\ Z.$$

Lemma 6.2 Suppose G is not A-periodic. Let $0\to(E_2)^1\to F\oplus G^1\to E_2\to 0$ be an almost split sequence in mod Λ^K. Then, ${}_\Lambda F\cong{}_\Lambda G\oplus{}_\Lambda G^1\oplus{}_\Lambda G^2$ and if F is decomposable $F\cong Z_1\oplus Z_2$ with Z_1 and Z_2 indecomposables,

$$ {}_\Lambda Z_2\cong{}_\Lambda G^1,\quad {}_\Lambda Z_1\cong{}_\Lambda G\oplus{}_\Lambda G^2. $$

Proof From Proposition 6.1 we obtain ${}_\Lambda F\cong{}_\Lambda G\oplus{}_\Lambda G^1\oplus{}_\Lambda G^2$. Thus $s(F)=3$. Suppose $F=Z_1\oplus Z_2\oplus Z_3$, with $Z_i\not\cong 0$ for $i=1,2,3$. Then for $i=1,2,3$, $s(Z_i)=1$. Thus, for some i, ${}_\Lambda Z_i\cong{}_\Lambda G^2$. There is an almost split sequence $0\to Z_i^1\to(E_2)^1\to Z_i\to 0$. But then

$$ {}_\Lambda(E_2)^1\cong{}_\Lambda G^1\oplus{}_\Lambda G^2\cong{}_\Lambda(Z_i\oplus Z_i^1)\cong{}_\Lambda G^2\oplus G^3. $$

Thus, ${}_\Lambda G^1\cong{}_\Lambda G^3$, which is not the case.

Thus, if F is decomposable, $F=Z_1\oplus Z_2$ with $s(Z_1)=2$, $s(Z_2)=1$. There is an almost split sequence $0\to Z_2^1\to E_2^1\to Z_2\to 0$. Then ${}_\Lambda(Z_2\oplus Z_2^1)\cong{}_\Lambda G^1\oplus{}_\Lambda G^2$. Thus, ${}_\Lambda Z_2\cong{}_\Lambda G^1$. Consequently, ${}_\Lambda Z_1\cong{}_\Lambda G\oplus{}_\Lambda G^2$. □

Now we will use the notation and definitions of [7]. We will say that the translation quiver $\mathcal{C}$ has shape $\mathbf{Z}\Gamma$, with Γ a valued graph if $\mathcal{C}$ is isomorphic as translation quiver with $\mathbf{Z}\bar{\Gamma}$ where $\bar{\Gamma}$ is an orientation for Γ.

We will consider the following valued graphs (fig. 6.1):

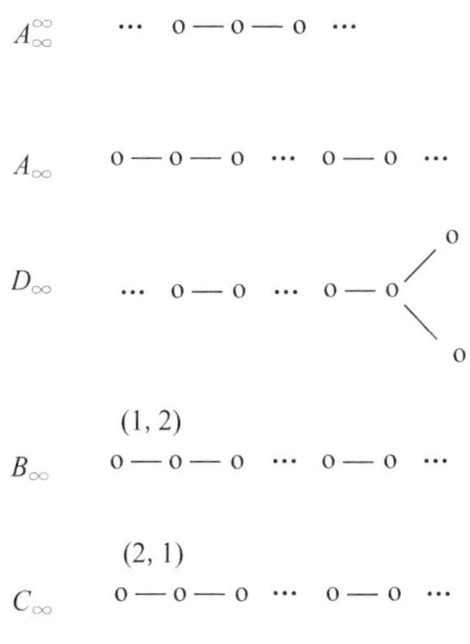

Fig. 6.1

Theorem 6.1 Let G be in mod Λ^K such that ${}_\Lambda G$ is indecomposable. Then $\mathcal{C}$, the component of the Auslander-Reiten quiver of Λ^K containing

$[G]$ is a tube or it has the shape $\mathbf{Z}A_\infty$, $\mathbf{Z}B_\infty$, or $\mathbf{Z}D_\infty$. In the last two cases $s(Z)\leqslant 2$ for all $Z\in\mathcal{C}$.

Proof By Proposition 6.2, $\mathcal{C}$ is a regular component and $s:\mathcal{C}\to\mathbf{N}$ is a $D\ \mathrm{tr}_{\Lambda^K}$-invariant additive function. By [15, Corollary 1], $\mathcal{C}$ is isomorphic as a valued translation quiver with some $\mathbf{Z}\bar{\Gamma}$. In particular, the valued subquiver $\mathcal{S}$ of $\mathcal{C}$ corresponding to $(0,\bar{\Gamma})$ has the following properties:

(1) if $[X]\to[Y]$ is an arrow in $\mathcal{C}$ with $[X]\in\mathcal{S}$ then either $[X]\to[Y]$ is in $\mathcal{S}$ or $[D\ \mathrm{tr}_{\Lambda^K}\ Y]\to[X]$ is an arrow in $\mathcal{S}$;

(2) if $[X]\in\mathcal{S}$ there is not a pair of arrows in $\mathcal{S}$ of the form

$$[D\ \mathrm{tr}_{\Lambda^K}\ Y]\to[X]\to[Y].$$

For the valued quiver $\mathcal{S}$ for any pair $[X]$, $[Y]$ of vertices of $\mathcal{S}$ we put $a(X,Y)=a(\alpha)$ if there is an arrow $\alpha:[X]\to[Y]$ in $\mathcal{S}$ and zero otherwise. We also put $a'(X,Y)=a'(\alpha)$ in case there is an arrow $\alpha:[X]\to[\tau^{-1}Y]$, otherwise we put zero. We denote by $\mathcal{S}_0$ the set of vertices of $\mathcal{S}$.

Following [7], we define the Cartan matrix C of $\mathcal{S}$ as a function from $\mathcal{S}_0\times\mathcal{S}_0\to\mathbf{N}$ by $C([X],[X])=2$ and for

$$[X]\neq[Y],\quad C([X],[Y])=-a(X,Y)-a'(Y,X).$$

We can now consider the restriction of s to $\mathcal{S}_0$, $s:\mathcal{S}_0\to N$. Now for $[X]\in\mathcal{S}_0$ by the properties (1) and (2) of $\mathcal{S}$, the almost split sequence starting in X has the form:

$$0\to X\to\bigoplus_{[Y]\in\mathcal{S}}a(X,Y)Y\bigoplus_{[Y]\in\mathcal{S}}a(X,\tau^{-1}Y)\tau^{-1}Y\to Z\to 0,$$

where

$$\tau Y=D\ \mathrm{tr}_{\Lambda^K}Y,\quad \tau^{-1}Y=\mathrm{tr}\ D_{\Lambda^K}Y.$$

But s is $D\ \mathrm{tr}_{\Lambda^K}$-invariant additive, then

$$\begin{aligned}2s([X])&=\sum_{[Y]\in\mathcal{S}}a(X,Y)s([Y])+\sum_{[Y]\in\mathcal{S}}a(X,\tau^{-1}Y)s(Y)\\&=\sum_{[Y]\in\mathcal{S}}(a(X,Y)+a'(Y,X)s[Y]).\end{aligned}$$

Therefore, for all $[X]\in\mathcal{S}$ we have

$$\sum_{[Y]\in\mathcal{S}}C(X,Y)s([Y])=0.$$

Consequently, the function $s: \mathcal{S}_0 \to \mathbf{N}$ is an additive function for the Cartan matrix C. Therefore, by [7, Theorem 1] $\Gamma \cong A_\infty, A_\infty^\infty, B_\infty, D_\infty$. By Lemma 6.1, A_∞^∞ and C_∞ are excluded. This proves the first part of our result.

For the second part assume $\Gamma = D_\infty$, then we may take $\mathcal{S}$ as fig. 6.2.

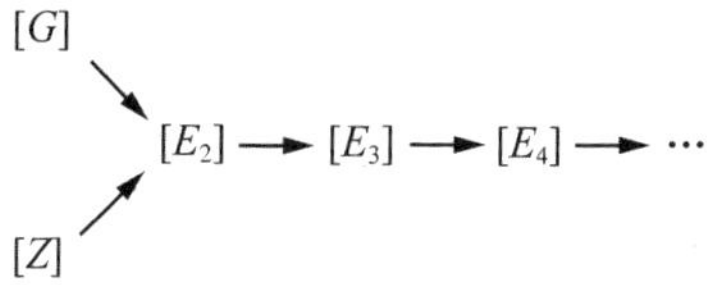

Fig. 6.2

By Proposition 5.1 G is not A-periodic, thus by Lemma 6.2, $s([G]) = s([Z]) = 1$, then we must have $s([E_2]) = 2$ and then $2 = s([E_3]) = s([E_4]) = \cdots$. The same argument works for B_∞. □

Proposition 6.3 Let $L \in \text{mod}\ \Lambda^K$. Then there is a rational algebra Γ and L_0, L_1, Λ^Γ-modules, free finitely generated over Γ, such that

(1) $L_0 \otimes_\Gamma K \cong L, L_1 \otimes_\Gamma K \cong D\ \text{tr}_{\Lambda^K} L$;

(2) for $\lambda \in S(\Gamma), L_1 \otimes_\Gamma S_\lambda \cong D\ \text{tr}_\Lambda (L_0 \otimes_\Gamma S_\lambda)$.

Proof Take a minimal projective presentation for L, $P^K \xrightarrow{f} Q^K \xrightarrow{g} L \to 0$. We have the exact sequence: $0 \to D\ \text{tr}_{\Lambda^K} L \xrightarrow{u} F(P)^K \xrightarrow{F^{k(x)}(f)} F(Q)^K$. Now $f = \sum_j f_j \otimes a_j$ with $f_j: P \to Q$. Then

$$F^K(f) = \sum_j F(f_j) \otimes a_j.$$

There is a rational algebra Γ with $a_j \in \Gamma$ and exact sequences:

$$P^\Gamma \xrightarrow{f_0} Q^\Gamma \xrightarrow{g_0} L_0 \to 0, \quad 0 \to L_1 \xrightarrow{u_0} F(P)^\Gamma \xrightarrow{F^\Gamma(f_0)} F(Q)^\Gamma$$

with L_0, L_1, Λ^Γ-modules free finitely generated over Γ, $f_0 = \sum_j f_j \otimes a_j$, $F^K(f_0) = \sum_j F(f_j) \otimes a_j$. Then $L_0 \otimes_\Gamma K \cong L, L_1 \otimes_\Gamma K \cong D\ \text{tr}_{\Lambda^K} L$. For $\lambda \in S(\Gamma)$ we obtain the exact sequences:

$$P^\Gamma \otimes_\Gamma S_\lambda \xrightarrow{f_0 \otimes 1} Q^\Gamma \otimes_\Gamma S_\lambda \xrightarrow{g_0 \otimes 1} L_0 \otimes_\Gamma S_\lambda \to 0,$$

$$0 \to L_1 \otimes_\Gamma S_\lambda \xrightarrow{u_0 \otimes 1} F(P)^\Gamma \otimes_\Gamma S_\lambda \xrightarrow{F^\Gamma(f_0)\otimes 1} F(Q)^\Gamma \otimes_\Gamma S_\lambda.$$

The morphism $f_0 \otimes 1$ is isomorphic to $\mathrm{f}(\lambda) = \sum_j f_j a_j(\lambda)$. The morphism $F^\Gamma \otimes 1$ is isomorphic to $F^\Gamma(f_0)(\lambda) = \sum_j F(f_j) a_j(\lambda) = F(f_0(\lambda))$. Therefore, $L_1 \otimes_\Gamma S_\lambda \cong D\ \mathrm{tr}_\Lambda(L_0 \otimes_\Gamma S_\lambda)$. □

Definition 6.1 Suppose $H \in \mathrm{mod}\ \Lambda^K$ and Γ is a rational k-algebra, we will say that H is Γ-realizable if there is a Λ-Γ-bimodule H_0, free finitely generated over Γ such that

(1) $H_0 \otimes_\Gamma K \cong H$ as Λ-K-bimodules;

(2) for all $\lambda \in S(\Gamma)$, $H_0 \otimes_\Gamma S_\lambda$ is indecomposable;

(3) if $\lambda \neq \mu$ are in $S(\Gamma)$ then $H_0 \otimes_\Gamma S_\lambda \not\cong H_0 \otimes_\Gamma S_\mu$. The bimodule H_0 is said to be a Γ-realization of H. If H is Γ-realizable for some Γ we will say that H is realizable.

Lemma 6.3 Suppose $\Gamma = k[x]_{h(x)}$ and $\Gamma' = k[x]_{h(x)g(x)}$. Then if H_0 is a Γ-realization of $H \in \mathrm{mod}\ \Lambda^K$, $H_1 = H_0 \otimes_\Gamma \Gamma'$ is a Γ'-realization of H.

Proof H_1 is free finitely generated over Γ'. For $\lambda \in S(\Gamma') \subset S(\Gamma)$ we have $H_1 \otimes_{\Gamma'} S_\lambda \cong H_0 \otimes_\Gamma S_\lambda$, and then conditions (1) ~ (3) of definition hold for H_1. □

Proposition 6.4 If $H \in \mathrm{mod}\ \Lambda^K$ is realizable, then $D\ \mathrm{tr}_{\Lambda^K} H$ is also realizable. Moreover, there are Γ-realizations H_0 and H_1 of H and $D\ \mathrm{tr}_{\Lambda^K} H$, respectively, for some rational algebra Γ, such that for all $\lambda \in S(\Gamma)$

$$D\ \mathrm{tr}_{\Lambda^K}(H_0 \otimes_\Gamma S_\lambda) \cong H_1 \otimes_\Gamma S_\lambda.$$

Proof By definition there is a Λ-Γ_1-bimodule M free finitely generated over Γ_1 with conditions (1), (2) and (3) of Definition 6.2. On the other hand, there is a rational k-algebra Γ_2 and Λ-Γ_2-bimodules N_1, N_2 free finitely generated over Γ_2 realizations of H and $D\ \mathrm{tr}_{\Lambda^K} H$, respectively such that the condition (2) of Proposition 6.2 holds. Suppose $\Gamma_1 = k[x]_{h(x)}$, $\Gamma_2 = k[x]_{g(x)}$, consider $\Gamma_3 = k[x]_{h(x)g(x)}$. Then

$$(N_1 \otimes_{\Gamma_2} \Gamma_3) \otimes_{\Gamma_3} K \cong H \cong M \otimes_{\Gamma_1} K \cong (M \otimes_{\Gamma_1} \Gamma_3) \otimes_{\Gamma_3} K.$$

Thus, there exists a $\Gamma=k[x]_{h(x)g(x)t(x)}$ such that

$$(N_1\otimes_{\Gamma_2}\Gamma_3)\otimes_{\Gamma_3}\Gamma\cong(M\otimes_{\Gamma_1}\Gamma_3)\otimes_{\Gamma_3}\Gamma\cong M\otimes_{\Gamma_1}\Gamma.$$

Take now $H_0=M\otimes_{\Gamma_1}\Gamma$, $H_1=N_2\otimes_{\Gamma_2}\Gamma$. By our previous lemma H_0 is a Γ-realization of H. Moreover, for $\lambda\in S(\Gamma)\subset S(\Gamma_2)$,

$$H_1\otimes_\Gamma S_\lambda\cong N_2\otimes_{\Gamma_2}S_\lambda\cong D\ \mathrm{tr}_\Lambda(N_1\otimes_{\Gamma_2}S_\lambda),$$

now,

$$N_1\otimes_{\Gamma_2}S_\lambda\cong(N_1\otimes_{\Gamma_2}\Gamma_3)\otimes_{\Gamma_3}\Gamma\otimes_\Gamma S_\lambda\cong H_0\otimes_\Gamma S_\lambda.$$

Therefore, for all $\lambda\in S(\Gamma)$, $H_1\otimes_\Gamma S_\lambda\cong D\ \mathrm{tr}_\Lambda(H_0\otimes_\Gamma S_\lambda)$. From here we can see that conditions (2) and (3) of Definition 6.1 are hold for H_1, therefore H_1 is a Γ-realization of $D\mathrm{tr}_{\Lambda^K}H$. □

Let Λ be a finite-dimension algebra of tame representation type over an algebraically closed field k and G a generic Λ-module. We recall that in this case a realization in the sense of [5] of G over R a commutative principal ideal domain which is finitely generated over k, is a finitely generated Λ-R-bimodule M such that if F is the quotient field of R, then $G\cong M\otimes_R F$ and $\dim_F(M\otimes_R F)=\mathrm{el}(G)$.

Proposition 6.5 Let Λ be a finite-dimension algebra of tame representation type over an algebraically closed field k. Suppose $L\in\mathrm{mod}\ \Lambda^K$ with ${}_\Lambda L$ indecomposable and $\dim_K\ \mathrm{End}_\Lambda(L)/\mathrm{rad}\ \mathrm{End}_\Lambda(L)=1$, then L is realizable and $D\ \mathrm{tr}_{\Lambda^K}L\cong L$.

Proof Since $\dim_K L$ and $\dim_K\ \Lambda^K$ are finite, there is a rational algebra $\Gamma_0\subset K$ and a Λ-Γ_0-bimodule M free finitely generated over Γ_0 such that $L\cong M\otimes_{\Gamma_0}K$. From Lemma 5.1 we obtain $\mathrm{el}({}_\Lambda L)=\dim_K L$, then M is a realization in the sense of [5] of ${}_\Lambda L$ over Γ.

On the other hand, by (i) and (ii) of [5, Theorem 5.4] there is a rational algebra $\Gamma_1\subset K$ and a realization in the sense of [5] N of ${}_\Lambda L$ over Γ_1 such that N is free over Γ_1 and the functor $N\otimes_{\Gamma_1}$-preserves isomorphism classes and indecomposability.

By [5, Lemma 5.2(3)] there are rational algebras Γ_2, Γ_3, and an isomorphism $\phi:\Gamma_2\to\Gamma_3$ such that $\Gamma_0\subset\Gamma_2$, $\Gamma_1\subset\Gamma_3$, and $M\otimes_{\Gamma_0}\Gamma_2\cong(N\otimes_{\Gamma_1}\Gamma_3)_\phi$. Then for $\lambda\in S(\Gamma_2)$ we have

$$(M\otimes_{\Gamma_0}\Gamma_2)\otimes_{\Gamma_2}S_\lambda\cong((N\otimes_{\Gamma_1}\Gamma_3)\otimes_{\Gamma_3}(S_\lambda)_{\phi^{-1}})_\phi.$$

Therefore, M is a Γ_2-realization of L. By Proposition 6.4 there are Γ-realizations L_0 and L_1 of L and D tr $_{\Lambda^K}L$, respectively, with $D\ \mathrm{tr}_\Lambda(L_0\otimes_\Gamma S_\lambda)\cong L_1\otimes_\Gamma S_\lambda$, for $\lambda\in S(\Gamma)$. Here by Lemma 5.2, $\dim_K F({}_\Lambda D\ \mathrm{tr}_{\Lambda^K}L)=\dim_K F({}_\Lambda L)$, therefore L_0 and L_1 are realizations in the sense of [5] of ${}_\Lambda L$ and ${}_\Lambda D\ \mathrm{tr}_\Lambda L$. The infinitely many nonisomorphic objects $L_0\otimes_\Gamma S_\lambda$ have the same k-dimension, therefore by [4], for almost all $\lambda\in S(\Gamma)$, $L_0\otimes_\Gamma S_\lambda\cong D\ \mathrm{tr}_\Lambda(L_0\otimes_\Gamma S_\lambda)\cong L_1\otimes_\Gamma S_\lambda$, but by [5, Lemma 5.2 (4)] this implies our result. □

As a consequence of the above proposition we obtain a new proof of the following result in [11].

Theorem 6.2 Let Λ be a finite-dimension algebra of tame representation type, over an algebraically closed field k. Then if H is a generic module in Mod Λ, $A\ (H)\cong H$.

Proof By [5] there is a decomposition

$$\mathrm{End}_\Lambda(G)=k(x)\oplus\mathrm{radEnd}_\Lambda(G),$$

using this decomposition H has a structure of Λ^K-module G, with $\dim_K F({}_\Lambda H)=1$, then from the previous proposition we have:

$$A(H)\cong{}_\Lambda D\ \mathrm{tr}\ _{\Lambda^K}G\cong{}_\Lambda G=H.\quad\square$$

References

[1] Auslander M, Reiten I, Smalo S. Representation Theory of Artin Algebras. Cambridge Univ. Press, Cambridge, 1995.

[2] Bautista R. On algebras of strongly unbounded representation type, Comment. Math. Helv., 1985, 60(3): 392-399.

[3] Bongartz K. Indecomposables are standard. Comment. Math. Helv. 1985, 60(3): 400-410.

[4] Crawley-Boevey W W. On tame algebras and bocses. Proc. London Math. Soc., 1988, 56(3): 451-483.

[5] Crawley-Boevey W W. Tame algebras and generic modules. Proc. London Math. Soc., 1991, 63: 241-264.

[6] Crawley-Boevey W W. Modules of finite length over their endomorphism

rings. in: Brenner S, Tachikawa H (Eds.). Representations of Algebras and Related Topics. in: London Math. Soc. Lecture Note Ser., Vol. 168, 1992, 127-184.

[7] Happel D, Preiser U, Ringel C M. Vinbergs characterization of Dynkin diagrams using subadditive functions with applications to Dtr-periodic modules. in: Lecture Notes in Math., Vol. 832, Springer-Verlag, Berlin, 1980: 280-294.

[8] Jacobson N. Basic Algebra Ⅱ. Freeman, New York, 1980.

[9] Jensen Ch, Lenzing H. Homological dimension and representation type of algebras under base extension. Manuscripts Maths, 1982, 39: 1-13.

[10] Kasjan S. Auslander-Reiten sequences under base field extensions. Proc. Amer. Math. Soc., 2000, 128: 2 885-2 896.

[11] Krause H. Stable equivalence preserves representation type. Comment. Math. Helv. 1997, 72: 266-284.

[12] Krause H. Generic modules over Artin algebras. Proc. London. Math. Soc., 1998, 76: 276-306.

[13] Liu S. Shapes of connected components of the Auslander-Reiten quivers of Artin algebras. in: Representation Theory of Algebras and Related Topics. in: Canad. Math. Soc. Conf. Proc., Vol. 19, 1996: 109-137.

[14] Todorov G. Almost Split Sequences for Dtr-periodic modules. in: Lecture Notes in Math., Vol. 832, Springer-Verlag, Berlin, 1980: 600-631.

[15] Zhang Y. The structure of stable components. Canadian J. Math., 1991, 43: 652-672.

Science in China: Mathematics, 2005, 48A(4): 456—468.

驯顺型和野型 Bocses①

On Tame and Wild Bocses

Abstract　We first give an alternative proof of the well-known Drozd's wild Theorem, which lowers down the dimension 43 to 20. Then we list more minimally wild bocses and discuss the possible differentials of the first arrow of a tame bocs, which is useful for reductions of bocses.

Keywords　Bocs; tameness; wildness; differential.

§ 1.　Introduction

Let k be an algebraically closed field and Λ a finite dimensional k-algebra (associative, with 1). The Yu. A. Drozd's remarkable "Tame and Wild Theorem"[1], which confirmed a conjecture by P. Donovan and M. R. Freislich, states that every algebra is either tame or wild. The proof of the Tame and Wild Theorem relies on the notion of a bocs, an idea which was introduced by A. V. Roiter[2] to formalize the matrix methods and differential graded category approach to representation theory.

① This paper is dedicated to Prof. R. Bautista for his 60's birthday.

Received: 2003-11-23; Revised: 2004-11-21.

本文与徐运阁合作.

Again based on the method of bocses and their reduction techniques, Crawley-Boevey proved that for a tame algebra Λ, and for each dimension d, almost all indecomposable Λ-modules of dimension at most d are isomorphic to their AR-translations, thus lying in homogeneous tubes, in terminology used by C. M. Ringel to describe certain components of the Auslander-Reiten quiver of Λ.

$\mathcal{A}=(A,V)$ is said to be a bocs if A is a skeletal small category and V is a coalgebra. Thus V is an A-A-bimodule with coalgebra structure, i. e. counite $\varepsilon: V\to A$ and comultiplication $\mu: V\to V\otimes_A V$. Denote by $\bar{V}$ the kernel of ε. The category $R(\mathcal{A})$ of representations of $\mathcal{A}$ has the same objects as $R(\mathcal{A})$, the contravariant functors $A\to \mathrm{mod}(k)$, and if M and N are two representations, then a morphism from M to N is given by an A-module map from $V\otimes_A M$ to N. The composition $\theta\phi$ of $\theta: V\otimes_A M\to N$ and $\phi: V\otimes_A N\to L$ is given by

$$V\otimes_A M \xrightarrow{\mu\otimes 1} V\otimes_A V\otimes_A M \xrightarrow{1\otimes\theta} V\otimes_A N \xrightarrow{\phi} L.$$

Throughout we consider only layered bocs (see ref. [3], Definition 3. 6 for details).

There are several equivalent formulations of matrix problems[4] such as bocses, differential graded categories[2], differential biquivers, bimodule problems[5], relatively projective categories[6]. Among their reduction techniques, the reductions of bocses, which provide the inductive step in the proof of Drozd's Teorem, seem to be very elegant formulation for some theoretical purposes. Theorem 3. 2 (see refs. [1], Prop. 9 or [3], Prop. 3. 10) is essential in the proof of Drozd's Theorem to ensure that certain bad configurations do not occur in non-wild bocses. It is a remarkable fact that if $\mathcal{A}$ is wild, then we are bound to meet one of these so-called minimally wild bocses at some stage of reductions. In this paper we first give an alternative proof of the theorem, which may not be an essential improvement, since the idea of the orginal proof is perfect. But the new proof lowers down the dimension 43 to 20, which makes the matrices used in the proof simpler. Also, we list more minimally wild bocses in order to describe the differential of the first arrow of a tame bocs. All this is very useful for the study of algebras of tame representation type and for the possible calculation of the wild rank of finite-dimensional algebras, and the latter is helpful to Tame-open conjecture①.

Throughout this paper k is an algebraically closed field. We write the composition of arrows or morphisms from left to right.

§ 2. The differential of the first arrow

From now on we fix a bocs $\mathcal{A}=(A,V)$ with a layer $L=(A';\omega;a_1,a_2,\cdots,a_n;v_1,v_2,\cdots,v_m)$. One of the advantages of bocs methods is the reduction technique, which provide the inductive steps in the proof of Drozd's Teorem. Thus we have to discuss the differential of the first arrow. There are 4 possibilities of the position of a_1.

(1) $\mathcal{P}=\mathcal{Q}$, 

. $A'(\mathcal{P},\mathcal{P})=k[\lambda,f(\lambda)^{-1}]$ for some $f(\lambda)\in k[\lambda]$.

(2) $\mathcal{P}\neq\mathcal{Q}$, 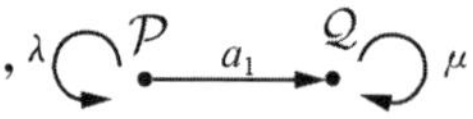

. $A'(\mathcal{P},\mathcal{P})=k[\lambda,f(\lambda)^{-1}]$ and $A'(\mathcal{Q},\mathcal{Q})=k[\mu,g(\mu)^{-1}]$ for some $f(\lambda)\in k[\lambda]$, $g(\mu)\in k[\mu]$.

(3) $\mathcal{P}\neq\mathcal{Q}$, 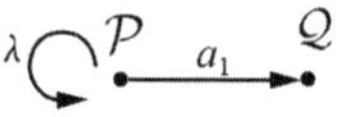

(resp. 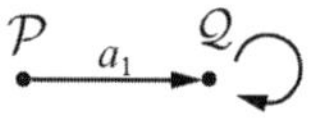

). $A'(\mathcal{P},\mathcal{P})=k[\lambda,f(\lambda)^{-1}]$(resp. $A'(\mathcal{Q},\mathcal{Q})=k[\mu,g(\mu)^{-1}]$) for some $f(\lambda)\in k[\lambda]$ (resp. $g(\mu)\in k[\mu]$).

(4) 

or 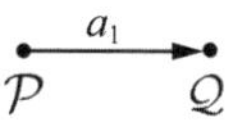

. Both $A'(\mathcal{P},\mathcal{P})$ and $A'(\mathcal{Q},\mathcal{Q})$ are trivial.

Then in possibilities (1) and (2)

$$\delta(a_1)=\sum_{j=1}^{m}h_j(\lambda,\mu)v_j,$$

where λv_j stands for the left multiplication by λ and μv_j for the right multiplication by λ in case (1) and by μ in case (2).

---

① Han Yang. Is tame open? J. Alg., 2005, 284: 801-810.

Recalling from ref. [3], let $h(\lambda,\mu)$ be the highest common factor of the $h_j(\lambda,\mu)$ and let $q_j(\lambda,\mu)=h_j(\lambda,\mu)/h(\lambda,\mu)$. Since $q_j(\lambda,\mu)$ are coprime, there are polynomials $s_j(\lambda,\mu)$ and a non-zero polynomial $c(\lambda)$, such that

$$c(\lambda)=\sum_{j=1}^{m}s_j(\lambda,\mu)q_j(\lambda,\mu)$$

in ring $k[\lambda,\mu]$, thus

$$1=\sum_{j=1}^{m}c(\lambda)^{-1}s_j(\lambda,\mu)q_j(\lambda,\mu)$$

in ring $S=k[\lambda,\mu,f(\lambda)^{-1},g(\mu)^{-1},c(\lambda)^{-1}]$ which is a Hermite ring, so there is an invertible matrix Q in $M_{m\times m}(S)$ with the first row $(q_j(\lambda,\mu))_{j=1,2,\cdots,m}$, and we can make a change of the basis of the form

$$(w_1,w_2,\cdots,w_m)^{\mathrm{T}}=Q(v_1,v_2,\cdots,v_m)^{\mathrm{T}}$$

so that

$$\delta(a_1)=h(\lambda,\mu)w_1.$$

In possibility (3), $\delta(a_1)=\sum_{j=1}^{m}h_j(\lambda)v_j$. If $h(\lambda)$ is the highest common factor of $h_j(\lambda)$, then $q_j(\lambda)=h_j(\lambda)/h(\lambda)$ are coprime and $1=\sum_{j=1}^{m}s_j(\lambda)q_j(\lambda)$ for some polynomials $s_j(\lambda)\in k[\lambda]$. Make a similar basis change given by a transformation matrix over $S=k[\lambda,f(\lambda)^{-1}]$, so that

$$\delta(a_1)=h(\lambda)w_1.$$

In possibility (4), $\delta(a_1)=w_1$, or $\delta(a_1)=0$ after some basis changes given by a transformation matrix over $S=k$.

§ 3. Minimally wild bocses

This section is devoted to an alternative proof of Drozd's theorem, which lowers down the dimension 43 to 20. Our proof is based on the Belitskii algorithm, which is very effective in reducing an individual matrix into a canonical form under some admissible transformations which may be considered as a generalized Jordan form[7].

Definition 3.1[3] We say a bocs $\mathcal{A}=(A,V)$ is wild if there is a functor $F: k\langle x,y\rangle$-mod $\to R(\mathcal{A})$ which preserves iso-classes and indecomposability. Otherwise, $\mathcal{A}$ is said to be tame.

We restate the Drozd's theorem on minimally wild bocses as follows:

Theorem 3.2[1,3]　Let $\mathcal{A}=(A,V)$ be a bocs given in 2. Then $\mathcal{A}$ must be wild if one of the following two cases appears.

Case 1　$\delta(a_1)=h(\lambda,\mu)w_1$, $h(\lambda,\mu)\in k[\lambda,\mu,f(\lambda)^{-1},g(\mu)^{-1}]$ is non-invertible in cases 2(1) and 2(2).

(1)

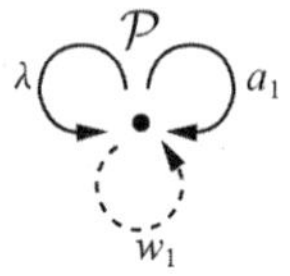

when $\mathcal{P}=\mathcal{Q}$;

Fig. 3.1

(2)

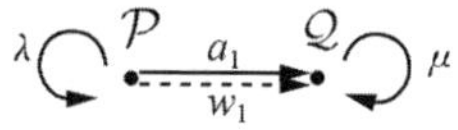

when $\mathcal{P}\neq\mathcal{Q}$.

Fig. 3.2

Case 2　$\delta(a_1)=0$, in case 2(3).

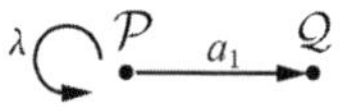

(resp.

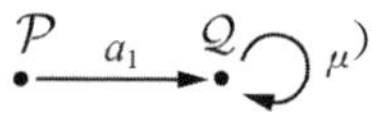

).

Fig. 3.3(a)　　**Fig. 3.3(b)**

Proof　Let $h(\lambda,\mu)=\alpha(\lambda-\lambda_0)+\beta(\mu-\mu_0)+\gamma_1(\lambda-\lambda_0)^2+\gamma_2(\lambda-\lambda_0)\cdot(\mu-\mu_0)+\gamma_3(\mu-\mu_0)^2+\cdots$ and denote by D the domain of the parameters $\{\lambda,\mu\}$. In case 1(2), we define a dimension vector $\underline{m}=(m_{\mathcal{P}},m_{\mathcal{Q}})$ and a representation M of $\mathcal{A}$ with $M(\mathcal{P})=k^{m_{\mathcal{P}}}$, $M(\mathcal{Q})=k^{m_{\mathcal{Q}}}$. Choosing suitable k-bases of these vector spaces, we can write $\boldsymbol{M}(\lambda)=\boldsymbol{W}^{\lambda}$, $\boldsymbol{M}(\mu)=\boldsymbol{W}^{\mu}$ as Weyr matrices[7]. An automorphism of $\boldsymbol{M}$ can be given by invertible matrices $\boldsymbol{S}_{\mathcal{P}}$ satisfying $\boldsymbol{S}_{\mathcal{P}}\boldsymbol{M}(\lambda)-\boldsymbol{M}(\lambda)\boldsymbol{S}_{\mathcal{P}}=\boldsymbol{0}$, $\boldsymbol{S}_{\mathcal{Q}}$ satisfying $\boldsymbol{S}_{\mathcal{Q}}\boldsymbol{M}(\mu)-\boldsymbol{M}(\mu)\boldsymbol{S}_{\mathcal{Q}}=\boldsymbol{0}$, and matrix $\boldsymbol{S}(w_1)=(v_{ij})_{m_{\mathcal{P}}\times m_{\mathcal{Q}}}$. Then $M(a_1)$ can be taken as a canonical form (see ref. [7]) calculated according to

$$\boldsymbol{M}(a_1)\boldsymbol{S}_{\mathcal{Q}}-\boldsymbol{S}_{\mathcal{P}}\boldsymbol{M}(a_1)=\boldsymbol{S}(\delta(a_1))=\boldsymbol{M}(h(\lambda,\mu))\boldsymbol{S}(w_1).$$

Case 1(1)　It's obvious.

Case 1(2)　$\alpha\beta\neq 0$.

Set $m_{\mathcal{P}}=9$, $m_{\mathcal{Q}}=11$, $(\lambda_0,\mu_0)\in D$ with $h(\lambda_0,\mu_0)=0$, $f(\lambda_0)\neq 0$, $g(\mu_0)\neq 0$, $\boldsymbol{M}(\lambda)\simeq\boldsymbol{J}_1(\lambda_0)\oplus\boldsymbol{J}_3(\lambda_0)\oplus\boldsymbol{J}_5(\lambda_0)$, $\boldsymbol{M}(\mu)\simeq\boldsymbol{J}_2(\mu_0)\oplus\boldsymbol{J}_4(\mu_0)\oplus\boldsymbol{J}_5(\mu_0)$. Then

$$
\boldsymbol{M}(\lambda)=\left(\begin{array}{ccc|cc|cc|c|c}
\lambda_0 & 0 & 0 & 1 & 0 & & & & \\
 & \lambda_0 & 0 & 0 & 1 & & & & \\
 & & \lambda_0 & 0 & 0 & & & & \\
\hline
 & & & \lambda_0 & 0 & 1 & 0 & & \\
 & & & & \lambda_0 & 0 & 1 & & \\
\hline
 & & & & & \lambda_0 & 0 & 1 & \\
 & & & & & & \lambda_0 & 0 & \\
\hline
 & & & & & & & \lambda_0 & 1 \\
\hline
 & & & & & & & & \lambda_0
\end{array}\right) \text{①}
$$

and

$$
\boldsymbol{S}_{\mathcal{P}}=\left(\begin{array}{ccc|cc|cc|c|c}
s_1 & s^{1}_{12} & s^{1}_{13} & s^{2}_{11} & s^{2}_{12} & s^{3}_{11} & s^{3}_{12} & s^{4}_{11} & s^{5}_{11} \\
 & s_2 & s^{1}_{23} & 0 & s^{2}_{22} & s^{3}_{21} & s^{3}_{22} & s^{4}_{21} & s^{5}_{21} \\
 & & s_3 & 0 & 0 & 0 & s^{3}_{32} & 0 & s^{5}_{31} \\
\hline
 & & & s_1 & s^{1}_{12} & s^{2}_{11} & s^{2}_{12} & s^{3}_{11} & s^{4}_{11} \\
 & & & & s_2 & 0 & s^{2}_{22} & s^{3}_{21} & s^{4}_{21} \\
\hline
 & & & & & s_1 & s^{1}_{12} & s^{2}_{11} & s^{3}_{11} \\
 & & & & & & s_2 & 0 & s^{3}_{21} \\
\hline
 & & & & & & & s_1 & s^{2}_{11} \\
\hline
 & & & & & & & & s_1
\end{array}\right)
$$

is given by $\boldsymbol{S}_{\mathcal{P}}\boldsymbol{M}(\lambda)-\boldsymbol{M}(\lambda)\boldsymbol{S}_{\mathcal{P}}=\boldsymbol{0}$. Similarly,

$$
\boldsymbol{M}(\mu)=\left(\begin{array}{ccc|ccc|cc|cc|c}
\mu_0 & 0 & 0 & 1 & 0 & 0 & & & & & \\
 & \mu_0 & 0 & 0 & 1 & 0 & & & & & \\
 & & \mu_0 & 0 & 0 & 1 & & & & & \\
\hline
 & & & \mu_0 & 0 & 0 & 1 & 0 & & & \\
 & & & & \mu_0 & 0 & 0 & 1 & & & \\
 & & & & & \mu_0 & 0 & 0 & & & \\
\hline
 & & & & & & \mu_0 & 0 & 1 & 0 & \\
 & & & & & & & \mu_0 & 0 & 1 & \\
\hline
 & & & & & & & & \mu_0 & 0 & 1 \\
 & & & & & & & & & \mu_0 & 0 \\
\hline
 & & & & & & & & & & \mu_0
\end{array}\right),
$$

and

---

① Such a matrix is called Weyr matrix that guarantees the matrices commutated with it are upper triangular, It can be obtained from Jordan matrix by permuting some rows and columns simultaneously.

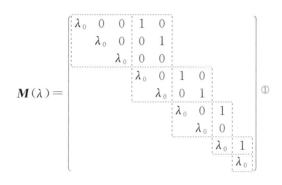

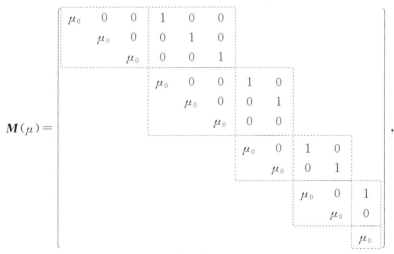

$$
\boldsymbol{S}_Q=\left(\begin{array}{ccc:ccc:c}
t_1 & t_{12}^1 & t_{13}^1 & t_{11}^2 & t_{12}^2 & t_{13}^2 & \\
 & t_2 & t_{23}^1 & t_{21}^2 & t_{22}^2 & t_{23}^2 & \\
 & & t_3 & 0 & 0 & t_{33}^2 & * \\
\hdashline
 & & & t_1 & t_{12}^1 & t_{13}^1 & \\
 & & & & t_2 & t_{23}^1 & \\
 & & & & & t_3 & \\
\hdashline
 & & & & & & *
\end{array}\right)
$$ ①

Then $\boldsymbol{S}(\delta(a_1))$ is a 9×11 matrix of the form

$$
\left(\begin{array}{cccccccccc}
* & * & * & & & & & & & \\
* & * & * & & & & & & & \\
0 & 0 & 0 & & & & & * & & \\
0 & 0 & 0 & & & & & & & \\
\alpha v_{71} & \alpha v_{72} & * & & & & & & & \\
\alpha v_{81}+\gamma_1 v_{91} & \alpha v_{82}+\gamma_1 v_{92} & * & & & & & & & \\
0 & 0 & 0 & \beta v_{71} & \beta v_{72} & * & & & & \\
\alpha v_{91} & \alpha v_{92} & \alpha v_{93} & \alpha v_{94}+\beta v_{81}+\gamma_2 v_{91} & \alpha v_{95}+\beta v_{82}+\gamma_2 v_{92} & * & & & & \\
0 & 0 & 0 & \beta v_{91} & \beta v_{92} & \beta v_{93} & \beta v_{94}+\gamma_3 v_{91} & \beta v_{95}+\gamma_3 v_{92} & \cdots &
\end{array}\right).
$$

Let

$$
\boldsymbol{M}(a_1)=\left(\begin{array}{cccc}
0 & 0 & 0 & \\
0 & 0 & 0 & \\
1 & \xi & 0 & \\
0 & 0 & 0 & \\
\eta & 1 & 0 & \mathbf{0} \\
1 & 0 & 0 & \\
0 & 0 & 1 & \\
0 & 0 & 1 & \\
0 & 0 & 0 &
\end{array}\right)_{9\times 11}.
$$

① We adopt the convention that only important entries in matrices are written. Zeros are replaced with blanks, and other nonzero entries with *.

Thus $\boldsymbol{S}_{\mathcal{P}} = s\boldsymbol{I}_9 + N_1$ and $\boldsymbol{S}_{\mathcal{Q}} = s\boldsymbol{I}_{11} + N_2$ for some upper triangular nilpotent matrices N_1, N_2, and $s \in k$.

Case 1(2) $\alpha \neq 0, \beta \neq 0, \gamma_3 \neq 0$.

$\boldsymbol{M}(\lambda)$ can be taken as the same as above, and $\boldsymbol{M}(\mu) \simeq \boldsymbol{J}_1(\mu_0) \oplus \boldsymbol{J}_3(\mu_0) \oplus \boldsymbol{J}_5(\mu_0)$, then $\boldsymbol{S}_{\mathcal{Q}}$ has the same form as $\boldsymbol{S}_{\mathcal{P}}$. And $\boldsymbol{S}(\delta(a_1)) =$

$$\begin{pmatrix} * & * & * & * & * & & & & \\ * & * & * & * & * & & & & \\ 0 & 0 & 0 & 0 & 0 & * & & & \\ * & * & * & * & * & & & & \\ \alpha v_{71} & * & * & * & * & & & & \\ * & * & * & * & * & & & & \\ 0 & 0 & 0 & 0 & 0 & \gamma_3 v_{71} & * & * & * \\ \alpha v_{91} & \alpha v_{92} & * & \alpha v_{94} + \gamma_2 v_{91} & * & * & * & * & * \\ 0 & 0 & 0 & 0 & 0 & \gamma_3 v_{91} & \gamma_3 v_{92} & \gamma_3 v_{94} + \rho v_{91} & * \end{pmatrix}_{9\times 9},$$

where ρ is the coefficient of $(\mu - \mu_0)^3$ in $h(\lambda, \mu)$, and

$$\boldsymbol{M}(a_1) = \begin{pmatrix} 0 & 0 & 0 & 0 & 0 & \\ 0 & 0 & 0 & 0 & 0 & \\ 0 & 1 & 0 & \xi & 0 & \\ 0 & 0 & 0 & 0 & 0 & \\ \eta & 0 & 0 & 0 & 0 & \boldsymbol{0} \\ 0 & 0 & 0 & 0 & 0 & \\ 0 & 0 & 1 & 0 & 1 & \\ 0 & 1 & 0 & 1 & 0 & \\ 0 & 0 & 0 & 0 & 0 & \end{pmatrix}_{9\times 9},$$

Case 1(2) $\alpha \neq 0, \beta = 0$ and $\gamma_3 = 0$; or $\alpha = \beta = 0$.

Set $m_{\mathcal{P}} = 9, m_{\mathcal{Q}} = 4, \boldsymbol{M}(\lambda) \simeq \boldsymbol{J}_1(\lambda_0) \oplus \boldsymbol{J}_3(\lambda_0) \oplus \boldsymbol{J}_5(\lambda_0)$ and $\boldsymbol{M}(\mu) \simeq \boldsymbol{J}_1(\mu_0) \oplus \boldsymbol{J}_3(\mu_0)$.

$$\boldsymbol{M}(\mu) = \begin{pmatrix} \mu_0 & 0 & 1 & \\ & \mu_0 & 0 & \\ & & \mu_0 & 1 \\ & & & \mu_0 \end{pmatrix}_{4\times 4},$$

$$\boldsymbol{S}_{\mathcal{Q}} = \begin{pmatrix} t_1 & t_{12}^1 & t_1^2 & t_1^3 \\ & t_2 & 0 & t_{21}^3 \\ & & t_1 & t_1^2 \\ & & & t_1 \end{pmatrix}_{4\times 4}$$

is given by $\boldsymbol{S}_{\mathcal{Q}}\boldsymbol{M}(\mu)=\boldsymbol{M}(\mu)\boldsymbol{S}_{\mathcal{Q}}$. Then $\delta(a_1)$ has the form

$$\begin{pmatrix} * & * & * & * \\ * & * & * & * \\ 0 & 0 & 0 & * \\ * & * & * & * \\ 0 & 0 & 0 & * \\ * & * & * & * \\ 0 & 0 & 0 & * \\ 0 & 0 & 0 & * \\ 0 & 0 & 0 & * \end{pmatrix}_{9\times 4}. \quad \text{Set } \boldsymbol{M}(a_1)=\begin{pmatrix} 0 & 0 & 0 & 0 \\ 0 & 0 & 0 & 0 \\ 1 & \eta & \xi & 0 \\ 0 & 0 & 0 & 0 \\ 0 & 1 & 1 & 0 \\ 0 & 0 & 0 & 0 \\ 0 & 0 & 0 & 0 \\ 1 & 0 & 0 & 0 \\ 0 & 0 & 0 & 0 \end{pmatrix}_{9\times 4}.$$

A similar calculation shows that $\boldsymbol{S}_{\mathcal{P}}=s\boldsymbol{I}_9+\boldsymbol{N}_1$, $\boldsymbol{S}_{\mathcal{Q}}=s\boldsymbol{I}_4+\boldsymbol{N}_2$, $s\in k$, $\boldsymbol{N}_1$, $\boldsymbol{N}_2$ are upper triangular nilpotent.

Case 1(1) $\alpha\beta\neq 0$.

If $h(\lambda,\lambda)=0$, or the degree $\deg(h(\lambda,\lambda))>0$, but there exists a $\lambda_0\in D$, such that $h(\lambda_0,\lambda_0)=0$, $f(\lambda_0)\neq 0$. Set $m_p=4$, $\boldsymbol{M}(\lambda)\simeq\boldsymbol{J}_1(\lambda_0)\oplus\boldsymbol{J}_3(\lambda_0)$,

$$\boldsymbol{M}(\lambda)=\begin{pmatrix} \lambda_0 & 0 & 1 & \\ & \lambda_0 & 0 & \\ & & \lambda_0 & 1 \\ & & & \lambda_0 \end{pmatrix}_{4\times 4}, \quad \boldsymbol{S}_{\mathcal{P}}=\begin{pmatrix} s_1 & s_{12}^1 & s_{12}^2 & s_{11}^3 \\ & s_2 & 0 & s_{21}^3 \\ & & s_1 & s_{12}^2 \\ & & & s_1 \end{pmatrix}_{4\times 4}$$

Then

$$\boldsymbol{S}(\delta(a_1))=\begin{pmatrix} * & * & * & * \\ 0 & * & * & * \\ \alpha v_{41} & \alpha v_{42} & * & * \\ 0 & 0 & -\beta v_{41} & * \end{pmatrix}_{4\times 4}, \quad \boldsymbol{M}(a_1)=\begin{pmatrix} 0 & 0 & 0 & 0 \\ \xi & 0 & 0 & 0 \\ \eta & 0 & 0 & 0 \\ 0 & 1 & 0 & 0 \end{pmatrix}_{4\times 4}.$$

Thus $\boldsymbol{S}_{\mathcal{P}}=s\boldsymbol{I}_4+\boldsymbol{N}$ for an upper triangular nilpotent matrix $\boldsymbol{N}$ and $s\in k$.

Case 1(1) $\alpha\neq 0$, $\beta=0$. $\boldsymbol{M}(\lambda)$ is the same as above.

$$\boldsymbol{S}(\delta(a_1))=\begin{pmatrix} * & * & * & * \\ 0 & 0 & 0 & * \\ * & * & * & * \\ 0 & 0 & 0 & * \end{pmatrix}_{4\times 4}. \quad \boldsymbol{M}(a_1)=\begin{pmatrix} 0 & 0 & 0 & 0 \\ \xi & 0 & 0 & 0 \\ 0 & 0 & 0 & 0 \\ 0 & 1 & \eta & 0 \end{pmatrix}_{4\times 4},$$

Case 1(1) $\alpha=\beta=0$.

$\boldsymbol{M}(\lambda)\simeq\boldsymbol{J}_1(\lambda_0)\oplus\boldsymbol{J}_3(\lambda_0)$,

$$\boldsymbol{S}(\delta(a_1))=\begin{pmatrix} * & * & * & * \\ 0 & 0 & 0 & * \\ 0 & 0 & * & * \\ 0 & 0 & 0 & * \end{pmatrix}_{4\times 4}.$$

In this case $\boldsymbol{M}(a_1)$ can be taken as the same as in the case $\beta=0,\alpha\neq 0$.

If for any $\lambda_0\in D, h(\lambda_0,\lambda_0)\neq 0$, there must exist some $\lambda_0\neq\mu_0$ with $h(\lambda_0,\mu_0)=0, f(\lambda_0)\neq 0, f(\mu_0)\neq 0$. We may use a rolled up version $\overline{\boldsymbol{M}}$ of the construction given in case 1(2). So $\overline{\boldsymbol{M}}(\lambda)=\boldsymbol{M}(\lambda)\oplus\boldsymbol{M}(\mu)$ and

$$\overline{\boldsymbol{M}}(a_1)=\begin{pmatrix}\boldsymbol{0} & \boldsymbol{M}(a_1)\\ \boldsymbol{0} & \boldsymbol{0}\end{pmatrix}.$$

Case 2 Set $m_{\mathcal{P}}=5, m_{\mathcal{Q}}=2, \boldsymbol{M}(\lambda)=\boldsymbol{J}_5(\lambda_0)$ with $f(\lambda_0)\neq 0$. Then

$$\boldsymbol{S}_{\mathcal{P}}=\begin{pmatrix} s_1 & s_2 & s_3 & s_4 & s_5 \\ & s_1 & s_2 & s_3 & s_4 \\ & & s_1 & s_2 & s_3 \\ & & & s_1 & s_2 \\ & & & & s_1 \end{pmatrix}, \boldsymbol{S}_{\mathcal{Q}}=\begin{pmatrix} t_{11} & t_{12} \\ t_{21} & t_{22}\end{pmatrix}, \text{and } \boldsymbol{M}(a_1)=\begin{pmatrix}\xi & 0\\ \eta & 0\\ 0 & 0\\ 1 & 0\\ 0 & 1\end{pmatrix},$$

thus $\boldsymbol{S}_{\mathcal{P}}=s\boldsymbol{I}_5, \boldsymbol{S}_{\mathcal{Q}}=s\boldsymbol{I}_2, s\in k$.

Let all the entries after ξ in M be equal to 0. Then $\boldsymbol{M}\in R(\mathcal{A})$ is a parameterized indecomposable representation since its endomorphism ring is local[7,8]. There are two algebraically independent parameters η and ξ in $\boldsymbol{M}$, the domain of η,ξ is k^2. $\boldsymbol{M}$ induces a functor

$$F_M: k(x,y)\text{-mod}\rightarrow R(\mathcal{A})$$

by sending (fig. 3. 4)

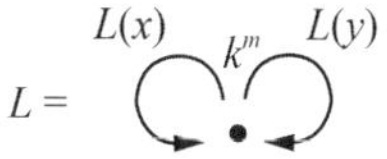

Fig. 3. 4

to $M^m(L(x),L(y))$ with each constant entry a enlarged into aI_m, and η,ξ changed into $L(x),L(y)$ respectively. The endomorphism ring of $M^m(L(x),L(y))$ consists of upper triangular partitioned matrices with diagonal blocks $\text{End}_{k\langle x,y\rangle}(L)$. Thus functor F preserves indecomposability and iso-classes as desired. □

§ 4. Differentials of tame bocses

This section is devoted to giving more minimally wild bocses in order to see the possibilities of the differential of the first arrow of a tame bocs.

Lemma 4. 1 Let $\mathcal{A}=(A,V)$ be a bocs with a layer $L=(A';\omega;a_1, a_2,\cdots,a_n;v_1,v_2,\cdots,v_m)$, where Obj. $A'\supseteq\{\mathcal{P},\mathcal{Q}\}$, $A'(\mathcal{P},\mathcal{P})=k[\lambda,f(\lambda)^{(-1)}]$, $A'(\mathcal{Q},\mathcal{Q})=k[\mu,g(\mu)^{-1}]$, $\delta(a_1)=h_1(\lambda)w_1+w_2h_2(\mu)$. (fig. 4. 1)

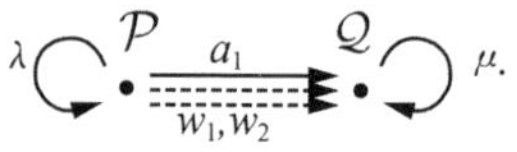

Fig. 4. 1

Then $\mathcal{A}$ must be wild if

(1) $h_1(\lambda)=(\lambda-\lambda_1)(\lambda-\lambda_2)$, $h_2(\mu)=(\mu-\mu_1)(\mu-\mu_2)$;

(2) $h_1(\lambda)=(\lambda-\lambda_1)(\lambda-\lambda_2)(\lambda-\lambda_3)$, $h_2(\mu)=\mu-\mu_1$; or dually,

$$h_1(\lambda)=\lambda-\lambda_1,\ h_2(\mu)=(\mu-\mu_1)(\mu-\mu_2)(\mu-\mu_3),$$

where $f(\lambda_i)\neq 0$, $g(\mu_j)\neq 0$.

Proof Construct a parameterized indecomposable representation M of $\mathcal{A}$ as the same as in the proof of Theorem 3. 2.

(1) In case $\lambda_1=\lambda_2$, $\mu_1=\mu_2$, set $\boldsymbol{M}(\lambda)$ similar to $\boldsymbol{J}_1(\lambda_1)\oplus\boldsymbol{J}_3(\lambda_1)$, $\boldsymbol{M}(\mu)$ to $\boldsymbol{J}_1(\mu_1)\oplus\boldsymbol{J}_3(\mu_1)$; in case $\lambda_1\neq\lambda_2$, $\mu_1=\mu_2$ (or dual) set

$$\boldsymbol{M}(\lambda)\simeq\boldsymbol{J}_1(\lambda_1)\oplus(\boldsymbol{J}_1(\lambda_2)\oplus\boldsymbol{J}_2(\lambda_2)),$$

$$\boldsymbol{M}(\mu)\simeq\boldsymbol{J}_1(\mu_1)\oplus\boldsymbol{J}_1(\mu_1)\quad\text{(or dual)};$$

in case $\lambda_1\neq\lambda_2$, $\mu_1\neq\mu_2$, set

$$\boldsymbol{M}(\lambda)\simeq\boldsymbol{J}_1(\lambda_1)\oplus(\boldsymbol{J}_1(\lambda_2)\oplus\boldsymbol{J}_2(\lambda_2)),$$

$$\boldsymbol{M}(\mu)\simeq\boldsymbol{J}_1(\mu_1)^2\oplus(\boldsymbol{J}_1(\mu_2)\oplus\boldsymbol{J}_2(\mu_2)).$$

Then $\boldsymbol{M}(a_1)$ equals

$$\begin{pmatrix}0&0&0&0\\0&1&\xi&0\\0&1&\eta&0\\1&0&0&0\end{pmatrix},\quad\begin{pmatrix}0&\eta&\xi&0\\0&0&0&0\\0&1&1&0\\1&0&0&0\end{pmatrix},\quad\begin{pmatrix}0&\eta&1&\xi&0\\0&0&0&0&0\\1&0&0&1&0\\0&1&1&0&0\end{pmatrix}$$

respectively.

(2) If $\lambda_1=\lambda_2=\lambda_3$, set

$$\boldsymbol{M}(\lambda)\simeq\boldsymbol{J}_1(\lambda)\oplus\boldsymbol{J}_4(\lambda)\oplus\boldsymbol{J}_7(\lambda_2),$$

$$\boldsymbol{M}(\mu)\simeq\boldsymbol{J}_1(\mu_1)^2\oplus\boldsymbol{J}_2(\mu_1);$$

if $\lambda_1=\lambda_2\neq\lambda_3$, set

$$\boldsymbol{M}(\lambda)\simeq(\boldsymbol{J}_1(\lambda_1)\oplus\boldsymbol{J}_3(\lambda_1))\oplus(\boldsymbol{J}_1(\lambda_2)\oplus\boldsymbol{J}_2(\lambda_2)),$$

$$\boldsymbol{M}(\mu)\simeq\boldsymbol{J}_1(\mu_1)^2\oplus\boldsymbol{J}_2(\mu_0);$$

if $\lambda_1,\lambda_2,\lambda_3$ are pairwise different, set

$$\boldsymbol{M}(\lambda)\simeq\boldsymbol{J}_1(\lambda_1)\oplus(\boldsymbol{J}_1(\lambda_2))\oplus(\boldsymbol{J}_2(\lambda_2))\oplus(\boldsymbol{J}_1(\lambda_3)\oplus\boldsymbol{J}_2(\lambda_3)),$$

$$\boldsymbol{M}(\mu)\simeq\boldsymbol{J}_1(\mu_0)^2\oplus\boldsymbol{J}_2(\mu_0).$$

Then $\boldsymbol{M}(a_1)$ equals

$$\begin{pmatrix}0&0&0&0\\0&0&0&0\\0&1&\xi&0\\0&0&0&0\\0&\eta&0&0\\0&0&0&0\\0&1&0&0\\0&0&0&0\\1&0&1&0\\1&0&0&0\\0&0&0&0\\0&0&0&0\end{pmatrix},\quad\begin{pmatrix}0&0&0&0\\0&1&\xi&0\\0&\eta&1&0\\1&1&0&0\\0&0&0&0\\0&0&1&0\\1&0&0&0\end{pmatrix},\quad\begin{pmatrix}1&\eta&\xi&0\\0&0&0&0\\0&1&1&0\\1&1&0&0\\0&0&0&0\\0&0&1&0\\1&0&0&0\end{pmatrix}$$

respectively. □

Lemma 4.2 Let $\mathcal{A}=(A,V)$ be a local bocs with a layer $L=(A';\omega;a_1,a_2,\cdots,a_n;v_1,v_2,\cdots,v_m)$, where Obj.

$A'\supseteq\{\mathcal{P}\}$, $A'(\mathcal{P},\mathcal{P})=k[\lambda,f(\lambda)^{-1}]$, $\delta(a_1)=h_1(\lambda)w_1+w_2h_2(\mu)$.

Fig. 4. 2

Then $\mathcal{A}$ must be wild if

(1) $h_1(\lambda)=(\lambda-\lambda_1)(\lambda-\lambda_2), h_2(\mu)=(\mu-\mu_1)(\mu-\mu_2)$ and

$$\{\lambda_1,\lambda_2\}\cap\{\mu_1,\mu_2\}=\varnothing;$$

(2) $h_1(\lambda)=(\lambda-\lambda_1)(\lambda-\lambda_2)(\lambda-\lambda_3), h_2(\mu)=\mu-\mu_0$ and

$$\{\lambda_1,\lambda_2,\lambda_3\}\cap\{\mu_0\}=\varnothing \text{ or dually;}$$

(3) $h_1(\lambda)=(\lambda-\lambda_1)(\lambda-\lambda_2), h_2(\mu)=(\mu-\lambda_1)$ or dually, where

$$f(\lambda_i)\neq 0, f(\mu_j)\neq 0.$$

Proof (1) and (2) follow from a rolled up version of 4. 1 Lemma.

(3) If $\lambda_1=\lambda_2$, set $\boldsymbol{M}(\lambda)\simeq\boldsymbol{J}_1(\lambda_1)\oplus\boldsymbol{J}_3(\lambda_1)$, if $\lambda_1\neq\lambda_2$, set $\boldsymbol{M}(\lambda)\simeq(\boldsymbol{J}_1(\lambda_1)\oplus\boldsymbol{J}_2(\lambda_1))\oplus\boldsymbol{J}_1(\lambda_2)$. Then $M(a_1)$ equals

$$\begin{pmatrix}0&0&0&0\\ \xi&0&0&0\\ \eta&0&0&0\\ 0&1&0&0\end{pmatrix},\quad \begin{pmatrix}0&0&0&0\\ 0&\xi&0&0\\ \eta&1&0&0\\ 1&0&0&0\end{pmatrix}$$

respectively. □

Lemma 4. 3 Let $\mathcal{A}=(A,V)$ be a bocs with a layer $L=(A';\omega;a_1,a_2,\cdots,a_n;v_1,v_2,\cdots,v_m)$, where Obj. $A'\supseteq\{\mathcal{P},\mathcal{Q}\}$, $A'(\mathcal{P},\mathcal{P})=k[\lambda,f(\lambda)^{-1}]$, $A'(\mathcal{Q},\mathcal{Q})=k$, $\delta(a_1)=h(\lambda)w_1$, or dually(fig. 4. 3)

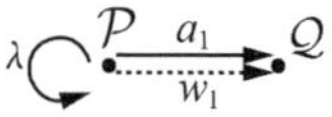

Fig. 4. 3

Then $\mathcal{A}$ must be wild if $h(\lambda)=(\lambda-\lambda_1)^{e_1}(\lambda-\lambda_2)^{e_2}\cdots(\lambda-\lambda_r)^{e_r}$ with $f(\lambda_i)\neq 0$, where $\lambda_1,\lambda_2,\cdots,\lambda_r$ are pairwise different and e_i's are non-negative integers with $e_1+e_2+\cdots+e_r\geqslant 4$.

Proof It suffices to consider the case $e_1+e_2+\cdots+e_r=4$. The proof is according to the partition $4=(e_1,e_2,\cdots,e_r)$.

(4): $\boldsymbol{M}(\lambda)\simeq\boldsymbol{J}_1(\lambda_1)\oplus\boldsymbol{J}_4(\lambda_1)\oplus\boldsymbol{J}_7(\lambda_1)$;

(3,1): $\boldsymbol{M}(\lambda)\simeq(\boldsymbol{J}_1(\lambda_1)\oplus\boldsymbol{J}_4(\lambda_1))\oplus(\boldsymbol{J}_1(\lambda_2)\oplus\boldsymbol{J}_2(\lambda_2))$;

$(2,2):\boldsymbol{M}(\lambda)\simeq(\boldsymbol{J}_1(\lambda_1)\oplus\boldsymbol{J}_3(\lambda_1))\oplus(\boldsymbol{J}_1(\lambda_2)\oplus\boldsymbol{J}_3(\lambda_2))$;

$(2,1,1):\boldsymbol{M}(\lambda)\simeq(\boldsymbol{J}_1(\lambda_1)\oplus\boldsymbol{J}_3(\lambda_1))\oplus(\boldsymbol{J}_1(\lambda_2)\oplus\boldsymbol{J}_2(\lambda_2))\oplus\boldsymbol{J}_1(\lambda_3)$;

$(1,1,1,1):\boldsymbol{M}(\lambda)\simeq(\boldsymbol{J}_1(\lambda_1)\oplus\boldsymbol{J}_2(\lambda_1))\oplus(\boldsymbol{J}_1(\lambda_2)\oplus\boldsymbol{J}_2(\lambda_2))\oplus \boldsymbol{J}_1(\lambda_3)\oplus\boldsymbol{J}_1(\lambda_4)$. Then $\boldsymbol{M}(a_1)$ equals

$$\begin{pmatrix}0&0&0\\0&0&0\\1&\xi&0\\0&0&0\\0&0&0\\0&0&0\\\eta&0&0\\1&0&0\\0&0&1\\1&1&0\\0&0&1\\0&0&0\end{pmatrix},\quad\begin{pmatrix}0&0&0\\0&1&\xi\\0&\eta&1\\0&1&0\\1&0&0\\0&0&0\\0&1&0\\0&0&1\end{pmatrix},\quad\begin{pmatrix}0&0&0\\0&1&\xi\\0&\eta&0\\1&1&0\\0&0&0\\1&0&0\\0&1&0\\0&0&1\end{pmatrix},\quad\begin{pmatrix}0&0&0\\0&1&\xi\\0&1&\eta\\1&0&1\\0&0&0\\1&0&0\\0&1&0\\0&0&1\end{pmatrix},\quad\begin{pmatrix}0&0&0\\0&1&\xi\\1&1&\eta\\0&0&0\\0&1&1\\1&0&0\\0&1&0\\0&0&1\end{pmatrix}$$

respectively. □

By the lemmas above, we obtain

Theorem 4.4 Let $\mathcal{A}=(A,V)$ be a tame bocs with a layer $L=(A';\omega;a_1,a_2,\cdots,a_n;v_1,v_2,\cdots,v_m)$ given in 2. Then $\delta(a_1)$ has the following possibilities.

(1) 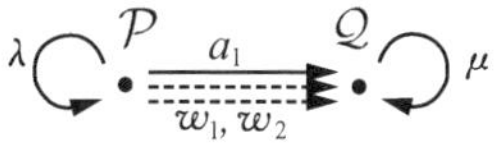

Fig 4.4 in case 2(2),

Fig. 4.4

$\delta(a_1)=(\lambda-\lambda_1)^{e_1}(\lambda-\lambda_2)^{e_2}w_1+w_2(\mu-\mu_1)^{f_1}(\mu-\mu_2)^{f_2}$, with $f(\lambda_i)\neq 0$, $g(\mu_j)\neq 0$, $e_1f_1\neq 0$, $e_1+e_2+f_1+f_2\leqslant 3$, w_1 and w_2 are linearly independent.

(2) 

Fig 4.5 in case 2(1), $\delta(a_1)=(\lambda-\lambda_0)w_1+$

Fig. 4.5

$w_2(\mu-\lambda_0)$, with $f(\lambda_0)\neq 0$ or $\delta(a_1)$ is given in (1) with $\{\lambda_1,\lambda_2\}\cap\{\mu_1,$

$\mu_2\}=\varnothing$, w_1 and w_2 are linearly independent.

(3) 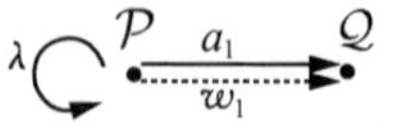

Fig 4. 6 in case 2(3),

Fig. 4. 6

$\delta(a_1)=(\lambda-\lambda_1)^{e_1}(\lambda-\lambda_2)^{e_2}(\lambda-\lambda_3)^{e_3}w_1$ with $f(\lambda_i)\neq 0$, where $\lambda_1,\lambda_2,\lambda_3$ are pairwise different and $1\leqslant e_1+e_2+e_3\leqslant 3$, or dually.

(4) 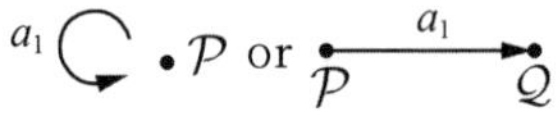

Fig 4. 7 in case 2(4), $\delta(a_1)=0$.

Fig. 4. 7

(5) $\delta(a_1)=w_1$ in all the above cases. □

Acknowledgements

The authors would like to thank the referee for their valuable suggestions which improve this note.

References

[1] Drozd Y A. Tame and wild matrix problems. in: Representations and Quadratic Forms (in Russian), Kiev: Inst. Math. Akad. Nauk Ukrain. SSR, 1979:39-74.

[2] Rojter A V. Matrix problems and representations of BOCS's. in: Representation Theory I (Dlab V, Gabriel P. eds.), LNM 831. Berlin: Springer-Verlag, 1980: 288-324.

[3] Crawley-Boevey W W. On tame algebras and bocses. Proc. London Math. Soc. , 1988, 56(3): 451-483.

[4] Zeng X Y, Zhang Y B. A correspondence of almost split sequences between some categories, Comm. Alg. , 2001, 29(2): 1-26.

[5] Crawley-Boevey W W. Matrix problems and Drozd's theorem. in: Topics in Algebra (eds. Balcerzyk, S. et al.). Warsaw: Banach Center publications, PWN-Polish Scientific Publishers, 1990, 26(1): 199-222.

[6] Bautista R, Kleiner M. Almost split sequences for relatively projective modules. J. Alg. , 1990, 135(1): 19-56.

[7] Sergeichuk V V. Canonical matrices for basic matrix problems. Linear Algebra and Its Applications, 2000, 317(1/3): 53-102.

[8] Xu Y G, Zhang Y B. Indecomposability and the number of links. Science in China, 2001, 44A(12): 1 515-1 522.

Central European Journal of Mathematics, 2007, 5(2): 215-263.

驯顺代数模范畴的态射空间①

On Hom-Spaces of Tame Algebras*

Abstract Let Λ be a finite dimensional algebra over an algebraically closed field k and Λ has tame representation type. In this paper, the structure of Hom-spaces of all pairs of indecomposable Λ-modules having dimension smaller than or equal to a fixed natural number is described, and their dimensions are calculated in terms of a finite number of finitely generated Λ-modules and generic Λ-modules. In particular, such spaces are essentially controlled by those of the corresponding generic modules.

Keywords Generic module; infinite radical; bocs.

§ 1. Introduction

Let Λ be a finite-dimensional k-algebra of tame representation type, k an algebraically closed field. We recall that Λ is of tame representation type if for all natural numbers d, there is a finite number of Λ-$k[x]$-bimodules $M_1, M_2, \cdots, M_n$ which are free of finite rank as right $k[x]$-

① Bautista R thanks the support of project "43374F" of Fondo Sectorial SEP-Conacyt. Y. Zhang thanks the support of Important project 10331030 of Natural Science Foundation of China.

Received: 2006-07-11; Accepted: 2006-11-09; 本文与 Bautista R, Drozd Y. A, 曾祥勇合作.

modules and such that if M is an indecomposable Λ-module of k-dimension equal to d, then $M \cong M_i \otimes_{k[x]} k[x]/(x-\lambda)$ for some $1 \leqslant i \leqslant n$ and $\lambda \in k$.

It is known from [6] that for each dimension d, almost all Λ-modules of dimension at most d are controlled by finitely many isomorphism classes of generic modules in the sense of (i) of Theorem 1.2. A question arises naturally: are Hom-spaces of Λ-modules also controlled by those of generic modules? In this paper, we will give a positive answer.

If G is a left Λ-module then G can be regarded as a left $\mathrm{End}_\Lambda(G)$-module, and we call its length as $\mathrm{End}_\Lambda(G)$-module, the endolength of G. We say that G is a generic module if it is indecomposable, of infinite dimension over k but finite endolength. We recall that if G is a generic Λ-module and R a commutative principal ideal domain which is finitely generated over k, then a realization of G over R is a finitely generated Λ-R-bimodule T such that if K is the quotient field of R, then $G \cong T \otimes_R K$ and $\dim_K (T \otimes_R K)$ is equal to the endolength of G.

As an example consider, $\Lambda = kQ$, the Kronecker algebra defined by quiver Q, then G is a generic module, and T is a realization of G over $R = k[x]$ (fig. 1.1).

$$Q =: \cdot \underset{b}{\overset{a}{\rightrightarrows}} \cdot, \qquad G =: k(x) \underset{id}{\overset{x}{\rightrightarrows}} k(x), \qquad T =: k[x] \underset{id}{\overset{x}{\rightrightarrows}} k[x].$$

Fig. 1.1

We denote by Λ-Mod the category of left Λ-modules, by Λ-mod the full subcategory of Λ-Mod consisting of the finite-dimensional Λ-modules, and by Λ-ind the full subcategory of Λ-mod consisting of the indecomposable Λ-modules.

We recall from Theorem 5.4 of [6] that if Λ is of tame representation type then given any generic Λ-module there is a good realization of G over some R in the sense of the following definition:

Definition 1.1 Let T be a realization of a generic module G over some R, then T is called a good realization if:

(1) T is free as right R-module;

(2) the functor $T\otimes_R -: R\text{-Mod}\to\Lambda\text{-Mod}$ preserves isomorphism classes and indecomposability;

(3) if $p\in R$ is a prime, $n\geqslant 1$ and $S_{p,n}$ denotes the exact sequence

$$0\to R/(p^n)\xrightarrow{(p,\pi)}R/(p^{n+1})\oplus R/(p^{n-1})\xrightarrow{\binom{\pi}{-p}}R/(p^n)\to 0$$

where π is the canonical projection, then $T\otimes_R S_{p,n}$ is an almost split sequence in Λ-mod.

We know from Theorem 4.6 of [6] that if G is a generic Λ-module then there is a splitting $\mathrm{End}_\Lambda(G)=k(x)\oplus\mathrm{radEnd}_\Lambda(G)$. This splitting induces a structure of left $\Lambda^{k(x)}=\Lambda\otimes_k k(x)$-module for G and such structure is called an admissible structure. The main aim of this paper is to prove of the following theorem:

Theorem 1.2 Let Λ be a finite-dimensional k-algebra of tame representation type, k an algebraically closed field. Let d be an integer greater than the dimension of Λ over k. Then there are generic Λ-modules $G_1, G_2, \cdots, G_s$ with admissible structures of left $\Lambda^{k(x)}$-modules and good realizations T_i over some R_i, finitely generated localization of $k[x]$, of each G_i and indecomposable Λ-modules $L_1, L_2, \cdots, L_t$ with $\dim_k L_j\leqslant d$ for $j=1,2,\cdots,t$ with the following properties:

(1) If M is an indecomposable left Λ-module with $\dim_k M\leqslant d$, then either $M\cong L_j$ for some $j\in\{1,2,\cdots,t\}$ or $M\cong T_i\otimes_{R_i}R_i/(p^m)$ for some $i\in\{1,2,\cdots,s\}$ some prime element $p\in R_i$ and some natural number m. If M is an indecomposable which is simple, projective or injective left Λ-module, then $M\cong L_j$ for some $j\in\{1,2,\cdots,t\}$.

(2) If $M=T_i\otimes_{R_i}R_i/(p^m)$, $N=T_j\otimes_{R_j}R_j/(q^n)$, $L_u^{k(x)}=L_u\otimes_k k(x)$ with $i,j\in\{1,2,\cdots,s\}$, $u\in\{1,2,\cdots,t\}$, p a prime in R_i, q a prime in

R_j, then

$$\dim_k \mathrm{rad}_\Lambda^\infty(M,N)=mn\dim_{k(x)}\mathrm{rad}_\Lambda k(x)(G_i,G_j),$$
$$\dim_k \mathrm{rad}_\Lambda^\infty(L_u,M)=m\dim_{k(x)}\mathrm{rad}_\Lambda k(x)(L_u^{k(x)},G_i),$$
$$\dim_k \mathrm{rad}_\Lambda^\infty(M,L_u)=m\dim_{k(x)}\mathrm{rad}_\Lambda k(x)(G_i,L_u^{k(x)}).$$

(3) Suppose $M=T_i\otimes_{R_i}R_i/(p^m)$, $N=T_j\otimes_{R_j}R_j/(q^n)$, then if $i=j$, $p=q$,

$$\mathrm{Hom}_\Lambda(M,N)\cong\mathrm{Hom}_{R_i}(R_i/(p^m),R_i/(p^n))\oplus\mathrm{rad}_\Lambda^\infty(M,N).$$

And if $i\neq j$ or $(p)\neq(q)$:

$$\mathrm{Hom}_\Lambda(M,N)=\mathrm{rad}_\Lambda^\infty(M,N).$$

Moreover,

$$\mathrm{Hom}_\Lambda(L_u,M)=\mathrm{rad}_\Lambda^\infty(L_u,M),\quad \mathrm{Hom}_\Lambda(M,L_u)=\mathrm{rad}_\Lambda^\infty(M,L_u).$$

For the proof of our main result we first study layered bocses of tame representation type (see Theorem 9.2). For this we use the method of reduction functors $F: \mathcal{B}_1$-Mod $\to \mathcal{B}_2$-Mod between the representation categories of two layered bocses $\mathcal{B}_1$ and $\mathcal{B}_2$ (see [5][7] and section 7 of this paper). We prove that given a layered bocs $\mathcal{A}$ of tame representation type and a dimension vector $\mathbf{d}$ of $\mathcal{A}$ there is a composition of reduction functors $F:\mathcal{B}$-Mod$\to\mathcal{A}$-Mod with $\mathcal{B}$ a minimal bocs such that if $M\in\mathcal{A}$-Mod with $\dim M\leqslant\mathbf{d}$, then there is a $N\in\mathcal{B}$-Mod with $F(N)\cong M$. Observe that in Theorem A of [5] several minimal bocses are needed. In section 6 we study the Hom-spaces for minimal bocses. Consider now the category $P^1(\Lambda)$ of morphisms $f:P\to Q$ with P,Q projective Λ-modules and $f(P)\subset \mathrm{rad}Q$. There is a layered bocs $\mathcal{D}(\Lambda)$, the Drozd's bocs, such that $\mathcal{D}(\Lambda)$-Mod is equivalent to $P^1(\Lambda)$. Using our results on Hom-spaces for minimal layered bocses we study the Hom-spaces in $P^1(\Lambda)$ obtaining a version of Theorem 1.2 for $P^1(\Lambda)$ (see Theorem 9.5). Finally, we use the relations between Hom-spaces in $P^1(\Lambda)$ and Λ-Mod collected in the results of sections 2 and 3.

§ 2. Generalities

Here we state the general results needed in our work. We recall that an additive k-category $\mathcal{R}$ is a Krull-Schmidt category if each object is a finite direct sum of indecomposable objects with local endomorphism rings. In this case, the indecomposable objects coincide with those having local endomorphism rings.

Let $\mathcal{R}$ be a Krull-Schmidt category. A morphism $f: E \to M$ in $\mathcal{R}$ is called irreducible if it is neither a retraction nor a section and for any factorization $f = vu$, either u is a section or v is a retraction.

A morphism $f: E \to M$ in $\mathcal{R}$ is called right almost split if

(1) f is not a retraction;

(2) if $g: X \to M$ is not a retraction, there is a $s: X \to E$ with $fs = g$.

Moreover, $f: E \to M$ a right almost split morphism is said to be minimal if $fu = f$ with $u \in \operatorname{End}_{\mathcal{R}}(E)$ implies u is an isomorphism.

One has the dual concepts for left almost split morphisms and minimal left almost split morphisms.

Remark Any minimal right almost split morphism $f: E \to M$ is an irreducible morphism. Moreover if $X \neq 0$, $g: X \to M$ is an irreducible morphism iff there is a section $\sigma: X \to E$ with $f\sigma = g$.

In particular if $h: F \to M$ is also a minimal right almost split morphism there is an isomorphism $u: F \to E$ with $fu = h$.

Similar properties hold for minimal left almost split morphisms.

Definition 2.1 A pair of composable morphisms in $\mathcal{R}$,

$$M \xrightarrow{f} E \xrightarrow{g} N$$

is said to be almost split if

(1) g is a minimal right almost split morphism;

(2) f is a minimal left almost split morphism, and;

(3) $g f = 0$

In the following, we use the following notation. If $f: E \to M$ and $f': E' \to M'$ are morphisms in $\mathcal{R}$, a morphism from f to f' is a pair (u, v) where $u: E \to E'$ and $v: M \to M'$ are morphisms such that $f'u = vf$. If u, v are isomorphisms, we say that f and g are isomorphic. Similarly if $M \xrightarrow{f} E \xrightarrow{g} N$, $M' \xrightarrow{f'} E' \xrightarrow{g'} N'$ are pairs of composable morphisms, a morphism from (f, g) into (f', g') is a triple (u_1, u_2, u_3) where $u_1: M \to M'$, $u_2: E \to E'$, $u_3: N \to N'$ are morphisms such that $u_2 f = f' u_1$, $u_3 g = g' u_2$. If u_1, u_2, u_3 are isomorphisms we say that the pair (f, g) is isomorphic to the pair (f', g'). The pairs (f, g) and (f', g') are equivalent if $M = M'$, $N = N'$ and there is an isomorphism from the first pair into the second one of the form $(1_M, u, 1_N)$.

If $\mathcal{A}$ is an additive category with split idempotents a pair (i, d) of composable morphisms $X \xrightarrow{i} Y \xrightarrow{d} Z$ in $\mathcal{A}$ is said to be exact if i is a kernel of d, and d is a cokernel of i. Let $\mathcal{E}$ be a class of exact pairs closed under isomorphisms. The morphisms i and d appearing in a pair of $\mathcal{E}$ are called an inflation and a deflation of $\mathcal{E}$, respectively.

We recall from [9] that the class $\mathcal{E}$ is an exact structure for $\mathcal{E}$ if the following axioms are satisfied:

E. 1 The composition of two deflations is a deflation.

E. 2 If $f: Z' \to Z$ is a morphism in $\mathcal{A}$ for each deflation $d: Y \to Z$ there is a morphism $f': Y' \to Y$ and a deflation $d': Y' \to Z'$ such that $df' = fd'$.

E. 3 Identities are deflations. If de is deflation, then so is d.

E. 3^{op} Identities are inflations. If ji is a inflation, then so is i.

If $\mathcal{E}$ is an exact structure for $\mathcal{A}$ then we denote by $\mathrm{Ext}_{\mathcal{A}}(X, Y)$ the equivalence class of the pairs $Y \xrightarrow{i} E \xrightarrow{d} X$ in $\mathcal{E}$. If $\mathcal{A}$ is a k-category, $\mathrm{Ext}_{\mathcal{A}}(?, -)$ is a bifunctor from $\mathcal{A}$ into the category of k-vector spaces, contravariant in the first variable and covariant in the second variable.

An object $X \in \mathcal{A}$ is called $\mathcal{E}$-projective if $\mathrm{Ext}_{\mathcal{A}}(X, -) = 0$, and it is called $\mathcal{E}$-injective if $\mathrm{Ext}_{\mathcal{A}}(-, X) = 0$.

Definition 2. 2 An almost split pair $X \to Y \to Z$ in $\mathcal{A}$ which is in $\mathcal{E}$ is called an almost split $\mathcal{E}$-sequence.

As in the case of modules, one can prove that in the above definition, X and Z are indecomposables.

Now, consider $(\mathcal{A}, \mathcal{E})$ an exact category with $\mathcal{A}$ a Krull-Schmidt k-category such that for $X, Y \in \mathcal{A}$, $\dim_k \mathrm{Hom}_{\mathcal{A}}(X, Y)$ is finite. Let $\mathcal{C}$ be a full subcategory of $\mathcal{A}$ having the following property:

(A) If X is an indecomposable object in $\mathcal{C}$ there is a minimal left almost split morphism in $\mathcal{A}$, $f: X \to Y_1 \oplus \cdots \oplus Y_t$ with $Y_i \in \mathcal{C}$.

We recall that a morphism $f: M \to N$ with M, N indecomposable objects in $\mathcal{A}$ is called a radical morphism if f is not an isomorphism.

Proposition 2.3 Let $\mathcal{C}$ be a full subcategory of $\mathcal{A}$ with condition (A).

Suppose $h: M \to N$ is a morphism in $\mathcal{A}$ with M, N indecomposable objects in $\mathcal{C}$ such that $h = \sum h_i$, where each h_i is a composition of m radical morphisms between indecomposables in $\mathcal{A}$, then $h = \sum g_j$ with each g_j composition of m radical morphisms between indecomposables in $\mathcal{C}$.

Proof By induction on m. If $m=1$ our assertion is trivial. Assume our assertion is true for $m-1$. We may assume $h = s_m, s_{m-1}, \cdots, s_1$ with $s_i: M_i \to M_{i+1}$, M_j indecomposable object of $\mathcal{A}$ for $j = 1, 2, \cdots, m+1$, $M_1 = M$, $M_{m+1} = N$. By (A), there is a left almost split morphism $M = M_1 \xrightarrow{u} Y_1 \oplus Y_2 \oplus \cdots \oplus Y_t$ with $Y_1, Y_2, \cdots, Y_t \in \mathcal{C}$. We have $u = \begin{bmatrix} u_1 \\ u_2 \\ \vdots \\ u_t \end{bmatrix}$.

Then there is $v = (v_1, v_2, \cdots, v_t): Y_1 \oplus Y_2 \oplus \cdots \oplus Y_t \to M_2$ with $vu = s_1 = \sum_{i=1}^{t} v_i u_i$. Therefore,

$$h = s_m s_{m-1} \cdots s_2 s_1 = \sum_{i=1}^{t} s_m s_{m-1} \cdots s_2 v_i u_i.$$

Now, consider $g_i = s_m s_{m-1} \cdots s_2 v_i: Y_i \to N$ which is a composition of $m-1$ radical morphisms. Then, by induction hypothesis, each g_i is a sum of $m-1$ radical morphisms between indecomposables in $\mathcal{C}$. Consequently, h is a sum of compositions of m radical morphisms between objects in $\mathcal{C}$. This proves our claim. □

We recall that an ideal of a k-category $\mathcal{R}$ is a subfunctor of $\mathrm{Hom}_{\mathcal{R}}(-, ?)$. If I, J are ideals of $\mathcal{R}$, IJ is the ideal such that for $X, Y \in$

$\mathcal{R}$, $IJ(X,Y)$ consists of sums of compositions gf with $f\in J(X,Z)$, $g\in I(Z,Y)$ for some $Z\in\mathcal{R}$. We denote by I^2 the ideal II and, by induction, $I^n=I^{n-1}I$. For $\mathcal{R}$ a Krull-Schmidt k-category we define the ideal $\mathrm{rad}_{\mathcal{R}}$ such that for X and Y indecomposable objects of $\mathcal{R}$, $\mathrm{rad}_{\mathcal{R}}(X,Y)=$the morphisms which are not isomorphisms. The infinity radical is defined by

$$\mathrm{rad}_{\mathcal{R}}^{\infty}=\bigcap_{n}\mathrm{rad}_{\mathcal{R}}^{n}.$$

Corollary 2.4 With the hypothesis of proposition 2.3, for $X,Y\in\mathcal{C}$,

$$\mathrm{rad}_{\mathcal{C}}^{\infty}(X,Y)=\mathrm{rad}_{\mathcal{A}}^{\infty}(X,Y).$$

Proof We may assume X and Y are indecomposables. It follows from Proposition 2.3 that $\mathrm{rad}_{\mathcal{C}}^{m}(X,Y)=\mathrm{rad}_{\mathcal{C}}^{m}(X,Y)$ for all m. Hence,

$$\mathrm{rad}_{\mathcal{C}}^{\infty}(X,Y)=\bigcap_{m}\mathrm{rad}_{\mathcal{C}}^{m}(X,Y)=\bigcap_{m}\mathrm{rad}_{\mathcal{A}}^{m}(X,Y)=\mathrm{rad}_{A}^{\infty}(X,Y).\quad\square$$

Now, we recall the following definition of [5], section 2:

Definition 2.5 If $(\mathcal{A},\mathcal{E})$ is an exact category with $\mathcal{A}$ a Krull-Schmidt category, we say that it has almost split sequences if

(1) for any indecomposable Z in $\mathcal{A}$ there is a right almost split morphism $Y\to Z$ and a left almost split morphism $Z\to X$;

(2) for each indecomposable Z in $\mathcal{A}$ which is not $\mathcal{E}$-projective, there is an almost split $\mathcal{E}$-sequence ending in Z, and for each indecomposable Z in $\mathcal{A}$ which is not $\mathcal{E}$-injective, there is an almost split $\mathcal{E}$-sequence starting in Z.

Remark If the exact category $(\mathcal{A},\mathcal{E})$ has almost split sequences one can consider the valued Auslander—Reiten quiver of $\mathcal{A}$ as in the case of the category of finitely generated modules over an artin algebra.

Proposition 2.6 Suppose $(\mathcal{A},\mathcal{E}_{\mathcal{A}})$ and $(\mathcal{B},\mathcal{E}_{\mathcal{B}})$ are two exact categories such that the first category has almost split sequences and $F:\mathcal{B}\to\mathcal{A}$ is a full and faithful functor sending $\mathcal{E}_{\mathcal{B}}$-sequences into $\mathcal{E}_{\mathcal{A}}$-sequences. Let $\{E_i\}_{i\in\mathbf{N}}$ be a set of pairwise non-isomorphic objects in $\mathcal{B}$ which are not $\mathcal{E}_{\mathcal{B}}$-projectives, and almost split $\mathcal{E}_{\mathcal{B}}$-sequences:

$$(e_1): E_1\xrightarrow{f_1}E_2\xrightarrow{g_1}E_1,$$

$$(e_i): E_i \xrightarrow{\binom{f_i}{g_{i-1}}} E_{i+1} \oplus E_i \xrightarrow{(g_i, f_{i-1})} E_i,$$

for $i>1$. Then, if there is an almost split $\mathcal{E}_{\mathcal{A}}$-sequence ending in $F(E_1)$ which is the image under F of a sequence in $\mathcal{E}_{\mathcal{B}}$, then the image $F(e_i)$ of the sequence e_i is an $\mathcal{E}_{\mathcal{A}}$-almost split sequence for all $i \in \mathbf{N}$.

Proof There is a sequence in $\mathcal{E}_{\mathcal{B}}$, $(a): M \xrightarrow{u} E \xrightarrow{v} E_1$ whose image under F is an almost split $\mathcal{E}_{\mathcal{A}}$-sequence. Since F is a full and faithful functor, then (a) is an almost split sequence. This implies that (a) is isomorphic to (e_1). Therefore, the image under F of (e_1) is isomorphic to the image under F of (a) which is an almost split sequence, and so, the image of (e_1) under F is an almost split sequence.

Suppose that $F(e_l)$ is an almost split sequence for all $l \leqslant i$. By hypothesis, (e_{i+1}) is a non-trivial $\mathcal{E}_{\mathcal{B}}$-sequence, since F is a full and faithful functor. Then $F(e_{i+1})$ is a non-trivial $\mathcal{E}_{\mathcal{A}}$-sequence. Thus, $F(E_{i+1})$ is not $\mathcal{E}_{\mathcal{A}}$-projective. Then there is an almost split sequence

$$L_{i+1} \to M_{i+1} \to F(E_{i+1}).$$

Here $F(e_i)$ is an almost split sequence. Then we have an almost split sequence:

$$F(E_i) \to F(E_{i+1}) \oplus F(E_{i-1}) \to F(E_i),$$

and so, we have an irreducible morphism $F(E_i) \to F(E_{i+1})$. Therefore, $M_{i+1} \cong F(E_i) \oplus Y$. Thus, we have an irreducible morphism $L_{i+1} \to F(E_i)$. This implies that $L_{i+1} \cong F(E_{i+1})$ or $L_{i+1} \cong F(E_{i-1})$. But we have an almost split sequence starting and ending in $F(E_{i-1})$. Therefore, if $L_{i+1} \cong F(E_{i+1})$, then $F(E_{i+1}) \cong F(E_{i-1})$ implies $E_{i+1} \cong E_{i-1}$, which is not the case, therefore $L_{i+1} \cong F(E_{i+1})$. Then the socle of $\mathrm{Ext}_{\mathcal{A}}(F(E_{i+1}), F(E_{i+1}))$ as $\mathrm{End}_{\mathcal{A}}(F(E_{i+1}))$-module is simple. As previously stated, $F(e_{i+1})$ is a non-zero element of the above socle, and; therefore, $F(e_{i+1})$ is an almost split sequence. □

§ 3. The categories $P(\mathrm{A})$ and $P^1(\Lambda)$

Let Λ be a finite-dimensional algebra over an arbitrary field k. We denote by Λ-Proj the full subcategory of Λ-Mod whose objects are projective Λ-modules, and by Λ-proj, the full subcategory of Λ-mod whose objects are projective Λ-modules.

Here Λ-proj has only a finite number of isoclasses of indecomposable objects, then for any indecomposable projective Λ-module P there are morphisms

$$\rho(P):r(P)\to P,\quad \lambda(P):P\to l(P)$$

such that they are a minimal right almost split in Λ-proj and a minimal left almost split in Λ-proj, respectively. Observe that $\rho(P)$ and $\lambda(P)$ are also a minimal right almost split and a minimal left almost split morphism, respectively, in the category Λ-Proj.

Denote by $P(\Lambda)$ the category whose objects are morphisms $X=f_X:P_X\to Q_X$, with $P_X, Q_X\in\Lambda$-Proj. The morphisms from X to Y, objects of $P(\Lambda)$, are pairs $u=(u_1,u_2)$ with $u_1:P_X\to P_Y$, $u_2:Q_X\to Q_Y$ such that $u_2f_X=f_Yu_1$. If $u=(u_1,u_2):X\to Y$ and $\boldsymbol{v}=(v_1,v_2):Y\to Z$ are morphisms, its composition is defined by $\boldsymbol{vu}=(v_1u_1,v_2u_2)$.

We denote by $\mathcal{E}$ the class of pairs of composable morphisms $X\xrightarrow{u}Y\xrightarrow{v}Z$ such that the sequences of Λ-modules:

$$0\to P_X\xrightarrow{u_1}P_Y\xrightarrow{v_1}P_Z\to 0$$

$$0\to Q_X\xrightarrow{u_2}Q_Y\xrightarrow{v_2}Q_Z\to 0$$

are exact and then split exact.

Proposition 3.1 The pair $(P(\Lambda),\mathcal{E})$ is an exact category.

Proof See [1]. □

For P any projective Λ-module consider $J(P)=(P\xrightarrow{id_P}P)$, $Z(P)=(P\xrightarrow{0}0)$, $T(P)=(0\xrightarrow{0}P)$. It is easy to see that the objects $J(P)$ and $T(P)$ are $\mathcal{E}$-projectives and the objects $J(P)$, $Z(P)$ are $\mathcal{E}$-injectives. One can see without difficulty that the exact category $(P(\Lambda),\mathcal{E})$ has enough projectives and enough injectives.

Proposition 3.2 The indecomposable $\mathcal{E}$-projectives in $P(\Lambda)$ are the objects $J(P)$ and $T(P)$ for P indecomposable projective Λ-module.

The indecomposable $\mathcal{E}$-injectives in $P(\Lambda)$, are the objects $J(P)$ and $Z(P)$ for P indecomposable projective Λ-module.

We denote by $\overline{P(\Lambda)}$ the category having the same objects as $P(\Lambda)$ and morphisms those of $P(\Lambda)$ modulo the morphisms which factorizes through $\mathcal{E}$-injective objects.

We have a full and dense functor $Cok: P(\Lambda) \to \Lambda\text{-Mod}$ which in objects is given by $Cok(f_X: P_X \to Q_X) = \text{Coker } f_X$.

Proposition 3.3 The functor $Cok: P(\Lambda) \to \Lambda\text{-Mod}$ induces an equivalence $\overline{Cok}: \overline{P(\Lambda)} \to \Lambda\text{-Mod}$.

Proof One can prove (see [1]) that if $f: X \to Y$ is a morphism in $P(\Lambda)$ then $Cok(f) = 0$ iff f factorizes through some $\mathcal{E}$-injective object in $P(\Lambda)$. □

We consider now $p(\Lambda)$, the full subcategory of $P(\Lambda)$ whose objects are morphisms between finitely generated Λ-modules.

Proposition 3.4 The exact category $(p(\Lambda), \mathcal{E})$ has almost split $\mathcal{E}$-sequences.

Proof See [1]. □

Now consider $P^1(\Lambda)$ the full subcategory of $P(\Lambda)$ whose objects are those $X = f_X: P_X \to Q_X$ with $\text{Im}(f_X) \subset \text{rad}(Q_X)$. We denote by $\mathcal{E}_1$ the class of composable morphisms in $P^1(\Lambda)$ which are in $\mathcal{E}$. By $p^1(\Lambda)$ we denote the full subcategory of $P^1(\Lambda)$, whose objects are morphisms between finitely generated projective Λ-modules.

Proposition 3.5 The pair $(P^1(\Lambda), \mathcal{E}_1)$ is an exact category.

Proof See [1]. □

For an indecomposable projective Λ-module P denote by $R(P)$ the object $\rho(P): r(P) \to P$ and by $L(P)$ the object $\lambda(P): P \to l(P)$. Observe that P a left Λ-module is in Λ-proj if P is indecomposable and projective.

Lemma 3.6 The morphism

$$\sigma(P) = (\rho(P), id_P): R(P) \to J(P)$$

is a minimal right almost split morphism in $P(\Lambda)$, the morphism

$$\tau(P)=(id_P,\lambda(P)):J(P)\to L(P)$$

is a minimal left almost split morphism in $P(\Lambda)$.

Proposition 3.7 Suppose $u:X\to Y$ is a morphism in $P^1(\Lambda)$ such that $Cok(u)=0$, then $u=gh$ with $h:X\to W$, $g:W\to Y$ and W a sum of objects of the form $Z(P)$ and $R(Q)$.

Proof It follows from Proposition 3.3 and Lemma 3.6. □

Proposition 3.8 The indecomposable $\mathcal{E}_1$-projectives in $P^1(\Lambda)$ are the objects $T(P)$ and $L(P)$ with P indecomposable projective Λ-module. The indecomposable $\mathcal{E}_1$-injectives are the objects $Z(P)$ and $R(P)$ with P an indecomposable projective Λ-module.

Proof It follows from Proposition 3.2 and Lemma 3.6. □

Proposition 3.9 For $X, Y\in P^1(\Lambda)$, there is an exact sequence

$$0\to \mathrm{Hom}_{P^1(\Lambda)}(X,Y)\xrightarrow{i}\mathrm{Hom}_\Lambda(P_X,P_Y)\oplus\mathrm{Hom}_\Lambda(Q_X,Q_Y)$$
$$\xrightarrow{\delta}\mathrm{rad}_\Lambda(P_X,Q_Y)\xrightarrow{\eta}\mathrm{Ext}_{P^1(\Lambda)}(X,Y)\to 0$$

Proof See Proposition 5.1 of [1]. □

Now, if $X=(P_X\xrightarrow{f_X}Q_X)\in P(\Lambda)$ choose some minimal projective cover $P_2\xrightarrow{g}P_1\xrightarrow{\eta}\mathrm{Ker}h\to 0$ with

$$h=D(\Lambda)\otimes f_X:D(\Lambda)\otimes_\Lambda P_X\to D(\Lambda)\otimes_\Lambda Q_X.$$

We put $\tau X=(P_2\xrightarrow{g}P_1)$.

Proposition 3.10 If X is an indecomposable which is not $\mathcal{E}_1$-projective in $p^1(\Lambda)$, then there is an almost split $\mathcal{E}_1$-sequence:

(1) $$Y\to E\to X$$

with $Y\cong_\tau X$. Dually if Y is indecomposable non $\mathcal{E}_1$-injective, then there is an almost split $\mathcal{E}_1$-sequence (1).

Proof See [10] for k a perfect field and [1] for the general case. □

Proposition 3.11 For $X, Y\in p^1(\Lambda)$, there is an isomorphism of k-modules

$$\mathrm{Ext}_{p^1(\Lambda)}(X,Y)\cong D\overline{\mathrm{Hom}}_{p^1(\Lambda)}(Y,\tau(X)).$$

Here $\overline{\mathrm{Hom}}_{p^1(\Lambda)}(Z,W)$ stands for the morphisms from Z to W modulo those morphisms which are factorized through $\mathcal{E}_1$-injectives objects.

Proof It follows from Corollary 9.4 of [9]. □

As a consequence we obtain:

Proposition 3.12 (See [3] and [1]) For $X,Y\in p^1(\Lambda)$, there is an isomorphism of k-modules:

$$\mathrm{Ext}_{p^1(\Lambda)}(X,Y)\cong$$
$$D(\mathrm{Hom}_\Lambda(Cok(Y),DtrCok(X))/\mathcal{S}(Cok(Y),Dtr(Cok(X)))$$

where $\mathcal{S}(M,N)$ are the morphisms which factorizes through semisimple Λ-modules.

Proposition 3.13 If $Y\xrightarrow{v}E\xrightarrow{u}X$ is an almost split sequence in $p(\Lambda)$ with $Cok(Y)\neq 0$ and $Cok(X)\neq 0$, then

$$0\to Cok(Y)\xrightarrow{Cok(v)}Cok(E)\xrightarrow{Cok(u)}Cok(X)\to 0$$

is an almost split sequence in Λ-mod. Moreover, if $Cok(Y)$ is not a simple Λ-module, then the sequence $Y\xrightarrow{v}E\xrightarrow{u}X$ lies in $p^1(\Lambda)$.

Proof For the first part of our statement see Proposition 5.6 of [1], for the second part see Theorem 2.6 of [10] arid Proposition 5.7 of [1]. □

Suppose now that Λ is a basic finite-dimensional k-algebra, and $1_\Lambda=\sum_{i=1}^{n}e_i$ is a decomposition into pairwise orthogonal primitive idempotents. Moreover, assume that $\dim_k(\Lambda/\mathrm{rad}\Lambda)e_i=1$ for all $i=1,2,\cdots,n$. For $M\in\Lambda$-mod we put

$$\dim M=(\dim_k e_1M,\dim_k e_2M,\cdots,\dim_k e_nM).$$

For $X=f_X:P_X\to Q_X$ an object in $p^1(\Lambda)$ we put

$$\dim X=(\dim(P_X/\mathrm{rad}P_X),\quad \dim(Q_X/\mathrm{rad}Q_X))\in\mathbf{Z}^{2n}.$$

In the following, we consider three bilinear forms defined on $\mathbf{Z}^{2n}$:

For $\boldsymbol{x}=(x_1,x_2,\cdots,x_n;x_1',x_2',\cdots,x_n')$, $\boldsymbol{y}=(y_1,y_2,\cdots,y_n;y_1',y_2',\cdots,y_n')$, we put

$h_\Lambda(\boldsymbol{x},\boldsymbol{y}) = \sum_{i,j}(x_iy_j + x'_iy'_j)\dim_k(e_i\Lambda e_j) - \sum_{i,j}x_iy'_j\dim_k(e_i \mathrm{rad}\Lambda e_j)$,

$s_\Lambda(\boldsymbol{x},\boldsymbol{y}) = \sum_{i=1}^{n}x_iy'_i$, $g_\Lambda(x,y) = \sum_{i,j}(x_iy_j + x'_iy'_j - x_iy'_j)(\dim_k e_i\Lambda e_j)$.

Clearly $g_\Lambda(\boldsymbol{x},\boldsymbol{y}) = h_\Lambda(\boldsymbol{x},\boldsymbol{y}) - s_\Lambda(\boldsymbol{x},\boldsymbol{y})$.

Proposition 3.14 For $X,Y \in p^1(\Lambda)$ we have:

(1) $\dim_k \mathrm{Hom}_{p^1(\Lambda)}(X,Y) - \dim_k \mathrm{Ext}_{p^1(\Lambda)}(X,Y) = h_\Lambda(\dim X, \dim Y)$;

(2) $\dim_k \mathrm{Ext}_{p^1(\Lambda)}(X,Y) = \dim_k \mathrm{Hom}_\Lambda(Cok(Y), DtrCok(X)) - s_\Lambda(\dim X, \dim Y)$;

(3) $\dim_k \mathrm{Hom}_\Lambda(Cok(Y), DtrCok(X)) = \dim_k \mathrm{Hom}_{p^1(\Lambda)}(X,Y) - g_\Lambda(\dim X, \dim Y)$.

Proof The part (1) follows from Proposition 3.9, part (2) follows from Proposition 3.12 and from the equalities:

$$\dim_k \mathcal{S}(Cok(Y), DtrCok(X)) = \dim_k \mathrm{Hom}_\Lambda(\mathrm{top}Cok(Y), \mathrm{soc}DtrCok(X)) = s_\Lambda(\dim X, \dim Y).$$

Finally, (3) follows from (1) and (2). □

§ 4. Bocses

We recall that a coalgebra over a k-category A is an A-bimodule V endowed with two bimodule homomorphisms, a comultiplication $\mu: V \to V \otimes_A V$ and a counite $\epsilon: V \to A$, subject to the conditions

$$(\mu \otimes 1)\mu = (1 \otimes \mu)\mu,$$
$$(\epsilon \otimes 1)\mu = i_l, (1 \otimes \epsilon)\mu = i_r,$$

with $i_l: V \cong A \otimes_A V$ and $i_r: V \cong V \otimes_A A$ the natural isomorphisms. Observe that A is a coalgebra over A with comultiplication $A \cong A \otimes_A A$ the natural isomorphism and the counite the identity morphism $id_A: A \to A$.

A bocs is a pair $\mathcal{A} = (A, V)$ with A a skeletally small k-category and V a coalgebra over A.

The bocs (A, A) is called the principal bocs.

The category $\mathcal{A}$-Mod has the same objects as A-Mod, the covariant functors $A \to k$-Mod. Then, if M, N are in $\mathcal{A}$-Mod, a morphism in $\mathcal{A}$-Mod is given by an A-module morphism from $V \otimes_A M$ to N. The composition of $f: V\otimes_A M\to N$ and $g: V\otimes_A N\to L$ is given by the composition

$$V\otimes_A M\xrightarrow{\mu\otimes 1}V\otimes_A V\otimes_A M\xrightarrow{1\otimes f}V\otimes_A N\xrightarrow{g}L,$$

the identity morphism for M in $\mathcal{A}$-Mod is given by the composition:

$$V\otimes_A M\xrightarrow{\epsilon\otimes 1}A\otimes_A M\xrightarrow{\sigma}M,$$

where σ is given by $\sigma(a\otimes m)=am$ for $a\in A, m\in M$. We identify A-Mod with (A,A)-Mod.

Suppose now $\mathcal{A}=(A,V)$ and $\mathcal{B}=(B,W)$ are two bocses, denote by $\epsilon_V, \mu_V, \epsilon_W, \mu_W$ the corresponding counites and comultiplications. A morphism of bocses $\theta:\mathcal{A}\to\mathcal{B}$ is a pair (θ_0,θ_1) where $\theta_0: A\to B$ is a functor and $\theta_1: V\to {}_{\theta_0}W_{\theta_0}$ is a morphism of A-A bimodules such that

$$\epsilon_W\theta_1=\theta_0\epsilon_V, \quad \text{and} \quad \pi(\theta_1\otimes\theta_1)\mu_V=\mu_W\theta_1,$$

where π is the natural map $W\otimes_A W\to W\otimes_B W$. A morphism of bocses θ: $\mathcal{A}\to\mathcal{B}$ induces a functor $\theta^*:\mathcal{B}$-Mod$\to\mathcal{A}$-Mod. For $M\in\mathcal{B}$-Mod we put $\theta^* M={}_{\theta_0}M$ and if $f: W\otimes_B M\to N$ is a morphism in $\mathcal{B}$-Mod then $\theta^*(f)$ is the composition:

$$V\otimes_A({}_{\theta_0}M)\xrightarrow{\theta_1\otimes 1}W\otimes_A({}_{\theta_0}M)\xrightarrow{\pi}W\otimes_B M\xrightarrow{f}N$$

where π is the natural morphism.

Observe that if

$$\mathcal{A}\xrightarrow{(\theta_0,\theta_1)}\mathcal{B}\xrightarrow{(\phi_0,\phi_1)}\mathcal{C}$$

are morphisms of bocses then $(\phi_0\theta_0,\phi_1\theta_1)=\phi\theta:\mathcal{A}\to\mathcal{C}$ is a morphism of bocses. Clearly $(\phi\theta)^*=(\theta)^*(\phi)^*$.

Lemma 4.1 If $\theta=(\theta_0,\theta_1):\mathcal{A}=(A,V)\to\mathcal{B}=(B,W)$ is a morphism of bocses then

$$(\theta)^*(1,\epsilon_W)^*=(1,\epsilon_V)^*(\theta_0,\theta_0)^*.$$

Proof It follows from the definition of morphism of bocses and the above. □

Let $\mathcal{A}=(A,V)$ be a bocs and A' a subcategory of A with the same objects as A. A morphism $\omega: A'\to {}_{A'}V_{A'}$ of A'-A' bimodules is said to be a grouplike of $\mathcal{A}$ relative to A' if $(i,\omega):(A',A')\to\mathcal{A}$ is a morphism of bocses, where $i: A'\to A$ is the inclusion. If the induced functor $(i,\omega)^*$:

$\mathcal{A}$-Mod$\to A'$-Mod reflects isomorphisms we say that ω is a reflector. If $\omega: {}_{A'}A'_{A'} \to -{}_{A'}V_{A'}$ is a grouplike we have that ω is completely determined by the elements $\omega_X = \omega(id_X)$ for all $X \in \mathrm{ind}A'$ such that $\mu(\omega_X) = \omega_X \otimes \omega_X$.

If $\mathcal{A} = (A, V)$ is a bocs $\bar{V} = \mathrm{Ker}\,\epsilon$ is called the kernel of $\mathcal{A}$. Then there is the following exact sequence of A-A bimodules:

$$0 \to \bar{V} \xrightarrow{\sigma} V \xrightarrow{\epsilon} A \to 0$$

where σ is the inclusion.

We recall that if $\omega: A' \to {}_{A'}V_{A'}$ is a grouplike, it determines two morphisms $\delta_1: {}_{A'}A_{A'} \to {}_{A'}\bar{V}_{A'}$ and $\delta_2: {}_{A'}\bar{V}_{A'} \to {}_{A'}\bar{V} \otimes_A \bar{V}_{A'}$, given for $a \in \mathrm{Hom}_A(X, Y)$ and $v \in V(X, Y)$ by:

$$\delta_1(a) = a\omega_X - \omega_Y a, \quad \delta_2(v) = \mu(v) - \omega_Y \otimes v - v \otimes \omega_X.$$

Observe that $(id_A, \epsilon): \mathcal{A} \to (A, A)$ is a morphism of bocses. Therefore, it induces a functor $(id_A, \epsilon)^*: A$-Mod$\to\mathcal{A}$-Mod. For $M \in A$-Mod, $(id_A, \epsilon)^*(M) = M$, and for $h: M \to N$ a morphism of A-modules $(id_A, \epsilon)^* h: V \otimes_A M \to N$ is given by $(id_A, \epsilon)^*(h)(v \otimes m) = h(\epsilon(v)m)$ for $m \in M, v \in V$.

For $M \in \mathcal{A}$-Mod, $(i, \omega)^*(M) = {}_{A'}M$ and if $f: V \otimes_A M \to N$ is a morphism in $\mathcal{A}$-Mod, $f^0 = (i, \omega)^* f: {}_{A'}M \to {}_{A'}N$ is given by $f^0(m) = f(\omega_X \otimes m)$ for $m \in M(X)$.

Given $\mathcal{A} = (A, V)$ a bocs with a grouplike ω relative to some A' subcategory of A, for any morphism, $f: V \otimes_A M \to N$ we have the morphisms $f^0 = (i, \omega)^* f \in \mathrm{Hom}_{A'}(M, N)$, $f^1 = f(\sigma \otimes 1): \bar{V} \otimes_A M \to N$. The pair of morphisms (f^0, f^1) satisfies the following property:

$$\text{(A)} \qquad f^0(am) = af^0(m) + f^1(\delta_1(a) \otimes m).$$

Now, for any object $Y \in A$ we have:

$$(V \otimes_A M)(Y) = V(-, Y) \otimes_A M = \omega_Y \otimes M(Y) \oplus (\bar{V} \otimes_A M)(Y),$$

therefore, a pair of morphisms (f^0, f^1) with

$$f^0 \in \mathrm{Hom}_{A'}(M, N) \quad \text{and} \quad f^1 \to \mathrm{Hom}_A(\bar{V} \otimes_A M, N)$$

which satisfies the condition (A) determines a morphism of A-modules

$f: V\otimes_A M\to N$. Thus, any morphism $f: V\otimes_A M\to N$ is completely determined by the pair (f^0, f^1) satisfying property (A). In the rest of the paper, we put $f=(f^0, f^1)$.

Proposition 4.2 If $f=(f^0, f^1): M\to N, g=(g^0, g^1): N\to L$ are morphisms in $\mathcal{A}$-Mod ther, $g f=(g^0 f^0, (g f)^1)$ with

$$(gf)^1(v\otimes m)=g^1(v\otimes f^0(m))+g^0(f^1(v\otimes m))+\sum_i g^1(v_i^1\otimes f^1(v_i^2\otimes m)),$$

where $v\in V, m\in M$ and $\delta_2(v)=\sum_i v_i^1\otimes v_i^2$.

Proof It follows from the fact that $(i,\omega)^*$ is a functor and from the definitions. □

Following [5], if A is a k-category a morphism $a\in A(X,Y)$ is called indecomposable if both X and Y are indecomposable objects of A. Similarly, if W is an A-A bimodule an element of W is an element $w\in W(X,Y)$ for some X, Y. In case both X and Y are indecomposable, w will be called indecomposable. If X and Y are objects of A, then we denote by $F_{X,Y}$ the A-A bimodule given by

$$F_{X,Y}=\mathrm{Hom}_A(-,X)\otimes_k \mathrm{Hom}_A(Y,-).$$

We say that the A-A bimodule W is freely generated by the elements $w_i\in W(X_i,Y_i), i=1,2,\cdots,n$ if there is an isomorphism of A-A bimodules

$$\psi: F_{X_1,Y_1}\oplus F_{X_2,Y_2}\oplus\cdots\oplus F_{X_n,Y_n}\to W$$

such that $\psi(id_{X_i}\otimes id_{Y_i})=w_i$, for $i=1,2,\cdots,n$.

Now, suppose that A' has the same objects as A, and T is an A'-A'-subimodule of ${}_{A'}A_{A'}$, denote by $T^{\otimes n}$ the tensor product $T\otimes_{A'} T\otimes_{A'}\cdots\otimes_{A'} T$ of n copies of T and set $T^0=A'$. Then the direct sum of A'-A'-bimodules:

$$T^{\otimes}=\bigoplus_{n=0}^{+\infty} T^{\otimes n}$$

can be regarded as a category with the same objects as A and product given by the natural isomorphisms $T^{\otimes n}\otimes_A T^{\otimes m}\to T^{\otimes m+n}$.

We recall from Definition 2.5 of [5] that if A' has the same objects as A, we say that A is freely generated over A by morphisms $a_1, a_2, \cdots, a_n$ in A if the a_i freely generate an A-A' subimodule T of ${}_{A'}A_{A'}$ such that the functor $T^{\otimes} \to A$ induced by the inclusion of A' and T in A is an isomorphism.

Definition 4.3 A k-category A is called minimal if it is skeletal and is equivalent to

$$\mathrm{mod}(k) \times \mathrm{mod}(k) \times \cdots \times \mathrm{mod}(k) \times P(R_1) \times P(R_2) \times \cdots \times P(R_n)$$

where $R_i = k[x, f_i(x)^{-1}]$ with $f_i(x)$ is a nonzero element of $k[x]$ and $P(R)$ denotes the category of finitely generated projective left R-modules. We denote by $\mathrm{ind}A$ the set of indecomposable objects of a minimal category A.

Definition 4.4 Let $\mathcal{A} = (A, V)$ be a bocs with kernel $\bar{V}$. A collection $L = (A'; \omega; a_1, a_2, \cdots, a_n; v_1, v_2, \cdots, v_m)$, is a layer for $\mathcal{A}$, if

(L1) A' is a minimal category;

(L2) A is freely generated over A' by indecomposable elements $a_1, a_2, \cdots, a_n$;

(L3) ω is a reflector for $\mathcal{A}$ relative to A';

(L4) $\bar{V}$ is freely generated as an A-A bimodule by indecomposable elements $v_1, v_2, \cdots, v_m$;

(L5) let $\delta_1: A \to \bar{V}$ be the morphism induced by ω, $A_0 = A'$ and for $i \in \{1, 2, \cdots, n-1\}$, A_i the subcategory of A generated by A' and $a_1, a_2, \cdots, a_i$, then for any $0 \leqslant i < n$, $\delta_1(a_{i+1})$ is contained in the A_i-A_i subimodule of $\bar{V}$ generated by $v_1, v_2, \cdots, v_m$.

A bocs having a layer will be called layered.

Suppose $\mathcal{A} = (A, V)$ is a bocs with layer $L = (A'; \omega; a_1, a_2, \cdots, a_n; v_1, v_2, \cdots, v_m)$. Throughout this paper, we denote by $\mathcal{A}$-mod the full subcategory of $\mathcal{A}$-Mod whose objects are representations M such that

$$\sum_{X \in \mathrm{ind}A'} \dim_k M(X) < +\infty.$$

For $\mathcal{A}$ as before we have

$$\bar{V} \otimes_A M \cong \bigoplus_{v_i} A(-, Y_i) \otimes_k M(X_i)$$

for $M \in A$-Mod. Thus, for $M, N \in A$-Mod we have an isomorphism:

$$\phi_{M,N}: \bigoplus_{v_i} \mathrm{Hom}_k(M(X_i), N(Y_i)) \to \mathrm{Hom}_A(\bar{V} \otimes_A M, N).$$

Therefore, in this case a morphism $f: M \to N$ in $\mathcal{A}$-Mod is given by a pair of morphisms

$$(f^0, \phi_{M,N}(f_1^1, f_2^1, \cdots, f_m^1)), \quad f^0 \in \mathrm{Hom}_{A'}(M, N),$$
$$f_i^1 \in \mathrm{Hom}_k(M(X_i), N(Y_i)),$$

$i = 1, 2, \cdots, m$ such that for all $a_j: X_j \to Y_j$, $j = 1, 2, \cdots, n$ and $u \in M(X_j)$

$$f_{Y_j}^0(a_j u) = a_j f_{X_j}^0(u) + \phi_{M,N}(f_1^1, f_2^1, \cdots, f_m^1)(\delta_1(a_j) \otimes u).$$

Observe that $\phi_{M,N}(f_1^1, f_2^1, \cdots, f_m^1)(v_i \otimes u) = f_i^1(u)$ for $u \in M(X_i)$, $i = 1, 2, \cdots, m$.

Lemma 4.5 With the above notations, if $(f, 0): M \to N$ and $(h^0, \phi_{N,L}(h_1, h_2, \cdots, h_m)): N \to L$ are morphisms in $\mathcal{A}$-Mod then:

$$(h^0, \phi_{N,L}(h_1, h_2, \cdots, h_m))(f, 0) = (h^0 f, \phi_{M,L}(g_1, g_2, \cdots, g_m)) \text{ with}$$
$$g_i = h_i f_{X_i}.$$

Similarly, if $(h^0, \phi_{M,N}(h_1, h_2, \cdots, h_m)): M \to N$, $(f, 0): N \to L$ are morphisms in $\mathcal{A}$-Mod, then:

$$(f, 0)(h^0, \phi_{M,N}(h_1, h_2, \cdots, h_m)) = (fh^0, \phi_{M,N}(g_1, g_2, \cdots, g_m)),$$
$$\text{with } g_i = f_{Y_i} h_i.$$

In later sections we need the following.

Definition 4.6 Let $\mathcal{A} = (A, V)$ be a bocs with layer $(A'; \omega; a_1, a_2, \cdots, a_n; v_1, v_2, \cdots, v_m)$. Then a sequence of morphisms in $\mathcal{A}$-Mod,

$$M \xrightarrow{f} E \xrightarrow{g} N$$

is called proper exact if $gf = 0$ and the sequence of morphisms

$$0 \to M \xrightarrow{(i,\omega)^* f} E \xrightarrow{(i,\omega)^* g} N \to 0$$

in A'-Mod is exact. An almost split sequence in $\mathcal{A}$-mod which is also a proper exact sequence is called a proper almost split sequence.

Definition 4.7 With the notation of Definition 4.6 an indecomposable object $X \in A'$ is called marked if $A'(X, X) \neq k\mathrm{id}_X$.

§ 5. Hom-spaces of minimal bocses

We recall from [5] that a minimal bocs is a bocs $\mathcal{A} = (A, V)$ with layer

$$L=(A';\omega;a_1,a_2,\cdots,a_n;v_1,v_2,\cdots,v_m)$$

such that $A'=A$. Therefore in this case the $a_1,a_2,\cdots,a_n$ do not appear.

Throughout this section, $\mathcal{B}=(B,W)$ is a minimal bocs with layer

$$L=(B;\omega;w_1,w_2,\cdots,w_m),\text{where}\quad w_i\in\overline{W}(X_i,Y_i).$$

For $M,N\in\mathcal{B}$-Mod we put $\mathrm{Hom}_{\mathcal{B}}(M,N)^1=\{f:M\to N\mid(1,\omega)^*(f)=0\}$.

Proposition 5.1 Let $\mathcal{B}=(B,W)$ be a minimal bocs and $\epsilon:W\to B$ the counit of W. Then for $M,N\in\mathcal{B}$-Mod we have

$$\mathrm{Hom}_{\mathcal{B}}(M,N)=(1,\epsilon)^*(\mathrm{Hom}_B(M,N))\oplus\mathrm{Hom}_{\mathcal{B}}(M,N)^1.$$

Proof We have $(1,\epsilon)^*(1,\omega)^*\cong id_{B\text{-Mod}}$. □

Observe that if we have any pair of morphisms $(f,\phi_{M,N}(h_1,h_2,\cdots,h_m))$ with $f\in\mathrm{Hom}_B(M,N)$, $h_i\in\mathrm{Hom}_k(M(X_i),N(Y_i))$ where $w_i:X_i\to Y_i$, this pair is a morphism from M to N in $\mathcal{B}$-Mod, because in a minimal bocs $\delta_1=0$ and condition (A) before Proposition 4.2 is trivially satisfied. Then we have:

Corollary 5.2 For $M,N\in\mathcal{B}$-mod:

$$\dim_k\mathrm{Hom}_{\mathcal{B}}^1(M,N)=\sum_{w_i}\dim_k\mathrm{Hom}_k(M(X_i),N(Y_i)).$$

The morphisms in the image of $(1,\mathcal{E})^*$ have the form $(f,0)$ where the morphism f is in $\mathrm{Hom}_B(M,N)$.

Lemma 5.3 (Compare Definition 3.8 in [5]) Let M,N be two objects in $\mathcal{B}$-Mod, then $M\cong N$ in $\mathcal{B}$-Mod iff $M\cong N$ in $\mathcal{B}$-Mod.

Proof If $h:M\to N$ is an isomorphism in $\mathcal{B}$-Mod then $(1,\omega)^*(h)$ is an isomorphism in B-Mod. Conversely, if $g:M\to N$ is an isomorphism in B-Mod then $(1,\epsilon)^*(g)$ is an isomorphism in $\mathcal{B}$-Mod. □

Clearly, Lemma 5.3 implies that indecomposable objects in B-Mod and $\mathcal{B}$-Mod coincide.

We have $B(Z,Z')=0$ for $Z\neq Z'\in\mathrm{ind}B$ and for $Z\in\mathrm{ind}B$, $B(Z,Z)=R_Z=k[x,h(x)^{-1}]id_Z$ with $h(x)\in k[x]$ or $B(Z,Z)=kid_Z$. Take M an indecomposable object in B-mod, then there is only one $Z\in\mathrm{ind}B$ such that $M(Z)\neq 0$. Here M is a covariant functor of B into k-Mod,

$M(Z)$ is a left R_Z- module. Therefore if $B(Z,Z)=R_Z\neq kid_Z$, $M(Z)\cong R_Z/(p^n)$ with $p=x-\lambda$ a prime element in R_Z, if $B(Z,Z)=kid_Z$, $M(Z)=k$.

For $Z\in \text{ind}B$ with $B(Z,Z)=R_Z\neq kid_Z$ and $p=x-\lambda$, a prime element in R_Z we define $M(Z,p,n)\in B$-Mod by

$$M(Z,p,n)(W)=0 \text{ for } W\neq Z, W\in \text{ind}B, M(Z,p,n)(Z)=R_Z/(p^n).$$

If $B(Z,Z)=kid_Z$ we define $S_Z\in B$-mod by

$$S_Z(W)=0 \text{ for } W\neq Z, W\in \text{ind}B, S_Z(Z)=k.$$

Lemma 5.4 If M is an indecomposable object in B-mod then $M\cong M(Z,p,n)$ or $M\cong S_Z$ for some $Z\in \text{ind}B$.

Lemma 5.5 Let $(f,0):M\to N$ be a morphism in $\mathcal{B}$-Mod such that for all $Z\in\text{ind}B$, $f_Z:M(Z)\to N(Z)$ is surjective. Then if $h:L\to N$ is a morphism in $\mathcal{B}$-Mod with $(1,\omega)^*(h)=0$, there is a morphism $g:L\to M$ in $\mathcal{B}$-Mod with $(f,0)g=h$.

Proof Take $h:L\to N$ with $(1,\omega)^*(h)=0$, then $h=(0,\phi_{L,N}(h_1, h_2,\cdots,h_m))$. We may assume that there is a j with $0\neq h_j\in \text{Hom}_k(M(X_j), N(Y_j))$ and $h_i=0$ for $i\neq j$.

We have that $f_{Y_j}:M(Y_j)\to N(Y_j)$ is an epimorphism. Consequently, there is a k-linear map $\sigma:N(Y_j)\to M(Y_j)$ with $f_{Y_j}\sigma= id_{N(Y_j)}$. Take now

$g_j=\sigma h_j\in \text{Hom}_k(L(X_j),M(Y_j))$, and $0=g_i\in \text{Hom}_k(L(X_i),M(Y_i))$, for $i\neq j$. Take now the morphism

$$g=(0,\phi_{L,M}(g_1,g_2,\cdots,g_m)):L\to M$$

then by Lemma 4.5 $(f,0)g=(0,\phi_{L,N}(\lambda_1,\lambda_2,\cdots,\lambda_m))$ with $\lambda_i=f_{Y_i}\ g_i$. Therefore, $\lambda_i=0$ for $i\neq j$ and $\lambda_i=f_{Y_j}\ g_j=f_{Y_j}\sigma h_j=h_j$. Consequently,

$$(f,0)g=(0,\phi_{L,N}(\lambda_1,\lambda_2,\cdots,\lambda_m))=(0,\phi_{L,N}(h_1,h_2,\cdots,h_m))=h. \quad \square$$

Similarly, we have the dual version of the above result.

Lemma 5.6 Let $(f,0):M\to N$ be a morphism in $\mathcal{B}$-Mod such that for all $Z\in\text{ind}B$, $f_Z:M(Z)\to N(Z)$ is an injection. Then if $u:M\to L$ is a morphism with $(1,\omega)^*(u)=0$ there is a morphism $v:N\to L$ with $v(f,0)=u$.

For $Z, Z' \in \operatorname{ind}B$ we denote by $t(Z,Z')$ the number of $w_i \in \overline{W}(Z,Z')$.

Lemma 5.7 Suppose M, N are indecomposable objects in $\mathcal{B}$-mod with $M(Z) \neq 0, N(Z') \neq 0, Z, Z' \in \operatorname{ind}B$. Then

$$\dim_k \operatorname{Hom}_{\mathcal{B}}(M,N)^1 = t(Z,Z')\dim_k M(Z)\dim_k N(Z').$$

Proof It follows from Corollary 5.2 □

Lemma 5.8 If M, N are indecomposable objects in $\mathcal{B}$-mod, then

$$\operatorname{rad}_{\mathcal{B}}^{\infty}(M,N) \subset \operatorname{Hom}_{\mathcal{B}}(M,N)^1.$$

Proof Suppose there is a $h \in \operatorname{rad}_{\mathcal{B}}^{\infty}(M,N)$ with $(1,\omega)^*(h) \neq 0$. Then there is a $Z \in \operatorname{ind}B$ with $M(Z) \neq 0, N(Z) \neq 0$. Since $(1,\omega)^*$ reflects isomorphisms, then $(1,\omega)^*(h)$ is not an isomorphism. Consequently, $B(Z,Z) = R_Z \neq k id_Z$ and $M \cong M(Z,p,m), N \cong M(z,p,n)$.

Here $\operatorname{rad}_B^{\infty}(M,N) \cong \operatorname{rad}_{R_Z}^{\infty}(R_Z/(p^m), R_Z/(p^n)) = 0$. Then there is a s with $\operatorname{rad}_B^s(M,N) = 0$.

On the other hand, there is a chain of non-isomorphisms between indecomposables:

$$M \xrightarrow{f_1} X_1 \xrightarrow{f_2} X_2 \to \cdots \to X_{s-1} \xrightarrow{f_s} N$$

with $g = (1,\omega)^*(f_s f_{s-1} \cdots f_2 f_1) \neq 0$.

But $g = (1,\omega)^*(f_s)(1,\omega)^*(f_{s-1}) \cdots (1,\omega)^*(f_1) \in \operatorname{rad}_B^s(M,N) = 0$, a contradiction. This proves our claim. □

Consider $M = M(Z,p,m), N = M(Z,p,n)$ indecomposables in B-mod. If $f: R_Z/(p^m) \to R_Z/(p^n)$ is a morphism of R_Z-modules, we put $u(f): M \to N$ given by $u(f)_Z = f$ and $u(f)_W = 0$ for $W \neq Z$.

Proposition 5.9 Let M, N be indecomposables in $\mathcal{B}$-mod with $M(Z) \neq 0$ or $N(Z) \neq 0$ for some $Z \in \operatorname{ind}B$ with $B(Z,Z) \neq k id_Z$, then

$$\operatorname{rad}_{\mathcal{B}}^{\infty}(M,N) = \operatorname{Hom}_{\mathcal{B}}(M,N)^1.$$

Proof By Lemma 5.8, it is enough to prove that if $f: M \to N$ is a morphism in $\mathcal{B}$-mod with $(1,\omega)^*(f) = 0$ then $f \in \operatorname{rad}_{\mathcal{B}}^{\infty}(M,N)$. Suppose $M(Z) \neq 0$ with $B(Z,Z) = R_Z \neq id_Z k$. Then we may assume $M = M(Z,p,m)$. Take any natural number n. Consider the monomorphism $i_l: R_Z/(p^l) \to R_Z/(p^{l+1})$ given by $i_l(\eta_l(a)) = \eta_{l+1}(pa)$ for $a \in R_Z$ and

$\eta_j : R_Z \to R_Z/(p^j)$ the quotient map. Take

$$(u,0)=(u(i_{n+m-1}),0)\cdots(u(i_{m+1}),0)(u(i_m),0):$$
$$M(Z,p,m)\to M(Z,p,m+n).$$

Here $u_Z : M(Z,p,m)(Z)\to M(Z,p,m+n)(Z)$ is a monomorphism. By Lemma 5.6, there is a morphism $t : M(Z,p,m+n)\to N$ in $\mathcal{B}$-Mod such that $t(u,0)=f$.

Now, $(u,0)\in \mathrm{rad}_{\mathcal{B}}^{\infty}(M,M(Z,p,m+n))$, and, therefore, $f=t(u,0)\in \mathrm{rad}_{\mathcal{B}}^{\infty}(M,N)$ for all n, then $f\in \mathrm{rad}_{\mathcal{B}}^{\infty}(M,N)$.

For the case in which $N(Z)\neq 0$ with $B(Z,Z)\neq kid_Z$ one proceeds in a similar way. □

Corollary 5.10 If M,N are indecomposable objects in $\mathcal{B}$-mod, and $Z,Z'\in \mathrm{ind}B$ with $M(Z)\neq 0, N(Z')\neq 0$, and $B(Z,Z)\neq kid_Z$ or $B(Z',Z')\neq kid_{Z'}$, then

$$\dim_k \mathrm{rad}_{\mathcal{B}}^{\infty}(M,N)=\dim_k M(Z)\dim_k N(Z')t(Z,Z').$$

Corollary 5.11 Let $M=M(Z,p,m)$, $N=M(Z',q,n)$, $S=S_W$ be indecomposables in $\mathcal{B}$-mod, with $B(Z,Z)\neq kid_Z$, $B(Z',Z')\neq kid_{Z'}$, $B(W,W)=kid_W$. Then, if $Z=Z'$, $p=q$,

$$\mathrm{Hom}_{\mathcal{B}}(M,N)\cong \mathrm{Hom}_B(M,N)\oplus \mathrm{rad}_{\mathcal{B}}^{\infty}(M,N),$$

with $\dim_k(\mathrm{Hom}_B(M,N))=\min\{m,n\}$.

And if $Z\neq Z'$ or Z=Z', and $(p)\neq(q)$

$$\mathrm{Hom}_{\mathcal{B}}(M,N)=\mathrm{rad}_{\mathcal{B}}^{\infty}(M,N).$$

Moreover,

$$\mathrm{Hom}_{\mathcal{B}}(M,S)=\mathrm{rad}_{\mathcal{B}}^{\infty}(M,S)\quad \text{and}\quad \mathrm{Hom}_{\mathcal{B}}(S,M)=\mathrm{rad}_{\mathcal{B}}^{\infty}(S,M).$$

Lemma 5.12 If $0\to M\xrightarrow{f^0}E\xrightarrow{g^0}N\to 0$ is a short exact sequence in B-Mod, then the pair of morphisms in $\mathcal{B}$-Mod, $M\xrightarrow{(f^0,0)}E\xrightarrow{(g^0,0)}N$ is an exact pair of morphisms.

Proof We claim that $f=(f^0,0)$ is a kernel of $(g^0,0)$. Assume there is a morphism $u=(u^0,u^1)=(u^0,0)+(0,u^1): L\to E$ such that $gu=(g^0u^0,(gu)^1)=0$. Here $g^0u^0=0$, then there is a unique morphism in B-Mod, $v^0: L\to M$ with $f^0v^0=u^0$. Now, $u^1=\phi_{L,E}(u_1,u_2,\cdots,u_m)$, with $u_i: L(X_i) \to E(Y_i)$ where $w_i \in \overline{W}(X_i, Y_i)$. Then $(gu)^1 = \phi_{L,N}(g^0_{Y_1}u_1, g^0_{Y_2}u_2, \cdots, g^0_{Y_m}u_m)$. Therefore, for $i=1,2,\cdots,m$, $g^0_{Y_i}u_i=0$. Thus, there are linear maps $v_i: L(X_i) \to M(Y_i)$ with $f^0_{Y_i}v_i = u_i$ for $i = 1,2,\cdots,m$. Then taking $v=(v^0, \phi_{L,M}(v_1, v_2, \cdots, v_m))$ we have $fv=u$. Clearly v is unique with this property. This proves our claim. In a similar way one can prove that g is a cokernel of f. □

Lemma 5.13 Suppose (1): $M \xrightarrow{f} E \xrightarrow{g} N$ is a proper exact sequence in $\mathcal{B}$-Mod. Then(1) is isomorphic to the sequence:

$$M \xrightarrow{(f^0, 0)} E \xrightarrow{(g^0, 0)} N.$$

Proof By Lemma 5.5 and its proof, there is a morphism $u=(0, u^1): E \to E$ such that $(g^0, 0)u = (0, g^1)$. Then $(g^0, 0)(1_E, u^1) = g$, with $\sigma = (1_E, u^1)$ an isomorphism. Thus, $(g^0, 0)\sigma f = g f = 0$. But by the above lemma, $(f^0, 0)$ is a kernel of $(g^0, 0)$, then there is a morphism $\lambda = (\lambda^0, \lambda^1): M \to M$ with $(f^0, 0)\lambda = \sigma f$. Here $f^0\lambda^0 = f^0$, since f^0 is a monomorphism then $\lambda^0 = 1_M$. Therefore, $\lambda: M \to M$ is an isomorphism. This proves our claim. □

From Lemma 5.12 and Lemma 5.13, we deduce that proper exact sequences are exact pairs of morphisms. Denote by $\mathcal{E}_p$ the class of proper exact sequences in $\mathcal{B}$-Mod, then we have the following.

Proposition 5.14 The pair $(\mathcal{B}\text{-Mod}, \mathcal{E}_p)$ is an exact category.

Proof Observe first that $g=(g^0, g^1): E \to M$ is a deflation if and only if g^0 is an epimorphism. In fact, if g is a deflation, by definition of proper exact sequence g^0 is an epimorphism. Conversely, suppose g^0 is an epimorphism, then as in the proof of Lemma 5.5 there is an isomorphism $\tau: E \to E$ such that $(g^0, 0) = g\tau$. Taking $f^0: N \to E$ the kernel of g^0 in B-Mod, we see that $(g^0, 0)$ is a deflation, thus g is a deflation too. Similarly, one can prove that $f: N \to E$ is an inflation if and only if f^0 is a monomorphism. From this, it is clear that conditions E.1, E.3 and E.3^{op} hold. For proving E.2, assume $g: E \to N$ is a deflation and $h: L \to N$ is an arbitrary morphism. Then we have the morphism (g, h):

$E \oplus L \to N$. Now, $(g, h) = ((g^0, h^0), (g^1, h^1))$, here g^0 is an epimorphism, then (g^0, h^0) is also an epimorphism, thus (g, h) is a deflation, therefore it has a kernel, $M \xrightarrow{u} E \oplus L$. Take $u_1: M \to E$ equal to u composed with the projection on E and $-u_2: M \to L$, the composition of u with the projection on L. Now, one can see that u_2 is a deflation and $g u_1 = h u_2$. Therefore, E. 2 holds. □

Let $Z_1, Z_2, \cdots, Z_s$ be all marked objects in ind$B$. For $i = 1, 2, \cdots, s$ take $R_i = B(Z_i, Z_i)$ and the B-R_i-bimodule $B_i = B(Z_i, -)$. Then if p is a prime element of R_i and n a positive integer, $M(Z_i, p, n) \cong B_i \otimes_{R_i} R_i/(p^n)$. We denote by $S^i_{p,n}$ the exact sequence in R_i-mod:

$$0 \to R_i/(p^n) \xrightarrow{(p,\pi)} (R_i/(p^{n+1}) \oplus R_i/(p^{n-1})) \xrightarrow{\binom{\pi}{-p}} R/(p^n) \to 0.$$

Proposition 5.15 The sequence $B_i \otimes_{R_i} S^i_{p,n}$:

$$B_i \otimes_{R_i} R_i/(p^n) \xrightarrow{id \otimes (p,\pi)} B_i \otimes_{R_i} (R_i/(p^{n+1}) \oplus R_i/(p^{n-1}))$$

$$\xrightarrow{id \otimes \binom{\pi}{-p}} B_i \otimes_{R_i} R_i/(p^n)$$

is a proper almost split sequence in $\mathcal{B}$-mod.

Proof The sequence $S^i_{p,n}$ is an almost split sequence in R_i-mod. Now, using Lemma 5.5 and Lemma 5.6 one can prove that $B_i \otimes_{R_i} S^i_{p,n}$ is a proper almost split sequence. □

§ 6. Hom-spaces between $\mathcal{A}$-$k(x)$-bimodules

Let $\mathcal{A} = (A, V)$ be a bocs with layer $(A'; \omega; a_1, a_2, \cdots, a_n; v_1, v_2, \cdots, v_m)$. We recall from [6] that an $\mathcal{A}$-$k(x)$-bimodule is an object $M \in \mathcal{A}$-Mod with a morphism $\alpha_M: k(x) \to \mathrm{End}_{\mathcal{A}}(M)$. If M and N are $\mathcal{A}$-$k(x)$-bimodules, a morphism $f: M \to N$ in $\mathcal{A}$-Mod is a morphism of $\mathcal{A}$-$k(x)$-bimodules if for all $q \in k(x)$, $f\alpha_M(q) = \alpha_N(q) f$.

We denote by $\mathcal{A}$-$k(x)$-Mod the category whose objects are the $\mathcal{A}$-$k(x)$-bimodules and the morphisms are morphisms of $\mathcal{A}$-$k(x)$-bimodules. If $F: \mathcal{B}$-Mod $\to \mathcal{A}$-Mod is a functor with $\mathcal{A}, \mathcal{B}$ layered bocses, then F induces a functor $F^{k(x)}: \mathcal{B}$-$k(x)$-Mod $\to \mathcal{A}$-$k(x)$-Mod. If M is a

$\mathcal{B}$-$k(x)$-bimodule, with $\alpha_M : k(x) \to \mathrm{End}_{\mathcal{B}}(M)$ then $F(M)$ is an $\mathcal{A}$-$k(x)$-bimodule with $\alpha_{F(M)} = F\alpha_M : k(x) \to \mathrm{End}_{\mathcal{A}}(F(M))$. Observe that if $f: M \to N$ is a morphism of $\mathcal{B}$-$k(x)$-bimodules, then $F(f)$ is a morphism of $\mathcal{A}$-$k(x)$-bimodules. Now, if F is full and faithful then $F(f): F(M) \to F(N)$ is a morphism of $\mathcal{A}$-$k(x)$-bimodules if and only if for all $q \in k(x)$, $F(f)F(\alpha_M(q)) = F(\alpha_N(q))F(f)$ and this is true if and only if $f\alpha_M(q) = \alpha_N(q)f$ for all $q \in k(x)$. Thus, F induces a full and faithful functor

$$F^{k(x)} : \mathcal{B}\text{-}k(x)\text{-Mod} \to \mathcal{A}\text{-}k(x)\text{-Mod}.$$

The $\mathcal{A}$-$k(x)$-bimodule M is called proper if there is a $\beta_M : k(x) \to \mathrm{End}_A(M)$ such that $\alpha_M = (1, \epsilon)^* \beta_M$, thus $\alpha_M(q) = (\beta_M(q), 0)$ for all $q \in k(x)$. Observe that if M is a proper $\mathcal{A}$-$k(x)$-bimodule then M is an $\mathcal{A}$-$k(x)$-bimodule. We denote by $\mathcal{A}$-$k(x)$-Mod^p, the full subcategory of $\mathcal{A}$-$k(x)$-Mod whose objects are the proper bimodules. Suppose $\theta : \mathcal{A} \to \mathcal{B}$ is a morphism of bocses with $\epsilon_{\mathcal{B}}$ the counit of $\mathcal{B}$ and $\epsilon_{\mathcal{A}}$ the counit of $\mathcal{A}$, then $\theta^* : \mathcal{B}$-Mod$\to \mathcal{A}$-Mod is a full and faithful functor. Observe that if M is a proper $\mathcal{B}$-$k(x)$-bimodule then $\alpha_M = (1, \epsilon_{\mathcal{B}})^* \beta_M$ with $\beta_M : k(x) \to \mathrm{End}_B(M)$. Then $\theta^*(M)$ is a $\mathcal{A}$-$k(x)$-bimodule, using Lemma 4.1 we have

$$\alpha_{\theta^*(M)} = (\theta_0, \theta_1)^* (1, \epsilon_{\mathcal{B}})^* \beta_M = (1, \epsilon_{\mathcal{A}})^* (\theta_0, \theta_0)^* \beta_M,$$

thus $\theta^*(M)$ is a proper $\mathcal{B}$-$k(x)$-bimodule, consequently θ^* induces a full and faithful functor $(\theta^*)^{k(x)} : \mathcal{B}$-$k(x)$-$\mathrm{Mod}^p \to \mathcal{A}$-$k(x)$-$\mathrm{Mod}^p$.

Proposition 6.1 Let M, N be proper $\mathcal{A}$-$k(x)$-bimodules. Then $f = (f^0, \phi_{M,N}(f_1, f_2, \cdots, f_m)) : M \to N$ is a morphism of $\mathcal{A}$-$k(x)$-bimodules if and only if f^0 is a morphism of A'-$k(x)$-bimodules and $f_i \in \mathrm{Hom}_{k(x)}(M(X_i), N(Y_i))$ for all $v_i \in \overline{V}(X_i, Y_i)$.

Proof We have that M and N are proper bimodules so, $\alpha_M(q) = (\beta_M(q), 0)$ and $\alpha_N(q) = (\beta_N(q), 0)$ with morphisms of k-algebras $\beta_M : k(x) \to \mathrm{End}_A(M)$ and $\beta_N : k(x) \to \mathrm{End}_A(N)$. Then a morphism $f: M \to N$ in $\mathcal{A}$-Mod is a morphism of $\mathcal{A}$-$k(x)$-bimodules if and only if $f\alpha_M(q) = \alpha_N(q)f$ for all $q \in k(x)$. Then, by Proposition 4.2, the above holds if and only if $f^0\beta_M(q)=\beta_N(q)f^0$ for all $q\in k(x)$, and for all v_i and all $q\in k(x), u\in M(X_i)$: $\beta_N(q)\phi_{M,N}(f_1, f_2, \cdots, f_m)(v_i\otimes u)=\phi_{M,N}(f_1, f_2, \cdots, f_m)(v_i\otimes\beta_M(q)(u))$. Using the relations given in Lemma 4.5, we obtain that the latter equality is equivalent to $\beta_N(q)f_i(u)=f_i(\beta_M(q)(u))$. From here we obtain our result. □

Corollary 6.2 Let $\mathcal{B}=(B, W)$ be a minimal bocs with layer $(B; \omega_B; w_1, w_2, \cdots, w_m)$, with $w_i\in\overline{W}(X_i, Y_i)$. Then if M and N are proper $\mathcal{B}$-$k(x)$-bimodules we have:

$$\mathrm{Hom}_{\mathcal{B}\text{-}k(x)}(M, N)\cong\mathrm{Hom}_{B\text{-}k(x)}(M, N)\oplus\bigoplus_i\mathrm{Hom}_{k(x)}(M(X_i), N(Y_i)).$$

Let $\mathcal{B}=(B, W)$ be a minimal bocs with layer $(B; \omega; w_1, w_2, \cdots, w_m)$, for Z a marked object in ind$B$ we define $Q_Z\in\mathcal{B}$-Mod as follows: $Q_Z(Z)=k(x)$ where $B(Z, Z)=k[x, f(x)^{-1}]id_Z$ and the action of x on $Q_Z(Z)$ is the multiplication by x, $Q_Z(W)=0$ for $Z\neq W$. The action of $k(x)$ is the multiplication on the right by the elements of $k(x)$. Here Q_Z is a proper $\mathcal{B}$-$k(x)$-bimodule. Using the notation of section 5, we have as a consequence of the above corollary:

Corollary 6.3 If Z, Z' are marked objects and W is a non-marked object in, ind$B$, write $S_W^{k(x)}=S_W\otimes_k k(x)$. We have

$$\dim_{k(x)}\mathrm{Hom}_{\mathcal{B}\text{-}k(x)}(Q_Z, Q_{Z'})=\delta(Z, Z')+t(Z, Z')$$

where $\delta(Z, Z')=1$ if $Z=Z'$ and zero otherwise. Moreover

$$\dim_{k(x)}(\mathrm{rad}_{\mathcal{B}\text{-}k(x)}(Q_Z, S_W^{k(x)}))=t(Z, W),$$

$$\dim_{k(x)}(\mathrm{rad}_{\mathcal{B}\text{-}k(x)}(S_W^{k(x)}, Q_Z))=t(W, Z).$$

Corollary 6.4 With the notations in Corollary 6.3 we have

$$\mathrm{Hom}_{\mathcal{B}\text{-}k(x)}(Q_Z, Q_{Z'})=k(x)\oplus\mathrm{rad}_{\mathcal{B}\text{-}k(x)}(Q_Z, Q_{Z'})\text{ when }Z=Z',$$

$$\mathrm{Hom}_{\mathcal{B}\text{-}k(x)}(Q_Z, Q_{Z'})=\mathrm{rad}_{\mathcal{B}\text{-}k(x)}(Q_Z, Q_{Z'})\text{ when }Z\neq Z'.$$

Moreover

$$\mathrm{Hom}_{\mathcal{B}\text{-}k(x)}(Q_Z, S_W^{k(x)})=\mathrm{rad}_{\mathcal{B}\text{-}k(x)}(Q_Z, S_W^{k(x)}),$$

$$\mathrm{Hom}_{\mathcal{B}\text{-}k(x)}(S_W^{k(x)}, Q_Z)=\mathrm{rad}_{\mathcal{B}\text{-}k(x)}(S_W^{k(x)}, Q_Z).$$

From the above corollaries, we obtain the next proposition.

Proposition 6.5 Let $\mathcal{B}=(B, W)$ be a minimal bocs with layer $(B; \omega; w_1, w_2, \cdots, w_m)$. Suppose Z, Z', and W are objects in ind B with

$B(W,W)=id_W k$, $B(Z,Z)\neq id_Z k$, $B(Z',Z')\neq id_{Z'}k$. Take $M=M(Z,p,m)$, $N=M(Z',q,n)$, $L=S_W$ with p,q prime elements in $B(Z,Z)$ and $B(Z',Z')$, respectively. Then

$$\dim_k \mathrm{rad}_{\mathcal{B}}^{\infty}(M,N)=mn(\dim_{k(x)}\mathrm{Hom}_{\mathcal{B}\text{-}k(x)}(Q_Z,Q_{Z'})-\delta(Z,Z'));$$

$$\dim_k \mathrm{rad}_{\mathcal{B}}^{\infty}(M,L)=m\dim_{k(x)}\mathrm{rad}_{\mathcal{B}\text{-}k(x)}(Q_Z,L^{k(x)});$$

$$\dim_k \mathrm{rad}_{\mathcal{B}}^{\infty}(L,M)=m\dim_{k(x)}\mathrm{rad}_{\mathcal{B}\text{-}k(x)}(L^{k(x)},Q_Z).$$

§ 7.　$\mathcal{D}$-isolated Objects

Let $\mathcal{A}=(A,V)$ be a bocs with layer $L=(A';\omega;a_1,a_2,\cdots,a_n;v_1,v_2,\cdots,v_m)$. We recall that an object $X\in \mathrm{ind}A'$ is called marked if $A'(X,X)\neq kid_X$, we denote by $m(A')$, the set of marked objects of A'. For $M\in\mathcal{A}$-mod we define its dimension vector

$$\mathbf{dim}M:\mathrm{ind}A'\to\mathbb{N} \text{ by } \mathbf{dim}M(X)=\dim_k M(X).$$

By Dim $\mathcal{A}$ we denote the set of functions $\mathbf{d}:\mathrm{ind}A'\to\mathbb{N}$. If $\mathbf{d},\mathbf{d}'\in$ Dim $\mathcal{A}$ we have $\mathbf{d}+\mathbf{d}'$, defined by $(\mathbf{d}+\mathbf{d}')(X)=\mathbf{d}(X)+\mathbf{d}'(X)$ for all $X\in \mathrm{ind}A'$. The norm of $\mathbf{d}\in$ Dim $\mathcal{A}$ is defined by $\|\mathbf{d}\|=\sum_{i=1}^{n}\mathbf{d}(X_i)\mathbf{d}(Y_i)+\sum_{X\in m(A')}\mathbf{d}(X)^2$, where $a_i:X_i\to Y_i$. For $M\in\mathcal{A}$-mod we define the norm of M, $\|M\|=\|\mathbf{dim}M\|$.

If $\mathbf{d}\in$ Dim$(\mathcal{A})$ we define $|\mathbf{d}|=\sum_{X\in\mathrm{ind}A'}\mathbf{d}(X)$. For $M\in\mathcal{A}$-mod, we put $|M|=|\mathbf{dim}M|$ which is called the dimension of M.

Take $\theta:A\to B$ a functor with B a skeletally small category, the induced bocs $\mathcal{A}^B=(B,W)$ is given as follows: $W=B\otimes_A V\otimes_A B$ with counit

$$\epsilon_B:W\to B$$

given by $\epsilon_B(b_1\otimes v\otimes b_2)=b_1\theta(\epsilon(v))b_2$ for b_1,b_2 morphisms in B, $v\in V$. The coproduct

$$\mu_B:W\to W\otimes_B W$$

is given by $\mu_B(b_1\otimes v\otimes b_2)=\sum_i b_1\otimes v_i^1\otimes 1\otimes 1\otimes v_i^2\otimes b_2$, where b_1,b_2 are morphisms in B and $v\in V$ with $\delta(v)=\sum_i v_i^1\otimes v_i^2$.

There is a morphism of A-A-bimodules

$$\theta_1: V \to W$$

given by $\theta_1(v) = 1 \otimes v \otimes 1$, for $v \in V$. Then we obtain a morphism of bocses $(\theta, \theta_1): \mathcal{A} \to \mathcal{A}^B$ which induces a full and faithful functor $\theta^*: \mathcal{A}^B$-Mod $\to \mathcal{A}$-Mod.

Assume $\mathcal{A}^B$ has layer

$$L^\theta = (B'; \omega'; b_1, b_2, \cdots, b_{n'}; w_1, w_2, \cdots, w_{m'}).$$

There is an additive function $t^\theta: \mathrm{Dim}(\mathcal{A}^B) \to \mathrm{Dim}(\mathcal{A})$, given by $t^\theta(\mathbf{d})(X) = \sum_j \mathbf{d}(Y_j)$ with $\theta(X) = \bigoplus_j Y_j$, $Y_j \in \mathrm{ind}B'$. We have

$$\dim\theta^*(M) = t^\theta(\dim M), \text{for } M \in \mathcal{A}^B\text{-mod}.$$

Following [6], we say that the bocs $\mathcal{A} = (A, V)$ with counite $\epsilon: V \to A$ and layer $L = (A'; \omega; a_1, a_2, \cdots, a_n; v_1, v_2, \cdots, v_m)$ is of wild representation type or simply wild if there is a functor $F: A \to \Sigma$, where Σ are the finitely generated free $k\langle x, y\rangle$-modules such that the induced functor

$$(F, F\epsilon)^*: \Sigma\text{-Mod} \to \mathcal{A}\text{-Mod}$$

preserves isomorphism classes and indecomposables.

From [7], we know that a layered bocs $\mathcal{A} = (A, V)$ which is not of wild representation type is of tame representation type. This is, for each natural number d, there are a finite number of A-$k[x]$-bimodules M_1, $M_2, \cdots, M_s$ free of finite rank as right $k[x]$-modules, and such that every indecomposable M in $\mathcal{A}$-Mod with $|\dim M| \leqslant d$ is isomorphic to

$$M_i \otimes_{k[x]} k[x]/(x-\lambda) \text{ for some } 1 \leqslant i \leqslant s \text{ and } \lambda \in k.$$

This section is devoted to find some subset $\mathcal{D}$ of Dim $\mathcal{A}$ with $\mathcal{A}$ a bocs of tame representation type such that the marked indecomposable objects of $\mathcal{A}$ become $\mathcal{D}$-isolated objects in the sense of Definition 7.4. For this we need the following specific functors (see section 4 of [5]):

7.1 Regularization Suppose $a_1: X_1 \to Y_1$ with $\delta(a_1) = v_1$. Then B is freely generated by A' and $a_2, a_3, \cdots, a_n$. The functor $\theta: A \to B$ is the identity on A', $\theta(a_1) = 0$, $\theta(a_i) = a_i$ for $i = 2, 3, \cdots, n$. The bocs $\mathcal{A}^B =$

(B,W) has layer $(A';\omega_B;a_2,a_3,\cdots,a_n;\theta_1(v_2),\theta_1(v_3),\cdots,\theta_1(v_m))$.

The functor $\theta^*:\mathcal{A}^B\text{-Mod}\to\mathcal{A}\text{-Mod}$ is an equivalence of categories, $\mathrm{Dim}(\mathcal{A}^B)=\mathrm{Dim}(\mathcal{A})$ and $t^\theta=id$. In this case $\|t^\theta(\mathbf{d})\|\geqslant\|\mathbf{d}\|$, and one has the equality if and only if $\mathbf{d}(X_1)\mathbf{d}(Y_1)=0$.

7.2 Deletion of objects Let C be a subcategory of A. Let B' be the full subcategory of A' whose objects have no non-zero direct summand isomorphic to a direct summand of an object of C. Take I_0 the set of $i=\{1,2,\cdots,n\}$ such that $a_i\in A(X_i,Y_i)$ with X_i,Y_i in B', and I_1 the set of $j\in\{1,2,\cdots,m\}$ such that $v_j\in V(X_j,Y_j)$ with X_j,Y_j in B'. Then B is freely generated by B' and the a_i with $i\in I_0$. The functor θ: $A\to B$ is the identity on B' and $\theta(X)=0$ for all $X\in C$. The bocs $\mathcal{A}^B$ has layer $(B';\omega_B;(a_i)_{i\in I_0};(\theta_1(v_j))_{j\in I_1})$. Here $M\in\mathcal{A}$-Mod is isomorphic to some $\theta^*(N)$ if and only if $M(X)=0$ for all X indecomposable objects of C. The function $t^\theta:\mathrm{Dim}(\mathcal{A}^B)\to\mathrm{Dim}(\mathcal{A})$ is an inclusion, $\mathbf{d}\in\mathrm{Dim}(\mathcal{A})$ is in the image of t^θ if and only if $\mathbf{d}(X)=0$ for all X indecomposable objects of C. In this case $\|t^\theta\mathbf{d}\|=\|\mathbf{d}\|$.

7.3 Edge reduction Suppose $a_1:X_1\to Y_1$ with $X_1\neq Y_1$ is such that $\delta(a_1)=0$, and $A'(X_1,X_1)=kid_{X_1}$, $A'(Y_1,Y_1)=kid_{Y_1}$. Let C be the full subcategory of A' whose objects have no direct summands isomorphic to X_1 or Y_1. Now denote by D a minimal category with three indecomposable objects Z_1,Z_2,Z_3, $D(Z_i,Z_i)=kid_{Z_i}$ for $i=1,2,3$. Take $B'=C\times D$. The category B is freely generated by B' and elements b_1, $b_2,\cdots,b_s$. The number of arrows $b_j:W_j\to W_j'$ with W_j and W_j' different from Z_2 is $n-1$, where n is the number of a_i.

The functor $\theta:A\to B$ is the identity on C and

$$\theta(X_1)=Z_1\oplus Z_2,\quad \theta(Y_1)=Z_2\oplus Z_3.$$

The bocs $\mathcal{A}^B=(B,W)$ has a layer of the form $(B',\omega_B;b_1,b_2,\cdots,b_s;w_1,w_2,\cdots,w_u)$. Moreover, if $M\in\mathcal{A}^B$-Mod, $\theta^*(M)(a_i)=0$ for all $i\in\{1,2,\cdots,n\}$ if and only if $M(b_j)=0$ for all $j\in\{1,2,\cdots,s\}$ and $M(Z_2)=0$. The functor θ^* is an equivalence of categories. Moreover $\|t^\theta\mathbf{d}\|>\|\mathbf{d}\|$ if and only if $(t^\theta(\mathbf{d}))(X_1)(t^\theta(\mathbf{d})(Y_1))\neq0$. If $\|t^\theta(\mathbf{d})\|=\|\mathbf{d}\|$

and $\| t^{\theta}(\mathbf{d}') \| = \| \mathbf{d}' \|$, then $t^{\theta}(\mathbf{d})=t^{\theta}(\mathbf{d}')$ implies $\mathbf{d}=\mathbf{d}'$.

7.4 Unraveling Let X be an indecomposable object in A' with $A'(X,X)=k[x,f(x)^{-1}]id_X$. Suppose $S=\{\lambda_1,\lambda_2,\cdots,\lambda_t\}$ is a set of elements of k which are not roots of $f(x)$. For r a positive integer there is a functor $\theta: A\to B$, where B is freely generated by B' and elements $b_1,b_2,\cdots,b_s$, $B'=C\times D$, where C is the full subcategory of A' whose objects have no direct summands isomorphic to X. The category D is the minimal category with indecomposable objects $Y, Z_{i,j}$ with $i\in\{1,2,\cdots,r\}$, $j\in\{1,2,\cdots,t\}$, $D(Z_{i,j},Z_{i,j})=kid_{Z_{i,j}}$, $D(Y,Y)=k[x,f(x)^{-1},g(x)^{-1}]id_Y$, where $g(x)=(x-\lambda_1)(x-\lambda_2)\cdots(x-\lambda_t)$. The functor θ: $A\to B$ acts as the identity on C and $\theta(X)=Y\oplus\bigoplus_{j=1}^{t}\bigoplus_{i=1}^{r}Z_{i,j}^{i}$, where $Z_{i,j}^{i}$ is the direct sum of i copies of $Z_{i,j}$.

The bocs $\mathcal{A}^B=(B,W)$ has a layer of the form $(B';\omega_B;b_1,b_2,\cdots,b_s;w_1,w_2,\cdots,w_u)$.

Moreover for $N\in\mathcal{A}^B$-mod we have the following:

(1) $\| N \| \leqslant \| \theta^*(N) \|$, with strict inequality if $\theta^*(N)(g(x))$ is not invertible.

(2) If $M\in\mathcal{A}$-mod and for all $Z\in \mathrm{ind}A'$, $\dim_k M(Z)\leqslant r$ then there is a $N\in\mathcal{A}^B$-mod such that $\theta^*(N)\cong M$.

(3) $\theta^*(N)(x)=N(x)\oplus\bigoplus_{j=1}^{s}\bigoplus_{i=1}^{r}N(Z_{i,j}^{i})(x)$ with eigenvalues of $N(x)$ not in S, and $N(Z_{i,j}^{i})(x)=J_i(\lambda_j)$, the Jordan block of size i and eigenvalue λ_j.

(4) Suppose $M\in\mathcal{A}$-mod is an indecomposable with $M(X)\neq 0$ and $M(W)=0$ for all $W\neq X$, $W\in \mathrm{ind}A'$, $M(a_i)=0$ for $i\in\{1,2,\cdots,n\}$. Then if the unique eigenvalue of $M(x)$ is not in the set S, there is a $N\in\mathcal{A}^B$-mod with $N(W)=0$ for all $W\in \mathrm{ind}B'$, with $W\neq Y$, $N(b_j)=0$ for all $j\in\{1,2,\cdots,s\}$ and $\theta^*(N)\cong M$.

(5) The number of $b_j: Y_1\to Y_2$ with Y_1, Y_2 non isomorphic to $Z_{i,j}$ is equal to n, the number of a_i.

Definition 7.1 Let $\mathcal{A}=(A,V)$ be a bocs with layer $(A';\omega;a_1,$

$a_2,\cdots,a_n;v_1,v_2,\cdots,v_m$). We say that $M\in\mathcal{A}$-Mod is concentrated in the indecomposable $X\in A'$ if $M(X)\neq 0$, $M(Y)=0$ for Y indecomposable in A', $Y\neq X$ and $M(a_i)=0$ for all $i\in\{1,2,\cdots,n\}$.

Proposition 7.2 Let $\mathcal{A}=(A,V)$ be a bocs which is not wild, with layer $(A';\omega;a_1,a_2,\cdots,a_n;v_1,v_2,\cdots,v_m)$. Let X be an indecomposable object in A' with $A'(X,X)=k[x,f(x)^{-1}]$. Then given a fixed dimension vector $\mathbf{d}$ with $\mathbf{d}(X)\neq 0$, there is a finite subset $S(X,\mathbf{d})$ of k such that if M is indecomposable in $\mathcal{A}$-mod with $\dim M=\mathbf{d}$ and λ in k but not in $S(X,\mathbf{d})$ is an eigenvalue of $M(x)$, then $M\cong M'$, with M' concentrated in X.

Proof We may assume $\mathbf{d}$ is sincere. We prove our assertion by induction on $\|\mathbf{d}\|$. If $\|\mathbf{d}\|=1$, take $S(X,\mathbf{d})$ the set of roots of $f(x)$. Then if M is an indecomposable in $\mathcal{A}$-mod, $M(X)\neq 0$, $\dim M=\mathbf{d}$, clearly M is concentrated in X.

Suppose our result proved for all non-wild layered bocses and dimension vectors with norm smaller than r. We may assume that for all $a_i:X_i\to Y_i$ with $\delta(a_i)=0$, Y_i is not equal to X_i, since if $X_i=Y_i$, then because $\mathcal{A}$ is not wild and by Proposition 9 of [7] we have $A'(X_i,X_i)=k\,id_{X_i}$, so we may move a_i into A', such that $A'(X_i,X_i)=k[z]$, with

$$z=a_i.$$

Take $a_1:X_1\to Y_1$ the first arrow. By condition $L.5$ of a layered bocs we have

$$\delta(a_1)=\sum_{j\in T}c_jv_jd_j,$$

where $c_j\in A'(Y_1,Y_1)$, $d_j\in A'(X_1,X_1)$ and T is the set of all $j\in\{1,2,\cdots,m\}$ such that $v_j:\bar{V}(X_1,Y_1)$. We have then the following possibilities: $\delta(a_1)=0$ or $\delta(a_1)=\sum_j c_jv_jd_j$ with some $c_jv_jd_j\neq 0$. If all $c_i,d_i\in k$, we may assume $d_i=1$ for all $i\in T$. In this case we put $v_i'=v_i$ for $i\neq j$ and $v_j'=\sum_j c_jv_j$. Taking $\{v_j',v_1',v_2',\cdots,v_m'\}$ instead of $\{v_1,v_2,\cdots,v_m\}$ we have again a layer for $\mathcal{A}$, thus in this case we may assume $\delta(a_1)=v_1$. In case that for some $j\in T$, c_j is not in k or d_j is not in k, we have $A'(Y_1, Y_1) \neq k id_{Y_1}$ or $A'(X_1, X_1) \neq k id_{X_1}$.

Case 1 $\delta(a_1) = v_1$. Take $\theta^*: \mathcal{A}^B\text{-Mod} \to \mathcal{A}\text{-Mod}$ the regularization of a_1. Here θ^* is an equivalence and the norm of $\mathbf{d}$ in $\mathcal{A}^B$ is smaller than r. Our claim is true for X and the norm r' of $\mathbf{d}$ in $\mathcal{A}^B$. Take $S(X, \mathbf{d}) = S'(X, \mathbf{d})$, with $S'(X, \mathbf{d})$ the subset of k for which our claim is true in $\mathcal{A}^B$.

Then if $M \in \mathcal{A}$-mod is indecomposable with $\dim M = \mathbf{d}$ and λ is an eigenvalue of $M(x)$ which is not in $S(X, \mathbf{d})$, we may assume $M = \theta^*(N)$. Here $M(x) = N(x)$, thus $N \cong N'$, with N' concentrated in X, but this implies that $\theta^*(N')$ is concentrated in X, thus $\theta^*(N') \cong \theta^*(N) = M$, proving our claim.

Case 2 $\delta(a_1) = 0$. Since $\mathcal{A}$ is not wild, by Proposition 9 of [7], $A'(X_1, X_1) = k id_{X_1}$ and $A'(Y_1, Y_1) = k id_{Y_1}$. Here X_1 is not equal to Y_1. We have the edge reduction of a_1, $\theta^*: \mathcal{A}^B\text{-Mod} \to \mathcal{A}\text{-Mod}$, with $\mathcal{A}^B = (B, W)$. Consider the dimension vectors $\mathbf{d}_1, \mathbf{d}_2, \cdots, \mathbf{d}_l$ of those $N \in \mathcal{A}^B$-mod such that $\dim\theta^*(N) = \mathbf{d}$.

The norms of the $\mathbf{d}_i$ are smaller than r. Here X is not equal to X_1 and to Y_1. Therefore X is an indecomposable object of B'. We may consider the subsets $S(X, \mathbf{d}_1), S(X, \mathbf{d}_2), \cdots, S(X, \mathbf{d}_l)$. Take

$$S(X, \mathbf{d}) = S(X, \mathbf{d}_1) \cup S(X, \mathbf{d}_2) \cup \cdots \cup S(X, \mathbf{d}_l).$$

Let M be an indecomposable in $\mathcal{A}$-mod with $\dim M = \mathbf{d}$. Suppose λ is an eigenvalue of $M(x)$ which is not in $S(X, \mathbf{d})$. Since θ^* is an equivalence there is a $N \in \mathcal{A}^B$-mod such that $\theta^*(N) \cong M$. We may assume $\theta^*(N) = M$, then $M(X) = N(X)$ and $M(x) = N(x)$. Here $\dim N = \mathbf{d}_i$ for some $i \in [1, l]$. Therefore, since λ is an eigenvalue of $N(x)$ which is not in $S(X, \mathbf{d}_i)$, $N \cong N'$, with N' concentrated in X, consequently $\theta^*(N')$ is concentrated in X and $\theta^*(N') \cong M$.

Case 3 $a_1: X_1 \to Y_1$ with $A'(X_1, X_1) \neq k id_{X_1}$ or $A'(Y_1, Y_1) \neq k id_{Y_1}$.

Using the notation of [5], we have an unraveling in X_1 or in Y_1, for r and some elements of $k, \lambda_1, \lambda_2, \cdots, \lambda_s$ followed by regularization of b:

$Y \to Y_1$ or of $b: X_1 \to Y$, with b the generator corresponding to a_1. Let θ^*: $\mathcal{A}^B$-Mod $\to$ $\mathcal{A}$-Mod be the unraveling functor followed by the corresponding regularization, with $\mathcal{A}^B = (B, W)$ and layer

$$(B', \omega_B; b_1, b_2, \cdots, b_v; w_1, w_2, \cdots, w_u).$$

In case X is not equal to X_1 and to Y_1 we proceed as in Case 2.

Suppose now that the unraveling is in X with $X = X_1$ or $X = Y_1$, such that $\theta(X) = Y \oplus (\bigoplus_{i,j} Z_{i,j}^i)$. Take all dimension vectors $\mathbf{d}_1, \mathbf{d}_2, \cdots, \mathbf{d}_l$ of those $N \in \mathcal{A}^B$-mod with $\dim\theta^*(N) = \mathbf{d}$.

The norms of all $\mathbf{d}_i$ are smaller than r. Then we may take $S(Y, \mathbf{d}_i)$. We put $S(X, \mathbf{d}) = S(Y, \mathbf{d}_1) \cup S(Y, \mathbf{d}_2) \cup \cdots \cup S(Y, \mathbf{d}_l) \cup \{\lambda_1, \lambda_2, \cdots, \lambda_s\}$.

Let M be an indecomposable in $\mathcal{A}$-mod with $\dim M = \mathbf{d}$, $M(X) \neq 0$ and λ an eigenvalue of $M(x)$ which is not in $S(X, \mathbf{d})$.

There is a $N \in \mathcal{A}^B$ with $\theta^*(N) \cong M$. We may assume $\theta^*(N) = M$. There is a $\mathbf{d}_i$ with $i \in [1, l]$ such that $\dim N = \mathbf{d}_i$.

Here $M(x) = N(x) \oplus M'(x)$ with eigenvalues of $M'(x)$ contained in $\{\lambda_1, \lambda_2, \cdots, \lambda_s\}$. The eigenvalue λ of $M(x)$ is not in $S(X, \mathbf{d})$, therefore, λ is an eigenvalue of $N(x)$. But λ is not in $S(Y, \mathbf{d}_i)$, then $N \cong N'$, with N' concentrated in Y. This implies that $\theta^*(N')$ is concentrated in X and $M \cong \theta^*(N')$. □

Notation 7.3 We recall that if $\mathbf{d}$ and $\mathbf{d}'$ are dimension vectors of the bocs $\mathcal{A} = (A, V)$ we say that $\mathbf{d} \leqslant \mathbf{d}'$ if for all indecomposable objects X of A', $\mathbf{d}(X) \leqslant \mathbf{d}'(X)$. Then if $\mathcal{D}$ is a finite set of dimension vectors of $\mathcal{A}$, we denote by $s(\mathcal{D})$ the set consisting of all vectors in $\mathcal{D}$, all sums $\mathbf{d} + \mathbf{d}'$ with $\mathbf{d}, \mathbf{d}' \in \mathcal{D}$, and all vectors $\mathbf{e}$ with $\mathbf{e} \leqslant \mathbf{f}$ with $\mathbf{f}$ one of the above dimension vectors. Clearly $s(\mathcal{D})$ is also a finite set.

Definition 7.4 Let $\mathcal{A} = (A, V)$ be a bocs with layer $(A'; \omega; a_1, a_2, \cdots, a_n; v_1, v_2, \cdots, v_m)$ and $\mathcal{D}$ be a finite set of dimension vectors of $\mathcal{A}$. We say that X, an indecomposable object in A', with $A'(X, X) = k[x, f(x)^{-1}]id_X$ is $\mathcal{D}$-isolated if for any indecomposable $M \in \mathcal{A}$-mod with $\dim M \in s(\mathcal{D})$ and $M(X) \neq 0$, there is a $M' \in \mathcal{A}$-mod, concentrated in X with $M \cong M'$.

Lemma 7.5 Let $\mathcal{A}=(A,V)$ be a layered bocs as above, which is not of wild representation type, and $\mathcal{D}$ be a finite set of dimension vectors of $\mathcal{A}$ such that for all indecomposable $X\in A'$ there is a $\mathbf{d}\in\mathcal{D}$ with $\mathbf{d}(X)\neq 0$, and $a_1: X_1\to Y_1$. Then

(1) if X_1 and Y_1 are both $\mathcal{D}$-isolated and $\delta(a_1)\in\mathcal{I}_2\bar{V}+\bar{V}\mathcal{I}_1$ with $\mathcal{I}_1$ an ideal of $A'(X_1,X_1)$, $\mathcal{I}_2$ an ideal of $A'(Y_1,Y_1)$, then $\mathcal{I}_1=A'(X_1,X_1)$ or $\mathcal{I}_2=A'(Y_1,Y_1)$;

(2) if X_1 is $\mathcal{D}$-isolated, $A'(Y_1,Y_1)=k\,id_{Y_1}$, $\delta(a_1)\in\bar{V}\mathcal{I}_1$ with $\mathcal{I}_1$ an ideal of $A'(X_1,X_1)$, then $\mathcal{I}_1=A'(X_1,X_1)$;

(3) if Y_1 is $\mathcal{D}$-isolated, $A'(X_1,X_1)=k\,id_{X_1}$, $\delta(a_1)\in\mathcal{I}_2\bar{V}$ with $\mathcal{I}_2$ an ideal of $A'(Y_1,Y_1)$, then $\mathcal{I}_2=A'(Y_1,Y_1)$.

Proof We have

$$\delta(a_1)=\sum_{s\in T_1}h_s v_s+\sum_{s\in T_2}v_s\,g_s \tag{7.1}$$

with $h_s\in\mathcal{I}_2$, $g_s\in\mathcal{I}_1$.

(1) Suppose our claim is not true, then we may assume $\mathcal{I}_1$ and $\mathcal{I}_2$ are maximal ideals. Then $A'(X_1,X_1)/\mathcal{I}_1\cong k$ and $A'(Y_1,Y_1)/\mathcal{I}_2\cong k$. First assume $X_1=Y_1$. Take the representation M of A such that $M(X_1)=M_1\oplus M_2$ with $M_i=A'(X_1,X_1)/\mathcal{I}_i$ for $i=1,2$, $M(W)=0$ for $W\neq X_1$. Take $M(a_1)$ such that $0\neq M(a_1)(M_1)\subset M_2$, $M(a_1)(M_2)=0$ and $M(a_j)=0$ for $j>1$. Here $\dim M\in s(\mathcal{D})$, then if M is indecomposable, $M\cong M'$ with M' concentrated in X_1, but this implies that M' is indecomposable as A'-module, which is not the case because as A'-modules, we have $M'\cong M\cong M_1\oplus M_2$. Therefore, $M\cong L_1\oplus L_2$, with L_1, L_2 indecomposables, and $\dim L_1$, $\dim L_2$ are in $s(\mathcal{D})$. Then $L_1\cong L_1'$, $L_2\cong L_2'$, with L_1', L_2' concentrated in X_1, thus $M\cong L=L_1'\oplus L_2'$, and $L(a_1)=0$. There is an isomorphism $f=(f^0,f^1): M\to L$. Then from (7.1) we obtain

$$L(a_1)f^0_{X_1}-f^0_{Y_1}M(a_1)=\sum_{s\in T_1}L(h_s)f^1(v_s)+\sum_{s\in T_2}f^1(v_s)M(g_s),$$

then, since $L(a_1)=0$ and $\mathcal{I}_1 M_1=0$, from the above formula we obtain

$$f^0_{Y_1} M(a_1)(M) = f^0_{Y_1} M(a_1)(M_1) \subset \mathcal{I}_2 L,$$

then if $\mathcal{I}_1 = \mathcal{I}_2$, $\mathcal{I}_2 L = 0$, so $f^0_{Y_1} M(a_1)(M) = 0$. If $\mathcal{I}_1 \neq \mathcal{I}_2$, $A'(X_1, X_1) = \mathcal{I}_1 + \mathcal{I}_2$. We have

$$\mathcal{I}_1 f^0_{Y_1} M(a_1)(M) \subset \mathcal{I}_1 \mathcal{I}_2 L = 0,$$

$$\mathcal{I}_2 f^0_{Y_1} M(a_1)(M) \subset f^0_{Y_1}(\mathcal{I}_2 M_2) = 0.$$

Consequently, $f^0_{Y_1} M(a_1) = 0$, a contradiction to $M(a_1) \neq 0$. Thus we obtain our statement in this case.

Now, assume $X_1 \neq Y_1$, take M the representation of A such that $M(X_1) = A'(X_1, X_1)/\mathcal{I}_1$, $M(Y_1) = A'(Y_1, Y_1)/\mathcal{I}_2$, $M(Z) = 0$ for Z indecomposable non-isomorphic to X_1 or Y_1; $M(a_1) \neq 0$ and $M(a_j) = 0$ for all $j > 1$. Clearly $\dim M \in s(\mathcal{D})$. We claim that $M \cong L$ with $L(a_1) = 0$. In fact if M is indecomposable then $M \cong M'$ with M' concentrated in X_1 since $M(X_1) \neq 0$, and $M \cong M''$ with M'' concentrated in Y_1, since $M(Y_1) \neq 0$. Thus $X_1 = Y_1$ a contradiction, therefore M is decomposable $M \cong L = L_1 \oplus L_2$ with $L_1(X_1) \cong M(X_1)$, $L_1(Y_1) = 0$ and $L_2(X_1) = 0$, $L_2(Y_1) \cong M(Y_1)$, consequently, $L_1(a_1) = 0$ and $L_2(a_1) = 0$, and, therefore $L(a_1) = 0$, proving our claim.

Then there is an isomorphism $(f^0, f^1): M \to L$. Here $f^0_{X_1}: M(X_1) \to L(X_1)$ and $f^0_{Y_1}: M(Y_1) \to L(Y_1)$ are isomorphisms. From (7.1) we obtain

$$L(a_1) f^0_{X_1} - f^0_{Y_1} M(a_1) = \sum_{s \in T_1} L(h_s) f^1(v_s) + \sum_{s \in T_2} f^1(v_s) M(g_s) = 0,$$

consequently, $f^0_{Y_1} M(a_1) = 0$, so $M(a_1) = 0$, a contradiction.

(2) We are assuming that X_1 is $\mathcal{D}$-isolated, by Definition 7.4, $A'(X_1, X_1) \neq k\,id_{X_1}$. Here we suppose $A'(Y_1, Y_1) = k\,id_{Y_1}$, then $X_1 \neq Y_1$. If our claim is not true, we may assume that $\mathcal{I}_1$ is a maximal ideal and $A'(X_1, X_1)/\mathcal{I}_1 = k$. Consider now M, the representation of A, such that $M(X_1) = A'(X_1, X_1)/\mathcal{I}_1$, $M(Y_1) = k$, $M(Z) = 0$ for Z indecomposable non-isomorphic to X_1 and to Y_1, $M(a_1) \neq 0$, $M(a_j) = 0$ for all $j \geqslant 2$. If M is indecomposable, then $M \cong M'$ with M' concentrated in X_1, since $M(X_1) \neq 0$, a contradiction to $M(Y_1) \neq 0$. If M is

decomposable, we may construct a module $L=L_1\oplus L_2$ and lead to a contradiction similar to (1).

(3) The proof is similar to (2). □

Remark 7.6 Let $\mathcal{A}$ be a non wild bocs and $\theta: A\to B$ any of our reduction functors such that it does not delete marked indecomposable objects. If $\mathcal{A}$ has layer $(A';\omega;a_1,a_2,\cdots,a_n;v_1,v_2,\cdots,v_m)$ and $\mathcal{A}^B$ has layer $(B';\omega_B;b_1,b_2,\cdots,b_{n'};w_1,w_2,\cdots,w_{m'})$, then to each marked $X\in \mathrm{ind}A'$ corresponds a marked $X^m\in B'$ such that $\theta(X)=X^m\oplus Y$ with Y either 0 or a sum of non-marked indecomposables. Conversely each marked object in B' is equal to some X^m. Moreover,

1) if $N\in\mathcal{A}^B$-Mod is concentrated in X^m then $\theta^*(N)$ is concentrated in X.

2) Suppose $N\in\mathcal{A}^B$-Mod is indecomposable with $N(X^m)\neq 0$ and $\theta^*(N)\cong M$ with M concentrated in X, then there exists $N'\in\mathcal{A}^B$-Mod concentrated in X^m such that $N'\cong N$.

Lemma 7.7 If $\theta: A\to B$ is a reduction functor and $(e): M\xrightarrow{f} E\xrightarrow{g} N$ is a proper exact sequence in $\mathcal{A}^B$-mod, then

$$\theta^*(e):\theta^*(M)\xrightarrow{\theta^*(f)}\theta^*(E)\xrightarrow{\theta^*(g)}\theta^*(N)$$

is a proper exact sequence in $\mathcal{A}$-mod (see Definition 4.6).

Proof Let $f: L\to H$ be a morphism in $\mathcal{A}^B$-Mod. From the explicit description of θ^* for each of the reduction functors given in section 4 of [5] one can see that if $(i,\omega_B)^*(f)$ is a monomorphism (respectively an epimorphism), then $(i,\omega)^*\theta^*(f)$ is a monomorphism (respectively an epimorphism). We have $\dim E=\dim M+\dim N$, then $\dim\theta^*(E)=t^\theta(\dim E)=\dim\theta^*(M)+\dim\theta^*(N)$. Therefore,

$$\dim_k\theta^*(E)(X)=\dim_k\theta^*(M)(X)+\dim_k\theta^*(N)(X),$$

for each $X\in\mathrm{ind}A'$. From this and our first observation we may conclude that $\theta^*(e)$ is a proper exact sequence, proving our claim. □

§ 8. An improvement of the tame theorem

In this section, we prove in Theorem 8.5 that given a tame layered bocs $\mathcal{A}$ and a positive integer r, then there is a minimal layered bocs $\mathcal{B}$ and a functor $F: \mathcal{B}\text{-Mod} \to \mathcal{A}\text{-Mod}$, which is a composition of the reduction functors of section 7, such that for any M representation of $\mathcal{A}$, with dimension smaller than or equal to r there is a representation N of $\mathcal{B}$ with $F(N) \cong M$. This is an improvement of Theorem A in [5] which needs several minimal bocses.

We recall that if $\mathcal{A} = (A, V)$ is a bocs, then a family $\mathcal{F}$ of non-isomorphic indecomposable objects in $\mathcal{A}$-mod is called a one-parameter family if there is T an A-$k[x, f(x)^{-1}]$-bimodule free of finite rank as right $k[x, f(x)^{-1}]$-module, such that for all $\lambda \in k$ which is not a root of $f(x)$, there is a $N \in \mathcal{F}$ with $T \otimes_{k[x, f(x)^{-1}]} k[x]/(x-\lambda) \cong N$ and for each $N \in \mathcal{F}$ there is an unique $\lambda \in k$ which is not a root of $f(x)$ with

$$N \cong T \otimes_{k[x, f(x)^{-1}]} k[x]/(x-\lambda).$$

Two one-parameter families $\mathcal{F}_1$ and $\mathcal{F}_2$ are said to be equivalent if there is only a finite number of elements in $\mathcal{F}_1$ which are not isomorphic to objects in $\mathcal{F}_2$. It follows from Theorem 5.6 of [6] that if $\mathcal{A}$ is not of wild representation type and $\mathcal{D}$ is a finite set of dimension vectors there is only a finite number $m(\mathcal{A}, \mathcal{D})$ of non-equivalent one-parameter families of objects in $\mathcal{A}$-mod having dimension vectors in $s(\mathcal{D})$. Observe that the number of $\mathcal{D}$-isolated objects X in A' is smaller than or equal to $m(\mathcal{A}, \mathcal{D})$.

In the following, $\mathcal{A}_0 = (A_0, V_0)$ is a fixed layered bocs which is not of wild representation type and $\mathcal{D}_0$ a fixed finite set of dimension vectors of $\mathcal{A}_0$. Consider the family $\mathcal{P}$ of pairs $(\mathcal{A}, \mathcal{D})$ with $\mathcal{A}$ a bocs with layer $(A'; \omega; a_1, a_2, \cdots, a_n; v_1, v_2, \cdots, v_m)$, $\mathcal{D}$ a finite set of dimension vectors of $\mathcal{A}$ such that there exists $\theta: A_0 \to A$ a composition of reduction functors with $\mathcal{A}_0^A = \mathcal{A}$ and $t^\theta(\mathcal{D}) \subset \mathcal{D}_0$. We denote by m_0 the number $m(\mathcal{A}_0, s(\mathcal{D}_0))$. Observe that since θ^* is a full and faithful functor and $\mathcal{A}_0$ is not of wild representation type, then $\mathcal{A}$ is not of wild representation type.

If $(\mathcal{A}, \mathcal{D}) \in \mathcal{P}$, for each $X \in \text{ind}A'$ which is $\mathcal{D}$-isolated we have a one-parameter family of representations of $\mathcal{A}$. To different $\mathcal{D}$-isolated indecomposables in indA' correspond non-equivalent one-parameter families of representations of $\mathcal{A}$. By the definition of $\mathcal{P}$, there exists a composition of reduction functors $\theta: A_0 \to A$ with $t^\theta(\mathcal{D}) \subset \mathcal{D}_0$. Therefore, the image under θ^* of the one-parametric family corresponding to a $\mathcal{D}$-isolated indecomposable in A' is a one-parametric family of $\mathcal{A}_0$ with dimension vector in $s(\mathcal{D}_0)$. Therefore, the number of $\mathcal{D}$-isolated indecomposables in A' is smaller or equal to m_0.

Notation Suppose $\mathcal{A}$ is a layered bocs which is not of wild representation type and $\mathcal{D}$ is a finite set of dimension vectors of $\mathcal{A}$. For j a non-negative integer, we denote by $\mathcal{S}(\mathcal{A},\mathcal{D})(j)$ the subset of $\mathcal{D}$ consisting of the $\mathbf{d}$ in $\mathcal{D}$ with $\|\mathbf{d}\|=j$.

Take $(\mathcal{A},\mathcal{D})$ a pair in $\mathcal{P}$, we define a function $c(\mathcal{A},\mathcal{D}):\{-1,0,1,2,\cdots,+\infty\}\to\{0,1,2,\cdots\}$ in the following way:

$$c(\mathcal{A},\mathcal{D})(+\infty)=m_0-i(\mathcal{A},\mathcal{D})$$

with $i(\mathcal{A},\mathcal{D})$ the number of indecomposables in A' which are $\mathcal{D}$-isolated.

$$c(\mathcal{A},\mathcal{D})(-1)=n$$

where n is the number of a_i in the layer of $\mathcal{A}$. For j a non-negative integer we put

$$c(\mathcal{A},\mathcal{D})(j)=\text{Card}\,\mathcal{S}(\mathcal{A},\mathcal{D})(j).$$

The functions $c(\mathcal{A},\mathcal{D})$ belong to $\mathcal{H}$, the set of functions

$$f:\{-1,0,1,\cdots,+\infty\}\to\{0,1,2,\cdots,\}$$

with $f(x)=0$ for almost all $x\in\{-1,0,1,\cdots,+\infty\}$.

If f,g are elements in $\mathcal{H}$ we put $f<g$ if there is a s in $\{-1,0,1,\cdots,+\infty\}$ such that $f(s)<g(s)$ and $f(u)=g(u)$ for $u\in\{-1,0,1,\cdots,+\infty\}$, $u>s$. Clearly if we have an infinite sequence of elements in $\mathcal{H}$ with:

$$f_1\geqslant f_2\geqslant\cdots\geqslant f_m\geqslant f_{m+1}\geqslant\cdots$$

then there exists l such that for all $m>l$, $f_m=f_l$.

Notation If $\theta: A\to B$ is any of our reduction functors and $\mathcal{D}$ is a finite set of dimension vectors of $\mathcal{A}$, we say that θ^* is $\mathcal{D}$-covering if for each $M \in \mathcal{A}$-mod with $\dim M \in \mathcal{D}$ there exists a $N \in \mathcal{A}^B$-mod with $\theta^*(N) \cong M$. If $\theta: A \to B$ is a composition of our reduction functors, we denote by $\mathcal{D}^B$ the set of $\mathbf{d} \in \mathrm{Dim}(\mathcal{A}^B)$ such that $t^\theta(\mathbf{d}') \in \mathcal{D}$.

In the statement of the following Lemma, we use the notation of Remark 7.6.

Lemma 8.1 Let $\theta: A \to B$ be any of our reduction functors such that it does not delete marked objects. Then if X is $\mathcal{D}$-isolated, one has that X^m is $\mathcal{D}^B$-isolated. Conversely if θ is a regularization or the deletion of an object W such that $\mathbf{d}(W)=0$ for all $\mathbf{d} \in \mathcal{D}$ and X^m is $\mathcal{D}^B$-isolated then X is $\mathcal{D}$-isolated.

Proof Suppose X is $\mathcal{D}$-isolated in $\mathcal{A}$. We shall prove that X^m is $\mathcal{D}^B$-isolated in $\mathcal{A}^B$. For this take an indecomposable $N \in \mathcal{A}^B$-mod, with $\dim N \in s(\mathcal{D}^B)$ and $N(X^m) \neq 0$. Consider $M = \theta^*(N)$, then following the notation of Remark 7.6, $M(X) = N(X^m) \oplus N(Y)$, thus $M(X) \neq 0$, moreover $\dim M \in s(\mathcal{D})$. Since X is $\mathcal{D}$-isolated, then there exists $M' \in \mathcal{A}$-mod, with $M \cong M'$ and M' concentrated in X. Therefore, by Remark 7.6 there is a N' concentrated in X^m such that $N \cong N'$. From here we conclude that X^m is $\mathcal{D}^B$-isolated. This proves the first part of our claim.

Suppose now that θ is a regularization. In this case $t^\theta = id$ and $\mathcal{D}^B = \mathcal{D}$. Suppose X^m is $\mathcal{D}^B$-isolated, let us prove that X is $\mathcal{D}$-isolated. Let M be an indecomposable in $\mathcal{A}$-mod, with $\dim M \in s(\mathcal{D})$ and $M(X) \neq 0$. Since θ^* is an equivalence of categories, there is a $N \in \mathcal{A}^B$-mod with $\theta^*(N) \cong M$. We have $N(X^m) = M(X)$, and, therefore, $N(X^m) \neq 0$. Moreover, $\dim N \in s(\mathcal{D}^B)$. Since X^m is $\mathcal{D}^B$-isolated, there is a $N' \in \mathcal{A}^B$-mod, concentrated in X^m such that $N' \cong N$. We have $M' = \theta^*(N')$ is concentrated in X, clearly $M \cong M'$, proving our claim.

A similar proof is done for the case θ is the deletion of an indecomposable W with $\mathbf{d}(W)=0$ for all $\mathbf{d} \in \mathcal{D}$. □

Lemma 8.2 Let $\theta: A \to B$ be a reduction functor which is not an unraveling or the deletion of some X for which there is a $\mathbf{d} \in \mathcal{D}$ with $\mathbf{d}(X) \neq 0$. Suppose there is a $\mathbf{d}'$ with $t^\theta(\mathbf{d}') \in \mathcal{D}$ and $\| t^\theta \mathbf{d}' \| > \| \mathbf{d}' \|$.

Let

$$r=\max\{\|t^\theta(\mathbf{d}')\| \mid t^\theta(\mathbf{d}')\in\mathcal{D}, \text{and } \|t^\theta(\mathbf{d}')\|>\|\mathbf{d}'\|\}.$$

Then for $j>r$.

$$c(\mathcal{A}^B,\mathcal{D}^B)(j)=c(\mathcal{A},\mathcal{D})(j) \quad \text{and} \quad c(\mathcal{A}^B,\mathcal{D}^B)(r)<c(\mathcal{A},\mathcal{D})(r).$$

Proof Let us prove first that for $j\geqslant r$, t^θ induces an injective function

$$t_j^\theta:\mathcal{S}(\mathcal{A}^B,\mathcal{D}^B)(j)\to\mathcal{S}(\mathcal{A},\mathcal{D})(j).$$

Take $\mathbf{d}'\in\mathcal{S}(\mathcal{A}^B,\mathcal{D}^B)(j)$, then $\|t^\theta(\mathbf{d}')\|>\|\mathbf{d}'\|=j\geqslant r$. By definition of r, $\|t^\theta(\mathbf{d}')\|=\|\mathbf{d}'\|=j$. Thus, t^θ induces a function t_j^θ. If $t_j^\theta(\mathbf{d}')=t_j^\theta(\mathbf{d}'')$, we have $\|t^\theta(\mathbf{d}')\|=\|\mathbf{d}'\|$ and $\|t^\theta(\mathbf{d}'')\|=\|\mathbf{d}''\|$, therefore $\mathbf{d}'=\mathbf{d}''$. Consequently, t_j^θ is an injective function.

Suppose $j>r$. Take $\mathbf{d}\in\mathcal{S}(\mathcal{A},\mathcal{D})(j)$, since θ^* does not delete indecomposable objects $X\in \operatorname{ind}A'$ for which there is a $\mathbf{f}\in\mathcal{D}$ with $f(X)\neq 0$ then there is a $\mathbf{d}'\in\mathcal{S}(\mathcal{A}^B,\mathcal{D}^B)$ with $t^\theta(\mathbf{d}')=\mathbf{d}$. We have $r<\|\mathbf{d}\|=\|t^\theta(\mathbf{d}')\|\geqslant\|\mathbf{d}'\|$. By definition of r, $\|t^\theta(\mathbf{d}')\|=\|\mathbf{d}'\|=j$. Thus $\mathbf{d}'\in\mathcal{S}(\mathcal{A}^B,\mathcal{D}^B)(j)$. Consequently, t_j^θ is a bijective function and we have proved the first part of our claim.

For the second part of our claim, take $\mathbf{d}'\in\mathcal{D}^B$ such that $r=\|t^\theta(\mathbf{d}')\|>\|\mathbf{d}'\|$. We have $\mathbf{d}=t^\theta(\mathbf{d}')$ in $\mathcal{S}(\mathcal{A},\mathcal{D})(r)$. Let us prove that $\mathbf{d}$ is not in the image of $t_r^\theta:\mathcal{S}(\mathcal{A}^B,\mathcal{D}^B)(r)\to\mathcal{S}(\mathcal{A},\mathcal{D})(r)$. If θ is a regularization or deletion of objects, t^θ is an injective function and if $\mathbf{d}=t_r^\theta(\mathbf{d}'')$, with $\|\mathbf{d}''\|=r$, since t^θ is injective we have $\mathbf{d}'=\mathbf{d}''$, a contradiction. We only need consider the case in which θ is an edge reduction of $a_1:X_1\to Y_1$. Since $\|\mathbf{d}\|=\|t^\theta(\mathbf{d}')\|>\|\mathbf{d}'\|$, $\mathbf{d}(X_1)\mathbf{d}(Y_1)\neq 0$ and if $\mathbf{d}=t^\theta(\mathbf{d}'')$ then $r=\|t^\theta(\mathbf{d}'')\|>\|\mathbf{d}''\|$, proving our claim. □

Lemma 8.3 Suppose $(\mathcal{A},\mathcal{D})$ is a pair in $\mathcal{P}$. Let $\theta:A\to B$ be the deletion of a non-marked indecomposable $X\in A'$, such that for all $\mathbf{d}\in\mathcal{D}$, $\mathbf{d}(X)=0$, then $c(\mathcal{A}^B,\mathcal{D}^B)(u)=c(\mathcal{A},\mathcal{D})(u)$ for all $u\in\{0,1,\cdots,+\infty\}$.

Proof By Lemma 8.1 $c(\mathcal{A}^B,\mathcal{D}^B)(+\infty)=c(\mathcal{A},\mathcal{D})(+\infty)$. On the other hand, by our hypothesis, t^θ induces a bijective function $t^\theta:\mathcal{D}^B\to\mathcal{D}$ and $\|t^\theta(\mathbf{d})\|=\|\mathbf{d}\|$, for all $\mathbf{d}\in\mathcal{D}^\theta$.

Therefore, $c(\mathcal{A}^B, \mathcal{D}^B)(j) = c(\mathcal{A}, \mathcal{D})(j)$ for all non-negative integers j. This proves our claim. □

Lemma 8.4 Let $(\mathcal{A}, \mathcal{D})$ be a pair in $\mathcal{P}$. Suppose that for each $X \in \mathrm{ind}A'$ there exists $\mathbf{d} \in \mathcal{D}$ with $\mathbf{d}(X) \neq 0$. Then, if $\mathcal{A}$ is not a minimal bocs, there is a composition of reduction functors $\theta: A \to B$, with θ^* a $s(\mathcal{D})$-covering functor, such that $c(\mathcal{A}^B, \mathcal{D}^B) < c(\mathcal{A}, \mathcal{D})$, or there is a change of layer of $\mathcal{A}$ such that if $c'(\mathcal{A}, \mathcal{D})$ is the corresponding function we have $c'(\mathcal{A}, \mathcal{D}) < c(\mathcal{A}, \mathcal{D})$.

Proof (1) Suppose $a_1: X_1 \to X_1$ and $\delta(a_1) = 0$. Since $\mathcal{A}$ is not of wild representation type, then by Proposition 9 of [7] we have $A'(X_1, X_1)\ kid_{X_1}$. Take $B' = A'(a_1)$ and change the layer $(A'; \omega; a_1, a_2, \cdots, a_n; v_1, v_2, \cdots, v_m)$ by the layer $(B'; \omega; a_2, a_3, \cdots, a_n; v_1, v_2, \cdots, v_m)$. We have $B'(X_1, X_1) = k[a_1] id_{X_1}$. Clearly if W is an object non isomorphic to X_1 in $\mathrm{ind}A'$, this object is $\mathcal{D}$-isolated with respect to the original layer of $\mathcal{A}$ if and only if it is $\mathcal{D}$-isolated with respect to the new layer. Here it is possible that X_1, which is not marked with respect to the original layer of $\mathcal{A}$, becomes a $\mathcal{D}$-isolated object with respect to the new layer. Therefore, if we denote by $c'(\mathcal{A}, \mathcal{D})$ the corresponding function with respect to the new layer we have $c'(\mathcal{A}, \mathcal{D})(+\infty) \leqslant c(\mathcal{A}, \mathcal{D})(+\infty)$.

The norm of a dimension vector does not depend of the choice of the layer, therefore, $c'(\mathcal{A}, \mathcal{D})(j) = c(\mathcal{A}, \mathcal{D})(j)$ for all non-negative integers j. Moreover,

$$c'(\mathcal{A}, \mathcal{D})(-1) = c(\mathcal{A}, \mathcal{D})(-1) - 1.$$

Therefore, $c'(\mathcal{A}, \mathcal{D}) < c(\mathcal{A}, \mathcal{D})$.

(2) Suppose there is a marked $X \in \mathrm{ind}A'$ which is not $\mathcal{D}$-isolated. Take $S = \bigcup_{\mathbf{d} \in s(\mathcal{D})} S(X, \mathbf{d})$, with $S(X, \mathbf{d})$ the sets of Proposition 7.2. Take r the maximal of the numbers $\mathbf{d}(X)$ with $\mathbf{d} \in s(\mathcal{D})$. Consider now the unraveling $\theta: A \to B$ in X with respect to r and S. Clearly, the functor $\theta^*: \mathcal{A}^B\text{-Mod} \to \mathcal{A}\text{-Mod}$ is a $s(\mathcal{D})$-covering functor. We have $\theta(X) = X^m \oplus \bigoplus_{i,j} Z_{i,j}^i$. We shall see that X^m is $\mathcal{D}^B$-isolated. Take N an indecomposable in $\mathcal{A}^B$-mod with $N(X^m) \neq 0$ and $\dim N \in s(\mathcal{D}^B)$, then

$\dim\theta^*(N)\in s(\mathcal{D})$. We have $\theta^*(N)(X)=N(X^m)\oplus\bigoplus_{i,j}N(Z_{i,j})^i\neq 0$. Take any eigenvalue of $N(x)$, this is an eigenvalue of $\theta^*(N)(x)$ which is not in S, therefore, it is not in $S(X,\mathbf{d})$ with $\mathbf{d}=\dim\theta^*(N)$. Therefore, by Proposition 7.2, $\theta^*(N)\cong M$, with M concentrated in X. But this implies that $M(x)$ has only one eigenvalue which is not in S. Therefore, $M\cong\theta^*(N')$ with N' concentrated in X^m. But $N\cong N'$, this proves that X^m is $\mathcal{D}^B$-isolated. We have

$$c(\mathcal{A}^B,\mathcal{D}^B)(+\infty)\leqslant c(\mathcal{A},\mathcal{D})(+\infty)-1.$$

Therefore,
$$c(\mathcal{A}^B,\mathcal{D}^B)<c(\mathcal{A},\mathcal{D}).$$

(3) Suppose $a_1:X_1\to Y_1$ with $\delta(a_1)=0$ and $X_1\neq Y_1$. Take $\theta:A\to B$ the reduction of a_1. By Lemma 8.1, $c(\mathcal{A}^B,\mathcal{D}^B)(+\infty)\leqslant c(\mathcal{A},\mathcal{D})(+\infty)$. If there is a $\mathbf{d}'\in\mathcal{D}^B$ such that $\|t^\theta(\mathbf{d}')\|>\|\mathbf{d}'\|$, by Lemma 8.2, $c(\mathcal{A}^B,\mathcal{D}^B)<c(\mathcal{A},\mathcal{D})$. On the other hand if for all $\mathbf{d}'\in\mathcal{D}^B$, $\|t^\theta(\mathbf{d}')\|=\|\mathbf{d}'\|$, then again by Lemma 8.2, $c(\mathcal{A}^B,\mathcal{D}^B)(j)=c(\mathcal{A},\mathcal{D})(j)$ for all non-negative integers j. We have that for all $\mathbf{d}\in\mathcal{D}$, $\mathbf{d}(X_1)\mathbf{d}(Y_1)=0$. This implies that for all $\mathbf{d}'\in\mathcal{D}^B$, $\mathbf{d}'(Z_2)=0$. Take $\theta:B\to C$ the deletion of Z_2. By Lemma8.3 we have $c((\mathcal{A})^B)^C,(\mathcal{D}^B)^C)(u)=c(\mathcal{A}^B,\mathcal{D}^B)(u)=c(\mathcal{A},\mathcal{D})(u)$ for all $u\neq-1$. Moreover, $c((\mathcal{A})^B)^C,(\mathcal{D}^B)^C)(-1)=c(\mathcal{A},\mathcal{D})(-1)-1$, therefore, $c((\mathcal{A}^B)^C,(\mathcal{D}^B)^C)<c(\mathcal{A},\mathcal{D})$.

(4) $\delta(a_1)=v_1$. In this case take $\theta:A\to B$ the regularization of a_1. As in the above case if there is a $\mathbf{d}'\in\mathcal{D}^B$ with $\|t^\theta(\mathbf{d}')\|>\|\mathbf{d}'\|$, then $c(\mathcal{A}^B,\mathcal{D}^B)<c(\mathcal{A},\mathcal{D})$. On the other hand if for all $\mathbf{d}'\in\mathcal{D}^B$, $\|t^\theta(\mathbf{d}')\|=\|\mathbf{d}'\|$, by Lemma 8.1 $c(\mathcal{A}^B,\mathcal{D}^B)(+\infty)=c(\mathcal{A},\mathcal{D})(+\infty)$. By Lemma 8.2, $c(\mathcal{A}^B,\mathcal{D}^B)(j)=c(\mathcal{A},\mathcal{D})(j)$ for all non-negative integers j. Moreover, $c(\mathcal{A}^B,\mathcal{D}^B)(-1)=c(\mathcal{A},\mathcal{D})(-1)-1$. Therefore,

$$c(\mathcal{A}^B,\mathcal{D}^B)<c(\mathcal{A},\mathcal{D}).$$

(5) $\delta(a_1)=\sum_{s\in T}r_sv_s$ with $a_1:X_1\to Y_1$, T the set of s such that $v_s\in\bar{V}(X_1,Y_1)$ and $r_s\in A'(Y_1,Y_1)\otimes_k(A'(X_1,X_1))^{op}=H$. If there is a marked object in indA' which is not $\mathcal{D}$-isolated we may proceed as in (2). Therefore, we may assume that all marked objects in indA' are $\mathcal{D}$-

isolated. The ring H is isomorphic either to k, or to $k[x, f(x)^{-1}]$, or to $k[x, y, f(x)^{-1}, g(y)^{-1}]$. Let $\mathcal{I}$ be the ideal of H generated by the elements $\{r_s\}_{s\in T}$. If $\mathcal{I}\neq H$, then $A'(X_1, X_1)\neq kid_{X_1}$ or $A'(Y_1, Y_1)\neq id_{Y_1}$. Moreover there are ideals $\mathcal{I}_2\subset A'(Y_1, Y_1)$ and $\mathcal{I}_1\subset A'(X_1, X_1)$ with $\mathcal{I}\subset \mathcal{I}_2\otimes_k (A'(X_1, X_0))^{op}+A'(Y_1, Y_1)\otimes_k \mathcal{I}_1$, $\mathcal{I}_2\neq A'(Y_1, Y_1)$ and $\mathcal{I}_1\neq A'(X_1, X_1)$. Thus, $\delta(a_1)\in \mathcal{I}_2\bar{V}(X_1, Y_1)+\bar{V}(X_1, Y_1)\mathcal{I}_1$ with $\mathcal{I}_2\neq A'(Y_1, Y_1)$ and $\mathcal{I}_1\neq A'(X_1, X_1)$.

Then if $A'(X_1, X_1)\neq kid_{X_1}$ and $A'(Y_1, Y_1)\neq kid_{Y_1}$, both X_1 and Y_1 are $\mathcal{D}$-isolated. But this contradicts (1) of Lemma 7. 5 (recall that $\mathcal{A}$ is not of wild representation type).

If $A'(X_1, X_1)\neq kid_{X_1}$ and $A'(Y_1, Y_1)=kid_{Y_1}$, then X_1 is marked, so it is $\mathcal{D}$-isolated, we have $\mathcal{I}_1\neq A'(X_1, X_1)$, and $\mathcal{I}_2=0$, but this contradicts (2) of Lemma 7. 5. In case $A'(X_1, X_1)=kid_{X_1}$, then Y_1 is a marked object in indA', so it is $\mathcal{D}$-isolated and this contradicts (3) of Lemma 7. 5.

Therefore, $\mathcal{I}=H$ and $1=\sum_{s\in T} u_i r_i$. This implies that there is a free basis of $\bar{V}(X_1, Y_1)$, with one of their elements equal to $\delta(a_1)$, then we may apply case (4). □

Theorem 8. 5 Let $\mathcal{A}_0=(A_0, V_0)$ be a layered bocs which is not of wild representation type. Then given a positive integer r there is a composition of reduction functors $\theta: A_0\to B$ with $\mathcal{A}^B$ a minimal layered bocs such that for all $M\in \mathcal{A}_0$-mod with $|M|\leqslant r$ there exists $N\in \mathcal{B}$-Mod with $\theta^*(N)\cong M$.

Proof Take $\mathcal{D}_0$ the set of $\mathbf{d}\in \mathrm{Dim}(\mathcal{A}_0)$ such that $\sum_{X\in \mathrm{ind}A'_0}\mathbf{d}(X)\leqslant r$, $\mathcal{D}_0$ is a finite set. Denote by $\mathcal{P}$ the family of pairs $(\mathcal{A}, \mathcal{D})$, with $\mathcal{A}$ a layered bocs, $\mathcal{D}$ a finite subset of $\mathrm{Dim}(\mathcal{A})$ such that there is a functor, composition of reduction functors $\theta: A_0\to B$ with $t^\theta(\mathcal{D})\subset \mathcal{D}_0$ and θ^* a $s(\mathcal{D}_0)$-covering functor.

Let $\mathcal{A}=(A, V)$ be a bocs with layer $(A'; \omega; a_1, a_2, \cdots, a_n; v_1, v_2, \cdots, v_m)$ and $\mathcal{D}$ be a set of dimension vectors of $\mathcal{A}$, such that $(\mathcal{A}, \mathcal{D})$ is in $\mathcal{P}$.

For $X \in \mathrm{ind}A'$ we denote by $\mathbf{d}_X$ the dimension vector of $\mathcal{A}$ such that $\mathbf{d}_X(X)=1$ and $\mathbf{d}_X(Z)=0$ for $Z \in \mathrm{ind}A'$ with $Z \neq X$.

We will consider non-empty sets $\mathcal{D}$ of dimension vectors of $\mathcal{A}$ with the following two conditions:

(1) If $\mathbf{d} \in \mathcal{D}$ and $\mathbf{d}' < \mathbf{d}$, then $\mathbf{d}' \in \mathcal{D}$.

(2) If X is a marked object in $\mathrm{ind}A'$ then $\mathbf{d}_X \in \mathcal{D}$.

Let $\theta: A \to B$ be a reduction functor which does not delete marked objects of $\mathrm{ind}A'$ and such that $\theta^*: \mathrm{Mod}\text{-}\mathcal{A}^B \to \mathrm{Mod}\text{-}\mathcal{A}$ is a $s(\mathcal{D})$-covering functor, we claim that if $\mathcal{D}$ satisfies properties (a) and (b), then $\mathcal{D}^B$ also satisfies these properties. Let $(B'; \omega; b_1, b_2, \cdots, b_t; w_1, w_2, \cdots, w_s)$ be a layer for $\mathcal{A}^B$.

Here θ^* is a $s(\mathcal{D})$-covering functor, then $\mathcal{D}^B$ is a non-empty set. Suppose now that $\mathcal{D}$ satisfies properties (a) and (b). Property (a) for $\mathcal{D}^B$, follows from the fact that $\mathbf{d}' < \mathbf{d}$ in $\mathcal{D}$ implies $t^\theta(\mathbf{d}') \leqslant t^\theta(\mathbf{d})$.

For proving property (b) of $\mathcal{D}^B$, suppose W is a marked object in B'. Then following the notation of Lemma 7.6, $W = X^m$ for some marked object $X \in \mathrm{ind}A'$. Consider $\mathbf{d}_{X^m}$, dimension vector of $\mathcal{A}^B$. Then for $Z \in \mathrm{ind}A'$, $Z \neq X$ we have $\theta(Z) = \bigoplus\limits_i Z_i$ with $Z_i \in \mathrm{ind}B'$, $Z_i \neq X^m$. Then $t^\theta(\mathbf{d}_{X^m})(Z) = \sum\limits_i \mathbf{d}_{X^m}(Z_i) = 0$. We have $\theta(X) = X^m \oplus \bigoplus\limits_j Y_j$ with $Y_j \in \mathrm{ind}B'$, $Y_j \neq X^m$, then $t^\theta(\mathbf{d}_{X^m})(X) = \mathbf{d}_{X^m}(X^m) = 1$. Consequently, $t^\theta(\mathbf{d}_{X^m}) = \mathbf{d}_X \in \mathcal{D}$, thus $\mathbf{d}_{X^m} \in \mathcal{D}^B$, proving our claim.

Now, suppose $\mathcal{D}$ satisfies properties (1) and (2), and $\theta: A \to B$ is the deletion of all objects $Z \in \mathrm{ind}A'$ such that $\mathbf{d}(Z)=0$ for all $\mathbf{d} \in \mathcal{D}$. Since $\mathcal{D}$ satisfies property (2), then θ does not delete marked objects. Therefore, $\mathcal{D}^B$ satisfies properties (1) and (2).

Now, if $\mathcal{A}^B$ is not a minimal bocs, by Lemma 8.4 there is a reduction functor $\rho: B \to A_1$ such that ρ^* is a $s(\mathcal{D}^B)$-covering functor with

$$c((\mathcal{A}^B)^{A_1}, (\mathcal{D}^B)^{A_1}) < c(\mathcal{A}^B, \mathcal{D}^B),$$

or there exists a new layer for $\mathcal{A}^B$ such that

$$c'(\mathcal{A}^B, \mathcal{D}^B) < c(\mathcal{A}^B, \mathcal{D}^B).$$

By the proof of Lemma 8.4, we know that ρ does not delete marked objects, then $(\mathcal{D}^B)^{A_1}$ satisfies properties (1) and (2). Now for any $Z \in$ indB' there exists some $\mathbf{d} \in \mathcal{D}^B$ with $\mathbf{d}(Z) \neq 0$, thus $\mathbf{d}_Z \leqslant \mathbf{d}$, so by property (1), $\mathbf{d}_Z \in \mathcal{D}^B$, then $\mathcal{D}^B$ also satisfies property (2) with respect to the new layer.

Then starting from $(\mathcal{A}_0, \mathcal{D}_0)$, we can construct a sequence of composition of reduction functors:

$$A_0 \xrightarrow{\theta_0} A_1 \xrightarrow{\theta_1} A_2 \rightarrow \cdots \xrightarrow{\theta_{l-1}} A_l,$$

with sets of dimension vectors $\mathcal{D}_i = (\mathcal{D}_{i-1})^{A_i}$ of $\mathcal{A}_i = (\mathcal{A}_{i-1})^{A_i}$ having conditions (1) and (2), such that all functors θ_i^* are $s(\mathcal{D}_i)$-covering functors. Moreover, we have a strictly decreasing sequence in $\mathcal{H}$,

$$c(\mathcal{A}_0, \mathcal{D}_0) > c(\mathcal{A}_1, \mathcal{D}_1) > \cdots > c(\mathcal{A}_l, \mathcal{D}_l).$$

In $\mathcal{H}$ we can not have infinite strictly decreasing sequences, so there is a sequence of reduction functors as before with $\mathcal{A}_l$ a minimal bocs, proving our result. □

§ 9. Hom-spaces in $\mathcal{D}(\Lambda)$-mod and in $P(\Lambda)$

We may observe that if Λ_1 and Λ_2 are two Morita-equivalent finite-dimensional k-algebras, then Theorem 1.2 is valid for Λ_1 if and only if it is valid for Λ_2. Therefore, without loss of generality, we assume in the rest of the paper that Λ is a basic algebra.

Assume k is an algebraically closed field and $1 = \sum_{i=1}^{n} e_i$ is a decomposition of the unit element of Λ as a sum of pairwise orthogonal primitive idempotents. Then we have ${}_\Lambda\Lambda = \bigoplus_{i=1}^{n} \Lambda e_i$ a decomposition as sum of indecomposable projective Λ-modules and $\Lambda = S \oplus J$ a decomposition as a direct sum of S-S-bimodules, with $J = \mathrm{rad}(\Lambda)$, $S = ke_1 \oplus ke_2 \oplus \cdots \oplus ke_n$ a basic semisimple algebra. We can construct a basis

$T=\{\alpha_1,\alpha_2,\cdots,\alpha_m\}$ of J with $\alpha_j\in e_{s(j)}\mathrm{rad}\Lambda e_{t(j)}$, inductively extending a basis of J^i to J^{i-1} by adding elements each of which lies in e_sJe_t for some s and t. In the following, if L is a right S-modulo we denote its dual with respect to S by $L^*=\mathrm{Hom}_S(L,S)$. For each element $\alpha_j\in e_{s(j)}Te_{t(j)}$ we define the element $\alpha_j^*\in J^*$, by $\alpha_j^*(\alpha_i)=0$ for $\alpha_i\neq\alpha_j$ and $\alpha_j^*(\alpha_j)=e_{t(j)}$, clearly $\alpha_j^*\in e_{t(j)}J^*e_{s(j)}$ the elements α_j^* form a basis for J^*.

In the following, if $\boldsymbol{U}_1,\boldsymbol{U}_2,\boldsymbol{U}_3$ are k-vector spaces we denote by $\begin{pmatrix}\boldsymbol{U}_1 & 0\\ \boldsymbol{U}_2 & \boldsymbol{U}_3\end{pmatrix}$, the set of matrices of the form $\begin{pmatrix}u_1 & 0\\ u_2 & u_3\end{pmatrix}$, with $u_i\in U_i$, $i=1,2,3$. With the usual sum of matrices and multiplication of scalars in k by matrices, the above set is a k-vector space.

In order to define the Drozd's bocs of Λ we need to consider the following two matrix algebras $\boldsymbol{A}=\begin{pmatrix}\boldsymbol{S} & \boldsymbol{0}\\ \boldsymbol{J}^* & \boldsymbol{S}\end{pmatrix}$, and $\boldsymbol{A}^{\mathrm{T}}=\begin{pmatrix}\boldsymbol{S} & \boldsymbol{J}^*\\ \boldsymbol{0} & \boldsymbol{S}\end{pmatrix}$. We are going to define a coalgebra V over $\boldsymbol{A}$ which is isomorphic to the coalgebra given in Proposition 6.1 of [5]. First consider the morphism of S-S-bimodules:

$$m:\boldsymbol{J}^*\xrightarrow{\nu^*}(\boldsymbol{J}\otimes_S\boldsymbol{J})^*\cong\boldsymbol{J}^*\otimes_S\boldsymbol{J}^*$$

where $\nu:\boldsymbol{J}\otimes_S\boldsymbol{J}\to\boldsymbol{J}$ is the multiplication. We have the k-vector spaces $\boldsymbol{W}_0=\begin{pmatrix}\boldsymbol{0} & \boldsymbol{0}\\ \boldsymbol{J}^* & \boldsymbol{0}\end{pmatrix}$, and $\boldsymbol{W}_1=\begin{pmatrix}\boldsymbol{J}^* & \boldsymbol{0}\\ \boldsymbol{0} & \boldsymbol{J}^*\end{pmatrix}$, the elements of both vector spaces can be multiplied as matrices by the right and the left by elements of A^{T}, thus $\boldsymbol{W}_0$ and $\boldsymbol{W}_1$ are A^{T}-A^{T}-bimodules.

We have a morphism of A^{T}-A^{T}-bimodules,

$$\underline{m}:\boldsymbol{W}_1\to\boldsymbol{W}_1\otimes_{A^{\mathrm{T}}}\boldsymbol{W}_1$$

such that its composition with the isomorphism

$$\boldsymbol{W}_1\otimes_{A^{\mathrm{T}}}\boldsymbol{W}_1\cong\begin{pmatrix}\boldsymbol{J}^*\otimes_S\boldsymbol{J}^* & \boldsymbol{0}\\ \boldsymbol{0} & \boldsymbol{J}^*\otimes_S\boldsymbol{J}^*\end{pmatrix},$$

is the map that sends $\begin{pmatrix}h & 0\\ 0 & g\end{pmatrix}$ to $\begin{pmatrix}m(h) & 0\\ 0 & m(g)\end{pmatrix}$.

Now, consider the k-vector space $\bar{\mathbf{V}}=\begin{pmatrix}\boldsymbol{J}^{*} & \mathbf{0}\\ \boldsymbol{M}\oplus\boldsymbol{M} & \boldsymbol{J}^{*}\end{pmatrix}$, with $\boldsymbol{M}=\boldsymbol{J}^{*}\otimes_{S}\boldsymbol{J}^{*}$, this is an A-A-bimodule with the following actions of A over $\bar{\mathbf{V}}$:

$$\begin{pmatrix}s_1 & 0\\ g & s_2\end{pmatrix}\begin{pmatrix}h_1 & 0\\ (w_1,w_2) & h_2\end{pmatrix}=\begin{pmatrix}s_1h_1 & 0\\ (s_2w_1+g\otimes h_1,s_2w_2) & s_2h_2\end{pmatrix},$$

$$\begin{pmatrix}h_1 & 0\\ (w_1,w_2) & h_2\end{pmatrix}\begin{pmatrix}s_1 & 0\\ g & s_2\end{pmatrix}=\begin{pmatrix}h_1s_1 & 0\\ (w_1s_1,w_2s_1+h_2\otimes g) & h_2s_2\end{pmatrix}.$$

The k-linear map $\delta:A\to\bar{\mathbf{V}}$ given by

$$\delta\left(\begin{pmatrix}s_1 & 0\\ h & s_2\end{pmatrix}\right)=\begin{pmatrix}0 & 0\\ (m(h),-m(h)) & s_2\end{pmatrix},$$

is a derivation, thus it gives an extension of A-A-bimodules:

$$0\to\bar{\mathbf{V}}\xrightarrow{i}V\xrightarrow{\epsilon}A\to 0$$

where $V=\bar{\mathbf{V}}\oplus A$ as right A-modules, and putting $\omega=(0,1)$, the left action of A over V is given by $a(v+\omega b)=av+\delta(a)b+\omega ab$, for $a,b\in A$, $v\in\bar{\mathbf{V}}$. Here $\bar{\mathbf{V}}$ is generated by W_1 as $A^{\mathrm{T}}-A^{\mathrm{T}}$-bimodule. We have:

(1) $A\cong W_0^{\otimes}=A'\oplus W_0$.

(2) The multiplication map $A\otimes_{A'}W_1\otimes_{A'}A\to\bar{\mathbf{V}}$ is an isomorphism.

We have a morphism of A-A-bimodules $\mu:V\to V\otimes_A V$, with $\mu(\omega)=\omega\otimes\omega$ and for $v\in W_1$, $\mu(v)=v\otimes\omega+\omega\otimes v+\lambda(v)$, where λ is the composition of morphisms:

$$W_1\xrightarrow{m}W_1\otimes_{A^{\mathrm{T}}}W_1\to\bar{\mathbf{V}}\otimes_A\bar{\mathbf{V}}\to V\otimes_A V.$$

The A-A-bimodule V is a coalgebra over A with counit ϵ and comultiplication μ.

We have $1=\sum_{i=1,j=1}^{n,2}f_{i,j}$ a decomposition of the unit of A as a sum of pairwise orthogonal primitive idempotents, where

$$\boldsymbol{f}_{i,2}=\begin{pmatrix}e_i & 0\\ 0 & 0\end{pmatrix}\text{ and }\boldsymbol{f}_{i,1}=\begin{pmatrix}0 & 0\\ 0 & e_i\end{pmatrix}.$$

Denote by D the full subcategory of A-proj whose objects are all finite direct sums of objects $Af_{i,j}$. By D' we denote the subcategory of D with the same objects as D and such that $D'(X,X)=kid_X$ for all $X\in \text{ind}D$ and $D'(X,Y)=0$ for $X,Y\in\text{ind}D$ with $X\neq Y$. If Af and Ag are in $\text{ind}D$, and $x\in fAg$ we denote by $\nu_x: Af\to Ag$ the right multiplication by x.

Now, if W is an A-A-bimodule we denote by $\vartheta(W)$ the D-D bimodule given by $\vartheta(W)(Af,Ag)=fWg$ and if $\nu_x:Af'\to Af$, $\nu_y: Ag\to Ag'$ are morphisms then $\vartheta(W)(\nu_x,\nu_y):\vartheta(W)(Af,Ag)\to\vartheta(W)(Af',Ag')$ is given by $\vartheta(W)(\nu_x,\nu_y)(w)=xwy$ for $w\in\vartheta(W)(Af,Ag)$. Similarly, for L a right A-module and M a left A-module we define functors, $\vartheta(L):D\to\text{Mod-}k$ and $\vartheta(M):D^{op}\to\text{Mod-}k$. If $f:W_1\to W_2$ is a morphism of A-A-bimodules we have an induced morphism $\vartheta(f):\vartheta(W_1)\to\vartheta(W_2)$. If $g:W_2\to W_3$ is a morphism of A-A-bimodules then $\vartheta(f_2f_1)=\vartheta(f_2)\vartheta(f_1)$. The morphisms between left A-modules and right A-modules induce also morphisms between the corresponding functors.

Fixed L a right A-module we have $F: A\text{-mod}\to\text{Mod-}k$, given in objects by $F(M)=\vartheta(L)\otimes_D\vartheta(M)$ and if $f:M_1\to M_2$ is a morphism of left A-modules, then $F(f)=1\otimes\vartheta(f)$. The functor F is right exact and commutes with direct sums. Consequently, $F\cong W\otimes_A M$, with W the right A-module $\vartheta(L)(A)\cong L$, therefore $\vartheta(L)\otimes_D\vartheta(M)\cong L\otimes_A M$ an isomorphism natural in L and M.

Now, suppose V_1 and V_2 are A-A-bimodules then for $Af, Ag\in \text{ind}D$ we have $(\vartheta(V_1)\otimes_D\vartheta(V_2))(Af,Ag)=\vartheta(V_1)(Af,-)\otimes_D\vartheta(V_2)(-,Ag)\cong\vartheta(fV_1)\otimes_D\vartheta(V_2g)\cong fV\otimes_A Vg$. Now, it is easy to see that in fact we have:

(3) $$\vartheta(V_1)\otimes_D\vartheta(V_2)\cong\vartheta(V_1\otimes_A V_2)$$

The morphism of A-bimodules $\mu: V\to V\otimes_A V$ induces a morphism of D-D-bimodules $\vartheta(\mu):\vartheta(V)\to\vartheta(V)\otimes_D\vartheta(V)$. In a similar way the morphism of A-A bimodules $\epsilon: V\to A$ induces a morphism of D-D-bimodules $\vartheta(\epsilon):\vartheta(V)\to\vartheta({}_AA_A)\cong D$. Now it is clear that $\mathcal{D}(\Lambda)=(D,$

V_D) with $V_D=\vartheta(V)$ is a bocs, the Drozd's bocs of Λ.

The bocs $\mathcal{D}(\Lambda)$ is isomorphic to the one given in Theorem 4.1 of [8] (see also the bocs given in the proof of Theorem 11 in [7]). We have now a grouplike ω_D relative to D', given by $\omega_{Af}=f\omega f\in\vartheta(V)(Af, Af)$. Observe that we have $\vartheta(\mu)(\omega_{Af})=\omega_{Af}\otimes\omega_{Af}$. The set of elements ω_{Af} is called a normal section in [8].

We are now going to construct a layer for $\mathcal{D}(\Lambda)$, with this purpose for each $i=1,2,\cdots,n$, consider the following elements of D and $V_D=\vartheta(V)$, $b_i=\nu_{x(i)}\in D(Af_{t(i),1}, Af_{s(i),2})=\mathrm{Hom}_A(Af_{t(i),1}, Af_{s(i),2})$, $\boldsymbol{x}(i)=\begin{pmatrix}0&0\\ \alpha_i^*&0\end{pmatrix}$; $\boldsymbol{v}_{i,1}=\begin{pmatrix}0&0\\0&\alpha_i^*\end{pmatrix}\in\vartheta(V)(Af_{t(i),1}, Af_{s(i),1})=f_{t(i),1}Vf_{s(i),1}$, $\boldsymbol{v}_{i,2}=\begin{pmatrix}\alpha_i^*&0\\0&0\end{pmatrix}$, an element in

$$\vartheta(V)(Af_{t(i),2},Af_{s(i),2})=f_{t(i),2}Vf_{s(i),2}.$$

Consider the set $L=(D';\omega_D;b_1,b_2,\cdots,b_n;v_{1,1},v_{2,1},\cdots,v_{n,1},v_{1,2},v_{2,2},\cdots,v_{n,2})$. We will see that L is a layer for $\mathcal{D}(\Lambda)$. Here D' is a minimal category, so $L.1$ is satisfied. Properties (1)~(3) imply $L.2$ and $L.4$. By (1) of Proposition 3.1 of [8], we have $L.3$.

For proving $L.5$ observe that $m(\alpha_i^*)=\sum_{s,t}\alpha_i^*(\alpha_s\alpha_t)\alpha_t^*\otimes\alpha_s^*$, then

$$\delta_1(b_i)=V(1,b_i)\omega_{X_{t(i),1}}-V(b_i,1)\omega_{X_{s(i),2}}=-\delta(x_i)=$$

$$\sum_{s,t}\alpha_i^*(\alpha_s\alpha_t)(v_{t,1}x_s-x_tv_{s,2})=\sum_{s,t}\alpha_i^*(\alpha_s\alpha_t)(b_sv_{t,1}-v_{s,2}b_t).$$

Then by our choice of the α_i, we have $\alpha_i^*(\alpha_s\alpha_t)=0$ for $s\geqslant i$ or $t\geqslant i$. This proves $L.5$, therefore L is a layer for $\mathcal{D}(\Lambda)$.

In the following we put $\mathcal{D}(\Lambda)=\mathcal{D}$ and $X_{i,j}=Af_{i,j}$ for $i=1,2,\cdots,n$; $j=1,2$.

There is an equivalence of categories $\Xi:\mathcal{D}\text{-Mod}\to P^1(\Lambda)$. If $M\in\mathcal{D}$-Mod then,

$$\Xi(M):\bigoplus_{i=1}^n\Lambda e_i\otimes_k M(X_{1,i})\to\bigoplus_{i=1}^n\Lambda e_i\otimes_k M(X_{2,i}).$$

such that for $m_i\in M(X_{1,i})$, and $c_i\in\Lambda e_i$,

$$\Xi(M)(\sum_{i=1}^{n}c_i\otimes m_i)=\sum_{j=1}^{n}c_{s(j)}\alpha_j\otimes M(b_j)(m_{s(j)}).$$

For a morphism of the form $f=(f^0,f^1):M\to N$ in $\mathcal{D}$-Mod, $\Xi(f)$ is given by the pair of morphisms:

$$\Xi(f)_u:\bigoplus_{i=1}^{n}\Lambda e_i\otimes_k M(X_{u,i})\to\bigoplus_{i=1}^{n}\Lambda e_i\otimes_k N(X_{u,i}),u=1,2$$

such that for $m_i\in M(X_{i,u})$ and $c_i\in\Lambda e_i$ we have

$$\Xi(f)_u(\sum_{i=1}^{n}c_i\otimes m_i)=\sum_{i=1}^{n}c_i\otimes f^0_{X_{i,u}}(m_i)+\sum_{j=1}^{n}c_{s(j)}\alpha_j\otimes f^1(v_{j,u})(m_{s(j)}).$$

Observe that if M is a proper $\mathcal{D}$-$k(x)$-bimodule then $\Xi(M)$ is an object in $P^1(\Lambda^{k(x)})$, and if $f:M\to N$ is a morphism between proper $\mathcal{D}$-$k(x)$-bimodules then $\Xi(f)$ is a morphism in $P^1(\Lambda^{k(x)})$. Therefore Ξ induces an equivalence:

$$\Xi^{k(x)}:\mathcal{D}\text{-}k(x)\text{-Mod}^p\to P^1(\Lambda^{k(x)}).$$

Lemma 9.1 There are constants l_1 and l_2 such that if we have an almost split sequence in $\mathcal{D}(\Lambda)$-mod starting in H' and ending in H such that ΞH is not $\mathcal{E}$-injective, then $|H'|\leqslant l_1|H|$ and $|H|\leqslant l_2|H'|$.

Proof We put $l=\dim_k\Lambda$. Suppose $\Xi H=f:P_1\to P_2$, here ΞH is indecomposable and it is not $\mathcal{E}$-injective. Therefore, ΞH has not direct summands of the form $P\to 0$, this implies that $\ker f$ is contained in $\mathrm{rad}P_1$, then f induces a monomorphism $P_1/\mathrm{rad}P_1\to \mathrm{Im}f/\mathrm{radIm}f$, consequently $\dim_k(P_1/\mathrm{rad}P_1)\leqslant\dim_k\mathrm{Im}f\leqslant\dim_k P_2$. Then we have

$$\dim_k Cok(\Xi H)\leqslant\dim_k P_2\leqslant\dim_k P_1+\dim_k P_2\leqslant|H|l.$$

Moreover

$$\dim_k P_2\leqslant l\dim_k(P_2/\mathrm{rad}P_2)\leqslant l\dim_k Cok(\Xi H)$$

and $|H|=\dim_k(P_1/\mathrm{rad}P_1)+\dim_k(P_2/\mathrm{rad}P_2)\leqslant\dim_k P_2+\dim_k Cok(\Xi H)\leqslant(1+l)\dim_k Cok(\Xi H)$.

On the other hand, there is a constant l_0 such that for all non projective indecomposable $M\in\Lambda$-mod, $\dim_k M\leqslant l_0\dim_k DtrM$ (see proof of Theorem D in [5]). By Propositions 3.10 and 3.13, $Cok(\Xi H')\cong DtrCok(\Xi H)$. Then $\dim_k Cok(\Xi H')\leqslant l_0\dim_k Cok(\Xi H)$. Therefore

$$|H'|\leqslant\dim_k(Cok(\Xi H'))(1+l)\leqslant l_0\dim_k(Cok(\Xi H))(1+l)\leqslant$$

$$l_0|H|l(1+l)=l_1|H|.$$

The second part of our statement is proved in a similar way. □

Theorem 9.2 Let $\mathcal{D}=(D,V)$ be the Drozd's bocs of a tame algebra Λ. Then $(\mathcal{D}\text{-Mod},\mathcal{E}_{\mathcal{D}})$ is an exact category, with $\mathcal{E}_{\mathcal{D}}$ the class of proper exact sequences. This exact category restricted to $\mathcal{D}$-mod has almost split sequences in the sense of Definition 2.5. Given a positive integer r, there is a composition of reduction functors $\theta: D \to B$ with $\mathcal{B}=(B,V_B)=\mathcal{D}^B$ a minimal layered bocs having the following properties.

(1) For any indecomposable $M\in\mathcal{D}$-mod with $|M|\leqslant r$ there is a $N\in\mathcal{B}$-mod with $M\cong\theta^*(N)$. Moreover any proper almost split sequence in $\mathcal{D}$-mod starting or ending in an indecomposable M with $|M|\leqslant r$ is the image under θ^* of an almost split sequence (in the sense of Definition 2.1) in $\mathcal{B}$-mod.

(2) The image under θ^* of a proper exact sequence in $\mathcal{B}$-mod is a proper exact sequence in $\mathcal{D}$-mod.

(3) The image under θ^* of a proper almost split sequence in $\mathcal{B}$-mod is an almost split sequence in $\mathcal{D}$-mod.

(4) Let $Z_1,Z_2,\cdots,Z_s$ be all the marked objects of ind$B$ with

$$R_i=B(Z_i,Z_i)=k[x,h_i(x)^{-1}],\quad h_i(x)\in k[x],$$

and $M(Z_i,p,m)$, Q_{Z_i}, the indecomposable objects in $\mathcal{B}$-Mod defined in section 5 and section 6 respectively. Then $B_i=\mathrm{Hom}_B(Z_i,-)$ is a B-R_i-bimodule such that

$$Q_{Z_i}\cong B_i\otimes_{R_i}k(x)\text{ and }M(Z_i,p,m)\cong B_i\otimes_{R_i}R_i/(p^m).$$

Take the D-R_i-bimodule $D_i=\theta^*(B_i)$, then

$$\theta*(Q_{Z_i})\cong D_i\otimes_{R_i}k(x),\quad\text{and}\quad\theta^*(M(Z_i,p,n))\cong D_i\otimes_{R_i}R_i/(p^m).$$

Moreove, $\dim(D_i\otimes_{R_i}R_i/(p^m))=m\dim_{k(x)}(D_i\otimes_{R_i}k(x))$.

Proof There is an equivalence $\Xi:\mathcal{D}\text{-Mod}\to P^1(\Lambda)$, observe that if (a) is a pair of composable morphisms $X\to E\to Y$ in $\mathcal{D}$-Mod, $\Xi(a)$ is a sequence in the class $\mathcal{E}$ in $P^1(\Lambda)$ if and only if (a) is a proper exact sequence. Therefore if $\mathcal{E}_1$ is the class of proper exact sequences in $\mathcal{D}$-mod, the pair $(\mathcal{D}\text{-mod},\mathcal{E}_1)$ is an exact category with almost split sequences, moreover if (a) is a pair of composable morphisms in $\mathcal{D}$-mod, $\Xi(a)$ is an almost split $\mathcal{E}$-sequence if and only if (a) is an almost split $\mathcal{E}_1$-sequence.

Take the number $r(1+l)$, with $l=\max\{l_1,l_2\}$, l_1,l_2 the constants of Lemma 9.1. Then by Theorem 8.5 there is a composition of reduction functors $\theta_1:D\to C$ with $\mathcal{C}=(C,V_C)=\mathcal{D}^C$ a minimal bocs with layer $(C';\omega;w_1,w_2,\cdots,w_s)$ such that the full and faithful functor θ_1^*: $\mathcal{C}$-Mod$\to\mathcal{D}$-Mod has the property that for all $M\in\mathcal{D}$-Mod with $|M|\leqslant r$, there is a $N\in\mathcal{C}$-Mod with $(\theta_1)^*(N)\cong M$. Take now $\theta_2:C\to B$ the deletion of all marked indecomposable objects $Z\in$ ind$C$ with $|t^{\theta_1}(\mathbf{d}_Z)|>r$, where $\mathbf{d}_Z\in$ Dim$(\mathcal{C})$ with $\mathbf{d}_Z(Z)=1$, and $\mathbf{d}_Z(Z')=0$ for $Z'\neq Z$, $Z'\in$ ind$C$. Then we have $\theta=\theta_2\theta_1:D\to B$ and

$$\mathcal{B}=(B,V_B)=((\mathcal{D})^C)^B=\mathcal{D}^B$$

is a minimal layered bocs.

(1) Take an indecomposable object $M\in\mathcal{D}$-mod with $|M|\leqslant r$, then there is a $N_1\in\mathcal{C}$-mod with $(\theta_1)^*(N_1)\cong M$. Since N_1 is an indecomposable object in the minimal bocs $\mathcal{C}$, then either $M\cong M(Z,p,m)$ for some marked $Z\in$ ind$C$ or $M\cong S_Z$ for some non-marked $Z\in$ ind$C$. In the first case $|t^{\theta_1}(\mathrm{dim}N_1)|=m\,|t^{\theta_1}(\mathbf{d}_Z)|=|\mathrm{dim}M|\leqslant r$. Thus, $|t^{\theta_1}(\mathbf{d}_Z)|\leqslant r$. Consequently, in both cases $N_1(W)=0$ for W a marked object in ind$C$ with $|t^{\theta_1}(\mathbf{d}_W)|>r$, then there is a $N\in\mathcal{B}$-mod with $N_1\cong(\theta_2)^*(N)$. Therefore $M\cong\theta^*(N)$ proving the first part of (1). For the second part take $M\to E\to L$ a proper almost split sequence in $\mathcal{D}$-mod, then if either M or L have dimension equal or smaller than r, all indecomposable summands of the other terms of the sequence have dimension equal or smaller than $(l+1)r$, consequently our proper almost split sequence is isomorphic to the image under $(\theta_1)^*$ of an almost split sequence (in the sense of Definition 2.1) (a_1): $M_1\to E_1\to L_1$ in $\mathcal{C}$-mod. Then if M_1 or L_1 is an object of the form $M(Z,p,m)$, with Z a marked object in ind$C$, we have $M_1\cong L_1$ and $E_1=M(Z,p,m-1)\oplus$

$M(Z,p,m+1)$. Here $|M(Z,p,m)|\leqslant r$ implies $|t^{\theta_1}(\mathbf{d}_Z)|\leqslant r$, then the sequence (a_1) is the image under $(\theta_2)^*$ of an almost split sequence in $\mathcal{B}$-mod. In case that M_1 or L_1 is an object of the form S_Z for a non marked object in ind$C$, then all other terms of (a_1) are sums of objects of the form S_W with W a non-marked object in ind$C$. Therefore, again (a_1) is the image under $(\theta_2)^*$ of an almost split sequence in $\mathcal{B}$-mod. This proves the second part of (1).

(2) Follows from Lemma 7.7.

(3) Take now Z a marked indecomposable in B and $M(Z,p,1)\in\mathcal{B}$-mod with p a fixed prime element in $R_Z=B(Z,Z)$. By definition of B we have $|t^\theta(\mathbf{d}_Z)|\leqslant r$ and $\theta_2(Z)=Z\in C$. There is a non-trivial proper sequence ending and starting in $M(Z,p,1)$, since θ^* is a full and faithful functor, there is a non-trivial proper exact sequence ending and starting in $\theta^*(M(Z,p,1))$. Then $H=\theta^*(M(Z,p,1))$ is not $\mathcal{E}_1$-projective. Therefore, there is an almost split sequence $(a): H'\to H_0\to H$. By the second part of (1) the sequence (a) is the image under θ^* of an almost split sequence (b) in $\mathcal{B}$-mod. Then using Proposition 2.6 we obtain (3).

(4) The first part follows from the definition of θ^*. For proving the second part take X an indecomposable object in D and assume $\theta(X)=\bigoplus_{j=1}^{t} n_jZ_j$, where $Z_1,Z_2,\cdots,Z_j$ are all indecomposable objects of B. Then for each $i\in\{1,2,\cdots,s\}$:

$$\dim_{k(x)}(\theta^* B_i\otimes_{R_i}k(x))(X)=\dim_{k(x)}(B(Z_i,\theta(X))\otimes_{R_i}k(x))=$$
$$\dim_{k(x)}(R_i^{n_i}\otimes_{R_i}k(x))=n_i.$$

On the other hand:

$$t^\theta(\mathbf{d}_{Z_i})(X)=\mathbf{d}_{Z_i}(\theta(X))=n_i.$$

Therefore $t^\theta(\mathbf{d}_{Z_i})=\dim(\theta^* B_i\otimes_{R_i}k(x))$. Then

$$\dim(D_i\otimes_{R_i}R_i/(p^m))=\dim(\theta^* M(Z_i,p,m))=mt^\theta(\mathbf{d}_{Z_i}),$$

proving (4). □

In the following we put $\Lambda^{k(x)}=\Lambda\otimes_k k(x)$.

Definition 9.3 If R is a k-algebra a $P(\Lambda)$-R-bimodule is a morphism $X=f_X: P_X\to Q_X$, where P_X and Q_X are Λ-R-bimodules which are projectives as left Λ-modules and f_X is a morphism of Λ-R-bimodules. If Z is a left R-module, $X\otimes_R Z=f\otimes 1: P_X\otimes_R Z\to Q_X\otimes_R Z$.

We recall from section 3 that if $X: P_X\to Q_X$ is an object in $p^1(\Lambda)$, then $\dim X=(\dim(\mathrm{top}P_X), \dim(\mathrm{top}Q_X))$. Then if $H'\in\mathcal{D}$-mod, $\dim(\Xi H')=\dim H'$. In case $X\in p^1(\Lambda^{k(x)})$ we put

$$\dim_{k(x)}X=(\dim_{k(x)}(\mathrm{top}P_X),\quad \dim_{k(x)}(\mathrm{top}Q_X)),$$

then if $H'\in\mathcal{D}$-$k(x)$-mod, we have $\dim_{k(x)}(\Xi H')=\dim_{k(x)}H'$.

An indecomposable object $H=f_H: P_H\to Q_H$ in $P(\Lambda)$ which is not in $p(\Lambda)$ is called generic if P_H and Q_H have finite length as $\mathrm{End}_{P(\Lambda)}(H)$-modules. A structure of $P(\Lambda)$-$k(x)$-bimodule for H is called admissible in case $\mathrm{End}_{P(\Lambda)}(H)=k(x)_m\oplus\mathcal{R}$, where $\mathcal{R}=\mathrm{radEnd}_{P(\Lambda)}(H)$ and $k(x)_m$ denotes the set of morphisms $h: H\to H$ of the form $h=(m(x)id_{P_H}, m(x)id_{Q_H})$ with $m(x)\in k(x)$.

Definition 9.4 Suppose $\hat{T}=f_{\hat{T}}: P_{\hat{T}}\to Q_{\hat{T}}$ is a $P(\Lambda)$-R-bimodule with R a finitely generated localization of $k[x]$ and $P_{\hat{T}}, Q_{\hat{T}}$ finitely generated as right R-modules. We say that $\hat{T}$ is a realization of H if $\hat{T}\otimes_R k(x)\cong H$. The realization $\hat{T}$ of H over R is called good if:

(1) $P_{\hat{T}}$ and $Q_{\hat{T}}$ are free as right R-modules;

(2) the functor $\hat{T}\otimes_R -: R\text{-Mod}\to P(\Lambda)$ preserves isomorphism classes and indecomposable objects;

(3) for p a prime in R, and n a positive integer $\hat{T}\otimes_R S_{p,n}$ is an almost split sequence, where $S_{p,n}$ is the sequence given in (3) of Definition 1.1.

We are now ready for giving a version of Theorem 1.2 for $P(\Lambda)$.

Theorem 9.5 Let Λ be a finite-dimensional algebra over an algebraically closed field k of tame representation type. Let r be a positive integer. Then there are indecomposable objects in $p^1(\Lambda)$, $\hat{L}_1$, $\hat{L}_2,\cdots,\hat{L}_t$ with $|\hat{L}_j|\leqslant r$ for $j=1,2,\cdots,t$ and generic objects in $P^1(\Lambda)$

with admissible structure of $P(\Lambda)$-$k(x)$-bimodules, $H_1, H_2, \cdots, H_s$ such that for $j=1,2,\cdots,s$, H_j has a good realization $\hat{T}_j$ over R_j, a finitely generated localization of $k[x]$, with the following properties:

(1) If X is an indecomposable object in $p^1(\Lambda)$ with $|X| \leqslant r$, then either $X \cong \hat{L}_j$ for some $j \in \{1,2,\cdots,t\}$ or $X \cong \hat{T}_i \otimes_{R_i} R_i/(p^m)$ for some $i \in \{1,2,\cdots,s\}$, some prime element $p \in R_i$ and some natural number m.

(2) If $X = \hat{T}_i \otimes_{R_i} R_i/(p^m)$, $Y = \hat{Y}_j \otimes_{R_j} R_j/(q^n)$, with $i, j \in \{1, 2,\cdots,s\}$, p a prime in R_i, q a prime in R_j, and $\hat{L}_u$ with $u \in \{1,2,\cdots,t\}$, then

$$\dim_k \mathrm{rad}^{\infty}_{p^1(\Lambda)}(X,Y) = mn\dim_{k(x)} \mathrm{rad}_{p^1(\Lambda^{k(x)})}(H_i, H_j),$$
$$\dim_k \mathrm{rad}^{\infty}_{p^1(\Lambda)}(X,\hat{L}_u) = m\dim_{k(x)} \mathrm{rad}_{p^1(\Lambda^{k(x)})}(H_i, \hat{L}_u^{k(x)}),$$
$$\dim_k \mathrm{rad}^{\infty}_{p^1(\Lambda)}(\hat{L}_u, X) = m\dim_{k(x)} \mathrm{rad}_{p^1(\Lambda^{k(x)})}(\hat{L}_u^{k(x)}, H_i),$$

(3) If $X = \hat{T}_i \otimes_{R_i} R_i/(p^m)$, $Y = \hat{T}_j \otimes_{R_j} R_j/(q^n)$, then if $i=j$ and $p=q$,

$$\mathrm{Hom}_{p^1(\Lambda)}(X,Y) \cong \mathrm{Hom}_{R_i}(R_i/(p^n), R_i/(p^m)) \oplus \mathrm{rad}^{\infty}_{p^1(\Lambda)}(X,Y).$$

If $i \neq j$ or $i=j$ and $(p) \neq (q)$:

$$\mathrm{Hom}_{p^1(\Lambda)}(X,Y) = \mathrm{rad}^{\infty}_{p^1(\Lambda)}(X,Y).$$

Moreover

$$\mathrm{Hom}_{p^1(\Lambda)}(\hat{L}_u, X) = \mathrm{rad}^{\infty}_{p^1(\Lambda)}(\hat{L}_u, X), \mathrm{Hom}_{p^1(\Lambda)}(X, \hat{L}_u) = \mathrm{rad}^{\infty}_{p^1(\Lambda)}(X, \hat{L}_u).$$

Proof We apply Theorem 8. 5 for the Drozd's bocs $\mathcal{D} = (D, V_D)$ of Λ and the positive integer $r(l+1)$ with $l = \max\{l_1, l_2\}$ where l_1, l_2 are the integers given in Lemma 9. 1. Then we obtain a minimal layered bocs $\mathcal{B} = (B, V_B)$ having properties (1)～(4) of Theorem 9. 2. We have the reduction functor $\theta: D \to B$, suppose $\theta(X_{j,i}) = \bigoplus_l n^l_{j,i} Z_l$ with $j=1,2$ and $i=1,2,\cdots,n$ given in the beginning of this section.

Let $Z_1, Z_2, \cdots, Z_s$ be the marked objects of ind$B$ and $Z_{s+1}, Z_{s+2}, \cdots, Z_{s+t}$ be the non-marked objects. We have B_i, R_i and D_i given in (4) of Theorem 9. 2.

Consider $\hat{T}_i = \Xi D_i$. $\hat{T}_i = g_i : P_i \to Q_i$, then:

$$P_i=\bigoplus_v \Lambda e_v \otimes D_i(X_{1,v})=\bigoplus_v \Lambda e_v \otimes_k \mathrm{Hom}_B(Z_i,\theta(X_{1,v}))\cong\bigoplus_v \Lambda e_v \otimes_k n^i_{1,v} R_i.$$

Similarly $Q_i\cong\bigoplus_v \Lambda e_v \otimes_k n^i_{2,v} R_i$. If $\lambda\in\Lambda e_v$, and $m\in D_i(X_{1,v})$, then:

$$g_i(\lambda\otimes m)=\sum_{d_j:X_{1,s(j)}\to X_{2,t(j)},s(j)=v}\lambda\alpha_j\otimes \mathrm{Hom}_B(1,\theta(b_j))(m)$$

We have

$H_i=\Xi D_i\otimes_{R_i}k(x)=f_i:P_{H_i}\to Q_{H_i}$, $P_{H_i}=P_i\otimes_{R_i}k(x)$, $Q_{H_i}=Q_i\otimes_{R_i}k(x)$, with $f_i=g_i\otimes 1_{k(x)}$, therefore $H_i=\hat{T}_i\otimes_{R_i}k(x)$.

Moreover, $P_{H_i}\cong\bigoplus_v n^i_{1,v}\Lambda^{k(x)}(e_v\otimes 1)$ and $Q_{H_i}\cong\bigoplus_v n^i_{2,v}\Lambda^{k(x)}(e_v\otimes 1)$.

For $i=1,2,\cdots,s$, consider the objects $H_i\in P^1(\Lambda)$. For all $i=1,2,\cdots,s$ we have an isomorphism induced by the functor $\Xi\theta^*$:

$$\mathrm{End}_{\mathcal{B}}(Q_{Z_i})=\mathrm{End}_{\mathcal{B}}(Q_{Z_i})^0\oplus\mathrm{End}_{\mathcal{B}}(Q_{Z_i})^1\to\mathrm{End}_{P^1(\Lambda)}(H_i),$$

where $\mathrm{End}_{\mathcal{B}}(Q_{Z_i})^0$ denotes the morphisms of the form $(f^0,0)$ and $\mathrm{End}_{\mathcal{B}}(Q_{Z_i})^1$ denotes the morphisms of the form $(0,f^1)$. Here $\mathrm{End}_{\mathcal{B}}(Q_{Z_i})^0\cong\mathrm{End}_{R_i}(k(x))=k(x)_m$, where $k(x)_m$ denotes the right multiplication by elements of $k(x)$. Here $\mathcal{B}$ is a layered bocs, therefore a morphism (f^0,f^1) is an isomorphism if and only if f^0 is an isomorphism, thus the elements in $\mathrm{End}_{\mathcal{B}}(Q_{Z_i})^1$ are the non-units in $\mathrm{End}_{\mathcal{B}}(Q_{Z_i})$. Thus since the sum of non-units is again non-unit, $\mathrm{End}_{\mathcal{B}}(Q_{Z_i})$ is a local ring and its radical is $\mathrm{End}_{\mathcal{B}}(Q_{Z_i})^1$. The image under $\Xi\theta^*$ of an element in $\mathrm{End}_{\mathcal{B}}(Q_{Z_i})^0$ is of the form $(id_{P_{H_i}}m(x),id_{Q_{H_i}}m(x))$, with $m(x)\in k(x)$. From here we obtain that the $P(\Lambda)$-$k(x)$-structure of H_i is admissible. Clearly, $\hat{T}_i$ is a realization of H_i.

In order to prove that $\hat{T}_i$ is a good realization of H_i, we must prove conditions (1), (2) and (3) of Definition 9.4. Condition (1) is clear. For proving condition (2) take $\epsilon_{\mathcal{B}}:V_{\mathcal{B}}\to B$ the counit of the bocs $\mathcal{B}$. By Lemma 5.3 the functor $(id_B,\epsilon_B)^*:B\text{-Mod}\to\mathcal{B}\text{-Mod}$ preserves indecomposables and isomorphism classes. Consider $\hat{B}_i$ the full subcategory of B whose unique indecomposable object is Z_i, then we have the composition η_i of full and faithful functors:

$$R_i\text{-Mod}\to\hat{B}_i\text{-Mod}\to B\text{-Mod}.$$

The composition

$$R_i\text{-Mod}\xrightarrow{\eta_i}B\text{-Mod}\xrightarrow{(id_B,\epsilon_{\mathcal{B}})^*}\mathcal{B}\text{-Mod}\xrightarrow{\theta^*}\mathcal{D}\text{-Mod}\xrightarrow{\Xi}P^1(\Lambda)$$

is isomorphic to $\hat{T}_i\otimes_{R_i}-$. Therefore the functor $\hat{T}_i\otimes_{R_i}-$ preserves isomorphism classes and indecomposable modules. The condition (3) of Definition 9.4 is a consequence of (3) of Theorem 9.2.

Now, we may assume that $\hat{L}_j=\Xi\theta^*(S_{Z_{s+j}})$ for $j=1,2,\cdots,t$ is such that $|\hat{L}_j|\leqslant r$.

1) Take X an indecomposable object in $p^1(\Lambda)$ with $|X|\leqslant r$, then by (1) of Theorem 9.2 there is an indecomposable object N in $\mathcal{B}$-mod with $\Xi\theta^*(N)\cong X$. Since N is indecomposable, then $N\cong S_{Z_{s+j}}$ for some $j=1,2,\cdots,t$ and then either $X\cong\hat{L}_j$, or $N\cong M(Z_i,p,n)$ for some $i=1,2,\cdots,s$, some prime element $p\in R_i$ and some positive integer n, in this case by (4) of Theorem 9.2 we have

$$M(Z_i,p,n)\cong B_i\otimes_{R_i}R_i/(p^n).$$

Then $$X\cong\Xi\theta^*B_i\otimes_{R_i}R_i/(p^n)\cong\hat{T}_i\otimes_{R_i}R_i/(p^n).$$

Thus we have proved (1).

2) Consider $\mathcal{C}$ the full subcategory of $p^1(\Lambda)$ whose objects are the objects of the form $\hat{T}_i\otimes_{R_i}R_i/(p^m)$. We have already proved that $\hat{T}_i$ is a good realization of H_i, then by property (3) of Definition 9.4 the category $\mathcal{C}$ consists of whole Auslander-Reiten components of $p^1(\Lambda)$, thus $\mathcal{C}$ has property (A) of section 2, then by Corollary 2.4 for

$$X=\hat{T}_i\otimes_{R_i}R_i/(p^m),\quad Y=\hat{T}_i\otimes_{R_i}R_i/(q^n),$$

$$\begin{aligned}\dim_k\text{rad}^{\infty}_{p^1(\Lambda)}(X,Y)&=\dim_k\text{rad}^{\infty}_{\mathcal{C}}(X,Y)\\&=\dim_k\text{rad}^{\infty}_{\mathcal{B}}(M(Z,p,m),M(Z',q,n)).\end{aligned}$$

We recall from the discussion at the beginning of section 6 that the full and faithful functor $\theta^*:\mathcal{B}\text{-Mod}\to\mathcal{A}\text{-Mod}$ restricts to a full and faithful functor $(\theta^*)^{k(x)}:\mathcal{B}\text{-}k(x)\text{-Mod}^p\to\mathcal{D}\text{-}k(x)\text{-Mod}^{op}$. Then the first equality of (2) follows from that of Proposition 6.5.

Observe that $\hat{L}_u^{k(x)}=\Xi\theta^*(S_{Z_{s+u}})^{k(x)}\cong\Xi\theta^*(S^{k(x)}_{Z_{s+u}})$. The second and third equality of (2) follow from those of Proposition 6.5.

(3) Follows from Corollary 5.11 and from Corollary 2.4. □

§ 10. Hom-spaces in Λ-Mod

In this section we discuss the Hom-spaces in Λ-Mod for a tame algebra Λ and prove our main result, Theorem 1.2. For $X=f_x: P_X\to Q_X\in p(\Lambda)$ we define

$$|X|=|\dim X|=\dim_k(P_X/\mathrm{rad}P_X)+\dim_k(Q_X/\mathrm{rad}Q_X).$$

There is an integer l_0 such that for any indecomposable non-injective Λ-module M, $\dim_k trDM\leqslant l_0\dim_k M$. Let d be any positive integer greater than $\dim_k\Lambda$, consider $d_0=d(1+l_0)$ take $s(d_0)=(\dim_k(\Lambda)+1)d_0$. If $M\in\Lambda$-mod with $\dim_k M\leqslant d_0$ and $X=f_X: P_X\to Q_X$ is a minimal projective presentation of M, we have $\dim_k(Q_X/\mathrm{rad}Q_X)\leqslant d_0$ and $\dim_k(P_X/\mathrm{rad}P_X)\leqslant\dim_k(\mathrm{Im}f_X)\leqslant\dim_k Q_X\leqslant\dim_k(M/\mathrm{rad}M)\dim_k\Lambda\leqslant d_0\dim_k\Lambda$, so $|X|\leqslant s(d_0)$. Taking the number $r=s(d_0)(1+l)$ in Theorem 9.5 with $l=\max\{l_1,l_2\}$, where l_1 and l_2 are the constants of Lemma 9.1, we obtain the generic objects in $P(\Lambda)$, $H_1,H_2,\cdots,H_s$ with admissible Λ-$k(x)$ structures and the indecomposables in $p^1(\Lambda)$, $\hat{L}_1,\hat{L}_2,\cdots,\hat{L}_t$. For each $i=1,2,\cdots,s$ we have the realizations $\hat{T}_i$ over R_i of H_i. We have the generic Λ-modules $G_i=Cok(H_i)$ and the following isomorphism of Λ-$k(x)$-bimodules, $G_i=Cok(H_i)\cong Cok(\hat{T}_i\otimes_{R_i}k(x))\cong Cok(\hat{T}_i)\otimes_{R_i}k(x)$, with $T_i=Cok(\hat{T}_i)$ a Λ-R_i-bimodule finitely generated as right R_i-module. The Λ-$k(x)$ structure of H_i is admissible, then $\mathrm{End}_{P(\Lambda)}(H_i)=k(x)_m\oplus\mathcal{R}_i$ with $\mathcal{R}_i$ a nilpotent ideal. Then, $\mathrm{End}_\Lambda(G_i)=k(x)id_{G_i}\oplus\mathrm{radEnd}_\Lambda(G_i)$, therefore, the endolength of G_i coincides with $\dim_{k(x)}G_i$. Consequently, T_i is a realization of G_i.

Lemma 10.1 G_i and T_i satisfy the conditions (2) and (3) of Definition 1.1.

Proof Take $W\in R_i$-Mod, we claim that $\hat{T}_i\otimes_{R_i}W$ has not indecomposable direct summands of the form $Z(P)=P\to 0$. Suppose some indecomposable $Z(P)$ is a direct summand of $\hat{T}_i\otimes_{R_i}W=\Xi\theta^*(W')$, with $W'=(id_B,\epsilon_{\mathcal{B}})^*\eta_i(W)$. Here $Z(P)$ is injective in $P^1(\Lambda)$, then $Z(P)=\Xi\theta^*(S_{Z_u})$ for some non-marked indecomposable object $Z_u \in B$. Since the functor $\Xi\theta^*$ is full and faithful, we have that S_{Z_u} is direct summand of W', but this is impossible because $W'(Z_u)=0$. The above proves that $\hat{T}_i \otimes_{R_i} W$ is in $P^2(\Lambda)$, the full subcategory of $P^1(\Lambda)$ whose objects have not direct summands of the form $Z(P)$. Now the functor $Cok: P^2(\Lambda) \to \Lambda$-Mod preserves indecomposables and isomorphism classes (see (2) of Lemma 3.2 of [6]). Consequently, the functor $Cok(\hat{T}_i \otimes_{R_i} -) \cong T_i \otimes_{R_i} -$ preserves indecomposables and isomorphism classes. This proves that T_i has property (2) of Definition 1.1.

For proving condition (3) of Definition 1.1 take p a prime element in R_i. There is an almost split sequence in $p^1(\Lambda)$ starting in $\hat{T}_i \otimes_{R_i} R_i/(p^m)$, therefore this object is not injective in $p^1(\Lambda)$ and therefore its cokernel is not zero. By Proposition 3.13 the image under the functor Cok of the almost split sequence starting in $\hat{T}_i \otimes_{R_i} R_i/(p^m)$ is an almost split sequence in Λ-mod. This proves that the Λ-R_i-bimodule T_i satisfies condition 3) for all $i \in \{1, 2, \cdots, s\}$. □

Lemma 10.2 Let $L_j = Cok(\hat{L}_j)$ with $j=1,2,\cdots,t$. If M is an indecomposable Λ-module with $\dim_k M \leqslant d$, then M has the form given in (1) of Theorem 1.2.

Proof There is an indecomposable object $X \in p^1(\Lambda)$ with $M \cong Cok(X)$, since $|X| \leqslant s(d) \leqslant r$, $X \cong \hat{T}_i \otimes_{R_i} R_i/(p^m)$ or $X \cong \hat{L}_j$. But then either $M \cong Co\hat{k} T_i \otimes_{R_i} R_i/(p^m) \cong T_i \otimes_{R_i} R_i/(p^m)$, or $M \cong L_j$. This proves the first part of (1). For the second part of (1), by Proposition 5.9 of [1] we have that if X is an indecomposable object in $p^1(\Lambda)$ with $Cok(X)$ non-simple injective, then there is an almost split sequence in $p(\Lambda)$ starting in X and ending in an injective object with all its terms in $p^1(\Lambda)$, so this is an almost split sequence in $p^1(\Lambda)$. If $Cok(X)$ is simple then X is injective in $p^1(\Lambda)$, if $Cok(X)$ is projective, then X is projective in $p^1(\Lambda)$. Now if $X \cong \hat{T}_i \otimes_{R_i} R_i/(p^m)$, since $\hat{T}_i$ is a good realization of H_i, there is an almost split sequence starting and ending in X. Therefore, if M is an injective, projective or simple Λ-module, then

$M \cong L_j$ for some $j=1,2,\cdots,t$. □

Lemma 10.3 Let $X=\hat{T}_i \otimes_{R_i} R_i/(p^n)$, $Y=\hat{T}_i \otimes_{R_i} R_i/(p^m)$, $M=\mathit{Cok}X$, $N=\mathit{Cok}Y$, then the functor Cok induces an isomorphism:

$$\underline{\mathit{Cok}}: \mathrm{Hom}_{p^1(\Lambda)}(X,Y)/\mathrm{rad}^{\infty}(X,Y) \to \mathrm{Hom}_{\Lambda}(M,N)/\mathrm{rad}^{\infty}(M,N).$$

Proof In fact, take a morphism $u: X \to Y$ such that $\mathit{Cok}(u)=0$. Then by Proposition 3.3, u is a morphism which is a sum of compositions of the form $u_2 u_1$ with $u_1: X \to W$, $u_2: W \to Y$ and W an indecomposable injective in $P(\Lambda)$. Then either $W=Z(P)=(P \xrightarrow{0} 0)$ or $W=J(P)=(P \xrightarrow{id_P} P)$ for some indecomposable projective Λ-module P. In the first case W is also an injective object in $p^1(\Lambda)$, then W is not in the Auslander-Reiten component containing X, therefore $u_2 u_1 \in \mathrm{rad}^{\infty}(X,Y)$. Now, if $W=J(P)$, we recall (see Lemma 3.6) that there is a right minimal almost split morphism $\sigma(P): R(P) \to J(P)$, then $u_1=\sigma(P)u'_1$, with $u'_1: X \to R(P)$. Here $R(P)$ is injective in $p^1(\Lambda)$, then $u_2 u_1 = u_2 \sigma(P) u'_1$ is in $\mathrm{rad}^{\infty}(X,Y)$, therefore, $u \in \mathrm{rad}^{\infty}(X,Y)$, proving our Lemma. □

Lemma 10.4 If $M=T_i \otimes_{R_i} R_i/(p^m)$, $N=T_j \otimes_{R_j} R_j/(q^n)$, $L_u^{k(x)}=L_u^{k(x)}$ with $i,j \in \{1,2,\cdots,s\}$, $u \in \{1,2,\cdots,t\}$, p a prime element of R_i, q a prime element of R_j, then M,N,L_u satisfy (3) of Theorem 1.2.

Proof Let $M=\mathit{Cok}X$, $N=\mathit{Cok}Y$, $X,Y \in p^1(\Lambda)$. If $i=j$ and $p=q$ by the first formula in (3) of Theorem 9.5 and Lemma 10.3 we obtain our result. If $i \neq j$ or $(p) \neq (q)$ we have $\mathrm{Hom}_{p^1(\Lambda)}(X,Y)=\mathrm{rad}^{\infty}_{p^1(\Lambda)}(X,Y)$, thus $\mathrm{Hom}_{\Lambda}(M,N)=\mathrm{rad}^{\infty}_{\Lambda}(M,N)$. Moreover, the third and fourth formula of (3) of Theorem 9.5 gives $\mathrm{Hom}_{\Lambda}(L_u,M)=\mathrm{rad}^{\infty}_{\Lambda}(L_u,M)$ and $\mathrm{Hom}_{\Lambda}(M,L_u)=\mathrm{rad}^{\infty}_{\Lambda}(M,L_u)$ respectively. □

Lemma 10.5 Let $M=T_i \otimes_{R_i} R_i/(p^m)$, $N=T_j \otimes_{R_j} R_j/(q^n)$, for $i,j \in \{1,2,\cdots,s\}$, p a prime in R_i, q a prime in R_j. Then

$$\dim_k \mathrm{rad}^{\infty}_{\Lambda}(M,N)=mn\dim_{k(x)} \mathrm{rad}_{\Lambda k(x)}(G_i,G_j).$$

Proof Suppose $X=\hat{T}_i \otimes_{R_i} R_i/(p^m)$ and $Y=\hat{T}_j \otimes_{R_j} R_j/(q^n)$ are minimal projective presentations of M and N respectively. Then if $\mathbf{z}_u=$

$\dim_{k(x)} H_u$ for $u=1,2,\cdots,s$, by (4) of Theorem 9.2 we have

$$\dim_k X = m\mathbf{z}_i, \quad \dim_k Y = n\mathbf{z}_j.$$

Suppose now $i \neq j$ or $i=j$ and $(p) \neq (q)$. In this case $\mathrm{Hom}_\Lambda(M,N) = \mathrm{rad}_\Lambda^\infty(M,N)$ and $\mathrm{Hom}_{p^1(\Lambda)}(Y,X) = \mathrm{rad}_{p^1(\Lambda)}^\infty(Y,X)$. Here $DtrN \cong N$, then by (3) of Proposition 3.14 and the first equality in (2) of Theorem 9.5 we obtain

$$\dim_k \mathrm{Hom}_\Lambda(M,N) = mn(\dim_{k(x)} \mathrm{rad}_{p^1(\Lambda^{k(x)})}(H_j,H_i) - g_\Lambda(\mathbf{z}_j,\mathbf{z}_i)).$$

On the other hand, since $Dtr_{\Lambda^{k(x)}} G_j \cong G_j$ (see Proposition 6.5 of [2]) we have

$$\dim_{k(x)} \mathrm{Hom}_{\Lambda^{k(x)}}(G_i,G_j) = \dim_{k(x)} \mathrm{Hom}_{p^1(\Lambda^{k(x)})}(H_j,H_i) - g_{\Lambda^{k(x)}}(\mathbf{z}_j,\mathbf{z}_i).$$

We know from Corollary 2.3 of [2], that the indecomposable projective $\Lambda^{k(x)}$-modules are of the form $P \otimes_k k(x)$, with P indecomposable projective Λ-module, then $g_\Lambda = g_{\Lambda^{k(x)}}$. Observe that if $i \neq j$,

$$\mathrm{rad}_{p^1(\Lambda^{k(x)})}(H_j,H_i) = \mathrm{Hom}_{p^1(\Lambda^{k(x)})}(H_j,H_i)$$

and $\mathrm{rad}_{\Lambda^{k(x)}}(G_i,G_j) = \mathrm{Hom}_{\Lambda^{k(x)}}(G_i,G_j)$, moreover for $i=j$, $\dim_{k(x)} \mathrm{End}_{p^1(\Lambda^{k(x)})}(H_i) = 1 + \dim_{k(x)} \mathrm{radEnd}_{p^1(\Lambda^{k(x)})}(H_i)$ and $\dim_{k(x)} \mathrm{End}_{\Lambda^{k(x)}}(G_i) = 1 + \dim_{k(x)} \mathrm{radEnd}_{\Lambda^{k(x)}}(G_i)$. Thus we obtain:

$$\dim_{k(x)} \mathrm{rad}_{\Lambda^{k(x)}}(G_i,G_j) = \dim_{k(x)} \mathrm{rad}_{p^1(\Lambda^{k(x)})}(H_j,H_i) - g_\Lambda(\mathbf{z}_j,\mathbf{z}_i).$$

From here we obtain our equality for $i \neq j$ or $i=j$ and $(p) \neq (q)$.

For $i=j$ and $p=q$ and the first equality of (3) of Theorem 9.5 we obtain

$$\dim_k \mathrm{Hom}_{p^1(\Lambda)}(X,Y) = \min\{m,n\} + mn\dim_{k(x)} \mathrm{radHom}_{p^1(\Lambda^{k(x)})}(H_i,H_i),$$

therefore

$$\dim_k \mathrm{Hom}_\Lambda(M,N) = \min\{m,n\} + mn\dim_{k(x)} \mathrm{radHom}_{\Lambda^{k(x)}}(G_i,G_i).$$

By Lemma 10.4 the first equality of (3) Theorem 1.2 holds, then we have $\dim_k \mathrm{rad}_\Lambda^\infty(M,N) = mn\dim_{k(x)} \mathrm{radEnd}_{\Lambda^{k(x)}}(G_i)$, obtaining our result. □

Lemma 10.6 Let $M = T_i \otimes_{R_i} R_i/(p^m)$ for $i \in \{1,2,\cdots,s\}$, p a prime element in R_i, $L_u = Cok(\hat{L}_u)$, for some $u \in \{1,2,\cdots,t\}$. Then

$$\dim_k \mathrm{rad}_\Lambda^\infty(L_u,M) = m\dim_{k(x)} \mathrm{rad}_{\Lambda^{k(x)}}(G_i,L_u^{k(x)}).$$

In particular for Λe an indecomposable projective Λ-module there is a $u\in\{1,2,\cdots,t\}$ such that $\Lambda e\cong L_u$, then $\dim_k eM=m\dim_{k(x)}eG_i$.

Proof Consider $\mathbf{l}_u=\dim_k\hat{L}_u=\dim_{k(x)}\hat{L}_u^{k(x)}$. We have $DtrM\cong M$, then by (3) of Proposition 3.14 and the second equality of (2) of Theorem 9.5 we have

$$\dim_k \mathrm{Hom}_\Lambda(L_u,M)=m\dim_{k(x)}\mathrm{Hom}_{p^1(\Lambda^{k(x)})}(H_i,\hat{L}_u^{k(x)})-m\, g_\Lambda(\mathbf{z}_i,\mathbf{l}_u).$$

We have $Co\hat{k}L_u^{k(x)}\cong(Co\hat{k}L_u)^{k(x)}=L_u^{k(x)}$, thus again by (3) of Proposition 3.14, recalling that $Dtr_{\Lambda^{k(x)}}G_i\cong G_i$, we obtain

$$\dim_{k(x)}\mathrm{Hom}_{\Lambda^{k(x)}}(L_u^{k(x)},G_i)=\dim_{k(x)}\mathrm{Hom}_{p^1(\Lambda^{k(x)})}(H_i,\hat{L}_u^{k(x)})-g_\Lambda(\mathbf{z}_i,\mathbf{l}_u).$$

From here we obtain the first part of our Lemma. For the second part of the Lemma, observe that by assumption, $\dim_k\Lambda\leqslant d$, then by Lemma 10.4 we obtain our result. □

Lemma 10.7 Let $M=T_i\otimes_{R_i}R_i/(p^m)$ for $i\in(1,2,\cdots,s)$, p a prime in R_i, $L_u=Cok(\hat{L}_u)$ for $u\in\{1,2,\cdots,t\}$. Then

$$\dim_k \mathrm{rad}_\Lambda^\infty(M,L_u)=m\dim_{k(x)}\mathrm{rad}_{\Lambda^{k(x)}}(G_i,L_u^{k(x)}).$$

Proof Assume first L_u is injective, then we may suppose $L_u=D(e\Lambda)$. We have

$$\begin{aligned}\dim_k \mathrm{Hom}_\Lambda(M,D(e\Lambda))&=\dim_k\mathrm{Hom}_{\Lambda^{op}}(e\Lambda,D(M))=\dim_k D(M)e\\&=\dim_k(eM)\\&=m\dim_{k(x)}\mathrm{Hom}_{\Lambda^{k(x)}}(G_i,D_x((e\otimes 1)\Lambda^{k(x)}))\\&=m\dim_{k(x)}\mathrm{Hom}_{\Lambda^{k(x)}}(G_i,(D(e\Lambda)^{k(x)})).\end{aligned}$$

Where $D_x(-)=\mathrm{Hom}_{k(x)}(-,k(x))$.

Now assume L is not injective. Consider an almost split sequence starting in L:

$$0\to L\xrightarrow{f}\bigoplus_{s=1}^{m}E_s\xrightarrow{g}L'\to 0,$$

with E_s indecomposable for $s=1,2,\cdots,m$.

By the choice of the integer d_0, the objects E_s and L' are isomorphic to objects L_v or $T_j\otimes_{R_i}R_i/(p^m)$, but in this latter case L is in the component of an object of the form $T_j\otimes_{R_j}R_j/(p^m)$, which implies that $L\cong T_j\otimes_{R_j}R_j/(p^n)$ for some n, which is not the case therefore $L'\cong L_v$ for some $v=1,2,\cdots,t$. Then $L'\cong Cok\hat{L}_v$. Take

$$\mathbf{l}_v=\dim\hat{L}_v=\dim_{k(x)}\hat{L}_u^{k(x)}.$$

By (3) of Proposition 3.14 and the third equality of (3) of Theorem 9.5 we obtain

$$\dim_k \mathrm{Hom}_\Lambda(M,L)=m(\dim_{k(x)}\mathrm{Hom}_{p^1(\Lambda^{k(x)})}(\hat{L}_v^{k(x)},H_i)-g_\Lambda(\mathbf{l}_v,\mathbf{z}_i)).$$

On the other hand, by Corollary 2.2 of [2] we have

$$Dtr_{\Lambda_{k(x)}}(L_v^{k(x)})\cong(DtrL_v)^{k(x)}\cong L^{k(x)}.$$

Then

$$\dim_{k(x)}\mathrm{Hom}_{\Lambda_{k(x)}}(G_i,L^{k(x)})=\dim_{k(x)}\mathrm{Hom}_{p^1(\Lambda^{k(x)})}(\hat{L}_v^{k(x)},H_i)-g_\Lambda(\mathbf{l}_v,\mathbf{z}_i).$$

From here we obtain our Lemma. □

Lemma 10.8 T_i is a free right R_i-module, for $i=1,2,\cdots,s$.

Proof Since T_i is a finitely generated right R_i-module if it is not a free right R_i-module there is a primitive idempotent e of Λ such that $eT_i=C_0\oplus C_1$ with C_0 free and C_1 a torsion R_i-module, then we may assume $C_1=(\bigoplus_{j=1}^{a}R_i/(p^{m_j}))\oplus C_2$ with a prime element $p\in R_i$, positive integers m_j, and $C_2\cong\bigoplus_b R_i/(q_b^{n_b})$, where p, q_b are coprime in R_i. Suppose $m=\min\{m_1,m_2,\cdots,m_a\}$, $C_0\cong R_i^l$. Take $M=T_i\otimes_{R_i}R_i/(p^m)$, then by the second part of Lemma 10.6, $\dim_k eM=m\dim_{k(x)}eG_i=m\dim_{k(x)}eT_i\otimes_{k(x)}k(x)=m\dim C_0\otimes_{k(x)}k(x)=ml$. But $\dim_k eM=\dim_k eT_i\otimes_{R_i}R_i/(p^m)=\dim_k C_0\otimes_{R_i}R_i/(p^m)+\dim_k(R_i/(p^m))^a=ml+am$, a contradiction. Therefore, T_i is free as right R_i-module proving our result. □

Proof (of Theorem 1.2) The Λ-R_i-bimodule T_i is a good realization of G_i over R_i for $i=1,2,\cdots,s$ by Lemma 10.8 and Lemma 10.1.

(1) of Theorem 1.2 follows from Lemma 10.2, (2) follows from Lemma 10.5, Lemma 10.6 and Lemma 10.7. Finally (3) follows from Lemma 10.4. □

Acknowledgements

The authors thank the referee for several helpful comments, suggestions and corrections.

References

[1] Bautista R. The category of morphisms between projective modules. Comm. Algebra, 2004, 32(11): 4 303-4 331.

[2] Bautista R, Zhang Y. Representations of a k-algebra over the rational functions over k. J. Algebra, 2003, 267: 342-358.

[3] Bautista R, Boza J, Pérez E. Reduction functors and exact structures for bocses. Bol. Soc. Mat. Mexicana, 2003, 9(3): 21-60.

[4] Dräxler P, Reiten I, Smalø S O, Solberg O and with an appendix by Keller B. Exact categories and vector space categories. Trans. A. M. S., 1999, 351(2): 647-682.

[5] Crawley-Boevey W W. On tame algebras and bocses. Proc. London Math. Soc., 1988, 56: 451-483.

[6] Crawley-Boevey W W. Tame algebras and generic modules. Proc. London Math. Soc., 1991, 63: 241-265.

[7] Drozd Yu A. Tame and wild matrix problems. Amer. Math. Soc. Transl., 1986, 128(2): 31-55.

[8] Drozd Yu A. Reduction algorithm and representations of boxes and algebras. C. R. Math. Acad. Sci. Soc. R. Can., 2001, 23(4): 91-125.

[9] Gabriel P, Roiter A V. Representations of finite-dimensional algebras. In: Kostrikin A I, Shafarevich IV. Eds. Encyclopaedia of the Mathematical Sciences, Vol. 73, Algebra VIII, Springer, 1992.

[10] Zeng X, Zhang Y. A correspondence of almost split sequences between some categories. Comm. Algebra, 2001, 29(2): 557-582.

Science in China: Mathematics, 2009, 52A(9): 2 036-2 068.

统一化驯顺定理[①]

Unified Tame Theorem

Abstract The well-known tame theorem tells that for a given tame bocs and a positive integer n there exist, finitely many minimal bocses, such that any representation of the original bocs of dimension at most n is isomorphic to the image of a representation of some minimal bocses under a certain reduction functor. In the present paper we will give an alternative statement of the tame theorem in terms of matrix problem, by constructing a unified minimal matrix problem whose indecomposable matrices cover all the canonical forms of the indecomposable representations of dimension at most n for each non-negative integer n.

Keywords matrix problem; canonical form; representation tame type; indecomposable.

§ 0. Introduction

The notion of tame and wild problem is now rather popular in

---

① Received: 2007-03-03; accepted: 2008-08-20.

This work was supported by National Natural Science Foundation of China (Grant Nos. 10731070, 10501010)

本文与徐运阁合作.

various branches of representation theory and related topics, especially because of the so-called tame-wild dichotomy. The well-known tame theorem for bocses, which is used to prove the tame-wild dichotomy and Crawley-Boevey theorem, tells that for a given tame bocs and a positive integer n there exist finitely many minimal bocses, such that any indecomposable representation of the original bocs of dimension at most n is isomorphic to the image of a representation of some minimal bocses under a certain reduction functor[1,2].

The natural question is that for each fixed dimension n, whether there exits a unified minimal bocs, such that any indecomposable representation of the original bocs of dimension at most n is isomorphic to the image of an indecomposable representation of this minimal bocs under the reduction functor. Which is quite useful if we study the morphisms of representations and other problems. The question has been answered in "Hom-spaces of tame algebras"[3]. Now we give an alternative proof in terms of matrix problem which is relatively concrete and more convenient for calculation.

Matrix problems, which include as special cases the representation theory of finite dimensional algebras, subspace, problems and projective modules, gained their importance in studying questions about representation types. The matrix problems have nice module-theoretical interpretation by means of projective modules, and the interpretation in terms of matrix is convenient for calculation.

§ 1. The matrix problem $(\mathcal{K},\mathcal{M},\boldsymbol{H})$

We first recall the notion of matrix problems, which is a class of bimodule problems[4] given in the language of matrices and is also an algebraical version of general matrix problems[5]. Throughout the paper, we always assume that k is a fixed algebraically closed field.

Let $T=\{1,2,\cdots,t\}$ be a set of integers with an equivalent relation $\sim$. Denote by $\mathcal{T}$ the set of equivalent classes of $\mathcal{T}$. Suppose that $\mathcal{T}=$

$\mathcal{T}_1 \cup \mathcal{T}_2$, with $\mathcal{T}_1 = \{X_1, X_2, \cdots, X_r\}$ and $\mathcal{T}_2 = \{X_{r+1}, X_{r+2}, \cdots, X_s\}$. Each $X_i \in \mathcal{T}_1$ is called non-trivial, there is an indeterminate x_{X_i} or x_i for short, with domain $D_{X_i} = k \backslash \{$the roots of $g_{X_i}(x_i)\}$, attached to X_i and $x_1, x_2, \cdots, x_r$ are algebraically independent. On the other hand, each $X_i \in \mathcal{T}_2$ is called trivial.

Denoting by $\boldsymbol{E}_{ij}$ the matrix with (i, j)-entry 1 and others 0, we writed $E_X = \sum_{i \in X} E_{ii}$, for $X \in \mathcal{T}$. Then $\{\boldsymbol{E}_X \mid X \in \mathcal{T}\}$ is a set of idempotents of the upper triangular matrix algebra $\mathbf{T}_t(k)$, with $\sum_{X \in \mathcal{T}} \boldsymbol{E}_X = \mathbf{I}_t$, the identity matrix of size t.

For every pair $(X, Y) \in \mathcal{T} \times \mathcal{T}$, define a system of homogeneous linear equations

$$\mathbb{E}_{XY}: \sum_{(i,j) \in X \times Y} a_{ijl} y_{ij} = 0, a_{ijl} \in k, l = 1, 2, \cdots, m_{XY}.$$

Define $\mathbb{E} = \bigcup_{(X,Y) \in \mathcal{T} \times \mathcal{T}} \mathbb{E}_{XY}$. Let $\mathcal{M} = \{(m_{ij})_{t \times t}\}$, such that $\{m_{ij} \mid (i, j) \in X \times Y\}$ satisfies the system $\mathbb{E}_{XY}$, it implies that $E_X \mathcal{M} E_Y$ is the solution space of $\mathbb{E}_{XY}$. $\mathcal{M}$ is a subspace of the matrix algebra $\boldsymbol{M}_t(k)$, and

$$\mathcal{M} = \bigoplus_{(X,Y) \in \mathcal{T} \times \mathcal{T}} E_X \mathcal{M} E_Y.$$

Definition 1.1 An order on the indices of $\boldsymbol{M}_t(k)$ is defined as follows: $(i, j) \prec (i', j')$, provided $i > i'$, or $i = i'$ and $j < j'$. The order is also valid on the block-indices of partitioned matrices. Let $\mathbf{C} = (c_{ij})_{t \times t} \in \boldsymbol{M}_t(k)$ be a matrix, if $c_{ij} = 0$ for all $(i, j) \prec (p, q)$ and $c_{pq} \neq 0$, the (p, q) is called the leading position of C.

Let $\mathcal{S}$ be a subspace of $\boldsymbol{M}_t(k)$, a basis $\mathcal{U} = \{\boldsymbol{U}_1, \boldsymbol{U}_2, \cdots, \boldsymbol{U}_r\}$ with the leading positions $(p_1, q_1), (p_2, q_2), \cdots, (p_r, q_r)$ respectively is called a normalized basis of $\mathcal{S}$, if the $\boldsymbol{U}_i$ at (p_i, q_i) has value 1, and for any $j \neq i$, the entry of $\boldsymbol{U}_j$ at (p_i, q_i) has value 0. We say that $\boldsymbol{U}_i \prec \boldsymbol{U}_j$ provided

$$(p_i, q_i) \prec (p_j, q_j).$$

Lemma 1.2 The normalized basis $\mathcal{U}$ is uniquely determined by the subspace $\mathcal{S} \subseteq \boldsymbol{M}_t(k)$.

Proof (Given by Liu) Suppose that we have another normalized basis $\mathcal{W} = \{\boldsymbol{W}_1, \boldsymbol{W}_2, \cdots, \boldsymbol{W}_r\}$ of $\mathcal{S}$. Without loss of generality we may assume in addition that $\boldsymbol{U}_1 \prec \boldsymbol{U}_2 \prec \cdots \prec \boldsymbol{U}_r$ and $\boldsymbol{W}_1 \prec \boldsymbol{W}_2 \prec \cdots \prec \boldsymbol{W}_r$. Write (p_l, q_l) as the leading position of $\boldsymbol{U}_l$ and (s_l, t_l) as the leading position of $\boldsymbol{W}_l$ for $1 \leqslant l \leqslant r$. Thus

$$\boldsymbol{W}_i = \sum_{j=1}^{r} \lambda_{ij} \boldsymbol{U}_j, \quad \boldsymbol{U}_i = \sum_{j=1}^{r} \mu_{ij} \boldsymbol{W}_j.$$

It is clear that $(p_1, q_1) = (s_1, t_1)$. Since the entry at (p_1, q_1) of $\boldsymbol{W}_2$ is 0, but that of $\boldsymbol{U}_1$ is 1 and those of $\boldsymbol{U}_j$ for $j > 1$ are all 0, we deduce that $\lambda_{21} = 0$. Consequently $(s_2, t_2) \succeq (p_2, q_2)$. Dually $\mu_{21} = 0$ and $(p_2, q_2) \succeq (s_2, t_2)$, therefore $(s_2, t_2) = (p_2, q_2)$. By induction, we get $\lambda_{ij} = \mu_{ij} = 0$ for all $1 \leqslant j < i \leqslant m$, and $(p_i, q_i) = (s_i, t_i)$ for all $1 \leqslant i \leqslant r$. Now the (p_r, q_r)-entry of $\boldsymbol{W}_{r-1}$ and that of $\boldsymbol{U}_{r-1}$ are both zero. This gives rise to $\lambda_{r-1,r} = \mu_{r-1,r} = 0$. By induction again, we have $\lambda_{ij} = \mu_{ij} = 0$ for all $1 \leqslant i < j \leqslant r$. Finally, comparing the (p_i, q_i)-entries, we have $\boldsymbol{U}_i = \boldsymbol{W}_i$, for all $1 \leqslant i \leqslant r$. The proof of the lemma is completed.

The Gauss-Jacobi elimination method allows us to choose the normalized basis for $\mathcal{M}$ according to the order given in Definition 1. 1. Denote then by $\mathcal{A}_{XY}$ the normalized basis of the solution space $E_X \mathcal{M} E_Y$ of equation system $\mathbb{E}_{XY}$ for any fixed pair (X, Y). Moreover write

$$\mathcal{A} = \bigcup_{X, Y \in \mathcal{T}} \mathcal{A}_{XY} = \{\boldsymbol{A}_1, \boldsymbol{A}_2, \cdots, \boldsymbol{A}_n\},$$

which is the normalized basis of $\mathcal{M}$. Denote by (p_i, q_i) the leading position of A_i for $i = 1, 2, \cdots, n$, then $(p_1, q_1) \prec (p_2, q_2) \prec \cdots \prec (p_n, q_n)$ according to the order given in Definition 1. 1.

Definition 1. 3 Define a matrix

$$\boldsymbol{H} = \boldsymbol{H}^0 + x_1 \boldsymbol{A}_1^0 + \cdots + x_r \boldsymbol{A}_r^0,$$

where $\boldsymbol{H}^0 \in \boldsymbol{M}_t(k)$ such that $\left(\sum_{i=1}^{r} \boldsymbol{E}_{X_i}\right) \boldsymbol{H}^0 = 0 = \boldsymbol{H}^0 \left(\sum_{i=1}^{r} \boldsymbol{E}_{X_i}\right)$, and $\boldsymbol{A}_i^0 \in \boldsymbol{E}_{X_i} \boldsymbol{M}_t(k) \boldsymbol{E}_{X_i}$ with entry 1 at the leading position for $1 \leqslant i \leqslant r$, which satisfies the following conditions:

(1) The leading positions of $\boldsymbol{H}^0, \boldsymbol{A}_1^0, \cdots, \boldsymbol{A}_r^0$ are all smaller than (p_1, q_1), the leading position of $\boldsymbol{A}_1$. Write $\boldsymbol{H} = (h_{ij})_{t \times t}$, then $h_{p_i q_i} = 0$ for (p_i, q_i), the leading position of $\boldsymbol{A}_i \in \mathcal{A}$ and $1 \leqslant i \leqslant n$.

(2) $\boldsymbol{H}\boldsymbol{E}_X-\boldsymbol{E}_X\boldsymbol{H}=0$ for all $X\in\mathcal{T}$.

Remark (1) It is clear by Definition 1.3, that for any pair taken from $\{\boldsymbol{A}_1^0,\boldsymbol{A}_2^0,\cdots,\boldsymbol{A}_r^0\}$, they have no common non-zero (i,j)-entries for any $0\leqslant i,j\leqslant t$.

(2) If $\boldsymbol{H}=(h_{ij})_{t\times t}$, then $h_{ij}=0$ whenever $i\not\sim j$. In fact, if $i\in X, j\in Y$, then $0=\boldsymbol{E}_X\boldsymbol{H}-\boldsymbol{H}\boldsymbol{E}_X$, therefore $\boldsymbol{E}_X\boldsymbol{H}\boldsymbol{E}_Y=0$ in the case of $X\neq Y$, consequently $h_{ij}=0$.

Let $\boldsymbol{C},\boldsymbol{D}$ be two partitioned matrices. We write $\boldsymbol{C}\equiv_{\prec(p,q)}\boldsymbol{D}$ (resp. $\boldsymbol{C}\equiv_{\preceq(p,q)}\boldsymbol{D}$) if the (i,j)-blocks of $\boldsymbol{C},\boldsymbol{D}$ are the same for all $(i,j)\prec(p,q)$ (resp. $(i,j)\preceq(p,q)$).

Suppose that for each pair $(X,Y)\in\mathcal{T}\times\mathcal{T}$, there exists another homogeneous system

$$\mathbb{F}_{XY}:\sum_{(i,j)\in X\times Y;i<j} b_{ijl}(x,y)z_{ij}=0,$$

$$b_{ijl}(x,y)\in k[x,y,g_X(x)^{-1},g_Y(y)^{-1}],l=1,2,\cdots,n_{XY},$$

where x multiples z_{ij} from left in the case that X is non-trivial, and x does not occur if X is trivial; y multiples z_{ij} from right in the case that Y is non-trivial, and y does not occur if Y is trivial. The notation of left and right multiplications is also valid even in the case that $X=Y$ is non-trivial. Define an equation system:

$$\mathbb{F}=\bigcup_{(X,Y)\in\mathcal{T}\times\mathcal{T}}\mathbb{F}_{XY}.$$

Define a set of upper triangular matrices,

$$\mathcal{K}=\{(s_{ij})_{t\times t}\mid s_{ii}\in k, s_{ij}\in k[x,y,g_X(x)^{-1},g_Y(y)^{-1}] \text{ for } i<j, i\in X, j\in Y\},$$

such that $s_{ii}=s_{jj}\in k$ for $i\sim j$, and $\{s_{ij}\mid i<j,i\in X,j\in Y\}$ satisfies the equation system $\mathbb{F}_{XY}$ for each pair $(X,Y)\in\mathcal{T}\times\mathcal{T}$. We denote by rad $\mathcal{K}$ the subspace of $\mathcal{K}$ consisting of strictly upper triangular matrices, and give the following additional assumption on rad $\mathcal{K}$.

Assumption 1.4 (1) $\boldsymbol{E}_X(\operatorname{rad}\mathcal{K})\boldsymbol{E}_Y$ is a free module over $k[x,y,g_X(x)^{-1},g_Y(y)^{-1}]$, with a set of free generators (basis) $\mathcal{V}_{XY}$.

(2) For any $\mathbf{V}_i\in\mathcal{V}_{XY},\mathbf{V}_j\in\mathcal{V}_{YZ}$, $\mathbf{V}_i\mathbf{V}_j$ is a linear combination of the basis in $\mathcal{V}_{XZ}$ with the coefficients in

$$k[x, y, z, g_X(x)^{-1}, g_Y(y)^{-1}, g_Z(z)^{-1}].$$

(3) For any $\boldsymbol{A}_i \in \mathcal{A}_{XY}, \boldsymbol{V}_j \in \mathcal{V}_{YZ}, \boldsymbol{A}_i\boldsymbol{V}_j$ is a linear combination of the basis in $\mathcal{A}_{XZ}$ with the coefficients in $k[y, z, g_Y(y)^{-1}, g_Z(z)^{-1}]$. Similarly for any $\boldsymbol{V}_i \in \mathcal{V}_{XY}, \boldsymbol{A}_j \in \mathcal{A}_{YZ}, \boldsymbol{V}_i\boldsymbol{A}_j$ is a linear combination of the basis in $\mathcal{A}_{XZ}$ with the coefficients in $k[x, y, g_X(x)^{-1}, g_Y(y)^{-1}]$.

(4)For any $V_i \in \mathcal{V}_{XY}$, define $d_H(\boldsymbol{V}_i) = \boldsymbol{V}_i\boldsymbol{H} - \boldsymbol{H}\boldsymbol{V}_i$, which is a linear combination of the basis in $\mathcal{A}_{XY}$ with the coefficients in $k[x, y, g_X(x)^{-1}, g_Y(y)^{-1}]$. In particular, $\boldsymbol{SH} \equiv_{<(p,q)} \boldsymbol{HS}$. The basis of rad $\mathcal{K}$ is usually written as, $\mathcal{V} = \{\boldsymbol{V}_1, \boldsymbol{V}_2, \cdots, \boldsymbol{V}_m\}$. $\mathcal{K}$ is called an admissible transformation set, and

$$\mathcal{K} = \bigoplus_{X \in \mathcal{T}} k\boldsymbol{E}_X \oplus \bigoplus_{(X,Y) \in \mathcal{T} \times \mathcal{T}} \boldsymbol{E}_X \operatorname{rad} \mathcal{K} \boldsymbol{E}_Y.$$

Definition 1.5 Let $t \geqslant 1$. A matrix problem $\mathfrak{M} = (\mathcal{K}, \mathcal{M}, H)$ of size t consists of the following data:

(1) a set of integers $T = \{1, 2, \cdots, t\}$ with an equivalent relation $\sim$;

(2) a k-subspace $\mathcal{M}$ of $\boldsymbol{M}_t(k)$ determined by the equation system $\mathbb{E}$;

(3) an admissible transformation set $\mathcal{K}$ determined by the equation system $\mathbb{F}$;

(4)a matrix $\boldsymbol{H}$ and a map d_H.

§ 2. The representation category of $\mathfrak{M}$

Definition 2.1 Let $(T, \sim)$ be a set of integers with an equivalent relation $\sim$. A vector $\underline{\boldsymbol{m}} = (m_1, m_2, \cdots, m_t)$ is called a size vector over $(T, \sim)$, provided that $m_i, 1 \leqslant i \leqslant t$, are non-negative integers with $m_i = m_j$ whenever $i \sim j$. $m = \sum_{j=1}^{t} m_j$ is called the size of the size vector $\underline{\boldsymbol{m}}$. In the case $i \in X$ we write $m_X = m_i$ sometimes.

Definition 2.2[5] Let $\boldsymbol{J}(\lambda) = \boldsymbol{J}_d(\lambda)^{e_d} \oplus \boldsymbol{J}_{d-1}(\lambda)^{e_{d-1}} \oplus \cdots \oplus \boldsymbol{J}_1(\lambda)^{e_1}$ be a Jordan form. Set

$$m_j \rightarrow e_d + e_{d-1} + \cdots + e_j.$$

The following partitioned matrix $\boldsymbol{W}_\lambda$ is called a Weyr matrix of eigenvalue λ:

$$\boldsymbol{W}(\lambda)=\begin{pmatrix} \lambda\boldsymbol{I}_{m_1} & \boldsymbol{J}_{12} & \boldsymbol{0} & \cdots & \boldsymbol{0} & \boldsymbol{0} \\ & \lambda\boldsymbol{I}_{m_2} & \boldsymbol{J}_{23} & \cdots & \boldsymbol{0} & \boldsymbol{0} \\ & & \lambda\boldsymbol{I}_{m_3} & & \cdots & \boldsymbol{0} \\ & & & \ddots & \vdots & \vdots \\ & & & & \lambda\boldsymbol{I}_{m_{d-1}} & \boldsymbol{J}_{d-1,d} \\ & & & & & \lambda\boldsymbol{I}_{m_d} \end{pmatrix}_{d\times d}, \quad \boldsymbol{J}_{j,j+1}=\begin{pmatrix} \boldsymbol{I}_{m_{j+1}} \\ \boldsymbol{0} \end{pmatrix}_{m_j\times m_{j+1}}$$

A direct sum $\boldsymbol{W}=\boldsymbol{W}(\lambda_1)\oplus\boldsymbol{W}(\lambda_2)\oplus\cdots\oplus\boldsymbol{W}(\lambda_s)$ of finitely many Weyr matrices with certain eigenvalues is said to be a Weyr matrix. We may define an order $<$ on the base field k, so that each Weyr matrix has a unique form according to the order from up to bottom.

A matrix which commutes with $\boldsymbol{W}(\lambda)$ has a nice block upper triangular form, for example $d=3, e_1=e_2=e_3=1$:

$$\begin{pmatrix} s_{11}^{(1)} & s_{12}^{(1)} & s_{13}^{(1)} & \vdots & s_{11}^{(2)} & s_{12}^{(2)} & \vdots & s_{11}^{(3)} \\ 0 & s_{22}^{(1)} & s_{23}^{(1)} & \vdots & s_{21}^{(2)} & s_{22}^{(2)} & \vdots & s_{21}^{(3)} \\ 0 & 0 & s_{33}^{(1)} & \vdots & 0 & s_{32}^{(2)} & \vdots & s_{31}^{(3)} \\ \cdots & \cdots & \cdots & \cdots & \cdots & \cdots & \cdots & \cdots \\ & & & \vdots & s_{11}^{(1)} & s_{12}^{(1)} & \vdots & s_{11}^{(2)} \\ & & & \vdots & 0 & s_{22}^{(1)} & \vdots & s_{21}^{(2)} \\ & & & & \cdots & \cdots & \cdots & \cdots \\ & & & & & & \vdots & s_{11}^{(1)} \end{pmatrix}$$

Given size vectors $\underline{\boldsymbol{m}}$ and $\underline{\boldsymbol{n}}$ over $(T,\sim)$, a partitioned matrix G of size $\underline{\boldsymbol{m}}\times\underline{\boldsymbol{n}}$ means that $\boldsymbol{G}=(\boldsymbol{G}_{ij})_{t\times t}$ with $\boldsymbol{G}_{ij}$ $m_i\times n_j$-matrices for $1\leqslant i, j\leqslant t$. In particular if $\underline{\boldsymbol{m}}=\underline{\boldsymbol{n}}$, the matrix $\boldsymbol{G}$ is called a matrix of size $\underline{\boldsymbol{m}}$.

Suppose now $X, Y\in\mathcal{T}$, that a $t\times t$ matrix $\boldsymbol{D}=(d_{ij})_{t\times t}$ is said to be an (XY)-matrix if $d_{ij}=0$ when $i\notin X$ or $j\notin Y$. It is clear that the matrices in $\mathcal{A}_{XY}$ and $\mathcal{V}_{XY}$ are all (XY)-matrices.

Definition 2.3 Let $\underline{\boldsymbol{m}}, \underline{\boldsymbol{n}}$ be two size vectors over $(T,\sim)$, and $\boldsymbol{D}$ be an (XY)-matrix for some $X, Y\in\mathcal{T}$, let $\boldsymbol{C}$ be an $m_X\times n_Y$ matrix, then $\boldsymbol{C}*\boldsymbol{D}$ stands for a partitioned matrix of size $\underline{\boldsymbol{m}}\times\underline{\boldsymbol{n}}$, such that $(\boldsymbol{C}*\boldsymbol{D})_{ij}=\boldsymbol{C}d_{ij}$ for $i\in X$ and $j\in Y$, $(\boldsymbol{C}*\boldsymbol{D})_{ij}=\boldsymbol{0}$ for $i\notin X$ or $j\notin Y$.

Fix a size vector $\underline{m}=(m_1,m_2,\cdots,m_t)$, let $\underline{W}=(\boldsymbol{W}_{X_1},\boldsymbol{W}_{X_2},\cdots,\boldsymbol{W}_{X_r})$ be a sequence of Weyr matrices, such that the size of $\boldsymbol{W}_{X_j}$ is m_{X_j} and the eigenvalues of $\boldsymbol{W}_{X_j}$ are taken from $\boldsymbol{D}_{X_j}$ for $1\leqslant j\leqslant r$. Set

$$\boldsymbol{H}_{\underline{m}}(\underline{W})=\boldsymbol{H}_{\underline{m}}^0+\sum_{j=1}^{r}\boldsymbol{W}_{X_j}*\boldsymbol{A}_j^0,$$

where $\boldsymbol{H}_{ij}^0=0$ whenever $i\not\sim j$; and $\boldsymbol{H}_{ij}^0=h_{ij}\boldsymbol{I}_{m_i}$ whenever $i,j\in X$ and X is trivial, with $\boldsymbol{I}_{m_i}$ an identity matrix of size m_i. Let $\mathcal{M}_{\underline{m}}$ be a set of partitioned matrices $(M_{ij})_{t\times t}$ such that M_{ij}, $1\leqslant i,j\leqslant t$, satisfy the defining system $\mathbb{E}$, i. e. $\boldsymbol{M}=\sum_{i=1}^{n}\boldsymbol{C}_i*\boldsymbol{A}_i$ for some $m_{X_i}\times m_{Y_i}$ matrix $\boldsymbol{C}_i$, $1\leqslant i\leqslant n$. A pair $(\boldsymbol{M},\boldsymbol{H}_{\underline{m}}(\underline{W}))$ is called a matrix over $\mathfrak{M}$.

Lemma 2. 4[5] Let $f(x,y)=\sum_{i,j\geqslant 0}\beta_{ij}x^iy^j\in k[x,y]$. Given two positive integers m, n, suppose that $\boldsymbol{R}=(r_{pq})_{m\times n}$ with r_{pq} indeterminates, $\boldsymbol{L}$, $\boldsymbol{K}$ are upper triangular nilpotent matrices of size m, n respectively. Define a matrix

$$f(x\boldsymbol{I}_m+\boldsymbol{L},y\boldsymbol{I}_n+\boldsymbol{K})\circ\boldsymbol{R}=\sum_{i,j\geqslant 0}\beta_{ij}(x\boldsymbol{I}_m+\boldsymbol{L})^i\boldsymbol{R}(y\boldsymbol{I}_n+\boldsymbol{K})^j. \qquad (2.1)$$

Then the (p,q)-entry of the matrix is

$$f(x,y)r_{pq}+\sum_{(p'q')\prec(pq)}d_{p'q'}(x,y)r_{p'q'},$$

where the order $\prec$ is defined in Definition 1. 1 and $d_{p'q'}(x,y)$ are polynomials. In particular, if $f(\lambda,\mu)\neq 0$ for some $\lambda,\mu\in k$, then the (p,q)-entry of the matrix is non-zero in the case of $x=\lambda$, $y=\mu$.

Lemma 2. 5 Let

$$\sum_{j=1}^{s}b_{ij}(x,y)z_j=0,i=1,2,\cdots,\eta, \qquad (2.2)$$

where x multiples z_j from left; y multiples z_j from right, and $b_{ij}(x,y)\in k[x,y]$ for $1\leqslant i\leqslant\eta$ and $1\leqslant j\leqslant s$. Suppose that the solutions of Formulae (2. 2) form a free $k[x,y,g_1(x)^{-1}g_2(y)^{-1}]$-module with free generators (or basis): $\{\boldsymbol{V}_1(x,y),\boldsymbol{V}_2(x,y),\cdots,\boldsymbol{V}_l(x,y)\}$, i. e. the rank of the coefficient matrix of Formulae (2. 2) equals $(s-l)$ over $k[x,y,g_1(x)^{-1}g_2(y)^{-1}]$, thus we have a general solution

$$v_1\boldsymbol{V}_1+v_2\boldsymbol{V}_2+\cdots+v_l\boldsymbol{V}_l.$$

Given two positive integers m, n, write $\boldsymbol{Z}j=(z^j_{pq})_{m\times n}$ with z^j_{pq} variables, and let L, K be upper triangular nilpotent matrices of sizes m, n respectively. Define a system of matrix equations for any fixed $\lambda, \mu \in k$ with $g_1(\lambda)\neq 0$ and $g_2(\mu)\neq 0$:

$$\sum_{j=1}^{s} b_{ij}(\lambda \boldsymbol{I}_m+\boldsymbol{L}, \mu \boldsymbol{I}_n+\boldsymbol{K}) \circ \boldsymbol{Z}_j=0, \quad i=1,2,\cdots,\eta. \tag{2.3}$$

Then the system has a general solution:

$$\bar{\boldsymbol{v}}_1 \circ \mathbf{V}_1(\lambda \boldsymbol{I}_m+\boldsymbol{L}, \mu \boldsymbol{I}_n+\boldsymbol{K})+\cdots+\bar{\boldsymbol{v}}_l \circ \mathbf{V}_l(\lambda \boldsymbol{I}_m+\boldsymbol{L}, \mu \boldsymbol{I}_n+\boldsymbol{K}), \tag{2.4}$$

where $\bar{\boldsymbol{v}}_j=(v^j_{\zeta\eta})_{m\times n}$, if $\mathbf{V}=(f^j_1(x,y), f^j_2(x,y), \cdots, f^j_s(x,y))^{\mathrm{T}}$, T stands for the transpose of a matrix. Then for each $1\leqslant j\leqslant l$,

$$\bar{\boldsymbol{v}}_j \circ \mathbf{V}_j(\lambda \boldsymbol{I}_m+\boldsymbol{L}, \mu \boldsymbol{I}_n+\boldsymbol{K})=(f^j_1(\lambda \boldsymbol{I}_m+\boldsymbol{L}, \mu \boldsymbol{I}_n+\boldsymbol{K})\circ \bar{\boldsymbol{v}}_j, \cdots,$$
$$f^j_s(\lambda \boldsymbol{I}_m+\boldsymbol{L}, \mu \boldsymbol{I}_n+\boldsymbol{K})\circ \bar{\boldsymbol{v}}_j)^{\mathrm{T}}.$$

Furthermore, define an $s\times 1$ block matrix $\mathbf{V}^{\zeta\eta}_j$, such that $\bar{\boldsymbol{v}}_j \circ \mathbf{V}_j(\lambda \boldsymbol{I}_m+\boldsymbol{L}, \mu \boldsymbol{I}_n+\boldsymbol{K})=\sum_{\zeta\eta} v^j_{\zeta\eta}\mathbf{V}^{\zeta\eta}_j$, then

$$\{\mathbf{V}^{\zeta\eta}_j \mid 1\leqslant \zeta\leqslant m, 1\leqslant \eta\leqslant n, 1\leqslant j\leqslant l\}$$

is a basis of the solution space of the equations (2.3).

Proof The $s\times 1$ block matrix $\mathbf{V}^{\zeta\eta}_j$ for $1\leqslant \zeta\leqslant m, 1\leqslant \eta\leqslant n, 1\leqslant j\leqslant l$ has indeterminates $v^j_{\zeta\eta}$ as coefficients in Formula (2.4). By Formula (2.1), the (ζ,η) entries at all the blocks of the $s\times 1$ block-matrix $\bar{\boldsymbol{v}}_j \circ \mathbf{V}_j(\lambda \boldsymbol{I}_m+\boldsymbol{L}, \mu \boldsymbol{I}_n+\boldsymbol{K})$ are

$$\Big((f^j_1(\lambda,\mu)v^j_{\zeta\eta}+\sum_{(\zeta',\eta')\prec(\zeta,\eta)} d^j_{\zeta',\eta'}(\lambda,\mu)v^j_{\zeta',\eta'}), \cdots, (f^j_s(\lambda,\mu)v^j_{\zeta\eta}+$$
$$\sum_{(\zeta',\eta')\prec(\zeta,\eta)} d^j_{\zeta',\eta'}(\lambda,\mu)v^j_{\zeta',\eta'})\Big)^{\mathrm{T}}.$$

Therefore the $s\times 1$ block matrix $\mathbf{V}^{\zeta\eta}_j$ has the following form at (ζ,η)-position of each block:

$$(f^j_1(\lambda,\mu), f^j_2(\lambda,\mu), \cdots, f^j_s(\lambda,\mu))^{\mathrm{T}},$$

and all the other possible non-zero entries sit at $(\zeta',\eta')>(\zeta,\eta)$. Thus mnl matrices $\{\mathbf{V}^{\zeta\eta}_j\}$ are linearly independent by induction on the indices (ζ,η) according to the ordering of the indices of the matrices. Thus the dimension of the solution space of Formulae (2.3) is bigger than or equal to mnl.

On the other hand, denote by $\boldsymbol{C}(x, y)$ the coefficient matrix of Formulae (2.2), then rank$(\boldsymbol{C}(\lambda, \mu)) = s - l$. Define an order on the entries of $\boldsymbol{Z}^j = (z^j_{pq})_{m\times n}$, such that $z^j_{pq} < z^{j'}_{p'q'}$ if and only if $p > p'$, or $p = p'$ and $q < q'$, or $p = p'$, $q = q'$ and $j < j'$. Denote by $\boldsymbol{C}(\lambda \boldsymbol{I}_m + \boldsymbol{L}, \mu \boldsymbol{I}_n + \boldsymbol{K})$ the coefficient matrix of (2.3) under the above order, then

$$\begin{pmatrix} \boldsymbol{C}(\lambda,\mu) & * & \cdots & * \\ \boldsymbol{0} & \boldsymbol{C}(\lambda,\mu) & \cdots & * \\ \vdots & \vdots & & \vdots \\ \boldsymbol{0} & \boldsymbol{0} & \cdots & \boldsymbol{C}(\lambda,\mu) \end{pmatrix}_{mn\times mn}$$

with rank bigger than or equal to $mn(s-l)$. Therefore the dimension of the solution space of Formulae (2.3) is smaller than or equal to mnl, thus it equals mnl. The basic solution $\{\boldsymbol{V}_i \mid 1 \leqslant j \leqslant l\}$ of Formulae (2.2) yields the basic solution $\{\boldsymbol{V}_j^{\zeta\eta} \mid 1 \leqslant \zeta \leqslant m, 1 \leqslant \eta \leqslant n, 1 \leqslant j \leqslant l\}$ of the matrix Formulae (2.3). Consequently the general solution of (2.3) is given by Formula (2.4), the proof of the lemma is finished.

Corollary 2.6 With the hypothesis in Lemma 2.5, let $\boldsymbol{W} = \boldsymbol{W}(\lambda_1) \oplus \boldsymbol{W}(\lambda_2) \oplus \cdots \oplus \boldsymbol{W}(\lambda_s)$ and $\boldsymbol{U} = \boldsymbol{W}(\mu_1) \oplus \boldsymbol{W}(\mu_2) \oplus \cdots \oplus \boldsymbol{W}(\mu_t)$ be two Weyr matrices with $g_1(\lambda_p) \neq 0$, $g_2(\mu_q) \neq 0$, and write m_p as the size of $\boldsymbol{W}(\lambda_p)$, n_q as the size of $\boldsymbol{W}(\mu_q)$ respectively for $p = 1, 2, \cdots, s$, $q = 1, 2, \cdots, t$. Set $\boldsymbol{Z}^j = (\boldsymbol{Z}^j_{pq})_{s\times t}$, such that $\boldsymbol{Z}^j_{pq}$ is an $m_p \times n_q$ matrix with variable entries. Then the equation system

$$\sum_{j=1}^{s} b_{ij}(\boldsymbol{W}, \boldsymbol{U}) \boldsymbol{Z}^j = 0, i = 1, 2, \cdots, \eta, \tag{2.5}$$

has a general solution:

$$\bar{\boldsymbol{v}}_1 \circ \boldsymbol{V}_1(\boldsymbol{W}, \boldsymbol{U}) + \bar{\boldsymbol{v}}_2 \circ \boldsymbol{V}_2(\boldsymbol{W}, \boldsymbol{U}) + \cdots + \bar{\boldsymbol{v}}_l \circ \boldsymbol{V}_l(\boldsymbol{W}, \boldsymbol{U}),$$

where $\bar{\boldsymbol{v}}_j = (v^j_{\zeta\eta})_{m\times n}$ with $m = \sum_{p=1}^{s} m_p$, $n = \sum_{q=1}^{t} n_q$, the meanings of $\boldsymbol{V}_j$ and $\bar{\boldsymbol{v}}_j \circ \boldsymbol{V}_j$ are the same as Formula (2.4). Furthermore, define an $s \times 1$ block matrix $\boldsymbol{V}_j^{\zeta\eta}$ for $1 \leqslant \zeta \leqslant m$, $1 \leqslant \eta \leqslant n$, such that $\bar{\boldsymbol{v}}_j \circ \boldsymbol{V}_j(\boldsymbol{W}, \boldsymbol{U}) = \sum_{\zeta,\eta} v^j_{\zeta\eta} \boldsymbol{V}_j^{\zeta\eta}$, then $\{\boldsymbol{V}_j^{\zeta\eta} \mid 1 \leqslant \zeta \leqslant m, 1 \leqslant \eta \leqslant n, 1 \leqslant j \leqslant l\}$ is a basis of the solution space of Formulae (2.5).

Proof Consider the matrix equations $\sum_{j=1}^{s} b_{ij}(\boldsymbol{W}(\lambda_p),\boldsymbol{U}(\mu_q))\boldsymbol{Z}_{pq}^{j}=0, i=1,2,\cdots,\eta_{pq}$ for each $1\leqslant p\leqslant s, 1\leqslant q\leqslant t$, the assertion follows from Lemma 2.5, the corollary is proved.

Given two size vectors $\underline{\boldsymbol{m}}$ and $\underline{\boldsymbol{n}}$, two Weyr sequences $\underline{\boldsymbol{W}}=(\boldsymbol{W}_1, \boldsymbol{W}_2,\cdots,\boldsymbol{W}_r)$ of size $\underline{\boldsymbol{m}}$ and $\underline{\boldsymbol{U}}=(\boldsymbol{U}_1,\boldsymbol{U}_2,\cdots,\boldsymbol{U}_r)$ of size $\underline{\boldsymbol{n}}$ respectively. Denote by $\mathbb{F}(\underline{\boldsymbol{W}},\underline{\boldsymbol{U}})$ the equation system obtained from $\mathbb{F}$ by substituting $\boldsymbol{W}_1,\boldsymbol{W}_2,\cdots,\boldsymbol{W}_r$ for $x_1,x_2,\cdots,x_r$, and $\boldsymbol{U}_1,\boldsymbol{U}_2,\cdots,\boldsymbol{U}_r$ for $y_1,y_2,\cdots,y_r$ in the coefficients $b_{ijl}(x_p,y_q)$, moreover substituting the variables z_{ij} by some matrices of suitable sizes with the variable entries.

Let $\mathcal{K}_{\underline{m}\times\underline{n}}(\underline{\boldsymbol{W}},\underline{\boldsymbol{U}})$ be the set of upper triangular partitioned matrices $\boldsymbol{S}=(\boldsymbol{S}_{ij})_{t\times t}$ of size $\underline{\boldsymbol{m}}\times\underline{\boldsymbol{n}}$, such that $\boldsymbol{S}_{ii}=\boldsymbol{S}_{jj}$ if $i\sim j$, $\boldsymbol{S}_{ii}$ is any matrix over k of size m_X for $i\in X$ being trivial; or $\boldsymbol{S}_{ii}$ is a matrix over k of size m_X commuting with $\boldsymbol{W}_X$ for $i\in X$ being non-trivial. $\boldsymbol{S}_{ij}$ are $m_i\times n_j$ matrices satisfying the equation system $\mathbb{F}(\underline{\boldsymbol{W}},\underline{\boldsymbol{U}})$. We have seen the exact structure of the solutions of $\mathbb{F}(\underline{\boldsymbol{W}},\underline{\boldsymbol{U}})$ in Corollary 2.6.

Let $(\boldsymbol{M},\boldsymbol{H}_{\underline{m}}(\underline{\boldsymbol{W}}))$ and $(\boldsymbol{N},\boldsymbol{H}_{\underline{n}}(\underline{\boldsymbol{U}}))$ be matrices over $\mathfrak{M}$, a partitioned matrix $\boldsymbol{S}\in\mathcal{K}_{\underline{m}\times\underline{n}}(\underline{\boldsymbol{W}},\underline{\boldsymbol{U}})$ is called a morphism from $(\boldsymbol{M}, \boldsymbol{H}_{\underline{m}}(\underline{\boldsymbol{W}}))$ to $(\boldsymbol{N},\boldsymbol{H}_{\underline{n}}(\underline{\boldsymbol{U}}))$ provided

$$(\boldsymbol{M}+\boldsymbol{H}_{\underline{m}}(\underline{\boldsymbol{W}}))\boldsymbol{S}=\boldsymbol{S}(\boldsymbol{N}+\boldsymbol{H}_{\underline{n}}(\underline{\boldsymbol{U}})). \tag{2.6}$$

Finally if $(\boldsymbol{L},\boldsymbol{H}_{\underline{l}}(\underline{\boldsymbol{Q}}))$ is also a matrix over $\mathfrak{M}$, and $R:(\boldsymbol{N},\boldsymbol{H}_{\underline{n}}(\underline{\boldsymbol{U}}))\to(\boldsymbol{L}, \boldsymbol{H}_{\underline{l}}(\underline{\boldsymbol{Q}}))$ is a morphism, then Formula (2.6) and $(\boldsymbol{N}+\boldsymbol{H}_{\underline{n}}(\underline{\boldsymbol{U}}))\boldsymbol{R}=\boldsymbol{R}(\boldsymbol{L}+ \boldsymbol{H}_{\underline{l}}(\underline{\boldsymbol{Q}}))$ yield $(\boldsymbol{M}+\boldsymbol{H}_{\underline{m}}(\underline{\boldsymbol{W}}))\boldsymbol{SR}=\boldsymbol{SR}(\boldsymbol{L}+\boldsymbol{H}_{\underline{l}}(\underline{\boldsymbol{Q}}))$. Thus $\boldsymbol{SR}:(\boldsymbol{M}, \boldsymbol{H}_{\underline{m}}(\underline{\boldsymbol{W}}))\to(\boldsymbol{L},\boldsymbol{H}_{\underline{l}}(\underline{\boldsymbol{Q}}))$ is a morphism. We have the composition of two morphisms.

Proposition 2.7 The matrices over $\mathfrak{M}$ with the morphisms defined by Formula (2.6) form a representation category of $\mathfrak{M}$, which is denoted by Mat($\mathfrak{M}$).

Corollary 2.8 Let $\mathfrak{M}$ be a matrix problem. Given two size vectors $\underline{\boldsymbol{m}}$ and $\underline{\boldsymbol{n}}$ over $(T,\sim)$, we define two Weyr sequences $\underline{\boldsymbol{W}}$ of size $\underline{\boldsymbol{m}}$ and $\underline{\boldsymbol{U}}$

of size $\underline{n}$ respectively.

(1) A morphism S is invertible if and only if the block-diagonal part is invertible. Then $\underline{n}=\underline{m}$, and $S^{-1}\in\mathcal{K}_{\underline{m}\times\underline{m}}(\underline{W},\underline{W})$.

(2) If S is invertible, then $H_{\underline{m}}(\underline{W})=H_{\underline{n}}(\underline{U})$.

Proof (1) Write $S=D+N$ with D being the block-diagonal part and N the block-nilpotent part. Then S is invertible if and only if D is invertible. In this case $D^{-1}S=(I+D^{-1}N)$ has an inverse in $\mathcal{K}_{\underline{m}\times\underline{m}}(\underline{W},\underline{U})$: $I-D^{-1}N+(D^{-1}N)^2-\cdots$, because $D^{-1}N$ is nilpotent.

(2) Since $\underline{n}=\underline{m}$, $H^0_{\underline{m}}=H^0_{\underline{n}}$ in two matrices. Furthermore comparing the blocks at the leading position of $A^0_1, A^0_2, \cdots, A^0_r$ in Formula (2.6), we obtain according to Definition 1.3, that $W_{X_i}S_{X_i}=S_{X_i}U_{X_i}$ for each $1\leqslant i\leqslant r$, where S_{X_i} is an invertible m_{X_i} square matrix at the block-diagonal of S. But the Weyr matrix is uniquely determined up to similarity under the fixed order of the base field, we conclude that $W_{X_i}=U_{X_i}$. The lemma is proved.

If S is invertible, we have $(N+H_{\underline{m}}(\underline{W}))S^{-1}=S^{-1}(M+H_{\underline{m}}(\underline{W}))$. S is said to be an isomorphism, and $(M,H_{\underline{m}}(\underline{W}))$, $(N,H_{\underline{m}}(\underline{W}))$ are said to be isomorphic, which is written by $(M,H_{\underline{m}}(\underline{W}))\simeq(N,H_{\underline{m}}(\underline{W}))$.

§ 3. Reductions

Let $\mathfrak{M}=(\mathcal{K},\mathcal{M},H)$ be a matrix problem, and $(\mathcal{A},\mathcal{V})$ be the basis of $(\mathcal{M},\operatorname{rad}\mathcal{K})$ defined in Section 1. Suppose that e_X and a_i are indeterminates of domain k for $X\in\mathcal{T}$ and $A_i\in\mathcal{V}_{X_iY_i}$, $1\leqslant i\leqslant n$, v_j are indeterminates of domain $D_{X'_j}\otimes_k D_{Y'_j}$ for $V_j\in\mathcal{V}_{X'_jY'_j}$, $1\leqslant j\leqslant m$. We take

$$R=\sum_{X\in\mathcal{T}}e_X*E_X+\sum_{j=1}^{m}v_j*V_j,\ N=\sum_{i=1}^{n}a_i*A_i,$$

then R is called a generic matrix of $\mathcal{K}$ and N is called a generic matrix of $\mathcal{M}$. Consider the matrix equation

$$(N+H)R=R(N+H). \tag{3.1}$$

Write $R=(r_{ij})_{t\times t}$ and $N=(n_{ij})_{t\times t}$. Suppose that the leading position of A_1 is (p,q) and $(p,q)\in X\times Y$, then Assumption 1.4 (2.4) gives rise to $HR\equiv_{<(p,q)}RH$, and the equation (3.1) yields an equation at the (p,q)-entry:

$$h_{p1}r_{1q}+\cdots+h_{p,q-1}r_{q-1,q}+h_{pq}r_{qq}+n_{pq}r_{qq}$$
$$=r_{pp}n_{pq}+r_{pp}h_{pq}+r_{p,p+1}h_{p+1,q}+\cdots+r_{pt}h_{tq}.$$

Since $h_{pq}=0$, and $h_{ij}=0$ whenever $i\not\prec j$ by Definition 1.3, the equation is just as follows:

$$n_{pq}r_{qq}-r_{pp}n_{pq}=\sum_{l=p+1}^{t}r_{pl}h_{lq}-\sum_{l=1}^{q-1}h_{pl}r_{lq}. \tag{3.2}$$

Let us define a new equation

$$\sum_{l=p+1}^{t}z_{pl}h_{lq}-\sum_{l=1}^{q-1}h_{pl}z_{lq}=0. \tag{3.3}$$

In particular if $\mathcal{V}_{XY}=\{\mathbf{V}_1,\mathbf{V}_2,\cdots,\mathbf{V}_l\}$, and denote x_X by x, x_Y by y, in the case that X or Y is non-trivial. Then Formula (3.2) can also be written as

$$a_1e_Y-e_Xa_1=n_{pq}r_{qq}-r_{pp}n_{pq}=f_1(x,y)v_1+f_2(x,y)v_2+\cdots+f_l(x,y)v_l, \tag{3.4}$$

where x and y stand for the left and right multiplications respectively, x does not occur if X is trivial, and y does not occur if Y is trivial. Then we have the following 6 reductions recalling from [1,2,5,6].

Case 3.1 Loop reduction

When $X=Y$ is trivial, and Formula (3.4) can be written as $n_{pq}r_{qq}-r_{pp}n_{pq}=0$, we have $r_{pp}=r_{qq}$, let

$$X'_{r+1}=X,\quad x'_{X'_{r+1}}=x,\quad \bar{n}_{pq}=(x)_{1\times1}.$$

We define an induced matrix problem $\mathfrak{M}'=(\mathcal{K}',\mathcal{M}',H')$ as follows. Set $T'=T$ and $\mathcal{T}'=\mathcal{T}$; $\mathbf{D}'_{X'_1}=\mathbf{D}_{X_1},\cdots,\mathbf{D}'_{X'_r}=\mathbf{D}_{X_r},\mathbf{D}'_X=k$; $\mathcal{M}'=\{\sum_{i=2}^{n}\mathbf{C}_i*\mathbf{A}_i\}$ with $\mathbf{C}_i=(c_i)_{1\times1}$ and $c_i\in k$, thus $\mathbb{E}'=\mathbb{E}\cup\{y_{pq}=0\}$; $\mathcal{K}'=\mathcal{K}$, thus $\mathbb{F}'=\mathbb{F}$; finally $\mathbf{H}'=\mathbf{H}+\bar{n}_{pq}*\mathbf{A}_1=\mathbf{H}+x\mathbf{A}_1$.

Case 3.2 Base change

If X, Y are both trivial, Formula (2.10) is $n_{pq}r_{qq}-r_{pp}n_{pq}=\sum_{j=1}^{l}f_jv_j$, with $f_j\in k$ being not all zero. Then there exists an invertible matrix $\boldsymbol{Q}$ over k with the first row $(f_1, f_2, \cdots, f_l)$.

If X is non-trivial, Y is trivial (or dual), Formula (3.4) is $n_{pq}r_{qq} - r_{pp}n_{pq} = \sum_{j=1}^{l} f_j(x)v_j$ with $f_j(x) \in k[x, g_X(x)^{-1}]$ being not all zero. Without loss of generality we may divide v_j by some power of $g_X(x)$, such that $f_j(x) \in k[x]$, which is an Euclidean domain. Suppose that the highest common factor of $f_j(x)$ for $1 \leqslant j \leqslant l$ is $c(z) \in k[x]$. Let $D'_X = D_X \setminus \{$the roots of $c(x)\}$. There exist some $q_j(x) \in k[x]$, such that $\sum_{j=1}^{l} f_j(x)q_j(x) = c(x)$. Therefore we have an invertible matrix $\boldsymbol{Q}$ with the first row $(f_1(x), f_2(x), \cdots, f_l(x))$ over the Hermite ring $k[x, c(x)^{-1}]$, since $\sum_{j=1}^{l} f_j(x)(c(x)^{-1}q_j(x)) = 1$.

If X and Y are both non-trivial, Formula (2.10) is $n_{pq}r_{qq} - r_{pp}n_{pq} = \sum_{j=1}^{l} f_j(x,y)v_j$, with $f_j(x,y) \in k[x, y, g_X(x)^{-1}, g_Y(y)^{-1}]$ being not all zero. Without loss of generality we may divide v_j by some power of $g_X(x)g_Y(y)$, such that $f_j(x,y) \in k[x,y]$ which is a unique factorization domain. Assume that the highest common factor of $f_j(x, y)$, for $j = 1, 2, \cdots, l$, equals $d(x,y) \in k[x,y]$. We assume in addition that

$$d(x,y) = d_1(x)d_2(y). \tag{3.5}$$

Thus $f_j(x,y) = d(x,y)h_j(x,y)$ with $h_j(x,y)$ is co-prime for $1 \leqslant j \leqslant l$ in $k[x,y]$. Since $k(x)[y]$ is an Euclidean domain, there exist some $q_j(x,y) \in k[x,y]$, such that

$$c(x) = \sum_{j=1}^{l} h_j(x,y)q_j(x,y) \tag{3.6}$$

for some $c(x) \in k[x]$. Define $D'_X = D_X \setminus \{$the roots of $c(x)d_1(x)\}$ and $D'_Y = D_Y \setminus \{$the roots of $d_2(y)\}$. Since $k[x, y, (c(x)d_1(x))^{-1}, d_2(y)^{-1}]$ is a Hermit ring and

$$\sum_{j=1}^{l} f_j(x,y)(c(x)^{-1}d(x,y)^{-1}q_j(x,y)) = \sum_{j=1}^{l} h_j(x,y)(c(x)^{-1}q_j(x,y)) = 1, \tag{3.7}$$

there exists an invertible matrix Q with the first row $(f_1(x,y), f_2(x,y), \cdots, f_l(x,y))$ over

$$k[x,y,(c(x)d_1(x))^{-1},d_2(y)^{-1}].$$

In the above 3 cases, let $(v_1', v_2', \cdots, v_l')^{\mathrm{T}} = \boldsymbol{Q}(v_1, v_2, \cdots, v_l)^{\mathrm{T}}$, where $v_j' = v_j$ for $l < j \leqslant m$.

Define a matrix problem $\mathfrak{M}' = (\mathcal{M}', \mathcal{K}', H')$, such that $T' = T$, $\sim' = \sim$, $D_Z' = D_Z$ for $Z \neq X, Y$, and D_X', D_Y' are defined above. $\mathcal{M}' = \mathcal{M}$ thus $\mathbb{E}' = \mathbb{E}$, $\mathcal{K}'$ has a basis $\{\boldsymbol{V}_1', \boldsymbol{V}_2', \cdots, \boldsymbol{V}_l', \cdots, \boldsymbol{V}_m'\}$ with $(\boldsymbol{V}_1, \boldsymbol{V}_2, \cdots, \boldsymbol{V}_l) = (\boldsymbol{V}_1, \boldsymbol{V}_2, \cdots, \boldsymbol{V}_l)\boldsymbol{Q}^{-1}$, and $\boldsymbol{V}_j' = \boldsymbol{V}_j$ for $l < j \leqslant m$, thus $\mathbb{F}' = \mathbb{F}$. $\boldsymbol{H}' = \boldsymbol{H}$ with the different domains of x_X', x_Y'. In particular, $n_{pq}' r_{qq}' - r_{pp}' n_{pq}' = v_1'$.

Remark We stress that the assumption given in Formula (2.11) is necessary, otherwise we are not able to obtain any induced matrix problem.

Case 3.3 Regularization

If Formula (3.4) can be written as $n_{pq} r_{qq} - r_{pp} n_{pq} = v_1$ for some $V_1 \in \mathcal{V}_{XY}$, then let $\bar{n}_{pq} = 0$. And denote zero here by $\varnothing$, i.e. $\bar{n}_{pq} = \varnothing$.

We define an induced matrix problem $\mathfrak{M}' = (\mathcal{K}', \mathcal{M}', H')$ as follows. Let $T' = T$, and $\sim' = \sim$; $\mathbb{E}' = \mathbb{E} \cup \{y_{pq} = 0\}$, i.e. $\mathcal{M}' = \{\sum_{i=2}^{n} \boldsymbol{C}_i * \boldsymbol{A}_i\}$ with $\boldsymbol{C}_i = (c_i)$ of 1×1 matrices; $\mathbb{F} = \mathbb{F} \cup \{$Formula (3.3)$\}$, thus $\mathcal{K}'$ is a subset of $\mathcal{K}$; $\boldsymbol{H}' = \boldsymbol{H} + \bar{n}_{pq} * \boldsymbol{A}_1 = \boldsymbol{H}$.

Case 3.4 Deletion

Let $\mathfrak{M} = (\mathcal{K}, \mathcal{M}, \boldsymbol{H})$ be a matrix problem defined in Definition 1.5. Suppose that $\mathcal{T}_0 = \{Y_1, Y_2, \cdots, Y_r\} \subset \mathcal{T}$. Let $\mathcal{T}' = \mathcal{T} \backslash \mathcal{T}_0$, $T' = \{i' \mid i \in X \in \mathcal{T}'\}$, $D_X' = D_X$ for $X \in \mathcal{T}'$. Define a size vector $\underline{n}$ over $(T, \sim)$ with $n_i = 1$ if $i \in T'$ and $n_i = 0$ if $i \notin T'$. Let $\mathcal{M}' = \mathcal{M}_{\underline{n}}$, $\mathcal{K}' = \mathcal{K}_{\underline{n}}$ and $H' = H_{\underline{n}}$, thus $\mathbb{E}_{XY}' = \mathbb{E}_{XY}$, $\mathbb{F}_{XY}' = \mathbb{F}_{XY}$, whenever $X, Y \in \mathcal{T}'$. Then we obtain an induced matrix problem $\mathfrak{M}' = (\mathcal{K}', \mathcal{M}', H')$.

Case 3.5 Edge reduction

When Formula (3.4) is $n_{pq} r_{qq} - r_{pp} n_{pq} = 0$, and $X \neq Y$ is trivial, let $\bar{\boldsymbol{n}}_{pq} = \begin{pmatrix} 0 & 1 \\ 0 & 0 \end{pmatrix}_{2\times 2}$. We first claim that if $\bar{\boldsymbol{n}}_{pq} \boldsymbol{S}_{qq} = \boldsymbol{S}_{pp} \bar{\boldsymbol{n}}_{pq}$ with $\boldsymbol{S}_{pp}$ and $\boldsymbol{S}_{qq}$

2×2 matrices respectively, then

$$\boldsymbol{S}_{pp}=\begin{pmatrix} s_{p_1p_1} & s_{p_1p_2} \\ 0 & s_{p_2p_2} \end{pmatrix},\quad \boldsymbol{S}_{qq}=\begin{pmatrix} s_{q_1q_1} & s_{q_1q_2} \\ 0 & s_{q_2q_2} \end{pmatrix},\quad s_{p_1p_1}=s_{q_2q_2}.$$

Next we define an induced matrix problem as follows. Set $T'=\{1', 2',\cdots,t'\}$ with $t'=t+|X|+|Y|$. Moreover, the index i splits to two indices $\{i_1,i_2\}$ for any $i\in X$, similarly j splits to two indices $\{j_1,j_2\}$, $\forall j\in Y$, and l keeps unchanged, $\forall l\in Z$ for $Z\neq X,Y$. There is a one-to-one correspondence between T' and the set of the splitting indices preserving the orders. Let $\mathcal{T}'=(\mathcal{T}\setminus\{X,Y\})\cup\{X_2,Y_1,X_1\cup Y_2\}$ with $X_2=\{i_2\mid \forall i\in X\}$, $Y_1=\{j_1\mid \forall j\in Y\}$ and $X_1\cup Y_2=\{i_1,j_2\mid \forall i\in X, j\in Y\}$.

Take a size vector $\underline{\boldsymbol{n}}=(n_1,n_2,\cdots,n_t)$, such that $n_Z=1$ if $Z\neq X,Y$, and $n_X=n_Y=2$. Set $\mathcal{M}'=\{\sum_{i=2}^{n}\boldsymbol{C}_i*\boldsymbol{A}_i\}$, where $\boldsymbol{A}_i\in\mathcal{A}_{X_iY_i}$ for $X_i,Y_i\in\mathcal{T}$, $\boldsymbol{C}_i=(c_i)_{1\times 1}$ if $X_i,Y_i\notin\{X,Y\}$; $\boldsymbol{C}_i=(c_{11}\ c_{12})_{1\times 2}$ if $X_i\notin\{X,Y\}$, but $Y_i\in\{X,Y\}$, $\boldsymbol{C}_i=\begin{pmatrix} c_{11} \\ c_{21}\end{pmatrix}_{2\times 1}$ if $X_i\in\{X,Y\}$ but $Y_i\notin\{X,Y\}$; $\boldsymbol{C}_i=\begin{pmatrix} c_{11} & c_{12} \\ c_{21} & c_{22}\end{pmatrix}_{2\times 2}$ if $X_i,Y_i\in\{X,Y\}$, all the entries of the matrices are taken from k. The equation system $\mathbb{E}'$ is obtained from $\mathbb{E}$ by a suitable refinement plus $\{y_{p_iq_j}=0\mid 1\leqslant i,j\leqslant 2\}$.

Let $\mathcal{K}'=\{\sum_{Z}\boldsymbol{B}_Z*\boldsymbol{E}_Z+\sum_{j=2}^{m}\boldsymbol{B}_j*\boldsymbol{V}_j\}$, where $\boldsymbol{B}_Z=(d)_{1\times 1}$ if $Z\neq X$, Y and $\boldsymbol{B}_X=\begin{pmatrix} b_{11} & b_{12} \\ 0 & b_{22}\end{pmatrix}_{2\times 2}$, $\boldsymbol{B}_Y=\begin{pmatrix} d_{11} & d_{12} \\ 0 & b_{11}\end{pmatrix}_{2\times 2}$ with entries in k, the meaning of $\boldsymbol{B}_j$ is similar to that of $\boldsymbol{C}_i$, but the entries of $\boldsymbol{B}_j$ are taken from k, or $k[x,g(x)^{-1}]$, or $k[x,y,g(x)^{-1},g(y)^{-1}]$. $\mathbb{F}'$ is obtained from $\mathbb{F}$ by a suitable refinement plus $\{z_{i_1i_2}-z_{p_1p_2}=0, z_{j_1j_2}-z_{q_1q_2}=0\mid \forall i\in X, j\in Y\}$.

Remark The solution space of $\mathbb{F}'$ has a basis $\{\boldsymbol{V}_l^{\xi\eta}\mid l=1,2,\cdots,m\}$. Let $\boldsymbol{V}_l\in\mathcal{V}_{X_lY_l}\subset\mathcal{V}$, then

(1) $\zeta=1,\eta=1$ for $X_l,Y_l\notin\{X,Y\}$;

(2) $\zeta=1,\eta=1,2$ for $X_l\notin\{X,Y\},Y_l\in\{X,Y\}$;

(3) $\zeta=1,2,\eta=1$ for $X_l\in\{X,Y\},Y_l\notin\{X,Y\}$;

(4) $1\leqslant\zeta,\eta\leqslant 2$ for $X_l,Y_l\in\{X,Y\}$.

Finally $\boldsymbol{H}'=\boldsymbol{H}_{\underline{m}}+\bar{\boldsymbol{n}}_{pq}*\boldsymbol{A}_1$.

Case 3.6 Unraveling

Let $\mathfrak{M}=(\mathcal{K},\mathcal{M},H)$ be a matrix problem defined in Definition 1.5. Fix an integer $1\leqslant j\leqslant r$ given in Definition 1.3. Denote X_j by X, D_{X_j} by D_X, A_j^0 by A^0, x_j by x for short.

Suppose that $\boldsymbol{W}=\boldsymbol{W}(\lambda_1)\oplus\boldsymbol{W}(\lambda_2)\oplus\cdots\oplus\boldsymbol{W}(\lambda_s)$ is the Weylr matrix similar to a Jordan form $\sum\limits_{j=1}^{s}\sum\limits_{i=1}^{d}\boldsymbol{J}_i(\lambda_j)$ with $\lambda_j\in D_X$ for $j=1,2,\cdots,s$, and $\boldsymbol{U}=\boldsymbol{W}\oplus(x')$. Let $m_0=0$,

$$m_j=d+(d-1)+\cdots+(d-j+1),\quad j\geqslant 1,$$

thus $m_1=d,m_2=2d-1,\cdots,m_d=\frac{1}{2}d(d+1)$. Write $m=m_d$ and $n_X=sm+1$. We give a basis of the subspace of $\boldsymbol{M}_{n_X(k)}$, whose elements commute with $\boldsymbol{U}$:

$\{\boldsymbol{B}(l,p,i,j)\mid i-j\leqslant p-1,1\leqslant i\leqslant d,1\leqslant j\leqslant d-p+1;1\leqslant p\leqslant d;1\leqslant l\leqslant s\}\cup\{E_{n_Xn_X}\}$,

$$\boldsymbol{B}(l,p,i,j)=\sum_{\zeta=0}^{d-p-j+1}\boldsymbol{E}_{k_l+m_\zeta+i,k_l+m_{p+\zeta-1}+j}$$

with $k_l=(l-1)m$. In particular,

$$\{\boldsymbol{B}(l,i)=\boldsymbol{B}(l,1,i,i)\mid 1\leqslant i\leqslant d,1\leqslant l\leqslant s\}\cup\{\boldsymbol{E}_{n_Xn_X}\}$$

is the set of idempotent.

Next we define an induced matrix problem $\mathfrak{M}'=(\mathcal{K}',\mathcal{M},H')$ as follows.

Let $T'=\{1',2',\cdots,t'\}$ for $t'=t+|X|(n_X-1)$. On the other hand we define a set T'' as follows: the index u splits into a set of indices $\{u_1,u_2,\cdots,u_{n_X}\}$ for each $u\in X$, and the index u keeps unchanged for $u\in T\backslash\{X\}$. There is a one-to-one correspondence between T' and T'' preserving the orders

$$\mathcal{T}''=(\mathcal{T}\setminus\{X\})\cup\{X_{(l,i)}\mid 1\leqslant i\leqslant d,1\leqslant l\leqslant s\}\cup\{X'\},$$

where $X_{l,i}=\{u_{k_l+m_0+i},\cdots,u_{k_l+m_{d-i}+i}\mid \forall u\in X\}$ and $X'=\{u_{n_X}\mid \forall u\in X\}$ with the domain $D'_{X'}=D_X\setminus\{\lambda_1,\lambda_2,\cdots,\lambda_s\}$, but other parameters do not change their domains.

Define a size vector $\underline{\boldsymbol{n}}=(n_1,n_2,\cdots,n_t)$ such that $n_u=1$ for $u\notin X$ and $n_u=n_X$ for $u\in X$. Then $\mathcal{M}'=\{\sum\limits_{i=1}^{n}\boldsymbol{C}_i*\boldsymbol{A}_i\}$, where $\boldsymbol{A}_i\in\mathcal{A}_{X_i,Y_i}$, $C_i=(c_i)_{1\times1}$ if $X_i,Y_i\neq X$; $C_i=(c_{11},c_{12},\cdots,c_{1n_X})_{1\times n_X}$ if $X_i\neq X$, but $Y_i=X$; $\boldsymbol{C}_i=(c_{11},c_{21},\cdots,c_{n_X,1})^{\mathrm{T}}_{1\times n_X}$ with T standing for the transpose, if $X_i=X$ but $Y_i\neq X$; $\boldsymbol{C}_i=(\boldsymbol{C}_{rs})_{n_X\times n_X}$ if $X_i=Y_i=X$, all the entries of the matrices are taken from the base field k. Equation system $\mathbb{E}'$ is obtained from $\mathbb{E}$ by a suitable refinement plus $\{y_{p_iq_j}=0\mid 1\leqslant i,j\leqslant n_X\}$.

$\mathcal{K}'=\{(\boldsymbol{S}_{ij})\}$ where $(\boldsymbol{S}_{ij})$ is an upper triangular partitioned matrix of size $\underline{\boldsymbol{n}}\times\underline{\boldsymbol{n}}$, such that $\boldsymbol{S}_{ii}=\boldsymbol{S}_{jj}$ if $i\sim j$, $\boldsymbol{S}_{ii}$ is a 1×1 matrix over k if $i\in\mathbf{Z}\neq X$; $\boldsymbol{S}_{ii}$ is a matrix of size n_X over k commuting with $\boldsymbol{U}$ if $i\in X$. S_{ij} for $i<j$ satisfies the equation $\mathbb{F}(\boldsymbol{U},\boldsymbol{U}')$, which is obtained from $\mathbb{F}$ by substituting $\boldsymbol{U},\boldsymbol{U}'$ for x,y respectively, where $\boldsymbol{U}'=\boldsymbol{U}$.

Remark The solution space of $\mathbb{F}(\boldsymbol{U},\boldsymbol{U}')$ has a basis $\{\boldsymbol{V}_l^{\zeta\eta}\mid l=1,2,\cdots,m\}$. In fact, Corollary 2.6 tells the solution of the equations $\mathbb{F}(\boldsymbol{U},\boldsymbol{U}')$. Recall that for $\boldsymbol{V}_l\in\mathcal{V}_{X_lY_l}\subset\mathcal{V}$, we have $\zeta=1,\eta=1$ if $X_l,Y_l\neq X$; $\zeta=1,1\leqslant\eta\leqslant n_X$ if $X_l\neq X,Y_l=X$; $1\leqslant\zeta\leqslant n_X,\eta=1$ if $X_l=X,Y_l\neq X$; and $1\leqslant\zeta,\eta\leqslant n_X$ if $X_l=Y_l=X$. The entries of each base element are taken from k, or $k[y',g'_{Y'}(y')^{-1}]$, or $k[y',z',g'_{Y'}(y')^{-1},g'_{Z'}(z')^{-1}]$. And $\mathbb{F}'$ consists of $\mathbb{F}(U,U')$ plus the equations:

$$\{z_{u_{k_l}+m_\zeta+i,u_{k_l}+m_{p+\zeta-1}+j}-z_{p_{k_l}+i,p_{k_l}+m_p+j}=0\mid i-j\leqslant p-1\}\cup$$
$$\{z_{u_{k_l}+m_\zeta+i,u_{k_l}+m_{p+\zeta-1}+j}=0\mid i-j>p-1\},$$

for all $0\leqslant\zeta\leqslant d-p-j+1$; $1\leqslant i\leqslant d$, $1\leqslant j\leqslant d-p+1$; $1<p\leqslant d$; $1\leqslant l\leqslant s$; $u\in X$.

Finally, $\boldsymbol{H}'=(\boldsymbol{H}-x\boldsymbol{A}^0)_{\underline{n}}+\boldsymbol{W}*\boldsymbol{A}^0$.

Lemma 3.7 Let $\mathfrak{M}=(\mathcal{K},\mathcal{M},H)$ be a matrix problem. Suppose that $(p,q)\in X\times Y$ is the leading position of A_1 with X or Y being non-trivial, or X, Y being both non-trivial. It is possible to make a reduction, only in case what Formula (3.4) must satisfy:

$$n_{pq}r_{qq}-r_{pp}n_{pq}=f_1(x,y)v_1+f_2(x,y)v_2+\cdots+f_l(x,y)v_l\neq 0.$$

In this case, suppose that the solution space of equations $\mathbb{F}_{XY}$ has a basis

$$\mathbf{V}_j(x,y)=(f_1^j(x,y),f_2^j(x,y),\cdots,f_s^j(x,y))^{\mathrm{T}}$$

for $j=1,2,\cdots,l$ given in Case 3.2, and let $\boldsymbol{Q}^{-1}=\boldsymbol{P}(x,y)=(g_{ij}(x,y))_{l\times l}$ over $k[x,y,g_1(x)^{-1},g_2(y)^{-1}]$ be invertible, and $(\mathbf{V}_1',\mathbf{V}_2',\cdots,\mathbf{V}_l')=(\mathbf{V}_1,\mathbf{V}_2,\cdots,\mathbf{V}_l)\boldsymbol{P}$ be a base change. Recall from Corollary 2.6, if $\boldsymbol{W},\boldsymbol{U}$ of size m,n are Weyr matrices, such that the eigenvalues of $\boldsymbol{W}$ and $\boldsymbol{U}$ are not the roots of $g_1(x)$ and $g_2(y)$ respectively. Let

$$\left(\sum_j(f_1^j g_{ji})(\boldsymbol{W},\boldsymbol{U})\circ\boldsymbol{R}'_i,\cdots,\sum_j(f_s^j g_{ji})(\boldsymbol{W},\boldsymbol{U})\circ\boldsymbol{R}'_i\right)^{\mathrm{T}}=\sum_{\zeta,\eta}r'^{i}_{\zeta\eta}\mathbf{V}'^{\zeta\eta}_i.$$

Then the $s\times 1$ block-matrix $\{\mathbf{V}'^{\zeta\eta}_i\mid 1\leqslant\zeta\leqslant m,1\leqslant\eta\leqslant n,1\leqslant i\leqslant l\}$ as the coefficient of $r'^{i}_{\zeta\eta}$ in the above formula is also a basis of Formulae (2.5).

Proof Similar to the proof of Lemma 2.5, the mnl matrices $\{\mathbf{V}'^{\zeta\eta}_l\}$ are linearly independent. On the other hand the dimension of the solution space equals mnl, therefore the set of the matrices is also a basis of Formulae (2.5), the lemma is proved.

Theorem 3.8 Let $\mathfrak{M}$ be a matrix problem with $(p,q)\in X\times Y$ being the leading position of $\boldsymbol{A}_1$, $\mathfrak{M}'$ be the induced matrix problem given by reductions 3.1～3.6.

(1) There exists a reduction functor ϑ: Mat$(\mathfrak{M}')\to$ Mat$(\mathfrak{M})$ by sending any object $(\boldsymbol{M}',\boldsymbol{H}'_{\underline{m}'}(\underline{\boldsymbol{W}}'))\in$ Mat$(\mathfrak{M}')$ to

$$(\boldsymbol{M},H_{\underline{m}}(\underline{\boldsymbol{W}}))=(\boldsymbol{M}'+\boldsymbol{H}'_{\underline{m}'}(\underline{\boldsymbol{W}}')_{pq}*\boldsymbol{A}_1,\boldsymbol{H}'_{\underline{m}'}(\underline{\boldsymbol{W}}')-\boldsymbol{H}'_{\underline{m}'}(\underline{\boldsymbol{W}}')_{pq}*\boldsymbol{A}_1)$$

in Mat$(\mathfrak{M})$, which acts on the morphisms identically. Thus

$$\boldsymbol{M}+\boldsymbol{H}_{\underline{m}}(\underline{\boldsymbol{W}})=\boldsymbol{M}'+\boldsymbol{H}_{\underline{m}'}(\underline{\boldsymbol{W}}').$$

(2) The functor ϑ is fully faithful. In fact, for any $\boldsymbol{S}\in\mathcal{K}_{\underline{m}\times\underline{n}}(\underline{\boldsymbol{W}},\underline{\boldsymbol{U}})$, if $\boldsymbol{H}'_{\underline{m}'}(\underline{\boldsymbol{W}}')\boldsymbol{S}\equiv_{\leq(pq)}\boldsymbol{S}\boldsymbol{H}'_{\underline{n}'}(\underline{\boldsymbol{U}}')$, then $\boldsymbol{S}\in\mathcal{K}'_{\underline{m}'\times\underline{n}'}(\underline{\boldsymbol{W}}',\underline{\boldsymbol{U}}')$.

(3) The size vector $\underline{\boldsymbol{m}}=\underline{\boldsymbol{m}}'$ in Cases 3.1～3.4. And $m_Z=m'_Z$ for $Z\neq X,Y$; $m_X=m'_{X_2}+m'_{X_1\cup Y_2}$; $m_Y=m'_{Y_1}+m'_{X_1\cup Y_2}$, in Case 3.5. $m_Z=m'_Z$ for $Z\neq X$ and $m_X=\sum_{l,i}m'_{l,i}+m'_{X'}$ in Case 3.6.

Proof (1) and (3) are clear.

(2) Suppose that $(\boldsymbol{M}, \boldsymbol{H}_{\underline{m}}(\underline{\boldsymbol{W}})) = \vartheta(\boldsymbol{M}', \boldsymbol{H}'_{\underline{m}'}(\underline{\boldsymbol{W}}'))$ and $(\boldsymbol{N}, \boldsymbol{H}_{\underline{n}}(\underline{\boldsymbol{H}})) = \vartheta(\boldsymbol{N}', \boldsymbol{H}'_{\underline{n}'}(\underline{\boldsymbol{U}}'))$. If $\boldsymbol{S} \in \mathcal{K}_{\underline{m} \times \underline{n}}(\underline{\boldsymbol{W}}, \underline{\boldsymbol{U}})$ is a morphism in Mat$(\mathfrak{M})$, we have $(\boldsymbol{M} + \boldsymbol{H}_{\underline{m}}(\underline{\boldsymbol{W}}))\boldsymbol{S} = \boldsymbol{S}(\boldsymbol{N} + \boldsymbol{H}_{\underline{n}}(\underline{\boldsymbol{U}}))$, therefore $(\boldsymbol{M}' + \boldsymbol{H}'_{\underline{m}'}(\underline{\boldsymbol{W}}'))\boldsymbol{S} = \boldsymbol{S}(\boldsymbol{N}' + \boldsymbol{H}'_{\underline{n}'}(\underline{\boldsymbol{U}}'))$ by (2.1). Comparing the (p, q)-blocks in the two sides of the matrix equation, we have

$$\boldsymbol{H}'_{\underline{m}'}(\underline{\boldsymbol{W}}')_{pq}\boldsymbol{S}_{qq} - \boldsymbol{S}_{pp}\boldsymbol{H}'_{\underline{m}'}(\underline{\boldsymbol{W}}')_{pq} = \sum_{l=p+1}^{t} \boldsymbol{S}_{pl}h_{lq} - \sum_{l=1}^{q-1} h_{pl}\boldsymbol{S}_{lq}. \tag{3.8}$$

In Case 3.1, $\mathcal{K}' = \mathcal{K}$.

In Case 3.2, $\boldsymbol{S}$ can be expressed by the new basis $\mathcal{V}' = \{\boldsymbol{V}'_1, \boldsymbol{V}'_2, \cdots, \boldsymbol{V}'_m\}$ according to Lemma 3.7.

In Case 3.3, Formula (3.8) becomes $\sum_{l=p+1}^{t} \boldsymbol{S}_{pl}hl_q - \sum_{l=1}^{q-1} h_{pl}\boldsymbol{S}_{lq} = 0$, i.e. the matrix $\boldsymbol{S}$ satisfies Formula (3.3).

In Case 3.4,

$$\boldsymbol{M} = \boldsymbol{M}', \boldsymbol{H}_{\underline{m}}(\underline{\boldsymbol{W}}) = \boldsymbol{H}_{\underline{m}'}(\underline{\boldsymbol{W}}'), \text{and } \boldsymbol{N} = \boldsymbol{N}', \boldsymbol{H}_{\underline{n}}(\underline{\boldsymbol{U}}) = \boldsymbol{H}_{\underline{n}'}(\underline{\boldsymbol{U}}').$$

In Case 3.5, Formula (3.8) becomes $\begin{pmatrix} \boldsymbol{0} & \boldsymbol{I} \\ \boldsymbol{0} & \boldsymbol{0} \end{pmatrix} \boldsymbol{S}_{qq} = \boldsymbol{S}_{pp} \begin{pmatrix} \boldsymbol{0} & \boldsymbol{I} \\ \boldsymbol{0} & \boldsymbol{0} \end{pmatrix}$, where the identity matrix $\boldsymbol{I}$ has a size of $m'_{X_1 \cup Y_2}$.

In Case 3.6, Formula (3.8) becomes $\boldsymbol{U}\boldsymbol{S}_{qq} = \boldsymbol{S}_{pp}\boldsymbol{U}$.

Therefore after any reduction from Case 3.1 to 3.6, $\boldsymbol{S} \in \mathcal{K}'_{\underline{m}' \times \underline{n}'}(\underline{\boldsymbol{W}}', \underline{\boldsymbol{U}}')$, ϑ is full. ϑ is obviously faithful. The theorem is proved.

The composition of the reduction functors defined in Theorem 3.8 is also called a reduction functor.

§ 4. Canonical forms

Definition 4.1 A matrix problem $\mathfrak{M} = (\mathcal{M}, \mathcal{K}, H)$ is said to be minimal, provided $\mathcal{M} = \{(0)\}$.

A minimal matrix problem $\mathfrak{M}$ is said to be local, if $\mathfrak{M}$ has only one equivalent class in T.

Lemma 4.2 Let $\mathfrak{M} = (\mathcal{M}, \mathcal{K}, H)$ be a minimal matrix problem with $\mathcal{M}=\{(0)\}$ and $\boldsymbol{H}=\boldsymbol{H}^0+x_1\boldsymbol{A}_1^0+x_2\boldsymbol{A}_2^0+\cdots+x_r\boldsymbol{A}_r^0$ given in Definition 1.3. Then the indecomposable matrices of $\mathfrak{M}$ are given by the following formula:

$$\{(0,\boldsymbol{H}_{\underline{\boldsymbol{m}}_{(X,d)}}(\boldsymbol{J}_d(\lambda)))\mid\forall\lambda\in D_X,d\in\mathbf{N},\text{non-trivial }X\in\mathcal{T}\},$$
$$\{(0,\boldsymbol{H}_{\underline{\boldsymbol{m}}_Y})\mid\forall\ \text{trivial }Y\in\mathcal{T}\},$$

where the matrices of two types have respectively the size vectors $\underline{\boldsymbol{m}}_{(X,d)}$ with the component d at X and 0 at $Z\neq X$; $\underline{\boldsymbol{m}}_Y$ with the component 1 at Y and 0 at $Z\neq Y$.

Proposition 4.3 Let $\mathfrak{M}=(\mathcal{M},\mathcal{K},H)$ be a matrix problem with trivial index set $(T,\sim)$. Suppose that $(\boldsymbol{M},\boldsymbol{H}_{\underline{\boldsymbol{m}}})\in\mathrm{Mat}(\mathfrak{M})$, then there exists a unique sequence of matrix problems with trivial index sets:

$$\mathfrak{M},\mathfrak{M}^1,\cdots,\mathfrak{M}^s,\tag{4.1}$$

given by deletion from $\mathfrak{M}$ to $\mathfrak{M}^1$; by one of the reductions 3.5+3.4, 3.1+3.6+3.4, or 3.2+3.3 from $\mathfrak{M}^l$ to $\mathfrak{M}^{l+1}$ for $1\leqslant l<s$; and the end term $\mathfrak{M}^s$ is minimal. There also exists a sequence of matrices:

$$(\boldsymbol{M},\boldsymbol{H}_{\underline{\boldsymbol{m}}}),(\boldsymbol{M}^1,\boldsymbol{H}^1_{\underline{\boldsymbol{m}}^1}),\cdots,(\boldsymbol{M}^s,\boldsymbol{H}^s_{\underline{\boldsymbol{m}}^s}),\tag{4.2}$$

where $(\boldsymbol{M}^l,\boldsymbol{H}^l_{\underline{\boldsymbol{m}}^l})\in\mathrm{Mat}(\mathfrak{M}^l)$ is unique up to isomorphisms, $\underline{\boldsymbol{m}}_l$ is sincere over the index set of $\mathfrak{M}^l$, and $(\boldsymbol{M}^l,\boldsymbol{H}^l_{\underline{\boldsymbol{m}}^l})$ is isomorphic to $\vartheta^l(\boldsymbol{M}^{l+1},\boldsymbol{H}^l_{\underline{\boldsymbol{m}}^{l+1}})$ in $\mathrm{Mat}(\mathfrak{M}^l)$, with $\vartheta^l:\mathrm{Mat}(\mathfrak{M}^{l+1})\to\mathrm{Mat}(\mathfrak{M}^l)$ being the reduction functor for $0\leqslant l<s$.

Proof We first delete $\mathcal{T}_0=\{X\in\mathcal{T}\mid m_X=0\}$ to obtain a matrix problem $\mathfrak{M}^1$, then the size vector $\underline{\boldsymbol{m}}^1$ with $m_Y^1=m_Y$ for $Y\notin\mathcal{T}_0$ is sincere over the index set of $\mathfrak{M}^1$. It is clear that all the equivalent classes in $\mathcal{T}^1$ are trivial. Suppose that we have reached to $\mathfrak{M}^l$ and $(M^l,H^l_{\underline{\boldsymbol{m}}^l})$, where all the equivalent classes in $\mathcal{T}^l$ are trivial, and $\underline{\boldsymbol{m}}^l$ is sincere over $(T^l,\sim^l)$. Considering the reduction from $\mathfrak{M}^l$ to $\mathfrak{M}^{l+1}$, suppose the leading position of A_1 in $\mathfrak{M}^l$ is $(p,q)\in X\times Y$. Set

$$\boldsymbol{M}^l_{pq}\boldsymbol{S}_{qq}-\boldsymbol{S}_{pp}\boldsymbol{C}=\sum_{l=p+1}^{t}\boldsymbol{S}_{pl}h_{lq}-\sum_{l=1}^{q-1}h_{pl}\boldsymbol{S}_{\mathrm{lq}}.\tag{4.3}$$

If Formula (3.4) shows that $n_{pq}r_{qq}-r_{pp}n_{pq}=0$, then Formula (4.3) becomes $\boldsymbol{M}^l_{pq}\boldsymbol{S}_{qq}-\boldsymbol{S}_{pp}\boldsymbol{C}=0$.

In the case of $X \neq Y$, let $\boldsymbol{C}=\begin{pmatrix}\boldsymbol{0} & \boldsymbol{I}\\ \boldsymbol{0} & \boldsymbol{0}\end{pmatrix}$ with the size of $\boldsymbol{I}$ being equal to rank($\boldsymbol{M}^l_{pq}$). We make an edge reduction possibly following a deletion, then obtain $\mathfrak{M}^{l+1}$ with sincere $\underline{\boldsymbol{m}}^{l+1}$. Set

$$\boldsymbol{S}^l=\text{diag}\,(\boldsymbol{S}_{11},\boldsymbol{S}_{22},\cdots,\boldsymbol{S}_{tt})\in\mathcal{K}_{\underline{m}^l\times\underline{m}^l} \tag{4.4}$$

with $\boldsymbol{S}_{ii}=\boldsymbol{I}_{m_i}$ for $i\notin X\cup Y$, $\boldsymbol{S}_{ii}=\boldsymbol{S}_{pp}$ for $i\in X$ and $\boldsymbol{S}_{ii}=\boldsymbol{S}_{pp}$ for $i\in Y$, where $\boldsymbol{S}_{pp}$ and $\boldsymbol{S}_{qq}$ satisfy Formula (17).

In the case of $X=Y$, let $\boldsymbol{C}=\boldsymbol{W}_0$ be a Weyr matrix similar to $\boldsymbol{M}^l_{pq}$. We make a loop reduction following an unraveling, then a deletion is as follows. Let d be the largest size of the Jordan block appearing in the Jordan form of $\boldsymbol{M}^l_{pq}$, and let Weyr matrix $\boldsymbol{W}=\bigoplus\boldsymbol{W}(\lambda_j)$ have exactly the eigenvalues of $\boldsymbol{M}^l_{pq}$ given in Corollary 2.6. Then delete the parameter x' and all Jordan blocks which do not occur in the Jordan form of $\boldsymbol{M}^l_{pq}$. We obtain an induced matrix problem $\mathfrak{M}^{l+1}$ with domain $D_X=k\setminus\{$the eigenvalues of $\boldsymbol{M}^l_{pq}\}$ and a sincere size vector $\underline{\boldsymbol{m}}^{l+1}$. Recall that our base field equipped an order, the matrix problem $\mathfrak{M}^{l+1}$ here is uniquely determined. Set S^l as Formula (4.4) with $\boldsymbol{S}_{ii}=\boldsymbol{I}_{m_i}$ for $i\notin X$, $\boldsymbol{S}_{ii}=\boldsymbol{S}_{pp}$ for $i\in X$, where $\boldsymbol{S}_{pp}=\boldsymbol{S}_{qq}$ satisfies Formula (4.3).

If Formula (3.4) shows $n_{pq}r_{qq}-r_{pp}n_{pq}\neq 0$, let $\boldsymbol{C}=(\boldsymbol{0})_{m_p\times m_q}$. We make a base change then a regularization, since X, Y are both trivial. Set $\boldsymbol{S}^l=\boldsymbol{I}+\boldsymbol{S}'$ with $\boldsymbol{I}$ being a $t\times t$ identity matrix and $\boldsymbol{S}'$ being an upper triangular nilpotent $\underline{\boldsymbol{m}}^l\times\underline{\boldsymbol{m}}^l$-block matrix, such that the blocks $\boldsymbol{S}_{pl}$ and $\boldsymbol{S}_{lq}$ of $\boldsymbol{S}'$ satisfy the equation (4.3) and blocks $\boldsymbol{S}_{ij}=\boldsymbol{0}$ for any $i\neq p$, $j\neq q$.

Since the size of $\underline{\boldsymbol{m}}$ is fixed, we finally reach to a minimal matrix problem $\mathfrak{M}^s=(\mathcal{M}^s,\mathcal{K}^s,\boldsymbol{H}^s)$ by induction. The uniqueness of two sequences (4.1) and (4.2) is clear. For each $1\leqslant l\leqslant s$, we have constructed a matrix $\boldsymbol{S}^l\in\mathcal{K}^l_{\underline{m}^l}$ in the reduction from $\mathfrak{M}^l$ to $\mathfrak{M}^{l+1}$, such that

$$(\boldsymbol{M}^l+\boldsymbol{H}^l_{\underline{m}^l})\boldsymbol{S}^l=\boldsymbol{S}^l(\boldsymbol{M}^{l+1}+\boldsymbol{H}^{l+1}_{\underline{m}^{l+1}}). \tag{4.5}$$

□

Definition 4.4[5] Let $\mathfrak{M}=(\mathcal{K},\mathcal{M},\boldsymbol{H})$ with trivial $(T,\sim)$, and let $(\boldsymbol{M},\boldsymbol{H}_{\underline{m}})\in \mathrm{Mat}(\mathfrak{M})$. Then $\boldsymbol{H}^s_{\underline{m}^s}$ given in Proposition 4.3 is said to be the canonical form of $(\boldsymbol{M},\boldsymbol{H}_{\underline{m}})$, the pair of sequences (4.1) and (4.2) are said to be the pair of canonical sequences with respect to $(\boldsymbol{M},\boldsymbol{H}_{\underline{m}})$. Moreover

$$(\boldsymbol{M}+\boldsymbol{H}_{\underline{m}})\boldsymbol{S}=\boldsymbol{H}^s_{\underline{m}^s}S \text{ for } \boldsymbol{S}=\prod_{l=1}^{s-1}\boldsymbol{S}^l\in\mathcal{K}_{\underline{m}}$$

constructed in Proposition 4.3.

Our special interest goes to the case $\mathfrak{M}=(\mathcal{K},\mathcal{M},\boldsymbol{H}=\boldsymbol{0})$ and $\boldsymbol{M}\in\mathcal{M}_{\underline{m}}$, then $\boldsymbol{H}^s_{\underline{m}_s}$ is the canonical form of $(\boldsymbol{M},\boldsymbol{0})$ with $\boldsymbol{MS}=\boldsymbol{SH}^s_{m_s}$.

Corollary 4.5[5] Any isomorphic matrices in $\mathrm{Mat}(\mathfrak{M})$ have the same canonical form and the same sequence (4.1), also the same sequence (4.2) up to isomorphism.

Proof Suppose that $(\boldsymbol{M},\boldsymbol{H}_{\underline{m}})\simeq(\boldsymbol{N},\boldsymbol{H}_{\underline{n}})$ in $\mathrm{Mat}(\mathfrak{M})$. Then $\underline{\boldsymbol{m}}=\underline{\boldsymbol{n}}$ by Corollary 2.8, thus $(\boldsymbol{M}^1,\boldsymbol{H}^1_{\underline{m}^1})\simeq(\boldsymbol{N}^1,\boldsymbol{H}^1_{\underline{n}^1})$ in $\mathrm{Mat}(\mathfrak{M}^1)$, since, $\boldsymbol{M}=\boldsymbol{M}^1$, $\boldsymbol{N}=\boldsymbol{N}^1$ and $\boldsymbol{H}_{\underline{m}}=\boldsymbol{H}^1_{\underline{m}^1}$, $\boldsymbol{H}_{\underline{n}}=\boldsymbol{H}^1_{\underline{n}^1}$.

Now assume that we have reached to $\mathfrak{M}^l$ with $(p,q)\in X\times Y$ being the leading position of $\boldsymbol{A}_1$ in the normalized basis of $\mathcal{M}^l$, and $(\boldsymbol{M}^l,\boldsymbol{H}^l_{\underline{m}^l})\simeq(\boldsymbol{N}^l,\boldsymbol{H}^l_{\underline{n}^l})$ in $\mathrm{Mat}(\mathfrak{M}^l)$. Thus $\underline{\boldsymbol{m}}^l=\underline{\boldsymbol{n}}^l$, and there is some invertible $\boldsymbol{R}^l\in\mathcal{K}^l_{\underline{m}^l}$ with

$$(\boldsymbol{M}^l+\boldsymbol{H}^l_{\underline{m}^l})\boldsymbol{R}^l=\boldsymbol{R}^l(\boldsymbol{N}^l+\boldsymbol{H}^l_{\underline{n}^l}). \tag{4.6}$$

Since $\boldsymbol{H}^l_{\underline{m}^l}\boldsymbol{R}^l\equiv_{<(p,q)}\boldsymbol{R}^l\boldsymbol{H}^l_{\underline{n}^l}$), we have the (p,q)-block of the matrix equation (4.6):

$$\boldsymbol{M}^l_{pq}\boldsymbol{R}_{qq}-\boldsymbol{R}_{pp}\boldsymbol{N}^l_{pq}=\sum_{l=p+1}^{t}\boldsymbol{R}_{pl}h_{lq}-\sum_{l=1}^{q-1}h_{pl}\boldsymbol{R}_{lq}. \tag{4.7}$$

If Formula (3.4) shows that $n_{pq}r_{qq}-r_{pp}n_{pq}=0$, then (4.7) becomes $\boldsymbol{M}^l_{pq}\boldsymbol{R}_{qq}-\boldsymbol{R}_{pp}\boldsymbol{N}^l_{pq}=\boldsymbol{0}$ with $\boldsymbol{R}_{pp}$, $\boldsymbol{R}_{qq}$ invertible. In the case of $X\neq Y$, $\mathrm{rank}(\boldsymbol{M}^l_{pq})=\mathrm{rank}\boldsymbol{N}^l_{pq}$, thus after an edge reduction, we have the same deletion. In the case of $X=Y$, $\boldsymbol{M}^l_{pq}$ is similar to $\boldsymbol{N}^l_{pq}$, thus after a loop reduction and a same unraveling, we have the same deletion. If Formula (3.4) shows that $n_{pq}r_{qq}-r_{pp}n_{pq}\neq 0$, we make a base change, then a regularization. In the above 3 cases, $\mathfrak{M}^{l+1}$ is uniquely determined with respect to both $\boldsymbol{M}^l$ and $\boldsymbol{N}^l$.

Recalling (4.5) and (4.6), combined with $(\boldsymbol{N}^l+\boldsymbol{H}^l_{\underline{\boldsymbol{n}}^l})\boldsymbol{U}^l=\boldsymbol{U}^l(\boldsymbol{N}^{l+1}+\boldsymbol{H}^{l+1}_{\underline{\boldsymbol{n}}^{l+1}})$, we have

$$(\boldsymbol{M}^{l+1}+\boldsymbol{H}^{l+1}_{\underline{\boldsymbol{m}}^{l+1}})((\boldsymbol{S}^l)^{-1}\boldsymbol{R}^l\boldsymbol{U}^l)=((\boldsymbol{S}^l)^{-1}\boldsymbol{R}^l\boldsymbol{U}^l)(\boldsymbol{N}^{l+1}+\boldsymbol{H}^{l+1}_{\underline{\boldsymbol{n}}^{l+1}}).$$

Since $((\boldsymbol{S}^l)^{-1}\boldsymbol{R}^l\boldsymbol{U}^l)\in \mathrm{Mat}(\mathfrak{M}^l)$, the above equality shows that $(\boldsymbol{S}^l)^{-1}\boldsymbol{R}^l\boldsymbol{U}^l\in \mathrm{Mat}(\mathfrak{M}^{l+1})$ by Theorem 3.8(2). The corollary follows by induction.

Lemma 4.6 Let $\mathfrak{M}=(\mathcal{K},\mathcal{M},H(x))$ be a matrix problem with $\mathcal{T}=\{X\}$ non-trivial, let $D=k\setminus\{$the roots of $g(x)\}$ be the domain of x. Suppose that $\mathfrak{M},\mathfrak{M}^1,\cdots,\mathfrak{M}^s$ is a sequence given by base changes without any localization and then a regularization from $\mathfrak{M}^l$ to $\mathfrak{M}^{l+1}$ for $l=0,1,\cdots,s-1$. Let $\boldsymbol{L}=\boldsymbol{J}_d(\lambda)$ be a Weyr matrix with $g(\lambda)\neq 0$.

Then there is a sequence of reduction $\mathfrak{N},\mathfrak{N}^1,\cdots,\mathfrak{N}^{sd^2}$, such that $\mathfrak{N}$ is obtained by substituting L for x from $\mathfrak{M}$, and $\mathfrak{N}^{l+1}$ is obtained by a base change then a regularization from $\mathfrak{N}^l$ for $l=0,1,\cdots,sd^2-1$.

Proof Suppose that $\mathcal{M}$ has a normalized basis $\mathcal{A}=\{\boldsymbol{A}_1,\boldsymbol{A}_2,\cdots,\boldsymbol{A}_n\}$ with the leading position $(p_i,q_i)\in X\times X$ of A_i and $\mathcal{V}=(\boldsymbol{V}_1,\boldsymbol{V}_2,\cdots,\boldsymbol{V}_m)$ with $\boldsymbol{V}_j\in\mathcal{V}_{XX}$. Recall the general matrices and general equation (3.1): $(\mathcal{N}+\boldsymbol{H})R=R(\mathcal{N}+\boldsymbol{H})$ of $\mathfrak{M}$ given in the beginning of Section 3. Set the (p_i,q_i)-entries of (3.1):

$$a_ie_{Y_i}-e_{X_i}a_i=\sum_j h_{ij}(x,y)v_j+\sum_j h'_{ij}(a_1,\bar{a}_1,\cdots,a_{i-1},\bar{a}_{i-1})v_j,$$

where $h_{ij}(x,y)\in k[x,y,g_X(x)^{-1},g_Y(y)^{-1}]$; $h'_{ij}(a_1,\bar{a}_1,\cdots,a_{i-1},\bar{a}_{i-1})$ are polynomials on $a_1,\bar{a}_1,\cdots,a_{i-1},\bar{a}_{i-1}$ over $k[x,y,g(x)^{-1},g(y)^{-1}]$, whose monomials must involve some a_l or $\bar{a}_l$; a_l stands for the left multiplication and $\bar{a}_l$ stands for the right multiplication. In particular,

$$(a_ie_{Y_i}-e_{X_i}a_i)^0=\sum_j h_{ij}(x,y)v_j.$$

It is clear that the rank of the following matrix over $k[x,y,g(x)^{-1},g(y)^{-1}]$ equals s:

$$\boldsymbol{C}(x,y)=\begin{pmatrix} h_{11}(x,y) & \cdots & h_{1l}(x,y)\\ \vdots & & \vdots\\ h_{s1}(x,y) & \cdots & h_{sl}(x,y)\end{pmatrix}.$$

Define a size vector $\underline{n}$ with $n_X = d$. Let $\widetilde{\mathbf{N}} = \sum_{i=1}^{n} \tilde{\mathbf{a}}_i * \mathbf{A}_i$ with $\tilde{\mathbf{a}}_i = (a_{ipq})_{d\times d}$. Let $\widetilde{\mathbf{R}} = \tilde{\mathbf{e}} * \widetilde{\mathbf{E}} + \sum_{j=1}^{m} \tilde{\mathbf{v}}_j * \mathbf{V}_j$ with $\tilde{\mathbf{e}} = (e_{pq})_{d\times d}$ and $\tilde{\mathbf{v}}_i = (v_{ipq})_{d\times d}$. Then the equation (3.1) of $\mathfrak{N}$ has the form $(\widetilde{\mathbf{N}} + \mathbf{H}(\mathbf{L},\mathbf{L}))\widetilde{\mathbf{R}} = \widetilde{\mathbf{R}}(\widetilde{\mathbf{N}} + \mathbf{H}(\mathbf{L},\mathbf{L}))$. The (p_i, q_i)-block of the equation (3.1) gives rise to

$$(\tilde{\mathbf{a}}_i\tilde{\mathbf{e}} - \tilde{\mathbf{e}}\tilde{\mathbf{a}}_i)^0 = \sum_j h_{ij}(\mathbf{L},\mathbf{L})\tilde{\mathbf{v}}_j.$$

Given an order on the entries of the matrices $(v_{ipq})_{d\times d}$: $v_{ipq} \prec v_{i'p'q'}$ if and only if $p > p'$, or $p = p'$ and $q < q'$, or $p = p'$, $q = q'$ and $i < i'$. Then the coefficients of v_{ipq} under this order form an upper triangular block matrix with the block diagonal $\mathrm{diag}(\mathbf{C}(\mathbf{L},\mathbf{L}), \cdots, \mathbf{C}(\mathbf{L},\mathbf{L}))$. Thus the rank of $\widetilde{\mathbf{C}}(\mathbf{L},\mathbf{L})$ is at least sd^2. The lemma is proved.

Corollary 4.7 Let $\mathfrak{M} = (\mathcal{K}, \mathcal{M}, H = 0)$ be a matrix problem. Let $\mathfrak{M}, \mathfrak{N}^1, \cdots, \mathfrak{N}^r, \mathfrak{N}^{r+1}, \cdots, \mathfrak{N}^{r+l}$ be a sequence of reductions, such that the end term is minimal and local with a non-trivial index set $\mathcal{T}^{r+l} = \{X\}$ and a domain D^{r+l}. Set $\mathbf{M} = \mathbf{H}^{r+l}_{\underline{n}^{r+l}}(\mathbf{J}_d(\lambda))$, $(0, \mathbf{M}) \in \mathrm{Mat}(\mathfrak{N}^{r+l})$ of size

$$\underline{n}^{r+l} = \underline{n}^{r+l}_{(X,d)}.$$

Assume in addition that (a) $\mathfrak{N}^1$ is obtained by deletion from $\mathfrak{M}$ according to the size vector of $(M, 0) \in \mathrm{Mat}(\mathfrak{M})$; (b) $\mathfrak{N}^i$ is obtained by one of the reductions 3.5+3.4, 3.1+3.6+3.4, or 3.2+3.3 from $\mathfrak{N}^{i-1}$ for $1 < i < r$ with respect to $(\mathbf{M}, \mathbf{0})$ as shown in Proposition 4.3; (c) $\mathcal{T}^{r-1} = \{X\}$ is trivial in $\mathfrak{N}^{r-1}$ with Formula (3.4) $n_{pq}r_{qq} - r_{pp}n_{pq} = 0$, and $\mathfrak{N}^r$ is obtained by the loop reduction; (d) $\mathfrak{N}^i$ is obtained by reductions 3.2+3.3 from $\mathfrak{N}^{i-1}$ for $r < i \leqslant r+l$.

Define sequence (4.1) as follows: set $\mathfrak{M}^i = \mathfrak{N}^i$ for $i = 1, 2, \cdots, r-1$; $\mathfrak{M}^r$ is obtained from $\mathfrak{M}^{r-1}$ by a loop reduction, an unraveling and a deletion in turn such that $\bar{\mathbf{n}}^r = \mathbf{J}_d(\lambda)$ with $\lambda \in D^{r+l}$; then $\mathfrak{M}^{r+i}$ is obtained from $\mathfrak{M}^{r+i-1}$ by a base change plus a regularization for $i = 1, 2, \cdots, ld^2$, $s = r + ld^2$. Define sequence (4.2) as follows: set $\mathbf{M}^i + \mathbf{H}^i_{\underline{m}^i} = \mathbf{M}$ for $i = 1, 2, \cdots, r, r+1, \cdots, s$.

Then the pair of the sequences (4.1) and (4.2) is the pair of reduction sequences with respect to $(\boldsymbol{M},\boldsymbol{0})$, and $\boldsymbol{M}$ is the canonical form of $(\boldsymbol{M},\boldsymbol{0})$.

Proof By the conditions (a)~(c) in Corollary 4.7, the partial sequence $\mathfrak{M},\mathfrak{M}^1,\cdots,\mathfrak{M}^{r-1},\mathfrak{M}^r$ satisfy Proposition 4.3. The rest of the reductions starting from $\mathfrak{M}^r$ are all given by the base changes and the regularizations according to Lemma 4.6. The proof of the lemma is finished.

§ 5. Canonical minimal matrix problems

Wild Theorem 5.1[1,2,7] A matrix problem $\mathfrak{M}=(\mathcal{M},\mathcal{K},H)$ with the leading position $(p,q)\in X\times Y$ of A_1 must be wild, if one of the following two cases holds:

Case 1 X,Y are both non-trivial, there is non-invertible $f(x,y)\in k[x,y,g_X(x)^{-1},g_Y(y)^{-1}]$, with $n_{pq}r_{qq}-r_{pp}n_{pq}=f(x,y)v_1$.

Case 2 X is non-trivial, but Y is trivial (or dual), $n_{pq}r_{qq}-r_{pp}n_{pq}=0$.

A non-wild matrix problem is called the representation tame type.

Corollary 5.2 Let $\mathfrak{M}$ be a matrix problem of representation tame type. Suppose that $(p,q)\in X\times Y$ is the leading position of A_1.

(1) If the formula (3.5) shows $n_{pq}r_{qq}-r_{pp}n_{pq}=0$, then X,Y must be trivial.

(2) If the formula (3.5) shows $n_{pq}r_{qq}-r_{pp}n_{pq}\neq 0$, then in the case of X,Y both non-trivial, the highest common factor $d(x,y)$ given in Case 3.2 must be invertible in $k[x,y,g_X(x)^{-1},g_Y(y)^{-1}]$. Thus after a localization given by $c(x)$, we have an invertible matrix $\boldsymbol{Q}=\boldsymbol{Q}(x)$ over the Hermite ring $k[x,c(x)^{-1}]$ for a base change.

(3) The reduction can be made in any case for a matrix problem of tame type. In particular, if the size of the matrix problem is bounded, we are able to reach to a minimal matrix problem by finitely many steps of reductions.

Lemma 5.3 Let D be a principal ideal domain, then any finitely generated localization of D is still a principal ideal domain.

Proof (A result in commutative algebra) Take any $a \in D$ and $a \neq 0$, then $D_a = \{f/g\}$, where g divides a^e for some positive integer e depending on g. Suppose that J is an ideal of D_a, set $I = \{f \in D \mid f/g \in J\}$. We claim that I is an ideal of D. In fact if $f_1, f_2 \in I$, then there exist some g_1, g_2 dividing a^e for some $e \in \mathbb{N}$ with $f_1/g_1, f_2/g_2 \in J$. Thus $f_1/(g_1 g_2), f_2/(g_1 g_2) \in J$, and $(f_1 + f_2)/(g_1 g_2) \in J$, consequently $f_1 + f_2 \in I$. On the other hand, if $f \in I$, for any $h \in D$, we have $f/g \in J$ for some g dividing a^e, and $fh/g \in J$, therefore $fh \in I$. If $I = (b)$, then $J = (b)$. The lemma is proved.

Lemma 5.4[8, Theorem 3.8] Let D be a principal ideal domain, $\boldsymbol{M} \in \boldsymbol{M}_{m\times n}(D)$. Then there exist some invertible $\boldsymbol{P} \in \boldsymbol{M}_m(D)$ and $\boldsymbol{Q} \in \boldsymbol{M}_n(D)$, such that

$$\boldsymbol{PMQ} = \begin{pmatrix} d_1 & \cdots & 0 & 0 \\ \vdots & & \vdots & \vdots \\ 0 & \cdots & d_r & 0 \\ 0 & \cdots & 0 & 0 \end{pmatrix}_{m\times n}$$

where $r \leqslant m, n, d_i \neq 0$ and $d_i \mid d_j$ if $i \leqslant j$.

Furthermore, let $a = d_r$, there exist some invertible $\boldsymbol{P}^{\mathrm{T}} \in \boldsymbol{M}_m(D_a)$ and $\boldsymbol{Q}^{\mathrm{T}} \in \boldsymbol{M}_n(Da)$, such that $\boldsymbol{P}^{\mathrm{T}} \boldsymbol{MQ}^{\mathrm{T}} = \begin{pmatrix} \boldsymbol{0} & I_r \\ \boldsymbol{0} & \boldsymbol{0} \end{pmatrix}$.

Lemma 5.5[8, Sections 3.9 and 3.10] Let $K = k(x)$ be the field of rational functions, and let $A \in \boldsymbol{M}_m(K)$ be a linear transformation of the vector space K^m. Then K^m is a $K[\lambda]$-module, with $\lambda v = Av$ for any $v \in K^m$.

(1) Suppose that the invariant factors of $\lambda \boldsymbol{I}_m - \boldsymbol{A}$ are $d_1(\lambda), d_2(\lambda), \cdots, d_q(\lambda)$ and the elementary factors are $d_{ij}(\lambda)$ with $d_i(\lambda) = \prod_j^l d_{ij}(\lambda)$. Moreover $d_{ij}(\lambda) = f_j(\lambda)^{e_{ij}}$ for some irreducible $f_1(\lambda), f_2(\lambda), \cdots, f_l(\lambda) \in K[\lambda]$ with $e_{qj} > 0, e_{qj} \geqslant \cdots \geqslant e_{1j} \geqslant 0, j = 1, 2, \cdots, l$.

(2) There is an invertible $\boldsymbol{P} \in \boldsymbol{M}_m(K)$, such that $\boldsymbol{P}^{-1}\boldsymbol{AP}$ is a rational canonical form $\boldsymbol{B} = \mathrm{diag}(\boldsymbol{B}_{11}, \boldsymbol{B}_{12} \cdots, \boldsymbol{B}_{1l}, \cdots, \boldsymbol{B}_{q1}, \boldsymbol{B}_{q2} \cdots, \boldsymbol{B}_{ql})$, such that

$$|\lambda \boldsymbol{I} - \boldsymbol{B}_{ij}| = d_{ij}(\lambda).$$

Lemma 5.6 Let $D = k[x, g(x)^{-1}]$ with the field of fractions $K = k(x)$, and assume that $\boldsymbol{A} \in \boldsymbol{M}_m(D) \subseteq \boldsymbol{M}_m(K)$ given in Lemma 5.5 If $\boldsymbol{P} \in \boldsymbol{M}_m(K)$ given by Lemma 5.5(2), set $c(x)$ the least common multiple of the denominators of all the entries of $\boldsymbol{P}$ and $\boldsymbol{P}^{-1}$. Define $D' = k[x, g'(x)^{-1}]$ with $g'(x) = g(x)c(x)$, we have $\boldsymbol{P}, \boldsymbol{P}^{-1} \in \boldsymbol{M}_m(D')$.

(1) Write $\boldsymbol{A} = \boldsymbol{A}(x)$, then for any $\mu \in k, g'(\mu) \neq 0$ the matrix $\boldsymbol{A}(\mu)$ is similar to $\boldsymbol{B}(\mu) = \bigoplus_{\mathrm{i,j}} \boldsymbol{B}_{ij}(\mu)$. If $q > 1$ or $l > 1$, $\boldsymbol{B}(\mu)$ can be decomposed into a direct sum of indecomposable rational canonical forms.

(2) If $q = l = 1$, there exists only one invariant as well as elementary factor, say $d(\lambda) = f(\lambda)^e$, with $f(\lambda) \in \boldsymbol{M}_m(D')$. In the case of char$(k) = 0$, $(f(\lambda), f'(\lambda)) = 1$. If $\deg(f(\lambda)) = d$, $f(\lambda)$ has d different roots. In the case of char$(k) = p > 0$, there is some positive integer w, such that $f(\lambda) = h(\lambda)^{p^w}$ with $h(\lambda) \in K_1[\lambda]$, where $K_1 = K^{1/p^w} = \{a^{1/p^w} \mid a \in K\}$, then $(h(\lambda), h'(\lambda)) = 1$. If $\deg(h(\lambda)) = d$, $f(\lambda)$ has d different roots.

Proof of (2) In the case of char$(k) = 0$, there are some $u(\lambda), v(\lambda) \in K[\lambda]$ with $f(\lambda)u(\lambda) + f'(\lambda)v(\lambda) = 1$. Suppose that $f(\lambda) = \sum_{j=0}^{n} \frac{a_j(x)}{b_j(x)} \lambda^j$ with $a_j(x), b_j(x) \in k[x]$ and $a_n(x) \neq 0, b_j(x) \mid g'(x)^{l_j}$ for some positive integers l_j; $u(\lambda) = \sum_j \frac{c_j(x)}{d_j(x)} \lambda^j$; and $v(\lambda) = \sum_j \frac{c'_j(x)}{d'_j(x)} \lambda^j$ with $c(x), d(x), c'(x), d'(x) \in k[x]$. Set $C(x) = a_n(x) \prod_j d_j(x) \prod d'_j(x)$, then for any $\mu \in k, g'(\mu) \neq 0$ and $\boldsymbol{C}(\mu) \neq 0$, $\boldsymbol{B}(\mu)$ has d different eigenvalues, since the characteristic polynomial of $\lambda \boldsymbol{I} - \boldsymbol{B}(\mu)$ has d different roots.

In the case of char$(k) = p$, there are some $u(\lambda), v(\lambda) \in K_1[\lambda]$ with $h(\lambda)u(\lambda) + h'(\lambda)v(\lambda) = 1$. Suppose that $h(\lambda) = \sum_{j=0}^{n} \frac{a_j(x)}{b_j(x)} \lambda^j$ with $a_j(x)^{p^w}, b_j(x)^{p^w} \in k[x]$ and $a_n(x) \neq 0, b_j(x)^{p^w} \mid g'(x)^{l_j}$ for some positive integers l_j; $u(\lambda) = \sum_j \frac{c_j(x)}{d_j(x)} \lambda^j$; and $v(\lambda) = \sum_j \frac{c'_j(x)}{d'_j(x)} \lambda^j$ with $c(x)^{p^w}$,

$d(x)^{p^w}, c'(x)^{p^w}, d'(x)^{p^w} \in k[x]$. Set $C(x) = \{a_n(x)\prod_j d_j(x)\prod d'_j(x)\}^{p^w}$, then for any $\mu \in k$, $g'(\mu) \neq \mathbf{0}$ and $\mathbf{C}(\mu) \neq \mathbf{0}$, $\mathbf{B}(\mu)$ has d different eigenvalues, since the characteristic polynomial of $\lambda\mathbf{I} - \mathbf{B}(\mu)$ has d different roots. The lemma is proved.

Lemma 5.7 Define a linear map

$$\mathbf{J}_m(h(x)) : k[x, g(x)^{-1}]^m \to k[x, g(x)^{-1}]^m,$$

where $\mathbf{J}_m(h(x))$ is a Jordan block of size m with eigenvalue $h(x) \in k[x, g(x)^{-1}]$, then we obtain a $k[x, (x)^{-1}]$-module $k[x, g(x)^{-1}]^m$. The module is decomposable whenever $m > 1$.

Lemma 5.8 Let k be an algebraically closed field. If the inverse function of the rational function $y = \dfrac{f(x)}{g(x)}$ with $f(x), g(x) \in k[x]$ has the form of $x = \dfrac{p(y)}{q(y)}$, $p(y), q(y) \in k[y]$, then

$$y = \frac{ax+b}{cx+d} \text{ and } x = -\frac{dy-b}{cy-a} \text{ with } ad - bc \neq 0.$$

Proof (A result in algebraical geometry) Assume that $(p(y), q(y)) = 1$. Let $f(x) = \sum_{i=0}^{n} a_i x^i \ (a_n \neq 0)$, and $g(x) = \sum_{j=0}^{m} b_j x^j \ (b_m \neq 0)$, then a_0 and b_0 cannot be zero simultaneously. Set $l = \max(m, n)$ we have

$$y = \frac{f(x)}{g(x)} = \frac{f(p/q)}{g(p/q)} = \frac{\sum_{i=0}^{n} a_i \left(\frac{p}{q}\right)^i}{\sum_{j=0}^{m} b_i \left(\frac{p}{q}\right)^j} = \frac{\sum_{i=0}^{n} a_i p^i q^{l-i}}{\sum_{j=0}^{m} b_j p^j q^{l-j}},$$

so

$$y\left(\sum_{j=0}^{m} b_j p^j q^{l-j}\right) = \sum_{j=0}^{n} a_i p^i q^{l-i}. \tag{5.1}$$

Thus

$$y \cdot b_0 q^l(y) - a_0 q^l(y) = \sum_{i=0}^{n} a_i p^i q^{l-i} - y \sum_{j=0}^{m} b_j p^j q^{l-j},$$

and hence

$$q^l(y)(b_0 y - a_0) = p(y)\left[\sum_{i=1}^{n} a_i p^{i-1} q^{l-i} - y \sum_{j=1}^{m} b_j p^{j-1} q^{l-j}\right].$$

So $p(y)|q^n(y)(b_0y-a_0)$, but $(p(y),q(y))=1$, we have $p(y)|(b_0y-a_0)$. Therefore, $p(y)=c$ or $c(b_0y-a_0)$ for some $c\in k$. By Formula (5.1), we also have

$$y\cdot b_mp^m(y)q^{l-m}(y)-a_np^n(y)q^{l-n}(y)=\sum_{i=0}^{n-1}a_ip^iq^{l-i}-y\sum_{j=0}^{m-1}b_jp^jq^{l-j}$$

and in the case $m\leqslant n$,

$$p^m(y)(b_myq^{n-m}(y)-a_np^{n-m}(y))=q(y)\left[\sum_{i=0}^{n-1}a_ip^iq^{n-i-1}-y\sum_{j=0}^{m-1}b_jp^jq^{n-j-1}\right].$$

So $q(y)|p^m(y)(b_myq^{n-m}(y)-a_np^{n-m}(y))$, but $(q(y),p(y))=1$, we have $q(y)|(b_myq^{n-m}(y)-a_np^{n-m}(y))$. If $m<n$, then $q(y)|p^{n-m}(y)$, a contradiction. Thus $n=m$, and $q(y)|(b_my-a_n)$, therefore $q(y)=d$ or $d(b_my-a_n)$. The case $n\leqslant m$ is dual.

As a consequence, we have $x=\dfrac{sy+t}{uy+v}$, $s,t,u,v\in k$, and hence $y=\dfrac{ax+b}{cx+d}$, $a,b,c,d\in k$. The proof of the lemma is completed.

Proposition 5.9 Let $\mathfrak{M}=(\mathcal{K},\mathcal{M},H=0)$ be a matrix problem of tame type. And let $\mathfrak{N}=(\mathcal{K}_X,(0),\boldsymbol{M}(x))$ be a minimal matrix problem induced from $\mathfrak{M}$, with a unique equivalent class X and a domain $k\setminus\{$the roots of $g(x)\}$, $M(x)$ has the size vector $\underline{\boldsymbol{m}}$ over $(T,\sim)$. Then there exists a unique sequence of matrix problems:

$$\mathfrak{M},\mathfrak{M}^1,\cdots,\mathfrak{M}^{r-1},\mathfrak{M}^r,\cdots,\mathfrak{M}^s, \tag{5.2}$$

(1) given by a deletion according to the size vector of $M(x)$ from $\mathfrak{M}$ to $\mathfrak{M}^1$;

(2) given by reductions 3.5+3.4, or 3.1+3.6+3.4, or 3.2+3.3 from $\mathfrak{M}^{l-1}$ to $\mathfrak{M}^l$ for $1<l<r$, and their index sets are all trivial;

(3) given by a loop reduction 3.1 from $\mathfrak{M}^{r-1}$ to non-trivial $\mathfrak{M}^r$ with a parameter $x'=\dfrac{ax+b}{cx+d}$, where $ac-bd\neq 0$;

(4) given by reductions 3.2+3.3 from $\mathfrak{M}^{l-1}$ to $\mathfrak{M}^l$ for $r<l\leqslant s$;

(5) the end term is minimal.

The following sequence of localizations of $k[x,g(x)^{-1}]$ and $k[x']$

gives the domains of the matrix problems in the sequence (5.2):

$$k[x,g(x)^{-1}],k[x,g_1(x)^{-1}],\cdots,k[x,g_{r-1}(x)^{-1}],$$
$$k[x',h_r(x')^{-1}],\cdots,k[x',h_s(x')^{-1}]. \tag{5.3}$$

Furthermore, there also exists a sequence of matrices with a parameter, such that each matrix in the sequence is that over the corresponding matrix problem of the sequence (5.2),

$$(\boldsymbol{M}(x),\boldsymbol{H}_{\underline{m}}),(\boldsymbol{M}^1(x),\boldsymbol{H}^1_{\underline{m}^1}),\cdots,(\boldsymbol{M}^{r-1}(x),\boldsymbol{H}_{\underline{m}^{r-1}}),$$
$$(\boldsymbol{M}^r(x'),\boldsymbol{H}_{\underline{m}^r}(x')),\cdots,(\boldsymbol{0},\boldsymbol{H}^s_{\underline{m}^s}(x')), \tag{5.4}$$

where $\underline{m}_l$ is sincere over $(T^l,\sim^l)$ for $l=1,2,\cdots,s$, and there is some $(T^{l-1},\sim^{l-1})$-partitioned invertible matrix $\boldsymbol{S}^{l-1}(x)$ or $\boldsymbol{S}^{l-1}(x')$ satisfying the defining system of $\mathcal{K}^{l-1}$, such that

$$(\boldsymbol{M}^{l-1}(x)+\boldsymbol{H}^{l-1}_{\underline{m}^{l-1}})\boldsymbol{S}^{l-1}(x)=\boldsymbol{S}^{l-1}(x)(\boldsymbol{M}^l(x)+\boldsymbol{H}^l_{\underline{m}^l}),l=2,3,\cdots,r-1; \tag{5.5}$$

$$(\boldsymbol{M}^{r-1}(x)+\boldsymbol{H}^{r-1}_{\underline{m}^{r-1}})\boldsymbol{S}^{r-1}(x)=\boldsymbol{S}^{r-1}(x)\left(\boldsymbol{M}^r\left(\frac{ax+b}{cx+d}\right)+\boldsymbol{H}^r_{\underline{m}^r}\left(\frac{ax+b}{cx+d}\right)\right); \tag{5.6}$$

$$(\boldsymbol{M}^{l-1}(x')+\boldsymbol{H}^{l-1}_{\underline{m}^{l-1}}(x'))\boldsymbol{S}^{l-1}(x')=\boldsymbol{S}^{l-1}(x')(\boldsymbol{M}^l(x')+\boldsymbol{H}^l_{\underline{m}^l}(x')),$$
$$l=r+1,\cdots,s. \tag{5.7}$$

$\mathfrak{M}^s$, the end term of the sequence (5.2), is called the canonical minimal matrix problem of $\mathfrak{N}$.

Proof (1) The reduction from $\mathfrak{M}$ to $\mathfrak{M}^1$ is a deletion of

$$\mathcal{T}^0=\{Z\mid m_Z=0\}.$$

(2) Suppose that we have already the sequences (5.2)~(5.5) up to some $l-1<r-1$. Now we continue the reduction from $\mathfrak{M}^{l-1}$ to get the r-th term in the above sequences. Still write $(p,q)\in X\times Y$, the leading position of $\boldsymbol{A}_1$ in $\mathfrak{M}^{l-1}$.

If the equation (3.4) shows $n_{pq}r_{qq}-r_{pp}n_{pq}=0$, then X and Y are both trivial by Corollary 5.2. If $X\neq Y$ and $\operatorname{rank}(\boldsymbol{M}^{l-1}_{\underline{m}^{l-1}}(x)_{pq})=r$, we have $k[x,g_l(x)^{-1}]$, a localization of $k[x,g_{l-1}(x)^{-1}]$, such that there exist some invertible matrices $\boldsymbol{S}_{pp}$, $\boldsymbol{S}_{qq}$ over $k[x,g_l(x)^{-1}]$, with $(\boldsymbol{M}^{l-1}_{\underline{m}^{l-1}}(x))_{pq}\boldsymbol{S}_{qq}=\boldsymbol{S}_{pp}\begin{pmatrix}\boldsymbol{0}&\boldsymbol{I}_r\\\boldsymbol{0}&\boldsymbol{0}\end{pmatrix}$ by Lemma 5.4. We perform first an edge reduction then a deletion to ensure that $\underline{\boldsymbol{m}}^{l}$ is sincere over $(T^{l},\sim^{l})$. Set $\boldsymbol{S}^{l-1}$, a partitioned diagonal matrix according to $(T^{l-1},\sim^{l-1})$, with $\boldsymbol{S}_X^{l-1}=\boldsymbol{S}_{pp}$, $\boldsymbol{S}_Y^{l-1}=\boldsymbol{S}_{qq}$, and $\boldsymbol{S}_Z^{l-1}=\boldsymbol{I}_{m_Z^{l-1}}$ for all $Z\neq X,Y$.

If the equation (3.4) shows $n_{pq}r_{qq}-r_{pp}n_{pq}=0$ and $X=Y$, then the eigenvalues of $M^{l-1}_{\underline{m}^{l-1}}(x)_{pq}$ must be all in k according to the assumption for sequence (5.4). Then we have $k[x,g_l(x)^{-1}]$, a localization of $k[x,g_{l-1}(x)^{-1}]$, such that there is some invertible matrix $S_{pp}=S_{qq}$ over $k[x,g_l(x)^{-1}]$, with $(\boldsymbol{M}^{l-1}_{\underline{\boldsymbol{m}}^{l-1}}(x))_{pq}\boldsymbol{S}_{qq}=\boldsymbol{S}_{pp}\boldsymbol{W}$ by Lemma 5.6, where $\boldsymbol{W}$ is a Weyr matrix under the ordering of the base field. We perform first a loop reduction, then an unraveling according to the eigenvalues of $\boldsymbol{W}$ and the positive integer d, the maximal size of the Jordan blocks in W, finally a deletion to insure that $\underline{m}^{l}$ is sincere over $(T^{l},\sim^{l})$. Set $\boldsymbol{S}^{l-1}$, a partitioned diagonal matrix according to $(T^{l-1},\sim^{l-1})$, with $\boldsymbol{S}_X^{l-1}=\boldsymbol{S}_{pp}$, and $\boldsymbol{S}_Z^{l-1}=\boldsymbol{I}_{m_Z^{l-1}}$ for all $Z\neq X$.

If the equation (3.4), shows $n_{pq}r_{qq}-r_{pp}n_{pq}\neq 0$, we have a base change over k, since X,Y are both trivial by the assumption for the sequence (5.4). Then we perform a regularization. Define a matrix $\boldsymbol{S}^{l-1}=\boldsymbol{I}+(\boldsymbol{M}^{l-1}_{\underline{\boldsymbol{m}}^{l-1}})_{pq}*\boldsymbol{V}_1'$ with $\boldsymbol{V}_1'$ being the first base element of $\mathcal{K}^{l-1}$ after the base change.

Summing up the above discussion, we obtain the l-terms in the sequences (5.2)~(5.5).

(3) Since $\underline{\boldsymbol{m}}$ is fixed, and $\boldsymbol{M}(x)$ contains a parameter, we must reach an induced matrix problem $\mathfrak{M}^{r-1}=(\mathcal{K}^{r-1},\mathcal{M}^{r-1},\boldsymbol{H}^{r-1})$ and a pair $(\boldsymbol{M}^{l-1}_{\underline{\boldsymbol{m}}^{l-1}}(x),\boldsymbol{H}^{r-1}_{\underline{m}^{r-1}})$, such that the equation (3.4) in $\mathfrak{M}^{r-1}$ shows $n_{pq}r_{qq}-r_{pp}n_{pq}=0$ and $X=Y$, but the eigenvalues of $\boldsymbol{A}=\boldsymbol{M}^{l-1}_{\underline{\boldsymbol{m}}^{l-1}}(x)_{pq}$ are not all in k. And we know from Lemma 5.3 that $k[x,g_{r-1}(x)^{-1}]$ is a principal ideal domain with the field of the fractions $K=k(x)$. Suppose that the elementary factors of $\lambda\boldsymbol{I}-\boldsymbol{A}$ depend on the irreducible polynomials $f_1(\lambda)$, $f_2(\lambda),\cdots,f_l(\lambda)\in K[\lambda]$ given by Lemma 5.5, and $f_1(\lambda),f_2(\lambda),\cdots,f_u(\lambda)\in k[x]$, thus they have degree 1, since k is algebraically closed. Furthermore, $u<l$ and $f_{u+1}(\lambda),f_{u+2}(\lambda),\cdots,f_l(\lambda)\notin k[\lambda]$. By Lemma 5.6, there is a

localization $D'=k[x,g'(x)^{-1}]$ of $D=k[x,g_{l-1}(x)^{-1}]$ and an invertible $\boldsymbol{P}\in \boldsymbol{M}_m(D')$, such that $\boldsymbol{P}^{-1}\boldsymbol{A}\boldsymbol{P}$ is a rational canonical form $\boldsymbol{B}=\mathrm{diag}(\boldsymbol{B}_{11},\boldsymbol{B}_{12},\cdots,\boldsymbol{B}_{1l},\cdots,\boldsymbol{B}_{q1},\boldsymbol{B}_{q2},\cdots,\boldsymbol{B}_{ql})$ given in Lemma 5.5. After exchanging some rows and columns simultaneously, we may assume that $\boldsymbol{Q}^{-1}\boldsymbol{B}\boldsymbol{Q}=\boldsymbol{W}\oplus\boldsymbol{C}$, where $\boldsymbol{W}$ is a Weyr matrix, and $\boldsymbol{C}$ is a rational canonical form, whose eigenvalues are not in k.

We use a loop reduction, then an unraveling determined by $\boldsymbol{W}\oplus(x')$, and finally a deletion, to obtain $\mathfrak{M}^r=(\mathcal{K}^r,\mathcal{M}^r,\boldsymbol{H}^r(x'))$, such that $\mathcal{T}^r$ has a non-trivial equivalent class X'; and $\underline{\boldsymbol{m}}^r$, the induced size vector of $\underline{\boldsymbol{m}}^{r-1}$, is sincere over $(T^r,\sim^r)$.

By Corollary 5.2, any further reduction connecting X' is a base change followed by a regularization, since we do not make any unraveling. Therefore X' is preserved until reaching the minimal matrix problem in the sequence (5.2).

Case 1 If the size of $\boldsymbol{W}$ is non-zero, then any matrix of size vector $\underline{\boldsymbol{m}}^r$ is decomposable, a contradiction to $(\boldsymbol{M}(\mu),0)$, and thus $(\boldsymbol{M}^{r-1}_{\underline{\boldsymbol{m}}^{r-1}}(\mu),\boldsymbol{H}^{r-1}_{\underline{\boldsymbol{m}}^{r-1}})$ is indecomposable for any $\mu\in k$, $g'(\mu)\neq 0$.

Case 2 If the size of $\boldsymbol{W}$ is zero, but $q>1$, or $l>1$ given in Lemma 5.6, then any matrix of size vector $\underline{\boldsymbol{m}}^r$ is decomposable, a contradiction.

Case 3 If the size of $\boldsymbol{W}$ is zero and $q=l=1$, but $d>1$ given in Lemma 5.6, then for almost all $\mu\in k$, $g'(\mu)\neq 0$, $B(\mu)$ has d different eigenvalues. Thus almost all matrices of size vector $\underline{\boldsymbol{m}}^r$ are decomposable, again a contradiction.

Case 4 If the size of $\boldsymbol{W}$ is zero, $q=l=1$ and $d=1$, then $f(\lambda)=\lambda-\dfrac{u(x)}{v(x)}$ for some $u(x),v(x)\in k[x]$. Suppose in this case the invariant and the elementary factor equals $d(\lambda)=f(\lambda)^e$. Write $h(x)=\dfrac{u(x)}{v(x)}$ and set $g_r(x)=g'(x)v(x)$. If $e>1$, Lemma 5.7 shows that the $k[x,g_r(x)^{-1}]$-module $k[x,g_r(x)^{-1}]^e$ with a linear transformation

$$\boldsymbol{J}_e(h(x)):k[x,g_r(x)^{-1}]^e\rightarrow k[x,g_r(x)^{-1}]^e$$

is decomposable, we have a contradiction to $\boldsymbol{M}^{r-1}(x)+\boldsymbol{H}_{\underline{m}^{r-1}}$ being indecomposable.

Case 5 If the size of W is zero, $q=l=1$ and $d=e=1$, then Lemma 5.8 gives rise to $x'=\dfrac{ax+b}{cx+d}$ and $x=-\dfrac{dx'-b}{cx'-a}$ with $a,b,c,d\in k, ad-bc\neq 0$, since the reduction procedure is invertible. Letting

$$h_r(x')=g_r\left(-\frac{dx'-b}{cx'-a}\right)(cx'-a)^p\in k[x'], (cx'-a)\mid h_r(x'),$$

we obtain a matrix problem $\mathfrak{M}^r$ in sequence (5.2), and a ring $k[x', h_r(x')^{-1}]$ in sequence (5.3), a matrix $(\boldsymbol{M}^r(x'), \boldsymbol{H}_{\underline{m}^r}(x'))$ over $\mathfrak{M}^r$ with a size vector $\underline{\boldsymbol{m}}^r=(1,1,\cdots,1)$ in sequence (5.4). Setting S^{r-1} to be a partitioned diagonal matrix according to $(T^{r-1}, \sim^{r-1})$ with $\boldsymbol{S}_X^{r-1}=\boldsymbol{P}$ given in the beginning of (2.3), and $\boldsymbol{S}_Z^{r-1}=\boldsymbol{I}_{m_Z^{r-1}}$ for all $Z\neq X$, we have the formula (5.6).

(4) The further reductions are all 3.2+3.3, the sequence (5.2) is completed and the localizations $k[x', h_r(x')^{-1}], \cdots, k[x', h_s(x')^{-1}]$ in the sequence (5.3) are obtained. The rest terms of the sequence (5.4) and Formula (5.7) for $l=r+1, r+2, \cdots, s$ are obtained obviously. The proposition is proved.

The "generic modules" has been discussed intensively in [9~12]. The Proposition 5.9 gives an exact description of the generic modules in the tame case by language of matrix; and also gives the necessary and sufficient condition of two such modules being isomorphic.

Definition 5.10[9] Let $\mathfrak{M}=(\mathcal{K}, \mathcal{M}, 0)$ be a tame matrix problem, and $\mathfrak{M}_1=(\mathcal{K}, (0), H_1)$ and $\mathfrak{M}_2=(\mathcal{K}, (0), H_2)$ be two non-trivial local minimal matrix problems with domains D_1 and D_2 induced from $\mathfrak{M}$ respectively. If there exists an infinite subset $\overline{D}_1$, such that $\forall \lambda\in\overline{D}_1$, $(\boldsymbol{H}_1(J_1(\lambda)), 0)$ is isomorphic to $(\boldsymbol{H}_2(J_d(\mu)), 0)$ in Mat$(\mathfrak{M})$ for some positive integer d depending on λ, and $g_2(\mu)\neq 0$. Then $\mathfrak{M}_1$ and $\mathfrak{M}_2$ are said to be quasi-isomorphic.

Corollary 5.11 Let $\mathfrak{M}=(\mathcal{K}, \mathcal{M}, 0)$ be a tame matrix problem. If two local minimal matrix problems $\mathfrak{N}_1$ and $\mathfrak{N}_2$ induced from $\mathfrak{M}$ are quasi-isomorphic, then they have the same canonical minimal matrix problem up to domains.

§ 6. Direct sums

Definition 6.1 Let $(T,\sim)$ be a set of integers with an equivalent relation. Suppose that $\boldsymbol{C}=(\boldsymbol{C}_{ij})_{t\times t}$, $\boldsymbol{D}=(\boldsymbol{D}_{ij})_{t\times t}$ are two matrices partitioned by $(T,\sim)$. Then a matrix

$$\begin{pmatrix}\boldsymbol{C}_{ij} & \boldsymbol{0}\\ \boldsymbol{0} & \boldsymbol{D}_{ij}\end{pmatrix}_{t\times t}$$

is said to be a direct sum of $\boldsymbol{C}$ and $\boldsymbol{D}$, and is denoted by $\boldsymbol{C}\oplus_T\boldsymbol{D}$, or $\boldsymbol{C}\oplus\boldsymbol{D}$ for simplicity.

A matrix C is said to be indecomposable, if $\boldsymbol{C}=\boldsymbol{C}_1\oplus\boldsymbol{C}_2$, then the size of $\boldsymbol{C}_1$ or $\boldsymbol{C}_2$ is 0.

Lemma 6.2 (1) Let $m=m_1+m_2$, $n=n_1+n_2$, $r=r_1+r_2$, where m_i, n_i are positive integers and r_i are non-negative integers, with $r_i\leqslant m_i, n_i$. Then

$$\boldsymbol{P}\begin{pmatrix}\boldsymbol{0} & \boldsymbol{I}_r\\ \boldsymbol{0} & \boldsymbol{0}\end{pmatrix}_{m\times n}\boldsymbol{Q}\begin{pmatrix}\boldsymbol{0} & \boldsymbol{I}_{r_1}\\ \boldsymbol{0} & \boldsymbol{0}\end{pmatrix}_{m_1\times n_1}\oplus\begin{pmatrix}\boldsymbol{0} & \boldsymbol{I}_{r_2}\\ \boldsymbol{0} & \boldsymbol{0}\end{pmatrix}_{m_2\times n_2},$$

where

$$\boldsymbol{P}=\mathrm{diag}\left(\boldsymbol{I}_{r_1},\begin{pmatrix}\boldsymbol{0} & \boldsymbol{I}_{m_1-r_1}\\ \boldsymbol{I}_{r_2} & \boldsymbol{0}\end{pmatrix},\boldsymbol{I}_{m_2-r_2}\right),$$

and

$$\boldsymbol{Q}=\mathrm{diag}\left(\boldsymbol{I}_{n_1-r_1},\begin{pmatrix}\boldsymbol{0} & \boldsymbol{I}_{n_2-r_2}\\ \boldsymbol{I}_{r_1} & \boldsymbol{0}\end{pmatrix},\boldsymbol{I}_{r_2}\right)$$

are determined by a series of elementary transformations of exchanging rows and columns respectively.

(2) Let Weyr matrix $\boldsymbol{W}^i(\lambda)$ be similar to $\boldsymbol{J}_1(\lambda)^{e_{i1}}\oplus\boldsymbol{J}_2(\lambda)^{e_{i2}}\oplus\cdots\oplus\boldsymbol{J}_d(\lambda)^{e_{id}}$ with $e_{ij}\in\mathbb{Z}$, and $e_{ij}\geqslant 0$ for $i=1,2$, $j=1,2,\cdots,d$. Let $m=$

$\frac{1}{2}d(d+1)$, then $\boldsymbol{W}^i(\lambda)$ is an $m\times m$ block matrix with the block-sizes

$$\begin{aligned}(&e_{id}, e_{i,d-1}, \cdots, e_{i2} \quad e_{i1};\\ &e_{id}, e_{i,d-1}, \cdots, e_{i2};\\ &\cdots\\ &e_{id}, e_{i,d-1};\\ &e_{id}).\end{aligned}$$

Then the Weyr matrix $\boldsymbol{W}(\lambda)$ similar to $\boldsymbol{W}^1(\lambda)\oplus\boldsymbol{W}^2(\lambda)$ is obtained from $\boldsymbol{W}^1(\lambda)\oplus\boldsymbol{W}^2(\lambda)$ by moving $(m+j)$-block row to $(2j)$-block row, and $(m+j)$-block column to $(2j)$-block column simultaneously for $j=1, 2, \cdots, m$.

Moreover, if $\boldsymbol{W}^i=\boldsymbol{W}^i(\lambda_1)\oplus\boldsymbol{W}^i(\lambda_2)\oplus\cdots\oplus\boldsymbol{W}^i(\lambda_r)$ for $i=1,2$, and call $\boldsymbol{W}^1(\lambda_j)$ the j-diagonal block, and $\boldsymbol{W}^2(\lambda_j)$ the $(r+j)$-diagonal block for $j=1,2,\cdots,r$, then the Weyr matrix $\boldsymbol{W}$ similar to $\boldsymbol{W}^1\oplus\boldsymbol{W}^2$ is obtained by moving $(r+j)$-block to $(2j)$-block for each $1\leqslant j\leqslant r$, then using the conclusion above.

Thus there exists a matrix P determined by a series of elementary transformations of exchanging rows and columns, such that

$$\boldsymbol{P}^{-1}\boldsymbol{W}\boldsymbol{P}=\boldsymbol{W}^1\oplus\boldsymbol{W}^2.$$

(3) Let $\mathfrak{M}=(\mathcal{K},\mathcal{M},H)$ be a matrix problem with the index set $(T,\sim)$. If the size vectors $\underline{\boldsymbol{m}}=\underline{\boldsymbol{m}}_1+\underline{\boldsymbol{m}}_2$, then $\boldsymbol{H}_{\underline{\boldsymbol{m}}_1}\oplus_T\boldsymbol{H}_{\underline{\boldsymbol{m}}_2}=\boldsymbol{H}_{\underline{\boldsymbol{m}}}$.

Proof of (3) Write $\underline{\boldsymbol{m}}=\underline{\boldsymbol{m}}_0$, $(\boldsymbol{H}_{\underline{\boldsymbol{m}}_l})_{ij}=0$ for any $i\not\sim j$ and $(\boldsymbol{H}_{\underline{\boldsymbol{m}}_l})_{ij}=\boldsymbol{I}_{m_i}$ for any $i\sim j$, where $l=0,1,2$. The lemma is proved.

Theorem 6.3 Let $\mathfrak{M}$ be a matrix problem, and $\mathfrak{M}'$ be an induced matrix problem given by one of the reductions 3.1～3.5 or 3.1+3.6 under the ordering of the base field. Suppose that $\vartheta:\mathrm{Mat}(\mathfrak{M}')\to\mathrm{Mat}(\mathfrak{M})$ is the reduction functor or their composition defined in Theorem 3.8. If $\underline{m}'=\underline{n}'+\underline{l}'$ is a sum of the size vectors, then for each non-trivial $X\in\mathcal{T}'$, we have $n'_X=1, l'_X=0$, or $n'_X=0, l'_X=1$, or $n'_X=l'_X=0$. Let $\boldsymbol{N}'\in\mathcal{M}'_{\underline{n}'}$ and $\boldsymbol{L}'\in\mathcal{M}'_{\underline{l}'}$. If

$$\boldsymbol{M}'+\boldsymbol{H}_{\underline{\boldsymbol{m}}'}=(\boldsymbol{N}'+\boldsymbol{H}'_{\underline{n}'})\oplus_{T'}(\boldsymbol{L}'+\boldsymbol{H}'_{\underline{l}'}) \tag{6.1}$$

in $\mathrm{Mat}(\mathfrak{M}')$, then we have a direct sum in $\mathrm{Mat}(\mathfrak{M})$:

$$R^{-1}(M+H_{\underline{m}})R=(N+H_{\underline{n}})\oplus_T(L+H_{\underline{l}}) \tag{6.2}$$

where $(M,H_{\underline{m}})=\vartheta(M',H'_{\underline{m}'})$, $(N,H_{\underline{n}})=\vartheta(N',H'_{\underline{n}'})$, $(L,H_{\underline{l}})=\vartheta(L',H'_{\underline{l}'})$, and the invertible matrix $R\in\mathcal{K}_{\underline{m}}$ is given by a sequence of exchanging the rows and columns.

The theorem is still valid for any direct sum with finitely many summands.

Proof In Cases 3.1～3.3, $(T',\sim')=(T,\sim)$ ensures that the direct sum over T' is also over T. In the case 3.4, for any $Z\in\mathcal{T}_0$, we have obviously $m_Z=0$, $n_Z=0$ and $l_Z=0$, thus the direct sum over T' is also over T. Set $R=I\in\mathcal{K}_{\underline{m}}$ to be the identity matrix.

Now we only need to consider Cases 3.5 and 3.6. Let $\mathcal{A}=\{A_1, A_2,\cdots,A_n\}$ with (p_i,q_i) being the leading position of A_i. Assume that the blocks of M, N, and L at the position (p_i,q_i) are B_i, C_i, D_i respectively for $i=1,2,\cdots,n$. Still write $(p,q)\in X\times Y$, as the leading position of A_1.

In Case 3.5, assume that $C_1=\begin{pmatrix}0 & I_{r_1}\\ 0 & 0\end{pmatrix}$, $D_1=\begin{pmatrix}0 & I_{r_2}\\ 0 & 0\end{pmatrix}$, then

$$B_1=\begin{pmatrix}0 & I_{r_1+r_2}\\ 0 & 0\end{pmatrix}=C_1\oplus_{T'}D_1.$$

According to Lemma 6.2 (1), there are some invertible matrices P,Q determined by exchanging some rows and columns respectively, such that $PB_1Q=C_1\oplus_T D_1$.

If $p_i,q_i\in X\cup Y$, set C_i to be $\begin{pmatrix}C^{11} & C^{12}\\ C^{21} & C^{22}\end{pmatrix}$; D_i to be $\begin{pmatrix}D^{11} & D^{12}\\ D^{21} & D^{22}\end{pmatrix}$ with the sizes of C^{12} and D^{12} being $r_i\times r_j$, $i,j\in\{1,2\}$, then

$$P^{\mathrm T}\begin{bmatrix}C^{11} & 0 & C^{12} & 0\\ 0 & D^{11} & 0 & D^{12}\\ C^{21} & 0 & C^{22} & 0\\ 0 & D^{21} & 0 & D^{22}\end{bmatrix}Q^{\mathrm T}=\begin{bmatrix}C^{11} & C^{12} & 0 & 0\\ C^{21} & C^{22} & 0 & 0\\ 0 & 0 & D^{11} & D^{12}\\ 0 & 0 & D^{21} & D^{22}\end{bmatrix}$$

with $P^{\mathrm T},Q^{\mathrm T}\in\{P,Q\}$. If $q_i\in X\cup Y$ and $p_i\notin X\cup Y$, set C_i to be

$(\boldsymbol{C}^{11}\boldsymbol{C}^{12})$; $\boldsymbol{D}_i$ to be $(\boldsymbol{D}^{11}\boldsymbol{D}^{12})$ where $\boldsymbol{C}^{12}$ and $\boldsymbol{D}^{12}$ have r_1 or r_2 columns, then

$$\begin{pmatrix}\boldsymbol{C}^{11} & \boldsymbol{0} & \boldsymbol{C}^{12} & \boldsymbol{0}\\ \boldsymbol{0} & \boldsymbol{D}^{11} & \boldsymbol{0} & \boldsymbol{D}^{12}\end{pmatrix}\boldsymbol{Q}^{\mathrm{T}}=\begin{pmatrix}\boldsymbol{C}^{11} & \boldsymbol{C}^{12} & \boldsymbol{0} & \boldsymbol{0}\\ \boldsymbol{0} & \boldsymbol{0} & \boldsymbol{D}^{21} & \boldsymbol{D}^{22}\end{pmatrix}$$

with $\boldsymbol{Q}^{\mathrm{T}}\in\{\boldsymbol{P},\boldsymbol{Q}\}$. We have a dual statement for $q_i\in X\cup Y$ and $p_i\notin X\cup Y$. If $p_i, q_i\notin X\cup Y$, $\boldsymbol{C}_i\oplus_{T'}\boldsymbol{D}_i=\boldsymbol{C}_i\oplus_T\boldsymbol{D}_i$.

Let $\boldsymbol{R}=\mathrm{diag}(\boldsymbol{R}_1,\boldsymbol{R}_2,\cdots,\boldsymbol{R}_t)\in\mathcal{K}_{\underline{m}}$, where $\boldsymbol{R}_i=\boldsymbol{P}$ if $i\in X$, and $\boldsymbol{R}_i=\boldsymbol{Q}$ if $i\in Y$, $\boldsymbol{R}_i=\boldsymbol{I}_{m_i}$. if $i\notin X\cup Y$, then $\boldsymbol{R}^{-1}=\boldsymbol{R}$, and

$$\boldsymbol{R}^{-1}(\boldsymbol{M}+\boldsymbol{H}_{\underline{m}})\boldsymbol{R}=\boldsymbol{R}^{-1}((\boldsymbol{N}'+\boldsymbol{H}'_{\underline{n}'})\oplus_{T'}(\boldsymbol{L}'+\boldsymbol{H}'_{\underline{l}'}))\boldsymbol{R}=(\boldsymbol{N}+\boldsymbol{H}_{\underline{n}})\oplus_T(\boldsymbol{L}+\boldsymbol{H}_{\underline{l}}).\tag{6.3}$$

In Case 3.6, assume that $\boldsymbol{C}_1=\boldsymbol{W}^1$, $\boldsymbol{D}^1=\boldsymbol{W}^2$, then $\boldsymbol{B}_1=\boldsymbol{W}_1\oplus_{T'}\boldsymbol{W}_2=\boldsymbol{W}$. Lemma6.2 (2) tells that there is some invertible matrix $\boldsymbol{P}$ given by a sequence of exchanging the rows and columns, such that

$$\boldsymbol{P}^{-1}\boldsymbol{B}_1\boldsymbol{P}=\boldsymbol{C}_1\oplus_T\boldsymbol{D}_1.$$

If $p_i, q_i\in X$, $\boldsymbol{C}_i$, $\boldsymbol{D}_i$ are both $r\times r$ block-diagonal matrices, each block of them is an $m\times m$ partitioned matrix: $\boldsymbol{C}_i=(\boldsymbol{C}^{ulj})_{n\times n}$ and $\boldsymbol{D}_i=(\boldsymbol{D}^{ulj})_{n\times n}$, $1\leqslant i, l\leqslant m$, $1\leqslant u\leqslant r$, $n=mr$; the sizes of $\boldsymbol{C}^{uij}$, $\boldsymbol{D}^{uij}$ are $e^u_{1i}i\times e^u_{1j}$ and $e^u_{2i}\times e^u_{2j}$ respectively given in Lemma 6.2(2). For simplicity, we denote the blocks of $\boldsymbol{C}_i$, $\boldsymbol{D}_i$ by $\boldsymbol{C}_i^{\alpha\beta}$ and $\boldsymbol{D}_i^{\alpha\beta}$ respectively. Then

$$\boldsymbol{P}^{-1}\begin{pmatrix}\boldsymbol{C}_i^{11} & \boldsymbol{0} & \cdots & \boldsymbol{C}_i^{1n} & \boldsymbol{0}\\ \boldsymbol{0} & \boldsymbol{D}_i^{11} & \cdots & \boldsymbol{0} & \boldsymbol{D}_i^{1n}\\ \vdots & \vdots & & \vdots & \vdots\\ \boldsymbol{C}_i^{n1} & \boldsymbol{0} & \cdots & \boldsymbol{C}_i^{nn} & \boldsymbol{0}\\ \boldsymbol{0} & \boldsymbol{D}_i^{n1} & \cdots & \boldsymbol{0} & \boldsymbol{D}_i^{nn}\end{pmatrix}\boldsymbol{P}=\begin{pmatrix}\boldsymbol{C}_i^{11} & \cdots & \boldsymbol{C}_i^{1n} & \boldsymbol{0} & \cdots & \boldsymbol{0}\\ \vdots & & \vdots & \vdots & & \vdots\\ \boldsymbol{C}_i^{n1} & \cdots & \boldsymbol{C}_i^{nn} & \boldsymbol{0} & \cdots & \boldsymbol{0}\\ \boldsymbol{0} & \cdots & \boldsymbol{0} & \boldsymbol{D}_i^{11} & \cdots & \boldsymbol{D}_i^{1n}\\ \vdots & & \vdots & \vdots & & \vdots\\ \boldsymbol{0} & \cdots & \boldsymbol{0} & \boldsymbol{D}_i^{n1} & \cdots & \boldsymbol{D}_i^{nn}\end{pmatrix}$$

If $q_i\in X$ and $p_i\notin X$, set the partitioned matrices $\boldsymbol{C}_i=(\boldsymbol{C}^{uj})_{1\times n}$, $\boldsymbol{D}_i=(\boldsymbol{D}^{uj})_{1\times n}$ with $j=1,2,\cdots,m$ and $u=1,2,\cdots,r$, $\boldsymbol{C}^{uj}$ and $\boldsymbol{D}^{uj}$ have e^u_{1j} and e^u_{2j} columns given in Lemma 6.2 (2) respectively. For simplicity, we denote the blocks of $\boldsymbol{C}_i$, $\boldsymbol{D}_i$ by $\boldsymbol{C}_i^{1\beta}$ and $\boldsymbol{D}_i^{1\beta}$ respectively. Then

$$\begin{pmatrix}\boldsymbol{C}^{11} & \boldsymbol{0} & \cdots & \boldsymbol{C}^{1n} & \boldsymbol{0}\\ \boldsymbol{0} & \boldsymbol{D}^{11} & \cdots & \boldsymbol{0} & \boldsymbol{D}^{1n}\end{pmatrix}\boldsymbol{P}=\begin{pmatrix}\boldsymbol{C}^{11} & \cdots & \boldsymbol{C}^{12} & \boldsymbol{0} & \cdots & \boldsymbol{0}\\ \boldsymbol{0} & \cdots & \boldsymbol{0} & \boldsymbol{D}^{11} & \cdots & \boldsymbol{D}^{1n}\end{pmatrix}.$$

We have a dual statement for $p_i \in X$ and $q_i \notin X$. If $p_i, q_i \notin X$,

$$C_i \oplus_{T'} D_i = C_i \oplus_T D_i.$$

Let $\boldsymbol{R} = \mathrm{diag}(\boldsymbol{R}_1, \boldsymbol{R}_2, \cdots, \boldsymbol{R}_t) \in \mathcal{K}_{\underline{m}}$, where $\boldsymbol{R}_i = \boldsymbol{P}$ if $i \in X$, $\boldsymbol{R}_i = \boldsymbol{I}_{m_i}$ if $i \notin X$, then the formula (6.3) is valid. The proof of the theorem is completed.

Proposition 6.4 Let $\mathfrak{M}$ be a matrix problem and let $\mathfrak{M}'$ be an induced matrix problem obtained from $\mathfrak{M}$ by one of the 3.1 ~ 3.6 reductions. Suppose that $\mathfrak{N}'$ is obtained by a deletion from $\mathfrak{M}'$, then there exists a matrix problem $\mathfrak{N}$ with the minimal size induced from $\mathfrak{M}$ also by a deletion, such that $\mathfrak{N}'$ is either an induced matrix problem of $\mathfrak{N}$ given by the same reduction as that from $\mathfrak{M}$ to $\mathfrak{M}'$ (possibly followed by another deletion), or identical with $\mathfrak{N}$ (possibly having some restricted domain) (fig 6.1).

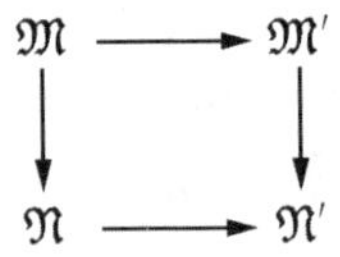

Fig. 6.1

Proof Set $\mathcal{T}_0' \subset \mathcal{T}'$ in $\mathfrak{M}'$ which is deleted for obtaining $\mathfrak{N}'$. Now we determine $\mathcal{T}_0$ in each reduction from 3.1 to 3.6 to obtain $\mathfrak{N}$ from $\mathfrak{M}$. Still write $(p, q) \in X \times Y$ as the leading position of A_1 in $\mathfrak{M}$.

3.1 Set $\mathcal{T}_0 = \mathcal{T}_0' \subset \mathcal{T}' = \mathcal{T}$. If $X \notin \mathcal{T}_0'$, we have also a loop reduction from $\mathfrak{N}$ to $\mathfrak{N}'$; if $X \in \mathcal{T}_0'$, we have an identity from $\mathfrak{N}$ to $\mathfrak{N}'$, the domain of $\mathfrak{N}'$ is preserved.

3.2 Since $(T, \sim) = (T', \sim')$, set $\mathcal{T}_0 = \mathcal{T}_0'$. (a) If $X, Y \notin \mathcal{T}_0'$, we have the same base change from $\mathfrak{N}$ to $\mathfrak{N}'$; (b) if $X \in \mathcal{T}_0'$, $Y \notin \mathcal{T}_0'$, or (c) if $X \notin \mathcal{T}_0'$, $Y \in \mathcal{T}_0'$, or (d) if $X, Y \in \mathcal{T}_0'$, we have an identity from $\mathfrak{N}$ to $\mathfrak{N}'$. In the cases (a) and (d), the domain of $\mathfrak{N}'$ is preserved. In the cases (b) and (c), the domain of the induced matrix problem of $\mathfrak{N}$ contains the domain of $\mathfrak{N}'$, i. e. the domain of $\mathfrak{N}'$ is restricted by the domain of $\mathfrak{M}'$.

3.3 Since $(T, \sim) = (T', \sim')$, set $\mathcal{T}_0 = \mathcal{T}_0'$. If $X, Y \notin \mathcal{T}_0'$, we have also a regularization from $\mathfrak{N}$ to $\mathfrak{N}'$. If X or Y or both of them are in $\mathcal{T}_0'$, we have an identity from $\mathfrak{N}$ to $\mathfrak{N}'$. The domain of $\mathfrak{N}'$ is preserved.

3.4 Set $\mathcal{T}=\mathcal{T}_0'$. If $\mathfrak{M}'$ is obtained from $\mathfrak{M}$ by deletion of $C\subset\mathcal{T}$, then $\mathfrak{N}'$ is obtained from $\mathfrak{N}$ also by deletion of C. The domain of $\mathfrak{N}'$ is preserved.

3.5 If $X_2, X_1\cup Y_2, Y_1\notin\mathcal{T}_0'$, then taking $\mathcal{T}_0=\mathcal{T}_0'$, we have an edge reduction from $\mathfrak{N}$ to $\mathfrak{N}'$. If only one of $X_2, X_1\cup Y_2, Y_1$ belongs to $\mathcal{T}_0'$, or $X_2, Y_1\in\mathcal{T}_0'$ but $X_1\cup Y_2\notin\mathcal{T}_0'$, then letting $\mathcal{T}_0=\mathcal{T}_0'\setminus\{X_2, X_1\cup Y_2, Y_1\}$, we have an edge reduction plus a suitable deletion from $\mathfrak{N}$ to $\mathfrak{N}'$. If X_2, $X_1\cup Y_2\in\mathcal{T}_0'$ but $Y_1\notin\mathcal{T}_0'$, or $X_1\cup Y_2, Y_1\in\mathcal{T}_0'$ but $X_2\notin\mathcal{T}_0'$, or $\{X_2, X_1\cup Y_2, Y_1\}\subseteq\mathcal{T}_0'$, then taking $\mathcal{T}=(\mathcal{T}_0'\setminus\{X_2, X_1\cup Y_2\})\cup\{X\}$, or $\mathcal{T}_0=(\mathcal{T}_0'\setminus\{X_1\cup Y_2, Y_1\})\cup\{Y\}$, or $\mathcal{T}_0=\mathcal{T}_0'$ respectively, we have an identity from $\mathfrak{N}$ to $\mathfrak{N}'$. The domain of $\mathfrak{N}'$ is preserved.

3.6 If some of the equivalent classes induced from X are not in $\mathcal{T}_0'$, taking $\mathcal{T}_0=(\mathcal{T}_0'\setminus\{\text{equivalent classes induced from } X\})$, we also have an unraveling from $\mathfrak{N}$ to $\mathfrak{N}'$ possibly followed by a suitable deletion. If the equivalent classes induced from X are all in $\mathcal{T}_0'$, taking $\mathcal{T}_0=(\mathcal{T}_0'\setminus\{\text{equivalent classes induced from } X\})\cup\{X\}$, we have an identity from $\mathfrak{N}$ to $\mathfrak{N}'$. The domain of $\mathfrak{N}'$ is preserved. The proof of the proposition is finished.

Corollary 6.5 Let $\mathfrak{M}=(\mathcal{K},\mathcal{M},H=0)$ be a matrix problem. Let $\mathfrak{M},\mathfrak{M}^1,\cdots,\mathfrak{M}^s$ be a sequence of reductions with the minimal end term. Suppose in addition that $\mathfrak{M}^1$ is obtained by deletion from $\mathfrak{M}$, and $\mathfrak{M}^l$ is obtained by reductions 3.5+3.4, 3.1+3.6+3.4, 3.2+3.3 from $\mathfrak{M}^{l-1}$ for $1<l\leqslant s$.

(1) Fix any non-trivial $X\in\mathcal{T}^s$, set $\mathcal{T}_0^s=\mathcal{T}^s\setminus\{X\}$. Deleting $\mathcal{T}_0^s$ from $\mathfrak{M}^s$, we have an induced matrix problem $\mathfrak{M}_X^s=(\mathcal{K}_X,(0),\boldsymbol{H}_X^s(x))$. Using Proposition 6.4 inductively, we obtain a sequence of Proposition 5.9 with some restricted domains (some terms may equal in the sequence). Therefore $\mathfrak{M}_X^s$ is a canonical minimal matrix problem.

(2) Fix any trivial $Y\in\mathcal{T}^s$, set $\mathcal{T}_0^s=\mathcal{T}^s\setminus\{Y\}$. Deleting $\mathcal{T}_0^s$ from $\mathfrak{M}^s$,

we have an induced matrix problem $\mathfrak{M}_Y^s=(\mathcal{K}_Y,(0),\boldsymbol{H}_Y^s)$. Using Proposition 6.4 inductively, we obtain a sequence of Proposition 4.3 (some terms may equal in the sequence). Therefore $\boldsymbol{H}_Y^s$ is the canonical form of $(\boldsymbol{H}_Y^s,0)\in \mathrm{Mat}(\mathfrak{M})$.

Lemma 6.6 Let $\mathfrak{M}$ be a matrix problem with trivial $\mathcal{T}$, and $(M,\boldsymbol{H}_{\underline{m}})\in \mathrm{Mat}(\mathfrak{M})$ with $\underline{\boldsymbol{m}}$ being a size vector over $(T,\sim)$. Let $\mathfrak{M}'$ be the induced matrix problem with trivial $\mathcal{T}'$ given by reductions 3.5+3.4, 3.1+3.6+3.4, or 3.2+3.3, such that $\underline{\boldsymbol{m}}$ has an induced size vector $\underline{\boldsymbol{m}}'$ over $(T',\sim')$. Suppose that we have a decomposition in $\mathrm{Mat}(\mathfrak{M})$:

$$\boldsymbol{M}+\boldsymbol{H}_{\underline{m}}=(\boldsymbol{N}+\boldsymbol{H}_{\underline{n}})\oplus_T(\boldsymbol{L}+\boldsymbol{H}_{\underline{l}}) \tag{6.4}$$

Then there are some $(\boldsymbol{N}',\boldsymbol{H}'_{\underline{n}'}),(\boldsymbol{L}',\boldsymbol{H}'_{\underline{l}'})\in \mathrm{Mat}(\mathfrak{M}')$, with $\vartheta(\boldsymbol{N}',\boldsymbol{H}'_{\underline{n}'})\simeq(\boldsymbol{N},\boldsymbol{H}_{\underline{n}})$ and $\vartheta(\boldsymbol{L}',\boldsymbol{H}'_{\underline{l}'})\simeq(\boldsymbol{L},\boldsymbol{H}_{\underline{l}})$ respectively. Let

$$\boldsymbol{M}'+\boldsymbol{H}'_{\underline{m}'}=(\boldsymbol{N}'+\boldsymbol{H}'_{\underline{n}'})\oplus_{T'}(\boldsymbol{L}'+\boldsymbol{H}'_{\underline{l}'})\in \mathrm{Mat}(\mathfrak{M}'). \tag{6.5}$$

Then $\vartheta(\boldsymbol{M}',\boldsymbol{H}'_{\underline{m}'})\simeq(\boldsymbol{M},\boldsymbol{H}_{\underline{m}})$.

The conclusion is still valid for a direct sum of finitely many summands.

Proof The existence of $(\boldsymbol{N}',\boldsymbol{H}'_{\underline{n}'})$ and $(\boldsymbol{L}',\boldsymbol{H}'_{\underline{l}'})$ in $\mathrm{Mat}(\mathfrak{M}')$ is clear, since $\underline{\boldsymbol{n}},\underline{\boldsymbol{l}}<\underline{\boldsymbol{m}}$. Write (p,q) as the leading position of $\boldsymbol{A}_1$ in $\mathfrak{M}$, then $\boldsymbol{M}_{pq}=\boldsymbol{N}_{pq}\oplus_T\boldsymbol{L}_{pq}$. Suppose that for some $\boldsymbol{S}_1=(\boldsymbol{S}_{ij}^1)\in\mathcal{K}_{\underline{n}}$ and $\boldsymbol{S}_2=(\boldsymbol{S}_{ij}^2)\in\mathcal{K}_{\underline{l}}$, we have

$$\boldsymbol{N}'+\boldsymbol{H}'_{\underline{n}'}=\boldsymbol{S}_1^{-1}(\boldsymbol{N}+\boldsymbol{H}_{\underline{n}})\boldsymbol{S}_1 \text{ and } \boldsymbol{L}'+\boldsymbol{H}'_{\underline{l}'}=\boldsymbol{S}_2^{-1}(\boldsymbol{L}+\boldsymbol{H}_{\underline{l}})\boldsymbol{S}_2.$$

Construct $\boldsymbol{S}_0=(\boldsymbol{S}_{ij}^0)\in\mathcal{K}_{\underline{m}}$, such that $\boldsymbol{S}_{ij}^0=\begin{pmatrix}\boldsymbol{S}_{ij}^1 & \boldsymbol{0}\\ \boldsymbol{0} & \boldsymbol{S}_{ij}^2\end{pmatrix}$. Thus

$$\boldsymbol{S}_0^{-1}(\boldsymbol{M}+\boldsymbol{H}_{\underline{m}})\boldsymbol{S}_0=\boldsymbol{S}_0^{-1}((\boldsymbol{N}+\boldsymbol{H}_{\underline{n}})\oplus_T(\boldsymbol{L}+\boldsymbol{H}_{\underline{l}}))\boldsymbol{S}_0=(\boldsymbol{N}'+\boldsymbol{H}'_{\underline{n}'})\oplus_T(\boldsymbol{L}'+\boldsymbol{H}'_{\underline{l}'}).$$

By Lemma 6.3, there is some invertible $R\in\mathcal{K}_{\underline{m}}$ given by exchanging rows and columns, such that

$$\boldsymbol{R}((\boldsymbol{N}'+\boldsymbol{H}'_{\underline{n}'})\oplus_T(\boldsymbol{L}'+\boldsymbol{H}'_{\underline{l}'}))\boldsymbol{R}^{-1}=(\boldsymbol{N}'+\boldsymbol{H}'_{\underline{n}'})\oplus_{T'}(\boldsymbol{L}'+\boldsymbol{H}'_{\underline{l}'})=\boldsymbol{M}'+\boldsymbol{H}'_{\underline{m}'}.$$

Thus $\boldsymbol{M}'+\boldsymbol{H}'_{\underline{m}'}=\boldsymbol{S}^{-1}(\boldsymbol{M}+\boldsymbol{H}_{\underline{m}})\boldsymbol{S}$ for $\boldsymbol{S}=\boldsymbol{S}_0\boldsymbol{R}^{-1}\in\mathcal{K}\,\underline{\boldsymbol{m}}$. □

Lemma 6.7 Let $\mathfrak{M}=(\mathcal{K},\mathcal{M},\boldsymbol{H}=\boldsymbol{0})$ be a matrix problem.

(1) If $(\boldsymbol{M},\boldsymbol{0})\in \mathrm{Mat}(\mathfrak{M})$ has the canonical form $\boldsymbol{M}$ and the canonical sequence (4.2) given in Proposition 4.3, then

$$\boldsymbol{M}^i+\boldsymbol{H}^i_{\underline{m}^i}=\boldsymbol{M}\ \text{for}\ i=1,2,\cdots,s.$$

(2) If $\mathfrak{M}$ is tame, and $\mathfrak{N}=(\mathcal{K}_X,0,\boldsymbol{M}(x))$ is a canonical minimal matrix problem induced from $\mathfrak{M}$ having the sequences (5.4) given in Proposition 5.9, then $x'=x$ and $\boldsymbol{M}^i(x)+\boldsymbol{H}^i_{\underline{m}^i}(x)=\boldsymbol{M}(x)$ for $i=1$, $2,\cdots,s$.

Key Corollary 6.8 Let $\mathfrak{M}=(\mathcal{K},\mathcal{M},\boldsymbol{H}=\boldsymbol{0})$ be a matrix problem. Suppose that $\mathfrak{L}_j=(\mathcal{K}_{X_j},(0),M_j(x_j))$ with the domain $k\backslash\{$the roots of $g_j(x)\}$ for $j=1,2,\cdots,r$ are canonical minimal matrix problems induced from $\mathfrak{M}$. And $\boldsymbol{N}_j$ are the canonical forms of $(\boldsymbol{N}_j,\boldsymbol{0})\in\mathrm{Mat}(\mathfrak{M})$ for $j=1$, $2,\cdots,l$. Define a parameterized matrix:

$$\overline{\boldsymbol{M}}=\boldsymbol{M}_1(x_1)\oplus_T\cdots\oplus_T\boldsymbol{M}_r(x_r)\oplus_T\boldsymbol{N}_1\cdots\oplus_T\boldsymbol{N}_l.$$

Then there is a sequence of reductions with respect to $\overline{\boldsymbol{M}}$: $\mathfrak{M}$, $\mathfrak{N}^1,\cdots,\mathfrak{N}^s$ with the index sets of $\mathfrak{N}^i$ being $(\overline{T}^i,\overline{\sim}^i)$, where $\mathfrak{N}^1$ is obtained by deletion from $\mathfrak{M}$; $\mathfrak{N}^i$ is obtained by one of the reductions 3.5+3.4, 3.1+3.6+3.4, 3.2+3.3 from $\mathfrak{N}^{i-1}$ for $1<i\leqslant s$; and the end term $\mathfrak{N}^s$ is minimal. Furthermore for $i=1,2,\cdots,s$, we have a parameterized matrix over $\mathfrak{N}^i$:

$$(\boldsymbol{M}_1^i(x_1)+\boldsymbol{H}^i_{\underline{m}_1^i}(x_1))\oplus_{\overline{T}^i}\cdots\oplus_{\overline{T}^i}(\boldsymbol{M}_r^i(x_r)+\boldsymbol{H}^i_{\underline{m}_r^i}(x_r))\oplus_{\overline{T}^i}(\boldsymbol{N}_1^i(x_r)+\boldsymbol{H}^i_{\underline{n}_1^i})\oplus_{\overline{T}^i}\cdots\oplus_{\overline{T}^i}(\boldsymbol{N}_l^i+\boldsymbol{H}^i_{\underline{n}_l^i}),\tag{6.6}$$

where its size vector $\underline{\boldsymbol{m}}^i$ over $(\overline{T}^i,\overline{\sim}^i)$ is sincere; $\boldsymbol{M}_j^i(x_j)+\boldsymbol{H}^i_{\underline{m}_r^i}(x_j)=\boldsymbol{M}_j(x_j)$ and $\boldsymbol{N}_j^i+\boldsymbol{H}^i_{\underline{n}_j^i}=\boldsymbol{N}_j$; i is obtained from $(i-1)$ by exchanging some rows and columns; in particular, the domain of x_j in $\mathfrak{N}^i$ is a restriction of that in $\mathfrak{L}_j$ for $1\leqslant j\leqslant r, 1\leqslant i\leqslant s$.

Proof Collect all the numbers in k which occur in $\boldsymbol{M}_j(x_j)$, $j=1$, $2,\cdots,r$, or in $\boldsymbol{N}_j$, $j=1,2,\cdots,l$, we obtain a finite set $\mathcal{S}$. Take some $\lambda_j\in k$, such that $g_j(\lambda_j)\neq 0$, and $\lambda_j>\alpha$ for any $\alpha\in\mathcal{S}$, define a matrix $(\boldsymbol{M},0)\in\mathrm{Mat}(\mathfrak{M})$ with

$$\boldsymbol{M}=\boldsymbol{M}_1(\lambda_1)\oplus_T\cdots\oplus_T\boldsymbol{M}_r(\lambda_r)\oplus_T\boldsymbol{N}_1\oplus\cdots\oplus_T\boldsymbol{N}_l,$$

where $\boldsymbol{M}_j(\lambda_j)$ is the canonical form of $(\boldsymbol{M}_j(\lambda_j),0)\in\mathrm{Mat}(\mathfrak{M})$ by Corollary 4.7. Then Proposition 4.3 gives a pair of reduction sequences

(4.1): $\mathfrak{M}, \mathfrak{M}^1, \cdots, \mathfrak{M}^s$ and (4.2) with respect to M. Write $(T^i, \sim^i)$ as the index set of $\mathfrak{M}^i$ in (4.1) for $i=1,2,\cdots,s$. Furthermore by an induction on the reduction steps, Lemma 6.6 ensures that $\boldsymbol{M}^i + \boldsymbol{H}^i_{\underline{\boldsymbol{m}}^i}$ is a direct sum:

$$(\boldsymbol{M}_1^i(\lambda_1) + \boldsymbol{H}^i_{\underline{\boldsymbol{m}}_1^i}(\lambda_1)) \oplus_{T^i} \cdots \oplus_{T^i} (\boldsymbol{M}_r^i(\lambda_r) + \boldsymbol{H}^i_{\underline{\boldsymbol{m}}_r^i}(\lambda_r)) \oplus_{T^i} (\boldsymbol{N}_1^i + \boldsymbol{H}^i_{\underline{\boldsymbol{n}}_1^i}) \oplus_{T^i} \cdots \oplus_{T^i} (\boldsymbol{N}_l^i + \boldsymbol{H}^i_{\underline{\boldsymbol{n}}_l^i}) \tag{6.7}$$

where $\boldsymbol{M}_j^i(\lambda_j) + \boldsymbol{H}^i_{\underline{\boldsymbol{m}}_r^i}(\lambda_j) = \boldsymbol{M}_j(\lambda_j)$ for $j=1,2,\cdots,r$ and $\boldsymbol{N}_j^i + \boldsymbol{H}^i_{\underline{\boldsymbol{n}}_j^i} = \boldsymbol{N}_j$ for $j=1,2,\cdots,l$ by Lemma 6.7, since $\boldsymbol{M}_j(\lambda)$ and $\boldsymbol{N}_j$ are both canonical forms.

By Theorem 6.3, the direct sum over $(T^{i-1}, \sim^{i-1})$ is obtained from the direct sum over $(T^i, \sim^i)$ by exchanging some rows and columns for $i=s, s-1, \cdots, 2$. Conversely the direct sum over $(T^i, \sim^i)$ is obtained from the direct sum over $(T^{i-1}, \sim^{i-1})$ by exchanging some rows and columns for $i=2,3,\cdots,s$.

Now we show that the sequence (4.1) yields a sequence $\mathfrak{M}, \mathfrak{N}_1, \mathfrak{N}_2, \cdots, \mathfrak{N}_s$, and the formula (6.7) yields the formula (6.6).

We first delete $\overline{\mathcal{T}}_0 = \{X \in \overline{\mathcal{T}} \mid m_X = 0\}$ with $\underline{\boldsymbol{m}}$ being the size vector of $\overline{\boldsymbol{M}}$, which is also the size vector of M to obtain a matrix problem $\mathfrak{N}^1$. Suppose that we have reached to $\mathfrak{N}^{i-1}$ and a direct sum over $(\overline{T}^{i-1}, \overline{\sim}^{i-1})$ given by (6.6). Consider the reduction from $\mathfrak{N}^{i-1}$ to $\mathfrak{N}^i$ suppose that the leading position of A_1 in $\mathfrak{N}^{i-1}$ is $(p,q) \in X \times Y$.

If the formula (3.4) shows that $n_{pq} r_{qq} - r_{pp} n_{pq} = 0$, we perform the reduction 3.5+3.4 in the case of $X \neq Y$, or 3.1+3.6+3.4 in the case of $X=Y$, being the same as that from $\mathfrak{M}^{i-1}$ to $\mathfrak{M}^i$. To obtain $\mathfrak{N}^i$ from $\mathfrak{N}^{i-1}$, the only difference is that we must substitute x for λ, whenever $X=Y$ and $\lambda \in \{\lambda_1, \lambda_2, \cdots, \lambda_r\}$ first occurs in $\mathfrak{M}^i$. By exchanging some rows and columns at $(i-1)$-th term of (6.6), (being the same as those of (6.7)), we obtain a direct sumover $(\overline{T}^i, \overline{\sim}^i)$ with $\boldsymbol{M}_j^i(x_j) + \boldsymbol{H}^i_{\underline{\boldsymbol{m}}_r^i}(x_j) = \boldsymbol{M}_j(x_j)$ for $j=1,2,\cdots,r$ and $\boldsymbol{N}_j^i + \boldsymbol{H}^i_{\underline{\boldsymbol{n}}_j^i} = \boldsymbol{N}_j$ for $j=1,2,\cdots,l$, i. e. the i-th term of (6.6).

If the formula (3.4) shows $n_{pq}r_{qq}-r_{pp}n_{pq}\neq 0$, we perform 3.2+3.3 to obtain $\mathfrak{N}^i$. If X, Y are both trivial, we have the same procedure as that from $\mathfrak{M}^{i-1}$ to $\mathfrak{M}^i$; if X or Y is non-trivial, (3.2) gives some localization of $k[x, g_X^{i-1}(x)^{-1}]$ or $k[y, g_Y^{i-1}(y)^{-1}]$, thus if $x=x_j$ or $y=x_j$, then the domain of x_j in $\mathfrak{N}^i$ is a restriction of that in $\mathfrak{L}_j$. The direct sum (6.6) over $(\overline{T}^{i-1}, \overline{\sim}^{i-1})$ is also a direct sum over $(\overline{T}^i, \overline{\sim}^i)$. The corollary is proved by induction.

Remark In the last step of the above proof, the following localization is useful when X and Y are both non-trivial If $X\neq Y, x=x_p$ and $y=x_q$, we make localization first for $k[x, g_X^{i-1}(x)]$ and then for $k[y, g_Y^{i-1}(y)]$, thus obtaining a restriction on the domain of $\mathfrak{L}_p$ and that of $\mathfrak{L}_q$ respectively. If $X=Y, x=y=x_p$, we make localization first for $k[x, g_X^{i-1}(x)]$ given by the left multiplication and then for $k[y, g_X^{i-1}(y)]$ given by the left multiplication, thus obtaining a restriction on the domain of $\mathfrak{L}_p$.

Proposition 6.9[13,14] Let $\mathfrak{M}$ be a matrix problem, then Mat($\mathfrak{M}$) is a Krull-Schmidt category.

Proof Let $(\boldsymbol{M}, \boldsymbol{H}_{\underline{m}})\in$ Mat($\mathfrak{M}$), we have

$$(\boldsymbol{M}, \boldsymbol{H}_{\underline{m}})\simeq(\boldsymbol{M}_1, \boldsymbol{H}_{\underline{m}_1})\oplus_T\cdots\oplus_T(\boldsymbol{M}_r, \boldsymbol{H}_{\underline{m}_r})$$

with $(\boldsymbol{M}_i, \boldsymbol{H}_{\underline{m}_i})\in$ Mat($\mathfrak{M}$) indecomposable for $i=1,2,\cdots,r$, since $m=\sum_{i\in T} m_i$ is finite.

Suppose that $(\boldsymbol{M}+\boldsymbol{H}_{\underline{m}})\simeq(\boldsymbol{N}_1+\boldsymbol{H}_{\underline{n}_1})\oplus_T\cdots\oplus_T(\boldsymbol{N}_{r'}+\boldsymbol{H}_{\underline{n}_{r'}})$ with $(\boldsymbol{N}_j, \boldsymbol{H}_{\underline{n}_j})\in$ Mat($\mathfrak{M}$) indecomposable is another decomposition. Let (4.1), (4.2) be the canonical sequences of M, then the canonical forms of $(\boldsymbol{M}_i, \boldsymbol{H}_{\underline{m}_i})$, $(\boldsymbol{N}_j, \boldsymbol{H}_{\underline{n}_j})$ for $1\leqslant i\leqslant r$, $1\leqslant j\leqslant r'$ are direct summands of $\boldsymbol{H}^s_{\underline{m}^s}$ by Lemma 6.6. But the sequence (4.1) and its end term are uniquely determined, and Mat($\mathfrak{M}^s$) is a Krull-Schmidt category. Therefore $r=r'$ and $(\boldsymbol{M}_i, \boldsymbol{H}_{\underline{m}_i})\simeq(\boldsymbol{N}_i, \boldsymbol{H}_{\underline{n}_i})$ under some reordering. The proposition is proved.

Remark The key point in Sections 4~6 is that the eigenvalues in any Weyr matrices have a fixed order. Otherwise any interchange of the eigenvalues causes the interchange of the leading positions of the normalized basis, consequently yielding some different reduction sequences, see Section 7 below.

§ 7. The unified tame theorem

Definition 7.1[2] Let $\mathfrak{M}$ be a matrix problem, and $\underline{m}=(m_1, m_2, \cdots, m_t)$ be a size vector over $(T, \sim)$. Then $\|\underline{m}\| = \sum_X m_{X^2} + \sum_{i=1}^{n} m_{X_i} m_{Y_i}$ is said to be the norm of $\underline{m}$, where X runs over the non-trivial equivalent classes of $\mathcal{T}$ and the normalized basis $A_i \in \mathcal{A}_{X_i Y_i}$ for $i = 1, 2, \cdots, n$.

Tame Theorem 7.2 [1,2] Let $\mathfrak{M}$ be a matrix problem of representation tame type. Then given a positive integer n,

(1) there exist finitely many minimal matrix problems $\mathfrak{M}_1', \mathfrak{M}_2', \cdots, \mathfrak{M}_s'$, each of them is obtained from $\mathfrak{M}$ by a sequence of reductions first by 3.4, then by 3.1, or 3.5+3.4, or 3.2+3.6+3.4+3.3 step by step;

(2) there are functors $\vartheta_i : \mathrm{Mat}(\mathfrak{M}_i') \to \mathrm{Mat}(\mathfrak{M})$, such that for any matrix $(\boldsymbol{M}, \boldsymbol{0}) \in \mathrm{Mat}(\mathfrak{M})$ of size at most n, there are some $1 \leqslant i \leqslant s$ and a matrix $(\boldsymbol{0}, \boldsymbol{N}) \in \mathrm{Mat}(\mathfrak{M}_i')$ with $\vartheta_i(\boldsymbol{0}, \boldsymbol{N}) \simeq (\boldsymbol{M}, \boldsymbol{0})$.

Proof The set of size vectors over $(T, \sim)$, $\{\underline{m} = (m_1, m_2, \cdots, m_t) \mid \sum_{j=1}^{t} m_j \leqslant n\}$, is finite. So we have a maximal norm d among those size vectors. If $d=0$, then $\mathfrak{M}$ is already trivial, we are done. Otherwise we will lower down the norms of the induced size vectors in finitely many induced matrix problems. Since d is finite, by induction we finally reach to some minimal matrix problems.

Suppose that $\mathcal{T}$ contains r equivalent classes, the number of the subsets of $\mathcal{T}$ equals 2^r. Then by deletion of each subset of $\mathcal{T}$, we obtain finitely many induced matrix problems. Moreover for any matrix $(\boldsymbol{M}, \boldsymbol{H}_{\underline{m}}) \in \mathrm{Mat}(\mathfrak{M})$ with $m \leqslant n$, set $\mathcal{T}_0 = \{X \mid m_X = 0\}$, and $\mathfrak{M}'$ is obtained by deletion of $\mathcal{T}_0$. Then $(\boldsymbol{M}, \boldsymbol{H}_{\underline{m}})$ is the image of some $(\boldsymbol{M}', \boldsymbol{H}_{\underline{m}'}) \in \mathrm{Mat}(\mathfrak{M}')$ under the functor given by deletion, and $\underline{\boldsymbol{m}}' = \underline{\boldsymbol{m}}$ is sincere over the index set of $\mathfrak{M}'$.

Suppose now we have obtained finitely many induced matrix problems after sequences of reductions given in (1) and for any matrix $(\boldsymbol{M}, \boldsymbol{H}_{\underline{m}}) \in \mathrm{Mat}(\mathfrak{M})$ with $m \leqslant n$, there exists an induced matrix problem, say $\mathfrak{N}$, such that $(\boldsymbol{M}, \boldsymbol{H}_{\underline{m}})$ is isomorphic to the image of some $(\boldsymbol{N}, \boldsymbol{H}_{\underline{n}}(\underline{U})) \in \mathrm{Mat}(\mathfrak{N})$ under the reduction functor, and $\underline{\boldsymbol{n}}$ is sincere over the index set of $\mathfrak{N}$. Denote by $(p, q) \in X \times Y$ the leading position of $\boldsymbol{A}_1$ in $\mathfrak{N}$.

If the formula (3.4) shows $n_{pq} r_{qq} - r_{pp} n_{pq} = 0$, then X, Y must be trivial by Corollary 5.2.

(1) In the case of $X = Y$, we perform 3.1 loop reduction and obtain a unique induced matrix problem $\mathfrak{N}'$, then $\underline{\boldsymbol{n}}' = \underline{\boldsymbol{n}}$ is sincere over the index set of $\mathfrak{N}'$ with $\| \underline{\boldsymbol{n}}' \| = \| \underline{\boldsymbol{n}} \|$. If we always meet the loop reduction until obtaining a minimal matrix problem, we are done. Otherwise we must meet some other reductions.

(2) In the case of $X \neq Y$, we perform 3.5 edge reduction, and then 3.4 deletions of $\varnothing$, $\{X_2\}$, $\{Y_1\}$, $\{X_1 \cup Y_2\}$, or $\{X_2, Y_1\}$ respectively, obtain 5 induced matrix problems. There must exist one of them, say $\mathfrak{N}'$, such that $\underline{\boldsymbol{n}}$ has an induced size vector $\underline{\boldsymbol{n}}'$, which is sincere over the index set of $\mathfrak{N}'$ with $\| \underline{\boldsymbol{n}}' \| = \| \underline{\boldsymbol{n}} \| - n_X n_Y < \| \underline{\boldsymbol{n}} \|$.

If the formula (3.4) shows $n_{pq} r_{qq} - r_{pp} n_{pq} \neq 0$,

(3) we perform the base change of 3.2 first, and obtain a new formula (3.4) $n_{pq} r_{qq} - r_{pp} n_{pq} = v_1'$. If X, Y are both trivial, or if X is non-trivial, and $c(x)$ given in 3.2 is invertible in $k[x, g_X(x)^{-1}]$, then after 3.3 regularization, we have an induced matrix problem $\mathfrak{N}'$. It is clear that $\underline{\boldsymbol{n}}' = \underline{\boldsymbol{n}}$ is sincere in $\mathfrak{N}'$ and $\| \underline{\boldsymbol{n}}' \| = \| \underline{\boldsymbol{n}} \| - n_X n_Y < \| \underline{\boldsymbol{n}} \|$.

If X is non-trivial, then there is a complete set $\{\lambda_1, \lambda_2, \cdots, \lambda_r\}$ with $g_X(\lambda_j) \neq 0$ but $c(\lambda_j) = 0$ for $j = 1, 2, \cdots, r$. Set d to be the largest integer with $t + d \leqslant n$. Let W be the Weyr matrix similar to

$$\bigoplus_{j=1}^{r}(\boldsymbol{J}_1(\lambda_j)\oplus\boldsymbol{J}_2(\lambda_j)\oplus\cdots\oplus\boldsymbol{J}_d(\lambda_j)) \text{ and } \boldsymbol{U}=\boldsymbol{W}\oplus(x')$$

defined in 3.6 with $g_{X'}(x')=g_X(x')c(x')$. Then we have an induced matrix problem $\mathfrak{N}$ given by 3.6 unraveling. Since the set of the equivalent classes determined by $\boldsymbol{U}$ is finite, the number of the subsets is also finite. Then after deletion of each subset given in Case 3.4, we have finitely many induced matrix problems of $\widetilde{\mathfrak{N}}$. There must exist one of them, say $\mathfrak{N}''$, such that $\underline{\boldsymbol{n}}$ has an induced size vector $\underline{\boldsymbol{n}}'$, which is sincere over the index set of $\mathfrak{N}''=(\mathcal{K}'',\mathcal{M}'',H'')$. Finally we use regularization 3.3 to take off the first base element of $\mathcal{M}''$, thus obtaining an induced matrix problem $\mathfrak{N}'$. Then $\underline{\boldsymbol{n}}'$ is still a size vector over the index set of $\mathfrak{N}'$ and

$$\|\boldsymbol{n}'\|=\|\boldsymbol{n}\|-n_X^2+{n'_X}^2-\begin{cases}n'_{X'}n_Y, & \text{if } X\neq Y.\\ n'_{X'}n'_{X'}, & \text{if } X=Y\end{cases}<\|\boldsymbol{n}\|.$$

Let ϑ: Mat$(\mathfrak{N}')\to$ Mat$(\mathfrak{N})$ be the reduction functor given in the above 3 cases, then there is a matrix $(\boldsymbol{N}',\boldsymbol{H}'_{\underline{n}'}(\underline{\boldsymbol{U}}'))\in$ Mat$(\mathfrak{N}')$ with $(\boldsymbol{N},\boldsymbol{H}_{\underline{n}}(\underline{\boldsymbol{U}}))=\vartheta(\boldsymbol{N}',\boldsymbol{H}'_{\underline{n}'}(\underline{\boldsymbol{U}}'))$. The tame theorem follows by induction.

Remark 7.3 Collection of local minimal matrix problems.

From now on, we assume that $\mathfrak{M}=(\mathcal{K},\mathcal{M},\boldsymbol{H}=\boldsymbol{0})$ is a tame matrix problem, n is a fixed positive integer. Then Theorem 7.2 gives rise to a set of finitely many minimal matrix problems $\{\mathfrak{M}'_1,\mathfrak{M}'_2,\cdots,\mathfrak{M}'_q\}$ of sizes at most n. If $\mathcal{T}'_i=\{X_1,X_2,\cdots,X_l\}$ in $\mathfrak{M}'_i$, by deletion of $\mathcal{T}'_i\setminus\{X_j\}$ we have l local minimal matrix problems. Let $\mathcal{S}=\{$minimal local matrix problems $\mathfrak{L}'_j$ obtained by deletion from $\mathfrak{M}'_i$ for $i=1,2,\cdots,q\}$. Set $\mathfrak{L}'_j=(\mathcal{K}'_j,(0),\boldsymbol{H}'_j)$ in the case of $\mathfrak{L}'_j$ trivial; and $\mathfrak{L}'_j=(\mathcal{K}'_j,(0),\boldsymbol{H}'_j(x'_j))$ in the case of $\mathfrak{L}'_j$ non-trivial.

Suppose $\mathfrak{L}'_j=(\mathcal{K}'_1,(\boldsymbol{0}),\boldsymbol{H}'_1)$ is trivial, then there is a complete subset of the trivial matrix problems of $\mathcal{S}$:

$$\mathfrak{L}'_2=(\mathcal{K}'_2,(\boldsymbol{0}),\boldsymbol{H}'_2),\cdots,\mathfrak{L}'_p=(\mathcal{K}'_p,(\boldsymbol{0}),\boldsymbol{H}'_p),$$

such that $(\boldsymbol{H}'_j,\boldsymbol{0})$ is isomorphic to $(\boldsymbol{H}'_1,\boldsymbol{0})$ in Mat$(\mathfrak{M})$ for $2\leqslant j\leqslant p$. Then by Corollary 4.5, $(\boldsymbol{H}'_1,\boldsymbol{0}),(\boldsymbol{H}'_2,\boldsymbol{0}),\cdots,(\boldsymbol{H}'_p,\boldsymbol{0})$ have the same canonical form say $\boldsymbol{N}_1$.

If $\mathfrak{L}_1'$ is non-trivial, then there is a complete subset of the non-trivial matrix problems of $\mathcal{S}$: $\mathfrak{L}_2', \mathfrak{L}_3', \cdots, \mathfrak{L}_p'$, such that $\mathfrak{L}_j'$ are quasi-isomorphic to $\mathfrak{L}_1'$ induced from $\mathfrak{M}$ for $2 \leqslant j \leqslant p$, see Definition 5.10. Then by Corollary 5.11, $\mathfrak{L}_1', \mathfrak{L}_2', \cdots, \mathfrak{L}_p'$ have a canonical minimal matrix problem in common up to domains. Let $D_1^0 = k \setminus \{\text{the roots of } g_1^0(x_1)\}$ be the intersection of all the domains of those canonical minimal matrix problems. Denote the common canonical minimal matrix problem by $\mathfrak{L}_1 = (\mathcal{K}_1, (0), \boldsymbol{M}_1(x_1))$ with the domain D_1^0.

Now consider the set $\mathcal{S} \setminus \{\mathfrak{N}_1', \mathfrak{N}_2', \cdots, \mathfrak{N}_p'\}$. If this set is non empty, we continue the procedure above, and obtain again some canonical form or canonical minimal matrix problem. Since $\mathcal{S}$ is a finite set, we are able to reach two sets: (1) $\{\mathfrak{L}_1, \mathfrak{L}_2, \cdots, \mathfrak{L}_r\}$, where $\mathfrak{L}_j = (\mathcal{K}_j, (0), \boldsymbol{M}_j(x_j))$ with the domains $D_j^0 = k \setminus \{\text{the roots of } g_j^0(x)\}$ is canonical minimal matrix problem induced from $\mathfrak{M}$, which is pairwise non quasi-isomorphic for $j = 1, 2, \cdots, r$; and (2) a set of canonical forms $\{\boldsymbol{N}_1, \boldsymbol{N}_2, \cdots, \boldsymbol{N}_l\}$ of $\{(\boldsymbol{N}_1, \boldsymbol{0}), (\boldsymbol{N}_2, \boldsymbol{0}), \cdots, (\boldsymbol{N}_l, \boldsymbol{0})\}$ in $\mathrm{Mat}(\mathfrak{M})$, which is pairwise non-isomorphic. Moreover we may assume that for any $1 \leqslant j \leqslant l$, $(\boldsymbol{N}_j, \boldsymbol{0})$ is not isomorphic to any $\boldsymbol{M}_i(\boldsymbol{J}_d(\lambda_i))$ for $g_i^0(\lambda_i) \neq 0$, $1 \leqslant i \leqslant r$ and $d \in \mathbf{N}$.

Define a parameterized matrix:

$$\overline{\boldsymbol{M}} = \boldsymbol{M}_1(x_1) \oplus_T \cdots \oplus_T \boldsymbol{M}_r(x_r) \oplus_T \boldsymbol{N}_1 \oplus_T \cdots \oplus_T \boldsymbol{N}_l. \tag{7.1}$$

By Proposition 6.9, there is a sequence of reductions with respected to the matrix (7.1)

$$\mathfrak{M}, \mathfrak{N}^1, \mathfrak{N}^2, \cdots, \mathfrak{N}^s \tag{7.2}$$

with a minimal end term, and the domains of x_j in $\mathfrak{M}^i$ are restrictions of those in $\mathfrak{L}_j$ for $1 \leqslant j \leqslant r$, $1 \leqslant i \leqslant s$. In particular, we determine the domains here step by step according to Remark given in Proposition 6.9. Let $D = D_1 \times D_2 \cdots \times D_r$ be the domain of $\mathfrak{M}^s$, define a set of matrices $\{\boldsymbol{M}_j(\boldsymbol{J}_d(\mu_j^u)) \mid \text{the size at most } n; \mu_j^u \in D_j^0 \setminus D_j, j = 1, 2, \cdots, r\}$, then $\boldsymbol{M}_j(\boldsymbol{J}_d(\mu_j^u))$ are canonical forms of $(\boldsymbol{M}_j(\boldsymbol{J}_d(\mu_j^u)), \boldsymbol{0}) \in \mathrm{Mat}(\mathfrak{M})$ by Corollary 4.7. Define the following direct sums over $(T, \sim)$:

$$\boldsymbol{L}_j = \oplus_{(u,d)} \boldsymbol{M}_j(\boldsymbol{J}_d(\mu_j^u)), \quad \boldsymbol{L} = \oplus_j \boldsymbol{L}_j.$$

Lemma 7.4 Let $\mathfrak{N}=(\mathcal{K},\mathcal{M},\boldsymbol{H}(x,y))$ be a matrix problem with $\mathcal{T}=\{X,Y\}$ being non-trivial (possible $X=Y$), let $D_X=k\backslash\{$the roots of $g_X(x)\}$, $D_Y=k\backslash\{$the roots of $g_Y(y)\}$ be the domains of x and y respectively.

Suppose that $\mathfrak{N},\mathfrak{N}^1,\mathfrak{N}^2,\cdots,\mathfrak{N}^s$ is a sequence, such that $\mathfrak{N}^{l+1}$ is obtained from $\mathfrak{N}^l$ by a base change, whose localization is made according to Remark of Corollary 6.9, and then a regularization for $l=0,1,\cdots,s-1$. Thus we have a sequence of localizations:

$$k[x,y,g_X(x)^{-1},g_Y(y)^{-1}],k[x,y,g_X^1(x)^{-1},$$
$$g_Y^1(y)^{-1}],\cdots,k[x,y,g_X^s(x)^{-1},g_Y^s(y)^{-1}].$$

Let $\boldsymbol{L}_j$ be a Weyr matrix similar to the Jordan form $\boldsymbol{J}_1(\lambda_j)\oplus\boldsymbol{J}_2(\lambda_j)\oplus\cdots\oplus\boldsymbol{J}_d(\lambda_j)$ with $g_Y^0(\lambda_j)\neq 0$ and $g_Y(\lambda_j)\neq 0$ for $j=1,2,\cdots,r$ and

$$\boldsymbol{L}=\boldsymbol{L}_1\oplus\boldsymbol{L}_2\oplus\cdots\oplus\boldsymbol{L}_r.$$

Assume that $\mathfrak{M}$ is given from $\mathfrak{N}$ by substituting $\boldsymbol{L}$ for y, and the domain of x is $k\backslash\{$the roots of $g_X^s(x)\}$, then there is a sequence of reductions $\mathfrak{M},\mathfrak{M}^1,\cdots,\mathfrak{M}^{s'}$, such that

(1) if $\boldsymbol{A}_1^l\in\mathcal{A}_{XZ}^l$, or $\mathcal{A}_{ZX}^l$, or $\mathcal{A}_{XX}^l$ for some trivial Z determined by L in $\mathfrak{M}^l$, then $\mathfrak{M}^{l+1}$ is obtained by a base change without any localization, followed by a regularization from $\mathfrak{M}^l$ for $l=0,1,\cdots,s'-1$;

(2) if $\boldsymbol{A}_1^l\in\mathcal{A}_{ZZ'}^l$ for some trivial Z,Z' determined by L, then $\mathfrak{M}^{l+1}$ is obtained from $\mathfrak{M}^l$ by 3.2+3.3, 3.5+3.4, or 3.1+3.6+3.4 with $\bar{n}_{pq}^l=(0)$ in the last two cases, see Section 3.

Proof Suppose $\mathcal{M}$ of $\mathfrak{N}$ has a normalized basis $\mathcal{A}=\{\boldsymbol{A}_1,\boldsymbol{A}_2,\cdots,\boldsymbol{A}_n\}$ with the leading position $(p_i,q_i)\in X_i\times Y_i$ of A_i and $\mathcal{V}=(\boldsymbol{V}_1,\boldsymbol{V}_2,\cdots,\boldsymbol{V}_m)$. Recall the general matrices and general equation (3.1) to be $(\mathcal{N}+\boldsymbol{H})R=R(\mathcal{N}+\boldsymbol{H})$ of $\mathfrak{M}$ given in Section 3. Set the (p_i,q_i)-entries of (3.1) to be

$$a_ie_{Y_i}-e_{X_i}a_i=\sum_j h_{ij}(x,y)v_j+\sum_j h'_{ij}(a_1,\bar{a}_1,\cdots,a_{i-1},\bar{a}_{i-1})v_j,$$

where according to 3.2 and the assumption that the localizations are made according to the Remark of Proposition 6.9, $h_{ij}(x,y)\in k[x,$

$g_X^s(x)^{-1}]$; and $h'_{ij}(a_1, \bar{a}_1, \cdots, a_{i-1}, \bar{a}_{i-1})$ are polynomials on $a_1, \bar{a}_1, \cdots, a_{i-1}, \bar{a}_{i-1}$ over $k[x, g_X^s(x)^{-1}]$, whose monomials must involve some a_l or $\bar{a}_l$. In particular, write

$$(a_i e_{Y_i} - e_{X_i} a_i)^0 = \sum_j h_{ij}(x, y) v_j.$$

Suppose that $\mathcal{A}_{XY} = \{\boldsymbol{B}_1, \boldsymbol{B}_2, \cdots, \boldsymbol{B}_{n_1}\}$, $\mathcal{V}_{XY} = \{\boldsymbol{U}_1, \boldsymbol{U}_2, \cdots, \boldsymbol{U}_{m_1}\}$; and $\mathcal{A}_{YX} = \{\boldsymbol{C}_1, \boldsymbol{C}_2, \cdots, \boldsymbol{C}_{n_2}\}$, $\mathcal{V}_{YX} = \{\boldsymbol{W}_1, \boldsymbol{W}_2, \cdots, \boldsymbol{W}_{m_1}\}$. It is clear that the following matrices over $k[x, g_X^s(x)^{-1}]$ have rank n_1 and n_2 respectively:

$$\boldsymbol{C}^1(x, y) = \begin{pmatrix} h^1_{11}(x, y) & \cdots & h^1_{1m_1}(x, y) \\ \vdots & & \vdots \\ h^1_{n_1 1}(x, y) & \cdots & h^1_{n_1 m_1}(x, y) \end{pmatrix},$$

$$\boldsymbol{C}^2(y, x) = \begin{pmatrix} h^2_{11}(y, x) & \cdots & h^2_{1m_2}(y, x) \\ \vdots & & \vdots \\ h^2_{n_2 1}(y, x) & \cdots & h^2_{n_2 m_2}(y, x) \end{pmatrix}.$$

Define a size vector $\underline{n}$ with $n_X = 1, n_Y = \frac{1}{2}d(d+1)r$. Let

$$\widetilde{\mathcal{N}} = \sum_{i=1}^{n} \tilde{a}_i * \boldsymbol{A}_i, \quad \widetilde{R} = \tilde{e}X * \boldsymbol{E}_X + \tilde{e}_Y * \boldsymbol{E}_Y + \sum_{j=1}^{m} \tilde{v}_j * \mathbf{V}_j,$$

where $\tilde{\boldsymbol{a}}_i = (a_{i1})$ when $\boldsymbol{A}_1 \in \mathcal{A}_{XX}$; $\tilde{\boldsymbol{a}}_i = (a_{i1}, a_{i2}, \cdots, a_{i,n_Y})$ when $\boldsymbol{A}_i \in \mathcal{A}_{XY}$; $\tilde{\boldsymbol{a}}_i = (a_{i1}, a_{i2}, \cdots, a_{i,n_Y})^{\mathrm{T}}$ when $\boldsymbol{A}_i \in \mathcal{A}_{YX}$; $\tilde{\boldsymbol{a}}_i = (a_{ipq})_{n_Y \times n_Y}$ when $\boldsymbol{A}_i \in \mathcal{A}_{YY}$; $\tilde{\boldsymbol{e}}_X = (e_X)$, $\tilde{\boldsymbol{e}}_Y = (e_{pq})_{n_Y \times n_Y}$; and the meaning of $\tilde{\boldsymbol{v}}_j$ is similar to that of $\tilde{\boldsymbol{a}}_i$. Then the equation (3.1) of $\mathfrak{M}$ and the (p_i, q_i)-block of the equation are

$$(\widetilde{\mathcal{N}} + \boldsymbol{H}(x, L))\widetilde{R} = \widetilde{R}(\widetilde{\mathcal{N}} + \boldsymbol{H}(x, L)), (\tilde{\boldsymbol{a}}_i \tilde{\boldsymbol{e}}_{Y_i} - \tilde{\boldsymbol{e}}_{X_i} \tilde{\boldsymbol{a}}_i)^0 = \sum_j h_{ij}(x, L) \tilde{\boldsymbol{v}}_j.$$

Given an order on the entries of the matrices $(u_{ipq})_{n_Y \times n_Y}$ for $i = 1, 2, \cdots, m_1$: $u_{ipq} < u_{i'p'q'}$ if and only if $p > p'$ or $p = p'$ but $q < q'$, or $p = p'$, $q = q'$ but $i < i'$. Then the coefficient matrix of u_{ipq} under this order forms an upper triangular block matrix with the block diagonal:

$$\mathrm{diag}(\boldsymbol{C}^1(x, L), \boldsymbol{C}^2(x, L), \cdots, \boldsymbol{C}^1(x, L)).$$

Thus the rank of $\widetilde{\boldsymbol{C}}^1(x, L)$ over $k[x, g_X^s(x)^{-1}]$ is at least $n_1 n_Y$.

Similarly, the coefficient matrix of w_{ipq} forms an upper triangular block matrix with the block diagonal: diag$(\boldsymbol{C}^2(L, x), \boldsymbol{C}^2(L, x), \cdots, \boldsymbol{C}^2(L, x))$. Thus the rank of $\widetilde{\boldsymbol{C}}^2(L, x)$, over $k[x, g_X^s(x)^{-1}]$ is at least $n_Y n_2$.

The conclusion for $\boldsymbol{A}_1^l \in \mathcal{A}_{XX}$ or $\mathcal{A}_{ZZ'}$ is trivial. The lemma is proved.

Proposition 7.5 Recall the matrix (7.1), the sequence (7.2), and the domain $D = D_1 \times D_2 \times \cdots \times D_r$ given in Remark 7.3. Define a parameterized matrix.

$$\boldsymbol{M}_1(x_1) \oplus_T \cdots \oplus_T \boldsymbol{M}_r(x_r) \oplus_T \boldsymbol{N}_1 \oplus_T \cdots \oplus_T \boldsymbol{N}_l \oplus_T \boldsymbol{L}, \tag{7.3}$$

then Corollary 6.9 gives rise to a sequence of reductions with respect to the matrix (7.3):

$$\mathfrak{M}, \overline{\mathfrak{N}}^1, \cdots, \overline{\mathfrak{N}}^s. \tag{7.4}$$

Then the domain of x_j in each $\mathfrak{N}^i$ containing x_j is D_j, for $1 \leqslant j \leqslant r$, $1 \leqslant i \leqslant s$. In particular, the domain of $\overline{\mathfrak{N}}^s$ is D.

If $(p_i, q_i) \in X \times Y$ is the leading position of $\boldsymbol{A}_1^i$ in $\overline{\mathfrak{N}}^i$ with X or Y non-trivial, then the reduction must be given by 3.2+3.3 without any localization.

Proof It is trivial that $\overline{\mathfrak{N}}^1$ is obtained from $\mathfrak{M}$ by deletion according to the size vector of matrix (7.4).

Assume that we have reached $\overline{\mathfrak{N}}^i$ with the index set $(\overline{\mathcal{T}}^i, \overline{\sim}^i)$ satisfying the hypothesis of the proposition, then consider reductions taken from 3.5+3.4, or 3.1+3.6+3.4, or 3.2+3.3 to obtain $\overline{\mathfrak{L}}^{i+1}$. Write the leading position of A_1^i by $(\bar{p}, \bar{q}) \in X \times Y$. Then we delete $\overline{\mathcal{T}}^i \setminus \{X, Y\}$, and obtain an induced matrix problem $\overline{\mathfrak{N}}_{XY}^i$, moreover we have a reduction sequence ending at $\overline{\mathfrak{N}}_{XY}^i$ by using Lemma 6.4 inductively.

If neither X nor Y is a equivalent class determined by L, then there exists some position $1 \leqslant i \leqslant s$ in the sequence (7.2) of Remark 7.3, such that the index set of $\mathfrak{N}^i$ contains X, Y, and the leading position of $\boldsymbol{A}_1^i$ in $\mathfrak{N}^i$ is $(p, q) \in X \times Y$, after deleting $\mathcal{T}^i \setminus \{X, Y\}$, we obtain an induced matrix problem $\mathfrak{N}_{XY}^i$ coinciding with $\overline{\mathfrak{N}}_{XY}^i$ except for the domains; moreover by using Lemma 6,4 inductively, we have a reduction sequence ending at $\mathfrak{N}^i_{XY}$ coinciding with the sequence ending at $\overline{\mathfrak{N}}^i_{XY}$ except for the domains.

(1) If $\mathfrak{N}^{i+1}$ is induced from $\mathfrak{N}^i$ by 3.5+3.4, or 3.1+3.6+3.4 with the non-trivial equivalent class deleted, or 3.2+3.3, then the reduction from $\overline{\mathfrak{N}}^i$ to $\overline{\mathfrak{N}}^{i+1}$ is completely the same.

(2) If $\mathfrak{N}^{i+1}$ is induced from $\mathfrak{N}^i$ by 3.1+3.6+3.4, such that the non-trivial equivalent class is preserved after the deletion, say X_j for a fixed $1 \leqslant j \leqslant r$, then the reduction from $\overline{\mathfrak{N}}^i$ to $\overline{\mathfrak{N}}^{i+1}$ is given by $\bar{n}_{\overline{pq}}$ (see Section 3), which is similar to $\bar{n}_{pq} \bigoplus_{(u,d)} \boldsymbol{J}_d(\mu_j^u)$, where μ_j^u runs over $D_j^0 \backslash D_j$, $d=1,2,\cdots$, such that the size of $M_j(\boldsymbol{J}_d(\mu_j^u))$ is at most n, see Remark 7.3. Therefore the domain of x_j in $\overline{\mathfrak{N}}^{i+1}$ is just D_j.

Now suppose that Y (or dually X) is determined by L. If $X = X_j$, $1 \leqslant j \leqslant r$, is non-trivial, Lemma 7.4 tells us that $\overline{\mathfrak{N}}^{i+1}$ is obtained from $\overline{\mathfrak{N}}^i$ by 3.2+3.3 without any localization; if X is trivial, $\overline{\mathfrak{N}}^{i+1}$ is obtained from $\overline{\mathfrak{N}}^i$ by 3.2+3.3, or 3.5+3.4, or 3.1+3.6+3.4, in the last two cases, $\bar{n}_{\overline{pq}} = (0)$. The proposition is proved.

Main Theorem 7.6 Let $\mathfrak{M} = (\mathcal{K}, \mathcal{M}, \boldsymbol{H} = \boldsymbol{0})$ be a matrix problem of representation tame type, and let n be any fixed positive integer. Then there is a minimal matrix problem $\mathfrak{N}$ induced from $\mathfrak{M}$ and a reduction functor $\vartheta: \mathrm{Mat}(\mathfrak{N}) \to \mathrm{Mat}(\mathfrak{M})$, such that any matrix $(\boldsymbol{M}, \boldsymbol{0}) \in \mathrm{Mat}(\mathfrak{M})$ of size at most n is isomorphic to $\vartheta(\boldsymbol{0}, \boldsymbol{N})$ for an unique matrix $(\boldsymbol{0}, \boldsymbol{N}) \in \mathrm{Mat}(\mathfrak{N})$. In particular $\boldsymbol{N}$ is the canonical form of $(\boldsymbol{M}, \boldsymbol{0})$.

Proof Let $\mathfrak{N} = \overline{\mathfrak{N}}^s$ given in Proposition 7.5, the theorem is proved.

References

[1] Drozd Y A. On tame and wild matrix problems. Matrix Problems. Institute of Mathematics, Academy of Sciences, Ukranian SSR. Kive: 1977: 104-114 .

[2] Crawley-Boevey W W. On tame algebras and bocses. Proc London Math Soc, 1988, 56(3): 451-483.

[3] Bautista R, Drozd Yu A, Zeng X Y, et al. On Hom-spaces of tame algebras. Central European J Math, 2007, 5(2): 215-263.

[4] Crawley-Boevey W W. Matrix problems and Drozd's theorem. Topics Algebra, 1990, 26: 199-222.

[5] Sergeichuk V V. Canonical matrices for basic matrix problems. Linear Algebra Appl, 2000, 317(1/3): 53-102.

[6] Gabriel P, Roiter A V. Representation of finite-dimensional algebras. Encyclopaedia of Math Sci, Vol 73, Algebra Ⅷ. New York: Springer-Verlag, 1992.

[7] Han Y. Controlled wild algebras. Proc London Math Soc, 2001, 83(3): 279-298.

[8] Jacobson N. Basic Algebra. San Francisco: WH Freemann and Company. 1974.

[9] Crawley-Boevev W W. Tame algebras and generic modules. Proc London Math Soc, 1991, 63(3): 241-265.

[10] Krause H. Generic modules over artin algebras. Proc London Math Soc, 1998, 76(3): 276-306.

[11] Ovsienko S A. Generic representations of free bocses, Preprint 93-100, Univ Bielefeld, 1993.

[12] Bautista R, Zhang Y B, Representatons of a k-algebra over the rational functions over k. J Algebra, 2003, 267: 342-358.

[13] Ringel C M. Tame Algebras and Integral Quadratic Forms. LNM 1 099. New York: Springer, 1984.

[14] Auslander M, Reiten I. Representation theory of artin algebras Ⅲ. Comm Algebra, 1975, 3(3): 239-294.

北京师范大学学报(自然科学版),2009,45(1):5-13.

Band-模的典范型①

Canonical Forms of Band Modules

Abstract Let $\mathscr{U}(\Lambda)=(\widetilde{\Lambda}\times\widetilde{\Lambda}, \mathrm{rad}\ \widetilde{\Lambda}, 0)$ the bimodule problem of the Gelfand-Ponomarev algebra $\Lambda := K[x,y]/(x^2, xy, y^3)$ over an algebraically closed field K. In the paper we determine the canonical forms of the indecomposable representations of $\mathfrak{A}(A)$ with dimension vectors (n,n) of and introduce the R-bands of $\mathfrak{A}(\Lambda)$, which enables us to give a bijection between the equivalence classes of bands (resp. band modules) of Λ and the R-bands (resp. indecomposable canonical forms) of $\mathfrak{A}(\Lambda)$.

Keywords Gelfand-Ponomarev algebra; band module; bimodule problem; canonical form.

§ 1. Λ-band modules

Let K be an algebraically closed field and let A be a finite-dimensional K-algebra given by a quiver with monomial relations. For an arrow α, we will denote by $s(\alpha)$ the starting point of α and by $e(\alpha)$

① 国家自然科学基金重点资助项目(10731070);教育部重点实验室资助项目.

收稿日期:2008-03-07.

本文与赵德科合作,由赵德科翻译为英文.

the ending point of α. Also, for every arrow α, we denote by α^- its formal inverse, that is, an arrow with $e(\alpha^-)=s(\alpha)$ and $s(\alpha^-)=e(\alpha)$.

1.1 Definition ([2]) The algebra A is said to be a string algebra if

(1) Every vertex is a starting point for at most two arrows and the ending point for at most two arrows;

(2) Given an arrow α, there is at most one arrow β such that $e(\beta)=s(\alpha)$, and the composition $\alpha\beta$ is not a relation in A (i. e., nonzero in A);

(3) Given an arrow α, there is at most one arrow γ such that $e(\alpha)=s(\gamma)$, and the composition $\gamma\alpha$ is not a relation in A.

According to [3], the Gelfand and Ponomarev algebra $G_{m,n}$ is the quotient algebra of $K[x,y]$ by modulo the ideal (x^m, xy, y^n). It is easy to see that $G_{m,n}$ is a local string algebra. In this paper we are only consider the Gelfand and Ponomarev algebra $\Lambda=G_{2,3}$, which can be realized as the path algebra of the quiver a  b with relations $a^2=0, ab=ba=0, b^3=0$.

1.2 Definition ([3,2]) A word $C=c_1c_2\cdots c_n$, where $c_i\in\{a, b, a^-, b^-\}$ is said to be a string of length n if it doesn't contains the subwords $a^-a, aa^-, bb^-, b^-b, a^2, b^3, ab$. The concatenation of strings $C=c_1c_2\cdots c_n$ and $D=d_1d_2\cdots d_m$ is defined to be $CD=c_1c_2\cdots c_nd_1d_2\cdots d_n$ whenever it is a string.

1.3 Definition A string $\mathcal{B}=b_1b_2\cdots b_n$ of length$\geqslant 1$ is called a band if all powers $\mathcal{B}^r$ is a string and $\mathcal{B}$ itself is not a power of a string of shorter length.

Let $\mathcal{B}=b_1b_2\cdots b_n$ be a band. We write $\mathcal{B}'\sim_r\mathcal{B}$ if $\mathcal{B}'=b_i\cdots b_nb_1\cdots b_{i-1}$ for some $1\leqslant i\leqslant n$. For bands $\mathcal{B}_1, \mathcal{B}_2$, we write $\mathcal{B}_1\sim\mathcal{B}_2$ if $\mathrm{B}_1\sim_r\mathcal{B}_2$ or $\mathcal{B}_1\sim_r\mathcal{B}_2^-$. Clearly $\sim$ is an equivalent relation on bands of Λ.

1.4 Definition ([3,2]) Given a band $\mathcal{B}=b_1b_2\cdots b_n$ and a positive integer r, we define a family of Λ-band module $\{M(\mathcal{B},\lambda,r)=\bigoplus_{i=1}^{n}V_i \mid \lambda\in K^*\}$, where $V_i=V$ is an r-dimensional K-space and $K^*=K\backslash\{0\}$, and the left Λ-action is defined as following

$$b_i = \mathrm{Id}_r : \begin{cases} V_i \to V_{i+1}, & b_i = a, b, \\ V_{i+1} \to V_i, & b_i = a^-, b^-, \end{cases} \quad 1 \leqslant i < n,$$

$$b_n = J_r(\lambda) : \begin{cases} V_n \to V_1, & b_i = a, b, \\ V_1 \to V_n, & b_i = a^-, b^-, \end{cases} \quad i = n,$$

and b_i acts on other K-space being zero.

1.5 Example It is easy to see that ab^-a is a string which is not a band and ab^-, ab^-b^-, $ab^-ab^-b^-$ are pair-wise non-equivalent bands. The band modules $M(ab^-, \lambda, r)$, $M(ab^-b^-, \lambda, r)$ and $M(ab^-ab^-b^-, \lambda, r)$ can be respectively described as fig. 1.1.

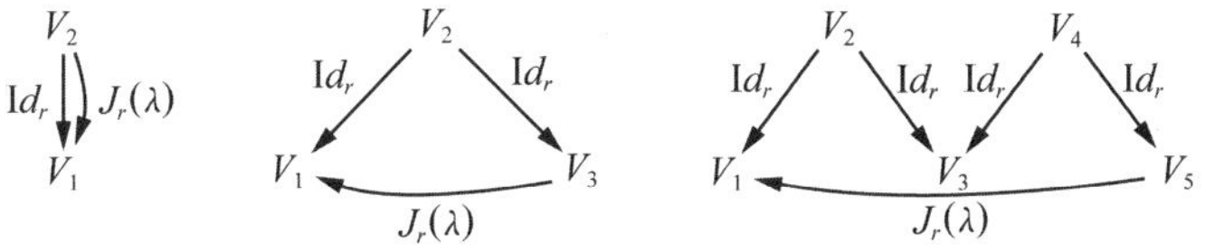

Fig. 1. 1

Let m be a positive integer. Denote by $\mathscr{P}_m$, the set of all m-tuples $\underline{\mathbf{p}} = (p_1, p_2, \cdots, p_m)$ of positive integers which is not of the form $(s_1, s_2, \cdots, s_t; \cdots; s_1, s_2, \cdots, s_t)$ for some $t < m$.

1.6 Lemma Denote by $\mathbb{B}$ the set of equivalent-classes or Λ-bands and by $M(\mathbb{B})$ the iso-classes of band modules. Then

(i) $\mathbb{B} = \{ab^-\} \cup_{m>0} \{(ab^-)^{p_1} b^- \cdots (ab^-)^{p_m} b^- \mid \underline{\mathbf{p}} \in \mathscr{P}_m\}$.

(ii) $M(\mathbb{B}) = \{M(\mathcal{B}, \lambda, r) \mid \mathcal{B} \in \mathbb{B}, r > 0, \lambda \in K^*\}$.

Proof Thanks to [2], we only need to prove (i). For any $B = b_1 b_2 \cdots b_n \in \mathbb{B}$, we may assume that $b_1 = a$. Then $b_2 = b^- = b_n$ due to $ab = 0 = ba = a^2$. If $n = 2$ then $B = ab^-$. If $n \geqslant 3$ then $B = ab^- b_3 \cdots b_{n-1} b^-$, which implies b_3 is either b^- or a. If $b_3 = b^-$ then $b_4 = a$, we yield $B = (ab^-) b^- (a \cdots b^-)$. If $b_3 = a$ then $b_4 = b^-$, we get $B = (ab^-)^2 (b_5 \cdots b^-)$, and b_5 is either b^- or a, therefore B is either $(ab^-)^2 b^- b_6 \cdots b^-$ or $(ab^-)^3 b_7 \cdots b^-$. By induction argument, we show that B is either $(ab^-)^{p_1} b^- \cdots (ab^-)^{p_m} b^-$ or $(ab^-)^{q_1} b^- \cdots (ab^-)^{q_m} b^- (ab^-)^{q_{m+1}}$ with p_i, $q_i > 0$. Note that $(ab^-)^{q_1} b^- \cdots (ab^-)^{q_m}$ is equivalent to $(ab^-)^{q_1 + q_m} b^- \cdots (ab^-)^{q_m} b^-$, that is, $B = (ab^-)^{p_1} b^- \cdots (ab^-)^{p_m} b^-$. Notice that B is not a power of a string with shorter length, which shows $(p_1, p_2, \cdots, p_m) \in \mathscr{P}_m$. □

§ 2. The bimodule problem of Λ

2.1 Definition A bimodule problem $\mathfrak{A} = (\mathcal{K}, \mathcal{M}, H)$ given by matrices over k consists of the following datum:

(1) A set of integers $T = \{1, 2, \cdots, t\}$ and an equivalent relation $\sim$ on T with $\mathscr{T} := T/\sim = \{X_1, X_2, \cdots, X_s\}$;

(2) An upper triangular matrix algebra $\mathcal{K} = \{(s_{ij})_{t\times t}\} \subset \boldsymbol{M}_t(k)$ where $s_{ii} = s_{jj}$ if $i \sim j$ and s_{ij} for $i < j$ satisfy the following k-linear equations

$$\sum_{X \ni i < j \in Y} c^l_{ij} x_{ij} = 0, c^l_{ij} \in K, 1 \leqslant l \leqslant q_{XY}, q_{XY} \in \mathbf{N}, \forall (X, Y) \in \mathscr{T} \times \mathscr{T}. \tag{2.1}$$

(3) A $\mathcal{K}$-bimodule $\mathcal{M} = \{(m_{ij})_{t\times t} \in \boldsymbol{M}_t(k)\}$ where m_{ij} satisfy the following k-linear equations:

$$\sum_{(i,j) \in X\times Y} d^l_{ij} z_{ij} = 0, d^l_{ij} \in K, 1 \leqslant l \leqslant q'_{XY}, q'_{XY} \in \mathbf{N}, \forall (X, Y) \in \mathscr{T} \times \mathscr{T}. \tag{2.2}$$

(4) A derivation $d: \mathcal{K} \to \mathcal{M}, d(\boldsymbol{S}) = \boldsymbol{SH} - \boldsymbol{HS}, \forall \boldsymbol{S} \in \mathcal{K}$, where $\boldsymbol{H} \in \boldsymbol{M}_{t\times t}(K)$ satisfying $h_{ij} = 0$ whenever $i \not\sim j$.

A t-tuple positive integers $\underline{\boldsymbol{n}} = (n_1, n_2, \cdots, n_t)$ is called a size vector of $(T, \sim)$ if $n_i = n_j$ whenever $i \sim j$. We write $n_X = n_i$, $\forall i \in \mathscr{T}$ and $\dim(\underline{\boldsymbol{n}}) = (n_{X_1}, n_{X_2}, \cdots, n_{X_s})$ is called a dimension vector of $(T, \sim)$ and $|\dim(\underline{\boldsymbol{n}})| = \sum_{X \in \mathscr{T}} n_X$ is the dimension of $\underline{\boldsymbol{n}}$.

Given a size vector $\underline{\boldsymbol{n}}$ of $(T, \sim)$, we let

$$\mathcal{M}_{\underline{n}} = \{(\boldsymbol{N}_{ij})_{t\times t} \mid \boldsymbol{N}_{ij} \in \boldsymbol{M}_{n_i \times n_j}(k) \text{ satisfy Equ. (2.2)}\};$$

$$\boldsymbol{H}_{\underline{n}} = (\boldsymbol{H}_{ij})_{t\times t} \text{ with } \boldsymbol{H}_{ij} = h_{ij}\boldsymbol{I}_{n_j} \text{ if } i \sim j \text{ and } \boldsymbol{H}_{ij} = \boldsymbol{0} \text{ for others.}$$

Then a pair of matrices $(\boldsymbol{M}, \boldsymbol{H})$ with $\boldsymbol{M} \in \mathcal{M}_{\underline{n}}$ is called a representation of $(\mathcal{K}, \mathcal{M})$ of dimension vector $\underline{\boldsymbol{n}}$. A representation of $(\mathcal{K}, \mathcal{M})$ is said to be decomposable when it is a direct sum of two nonzero representations of $(\mathcal{K},\mathcal{M})$.

Let $\underline{\boldsymbol{m}},\underline{\boldsymbol{n}}$ be size vectors of $(T,\sim)$. The set of morphisms from $(\boldsymbol{M},\boldsymbol{H}_{\underline{m}})$ to $(\boldsymbol{N},\boldsymbol{H}_{\underline{n}})$ with $\boldsymbol{M}\in\mathcal{M}_{\underline{m}}$, $\boldsymbol{N}\in\mathcal{M}_{\underline{n}}$, is defined to be

$\mathrm{Hom}_{\mathfrak{A}}(\boldsymbol{M},\boldsymbol{N}):=\{S\in\mathcal{K}_{\underline{m}\times\underline{n}}\mid(\boldsymbol{M}+\boldsymbol{H}_{\underline{m}})S=S(\boldsymbol{N}+\boldsymbol{H}_{\underline{n}})\}$, where

$\mathcal{K}_{\underline{m}\times\underline{n}}:=\{(\boldsymbol{S}_{ij})_{t\times t}\mid\boldsymbol{S}_{ij}\in\boldsymbol{M}_{m_i\times n_j}(K)\boldsymbol{S}_{ii}=\boldsymbol{S}_{jj}$ if $i\sim j$; $\boldsymbol{S}_{ij}$ satisfy Equ. (2.1) $\forall\, i<j\}$.

Denote by $\mathrm{Mat}(\mathfrak{A})$ the category of finite dimensional representations of $\mathfrak{A}$.

2.2 Definition For $\boldsymbol{J}(\lambda)=\bigoplus_{i=1}^{d}\boldsymbol{J}_i(\lambda)^{e_i}$, set $m_i=\sum_{j=i}^{d}e_j$. The Weyr matrix $\boldsymbol{W}(\lambda)$ is the following matrix similar to the $\boldsymbol{J}(\lambda)$

$$\begin{pmatrix}\lambda\boldsymbol{I}_{m_1} & \boldsymbol{W}_{12} & & & \\ & \lambda\boldsymbol{I}_{m_2} & \ddots & & \\ & & \ddots & \ddots & \\ & & & \ddots & \boldsymbol{W}_{d-1d} \\ & & & & \lambda\boldsymbol{I}_{m_d}\end{pmatrix},\quad \text{where } \boldsymbol{W}_{ii+1}=\begin{pmatrix}\boldsymbol{I}_{m_{i+1}}\\ \boldsymbol{0}\end{pmatrix}\in\boldsymbol{M}_{m_i\times m_{i+1}}(K). \tag{2.3}$$

If $\boldsymbol{J}=\bigoplus_{i=1}^{r}\boldsymbol{J}(\lambda_i)$ with λ_i pair-wisely different, then $\boldsymbol{W}=\bigoplus_{i=1}^{r}\boldsymbol{W}(\lambda_i)$ is called the Weyr matrix of J, which is similar to J.

2.3 Remark Thanks to [4], any matrix $\boldsymbol{X}$ commuting with $\boldsymbol{W}(\lambda)$ is a upper-triangular partitioned matrix $\boldsymbol{X}=(\boldsymbol{X}_{ij})_{d\times d}$ where for any $i\leqslant j$

$$\boldsymbol{X}_{ij}=\begin{pmatrix}\boldsymbol{X}^h_{11} & \boldsymbol{X}^h_{12} & \cdots & \boldsymbol{X}^h_{1\bar{j}}\\ \vdots & \vdots & \vdots & \vdots\\ \boldsymbol{X}^h_{h1} & \boldsymbol{X}^h_{h2} & \cdots & \boldsymbol{X}^h_{h\bar{j}}\\ & \boldsymbol{X}^h_{h+12} & \cdots & \boldsymbol{X}^h_{h+1\bar{j}}\\ & & \ddots & \vdots\\ & & & \boldsymbol{X}^h_{\bar{i}\bar{j}}\end{pmatrix} \tag{2.4}$$

where

$h=j-i+1$, $\bar{i}=d-i+1$, $\bar{j}=d-j+1$, $\boldsymbol{X}^h_{pq}\in\boldsymbol{M}_{e_{\bar{p}}\times e_{\bar{q}}}(K)$ for $p\leqslant\bar{i}$, $q\leqslant\bar{j}$.

Fix a size vector $\underline{\boldsymbol{n}}$, we may expand the Equations (2.1) and (2.2) into matrix linear systems of size $\underline{n}$. Now we define an order on the indices of the entries of any $m\times n$ matrix by setting $(i,j)\prec(p,q)$ provided $i>p$ or $i=p, j<q$. Let $(i,j):=\min\{(i',j')\mid m_{i'j'}\neq 0\}$. Clearly, if $M\in\mathcal{M}_{\underline{n}}$ then the first non-zero block of $\boldsymbol{M}$ is $\boldsymbol{M}_{ij}$. Now we can use the following algorithm to obtained a simpler new block $\overline{\boldsymbol{M}}_{ij}$.

2.4 Proposition ([4]) For $\boldsymbol{M}_{ij}$, we have the following three kinds of reductions (Bilitskii's) algorithms:

(1) If $\boldsymbol{M}_{ij}\boldsymbol{S}_{jj}-\boldsymbol{S}_{ii}\overline{\boldsymbol{M}}_{ij}=\sum \boldsymbol{S}_{il}h_{lj}-h_{il}\boldsymbol{S}_{lj}\neq\boldsymbol{0}$ then $\overline{\boldsymbol{M}}_{ij}=\boldsymbol{0}$ and denoted by $\varnothing$.

(2) If $\boldsymbol{M}_{ij}\boldsymbol{S}_{jj}-\boldsymbol{S}_{ii}\overline{\boldsymbol{M}}_{ij}=\boldsymbol{0}$ with $i\not\sim j$ then $\overline{\boldsymbol{M}}_{ij}=\begin{pmatrix}\boldsymbol{0} & \boldsymbol{I}_r\\ \boldsymbol{0} & \boldsymbol{0}\end{pmatrix}$, where $r=$ rank$(\boldsymbol{M}_{ij})$.

(3) If $\boldsymbol{M}_{ij}\boldsymbol{S}_{jj}-\boldsymbol{S}_{ii}\overline{\boldsymbol{M}}_{ij}=\boldsymbol{0}$ with $i\sim j$ then $\overline{\boldsymbol{M}}_{ij}=\boldsymbol{W}$, a Weyr matrix similar to $\boldsymbol{M}_{ij}$.

Let us remark that the Bilitskii's algorithms enable us to construct a new bimodule problem $\mathfrak{A}'=(\mathcal{K}',\mathcal{M}',H')$ with index set $(T',\sim')$, new equation systems $(2.1)'$, $(2.2)'$ and new size vector $\underline{n}'$ of $(T',\sim')$ (see [4,5] for more details).

2.5 Definition Let $\mathfrak{A}$ be a bimodule problem and $\underline{n}$ a fixed size vector. Suppose that

$$\begin{matrix}(\boldsymbol{M},\boldsymbol{H}_{\underline{n}}), & (\boldsymbol{M}^1,\boldsymbol{H}^1_{\underline{n}^1}), & \cdots, & (\boldsymbol{M}^s,\boldsymbol{H}^s_{\underline{n}^s})\\ \mathfrak{A}, & \mathfrak{A}_1, & \cdots, & \mathfrak{A}_s\end{matrix}$$

are sequences of reductions and the corresponding sequence of matrix problems respectively. If $\boldsymbol{M}^s=(0)$ then $H^s_{\underline{n}^s}$ is called the canonical form of $\boldsymbol{M}$. By a link, we means the integer 1 appearing in $\boldsymbol{I}_r$ or $\boldsymbol{W}_{ii+1}$ given by Proposition 2.4(2)(3) and we denote by $l(\boldsymbol{M})$ the number of the links in the matrix $\boldsymbol{M}$.

The following fact will be used frequently.

2.6 Proposition ([5]) Let $\boldsymbol{M}$ be d-dimensional representation of a bimodule problem $\mathfrak{A}=(\mathcal{K},\mathcal{M},0)$. Then $\boldsymbol{M}$ is indecomposable if and only if the number of the links of its canonical form is $d-1$.

Thanks to [5], the bimodule problem of Λ is $\mathfrak{A}(\Lambda)=(\widetilde{\boldsymbol{\Lambda}}\times\widetilde{\boldsymbol{\Lambda}},\operatorname{rad}\widetilde{\boldsymbol{\Lambda}},\mathbf{0})$, where

$$\widetilde{\boldsymbol{\Lambda}}=\left\{\begin{pmatrix} x_1 & x_3 & 0 & x_4 \\ 0 & x_1 & 0 & x_3 \\ 0 & 0 & x_1 & x_2 \\ 0 & 0 & 0 & x_1 \end{pmatrix} \middle| x_i\in K\right\}\quad \operatorname{rad}\widetilde{\boldsymbol{\Lambda}}=\left\{\begin{pmatrix} 0 & b & 0 & c \\ 0 & 0 & 0 & b \\ 0 & 0 & 0 & a \\ 0 & 0 & 0 & 0 \end{pmatrix} \middle| a,b,c\in K\right\}.$$

Denote by $\mathcal{K}=\begin{pmatrix}\widetilde{\boldsymbol{\Lambda}} & \mathbf{0}\\ \mathbf{0} & \widetilde{\boldsymbol{\Lambda}}\end{pmatrix}$ and $\mathcal{M}=\begin{pmatrix}\mathbf{0} & \operatorname{rad}\widetilde{\boldsymbol{\Lambda}}\\ \mathbf{0} & \mathbf{0}\end{pmatrix}$. Then the matrix problem of $\boldsymbol{\Lambda}$ is $\mathfrak{A}(\boldsymbol{\Lambda})=(\mathcal{K},\mathcal{M},\mathbf{0})$, where $T=\{1,2,\cdots,8\}$ with $1\sim2\sim3\sim4,5\sim6\sim7\sim8$.

Given a representation $\boldsymbol{M}$ of $\mathfrak{A}$ with dimension vector (m,n), the main aim of the paper is to determine the canonical form $\overline{\boldsymbol{M}}$ of $\boldsymbol{M}$ with respect to the following similar transformation:

$$\begin{pmatrix}\boldsymbol{X}^{-1} & \mathbf{0}\\ \mathbf{0} & \boldsymbol{Y}^{-1}\end{pmatrix}\begin{pmatrix}\mathbf{0} & \boldsymbol{M}\\ \mathbf{0} & \mathbf{0}\end{pmatrix}\begin{pmatrix}\boldsymbol{X} & \mathbf{0}\\ \mathbf{0} & \boldsymbol{Y}\end{pmatrix}\tag{2.5}$$

where

$$\boldsymbol{X}^{-1}=\begin{pmatrix}\boldsymbol{U} & \boldsymbol{U}_2 & \mathbf{0} & \boldsymbol{U}_3\\ \mathbf{0} & \boldsymbol{U} & \mathbf{0} & \boldsymbol{U}_2\\ \mathbf{0} & \mathbf{0} & \boldsymbol{U} & \boldsymbol{U}_1\\ \mathbf{0} & \mathbf{0} & \mathbf{0} & \boldsymbol{U}\end{pmatrix},\boldsymbol{M}=\begin{pmatrix}\mathbf{0} & \boldsymbol{B} & \mathbf{0} & \boldsymbol{C}\\ \mathbf{0} & \mathbf{0} & \mathbf{0} & \boldsymbol{B}\\ \mathbf{0} & \mathbf{0} & \mathbf{0} & \boldsymbol{A}\\ \mathbf{0} & \mathbf{0} & \mathbf{0} & \mathbf{0}\end{pmatrix},\boldsymbol{Y}=\begin{pmatrix}\boldsymbol{X} & \boldsymbol{X}_2 & \mathbf{0} & \boldsymbol{X}_3\\ \mathbf{0} & \boldsymbol{X} & \mathbf{0} & \boldsymbol{X}_2\\ \mathbf{0} & \mathbf{0} & \boldsymbol{X} & \boldsymbol{X}_1\\ \mathbf{0} & \mathbf{0} & \mathbf{0} & \boldsymbol{X}\end{pmatrix}$$

with $\boldsymbol{A},\boldsymbol{B},\boldsymbol{C}\in\boldsymbol{M}_{m\times n},\boldsymbol{U}\in\boldsymbol{GL}_m(K)$ and $\boldsymbol{X}\in\boldsymbol{GL}_n(K)$. For simplicity, we denote the representation by $(\boldsymbol{A},\boldsymbol{B},\boldsymbol{C})$, and by $(\overline{\boldsymbol{A}},\overline{\boldsymbol{B}},\overline{\boldsymbol{C}})$ its canonical form.

The following easy verified facts clarify the relationship between $(\boldsymbol{A},\boldsymbol{B},\boldsymbol{B})$ and $(\overline{\boldsymbol{A}},\overline{\boldsymbol{B}},\overline{\boldsymbol{C}})$.

2.7 Proposition Keeping notations as above. Then the $(\overline{\boldsymbol{A}},\overline{\boldsymbol{B}},\overline{\boldsymbol{C}})$ is determined by the following equations:

(1) $\boldsymbol{X}\overline{\boldsymbol{A}}=\boldsymbol{A}\boldsymbol{Y}$;

(2) $\boldsymbol{X}\overline{\boldsymbol{B}}=\boldsymbol{B}\boldsymbol{Y}$;

(3) $X\bar{C}=CY-X_2\bar{B}+\bar{B}Y_2$ where the entries in X_1,Y_2 are linearly independent.

§ 3. Reduction lemmas

3.1 Definition An R-bimodule problem $(\mathcal{K}_1\times\mathcal{K}_2,\mathcal{M},\mathbf{0})$ is a bimodule problem consisting of the following datum:

(1) $T=T_1\dot{\cup}T_2$, where $T_1=\{1,2,\cdots,t_1\}$, $T_2=\{1',2',\cdots,t_2'\}$ and the equivalent relation $\sim$ on T defined as following: let $T_0\subseteq T_1$ and σ: $T_0\rightarrow T_2$ is injective, then $i\sim\sigma(i)$.

(2) $\mathcal{K}_1=\{(s_{ij})_{t_1\times t_1}\mid s_{ij}$ is K-free if $i<j\}$, $\mathcal{K}_2=\{(s'_{ij})_{t_2\times t_2}\mid s'_{ij}$ is K-free if $i<j\}$;

(3) $\mathcal{M}=\{(m_{ij})_{t_1\times t_2}\mid m_{t_1j}=0$ if $(i,j)<(p,q)$, m_{ij} is K-free if $(i,j)\geq(p,q)\}$.

Let $\mathfrak{A}=(\mathcal{K}_1\times\mathcal{K}_2,\mathcal{M},\mathbf{0})$ be an R-bimodule problem and $\boldsymbol{M}\in\mathrm{Mat}(\mathfrak{A})$ with size vector $\underline{n}$. Then the Bilitskii's algorithms of the first non-zero block $\boldsymbol{M}_{pq}$ of $\boldsymbol{M}$ is either 2.4(1) or 2.4(2), which are called the edge reduction and unraveling reduction respectively, that is, we have the following two cases:

(1) If we reduce M by applying the edge reduction then $p\not\sim q'$ and

$$\bar{\boldsymbol{M}}_{pq}=\begin{pmatrix}\mathbf{0}&\boldsymbol{I}_r\\\mathbf{0}&\mathbf{0}\end{pmatrix},\quad \boldsymbol{X}_{pp}^{\mathrm{T}}=\begin{pmatrix}\boldsymbol{U}_1&\boldsymbol{U}_{12}\\\mathbf{0}&\boldsymbol{U}_2\end{pmatrix},\quad \boldsymbol{Y}_{qq}^{\mathrm{T}}=\begin{pmatrix}\boldsymbol{V}_1&\boldsymbol{V}_{12}\\\mathbf{0}&\boldsymbol{U}_1\end{pmatrix},$$

where $r=\mathrm{rank}(\boldsymbol{M}_{pq})$, $\boldsymbol{U}_1\in\boldsymbol{CL}_r(K)$. If $q\in\mathrm{Im}\sigma$ and let $i=\sigma^{-1}(q)$ then $\boldsymbol{X}_{ii}^{\mathrm{T}}=\boldsymbol{Y}_{qq}^{\mathrm{T}}$, and $\boldsymbol{X}^{\mathrm{T}}{}_{ll}=\boldsymbol{Y}_{ll}$ if $l\neq i,p$; if $j=\sigma(p)$ then $\boldsymbol{Y}_{jj}^{\mathrm{T}}=\boldsymbol{X}_{pp}^{\mathrm{T}}$ and $\boldsymbol{Y}_{ll}^{\mathrm{T}}=\boldsymbol{Y}_{ll}$ if $l\neq j,q$.

(2) If we reduce $\boldsymbol{M}$ by applying the unraveling reduction then $p\sim q'$ and $\bar{\boldsymbol{M}}_{pq}=\boldsymbol{W}_0\oplus\boldsymbol{W}$, which is similar to $\boldsymbol{M}_{pq}$, where $\boldsymbol{W}_0$ is the Weyr matrix of $\bigoplus_{i=1}^{r}\boldsymbol{J}_i(0)^{e_i}$ and $\boldsymbol{W}$ is an invertible Weyr matrix. Let

$$\boldsymbol{X}_{pp}^{\mathrm{T}}=\boldsymbol{Y}_{qq}^{\mathrm{T}}=\begin{pmatrix}\boldsymbol{Y}&\mathbf{0}\\\mathbf{0}&\boldsymbol{Z}\end{pmatrix},\quad \boldsymbol{W}_0\boldsymbol{Y}=\boldsymbol{Y}\boldsymbol{W}_0,\quad \boldsymbol{W}\boldsymbol{Z}=\boldsymbol{Z}\boldsymbol{W},\tag{3.1}$$

where $\boldsymbol{U}$ is determined by Equ. (2.4).

(1) and (2) enable us to obtain an induced bimodule problem $\mathfrak{A}'=(\mathcal{K}_1'\times\mathcal{K}_2',\mathcal{M}',H')$ of $\mathfrak{A}$ with the size vector $\underline{\boldsymbol{n}}^{\mathrm{T}}$ over $(T',\sim')$ induced from $\underline{\boldsymbol{n}}$, and the reduction for $\boldsymbol{M}^{\mathrm{T}}\in\mathrm{Mat}(\mathfrak{A}')$ can be written as $(\boldsymbol{X}^{\mathrm{T}},\boldsymbol{M}^{\mathrm{T}},\boldsymbol{Y}^{\mathrm{T}})$ where

$$\boldsymbol{X}^{\mathrm{T}}=\begin{pmatrix}\boldsymbol{X}_{11}' & \cdots & \boldsymbol{X}_{1p} & \cdots & \boldsymbol{X}_{1t_1}\\ & \ddots & \vdots & & \vdots\\ & & \boldsymbol{X}_{pp}' & \cdots & \boldsymbol{X}_{pt_1}\\ & & & \ddots & \vdots\\ & & & & \boldsymbol{X}_{t_1t_1}\end{pmatrix},\quad \boldsymbol{Y}^{\mathrm{T}}=\begin{pmatrix}\boldsymbol{Y}_{11}' & \cdots & \boldsymbol{Y}_{1p} & \cdots & \boldsymbol{Y}_{1t_2}\\ & \ddots & \vdots & & \vdots\\ & & \boldsymbol{Y}_{qq}' & \cdots & \boldsymbol{Y}_{qt_2}\\ & & & \ddots & \vdots\\ & & & & \boldsymbol{Y}_{t_2t_2}\end{pmatrix}, \tag{3.2}$$

where $\boldsymbol{X}_{ll}^{\mathrm{T}}=\boldsymbol{X}_{ll}$ if $l\neq p,i$; $\boldsymbol{Y}_{ll}^{\mathrm{T}}=\boldsymbol{Y}_{ll}$ if $l\neq q,j$, and the non-diagnoal entries of $\boldsymbol{X}^{\mathrm{T}}$ and $\boldsymbol{Y}^{\mathrm{T}}$ are the same as those of $\boldsymbol{X}$ and $\boldsymbol{Y}$ respectively, and

$$\boldsymbol{M}^{\mathrm{T}}=\begin{pmatrix}\boldsymbol{M}_{11} & \cdots & \boldsymbol{M}_{1q-1} & \cdots & \boldsymbol{M}_{1t_2}\\ \vdots & & \vdots & & \vdots\\ \boldsymbol{0} & \cdots & \boldsymbol{0} & \overline{\boldsymbol{M}}_{pq} & \cdots\\ \boldsymbol{0} & \cdots & \boldsymbol{0} & \cdots & \boldsymbol{0}\end{pmatrix}. \tag{3.3}$$

3.2 Lemma Keeping notations as above. Then we have the following reductions

(1) If we reduce $\boldsymbol{M}$ by applying the edge reduction then we remove those rows and columns of $\boldsymbol{X}^{\mathrm{T}}$ (resp. $\boldsymbol{Y}^{\mathrm{T}}$), which are intersect with that of the block $\boldsymbol{U}_1$ of $\boldsymbol{X}_{pp}^{\mathrm{T}}$ (resp. $\boldsymbol{Y}_{qq}^{\mathrm{T}}$), and remove those rows and columns of $\boldsymbol{M}^{\mathrm{T}}$, which ave intersect with that of the block $\boldsymbol{I}_r$ of $\overline{\boldsymbol{M}}_{t_1q}$.

(2) If we reduce $\boldsymbol{M}$ by applying the unraveling reduction then we remove those rows and columns of $\boldsymbol{X}^{\mathrm{T}}$ (resp. $\boldsymbol{Y}^{\mathrm{T}}$), which are intersect with that of the blocks $\boldsymbol{Z},\boldsymbol{U}_{jj}$, $j\leqslant d-i$ (resp. $j>1$), $i=1,2,\cdots,r$, of $\boldsymbol{X}_{pp}^{\mathrm{T}}$ (resp. $\boldsymbol{Y}_{qq}^{\mathrm{T}}$), and remove those rows and columns of $\boldsymbol{M}^{\mathrm{T}}$ which are intersect with that of $\boldsymbol{W}$ and the block $\boldsymbol{I}_{m_i+1}$ of $\boldsymbol{W}_{ii+1}$.

Denote by $(\widetilde{\boldsymbol{U}},\widetilde{\boldsymbol{M}},\widetilde{\boldsymbol{X}})$ the matrices obtained from $(\boldsymbol{X}^{\mathrm{T}},\boldsymbol{M}^{\mathrm{T}},\boldsymbol{Y}^{\mathrm{T}})$ by applying the above redutions. Then $(\widetilde{\boldsymbol{U}},\widetilde{\boldsymbol{M}},\widetilde{\boldsymbol{X}})$ is an R-bimodule problem $\widetilde{\mathfrak{A}}=(\widetilde{\mathcal{K}}_1\times\widetilde{\mathcal{K}}_2,\widetilde{\mathcal{M}},\boldsymbol{0})$ with $\dim(\widetilde{\boldsymbol{M}})<\dim(\boldsymbol{M}^{\mathrm{T}})$ and the reduction of $\boldsymbol{M}^{\mathrm{T}}$

over $(\mathcal{K}_1'\times\mathcal{K}_2',\mathcal{M}',H')$ is identity to the one of $\widetilde{\boldsymbol{M}}$ over $\widetilde{\mathfrak{A}}$.

Proof Since the equivalent relation $\sim$ on T of $(\mathcal{K}_1\times\mathcal{K}_2,\mathcal{M},\boldsymbol{0})$ is determined by the diagonal blocks of the elements of $\mathcal{K}_1\times\mathcal{K}_2$, we only need to determine the diagonal blocks of elements of $\widetilde{\mathcal{K}}_1\times\widetilde{\mathcal{K}}_2$ to determine the new equivalent relation $\sim'$.

(1) We have the following cases:

(a) if there exist i,j (we may assume that $j<q$), then the diagonal blocks of $\widetilde{\boldsymbol{X}}$ and $\widetilde{\boldsymbol{Y}}$ are respectively

$$(\boldsymbol{X}_{11},\cdots,\boldsymbol{X}_{i-1,i-1};\boldsymbol{V}_1,\boldsymbol{U}_1;\boldsymbol{X}_{i+1,i+1},\cdots,\boldsymbol{X}_{p-1,p-1};\widetilde{\boldsymbol{X}}_{pp};\boldsymbol{X}_{p+1,p+1},\cdots,\boldsymbol{X}_{t_1t_1})$$

and

$$(\boldsymbol{Y}_{11},\cdots,\boldsymbol{Y}_{j-1,j-1};\boldsymbol{U}_1,\boldsymbol{U}_2;\boldsymbol{Y}_{j+1,j+1},\cdots,\boldsymbol{Y}_{q-1,q-1};\widetilde{\boldsymbol{Y}}_{qq};\boldsymbol{Y}_{q+1,q+1},\cdots,\boldsymbol{Y}_{t_2t_2}),$$

where $\widetilde{\boldsymbol{X}}_{pp}=\boldsymbol{U}_2$ and $\widetilde{\boldsymbol{Y}}_{qq}=\boldsymbol{V}_1$. Then $\widetilde{T}_0=(T_0\backslash\{i\})\cup\{q\}$.

(b) if there don't exist i,j then there are no $\boldsymbol{V}_1,\boldsymbol{U}_1,\boldsymbol{U}_2$, which implies $\widetilde{T}_0=T_0,\tilde{\sigma}=\sigma$.

(c) if there only exists i then $\boldsymbol{U}_1,\boldsymbol{U}_2$ don't occur in $\widetilde{\boldsymbol{Y}}$, $\widetilde{T}_0=(T_0\backslash\{i\})\cup\{q\}$, and the block $\boldsymbol{V}_{12}$ of $\widetilde{\boldsymbol{X}}$ is removed in $\widetilde{\boldsymbol{Y}}$.

(d) if there only exists j then $\boldsymbol{V}_1,\boldsymbol{U}_1$ don't occur in $\widetilde{X}$, $\widetilde{T}_0=T_0$, and the block $\boldsymbol{U}_{12}$ of $\widetilde{Y}$ is removed.

(2) Note that the diagonal blocks of $\widetilde{\boldsymbol{X}}$ and $\widetilde{\boldsymbol{Y}}$ are respectively

$$(\boldsymbol{X}_{11},\cdots,\boldsymbol{X}_{p-1,p-1};\boldsymbol{U}_{dd}^1,\cdots,\boldsymbol{U}_{11}^1;\boldsymbol{X}_{p+1,p+1},\cdots,\boldsymbol{X}_{t_1t_1})$$

and

$$(\boldsymbol{Y}_{11},\cdots,\boldsymbol{Y}_{q-1,q-1};\boldsymbol{U}_{dd}^1,\cdots,\boldsymbol{U}_{11}^1;\boldsymbol{Y}_{q+1,q+1},\cdots,\boldsymbol{Y}_{t_2t_2}).$$

Furthermore, the new blocks occur in the nilpotent part of the upper triangle of $\widetilde{\boldsymbol{X}}$ (resp. $\widetilde{\boldsymbol{Y}}$) are $\boldsymbol{U}_{kl}^1,k<l$ (resp. $\boldsymbol{U}_{kl}^h,h\geqslant 2,l=h+k-1$).

Note that the strictly-upper entries of $(\widetilde{\boldsymbol{U}},\widetilde{\boldsymbol{X}})\in\widetilde{\mathcal{K}}_1\times\widetilde{\mathcal{K}}_2$ are linearly independent owing to the form of $(\boldsymbol{U},\boldsymbol{X})\in\mathcal{K}_1\times\mathcal{K}_2$ and the reductions, which means $(\widetilde{\boldsymbol{U}},\widetilde{\boldsymbol{M}},\widetilde{\boldsymbol{X}})$ is an R-bimodule problem. Finally, the freeness of strictly-upper entries of $(\boldsymbol{U},\boldsymbol{X})\in\mathcal{K}_1\times\mathcal{K}_2$ implies the reduction of those removing blocks is given by Proposition 2.4 (1). As a consequence, the reduction for those un-removing block of $\boldsymbol{M}$ in the R-

bimodule $(\mathcal{K}_1 \times \mathcal{K}_2, \mathcal{M}, \mathbf{0})$ is exactly the one for $\widetilde{\boldsymbol{M}}$ in $(\widetilde{\mathcal{K}}_1 \times \widetilde{\mathcal{K}}_2, \widetilde{\mathcal{M}}, \mathbf{0})$. The lemma is proved. □

3.3 Remark A point should be noted that the reductions 3.2(1) and 3.2(2) are not bimodule problem reductions, which can make the bimodule problem reductions simpler. The following picture is an example of the reduction 3.2(2)(Fig. 3.1):

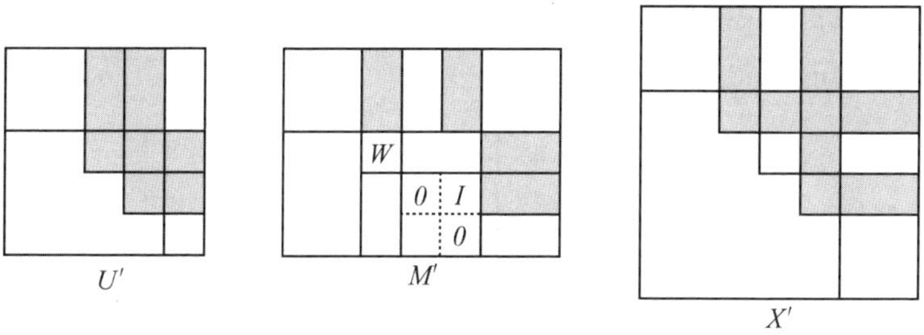

Fig. 3.1

3.4 Definition Let $\mathfrak{A}^0 = (\mathcal{K}_1^0 \times \mathcal{K}_2^0, \mathcal{M}^0, \mathbf{0})$ be an R-bimodule problem. Denote by

$$\mathcal{K}_1 \times \mathcal{K}_2 = \left\{ (\boldsymbol{X}, \boldsymbol{X}_1; \boldsymbol{Y}, \boldsymbol{Y}_1) := \begin{pmatrix} \boldsymbol{X} & \boldsymbol{X}_1 & & \\ & \boldsymbol{X} & & \\ & & \boldsymbol{Y} & \boldsymbol{Y}_1 \\ & & & \boldsymbol{Y} \end{pmatrix} \middle| \begin{matrix} (\boldsymbol{X}, \boldsymbol{Y}) \in \mathcal{K}_1' \times \mathcal{K}_2' \\ \boldsymbol{X}_1, \boldsymbol{Y}_1 \text{ are free blocks} \end{matrix} \right\},$$

$$\mathcal{M} = \left\{ (\boldsymbol{M}, \boldsymbol{N}) := \begin{pmatrix} \boldsymbol{M} & \boldsymbol{N} \\ \mathbf{0} & \boldsymbol{M} \end{pmatrix} \middle| \begin{matrix} \boldsymbol{M} \in \mathcal{M}' \\ \boldsymbol{N} \text{ is a free block} \end{matrix} \right\}.$$

The bimodule problem $\mathfrak{A} = (\mathcal{K}_1 \times \mathcal{K}_2, \mathcal{M}, \mathbf{0})$ is called the RR-bimodule problem induced by $\mathfrak{A}'$.

Let $(\boldsymbol{M}, \boldsymbol{N})$ be a representation of $\mathfrak{A}$ with size vector $\underline{\boldsymbol{n}}$. Definition 3.4 implies the reduction of $(\boldsymbol{M}, \boldsymbol{N})$ is determined by the following equations:

$$\begin{cases} \overline{\boldsymbol{M}} = \boldsymbol{X}^{-1} \boldsymbol{M} \boldsymbol{Y}, \\ \overline{\boldsymbol{N}} = \boldsymbol{X}^{-1} \boldsymbol{N} \boldsymbol{Y} + \boldsymbol{X}^{-1} \overline{\boldsymbol{M}} \boldsymbol{Y}_1 + \boldsymbol{X}^{-1} \boldsymbol{X}_1 \boldsymbol{X} \overline{\boldsymbol{M}} \boldsymbol{Y}. \end{cases}$$

Lemma 3.2 and Definition 3.4 implies the following fact.

3.5 Lemma Let $(\boldsymbol{M}, \boldsymbol{N})$ be a representation of the (RR)-bimodule problem $\mathfrak{A} = (\mathcal{K}_1 \times \mathcal{K}_2, \mathcal{M}, \mathbf{0})$. Denote by $(\boldsymbol{X}^{\mathrm{T}}, \boldsymbol{X}_1^{\mathrm{T}}; \boldsymbol{M}^{\mathrm{T}}, \boldsymbol{N}^{\mathrm{T}}; \boldsymbol{Y}^{\mathrm{T}}, \boldsymbol{Y}_1^{\mathrm{T}})$ the object obtained by applying the reductions 2.4 on $(\boldsymbol{M}, \boldsymbol{N})$

and by $\widetilde{\mathfrak{A}}=(\widetilde{\boldsymbol{X}},\widetilde{\boldsymbol{X}}_1;\widetilde{\boldsymbol{M}},\widetilde{\boldsymbol{N}};\widetilde{\boldsymbol{Y}},\widetilde{\boldsymbol{Y}}_1)$ the objects obtained by applying the reductions 3.2 on $(\boldsymbol{M}^{\mathrm{T}},\boldsymbol{N}^{\mathrm{T}})$. Then $\mathfrak{A}$ determines an RR-bimodule problem with $\dim(\widetilde{\boldsymbol{M}},\widetilde{\boldsymbol{N}})<\dim(\boldsymbol{M},\boldsymbol{N})$ and the reductions of $(\boldsymbol{M}^{\mathrm{T}}\boldsymbol{N}^{\mathrm{T}})$ and $(\widetilde{\boldsymbol{M}},\widetilde{\boldsymbol{N}})$ are same.

3.6 Example Assume that $(\boldsymbol{M}^{\mathrm{T}},\boldsymbol{N}^{\mathrm{T}})$ is obtained form $(\boldsymbol{M},\boldsymbol{N})$ by applying the unraveling reduction and $(\widetilde{\boldsymbol{M}},\widetilde{\boldsymbol{N}})$ is obtained form $(\boldsymbol{M}^{\mathrm{T}},\boldsymbol{N}^{\mathrm{T}})$ by applying the reduction 3.2. Then it can be illustrated by the following figure 3.2.

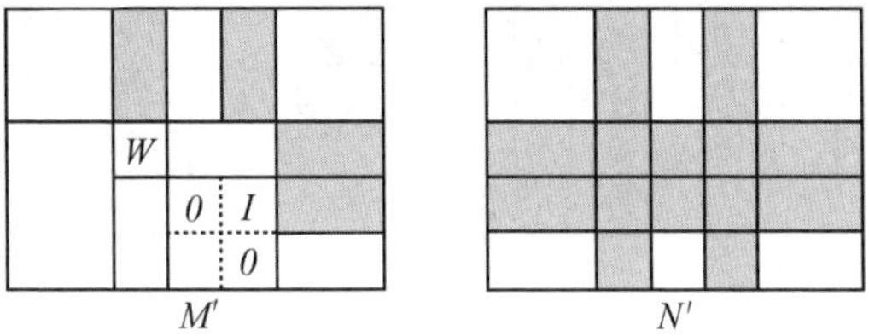

Fig. 3.2

where the shading rows and columns are those removed by applying the reduction 3.2.

3.7 Corollary Let $\mathfrak{A}=(\mathcal{K}_1\times\mathcal{K}_2,\mathcal{M},\mathbf{0})$ be an RR-bimodule problem. Let $(\overline{\boldsymbol{M}},\overline{\boldsymbol{N}})$ be the canonical form of the representation $(\boldsymbol{M},\boldsymbol{N})$ of $\mathfrak{A}$ with size vector $\underline{\boldsymbol{n}}=(n_1,n_2,\cdots,n_{t_1};n_{1'},n_{2'},\cdots,n_{t_2'})$. Put $k=\sum_{i=1}^{t_1}n_i$ and $l=\sum_{i=1}^{t_2}n_{i'}$. Then

(1) $(\overline{\boldsymbol{M}},\overline{\boldsymbol{N}})$ contains at most one link in each row and column and

$$l(\overline{\boldsymbol{M}},\overline{\boldsymbol{N}})\leqslant\min\{k,l\}.$$

(2) $(\boldsymbol{M},\boldsymbol{N})$ is indecomposable if and only if $|k-l|\leqslant 1$.

(3) $(\boldsymbol{M},\boldsymbol{N})$ is indecomposable and $(\overline{\boldsymbol{M}},\overline{\boldsymbol{N}})$ contains a Weyr matrix $\boldsymbol{W}$ with nonzero eigenvalue if and only if $k=l$. Moreover, $\boldsymbol{W}$ is unique and contains exactly one Jordan block.

(4) If a Weyr matrix occurs in the procedure of reduction with an succedent unraveling reduction, then $(\boldsymbol{M},\boldsymbol{N})$ is decomposable.

Proof (1) follows directly by Lemma 3.2. Note that the elements of $\{1,2,\cdots,t_1\}$ and of $\{1',2',\cdots,t_2'\}$ are pairwise non-equivalent. Thus

$\dim(\boldsymbol{M},\boldsymbol{N})\geqslant\max\{k,l\}$ and (2) follows by combing Proposition 2.6 and (1).

(3) Note that the number of links of a Weyr matrix is less than its degree. Therefore if $|k-l|=1$ then $l(\widetilde{\boldsymbol{M}},\widetilde{\boldsymbol{N}})<\dim(\boldsymbol{M},\boldsymbol{N})-1$, which is impossible; If $k=l$ then $l(\widetilde{\boldsymbol{M}},\widetilde{\boldsymbol{N}})$ equals the difference between $\dim(\boldsymbol{M},\boldsymbol{N})$ and the number of Weyr matrix with nonzero eigenvalue, which implies $(\widetilde{\boldsymbol{M}},\widetilde{\boldsymbol{N}})$ contains exactly one Weyr matrix with one Jordan block.

(4) Thanks to Equ. (3.1) and Lemma 3.2, the diagonal block $\boldsymbol{Z}$ determined by $\boldsymbol{W}$ will be removed, which will stable in the succedent reductions. Therefore $(\boldsymbol{M},\boldsymbol{N})$ can be decomposed into an nontrivial direct sum. □

§ 4. The canonical form $(\bar{A},\bar{B},\bar{C})$

Let $(\mathcal{K}_1\times\mathcal{K}_2,\mathcal{M},\boldsymbol{0})$ be an R-bimodule problem with $|T_1|=|T_2|=d$ and σ an bijective map form T_1 to T_2. Given a positive integer d, we denote by $\mathfrak{S}_d$ the symmetric group of degree d. For $\sigma\in\mathfrak{S}_d$, we set

$\mathcal{K}_1^\sigma\times\mathcal{K}_2^\sigma=\{(x_{ij})_{i\leqslant j}\times(y_{ij})_{i\leqslant j}\mid x_{ii}=y_{\sigma(i)\sigma(i)}\in K^*, x_{ij}, y_{\sigma(i)\sigma(j)}\in K, 1\leqslant i<j\leqslant d\}$,

$M^\sigma=\{(m_{i\sigma(j)})_{i\leqslant j}\mid m_{i\sigma(j)}\in K, 1\leqslant i,j\leqslant d\}$.

Then $\mathfrak{A}^\sigma=(\mathcal{K}^\sigma,\mathcal{M}^\sigma,0)$ is an bimodule problem with $|T_1|=|T_2|=d$.

For $\sigma\in\mathfrak{S}_d$ and $\underline{e}=(e_1,e_2,\cdots,e_d)\in\mathbf{N}^d$, we define $\boldsymbol{\sigma}\underline{\boldsymbol{e}}=(e_{\sigma(1)},e_{\sigma(2)},\cdots,e_{\sigma(d)})$, $e=\sum_{i=1}^{d}e_i$, and put

$$\bar{\sigma}=\begin{pmatrix}1 & 2 & \cdots & d\\ \bar{1} & \bar{2} & \cdots & \bar{d}\end{pmatrix}$$

where $\bar{i}=d-i+1, i=1,2,\cdots,d$.

The following facts is crucial to determine the canonical forms of the matrix problem $\mathfrak{A}(\Lambda)$.

4.1 Proposition Let Ω be an indecomposable canonical form of $\mathfrak{A}_\sigma$ with size vector $(\underline{e},\boldsymbol{\sigma}\underline{\boldsymbol{e}})$. Assume that Ω contains an Weyr matrix $\boldsymbol{W}$ with nonzero eigenvalue λ. Then

(1) $\boldsymbol{W}=\boldsymbol{J}_r(\lambda)$ and $r\mid \gcd(e_1,e_2,\cdots,e_d)$.

(2) there exists exactly one partition for Ω satisfying

(a) each subblock of Ω is one of $r\times r$ matrix $\mathbf{0}$, $\varnothing$, $\boldsymbol{I}_r$, and $\boldsymbol{J}_r(\lambda)$;

(b) there exists exactly one subblock in each row and colum-block of $\boldsymbol{\Omega}$, which is neither $\mathbf{0}$ nor $\varnothing$.

(3) If $\gcd(e_1,e_2,\cdots,e_d)=1$, then there exists exactly one link in each row and column of $\boldsymbol{\Omega}$ except the ones $\boldsymbol{W}=\lambda$ lying. Furthermore, assume that

$$\boldsymbol{\Omega}=((\omega_{i\sigma(j)}^{lm})_{(l,m)=(1,1)}^{(e_i,e_{\sigma(j)})})_{i,j=1}^{d},$$

then there exists a sequence $(i_1,l_1),(i_2,l_2),\cdots,(i_e,l_e)=(1,1)$ such that

$$\omega_{i_e i_1}^{l_e l_1}=\lambda \text{ and } \omega_{i_1 i_2}^{l_1 l_2}=\omega_{i_2 i_3}^{l_2 l_3}=\cdots=\omega_{i_e-1 i_e}^{l_e-1 l_e}=1. \tag{4.1}$$

Proof According to Corollary 3.7(4), $\boldsymbol{\Omega}$ contains exactly one Weyr matrix $\boldsymbol{J}_r(\lambda)$. We will prove the other assertions by applying the induction arguments on e, all the partitions $\underline{\boldsymbol{e}}^{\mathrm{T}}=(e_1^{\mathrm{T}},e_2^{\mathrm{T}},\cdots,e_{d'}^{\mathrm{T}})$ of $\underline{\boldsymbol{e}}$, and the permutations of $\underline{\boldsymbol{e}}^{\mathrm{T}}$. For our purpose, we will deal with all the bimodule problem $\mathfrak{A}_\sigma$, $\sigma\in\mathfrak{S}_d$. We begin with the case of $\sigma=\bar{\sigma}$. Assume that $\boldsymbol{\Omega}$ is the canonical form of an indecmposable $\boldsymbol{M}^{\mathrm{T}}$ of $\mathfrak{A}_{\bar{\sigma}}$ with size vector $(\underline{\boldsymbol{e}},\bar{\boldsymbol{\sigma}}\underline{\boldsymbol{e}})$.

Firstly, if $d=1$ then $\boldsymbol{\Omega}=\boldsymbol{J}_{e_1}(\lambda)$. We are done. If $d\geqslant 2$, $e=\sum_{i=1}^{d}e_i=1,2,3$, and all the partitions $\{(g_1,g_2,\cdots,g_f)\mid f\geqslant 2,\ \sum_{i=1}^{f}g_i\}$ of e and $\sigma\in\mathfrak{S}_f$, the proposition can be proved straightly.

Secondly, assume that $d\geqslant 2$, $e\geqslant 4$ and $\boldsymbol{\Omega}_{dd}\neq\mathbf{0}$. Then $\boldsymbol{\Omega}_{dd}$ is a singular Weyr matrix. Indeed, if $\boldsymbol{\Omega}_{dd}=\boldsymbol{W}\oplus\boldsymbol{W}_0$ where $\boldsymbol{W}$ is an invertible Weyr matrix of degree h and $\boldsymbol{W}_0$ is singular. Then $\boldsymbol{\Omega}=\begin{pmatrix}\varnothing & \boldsymbol{\Omega}_2 & \boldsymbol{\Omega}_3\\ \boldsymbol{W} & \mathbf{0} & \varnothing\\ \mathbf{0} & \boldsymbol{W}_0 & \boldsymbol{\Omega}_1\end{pmatrix}$, which implies $\widetilde{\boldsymbol{\Omega}}:=\begin{pmatrix}\boldsymbol{\Omega}_2 & \boldsymbol{\Omega}_3\\ \boldsymbol{W}_0 & \boldsymbol{\Omega}_1\end{pmatrix}$ is an indecomposable canonical form of dimension $e-h$ with $l(\widetilde{\Omega})=e-h-1$, therefore $l(\boldsymbol{\Omega})\leqslant e-2$, it is impossible.

Assume that $\boldsymbol{\Omega}_{d\sigma(1)}$ is the Weyr matrix of $\bigoplus_{i=1}^{s}\boldsymbol{J}_i(0)^{t_i}$. Thanks to Lemma 3.5, $\widetilde{\boldsymbol{\Omega}}$ is the canonical form of an indecomposable $(e-e_d+\sum_{i=1}^{s}t_i)$-dimensional representation of the bimodule problem $\mathfrak{A}^{\mu}$ with size vector $(\underline{e}^{\mathrm{T}},\boldsymbol{\mu}\underline{e}^{\mathrm{T}})$, where $\underline{e}^{\mathrm{T}}=(e_1,e_2,\cdots,e_{d-1},t_1,t_2,\cdots,t_s)$ and $\boldsymbol{\mu}\underline{e}^{\mathrm{T}}=(t_s,t_{s-1},\cdots,t_1,e_{d-1},e_{d-2},\cdots,e_1)$. The induction argument implies assertions (1)(2) holds for $\widetilde{\boldsymbol{\Omega}}$. As a consequence, $\boldsymbol{\Omega}$ satisfies (1)(2) by noticing that

$$e_d=\sum_{i=1}^{s}it_i.$$

For $\gcd(e_1,e_2,\cdots,e_d)=1$, we may assume $s=2$. Let $t=t_1+t_2$. Then ω_{dd}^{jt+j} $(j=1,2,\cdots,t_2)$ are links of $\boldsymbol{\Omega}_{dd}$. Owing to Lemma 3.2, $\widetilde{\boldsymbol{\Omega}}$ is obtained from $\boldsymbol{\Omega}_{dd}$ by reductions 3.2(1)(2), which means the double-indices of the links (or λ) of $\widetilde{\boldsymbol{\Omega}}$ is exactly the ones of $\boldsymbol{\Omega}$. By the induction argument, there is a sequence $\boldsymbol{S}_{\widetilde{\Omega}}$ of $\widetilde{\boldsymbol{\Omega}}$ satisfying Equ. (4.1). Thus the link $\omega_{i_jd}^{l_jj}$ in $\boldsymbol{S}_{\widetilde{\Omega}}$ is lying the left-side of $\omega_{da_j}^{t+jb_j}$ for $j=1,2,\cdots,t_2$. Clearly the links of $\boldsymbol{\Omega}$, which are not contained in $\boldsymbol{S}_{\widetilde{\Omega}}$ are ω_{dd}^{jt+j}, $j=1,2,\cdots,t_2$ of $\boldsymbol{\Omega}_{dd}$. As a consequence, we can get a sequence S_{Ω} for $\boldsymbol{\Omega}$ satisfying Equ. (4.1) by putting the links ω_{dd}^{jt+j} between $\omega_{i_jd}^{l_jj}$ and $\omega_{da_j}^{t+jb_j}$ for all $j=1,2,\cdots,t_2$.

Thirdly, suppose $\boldsymbol{\Omega}_{dd}=\boldsymbol{\Omega}_{dd-1}=\cdots=\boldsymbol{\Omega}_{dp-1}=\boldsymbol{0}$ and $q=\operatorname{rank}(\boldsymbol{\Omega}_{dp})>0$. Then $\sum_{i=1}^{p}e_i\geqslant e_d$ and $\boldsymbol{\Omega}_{dp}=\begin{pmatrix}\boldsymbol{0} & \boldsymbol{I}_q\\ \boldsymbol{0} & \boldsymbol{0}\end{pmatrix}$. By Lemma 3.2, $\widetilde{\boldsymbol{\Omega}}$ is the canonical forms of an $(e-q)$-dimensional indecomposable representation of the bimodule problem $\mathfrak{A}^{v}$ with size vector $(\underline{e}^{\mathrm{T}},\boldsymbol{v}\underline{e}^{\mathrm{T}})$, where

$$\underline{e}^{\mathrm{T}}=(e_1,e_2,\cdots,e_{p-1},e_p-q,q,e_{p+1},e_{p+2},\cdots,e_{d-1},e_d-q),$$
$$\boldsymbol{v}\underline{e}^{\mathrm{T}}=(q,e_d-q,e_{d-1},\cdots,e_{p+1},e_p-q,e_{p-1},\cdots,e_1).$$

By the induction argument, $\widetilde{\boldsymbol{\Omega}}$ satisfies (1)(2) and there is a sequence $\boldsymbol{S}_{\widetilde{\Omega}}$ satisfying Equ. (4.1). Therefore $\boldsymbol{\Omega}$ satisfies (1)(2) due to $r\mid\gcd(e_p-q,q,e_d-q)$ and the link $\omega_{i_jd}^{l_jj}$ in $\boldsymbol{S}_{\widetilde{\Omega}}$ is exactly in the left-side of $\omega_{pa_j}^{e'_p+jb_j}$

for $j=1,2,\cdots,q$, where $e'_p=e_p-q$. Clearly the only links of Ω, which are not contained in $\boldsymbol{S}_{\tilde{\Omega}}$ are the links $\omega_{dp}^{je'_p+j}$ $(j=1,2,\cdots,q)$ of $\boldsymbol{\Omega}_{dp}$. Thus there is a sequence $\boldsymbol{S}_{\Omega}$ for $\boldsymbol{\Omega}$ satisfying (3) by putting the links $\omega_{dp}^{je'_p+j}$ between $\omega_{i_j d}^{l_j j}$ and $\omega_{pa_j}^{e'_p+jb_j}$ for all $j=1,2,\cdots,p$.

Finally, note that the position of the Weyr matrix $\boldsymbol{W}=\lambda$ of $\boldsymbol{\Omega}$ can be determined by applying the induction argument in the first parts of the step 2 and step 3. Therefore we complete the proof of the proposition. □

4.2 Remark Proposition 4.1 implies that we may assume that the unique invertible Weyr matrix of an indecomposable canonical form of $\mathfrak{A}(\Lambda)$ is λ.

Now we are ready to determine the canonical forms of the bimodule problem $\mathfrak{A}(\Lambda)$ of the Gelfand-Ponomarev algebra Λ (see § 2).

4.3 Theorem Let $(\boldsymbol{A},\boldsymbol{B},\boldsymbol{C})$ be an indecomposable representation of $\mathfrak{A}$ with dimension vector (m,n) and let $(\bar{\boldsymbol{A}},\bar{\boldsymbol{B}},\bar{\boldsymbol{C}})$ be the canonical form of $(\boldsymbol{A},\boldsymbol{B},\boldsymbol{C})$. Then $|m-n|\leqslant 1$. Furthermore, if $(\bar{\boldsymbol{B}},\bar{\boldsymbol{C}})$ contains a Weyr matrix with nonzero eigenvalue λ, then $m=n$ and $(\bar{\boldsymbol{A}},\bar{\boldsymbol{B}},\bar{\boldsymbol{C}})$ is either $(\boldsymbol{I}_n,\boldsymbol{J}_n(\lambda),\varnothing)$ or $(\boldsymbol{I}_n,\boldsymbol{W}_0,\bar{\boldsymbol{C}})$, where $\boldsymbol{W}_0$ is the Weyr matrix of $\bigoplus_{i=1}^{d}\boldsymbol{J}_i(0)^{e_i}$ and

$$\bar{\boldsymbol{C}}=\begin{pmatrix} \varnothing & \cdots & \varnothing & \varnothing & \cdots & \varnothing \\ \boldsymbol{\Omega}_{1d} & \cdots & \boldsymbol{\Omega}_{11} & \varnothing & \cdots & \varnothing \\ \vdots & & \vdots & \vdots & & \vdots \\ \varnothing & \cdots & \varnothing & \varnothing & \cdots & \varnothing \\ \boldsymbol{\Omega}_{d-1,d} & \cdots & \boldsymbol{\Omega}_{d-1,1} & \varnothing & \cdots & \varnothing \\ \boldsymbol{\Omega}_{dd} & \cdots & \boldsymbol{\Omega}_{d1} & \varnothing & \cdots & \varnothing \end{pmatrix}$$

is a block matrix with row size vector $(m_2,e_1,m_3,e_2,\cdots,m_d,e_{d-1},e_d)$ and column size vector $(e_d,e_{d-1},\cdots,e_1,m_2,m_3,\cdots,m_d)$, where $m_i=\sum_{r=i}^{d}e_r$ for $i=2,3,\cdots,d$ and $\boldsymbol{\Omega}=(\boldsymbol{\Omega}_{id-j+1})_{i,j=1}^{d}$ is an indecomposable canonical forms of the bimodule problem $\mathfrak{A}_{\sigma}$ with size vector

$$\underline{e}=(e_1,e_2,\cdots,e_d),\bar{\sigma}(\underline{e}).$$

Proof Note that Proposition 2.7(1) shows $\bar{\boldsymbol{A}}=\begin{pmatrix}\boldsymbol{0} & \boldsymbol{I}_r \\ \boldsymbol{0} & \boldsymbol{0}\end{pmatrix}$ and

$$X=\begin{pmatrix} X_{11} & X_{12} \\ 0 & X_{22} \end{pmatrix},\quad Y=\begin{pmatrix} Y_{11} & Y_{12} \\ 0 & Y_{22} \end{pmatrix}, \tag{4.2}$$

where $Y_{22}=X_{11}\in GL_r(K)$, $X_{22}\in GL_{m-r}(K)$, and $Y_{11}\in GL_{n-r}(K)$. Therefore we obtain an RR-bimodule problem induced by $\mathfrak{A}(\Lambda)$ (see Definition 3.4) and the reductions of B and of C are determined by Equs. 2.7(2)(3) with X,Y being of the form Equ. (4.2). Now Corollary 3.7(2) implies $|m-n|\leqslant 1$ and $m=n$ whenever $(\bar{B},\bar{C})$ contains a Weyr matrix with nonzero eigenvalues.

Now assume that $(\bar{B},\bar{C})$ contains an invertible Weyr matrix. Then $l(\bar{B})+l(\bar{C})<n$ owing to Corollary 3.7(1). By applying Proposition 2.6, $l(\bar{A})+l(\bar{B})+l(\bar{C})=r+l(\bar{B})+l(\bar{C})=2n-1$, which means $r=n$, that is, $\bar{A}=I_n$. As a consequence, $\bar{B}$ is the Weyr matrix of the Jordan form of B. If $\bar{B}=J_n(\lambda)$ with $\lambda\in K^*$, then $\bar{C}=\varnothing$ by Equation 2.7(3). If $\bar{B}=W_0\oplus W$, where W_0 (resp. W) is an singular (resp. invertible) Weyr matrix, then (A,B,C) is decomposable according to Corollary 3.7(4). Therefore $\bar{B}=W_0$ is the Weyr matrix of $\bigoplus_{i=1}^{d} J_i(0)^{e_i}$.

If $\bar{B}=W_0$ then, by Proposition 2.7(3) and Lemma 3.2(2), the reduction for C is exactly the one for an indecomposable representation of dimension vector $(e_1,e_2,\cdots,e_d)$ of the bimodule problem $\mathfrak{A}_{\bar{\sigma}}$. We complete the proof by applying Proposition 4.1. □

From now on, we assume that $(A,B,C)=(I_n,W_0,C)$ is the canonical form of an indecomposable representation of $\mathfrak{A}(\Lambda)$ with dimension vector (n,n), where W_0 is the Weyr matrix of $\bigoplus_{i=1}^{d} J_i(0)^{e_i}$ and the invertible Weyr matrix of C is λ (see Remark 4.2) Let us remark that $e_d\neq 0$ while some e_i may be zero.

Following Definition 2.2, $B=(B_{j_1}^{i_1})_{j_1=1,2,\cdots,d}^{i_1=1,2,\cdots,d}$ is a block matrix where $B_{j_1}^{i_1}=(B_{j_1j_2}^{i_1i_2})_{j_2=d,d+1,\cdots,j_1}^{i_2=d,d+1,\cdots,i_1}$ is a block matrix with $B_{j_1j_2}^{i_1i_2}=(b_{j_1j_2j_3}^{i_1i_2i_3})_{j_3=1,2,\cdots,e_{j_2}}^{i_3=1,2,\cdots,e_{i_2}}$ being an $e_{i_2}\times e_{j_2}$-matrix.

The following definition will be crucial to our consideration.

4.4 Definition Keeping notations as above and let x be an entry of B. The row and column triple-indices of x is defined to be ($i_1 i_2 i_3$, $j_1 j_2 j_3$) if

(1) x is in the i_1-row-block (reps. j_1-column-block) $\boldsymbol{B}_{j_1}^{i_1}$ of $\boldsymbol{B}$;

(2) x is in the i_2-row-block (reps. j_2-column-block) $\boldsymbol{B}_{j_1 j_2}^{i_1 i_2}$ of $\boldsymbol{B}_{j_1}^{i_1}$;

(3) x is the (i_3, j_3)-entry of the matrix $\boldsymbol{B}_{j_1 j_2}^{i_1 i_2}$.

Let us remark that we may and will use the triple-indices of elements of $\boldsymbol{B}$ to index the elements of $\boldsymbol{A}$ and $\boldsymbol{C}$ in the same way.

4.5 Example Assume that $\boldsymbol{A} = \boldsymbol{I}_{13}$, $\boldsymbol{B}$ is the Weyr matrix of $\boldsymbol{J}_1(0)^3 \oplus \boldsymbol{J}_2(0)^3 \oplus \boldsymbol{J}_3(0)^2$ and that $\boldsymbol{C}$ is of form of the third fig 4.1:

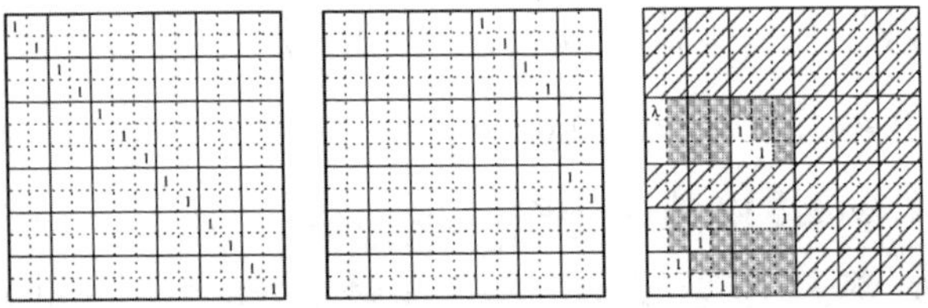

Fig. 4.1

Then the triple-indices of ($\boldsymbol{A}$, $\boldsymbol{B}$, $\boldsymbol{C}$) are fig. 4.2.

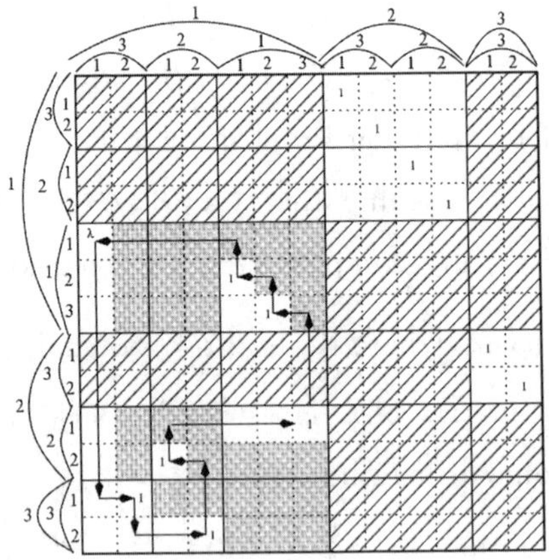

Fig. 4.2

where the order of row (resp. column) indices is from left (resp. top) to right (resp. bottom).

4.6 Remark The triple-indices of Definition 4.4 has the following properties:

(1) If the row triple-index of a link in $\boldsymbol{B}$ is (i,j,k), then its column triple-index is $(i+1,j,k)$.

(2) Let b_1 be a link of $\boldsymbol{B}$ with the row triple-index $(1,j,k)$. If $b_i\,(i=2,3,\cdots,j)$ are the link of $\boldsymbol{B}$ with row triple-index equaling the column triple-index of b_{i-1}. Then the row triple-indices of $b_1,b_2,\cdots,b_{j-1}$ are $(1,j,k),(2,j,k),\cdots,(j-1,j,k)$ and the column triple-index of b_{j-1} is (j,j,k), which is either the row triple-index of a link or the row triple-index of λ in C.

Theorem 4.3 and Remark 4.6 imply that the row triple-index of λ is $(i_0=1,j_0,1)$. Assume its column triple-index is $(1,j_1,k_1)$. Then there is a link $a_{1j_1k_1}$ of $\boldsymbol{A}$ with the same row and column triple-index $(1,j_1,k_1)$ and a link $b_{1j_1k_1}$ of $\boldsymbol{B}$ with the row triple-index $(1,j_1,k_1)$. Now we write the pair of links $(a_{1j_1k_1},b_{1j_1k_1})$ as , where the down-left (resp. down-right) arrow is indexed by the link $a_{1j_1k_1}$ (resp. $b_{1j_1k_1}$).

Repeat the above procedure, we obtain the following quiver

consisting of j_1-1 pair of arrows $(a_{ij_ik_i},b_{ij_ik_i})$ $(i=1,2,\cdots,j_1-1)$, where $\mathrm{a}_{ij_ik_i}$ and $b_{ij_ik_i}$ are links of $\boldsymbol{A}$ and $\boldsymbol{B}$ respectively.

Thanks to Remark 4.6, the column triple-index of $b_{j_1-1j_1k_1}$ is (j_1,j_1,k_1) and the link having the row triple-index (j_1,j_1,k_1) is a link of $c_{j_1j_1k_1}$. As a consequence, we can obtain a triple $(a_{j_1j_1k_1},\tilde{b}_{j_1j_1k_1},c_{j_1j_1k_1})$ and write it as the hook quiver 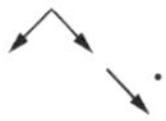.

Combing the above arguments, we obtain a sequence of arrows: $(a_{1j_1k_1},b_{1j_1k_1});(a_{2j_1k_1},b_{2j_1k_1});\cdots;(a_{j_1-1j_1k_1},b_{j_1-1j_1k_1});(a_{j_1j_1k_1},\tilde{b}_{j_1j_1k_1},c_{j_1j_1k_1})$ (note that $b'_{j_1j_1k_1}$ is not a link). In other words, we yield the following connected quiver:

(4.3)

Now if the column index of $c_{j_1j_1k_1}$ is (j_2,j_2,k_2), then we can repeat the above procedure. Using the induction argument, we obtain a connected quiver, which is called the R-band of $(\mathbf{A},\mathbf{B},\mathbf{C})$, consisting of m_1 pair-wisely different sub-quivers being of the form (4.3), where the i-th($i=1,2,\cdots,m_1$) sub-quiver is the quiver corresponding the sequence of arrows $(a_{1j_ik_i},b_{1j_ik_i});\cdots;(a_{j_i-1 1j_ik_i},b_{j_i-1j_ik_i});(a_{j_ij_ik_i},\tilde{b}_{j_ij_ik_i},c_{j_ij_ik_i})$. Note that $j_{m_1}=j_0$ and $k_{m_1}=1$ according to Proposition 4.1.

4.7 Example Let $(\mathbf{A},\mathbf{B},\mathbf{C})$ be the same as Example 4.5. Then the R-band of $(\mathbf{A},\mathbf{B},\mathbf{C})$ is the following quiver consisting of $m_1=e_1+e_2+e_3=7$ connected sub-quivers, where $a_{\dots},b_{\dots},c_{\dots}$ except $\tilde{b}_{\dots}$ are links and $c_{111}=\lambda$ (Fig. 4.3).

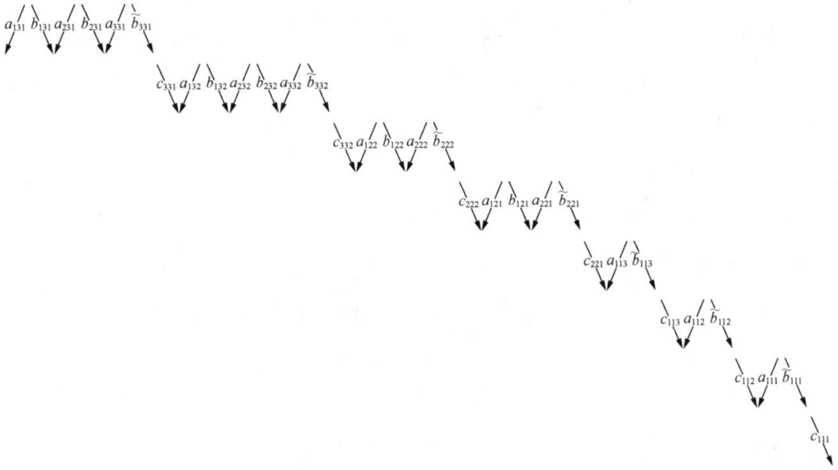

Fig. 4.3

4.8 Proposition Keep notations as above. If the R-band of $(\mathbf{A},\mathbf{B},\mathbf{C})$ is $(a_{1j_ik_i},b_{1j_ik_i});\cdots;(a_{j_i-1j_ik_i},b_{j_i-1j_ik_i});(a_{j_ij_ik_i},b_{j_ij_ik_i},c_{j_ij_ik_i})$, $i=1,2,\cdots,s$, then $s=m_1$ and

(1) the triple-index (i,j_t,k_t) $(1\leqslant i\leqslant j_t, 1\leqslant k_t\leqslant e_{j_t}, 1\leqslant j_t\leqslant d)$ runs through the triple-indices of the links and parameter λ of $(\mathbf{A},\mathbf{B},\mathbf{C})$. In particular, $c_{j_sj_sk_s}=c_{j_0j_01}$.

(2) The sequence $(j_1,k_1),(j_2,k_2),\cdots,(j_s,k_{m_1})$ consisting of the second and third indices of the R-band is the one of $\mathbf{C}$ satisfying Equ. (4.2).

(3) The vector $\underline{j}=(j_1,j_2,\cdots,j_{m_1})\in\mathscr{P}_{m_1}$ (see Lemma 1.6).

Proof (1) and (2) follows directly by applying Proposition 4.1.

Without look of generality we may assume that $e_i > 0$ for all $i=1,2,\cdots,d$. Thanks to Proposition 4.1 and (2), $k_{m_1} = k_1 = 1$, which implies $\underline{j}=(j_1,j_2,\cdots,j_{m_1}) \in \mathscr{P}_{m_1}$. Indeed, if $\underline{j}=(j_1,j_2,\cdots,j_{m_1}) \notin \mathscr{P}_{m_1}$, then there exists a sequence $\underline{q}=(j_1,j_2,\cdots,j_t)$ $(t<m_1)$ such that $\underline{j}=(\underline{q},\underline{q},\cdots,\underline{q})$, which implies $l(\boldsymbol{B},\boldsymbol{C})=n-\frac{m_1}{t}<n-1$ since the row and column triple-indices of the parameter λ, the links of $(\boldsymbol{B},\boldsymbol{C})$ are pairwise different. Then $(\boldsymbol{A},\boldsymbol{B},\boldsymbol{C})$ is decomposable according to [5], which is contradicts with the indecomposablity of $(\boldsymbol{A},\boldsymbol{B},\boldsymbol{C})$. □

§ 5. The correspondence between bands and *R*-bands

Recall that the algebra $\Lambda=K[x,y]/(x^2,xy,y^3)$, which is local and its indecomposable projective Λ-module is fig. 5.1.

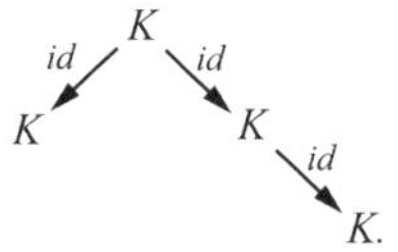

Fig. 5.1

We assume that ϵ,α,β and γ are a basis of the top space, the left space, the right middle space and the the bottom space. Furthermore, we let $(\epsilon_{ijk},\alpha_{ijk},\beta_{ijk},\gamma_{ijk})$ be a basis the projective module Λ^n of Λ, where $k=1,2,\cdots,e_j$, $j=d,\cdots,i,\cdots,d$, $i=1,2,\cdots,d$. Set

$$\underline{\boldsymbol{E}}=(\epsilon_{ijk}),\quad \underline{\boldsymbol{A}}=(\alpha_{ijk}),\quad \underline{\boldsymbol{B}}=(\beta_{ijk}),\quad \underline{\boldsymbol{C}}=(\gamma_{ijk}).$$

5.1 Lemma Let M be an indecomposable Λ-module. If the minimal projective presentation of M is

$$\Lambda^m \xrightarrow{\alpha} \Lambda^n \xrightarrow{\pi} M \to 0 \tag{5.1}$$

then $|n-m|\leqslant 1$, where n is the number of the starting points in the string of M. Furthermore, if M is an indecomposable band module then $m=n$.

Proof Note that Equ. (5.1) gives an indecomposable $(\boldsymbol{A},\boldsymbol{B},\boldsymbol{C})$ of $\mathfrak{A}(\Lambda)$ with dimensional vector (m,n). The lemma follows directly by applying Theorem 4.3. □

From now on, we may and will identify the morphism α in Equ. (5.1) with an indecomposable representation of $\mathfrak{A}(\Lambda)$ with dimension vector (n,n) and vice versa. We will use the two notations freely.

5.2 Lemma Assume that $(\boldsymbol{A}=\boldsymbol{I}_n,\boldsymbol{B},\boldsymbol{C})$ is an indecomposable representation of $\mathfrak{A}(\Lambda)$ with dimension vector (n,n), where B is the Weyr matrix of $\bigoplus_{i=1}^{d}\boldsymbol{J}_i(0)^{e_i}$ ($\sum_{i=1}^{d} ie_i=n$) and the invertible Weyr matrix of $\boldsymbol{C}$ is λ. Then

(1) The Λ-module structure of $\mathrm{Cok}(\alpha)$ is determined completely by the Λ-module structure of Λ^n and the following K-linear system

$$\begin{cases}\underline{\boldsymbol{C}}\boldsymbol{B}=\boldsymbol{0},\\ \underline{\boldsymbol{C}}\boldsymbol{C}+\underline{\boldsymbol{B}}\boldsymbol{B}+\underline{\boldsymbol{A}}=\boldsymbol{0}.\end{cases}\tag{5.2}$$

(2) $Cok(\alpha)$ is a band module. Moreover, the band associated to $Cok(\alpha)$ is the R-band of $(\boldsymbol{A},\boldsymbol{B},\boldsymbol{C})$ $w.r.t.$ the equivalent relation $\sim$.

(3) The eigenvalue of the Jordan block of $\mathrm{Cok}(\alpha)$ is $(-1)^n\lambda^{-1}$.

Proof (1) follows by noticing that $\mathrm{Cok}(\alpha)\simeq\Lambda^n/\mathrm{Im}(\alpha)$ and $\mathrm{Im}(\alpha)$ is generated by $\underline{\boldsymbol{C}}\boldsymbol{B}$ and $\underline{\boldsymbol{C}}\boldsymbol{C}+\underline{\boldsymbol{B}}\boldsymbol{B}+\boldsymbol{A}$.

(2) Thanks to Theorem 4.3 and Proposition 4.1(2): the second equation of linear system 5.2 is equivalent the linear system

$$\begin{cases}\underline{\boldsymbol{B}}\boldsymbol{B}+\underline{\boldsymbol{A}}=\boldsymbol{0},\\ \underline{\boldsymbol{C}}\boldsymbol{C}+\underline{\boldsymbol{A}}=\boldsymbol{0}.\end{cases}\tag{5.3}$$

For any (j,l), Equ. 5.3 implies $\gamma_{ijl}=0,\beta_{ijl}=-\alpha_{i+1jl}, i=1,2,\cdots,l-1$. As a consequence, we get the following quotient module of $\bigoplus_{i=1}^{j}\Lambda_{ijl}$ (fig. 5.2).

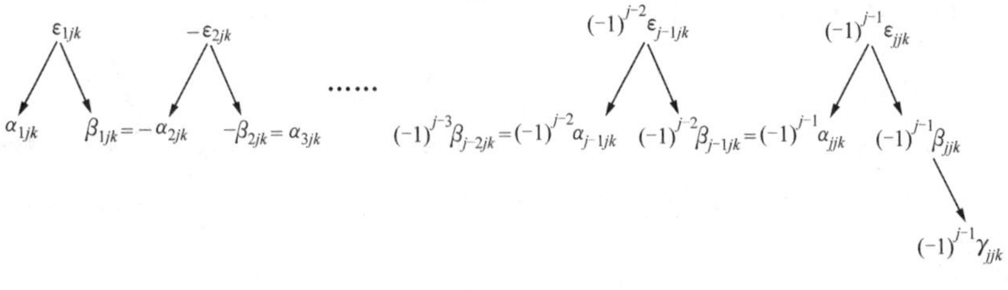

Fig. 5.2 (5.4)

If the column triple-index of the link $c_{1j'l'}$ of $\boldsymbol{C}$ with row triple index (jjl), then $\gamma_{ijl}=\alpha_{1j'l'}$. Repeating the above procedure for all $l=1,2,\cdots,e_j$ and $j=1,2,\cdots,d$, we yield m_1 connected quivers having the form of (5.4). Furthermore, Beginning with $c_{i_0 j_0 1}(i_0=1)$, we obtain an R-band of $(\boldsymbol{A},\boldsymbol{B},\boldsymbol{C})$. Note that the row triple-index of the parameter λ of $\boldsymbol{C}$ is $(1,j_0,j_0)$ and the sign of parameter of Cok(α) is determined by Equ. 5.3. Then (3) follows by applying (2). □

5.3 Example Let $(\boldsymbol{A},\boldsymbol{B},\boldsymbol{C})$ be the same as the one in Example 4.5. Then Cok$(\boldsymbol{A},\boldsymbol{B},\boldsymbol{C})$ is the following band module(fig. 5.3).

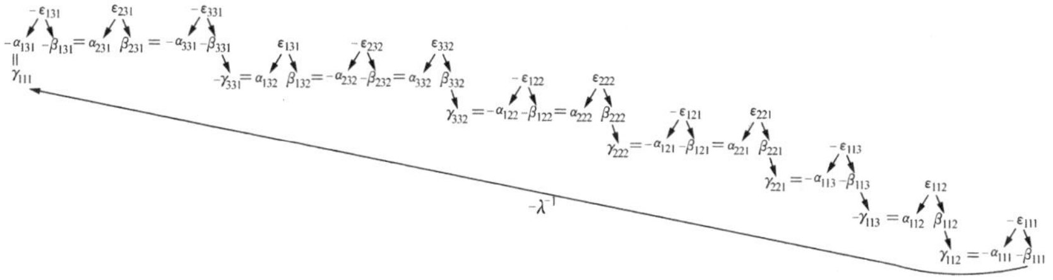

Fig. 5.3

5.4 Remark If $(\boldsymbol{A},\boldsymbol{B},\boldsymbol{C})=(1,\lambda,\varnothing)$ then $Cok(\boldsymbol{A},\boldsymbol{B},\boldsymbol{C})$ is the band module (fig. 5.4).

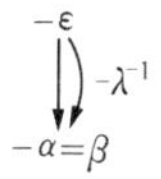

Fig. 5.4

Combining Lemmas 5.1, 5.2 and Proposition 4.1, we can prove the main result of this paper.

5.5 Theorem Let $\mathfrak{A}(\Lambda)=(\widetilde{\Lambda}\times\widetilde{\Lambda},\mathrm{rad}\widetilde{\Lambda},0)$ be the bimodule problem of $\Lambda=K[x,y]/(x^2,xy,y^3)$. Then there is a bijection between the equivalence classes of bands of Λ and R-bands $\mathfrak{A}(\Lambda)$. Furthermore, there is a bijection between the iso-classes of Λ-band modules and the canonical forms of $\mathfrak{A}(\Lambda)$ with parameters.

Let ϕ be the Euler's function and t a positive integer. Assume that $u_1,u_2,\cdots,u_t$ pair-wisely different symbols. Given non-negative integers m_i for $i=1,2,\cdots,t$ with $m=\sum_{i=1}^{t}m_i$, we denote by $\Omega(m_1,m_2,\cdots,m_t;t)$

the set of all words ω consisting of symbols $u_i s$ length m satisfying the symbol α_i occurs exactly m_i times in ω. Now we define an equivalent relation $\sim$ on $\Omega(m_1, m_2, \cdots, m_t; t)$ by: $\omega_1\omega_2\cdots\omega_m \sim \omega_i\omega_{i+1}\cdots\omega_m\omega_1\cdots\omega_{i-1}$, $\forall\, i=2,3,\cdots,m$.

The following fact is well-known (see e. g. [1, P_{226}]).

5. 6 Lemma Keep notations as above. Then

$$|\Omega(m_1, m_2, \cdots, m_t; t)| = \frac{1}{m}\sum_{d\,|\,\gcd(m_1, m_2, \cdots, m_t)} \phi(d)\frac{\left(\frac{m}{d}\right)!}{\prod_{i=1}^{t}\left(\frac{m_i}{d}\right)!}.$$

The following fact follows directly by applying Proposition 4. 8(3) and Theorem 5. 5, which is the main result of [?].

5. 7 Theorem Keep notations as Theorem 5. 5. Then

(1) If $(\boldsymbol{A}=\boldsymbol{I}_n, \boldsymbol{B}, \boldsymbol{C})$ is a canonical form of $\mathfrak{A}(\Lambda)$, where $\boldsymbol{B}$ is the Weyr matrix of $\bigoplus_{i=1}^{d}\boldsymbol{J}_i(0)^{e_i}$ and the Jordan block of $\boldsymbol{C}$ is λ, then there are $|\Omega(e_1, e_2, \cdots, e_d; d)|$ canonical forms of $\mathfrak{A}(\Lambda)$ being of the form.

(2) The number of indecomposable canonical forms $\mathfrak{A}$ with dimension vector (n, n) containing an invertible Weyr matrix equals

$$1+\sum_{(e_1, e_2, \cdots, e_d)\in Q_n}|\Omega(e_1, e_2, \cdots, e_d; d)|$$

where

$$Q_n=\{\underline{e}=(e_1, e_2, \cdots, e_d)\,|\,e_d>0, e_s\geqslant 0, 1\leqslant s\leqslant d-1, \sum_{i=1}^{d} ie_i=n\}.$$

References

[1] Aigner M. Cobinatorial theory. Springer, 1997.

[2] Bulter M C R, Ringel C M. Auslander-Rieten sequences with few middle terms and application to string algebras, Comm. Algebra, 1987, 15: 145-179.

[3] Gelfand I M. , Ponomarev V A. Indecomposable representations of the Lorentz group. Russian Math. Surveys, 1968, 23: 1C-58.

[4] Sergeichuk V V. Canonical matrices for linear matrix problems. Linear Algebra Appl. , 2000, 317: 53-102.

[5] Xu Yunge, Zhang Yingbo. Indecomposablity and the number of links (in Chinese). Science in China, 2001, 44A: 1 515-1 522.

Science in China: Mathematics, 2009, 52A(5): 949-958.

局部两点 Bocs 的表示型①

Representation Type of Local and Two-Vertex Bocses

Abstract Let k be an algebraically chosed field. It has been proved by Zhang and Xu that if a bocs is of tame representation type, then the degree of the differential of the first solid arrow must be less than or equal to 3. We will prove in the present paper that: The bocs is still wild when the degree of the differential of the first arrow is equal to 3. Especially, the bocs with only one solid arrow is of tame type if and only if the degree of the differential of the arrow is less than or equal to 2. Moreover, we classify in this case the growth problems of the representation category of the bocs and layout the sufficient and necessary conditions when the bocs is of finite representation type, tame domestic and tame exponential growth respectively.

Keywords bocs; reduction; tame representation type; wild representation type; domestic.

§ 1. The classification theorem

Let k be an algebraically closed field and Λ a finitely dimensional

---

① Received: 2008-07-13; Accepted: 2008-10-25; published online: 2009-04-04

This work was supported by National Natural Science Foundation of China (Grant No. 10731070)

本文与张学颖、赵双美合作.

algebra over k, which is associative with identity 1. The remarkable "Tame and Wild Theorem" of Drozd states that Λ is either tame or wild [2, 3], whose proof relies on transferring the algebra to the corresponding bocs which is representation equivalent to the algebra. It was proved in [1] that if a bocs is of tame representation type, the degree of the differential of the first solid arrow must be less than or equal to 3. We will improve the result and give some new ones.

A pair $\mathcal{A}=(A,V)$ is called a bocs provided that A is a skeletally small category and V is an A co-algebra with counit ϵ and co-multiplication μ. A collection $L=(A';\omega;a_1,a_2,\cdots,a_n;v_1,v_2,\cdots,v_m)$ is called a layer of bocs $\mathcal{A}$ whenever A' is a minimal category, A is freely generated over A' by indecomposable elements $a_1, a_2, \cdots, a_n$ (solid arrows), and $\ker(\epsilon)=\overline{V}$ is a free A-A bimodule generated by $v_1, v_2, \cdots$, v_m (dotted arrows), $V\simeq A\oplus\overline{V}$. $\omega: A'\to {}_{A'}V_{A'}$ is a reflector. Especially, if $A=A'$, then $\mathcal{A}=(A,V)$ is said to be a minimal bocs. And a minimal bocs $\mathcal{A}$ is trivial if

$$A'\simeq k\times k\times\cdots\times k.$$

If $\mathcal{A}=(A,V)$ is a bocs, and $\theta: A\to B$ is a functor from A to a skeletally small category B, then an induced bocs $\mathcal{A}^B=(B, {}^BV^B)$ can be constructed by setting ${}^BV^B=B\otimes_A V\otimes_A B$. In addition, we have a bocs morphism $\theta_I=(\theta,\theta_1):\mathcal{A}\to\mathcal{A}^B$, where θ_1 is the A-A bimodule map given by $\theta_1: V\cong A\otimes_A V\otimes_A V\xrightarrow{\theta\otimes\mathrm{id}\otimes\theta} {}_AB\otimes_A V\otimes_A B_A$. This induces a fully faithful functor $\theta_I^*: R(\mathcal{A}^B)\to R(\mathcal{A})$ between representation categories. The functor is generally obtained from 3 kinds of reductions: regularization, edge reduction and unraveling. We are able to turn a tame bocs into finitely many minimal bocses for any given dimension n through repeated reductions.

The following definitions are taken from [4, 5], while we make a parallel transformation from algebra Λ to bocs $\mathcal{A}$.

Definition 1.1 A bocs $\mathcal{A}$ is said to be of tame representation type, provided that for each dimension n, there are finitely many parameter-holding functors F_i: mod $k[x] \to R(\mathcal{A})$, $1 \leqslant i \leqslant r_n$, such that

(1) $F_i = M_i \otimes_{k[x]} -$, where M_i is a finitely generated $\mathcal{A}$-$k[x]$ bimodule and is free as a right $k[x]$ module, for $1 \leqslant i \leqslant r_n$;

(2) except finitely many iso-classes, every n dimensional indecomposable $\mathcal{A}$ module is of the form $F_i(S)$ up to isomorphism, where $\mathcal{S}$ is a simple $k[x]$ module.

Otherwise, $\mathcal{A}$ is said to be of representation wild type.

Denote by $\mu_{\mathcal{A}}(n)$ the minimal number of parameter-holding functors satisfying conditions (1) and (2) of Definition 1.1 for given bocs $\mathcal{A}$ and given dimension n.

Definition 1.2 The representation category $R(\mathcal{A})$ of a tame bocs $\mathcal{A}$ is said to be domestic if there exists a natural number m such that $\mu_{\mathcal{A}}(n) \leqslant m$ for each dimension n.

Definition 1.3 The representation category $R(\mathcal{A})$ of a tame bocs $\mathcal{A}$ is said to be polynomial growth if there exists a natural number m such that $\mu_{\Lambda}(n) \leqslant n^m$ for each dimension $n > 1$. Especially, if there exists a natural number p such that $\mu_{\Lambda}(n) \leqslant pn$ for each dimension n, then $\mathcal{A}$ is said to be linear growth.

Definition 1.4 The representation category $R(\mathcal{A})$ of a tame bocs $\mathcal{A}$ is said to be exponential growth if there exist some positive constants $p, q, \alpha, \beta, \alpha > 1$, such that $\mu_{\mathcal{A}}(n) \geqslant p\alpha^{qn^{\beta}}$ for each dimension n.

Theorem 1.5 Let $\mathcal{A} = (A, V)$ be a bocs with a layer $L = (A'; \omega; a_1, a_2, \cdots, a_n; v_1, v_2, \cdots, v_m)$. If $\mathcal{A}$ is of tame representation type, then the differential of the first solid arrow a_1 must be one of the following forms:

(1) $\delta(a_1) = (x-\lambda)u + v(y-\mu)$, where in local case, $A'(X, X) = k[x, f(x)^{-1}]$, $f(\lambda) \neq 0$, $f(\mu) \neq 0$ and in two-vertex case $A'(X, X) = k[x, f(x)^{-1}]$, $f(\lambda) \neq 0$, $A'(Y, Y) = k[y, g(y)^{-1}]$, $g(\mu) \neq 0$ (fig. 1.1).

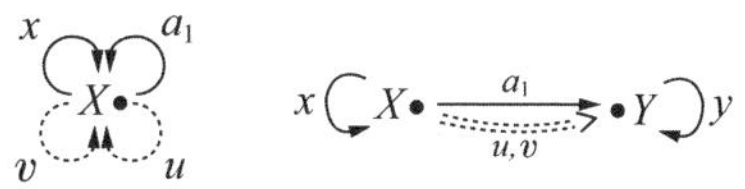

Fig. 1.1

(2) $\delta(a_1)=(x-\lambda)(x-\mu)v$ or $\delta(a_1)=(x-\lambda)v$, where $A'(X,X)=k[x,f(x)^{-1}]$, $f(\lambda)\neq 0$, $f(\mu)\neq 0$, $A'(Y,Y)=k$. (There is a dual case that X is trivial and Y nontrivial. fig. 1. 2)

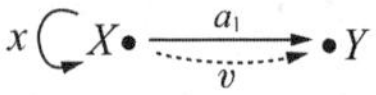

Fig. 1. 2

(3) Both the start and terminal vertices of a_1 are trivial.

Corollary 1. 6 Let $\mathcal{A}=(A,V)$ be a bocs with only one solid arrow and having a layer $L=(A';\omega;a;v_1,v_2,\cdots,v_m)$. Then $\mathcal{A}$ is tame if and only if the differential of a satisfies one of the three conditions given by Theorem 1. 5. In this case the growth of the representation categroy $R(\mathcal{A})$ of the bocs $\mathcal{A}$ is as follows:

(1) In local case $R(\mathcal{A})$ is exponential growth when $\lambda=\mu$, and is domestic when $\lambda\neq\mu$. In two-vertex case $R(\mathcal{A})$ is domestic.

(2) $R(\mathcal{A})$ is domestic.

(3) If bocs $\mathcal{A}$ is local and $\delta(a)=0$, then $R(\mathcal{A})$ is domestic. Otherwise bocs $\mathcal{A}$ is of finite representation type.

Proof of the above theorem and corollary

(1) In local case, Theorem 1. 5(1) is given by Lemma 4. 2(3) of [1] when $\lambda=\mu$, Corollary 1. 6(1) is proved in [6]. When $\lambda\neq\mu$, we can turn the bocs into two-vertex case by using rolled up technique. In two-vertex case, Theorem 1. 5(1) is given by Lemmas 2. 1 and 2. 2 below, Corollary 1. 6(1) is proved by Theorem 2. 1 of [7].

(2) Lemmas 3. 1～3. 3 below confirm Theorem 1. 5(2); Corollary 1. 5(2) is proved by [8] and Theorem 3. 1 of [7].

(3) Theorem 1. 5(3) is obvious. For Corollary 1. 6 (3), when $\delta(a)=0$ and the arrow a has the same start and terminal vertex, we make a loop reduction to turn the arrow a into a parameter, thus we obtain an induced local minimal bocs, whose representation category is domestic. If $\delta(a)=0$ and a starts and ends at different vertices, then after an edge reduction a new vertex appears, we obtain an induced trivial bocs with three vertices. When $\delta(a)=v$, a and v are vanished by regularization, hence we obtain a local or two-vertex trivial induced bocs.

§ 2. Minimally wild bocs of two nontrivial vertices

Lemma 2.1 Given a bocs by the following bi-quiver and differential with $A'(X,X)=k[x,f(x)^{-1}]$, $A'(Y,Y)=k[y,g(y)^{-1}]$ (fig. 2.1):

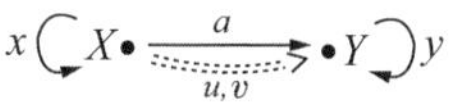

Fig. 2.1

$$\delta(a)=(x-\lambda_1)(x-\lambda_2)u+v(y-\mu),$$
$$\lambda,\mu\in k,\ f(\lambda_1)\neq 0,\ f(\lambda_2)\neq 0,\ g(\mu)\neq 0.$$

Set

$$x\to \boldsymbol{J}_1(\lambda_1)\oplus\boldsymbol{J}_2(\lambda_1)\oplus\boldsymbol{J}_1(\lambda_2)\oplus\boldsymbol{J}_2(\lambda_2)\oplus\boldsymbol{J}_3(\lambda_2),$$
$$y\to \boldsymbol{J}_1(\mu)\oplus\boldsymbol{J}_2(\mu)\oplus\boldsymbol{J}_3(\mu)\oplus\boldsymbol{J}_4(\mu).$$

Then after unraveling, we obtain an induced wild bocs by a sequence of edge reductions and regularizations (fig. 2.2):

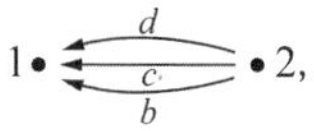

$\delta(b)=0,\ \delta(c)=0,\ \delta(d)=0.$

Fig. 2.2

Proof Using Belitskii's algorithm[9], we obtain the Weyr matrix $\boldsymbol{W}_1$ corresponding to x and $\boldsymbol{S}_1$ commuting with $\boldsymbol{W}_1$:

$$\begin{pmatrix}
\lambda_1 & 0 & 1 & & & & & & \\
0 & \lambda_1 & 0 & & & & & & \\
0 & 0 & \lambda_1 & & & & & & \\
 & & & \lambda_2 & 0 & 0 & 1 & 0 & \\
 & & & 0 & \lambda_2 & 0 & 0 & 1 & \\
 & & & 0 & 0 & \lambda_2 & 0 & 0 & \\
 & & & & & & \lambda_2 & 0 & 1 \\
 & & & & & & 0 & \lambda_2 & 0 \\
 & & & & & & & & \lambda_2
\end{pmatrix},\quad
\begin{pmatrix}
x_1 & x_{12} & x_{11}^1 & & & & & & \\
 & x_2 & x_{21}^1 & & & & & & \\
 & & x_1 & & & & & & \\
 & & & y_1 & y_{12} & y_{13} & y_{11}^1 & y_{12}^1 & y_{11}^2 \\
 & & & & y_2 & y_{23} & y_{21}^1 & y_{22}^1 & y_{21}^2 \\
 & & & & & y_3 & 0 & y_{32}^1 & y_{31}^2 \\
 & & & & & & y_1 & y_{12} & y_{11}^1 \\
 & & & & & & & y_2 & y_{21}^1 \\
 & & & & & & & & y_1
\end{pmatrix},$$

here, zeros are replaced by blanks, Similarly, we have $\boldsymbol{W}_2$ corresponding to y (omitted), and $\boldsymbol{S}_2$ commuting with $\boldsymbol{W}_2$:

$$\boldsymbol{S}_2=\begin{pmatrix}
z_1 & z_{12} & z_{13} & z_{14} & z_{11}^1 & z_{12}^1 & z_{13}^1 & z_{11}^2 & z_{12}^2 & z_{11}^3\\
 & z_2 & z_{23} & z_{24} & z_{21}^1 & z_{22}^1 & z_{23}^1 & z_{21}^2 & z_{22}^2 & z_{21}^3\\
 & & z_3 & z_{34} & 0 & z_{32}^1 & z_{33}^1 & z_{31}^2 & z_{32}^2 & z_{31}^3\\
 & & & z_4 & 0 & 0 & z_{43}^1 & 0 & z_{42}^2 & z_{41}^3\\
 & & & & z_1 & z_{12} & z_{13} & z_{11}^1 & z_{12}^1 & z_{11}^2\\
 & & & & & z_2 & z_{23} & z_{21}^1 & z_{22}^1 & z_{21}^2\\
 & & & & & & z_3 & 0 & z_{32}^1 & z_{31}^2\\
 & & & & & & & z_1 & z_{12} & z_{11}^1\\
 & & & & & & & & z_2 & z_{21}^1\\
 & & & & & & & & & z_1
\end{pmatrix}.$$

Denote by $\boldsymbol{U}, \boldsymbol{V}$. the two 9×10 full matrices corresponding to u and v whose entries are linearly independent, Then

$$\boldsymbol{W}_1(\boldsymbol{W}_1-1)\boldsymbol{U}+\boldsymbol{V}\boldsymbol{W}_2=\begin{pmatrix}
* & * & * & * & * & * & * & * & * & *\\
0 & 0 & 0 & 0 & * & * & * & * & * & *\\
0 & 0 & 0 & 0 & * & * & * & * & * & *\\
* & * & * & * & * & * & * & * & * & *\\
* & * & * & * & * & * & * & * & * & *\\
0 & 0 & 0 & 0 & * & * & * & * & * & *\\
* & * & * & * & * & * & * & * & * & *\\
0 & 0 & 0 & 0 & * & * & * & * & * & *\\
0 & 0 & 0 & 0 & * & * & * & * & * & *
\end{pmatrix}.$$

Here, $*$'s in the matrix represent linearly independent entries which are also independent to the entries of $\boldsymbol{S}_1$ and $\boldsymbol{S}_2$. Since

$$\boldsymbol{A}\boldsymbol{S}_2-\boldsymbol{S}_1\boldsymbol{A}=\boldsymbol{W}_1(\boldsymbol{W}_1-1)\boldsymbol{U}+\boldsymbol{V}\boldsymbol{W}_2,$$

the 9×10 matrix $\boldsymbol{A}$ corresponding to the solid arrow a is of the following form:

$$
\mathbf{A}=\begin{pmatrix}
\phi & \phi & \phi & \phi & \phi & \phi & \phi & \phi & \phi & \phi \\
a_{21} & a_{22} & a_{23} & a_{24} & \phi & \phi & \phi & \phi & \phi & \phi \\
a_{11} & a_{12} & a_{13} & a_{14} & \phi & \phi & \phi & \phi & \phi & \phi \\
\phi & \phi & \phi & \phi & \phi & \phi & \phi & \phi & \phi & \phi \\
\phi & \phi & \phi & \phi & \phi & \phi & \phi & \phi & \phi & \phi \\
b_{31} & b_{32} & b_{33} & b_{34} & \phi & \phi & \phi & \phi & \phi & \phi \\
\phi & \phi & \phi & \phi & \phi & \phi & \phi & \phi & \phi & \phi \\
b_{21} & b_{22} & b_{23} & b_{24} & \phi & \phi & \phi & \phi & \phi & \phi \\
b_{11} & b_{12} & b_{13} & b_{14} & \phi & \phi & \phi & \phi & \phi & \phi
\end{pmatrix}. \qquad (2.1)
$$

The elements denoted by ϕ are regularized by the $*$'s in matrix (2.1). For simpleness, we consider the following simplified reduction equation:

$$
\begin{pmatrix}
x_2 & x_{21} & & & \\
 & x_1 & & & \\
 & & y_3 & y_{32}^{1} & y_{31}^{2} \\
 & & & y_2 & y_{21}^{1} \\
 & & & & y_1
\end{pmatrix}
\begin{pmatrix}
a_{21} & a_{22} & a_{23} & a_{24} \\
a_{11} & a_{12} & a_{13} & a_{14} \\
b_{31} & b_{32} & b_{33} & b_{34} \\
b_{21} & b_{22} & b_{23} & b_{24} \\
b_{11} & b_{12} & b_{13} & b_{14}
\end{pmatrix}
=
\begin{pmatrix}
a_{21} & a_{22} & a_{23} & a_{24} \\
a_{11} & a_{12} & a_{13} & a_{14} \\
b_{31} & b_{32} & b_{33} & b_{34} \\
b_{21} & b_{22} & b_{23} & b_{24} \\
b_{11} & b_{12} & b_{13} & b_{14}
\end{pmatrix}
\begin{pmatrix}
z_1 & z_{12} & z_{13} & z_{14} \\
 & z_2 & z_{23} & z_{24} \\
 & & z_3 & z_{34} \\
 & & & z_4
\end{pmatrix}.
$$

The equation involving b_{11} is $y_1 b_{11} - b_{11} z_1 = 0$. Put $b_{11}=1$, then $y_1 = z_1$ and $b_{12}, b_{13}, b_{14}, b_{21}$ are regularized in which $z_{12}, z_{13}, z_{14}, y_{21}^{1}$ vanish. Then the equation involving b_{22} is $y_2 b_{22} - b_{22} z_2 = 0$. We also put $b_{22}=1$, then $y_2 = z_2$. Similarly, $b_{23}, b_{24}, b_{31}, b_{32}$ are regularized in which $y_{32}^{1}, y_{31}^{2}, z_{23}, z_{24}$ vanish. Equation $y_3 b_{33} - b_{33} z_3 = 0$ implies an edge reduction, so once again set $b_{33}=1$, then $y_3 = z_3$, b_{34} goes to ϕ and z_{34} to zero. Thus the reduction equation becomes

$$
\begin{pmatrix}
x_2 & x_{21} & & & \\
 & x_1 & & & \\
 & & y_3 & 0 & 0 \\
 & & & y_2 & 0 \\
 & & & & y_1
\end{pmatrix}
\begin{pmatrix}
a_{21} & a_{22} & a_{23} & a_{24} \\
a_{11} & a_{12} & a_{13} & a_{14} \\
\phi & \phi & 1 & \phi \\
\phi & 1 & \phi & \phi \\
1 & \phi & \phi & \phi
\end{pmatrix}
=
\begin{pmatrix}
a_{21} & a_{22} & a_{23} & a_{24} \\
a_{11} & a_{12} & a_{13} & a_{14} \\
\phi & \phi & 1 & \phi \\
\phi & 1 & \phi & \phi \\
1 & \phi & \phi & \phi
\end{pmatrix}
\begin{pmatrix}
y_1 & 0 & 0 & 0 \\
 & y_2 & 0 & 0 \\
 & & y_3 & 0 \\
 & & & z_4
\end{pmatrix}.
$$

Set $a_{11}=1, a_{12}=1, a_{13}=1, a_{14}=1$ then $x_1 = y_1 = y_2 = y_3 = z_4$. a_{21} is regularized, so x_{21} is forced to be zero. The above equation is changed into

$$\begin{pmatrix} x_2 & 0 & & & \\ & x_1 & & & \\ & & x_1 & 0 & 0 \\ & & & x_1 & 0 \\ & & & & x_1 \end{pmatrix} \begin{pmatrix} \phi & b & c & d \\ 1 & 1 & 1 & 1 \\ \phi & \phi & 1 & \phi \\ \phi & 1 & \phi & \phi \\ 1 & \phi & \phi & \phi \end{pmatrix} \begin{pmatrix} \phi & b & c & d \\ 1 & 1 & 1 & 1 \\ \phi & \phi & 1 & \phi \\ \phi & 1 & \phi & \phi \\ 1 & \phi & \phi & \phi \end{pmatrix} \begin{pmatrix} x_1 & 0 & 0 & 0 \\ & x_1 & 0 & 0 \\ & & x_1 & 0 \\ & & & x_1 \end{pmatrix},$$

where $b=a_{22}$, $c=a_{23}$, $d=a_{24}$. The induced bocs is given by the following quiver and differential (fig. 2. 2). This completes the proof.

Lemma 2. 2 Given a layered bocs by the following bi-quiver and differential with $A'(X,X)=k[x,f(x)^{-1}]$, $A'(Y,Y)=k[y,g(y)^{-1}]$ (fig. 2. 1),

$$\delta(a)=(x-\lambda)^2u+v(y-\mu),\lambda,\mu\in k,f(\lambda)\neq 0,g(\mu)\neq 0.$$

Let $x\to \boldsymbol{J}_2(\lambda)\oplus\boldsymbol{J}_3(\lambda)\oplus\boldsymbol{J}_4(\lambda)$, $y\to\boldsymbol{J}_1(\mu)\oplus\boldsymbol{J}_2(\mu)\oplus\boldsymbol{J}_3(\mu)\oplus\boldsymbol{J}_4(\mu)$. Then after unraveling, we obtain an induced wild bocs by a sequence of edge reductions and regularizations (fig. 2. 3).

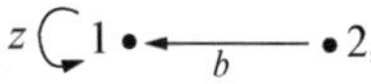

$\delta(b)=0.$

Fig. 2. 3

Proof Using Belitskii's reduction algorithm, we obtain two Weyr matrices $\boldsymbol{W}_1,\boldsymbol{W}_2$ corresponding to x,y and matrices $\boldsymbol{S}_1,\boldsymbol{S}_2$ commuting with $\boldsymbol{W}_1,\boldsymbol{W}_2$ respectively. Since it is similar to Lemma 2. 1, we do not write them explicitly here. The 9×10 matrices $\boldsymbol{A}$ and $\boldsymbol{U}$, $\boldsymbol{V}$ are corresponding to solid arrow a and dotted arrows u, v respectively.

$$\boldsymbol{AS}_2-\boldsymbol{S}_1\boldsymbol{A}=\boldsymbol{W}_1^2\boldsymbol{U}+\boldsymbol{VW}_2=\begin{pmatrix} * & * & * & * & * & * & * & * & * & * \\ * & * & * & * & * & * & * & * & * & * \\ 0 & 0 & 0 & 0 & * & * & * & * & * & * \\ * & * & * & * & * & * & * & * & * & * \\ 0 & 0 & 0 & 0 & * & * & * & * & * & * \\ 0 & 0 & 0 & 0 & * & * & * & * & * & * \\ 0 & 0 & 0 & 0 & * & * & * & * & * & * \\ 0 & 0 & 0 & 0 & * & * & * & * & * & * \\ 0 & 0 & 0 & 0 & * & * & * & * & * & * \end{pmatrix}. \quad (2.2)$$

Thus the reduction equation can be changed to the following simpler form:

$$\begin{pmatrix} x_3 & x_{32}^1 & x_{33}^1 & x_{31}^2 & x_{32}^2 & x_{31}^3 \\ & x_2 & x_{23}^1 & x_{21}^1 & x_{22}^2 & x_{21}^2 \\ & & x_3 & 0 & x_{32}^1 & x_{31}^2 \\ & & & x_1 & x_{12} & x_{11}^1 \\ & & & & x_2 & x_{21}^1 \\ & & & & & x_1 \end{pmatrix} \begin{pmatrix} a_{61} & a_{62} & a_{63} & a_{64} \\ a_{51} & a_{52} & a_{53} & a_{54} \\ a_{41} & a_{42} & a_{43} & a_{44} \\ a_{31} & a_{32} & a_{33} & a_{34} \\ a_{21} & a_{22} & a_{23} & a_{24} \\ a_{11} & a_{12} & a_{13} & a_{14} \end{pmatrix}$$

$$= \begin{pmatrix} a_{61} & a_{62} & a_{63} & a_{64} \\ a_{51} & a_{52} & a_{53} & a_{54} \\ a_{41} & a_{42} & a_{43} & a_{44} \\ a_{31} & a_{32} & a_{33} & a_{34} \\ a_{21} & a_{22} & a_{23} & a_{24} \\ a_{11} & a_{12} & a_{13} & a_{14} \end{pmatrix} \begin{pmatrix} y_1 & y_{12} & y_{13} & y_{14} \\ & y_2 & y_{23} & y_{24} \\ & & y_3 & y_{34} \\ & & & y_4 \end{pmatrix}. \tag{2.3}$$

Set $a_{11}=1$, then $a_{12}=a_{13}=a_{14}=a_{21}=\phi$, and $x_1=y_1$, $y_{12}=y_{13}=y_{14}=x_{21}^1=0$; set $a_{22}=1$, then $a_{23}=a_{24}=a_{31}=a_{32}=\phi$, and $x_2=y_2$, $y_{23}=y_{24}=x_{12}=x_{11}^1=0$; set $a_{33}=1$, then $a_{34}=a_{41}=a_{42}=\phi$, and $x_3=y_3$, $y_{34}=x_{32}^1=x_{31}^2=0$. Let $a_{43}=1$, $a_{44}=1$, we have $x_3=x_1$, $x_3=y_4$, and $a_{51}=a_{52}=a_{53}=\phi$, $x_{23}=x_{22}^2=x_{21}^2=0$, moreover $a_{61}=a_{62}=a_{63}=\phi$, and $x_{33}^1=x_{32}^2=x_{31}^3=0$. Thus the final equation is

$$\begin{pmatrix} x_1 & 0 & 0 & 0 & 0 & 0 \\ & x_2 & 0 & 0 & 0 & 0 \\ & & x_1 & 0 & 0 & 0 \\ & & & x_1 & 0 & 0 \\ & & & & x_2 & 0 \\ & & & & & x_1 \end{pmatrix} \begin{pmatrix} \phi & \phi & \phi & z \\ \phi & \phi & \phi & b \\ \phi & \phi & 1 & 1 \\ \phi & \phi & 1 & \phi \\ \phi & 1 & \phi & \phi \\ 1 & \phi & \phi & \phi \end{pmatrix} = \begin{pmatrix} \phi & \phi & \phi & z \\ \phi & \phi & \phi & b \\ \phi & \phi & 1 & 1 \\ \phi & \phi & 1 & \phi \\ \phi & 1 & \phi & \phi \\ 1 & \phi & \phi & \phi \end{pmatrix} \begin{pmatrix} x_1 & 0 & 0 & 0 \\ & x_2 & 0 & 0 \\ & & x_1 & 0 \\ & & & x_1 \end{pmatrix},$$

where $b=a_{54}$, $z=a_{64}$. The induced wild bocs is given by the following biquiver and differential (fig. 2. 3). □

§ 3. Minimally wild bocs with a nontrivial vertex and a trivial vertex

Lemma 3.1 Given a layered bocs by the following bi-quiver and differential with $A'(X,X)=k[x,f(x)]$, $A'(Y,Y)=k$ (fig. 3.1):

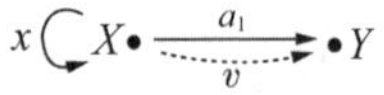

Fig. 3.1

$$\delta(a)=(x-\lambda_1)(x-\lambda_2)(x-\lambda_3)v, \lambda_i\in k,$$

$$f(\lambda_i)\neq 0, i=1,2,3.$$

Set $x\to(\boldsymbol{J}_1(\lambda_1)\oplus\boldsymbol{J}_2(\lambda_2))\oplus(\boldsymbol{J}_1(\lambda_2)\oplus\boldsymbol{J}_2(\lambda_2)\oplus\boldsymbol{J}_3(\lambda_2))\oplus(\boldsymbol{J}_1(\lambda_3)\oplus\boldsymbol{J}_2(\lambda_3)\oplus\boldsymbol{J}_3(\lambda_3))$. Then after unraveling and taking a 3 dimensional vector, space at vertex Y, we obtain an induced wild bocs by a sequence of edge reductions and regularizations(fig. 3.2):

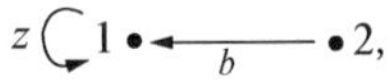

$\delta(b)=0.$

Fig. 3.2

Proof Using Belitskii's reduction algorithm, we obtain the Wyer matrix $\boldsymbol{W}$ corresponding to x and $\boldsymbol{S}_1$ the matrix commuting with $\boldsymbol{W}$:

$$\boldsymbol{W}=\begin{pmatrix}\boldsymbol{W}_1&&\\&\boldsymbol{W}_2&\\&&\boldsymbol{W}_3\end{pmatrix},\quad \boldsymbol{W}_1=\begin{pmatrix}\lambda_1&0&1\\0&\lambda_1&0\\&&\lambda_1\end{pmatrix},\quad \boldsymbol{W}_i=\begin{pmatrix}\lambda_i&0&0&1&0&\\0&\lambda_i&0&0&1&\\0&0&\lambda_i&0&0&\\&&&\lambda_i&0&1\\&&&&\lambda_i&0\\&&&&&\lambda_i\end{pmatrix},$$

$$\boldsymbol{S}_1=\begin{pmatrix}\boldsymbol{X}&&\\&\boldsymbol{Y}&\\&&\boldsymbol{Z}\end{pmatrix},\quad \boldsymbol{X}=\begin{pmatrix}x_1&x_{12}&x_{11}^1\\&x_2&x_{21}^1\\&&x_1\end{pmatrix},\quad \boldsymbol{Y}=\begin{pmatrix}y_1&y_{12}^1&y_{13}^1&y_{11}^2&y_{12}^2&y_{11}^3\\&y_2&y_{23}^1&y_{21}^2&y_{22}^2&y_{21}^3\\&&y_3&0&y_{32}^2&y_{31}^3\\&&&y_1&y_{12}^1&y_{11}^2\\&&&&y_2&y_{21}^2\\&&&&&y_1\end{pmatrix},$$

where $i=2,3$, $\boldsymbol{Z}$ has a similar form as $\boldsymbol{Y}$. Here we have a 3×3 matrix $\boldsymbol{S}_2=(w_{ij})$ at vertex Y, and 15×3 matrices $\boldsymbol{A},\boldsymbol{V}$ correspond to a, v respectively. Since

$$\boldsymbol{AS}_2=\boldsymbol{S}_1\boldsymbol{A}=(\boldsymbol{W}-\lambda_1)(\boldsymbol{W}-\lambda_2)(\boldsymbol{W}-\lambda_3)\boldsymbol{V}=$$

$$\begin{pmatrix} * & 0 & 0 & * & * & 0 & * & 0 & 0 & * & * & 0 & * & 0 & 0 \\ * & 0 & 0 & * & * & 0 & * & 0 & 0 & * & * & 0 & * & 0 & 0 \\ * & 0 & 0 & * & * & 0 & * & 0 & 0 & * & * & 0 & * & 0 & 0 \\ * & 0 & 0 & * & * & 0 & * & 0 & 0 & * & * & 0 & * & 0 & 0 \end{pmatrix}^{\mathrm{T}},$$

where T stands for transpose, it is clear that there are only 8 rows of the 15×4 matrix $\boldsymbol{A}$ valid, whereas the others are regularized. For briefness, we consider a simplified equation $\boldsymbol{A}^{\mathrm{T}}\boldsymbol{S}_2=\boldsymbol{S}_1^{\mathrm{T}}\boldsymbol{A}^{\mathrm{T}}$, with the following $\boldsymbol{S}_1^{\mathrm{T}}$, $\boldsymbol{A}^{\mathrm{T}}$, $\boldsymbol{S}_2$:

$$\begin{pmatrix} x_2 & x_{21}^1 & & & & & & \\ & x_1 & & & & & & \\ & & y_3 & y_{32}^1 & y_{31}^2 & & & \\ & & & y_2 & y_{21}^1 & & & \\ & & & & y_1 & & & \\ & & & & & z_3 & z_{32}^1 & z_{31}^2 \\ & & & & & & z_2 & z_{21}^1 \\ & & & & & & & z_1 \end{pmatrix}, \quad \begin{pmatrix} a_{21} & a_{22} & a_{23} \\ a_{11} & a_{12} & a_{13} \\ b_{31} & b_{32} & b_{33} \\ b_{21} & b_{22} & b_{23} \\ b_{11} & b_{12} & b_{13} \\ c_{31} & c_{32} & c_{33} \\ c_{21} & c_{22} & c_{23} \\ c_{11} & c_{12} & c_{13} \end{pmatrix}, \quad \begin{pmatrix} w_{11} & w_{12} & w_{13} \\ w_{21} & w_{22} & w_{23} \\ w_{31} & w_{32} & w_{33} \end{pmatrix}.$$

Reducing the last 3 rows of $\boldsymbol{A}^{\mathrm{T}}$, we have $\boldsymbol{I}_3$ and $\boldsymbol{S}_2=\boldsymbol{Z}$. The reductions of the middle 3 rows of $\boldsymbol{A}^{\mathrm{T}}$ result in

$$\begin{pmatrix} \phi & \phi & 1 \\ \phi & 1 & \phi \\ 1 & \phi & \phi \end{pmatrix}, \quad \boldsymbol{S}_2=\boldsymbol{Z}=\begin{pmatrix} y_3 & 0 & 0 \\ 0 & y_2 & 0 \\ 0 & 0 & y_1 \end{pmatrix}, \quad \boldsymbol{Y}=\begin{pmatrix} y_1 & 0 & 0 \\ 0 & y_2 & 0 \\ 0 & 0 & y_3 \end{pmatrix}.$$

Finally, by reducing the above two rows of $\boldsymbol{A}^{\mathrm{T}}$ we have the following equation:

$$\begin{pmatrix} x_2 & 0 \\ 0 & x_1 \end{pmatrix}\begin{pmatrix} \phi & b & z \\ 1 & 1 & 1 \end{pmatrix}=\begin{pmatrix} \phi & b & z \\ 1 & 1 & 1 \end{pmatrix}\begin{pmatrix} x_1 & 0 & 0 \\ 0 & x_1 & 0 \\ 0 & 0 & x_1 \end{pmatrix},$$

where $b=a_{12}$, z is obtained from a_{13} by a loop reduction (fig. 3.2). □

Lemma 3.2 Given a layered bocs by the following bi-quiver and differential with $A'(X,X)=k[x,f(x)^{-1}]$, $A'(Y,Y)=k$ (fig. 3.1).

$$\delta(a)=(x-\lambda_1)^2(x-\lambda_2)v,\lambda_i\in k,f(\lambda_i)\neq 0,i=1,2.$$

Set

$$x\rightarrow(\boldsymbol{J}_2(\lambda_1)\oplus\boldsymbol{J}_3(\lambda_1)\oplus\boldsymbol{J}_4(\lambda_1)\oplus\boldsymbol{J}_1(\lambda_2))\oplus(\boldsymbol{J}_2(\lambda)\oplus\boldsymbol{J}_3(\lambda_2)).$$

Then after unraveling and taken a 4 dimensional vector space at vertex Y, we obtain an induced wild bocs by a sequence of edge reductions and regularizations (fig. 3.2).

Proof Similar to the proof of the above lemmas, we first lower down the 15×4 matrix into 9×4 and consider the following reduction equation:

$$\begin{pmatrix} x_3 & x_{32}^{1} & x_{33}^{1} & x_{31}^{2} & x_{32}^{2} & x_{31}^{3} & & & \\ & x_2 & x_{23}^{1} & x_{21}^{1} & x_{22}^{2} & x_{21}^{2} & & & \\ & & x_3 & 0 & x_{32}^{1} & x_{31}^{2} & & & \\ & & & x_1 & x_{12} & x_{11}^{1} & & & \\ & & & & x_2 & x_{21}^{1} & & & \\ & & & & & x_1 & & & \\ & & & & & & y_3 & y_{32}^{1} & y_{31}^{2} \\ & & & & & & & y_2 & y_{21}^{1} \\ & & & & & & & & y_1 \end{pmatrix}\begin{pmatrix} a_{61} & a_{62} & a_{63} & a_{64} \\ a_{51} & a_{52} & a_{53} & a_{54} \\ a_{41} & a_{42} & a_{43} & a_{44} \\ a_{31} & a_{32} & a_{33} & a_{34} \\ a_{21} & a_{22} & a_{23} & a_{24} \\ a_{11} & a_{12} & a_{13} & a_{14} \\ b_{31} & b_{32} & b_{33} & b_{34} \\ b_{21} & b_{22} & b_{23} & b_{24} \\ b_{11} & b_{12} & b_{13} & b_{14} \end{pmatrix}$$

$$=\begin{pmatrix} a_{61} & a_{62} & a_{63} & a_{64} \\ a_{51} & a_{52} & a_{53} & a_{54} \\ a_{41} & a_{42} & a_{43} & a_{44} \\ a_{31} & a_{32} & a_{33} & a_{34} \\ a_{21} & a_{22} & a_{23} & a_{24} \\ a_{11} & a_{12} & a_{13} & a_{14} \\ b_{31} & b_{32} & b_{33} & b_{34} \\ b_{21} & b_{22} & b_{23} & b_{24} \\ b_{11} & b_{12} & b_{13} & b_{14} \end{pmatrix}\begin{pmatrix} z_{11} & z_{12} & y_{13} & y_{14} \\ z_{21} & z_{22} & y_{23} & y_{24} \\ z_{31} & z_{32} & y_{33} & y_{34} \\ z_{41} & z_{42} & y_{43} & y_{44} \end{pmatrix}.$$

After reducing the last three rows of matrix $\boldsymbol{A}$, we obtain a new equation:

$$
\begin{pmatrix}
x_3 & x_{32}^{1} & x_{33}^{1} & x_{31}^{2} & x_{32}^{2} & x_{31}^{3} & & & \\
 & x_2 & x_{23}^{1} & x_{21}^{1} & x_{22}^{2} & x_{21}^{2} & & & \\
 & & x_3 & 0 & x_{32}^{1} & x_{31}^{2} & & & \\
 & & & x_1 & x_{12} & x_{11}^{1} & & & \\
 & & & & x_2 & x_{21}^{1} & & & \\
 & & & & & x_1 & & & \\
 & & & & & & y_3 & y_{32}^{1} & y_{31}^{2} \\
 & & & & & & & y_2 & y_{21}^{1} \\
 & & & & & & & & y_1
\end{pmatrix}
\begin{pmatrix}
a_{61} & a_{62} & a_{63} & a_{64} \\
a_{51} & a_{52} & a_{53} & a_{54} \\
a_{41} & a_{42} & a_{43} & a_{44} \\
a_{31} & a_{32} & a_{33} & a_{34} \\
a_{21} & a_{22} & a_{23} & a_{24} \\
a_{11} & a_{12} & a_{13} & a_{14} \\
0 & 1 & \phi & \phi \\
0 & 0 & 1 & \phi \\
0 & 0 & 0 & 1
\end{pmatrix}
$$

$$
=\begin{pmatrix}
a_{61} & a_{62} & a_{63} & a_{64} \\
a_{51} & a_{52} & a_{53} & a_{54} \\
a_{41} & a_{42} & a_{43} & a_{44} \\
a_{31} & a_{32} & a_{33} & a_{34} \\
a_{21} & a_{22} & a_{23} & a_{24} \\
a_{11} & a_{12} & a_{13} & a_{14} \\
0 & 1 & \phi & \phi \\
0 & 0 & 1 & \phi \\
0 & 0 & 0 & 1
\end{pmatrix}
\begin{pmatrix}
z_{11} & z_{12} & y_{13} & y_{14} \\
 & y_3 & y_{32}^{1} & y_{31}^{2} \\
 & & y_2 & y_{21}^{1} \\
 & & & y_1
\end{pmatrix}.
$$

It is easy to see that the equation consisting of the first 6 rows is as the same as equation (3) in Lemma 2. 2, this completes the proof.

Lemma 3. 3 Given a layered bocs by the following bi-quiver and differential with $A'(X,X)=k[x,f(x)^{-1}]$, $A'(Y,Y)=k$ (fig. 3. 1).

$$\delta(a)=(x-\lambda)^3 v, \lambda\in k, f(\lambda)\neq 0.$$

Set $x\to \boldsymbol{J}_1(\lambda)\oplus \boldsymbol{J}_4(\lambda)\oplus \boldsymbol{J}_7(\lambda)$. Then after unraveling and taking a 3-dimensional vector space at vertex Y, we obtain an induced wild bocs by a sequence of edge reductions and regularizations (fig. 3. 2).

Proof In a similar way, we lower down the 12×3 matrix into 7×3 and consider the simplified reduction equation:

$$\begin{pmatrix} x_3 & 0 & 0 & x_{32}^3 & 0 & 0 & x_{31}^6 \\ & x_2 & x_{22}^1 & x_{22}^2 & x_{21}^3 & x_{21}^4 & x_{21}^5 \\ & & x_2 & x_{22}^1 & 0 & x_{21}^3 & x_{21}^4 \\ & & & x_2 & 0 & 0 & x_{21}^3 \\ & & & & x_1 & x_{11}^1 & x_{11}^2 \\ & & & & & x_1 & x_{11}^1 \\ & & & & & & x_1 \end{pmatrix} \begin{pmatrix} a_{11} & a_{12} & a_{13} \\ a_{21} & a_{22} & a_{23} \\ a_{31} & a_{32} & a_{33} \\ a_{41} & a_{42} & a_{43} \\ a_{51} & a_{52} & a_{53} \\ a_{61} & a_{62} & a_{63} \\ a_{71} & a_{72} & a_{73} \end{pmatrix} = \begin{pmatrix} a_{11} & a_{12} & a_{13} \\ a_{21} & a_{22} & a_{23} \\ a_{31} & a_{32} & a_{33} \\ a_{41} & a_{42} & a_{43} \\ a_{51} & a_{52} & a_{53} \\ a_{61} & a_{62} & a_{63} \\ a_{71} & a_{72} & a_{73} \end{pmatrix} \begin{pmatrix} y_{11} & y_{12} & y_{13} \\ y_{21} & y_{22} & y_{23} \\ y_{31} & y_{32} & y_{33} \end{pmatrix}.$$

Reducing the last three rows of $\boldsymbol{A}$ we obtain $\boldsymbol{I}_3$ and $\boldsymbol{S}_2=\boldsymbol{Y}$, where $\boldsymbol{Y}$ is given by the last 3 rows and columns of the first matrix in the above equation. Therefore the reduction equation becomes

$$\begin{pmatrix} x_3 & 0 & 0 & x_{32}^3 & 0 & 0 & x_{31}^6 \\ & x_2 & x_{22}^1 & x_{22}^2 & x_{21}^3 & x_{21}^4 & x_{21}^5 \\ & & x_2 & x_{22}^1 & 0 & x_{21}^3 & x_{21}^4 \\ & & & x_2 & 0 & 0 & x_{21}^3 \\ & & & & x_1 & x_{11}^1 & x_{11}^2 \\ & & & & & x_1 & x_{11}^1 \\ & & & & & & x_1 \end{pmatrix} \begin{pmatrix} a_{11} & a_{12} & a_{13} \\ a_{21} & a_{22} & a_{23} \\ a_{31} & a_{32} & a_{33} \\ a_{41} & a_{42} & a_{43} \\ 1 & \phi & \phi \\ 0 & 1 & \phi \\ 0 & 0 & 1 \end{pmatrix} = \begin{pmatrix} a_{11} & a_{12} & a_{13} \\ a_{21} & a_{22} & a_{23} \\ a_{31} & a_{32} & a_{33} \\ a_{41} & a_{42} & a_{43} \\ 1 & \phi & \phi \\ 0 & 1 & \phi \\ 0 & 0 & 1 \end{pmatrix} \begin{pmatrix} x_1 & x_{11}^1 & x_{11}^2 \\ & x_1 & x_{11}^1 \\ & & x_1 \end{pmatrix}.$$

Set $a_{41}=1$ we have $x_2=x_1$. The regularization of a_{42}, a_{43} makes $x_{11}^1=0$, $x_{21}^3=x_{11}^2$. Again $a_{31}=\phi, a_{32}=\phi, a_{33}=\phi, a_{21}=\phi, a_{11}=\phi$ force $x_{22}^1=0$, $x_{21}^3=0, x_{21}^4=0, x_{22}^2=0, x_{22}^3=0$. The loop a_{22} goes to a parameter z by a loop reduction. Subsequently, in $a_{23}=\phi, a_{13}=\phi, x_{21}^5$ and x_{31}^6 vanish. In the end, we have a solid arrow with zero differential and the following equation:

$$\begin{pmatrix} x_3 & 0 & 0 & 0 & 0 & 0 & 0 \\ & x_1 & 0 & 0 & 0 & 0 & 0 \\ & & x_1 & 0 & 0 & 0 & 0 \\ & & & x_1 & 0 & 0 & 0 \\ & & & & x_1 & 0 & 0 \\ & & & & & x_1 & 0 \\ & & & & & & x_1 \end{pmatrix} \begin{pmatrix} \phi & b & \phi \\ \phi & z & \phi \\ \phi & \phi & \phi \\ 1 & \phi & \phi \\ 1 & \phi & \phi \\ 0 & 1 & \phi \\ 0 & 0 & 1 \end{pmatrix} = \begin{pmatrix} \phi & b & \phi \\ \phi & z & \phi \\ \phi & \phi & \phi \\ 1 & \phi & \phi \\ 1 & \phi & \phi \\ 0 & 1 & \phi \\ 0 & 0 & 1 \end{pmatrix} \begin{pmatrix} x_1 & 0 & 0 \\ & x_1 & 0 \\ & & x_1 \end{pmatrix}$$

where $b = a_{12}$, and the induced wild bocs is of the following form (fig. 3. 2). □

References

[1] Zhang Y B, Xu Y G. On tame and wild bocses. Sci China, 2005, 48A(12): 456-468.

[2] Drozd Y A. Tame and wild matrix problems. Representations and Quadratic Forms (in Russian). Kiev: Inst Math Akad Nauk Ukrain SSR, 1979: 39-74.

[3] Crawley-Boevey W W. On tame algebras and bocses. Proc London Math Soc, 1998, 56(3): 451-483.

[4] Skowronski A. Algebras of polynomial growth. Topics in algebra. Banach Center Publication, 1990, 26(1).

[5] Norenberg R, Skowronski A. Tame minimal non-polynomial growth simply connected algebras. Colloq Math, 1997, 73(2): 301-330.

[6] Li L C. A linear matrix problem of exponential growth (in Chinese). J Beijing Normal Univ (Natur Sci), 2001, 37(4): 456-463.

[7] Zhang X Y, Zhang Y B. Two classes of domestic bocses. J Beijing Normal Univ (Natur Sci), 2009, 45(1): 22-25.

[8] Pan J, Zhang Y B. A class of domestic bocses (in Chinese). J Beijing Normal Univ (Natur Sci), 2006, 42(5): 467-482.

[9] Sergeichuk V V Canonical matrices for linear matrix problems. Linear Algebra Appl, 2000, 317: 53-102.

[10] Auslander M, Reiten I, Smalø S. Representatiou Theory of Artin Algebra. Cambridge: Cambridge University Press, 1995.

Algebra Colloquium, 2011, 18(3): 373-384.

$K[x,y]/(x^p,y^q,xy)$上不可分解模的典范型①

Canonical Forms of Indecomposable Modules over $K[x,y]/(x^p,y^q,xy)$

Abstract Let $\Lambda=K[x,y]/(x^p,y^q,xy)$ be a Gelfand-Ponomarev algebra, where $p,q\in\mathbf{N}$, and $\mathfrak{A}=(\widetilde{\Lambda}\times\widetilde{\Lambda},\text{rad }\widetilde{\Lambda})$ the matrix problem of Λ. In the present paper, we determine the canonical forms of indecomposable modules of Λ, and give a one-to-one correspondence between the iso-classes of the modules over Λ and the canonical forms over $\mathfrak{A}$. Finally, we discuss some properties of the module variety of Λ using the canonical forms.

Keywords band module; matrix problem; canonical form; module variety.

The well-known Drozd's theorem tells us that a finite-dimensional algebra Λ over an algebraically closed field K is either of tame representation type or of wild representation type (see [2, 3]). To determine the representation type of a finite-dimensional algebra is an important task in representation theory of algebras.

Sergeichuk [6] suggested a correspondence between the algebra Λ and the matrix problem $\mathfrak{A}=(\widetilde{\Lambda}\times\widetilde{\Lambda},\text{rad }\widetilde{\Lambda})$ so that the problem of

---

① Supported by the National Natural Science Foundation of China (10731070).

本文与刘根强合作.

classification of Λ-modules can be reduced to the problem of computing the canonical forms of indecomposable representations of $\mathfrak{A}=(\tilde{\Lambda}\times\tilde{\Lambda}, \operatorname{rad}\tilde{\Lambda})$. The computation is based on Belitskii's reduction algorithm which is very effective for reducing an individual matrix to a canonical form.

The class of string algebras is an interesting class of algebras. Butler and Ringel characterized the indecomposable modules over string algebras in [1]. The Gelfand-Ponomarev algebra $\Lambda=K[x,y]/(x^p,y^q,xy)$, where $p,q\in\mathbf{N}$ with $p\leqslant q$, is a special string algebra. The present paper describes the canonical forms corresponding to the indecomposable modules over Λ and the geometric properties of such canonical forms.

Throughout this paper, K stands for an algebraically closed field and $K^*=K\setminus\{0\}$. We always assume that the modules over an algebra Λ are left modules. Denote by $GL_n(K)$ the general linear group of size n, $M_{m\times n}(K)$ or $K^{m\times n}$ the set of $m\times n$ matrices, and $M_n(K)$ the set of square matrices of size n over K. Let $J_d(\lambda)$ be the Jordan block of eigenvalue λ of size d.

§ 1. Indecomposable modules over Λ

The indecomposable modules over a string algebra are characterized by Butler and Ringel [1]. Let $\Lambda:=K[x,y]/(x^p,y^q,xy)$ be a Gelfand-Ponomarev algebra, where $p,q\in\mathbf{N}$ with $p\leqslant q$. Denote the images of x, $y\in K[x,y]$ in Λ by a and b, respectively. Then a Λ-module is given by a triple $M=(V,a,b)$, where V is a K-vector space, and a,b are K-endomorphisms of V with $a^p=0, b^q=0$ and $ab=0=ba$.

Write a^{-1} and b^{-1} for the formal inverses of a and b, respectively. The elements a,b,a^{-1},b^{-1} are called letters. A word $\mathcal{C}$ is given by a sequence $c_1,c_2,\cdots,c_n$ of letters, where $c_i\in\{a,b,a^{-1},b^{-1}\}$. A string of length $n\geqslant 1$ is a word $\mathcal{C}=c_1,c_2,\cdots,c_n$ such that there is no any subword of $\mathcal{C}$ with the form a^ka^{-k} for $1\leqslant k<p$, b^lb^{-l} for $1\leqslant l<q$ or $a^p, b^q, a^{-p}, b^{-q}, ab, ba, a^{-1}b^{-1}, b^{-1}a^{-1}$. If $\mathcal{C}=c_1c_2\cdots c_n$ and $\mathcal{D}=d_1d_2\cdots d_m$ are

strings of length$\geqslant 1$, then the composition of $\mathcal{C}$ and $\mathcal{D}$ is defined by $\mathcal{CD}=c_1c_2\cdots c_nd_1d_2\cdots d_m$ provided that $c_1c_2\cdots c_nd_1d_2\cdots d_m$ is a string again. Let $\mathcal{C}^{-1}=c_n^{-1}c_{n-1}^{-1}\cdots c_1^{-1}$. We define $\mathcal{C}^n$ to be the n-fold composition $\mathcal{CC}\cdots\mathcal{C}$ of a string $\mathcal{C}$ whenever it is a string. Two strings $\mathcal{C}$ and $\mathcal{D}$ are equivalent if $\mathcal{C}=\mathcal{D}^{-1}$. Denote by **S** the set of the strings of the algebra Λ.

A string $\mathcal{B}=b_1b_2\cdots b_n$ of length$\geqslant 2$ is called a band if all powers $\mathcal{B}^r$ are defined and $\mathcal{B}$ itself is not a power of a string of shorter length. Let $\mathcal{B}=b_1b_2\cdots b_n$ be a band, if $\mathcal{B}'=b_ib_{i+1}\cdots b_nb_1b_2\cdots b_{i-1}$ for some $1\leqslant i\leqslant n$, we say $\mathcal{B}'\sim_r\mathcal{B}$. If $\mathcal{B}_1\sim_r\mathcal{B}_2$ or $\mathcal{B}_1\sim_r\mathcal{B}_2^{-1}$, we write $\mathcal{B}_1\sim\mathcal{B}_2$. Clearly, $\sim$ is an equivalent relation. Denote by $\mathbb{B}$ the set of the bands of the algebra Λ.

For any string $\mathcal{C}=c_1c_2\cdots c_n$, we have an indecomposable module $M(\mathcal{C})$ which is given by a K-vector space of dimension $n+1$ with basis $z_0, z_1, \cdots, z_n$ and linear maps $c_i=(1)$ for $1\leqslant i\leqslant n$ operating on the basis. $M(\mathcal{C})$ is called a string module. For example, in the case $q>2$, letting $\mathcal{C}=ab^{-1}ab^{-2}$, $M(\mathcal{C})$ can be expressed by the following schema:

$$Kz_0 \xleftarrow{a} Kz_1 \xrightarrow{b} Kz_2 \xleftarrow{a} Kz_3 \xrightarrow{b} Kz_4 \xrightarrow{b} Kz_5,$$

i. e., $a(z_1)=z_0, a(z_3)=z_2, b(z_1)=z_2, b(z_3)=z_4, b(z_4)=z_5$.

Similarly, for any band $\mathcal{B}=b_1b_2\cdots b_n$, we have an indecomposable module $M(\mathcal{B},\lambda,r)$. As a K-vector space, it is $\bigoplus_{i=0}^{n-1}V_i$ with V_i of dimension r, and linear maps $b_i=\mathrm{id}$ for $1\leqslant i\leqslant n-1$ and $b_n=J_r(\lambda)$. $M(\mathcal{B},\lambda,r)$ is called a band module. For example, in the case $q>2$, letting $\mathcal{B}=ab^{-1}\cdot ab^{-2}$, $M(\mathcal{B},\lambda,r)$ can be expressed by the following schema (fig. 1. 1):

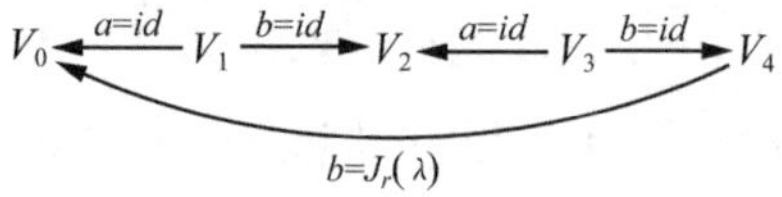

Fig. 1. 1

Theorem 1. 1 [1,4] The modules $M(\mathcal{C})$ with $\mathcal{C}\in\mathbb{S}$ and $M(\mathcal{B},\lambda, r)$ with $\mathcal{B}\in\mathbb{B}, \lambda\in K, r\in\mathbb{N}$ furnish a complete list of the indecomposable modules of Λ. There is no any string module isomorphic to a band module. Two modules $M(\mathcal{C})$ and $M(\mathcal{D})$ are isomorphic if and only if $\mathcal{C}$

and $\mathcal{D}$ belong to the same equivalence class in $\mathbb{S}$. Two modules $M(\mathcal{B},\lambda,r)$ and $M(\mathcal{B}',\lambda',r')$ are isomorphic if and only if $\mathcal{B}$ and $\mathcal{B}'$ belong to the same equivalence class in $\mathbb{B}$ and $\lambda=\lambda',r=r'$.

§ 2. Matrix problem of Λ

Definition 2.1[6] A matrix problem $\mathfrak{A}=(\mathcal{K},\mathcal{M},\boldsymbol{H})$ given by matrices over K consists of the following data:

(1) a set of integers $T=\{1,2,\cdots,t\}$ and an equivalent relation $\sim$ on T with $\mathcal{T}=T/\sim=\{X_1,X_2,\cdots,X_s\}$;

(2) an upper triangular matrix algebra $\mathcal{K}=\{(s_{ij})_{t\times t}\}\subset M_t(K)$, where $s_{ii}=s_{jj}$ if $i\sim j$, and those s_{ij} for $i<j$ satisfy the linear equations with $c_{ij}^l\in K$ for any $(X,Y)\in\mathcal{T}\times\mathcal{T}$:

$$\sum_{(i,j)\in X\times Y,i<j} c_{ij}^l x_{ij}=0; \tag{2.1}$$

(3) a $\mathcal{K}$-$\mathcal{K}$-bimodule $\mathcal{M}=\{(m_{ij})_{t\times t}\in M_t(K)\}$, where the m_{ij} satisfy the linear equations with $d_{ij}^l\in K$ for any $(X,Y)\in\mathcal{T}\times\mathcal{T}$:

$$\sum_{(i,j)\in X\times Y} d_{ij}^l z_{ij}=0; \tag{2.2}$$

(4) a matrix $\boldsymbol{H}\in M_{t\times t}(K)$ such that $h_{ij}=0$ for $i\nsim j$, and a derivation $d:\mathcal{K}\to\mathcal{M}$ given by $d(S)=SH-HS$ for any $S\in\mathcal{K}$.

Let $\underline{\boldsymbol{n}}=(n_1,n_2,\cdots,n_t)\in\mathbb{N}^t$. It is called a size vector of $(T,\sim)$ provided $n_i=n_j$ whenever $i\sim j$. Let $n_X=n_i$ for $i\in X$. Then $\underline{\boldsymbol{d}}=(n_X)_{X\in\mathcal{T}}$ is said to be a dimension vector of $\underline{\boldsymbol{n}}$, moreover, $d=\sum_{X\in\mathcal{T}} n_X$ is called the dimension of $\underline{\boldsymbol{n}}$. For any size vector $\underline{\boldsymbol{n}}$, let

$$\mathcal{M}_{\underline{n}}=\{(\boldsymbol{N}_{ij})_{t\times t}\mid \boldsymbol{N}_{ij}\in M_{n_i\times n_j}(K)\text{ satisfy (2.2)}\},$$

and set $\boldsymbol{H}_{\underline{n}}=(\boldsymbol{H}_{ij})_{t\times t}$, with $\boldsymbol{H}_{ij}=0$ for $i\nsim j$ and $\boldsymbol{H}_{ij}=h_{ij}\boldsymbol{I}_{n_j}$ for $i\sim j$. Then a pair of matrices $(\boldsymbol{M},\boldsymbol{H}_{\underline{n}})$ with $\boldsymbol{M}\in\mathcal{M}_{\underline{n}}$ is called a representation of $(\mathcal{K},\mathcal{M})$ of dimension vector $\underline{\boldsymbol{n}}$. If $\underline{\boldsymbol{m}}$ is also a size vector of $(T,\sim)$, let $\mathcal{K}_{\underline{m}\times\underline{n}}=\{(\boldsymbol{S}_{ij})_{t\times t}\mid \boldsymbol{S}_{ij}\in M_{m_i\times n_j}(K),\boldsymbol{S}_{ii}=\boldsymbol{S}_{jj}\text{ if } i\sim j,\boldsymbol{S}_{ij}(i<j)\text{ satisfy (2.1)}\}$.

The set of morphisms from $(\boldsymbol{M},\boldsymbol{H}_{\underline{m}})$ to $(\boldsymbol{N},\boldsymbol{H}_{\underline{n}})$ with $\boldsymbol{M}\in\mathcal{M}_{\underline{m}}$ and $\boldsymbol{N}\in\mathcal{M}_{\underline{n}}$ is

$$\{\boldsymbol{S}\in\mathcal{K}_{\underline{m}\times\underline{n}}\mid(\boldsymbol{M}+\boldsymbol{H}_{\underline{m}})\boldsymbol{S}=\boldsymbol{S}(\boldsymbol{N}+\boldsymbol{H}_{\underline{n}})\}.$$

Denote by Mat($\mathfrak{A}$) the category of finite-dimensional representations of the matrix problem $\mathfrak{A}$.

Given an order on the indices of the entries of any $m \times n$ matrix such that $(i,j) \prec (p,q)$ provided $i > p$ or $i = p$ and $j < q$. Let (i,j) be the minimal index such that there exists some $m_{ij} \neq 0$ in $\mathcal{M}$. Fix a size vector $\underline{\boldsymbol{n}}$, and expand the equations (2.1) and (2.2) into matrix equations of size $\underline{\boldsymbol{n}}$. Consider $\boldsymbol{M} \in \mathcal{M}_{\underline{n}}$ with the first non-zero block $\boldsymbol{M}_{ij}$, we will determine a new block $\overline{\boldsymbol{M}}_{ij}$ which is simpler.

Proposition 2.2[6]　There are three reductions called Belitskii's algorithms:

(1) If $\boldsymbol{M}_{ij}\boldsymbol{S}_{jj} - \boldsymbol{S}_{ii}\overline{\boldsymbol{M}}_{ij} = \sum \boldsymbol{S}_{il}h_{lj} - h_{il}\boldsymbol{S}_{lj} \neq \boldsymbol{0}$, set $\overline{\boldsymbol{M}}_{ij} = \boldsymbol{0}$, denoted by $\varnothing$.

(2) If $\boldsymbol{M}_{ij}\boldsymbol{S}_{jj} - \boldsymbol{S}_{ii}\overline{\boldsymbol{M}}_{ij} = \boldsymbol{0}$ and $i \nsim j$, set $\overline{\boldsymbol{M}}_{ij} = \begin{pmatrix} \boldsymbol{0} & \boldsymbol{I}_r \\ \boldsymbol{0} & \boldsymbol{0} \end{pmatrix}$ with

$$r = \operatorname{rank}(\boldsymbol{M}_{ij}).$$

(3) If $\boldsymbol{M}_{ij}\boldsymbol{S}_{jj} - \boldsymbol{S}_{ii}\overline{\boldsymbol{M}}_{ij} = \boldsymbol{0}$ and $i \sim j$, set $\overline{\boldsymbol{M}}_{ij} = \boldsymbol{W}$, called a Weyr matrix similar to $\boldsymbol{M}_{ij}$.

Thus, we are able to construct a new matrix problem $\mathfrak{A}' = (\mathcal{K}', \mathcal{M}', H')$ with a new index set $(T', \sim')$ and new equation systems (2.1′), (2.2′). Moreover, the size vector $\underline{\boldsymbol{n}}$ yields a new one $\underline{\boldsymbol{n}}'$; of $(T', \sim')$ (see [6]). Repeat the reduction procedure, we obtain a sequence of reduced matrices and a sequence of reduced matrix problems:

$$(\boldsymbol{M}, \boldsymbol{H}_{\underline{n}}), (\boldsymbol{M}^1, \boldsymbol{H}^1_{\underline{n}^1}), \cdots, (\boldsymbol{M}^s, \boldsymbol{H}^s_{\underline{n}^s}), \mathfrak{A}, \mathfrak{A}_1, \mathfrak{A}_2, \cdots, \mathfrak{A}_s. \tag{2.3}$$

In particular if $\boldsymbol{M}^s = (\boldsymbol{0})$, then $\boldsymbol{H}^s_{\underline{n}^s}$ is said to be the canonical form of $\boldsymbol{M}$. The number 1 is called a link of a canonical form if 1 appears from an edge reduction or sits at the nilpotent part of a Weyr matrix from an unraveling. Denote by $l(\boldsymbol{C})$ the number of the links appearing in the matrix $\boldsymbol{C}$.

Proposition 2.3[7]　Let $\mathfrak{A}$ be a matrix problem. An object $\boldsymbol{M} \in \operatorname{Mat}(\mathfrak{A})$ is indecomposable if and only if the number of links in the canonical form of $\boldsymbol{M}$ equals $\dim_K(\boldsymbol{M})$.

Proposition 2.4[6]　Let Λ be a finite-dimensional algebra. Then the matrix problem of Λ is given by $\mathfrak{A}=(\widetilde{\Lambda}\times\widetilde{\Lambda},\mathrm{rad}\,\widetilde{\Lambda},0)$ with Λ the left regular representation of Λ under some ordered basis. Moreover, there is a natural bijection between the set of iso-classes of indecomposable modules over Λ and the set of indecomposable matrix canonical forms of $\mathfrak{A}$.

Now we show an example for the algebra $\Lambda=K[x,y]/(x^5,y^5,xy)$. Write $a_1=a,b_1=b,a_i=a^i$ and $b_i=b^i$ for $i=1,2,3,4$. The left regular representation of Λ under the ordered basis $\{b_4,a_4,b_3,a_3,b_2,a_2,b_1,a_1,e\}$ with e the idempotent is given by

$$\widetilde{\Lambda}=\left\{\left.\begin{pmatrix} z & 0 & y_1 & 0 & y_2 & 0 & y_3 & 0 & y_4 \\ & z & 0 & x_1 & 0 & x_2 & 0 & x_3 & x_4 \\ & & z & 0 & y_1 & 0 & y_2 & 0 & y_3 \\ & & & z & 0 & x_1 & 0 & x_2 & x_3 \\ & & & & z & 0 & y_1 & 0 & y_2 \\ & & & & & z & 0 & x_1 & x_2 \\ & & & & & & z & 0 & y_1 \\ & & & & & & & z & x_1 \\ & & & & & & & & z \end{pmatrix}\right| \begin{matrix} x_i,y_i,z\in K \\ i\in\{1,2,3,4\} \end{matrix}\right\}.$$

Then $\mathfrak{A}=(\widetilde{\Lambda}\times\widetilde{\Lambda},\mathrm{rad}\,\widetilde{\Lambda},\mathbf{0})$ is the matrix problem of Λ, where $T=\{1,2,\cdots,18\}$ and $\mathcal{T}=\{X_1,X_2\}$ with $X_1=\{1,2,\cdots,9\}$ and $X_2=\{10,11,\cdots,18\}$.

Fix a size vector $(m,m,\cdots,m,n,n,\cdots,n)_{1\times 18}$, we have a reduction equation $\boldsymbol{ZM}=\overline{\boldsymbol{M}}\boldsymbol{Z}^{\mathrm{T}}$:

$$\boldsymbol{M}=\begin{pmatrix} \mathbf{0} & \mathbf{0} & \boldsymbol{B}_1 & \mathbf{0} & \boldsymbol{B}_2 & \mathbf{0} & \boldsymbol{B}_3 & \mathbf{0} & \boldsymbol{B}_4 \\ & \mathbf{0} & \mathbf{0} & \boldsymbol{A}_1 & \mathbf{0} & \boldsymbol{A}_2 & \mathbf{0} & \boldsymbol{A}_3 & \boldsymbol{A}_4 \\ & & \mathbf{0} & \mathbf{0} & \boldsymbol{B}_1 & \mathbf{0} & \boldsymbol{B}_2 & \mathbf{0} & \boldsymbol{B}_3 \\ & & & \mathbf{0} & \mathbf{0} & \boldsymbol{A}_1 & \mathbf{0} & \boldsymbol{A}_2 & \boldsymbol{A}_3 \\ & & & & \mathbf{0} & \mathbf{0} & \boldsymbol{B}_1 & \mathbf{0} & \boldsymbol{B}_2 \\ & & & & & \mathbf{0} & \mathbf{0} & \boldsymbol{A}_1 & \boldsymbol{A}_2 \\ & & & & & & \mathbf{0} & \mathbf{0} & \boldsymbol{B}_1 \\ & & & & & & & \mathbf{0} & \boldsymbol{A}_1 \\ & & & & & & & & \mathbf{0} \end{pmatrix} \qquad (2.4)$$

where $\boldsymbol{A}_1, \boldsymbol{B}_1, \boldsymbol{A}_2, \boldsymbol{B}_2 \in M_{m\times n}(K), \boldsymbol{Z}_0 \in GL_m(K), \boldsymbol{X}_i, \boldsymbol{Y}_i \in M_m(K); \boldsymbol{Z}'_0 \in GL_n(K), \boldsymbol{X}'_i, \boldsymbol{Y}'_i \in M_n(K)$, $\overline{\boldsymbol{M}}$ is the reduced matrix by reductions, we may assume that $\overline{\boldsymbol{M}}$ is the canonical form of $\boldsymbol{M}$. For simplicity, we denote $\boldsymbol{M}$ by $(\boldsymbol{A}, \boldsymbol{B})$ with $\boldsymbol{A} = (\boldsymbol{A}_1, \boldsymbol{A}_2, \boldsymbol{A}_3, \boldsymbol{A}_4), \boldsymbol{B} = (\boldsymbol{B}_1, \boldsymbol{B}_2, \boldsymbol{B}_3, \boldsymbol{B}_4)$, and $\overline{\boldsymbol{M}} = (\overline{\boldsymbol{A}}, \overline{\boldsymbol{B}})$ with $\overline{\boldsymbol{A}} = (\overline{\boldsymbol{A}}_1, \overline{\boldsymbol{A}}_2, \overline{\boldsymbol{A}}_3, \overline{\boldsymbol{A}}_4), \overline{\boldsymbol{B}} = (\overline{\boldsymbol{B}}_1, \overline{\boldsymbol{B}}_2, \overline{\boldsymbol{B}}_3, \overline{\boldsymbol{B}}_4)$, which is determined by the following equations:

$$\overline{\boldsymbol{A}}_1\boldsymbol{Z}'_0 - \boldsymbol{Z}_0\boldsymbol{A}_1 = \boldsymbol{0}, \quad \overline{\boldsymbol{B}}_1\boldsymbol{Z}'_0 - \boldsymbol{Z}_0\boldsymbol{B}_1 = \boldsymbol{0};$$

$$\overline{\boldsymbol{A}}_2\boldsymbol{Z}'_0 - \boldsymbol{Z}_0\boldsymbol{A}_2 = \boldsymbol{X}_1\overline{\boldsymbol{A}}_1 - \overline{\boldsymbol{A}}_1\boldsymbol{X}'_1, \quad \overline{\boldsymbol{B}}_2\boldsymbol{Z}'_0 - \boldsymbol{Z}_0\boldsymbol{B}_2 = \boldsymbol{Y}_1\overline{\boldsymbol{B}}_1 - \overline{\boldsymbol{B}}_1\boldsymbol{Y}'_1;$$

$$\overline{\boldsymbol{A}}_3\boldsymbol{Z}'_0 - \boldsymbol{Z}_0\boldsymbol{A}_3 = \sum_{k=1}^{2}\boldsymbol{X}_{3-k}\overline{\boldsymbol{A}}_k - \overline{\boldsymbol{A}}_k\boldsymbol{X}'_{3-k}, \quad \overline{\boldsymbol{B}}_3\boldsymbol{Z}'_0 - \boldsymbol{Z}_0\boldsymbol{B}_3 = \sum_{k=1}^{2}\boldsymbol{Y}_{3-k}\overline{\boldsymbol{B}}_k - \overline{\boldsymbol{B}}_k\boldsymbol{Y}'_{3-k},$$

$$\overline{\boldsymbol{A}}_4\boldsymbol{Z}'_0 - \boldsymbol{Z}_0\boldsymbol{A}_4 = \sum_{k=1}^{3}\boldsymbol{X}_{4-k}\overline{\boldsymbol{A}}_k - \overline{\boldsymbol{A}}_k\boldsymbol{X}'_{4-k}, \quad \overline{\boldsymbol{B}}_4\boldsymbol{Z}'_0 - \boldsymbol{Z}_0\boldsymbol{B}_4 = \sum_{k=1}^{3}\boldsymbol{Y}_{4-k}\overline{\boldsymbol{B}}_k - \overline{\boldsymbol{B}}_k\boldsymbol{Y}'_{4-k}.$$

Proposition 2.5 Let $\Lambda = K[x, y]/(x^p, y^q, xy)$ be a Gelfand-Ponomarev algebra, where $p, q \in \mathbb{N}$ with $q \geqslant p$, and let $\mathfrak{A} = (\widetilde{\Lambda} \times \widetilde{\Lambda}, \operatorname{rad} \widetilde{\Lambda}, \boldsymbol{0})$ be the corresponding matrix problem. A representation of $\mathfrak{A}$ can be expressed by $\boldsymbol{M} = (\boldsymbol{A}, \boldsymbol{B})$, where $\boldsymbol{A} = (\boldsymbol{A}_1, \boldsymbol{A}_2, \cdots, \boldsymbol{A}_{p-1})$ and $\boldsymbol{B} = (\boldsymbol{B}_1, \boldsymbol{B}_2, \cdots, \boldsymbol{B}_{q-1})$. The reduction equations are

$$\overline{\boldsymbol{A}}_i\boldsymbol{Z}'_0 - \boldsymbol{Z}_0\boldsymbol{A}_i = \sum_{k=1}^{i-1}\boldsymbol{X}_{i-k}\overline{\boldsymbol{A}}_k - \overline{\boldsymbol{A}}_k\boldsymbol{X}'_{i-k}, (1 \leqslant i \leqslant p-1); \quad (2.5)$$

$$\overline{\boldsymbol{B}}_j\boldsymbol{Z}'_0 - \boldsymbol{Z}_0\boldsymbol{B}_j = \sum_{k=1}^{j-1}\boldsymbol{Y}_{j-k}\overline{\boldsymbol{B}}_k - \overline{\boldsymbol{B}}_k\boldsymbol{Y}'_{j-k}, (1 \leqslant j \leqslant q-1). \quad (2.6)$$

§3. Canonical forms in Mat($\mathfrak{A}$) and projective presentations in mod(Λ)

Theorem 3.1 Let $\Lambda = K[x, y]/(x^p, y^q, xy)$ be a Gelfand-Ponomarev algebra, and let $\mathfrak{A} = (\widetilde{\Lambda} \times \widetilde{\Lambda}, \operatorname{rad} \widetilde{\Lambda}, \boldsymbol{0})$ be the matrix problem of Λ. Suppose that $(\boldsymbol{A}, \boldsymbol{B})$ in Mat($\mathfrak{A}$) is an indecomposable object with the canonical form $(\overline{\boldsymbol{A}}, \overline{\boldsymbol{B}})$.

(1) The links in $\overline{\boldsymbol{A}}$ (respectively, in $\overline{\boldsymbol{B}}$) locate in pairwise different rows and columns.

(2) The dimension vector $\underline{\boldsymbol{d}} = (m, n)$ of $(\boldsymbol{A}, \boldsymbol{B})$ has the property

$|m-n| \leqslant 1$.

(3) If $(\bar{\boldsymbol{A}}, \bar{\boldsymbol{B}})$ contains a Weyr matrix with some non-zero eigenvalues from an unraveling, then the dimension vector of $(\boldsymbol{A}, \boldsymbol{B})$ is (n, n). Moreover, the Weyr matrix is unique with only one non-zero eigenvalue and is sitting at the last non-empty block of $(\bar{\boldsymbol{A}}, \bar{\boldsymbol{B}})$.

(4) $l(\bar{\boldsymbol{A}})=l(\bar{\boldsymbol{B}})=\min\{m, n\}$ when $|m-n|=1$ in (2), and $l(\bar{\boldsymbol{A}})=n$ and $l(\bar{\boldsymbol{B}})=n-1$ or $l(\bar{\boldsymbol{A}})=n-1$ and $l(\bar{\boldsymbol{B}})=n$ when $m=n$ in (2) or in (3).

Proof (1) If there is a non-zero element at position (r, s) of $\bar{\boldsymbol{A}}_k$ with some $k<i$, then the rth row and sth column of $\boldsymbol{A}_i$ are deleted by regularization according to the reduction equation (2.5). The conclusion for $\boldsymbol{B}_j$ $(1 \leqslant j \leqslant q-1)$ is similar from the fomula (2.6). Thus, there are at most $\min\{m, n\}$ non-zero elements in $\bar{\boldsymbol{A}}$ and also in $\bar{\boldsymbol{B}}$ respectively.

(2) Proposition 2.3 tells us that the number of the links in any indecomposable object with dimension (m, n) of Mat($\mathfrak{A}$) equals $m+n-1$. If $|m-n| \geqslant 2$, then the number of the links is at most $2(\min\{m, n\})<m+n-1$ by (1), a contradiction. This ensures that $|m-n| \leqslant 1$.

(3) If $|m-n|=1$ and there is a Weyr matrix in $(\bar{\boldsymbol{A}}, \bar{\boldsymbol{B}})$ with some non-zero eigenvalue, then the number of the links equals $2\min\{m, n\}-1<m+n-1$ by (1), a contradiction to $(\boldsymbol{A}, \boldsymbol{B})$ being indecomposable. Thus, the dimension vector of $(\boldsymbol{A}, \boldsymbol{B})$ is (n, n). Moreover, if $(\bar{\boldsymbol{A}}, \bar{\boldsymbol{B}})$ has Weyr matrices with at least two non-zero eigenvalues, then the number of the links is less than or equal to $2n-2$, a contradiction to $(\boldsymbol{A}, \boldsymbol{B})$ being indecomposable. Finally, if the Weyr matrix $\boldsymbol{J}_d(\lambda)$ with $\lambda \neq 0$ is not sitting at the last non-empty block of $(\bar{\boldsymbol{A}}, \bar{\boldsymbol{B}})$, suppose that it appears at the k-th step in the reduction sequence (2.3). Then the equation $\boldsymbol{Z}^k \boldsymbol{M}^k=\bar{\boldsymbol{M}}^k \boldsymbol{Z}'^k$ yields a loop reduction with the first non-zero block $\boldsymbol{M}_{rs}^k$ in $\boldsymbol{A}^k$ (or in $\boldsymbol{B}^k$), thus $\boldsymbol{Z}_{rr}^k=\boldsymbol{Z}_{ss}'^k$. There must exist links sitting in the diagonal at the same position as $\boldsymbol{M}_{rs}^k$ in $\boldsymbol{B}^k$ (or in $\boldsymbol{A}^k$) before the k-th step. The vertex determined by $\boldsymbol{J}_d(\lambda)$ can not connect with other vertices at any further reductions, a contradiction to $(\boldsymbol{A}, \boldsymbol{B})$ being indecomposable.

(4) It is obvious. □

Given a canonical form $(\boldsymbol{A}, \boldsymbol{B})$ with dimension (m, n) and $|m-n|\leqslant 1$, then (A, B) corresponds to a morphism $\sigma: \Lambda^m \to \Lambda^n$. Writing $M=\mathrm{Coker}(\sigma)$. we have a minimal projective presentation

$$\Lambda^m \xrightarrow{\sigma} \Lambda^n \xrightarrow{\pi} M \to 0.$$

Now we give an exact expression of $\mathrm{Im}(\sigma)$ and $\mathrm{Coker}(\sigma)$ in the projective presentation $\Lambda^m \xrightarrow{\sigma} \Lambda^n$. Write $\Lambda^n = \Lambda_1 \oplus \Lambda_2 \oplus \cdots \oplus \Lambda_n$, a free module of n copies of Λ with $\Lambda_k = \Lambda$ for $k=1,2,\cdots,n$. Define a basis of Λ^n:

$\underline{E}=(e_1, e_2, \cdots, e_n)$ consists of the idempotents of the components;

$\underline{A}_i=(\alpha_{1i}, \alpha_{2i}, \cdots, \alpha_{ni})$ with $\alpha_{ki}=a^i e_k$ for $i=1,2,\cdots,p-1$;

$\underline{B}_j=(\beta_{1j}, \beta_{2j}, \cdots, \beta_{nj})$ with $\beta_{kj}=b^j e_k$ for $j=1,2,\cdots,q-1$.

Thus, the structure of Λ_k can be displayed as follows:

$$\alpha_{kp-1} \xleftarrow{a} \cdots \xleftarrow{a} \alpha_{k1} \xleftarrow{a} e_k \xrightarrow{b} \beta_{k1} \xrightarrow{b} \cdots \xrightarrow{b} \beta_{kq-1}. \qquad (3.1)$$

Lemma 3.2 Keep the notation as above.

(1) $\mathrm{Im}(\sigma)$ as a vector space is generated by

$$(\underline{B}_{q-1}, \underline{B}_{q-2}, \cdots, \underline{B}_{p-1}\underline{A}_{p-1}, \cdots, \underline{B}_1, \underline{A}_1, \underline{E})\overline{M}.$$

(2) $\mathrm{Coker}(\sigma)$ is determined by the relation

$$(\underline{B}_{q-1}, \underline{B}_{q-2}, \cdots, \underline{B}_{p-1}\underline{A}_{p-1}, \cdots, \underline{B}_1, \underline{A}_1, \underline{E})\overline{M}=0,$$

which is equivalent to

$$\sum_{i=1}^{p-1}\underline{A}_i\overline{A}_i + \sum_{j=1}^{q-1}\underline{B}_j\overline{B}_j = 0; \qquad (3.2)$$

$$\underline{A}_{i'}\overline{A}_i = 0\ (i' > i); \underline{B}_{j'}\overline{B}_j = 0\ (j' > j). \qquad (3.3)$$

Proof Referring to the formula (2.4), we obtain (1) of the lemma and the following formula directly:

$$\sum_{i=1}^{p-u-1}\underline{A}_{i+u}\overline{A}_i = 0\ (p-2 \geqslant u \geqslant 1); \quad \sum_{j=1}^{q-v-1}\underline{B}_{j+v}\overline{B}_j = 0\ (q-2 \geqslant v \geqslant 1).$$

But the non-zero entries in $\overline{A}_{i'}$, and $\overline{A}_{i''}$ are sitting in pairwise different rows and columns, (2) of the lemma follows. □

§ 4. Correspondence of canonical forms of $\mathfrak{A}$ and modules of Λ

Now we use the formulae (3. 2) and (3. 3) to calculate Coker(σ). Suppose that $(\boldsymbol{A},\boldsymbol{B})$ is an indecomposable canonical form of dimension vector $\underline{\boldsymbol{d}}=(m,n)$ with $|m-n|\leqslant 1$, furthermore if $(\boldsymbol{A},\boldsymbol{B})$ contains a Weyr matrix, then we may assume that the size r of the Weyr matrix equals 1, the proof is similar for any $r\in\mathbb{N}$.

For the purpose of simplicity, we write

$$y_i \xleftarrow{a} \cdots \xleftarrow{a} y_1 \xleftarrow{a} z = y_i \xleftarrow{a^i} z;$$

$$z \xrightarrow{b} y_1 \xrightarrow{b} \cdots \xrightarrow{b} y_j = z \xrightarrow{b^j} y_j.$$

For example, the module structure of $\Lambda\cong M(a^{p-1}b^{-(q-1)})$ can be displayed as

$$\alpha_{p-1} \xleftarrow{a^{p-1}} e \xrightarrow{b^{q-1}} \beta_{q-1} \quad \text{with} \quad a^{p-1}e=\alpha_{p-1},\quad b^{q-1}e=\beta_{q-1}.$$

Suppose that the last link or the Weyr matrix (λ) with $\lambda\neq 0$ is sitting in A_{i_1} and denoted by a^{i_1}, there must exist an index j_1 such that a link b^{j_1} in B_{j_1} has the same row index with a^{i_1} by (1) and (2) of Theorem 3. 1. Set $\Lambda_1=\alpha_{1p-1} \xleftarrow{a^{p-1}} e_1 \xrightarrow{b^{q-1}} \beta_{1q-1}$ (see the formula (3. 1)), we have $\alpha_{1i'}=0$ for any $i'>i_1$ from the formula (3. 2) and $\beta_{1j'}=0$ for any $j'>j_1$ from (3. 3). Next there exists a unique index i_2 such that the link a^{i_2} in $\boldsymbol{A}_{i_2}$ has the same column index with that of b^{j_1}. As the same as above, there is a unique index j_2, such that a link b^{j_2} in $\boldsymbol{B}_{j_2}$ has the same row index with a^{i_2}, set $\Lambda_2=\alpha_{2p-1} \xleftarrow{a^{p-1}} e_2 \xrightarrow{b^{q-1}} \beta_{2q-1}$. The equation (3. 2) and (1) and (4) of Theorem 3. 1 tell us that we have an equation $\beta_{1j_1}+\alpha_{2i_2}=0$. On the other hand, formulae (3. 2) and (3. 3) ensure that $\alpha_{2i'}=0$ for all $i'>i_2$ and $\beta_{2j'}=0$ for all $j'>J_2$. By induction, the above consideration suggests that Λ_{k-1} and $-\Lambda_k$ for $k=2,3,\cdots,n$ can be shown in the following fig. 4. 1,

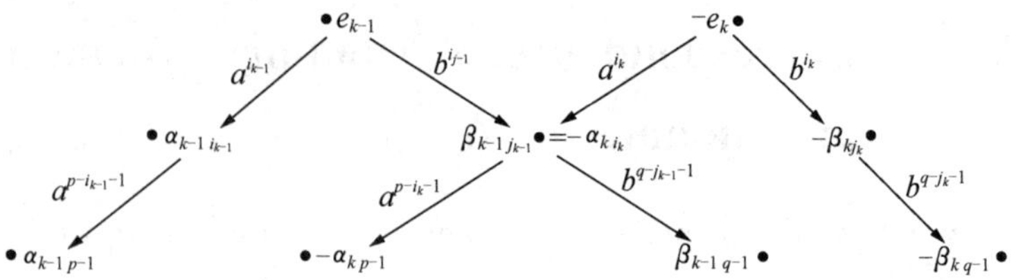

Fig. 4. 1

where the formula (3. 2) gives $\beta_{k-1j_{k-1}}+\alpha_{ki_k}=0$; and the formula (3. 3) ensures that $\alpha_{k-1i'}=0$ for all $i'>i_{k-1}$, $\beta_{k-1j'}=0$ for all $j'>j_{k-1}$, $\alpha_{ki'}=0$ for all $i'>i_k$ and $\beta_{kj'}=0$ for all $j'>j_k$.

Suppose that the last link or the Weyr matrix (λ) with $\lambda\neq 0$ is sitting in B_{j_n}, and denoted by b^{j_n}, there exists a unique index i_1 such that a link a^{i_1} in A_{i_1} has the same column index with that of b^{j_n}. Then we start the procedure given above.

Theorem 4. 1 Let $\Lambda=K[x,y]/(x^p,y^q,xy)$ be a Gelfand-Ponomarev algebra, and let $\mathfrak{A}=(\widetilde{\Lambda}\times\widetilde{\Lambda},\mathrm{rad}\,\widetilde{\Lambda},0)$ be the matrix problem of Λ. Suppose that a minimal projective presentation is given in Section 3, which determines an object of $\mathfrak{A}$ with the canonical form (A,B).

(1) If $|m-n|=1$ or $m=n$ but there is no any Weyr matrix with non-zero eigenvalue in (A,B), then the structure of Coker(σ) is given by fig. 4. 2,

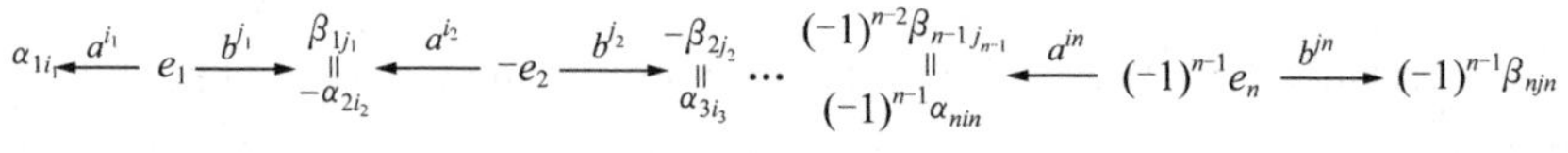

Fig. 4. 2

where $a=\mathrm{id}$ and $b=\mathrm{id}$. If $i_1=p-1$ and $j_n=q-1$, then $\underline{d}=(n-1,n)$; if $i_1=p-1$ and $j_n<q-1$ or $i_1<p-1$ and $j_n=q-1$, then $\underline{d}=(n,n)$; if $i_1<p-1$ and $j_n<q-1$, then $\underline{d}=(n+1,n)$. Furthermore,

$$\mathrm{Coker}(\sigma)\simeq M(a^{i_1}b^{-j_1}a^{i_2}b^{-j_2}\cdots a^{i_n}b^{-j_n})$$

is a string module over Λ.

(2) If $m=n$ and there is a Weyr matrix (λ) with $\lambda\neq 0$ in $(\boldsymbol{A},\boldsymbol{B})$, then the connection between Λ_n and Λ_1 is the following:

In the case of λ sitting in A, the structure of Coker(σ) is given by fig. 4. 3.

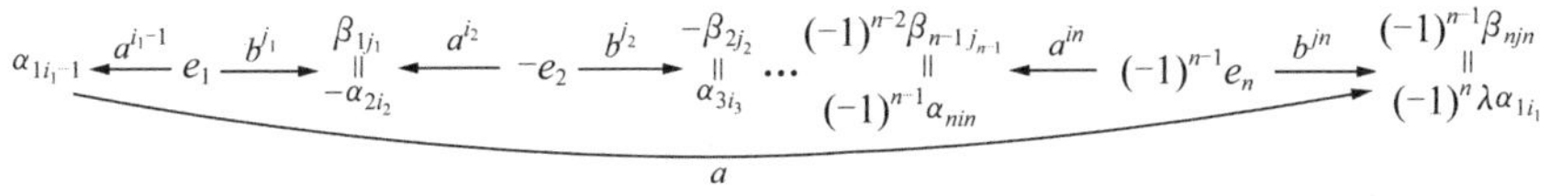

Fig. 4. 3

where $a=\mathrm{id}$ and $b=\mathrm{id}$, except one a equals λ. Furthermore

$$\mathrm{Coker}(\sigma)\simeq M((a^{i_1}b^{-j_1}a^{i_2}b^{-j_2}\cdots a^{i_n}b^{-j_n}),(-1)^n\lambda) \qquad (4.1)$$

is a band module over Λ by $b\beta_{nj_n-1}=\beta_{nj_n}=-\lambda\alpha_{1i_1}$.

In the case of λ sitting in B, the structure of Coker(σ) is given by fig. 4. 4,

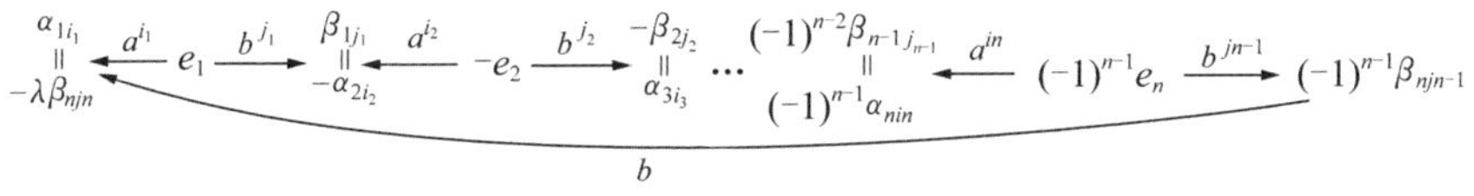

Fig. 4. 4

where $a=\mathrm{id}$ and $b=\mathrm{id}$, except one b equals λ. Furthermore,

$$\mathrm{Coker}(\sigma)\simeq M((a^{i_1}b^{-j_1}a^{i_2}b^{-j_2}\cdots a^{i_n}b^{-j_n}),(-1)^n\lambda^{-1},1)$$

is a band module over Λ by $b\beta_{nj_n-1}=\beta_{nj_n}=-\lambda^{-1}\alpha_{1i_1}$.

Corollary 4. 2 Let Λ be a Gelfand-Ponomarev algebra.

(1) If M is a string module of Λ, then a minimal projective presentation of M is given by

$$\Lambda^m \xrightarrow{\sigma} \Lambda^n \xrightarrow{\pi} M \to 0,\ |n-m|\leqslant 1.$$

(2) If M is a band module of Λ, then a minimal projective presentation of M is given by

$$\Lambda^n \xrightarrow{\sigma} \Lambda^n \xrightarrow{\pi} M \to 0.$$

Corollary 4. 3 Keep the notation as in Theorem 4. 1.

(1) There exists a one-to-one correspondence between the set of the iso-classes of the string modules and the set of the canonical forms:

$$\{M=M(\mathcal{C})\in \mathrm{mod}(\Lambda)\mid \dim_K \mathrm{top}(M)=n\},$$

$$\{(\boldsymbol{A},\boldsymbol{B})\in \mathrm{Mat}(\mathfrak{A})\mid \text{with size}(n-1,n),(n,n),(n+1,n)\}.$$

(2) There exists a one-to-one correspondence between the set of the iso-classes of the band modules and the set of the canonical forms:

$$\{M=M(\mathcal{B},\lambda,r)\in \operatorname{mod}(\Lambda)\mid \lambda\in K^*,\dim_K \operatorname{top}(M)=n\},$$

$$\{(\boldsymbol{A},\boldsymbol{B})\in \operatorname{Mat}(\mathfrak{A})\mid \text{with size } (nr,nr) \text{ and non-zero eigenvalues}\}.$$

Example 4.4 Consider the algebra $K[x,y]/(x^3,y^3,xy)$, i. e., $p=q=3$. Suppose that $(\boldsymbol{A}_1,\boldsymbol{B}_1,\boldsymbol{A}_2,\boldsymbol{B}_2)\in \operatorname{Mat}(\mathfrak{A})$ is a canonical form of dimension vector (2,2):

$$\boldsymbol{A}_1=\begin{pmatrix}0&1\\0&0\end{pmatrix},\quad \boldsymbol{B}_1=\begin{pmatrix}1&0\\0&1\end{pmatrix},\quad \boldsymbol{A}_2=\begin{pmatrix}0&0\\\lambda&0\end{pmatrix},\quad \boldsymbol{B}_2=\begin{pmatrix}0&0\\0&0\end{pmatrix},$$

where the index of λ in $\boldsymbol{A}_2$ is (2,1), consequently we have (2,2) in $\boldsymbol{B}_1$, (1,2) in $\boldsymbol{A}_1$, (1,1) in $\boldsymbol{B}_1$, (2,1) in $\boldsymbol{A}_2$. Thus, Coker(σ) is defined by the following equations:

$$\beta_{11}+\alpha_{21}=0,\quad \beta_{21}+\lambda\alpha_{12}=0.$$

And the module structure of Coker(σ) is fig. 4.5,

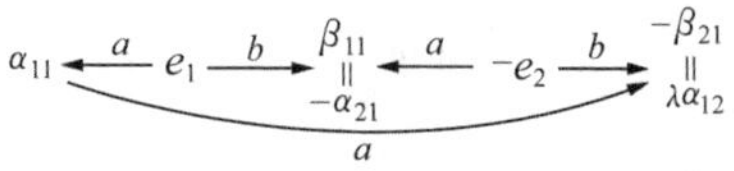

Fig. 4.5

where $be_2=\beta_{21}=-\lambda\alpha_{12}$. Therefore, $\operatorname{Coker}(\alpha)\cong M(a^2b^{-1}ab^{-1},\lambda,1)$ is a band module over Λ.

Example 4.5 With the same algebra as above. Suppose that $(\boldsymbol{A}_1,\boldsymbol{B}_1,\boldsymbol{A}_2,\boldsymbol{B}_2)\in \operatorname{Mat}(\mathfrak{A})$ is a canonical form of dimension vector (4,4):

$$\boldsymbol{A}_1=\begin{bmatrix}0&0&1&0\\0&0&0&1\\0&0&0&0\\0&0&0&0\end{bmatrix},\quad \boldsymbol{B}_1=\begin{bmatrix}0&1&0&0\\0&0&0&0\\0&0&0&1\\0&0&0&0\end{bmatrix},$$

$$\boldsymbol{A}_2=\begin{bmatrix}0&0&0&0\\0&0&0&0\\0&1&0&0\\1&0&0&0\end{bmatrix},\quad \boldsymbol{B}_2=\begin{bmatrix}0&0&0&0\\\lambda&0&0&0\\0&0&0&0\\0&0&1&0\end{bmatrix},$$

where the index of λ in $\boldsymbol{B}_2$ is (2,1), consequently we have (4,1) in $\boldsymbol{A}_2$,

(4,3) in $\boldsymbol{B}_2$, (1,3) in $\boldsymbol{A}_1$, (1,2) in $\boldsymbol{B}_1$, (3,2) in $\boldsymbol{A}_1$, (3,4) in $\boldsymbol{B}_1$, (2,4) in $\boldsymbol{A}_1$, (2,1) in $\boldsymbol{B}_2$. Thus, Coker(σ) is defined by the following equations:

$$\beta_{11}+\alpha_{22}=0, \quad \beta_{21}+\alpha_{31}=0, \quad \beta_{31}+\alpha_{41}=0, \quad \lambda\beta_{42}+\alpha_{12}=0.$$

And the module structure of Coker(σ) is fig. 4. 6,

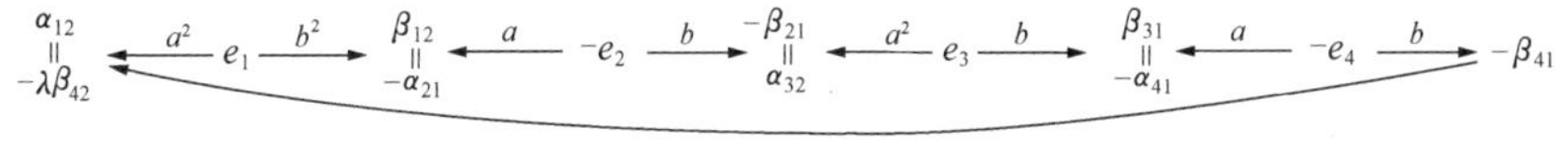

Fig. 4. 6

where $b\beta_{41}=-\frac{1}{\lambda}\alpha_{12}$. Therefore, Coker($\sigma$)$\cong M\left(a^2b^{-2}ab^{-1}a^2b^{-1}ab^{-2}, \frac{1}{\lambda}\right)$ is a band module over Λ.

§ 5. Some properties of a module variety proved by canonical forms

For a Celfand-Ponomarev algebra $\Lambda=K[x,y]/(x^p,y^q,xy)$ and any dimension n, the module variety $\mathrm{Mod}_\Lambda(n,K)$ can be described as follows under a fixed basis:

$$\mathrm{Mod}_\Lambda(n,K)=\{(X,Y)\mid X,Y\in M_n(K), X^p=0, Y^q=0, XY=YX=0\}.$$

We identify a module with its corresponding point in the module variety.

Definition 5. 1 A partition of n is a sequence $\mathbf{p}=(p_1,p_2,\cdots,p_t)$ of positive integers such that $\sum_{i=1}^{t}p_i=n$ and $p_i\geqslant p_{i+1}$ for all $1\leqslant i\leqslant t-1$. The length t of $\mathbf{p}$ is denoted by $l(\mathbf{p})$.

Let $\boldsymbol{N}\in M_n(K)$ be a matrix with only one eigenvalue. Denote by $\boldsymbol{J}(\boldsymbol{N})$ the Jordan form of $\boldsymbol{N}$ with the orders of Jordan blocks decreasing. Thus, we obtain a partition $\mathbf{p}(\boldsymbol{N})$ of n with the components being the orders of Jordan blocks.

Let $\mathcal{P}(n,p,q)=\{(\boldsymbol{a},\boldsymbol{b})\mid \boldsymbol{a},\boldsymbol{b}$ are partitions of n with $a_i\leqslant p, b_j\leqslant q$ for all $i,j\}$ and we define the stratum of the module variety $\mathrm{Mod}_\Lambda(n,K)$ as

$$\Delta(\boldsymbol{a},\boldsymbol{b})=\{(X,Y)\mid (X,Y)\in\mathrm{Mod}_\Lambda(n,K), \boldsymbol{p}(X)=\boldsymbol{a}, \boldsymbol{p}(Y)=\boldsymbol{b}\}.$$

Theorem 5. 2[5] If $\Delta(\boldsymbol{a},\boldsymbol{b})$ contains a band module, then $\Delta(\boldsymbol{a},\boldsymbol{b})$

is locally closed and irreducible in $\mathrm{Mod}_\Lambda(n, K)$ with regard to the Zariski topology.

It has been proved by Schröer that the Gelfand-Ponomarev algebra Λ is a subfinite algebra, i. e., any projective module has only finitely many isomorphism classes of submodules (see [5, 8]). The following lemma gives a direct proof using canonical forms.

Lemma 5.3 (1) Let $M=M(a^{i_1}b^{-j_1}\cdots a^{i_n}b^{-j_n})$ for $1\leqslant i_k<p$, $1\leqslant j_k<q$ and $1\leqslant k\leqslant n$ be a string module with $\dim_K \mathrm{top}(M)=n$, and $\Lambda^n \xrightarrow{\pi} M\to 0$ be a projective cover of M. Then $\mathrm{Ker}(\pi)$ is isomorphic to $M(a^{p-i_2-1}b^{-q+j_1+1})\oplus M(a^{p-i_3-1}b^{-q+j_2+1})\oplus\cdots\oplus M(a^{p-i_1-1})\oplus M(b^{-q+j_n+1})$ and $n-1\leqslant \dim_K \mathrm{top}(\mathrm{Ker}(\pi))\leqslant n+1$.

(2) Let $M=M(a^{i_1}b^{-j_1}\cdots a^{i_n}b^{-j_n}, \lambda, r)$ for $\lambda\in K$, $1\leqslant i_k<p$, $1\leqslant j_k<q$ and $1\leqslant k\leqslant n$ be a band module with $\dim_K \mathrm{top}(M)=nr$, and $\Lambda^{nr}\xrightarrow{\pi} M\to 0$ be a projective cover of M. Then $\mathrm{Ker}(\pi)$ is isomorphic to $M(a^{p-i_2-1}b^{-q+j_1+1})^r\oplus M(a^{p-i_3-1}b^{-q+j_2+1})^r\oplus\cdots\oplus M(a^{p-i_1-1}b^{-q+j_n+1})^r$ and $\dim_K \mathrm{top}(\mathrm{Ker}(\pi))=nr$.

Proof Suppose that we have the projective presentation

$$\Lambda^m \xrightarrow{\sigma} \Lambda^n \xrightarrow{\pi} M\to 0, \quad |n-m|\leqslant 1.$$

Then $\mathrm{Im}(\sigma)=\mathrm{Ker}(\pi)$. Denote by f_k the kth primitive idempotent in Λ^m. It is clear from the formula (4.1) that $\sigma(f_k)=\beta_{k-1 j_{k-1}}\alpha_{k i_k}$ for $k=2,3,\cdots,n$. In case (1), $\sigma(f_1)=\alpha_{1i_1+1}$ and $\sigma(f_{n+1})=\beta_{nj_n+1}$. In case (2), $\sigma(f_1)=\beta_{nj_n}+\lambda\alpha_{1i_1}$. □

Corollary 5.4 The indecomposable submodule of Λ^n $(n\in\mathbb{N})$ is of the form

$$M(a^i b^{-j}),\quad 0\leqslant i\leqslant p-1, 0\leqslant j\leqslant q-1.$$

Therefore, Λ is subfinite.

Proof Let N be an indecomposable submodule of Λ^n. Suppose that $M=\Lambda^n/N$ with $\dim_K \mathrm{top}(M)=n$. Then we have an exact sequence

$$0\longrightarrow N\longrightarrow \Lambda^n \xrightarrow{\pi} M\longrightarrow 0,$$

and $N \simeq \mathrm{Ker}(\pi)$. Lemma 5.3 tells us that $N \simeq M(a^i b^{-j})$ for some $0 \leqslant i \leqslant p-1$ and $0 \leqslant j \leqslant q-1$. □

Theorem 5.5 Let $\Lambda = K[x, y]/(x^p, y^q, xy)$ be a Gelfand-Ponomarev algebra, and $\mathfrak{A}$ the corresponding matrix problem. Let $M = M(\mathcal{B}, \lambda, r)$ and $M' = M(\mathcal{B}', \lambda', r')$ be two band modules of the same dimension in mod(Λ) with the canonical forms ($\boldsymbol{A}$, $\boldsymbol{B}$) and ($\boldsymbol{A}^{\mathrm{T}}$, $\boldsymbol{B}^{\mathrm{T}}$) in Mat($\mathfrak{A}$), respectively. Then M and M' belong to the same stratum in the module variety of Λ if and only if $\mathrm{rank}(\boldsymbol{A}_i) = \mathrm{rank}(\boldsymbol{A}_i^{\mathrm{T}})$ and $\mathrm{rank}(\boldsymbol{B}_j) = \mathrm{rank}(\boldsymbol{B}_j^{\mathrm{T}})$ for $1 \leqslant i \leqslant p-1$ and $1 \leqslant j \leqslant q-1$. Moreover, if it is the case, then M and M^{T} belong to the same irreducible component in the module variety of Λ.

Proof Firstly, we assume that $r = r' = 1$. In the module variety $\mathrm{Mod}_\Lambda(n, K)$, a module M can be expressed by a matrix pair ($\boldsymbol{X}$, $\boldsymbol{Y}$). By Theorem 4.1, $\mathrm{rank}(\boldsymbol{A}_i)$ equals the number of non-zero entries in $\boldsymbol{A}_i$ for $1 \leqslant i \leqslant p-1$, and $\mathrm{rank}(\boldsymbol{B}_j)$ equals the number of non-zero entries in $\boldsymbol{B}_j$ for $1 \leqslant j \leqslant q-1$. Therefore, $\mathrm{rank}(\boldsymbol{A}_i)$ equals the number of Jordan blocks $\boldsymbol{J}_{i+1}(0)$ in the Jordan normal form of X, and $\mathrm{rank}(\boldsymbol{B}_j)$ equals the number of Jordan blocks $\boldsymbol{J}_{j+1}(0)$ in the Jordan normal form of Y. The proof for $r = r' = 1$ is complete.

For the general case, the proof is similar. □

References

[1] Butler M C R, Ringel C M. Auslander-Reiten sequences with few middle terms and application to string algebras. Comm. Algebra, 1987, 15: 145-179.

[2] Crawley-Boevey W W, On tame algebras and bocses. Proc. London Math. Soc., 1998, 56(3): 451-483.

[3] Drozd Yu A. On tame and wild matrix problems. in: Matrix Problems. Kiev, 1977: 39-74.

[4] Gelfand I M, Ponomarev V A. Indecomposable representations of the Lorentz group. Russian Math. Surveys, 1968, 23: 1-58.

[5] Schröer J. Variety of pairs of nilpotent matrices annihilating each other. Commentarii Mathematici Helvetici, 2004, 79: 396-426.

[6] Sergeichuk V V. Canonical matrices for linear matrix problems. Linear Algebra and Its Applications, 2000, 317: 53-102.

[7] Xu Y G, Zhang Y B. Indecomposability and the number of links. Science in China 2001, 44A(12): 1 515-1 522 (in Chinese).

[8] Zimmermann-Huisgen B. Predicting syzygies over monomial relations algebras. Manuscripts Math., 1991, 70: 157-182.

Ring Theory, 2007, 15-18.

刘绍学教授的经历和工作简介

Professor Liu Shaoxue—His Live and Work

Professor Liu Shaoxue was born on November 6 , 1929 in Liaoyang, a city of Liaoning Province in North China. And his family moved to Beijing in 1937. Upon graduation from high school in 1946, he enrolled in Mathematics Department of National Peping Normal University (the predecessor of Beijing Normal University). In September 1953, he was sent to study in Mechanics and Mat hematics Department of Moscow University of Former Soviet Union.

His supervisor was A. G. Kurosh, a prestigious expert in Algebra. Under the guidance of Professor Kurosh, he completed his Ph. D thesis "On Decomposition of Infinite Algebras", and received Ph. D degree in 1956. He was the first student who received Ph. D degree out of the Chinese students, who studied Mathematics in Former Soviet Union. In his Ph. D thesis he proves the following theorems:

(1) An extension Jordan algebra of a locally finite Jordan algebra by a locally finite Jordan algebra is still locally finite.

(2) An extension Lie algebra of a locally finite Lie algebra by a locally finite Lie algebra is still locally finite in case it is an algebraic Lie algebra. (A Lie algebra is called algebraic, if for any two elements x, $y \in A$ there exists some positive integer $n = n(x, y)$, such that $x, xy, xy^2 = (xy)y, \cdots, xy^n = (xy^{n-1}y)$ are linearly dependent over k).

(3) Based on the result of (1), he proves the existence of the Livitzki radical of Jordan algebras independently of K. A. Zhevlakov.

(4) Based on the result of (2), it is obtained immediately that an algebraic solvable Lie algebra is locally finite.

Professor Liu's Ph. D thesis gives some nice and important results in ring theory, and many experts on ring theory quote his results in their papers. For example in the paper "On local finiteness in the sense of Shirshov" published in Algebra and Logical in 1973, the authors K. A. Zhevlakov and I. P. Shstakov quoted Professor Liu's Ph. D thesis 5 times. These results are the main contents of chapter 4 in the book "Rings that are nearly associative" published in 1978 in Russian and 1982 in English written by K. A. Zhevlakov, A. M. Slinko, I. P. Shestakov, the former students of Kurosh.

Professor Liu Shaoxue went back to Department of Mathematics of Beijing Normal University in 1956, and was promoted to be an associate Professor in 1961. From 1956 to 1966, because of some political policy, professors in China were asked to solve so called the problems of practice in farmland or factories, which was named the association of theory and practice. Professor Liu had no more chance to concentrate on his research work. But he still wrote more than ten papers, one of them was "On algebras in which every sub-algebra is an ideal" published in 1964, Chinese Mathematics Acta. An algebra which has the property that every sub-algebra is an ideal is called Hamilton algebra, Professor Liu gives a complete description of Hamilton algebra. The result is simple and beautiful, which is extended to the case of exponential associative algebras by Outcalt, and is quoted in chapter 9 of the book "Nilpotent Rings" by R. L. Kruse and D. T. Price in 1969.

During that time Professor Liu was teaching abstract algebra, calculus, partial differential equation and so on for undergraduate students. Besides his research experience: his lectures were made very clear, vivid and impressive. Many years have passed, still his former students remembered his lectures, and said that Professor Liu's lectures impressed them greatly, and guided them to do research on Algebras.

From 1966 to 1976, our country experienced a disaster called Culture Revolution. All universities were closed and professors were sent to countryside or factors. Research on basic sciences was stopped for more than ten years. Universites returned to normal in 1977, and graduate students enrolled into universities in 1978. Liu Shaoxue was promoted to be a professor in 1979. Since the year of 1978, he restarted his research work and his training of graduate students. He wrote a book "Rings and Algebras" in 1983, which was used widely as a textbook for graduate students on Algebra in China.

In 1988, Professor Liu visited Tsukuba University of Japan. One day Professor H. Tachikawa drove him to the Tsnkuba Mountain. During the trip, Professor Liu suggested to hold a symposium on ring theory hosted by China and Japan. Professor Tachikawa accepted the suggestion happily and said that, "We are able to compete with European experts on ring theory when we have good cooperation between Japan and China."

In autumn 1991, the First China-Japan international Symposium on ring theory was held in Guangxi Normal University in Guilin, the most beautiful city in South China. The main organizers were Liu and Tachikawa. Some algebraists from North America and Europe also attended the symposium, for example C. M. Ringel, B. J. Muller, and M. Beattie. Professor Li Baifei from Taiwan and Academician Wan Zhexian from Beijing also attended the symposium. The Proceedings were published in 1992 in Japan.

Around the year of 1985, after several visits to the United states and Europe, and having a lot of discussions with foreign algebraists, Professor Liu decided to find a new research area for his students pursing Ph. D degree. It should be new but related to ring theory. In May 1985, Professor Liu went to Belgium to visit Professor F. Van

Oystaeyen, and went to Germany to visit Professor Claus Ringel. Then he decided that his students should start the study on representation theory of algebras. Professor Liu is a kindhearted, honest and humorous person. Because of his personality Professor Liu made a lot of good friends.

Professor Liu was very lucky that God introduced a helpful and faithful friend, Claus Ringel, to him at the difficult time to change the research project. Claus visited China almost every year from 1987. He gave a series of lectures on representation theory of algebras for graduate students. There have been 3 students, who obtained Ph. D. degree under a program of joint training between China and Germany, and 4 students received Humboldt Research Award respectively to visit Germany and studied under his guidance for 2 years. Besides Claus, M. Auslande, I. Reiten, V. Dlab, P. Gabriel and many foreign algebraists also helped the algebra group in China. In the year of 2000, the Ninth International Conference on representation theory of algebras was held in Beijing Normal University. The Chinese group was merging more and more into the international family of representation theory of algebras.

There were altogether 17 Master degree students and 19 Ph. D degree students of Professor Liu. Now most of them are working in some of the best universities as Professors, some of them being outstanding mathematicians. Professor Liu has established a strong group on representation theory of algebras in China. Because of his efforts in educating graduate students, Professor Liu was awarded a first prize for excellent teaching in the City of Beijing in 1991. A person, who was already an expert on ring theory, but still determined to change with his students to representation theory of algebras at the age of 56, deserves to be respected.

Professor Liu has published more then 50 papers in mathematics and 40 of them after 1978. In the papers "Isomorphism problem of path algebras" and "Isomorphism problem For tensor algebras over valid quivers", Professor Liu and other authors proved that two path algebras are isomorphic if and only if their corresponding quivers are isomorphic. Based on this, P. A. Grillet started a systematical study of isomorphism problem of semi-group algebras. In the paper "Group graded rings, Smash Products and Additive categories", Professor Liu and F. Van. Oystaeyen have extended the concept of the smash product of group G and G-graded ring A to the case of G being an infinite group, and obtained the corresponding dual and co-dual theorems. In the paper "Comparing graded version of the prime radical" by Beattie, Liu and Stewart, the structure theorem of the graded rings is proved.

Professor Liu has written, translated and edited several books. He wrote the book "Rings and algebras" and the textbook "Basic Modern Algebra"; translated the books "Group Theory" and "General Algebra" by Kurosh, "Finite dimensional algebras" by Drozd and Gilchinko from Russian into Chinese. He edited "Ring and radical" with foreign algebraists, and edited "Rings, Groups and algebras in China" as a chief editor. He won a second prize of National Committee's prize for Scientific and Technological Development with the work of title "The structure and representation theory of rings" in 1988.

Professor Liu is a good professor and nice person.

论文和著作目录

Bibliography of Papers and Works

论文目录

[序号] 作者. 论文题目. 杂志名称, 年份, 卷(期): 起页-止页.

[1] 张英伯. 一类二秩无扭 Abel 群的结构. 数学学报, 1985, 28(1): 91-102.(摘要见: 张英伯. 一类二秩无扭 Abel 群的结构. 科学通报, 1982, 27(21): 1 285-1 288; Zhang Yingbo. Construction of a class of torsion-free Abelian groups of rank. Chinese Science Bulletin, 1983, 28(7): 869-873.)

[2] Zhang Yingbo. The modules in any component of the AR-guiver of a wild hereditary Artin algebra are uniquely determined by their composition factors. Archiv der Mathematik, 1989, 53(3): 250-251.

[3] Zhang Yingbo. Eigenvalue of coxeter transformation and the struture of regular components of an Auslander-Reiten. Communication in Algebra, 1989, 17(10): 2 347-2 362.

[4] Zhang Yingbo. The modules in any component of the AR-quiver of a wild hereditary Artin algebra are uniquely determined by their composition factors. Acta Mathematica Sinica, New Series, 1990, 6(2): 97-99.

[5] 肖杰, 郭晋云, 张英伯. A_n 类自入射代数的不可分解模由其 Loewy 因子唯一确定. 中国科学, 1989, 19A(7): 673-682. (Xiao Jie, Guo

Jinyun, Zhang Yingbo. Loewy factors of indecomposable modules over self-injective algebras of class A_n. Science in China, 1990, 33A (8): 897-908.)

[6] Zhang Yingbo. The Structure of stable components. Canadian Journal of Mathematics, 1991, 43(3): 652-672.

[7] Bautista R, Zhang Yingbo. A characterization of finite-dimensional algebras of tame representation type. Proceedings of the 1st China-Japan International Symposium on Ring Theory (Guilin, 1991), Okayama University, Okayama, 1992: 7-10.

[8] Zhang Yinbo, Lin Yanan. Some tame algebras with one parameters and their corresponding bocses. Proceedings of the 2nd Japan-China International Symposium on Ring Theory (Guilin, 1991), Okayama University, Okayama, 1992: 99-101.

[9] 张英伯. 只有一个不可分解么模的环. 北京师范大学学报(自然科学版), 1993, 29(1): 35-37.

[10] 林亚南, 张英伯. 对应于几个单参数变量 Tame 型代数的 Bocses. 北京师范大学学报(自然科学版), 1993, 29(3): 285-290.

[11] 张英伯, 肖杰. 代数表示论简介与综述. 数学进展, 1993, 22(6): 481-501.

[12] Zhang Yingbo, Lei Tiangang, Raymundo B. The representation category of a bocs (Ⅰ). 北京师范大学学报(自然科学版), 1995, 31(3): 313-316.

[13] Zhang Yingbo, Lei Tiangang, Raymundo B. The representation category of a bocs (Ⅱ). 北京师范大学学报(自然科学版), 1995, 31(4): 440-445.

[14] Zhang Yingbo, Li Size, Lei Tiangang, Bautista R. The representation category of a bocs (Ⅲ). 北京师范大学学报(自然科学版), 1996, 32(2): 143-148.

[15] Zhang Yingbo, Lei Tiangang, Bautista R. The representation category of a bocs (Ⅳ). 北京师范大学学报(自然科学版), 1996, 32(3): 289-295.

[16] Zhang Yingbo. The representation category of a wild bocs. Proceedings of the 2nd Japan-China International Symposium on Ring Theory and the 28th Symposium on Ring Theory (Okayma, 1995), Okayma University, 1996: 179-181.

[17] 林亚南,张英伯. 对应于 tame 遗传代数的 bocses (Ⅰ). 中国科学, 1996, 26A (2): 97-103. (Lin Yanan, Zhang Yingbo. Bocses corresponding to the hereditary algebras of tame type (Ⅰ). Science in China, 1996, 39A(5): 483-490.)

[18] 张英伯,林亚南. 对应于 tame 遗传代数的 bocses (Ⅱ). 中国科学, 1996, 26A (7): 595-603. (Zhang Yingbo, Lin Yanan. Bocses corresponding to the hereditary algebras of tame type (Ⅱ). Science in China, 1996, 39A(9): 909-918.)

[19] 张英伯,雷天刚. 具有强齐性条件的一个野 bocs. 科学通报, 1996, 41 (23): 2 119-2 122. (Zhang Yingbo, Lei Tiangang, Bautista R. A wild bocs having strong homogeneous property. Chinese Science Bulletin, 1997, 42(2): 108-112.)

[20] Lei Tiangang, Zhang Yingbo. A theorem on a class of special partitioned matrices. 北京师范大学学报(自然科学版), 1998, 34(3): 297-304.

[21] Assem I, Zhang Yingbo. Endomorphism algebras of exceptional sequences over path algebras of type A_n. Colloquium Mathematicum, 1998, 77(2): 271-292.

[22] Zhang Yingbo, Lei Tiangang, Bautista R. A matrix description of a wild category. Science in China, 1998, 41A(5): 461-475.

[23] Bautista R, Crawley-Boevey W, Lei Tiangang, Zhang Yingbo. On homogeneous exact categories. Journal of Algebra, 2000, 230: 665-675.

[24] Xu Yunge, Zhang Yingbo. Some geometrical properties on Bocses (Ⅰ). 北京师范大学学报(自然科学版), 2000, 36(3): 319-324.

[25] Xu Yunge, Zhang Yingbo. Some geometrical properties on Bocses (Ⅱ). 北京师范大学学报(自然科学版), 2000, 36(5): 604-606.

[26] 徐运阁，张英伯. 不可分解性与链环数. 中国科学，2001，31A(5)：385-391.（Xu Yunge，Zhang Yingbo. Indecomposability and number of links. Science in China，2001，44A(12)：1 515-1 522.）

[27] 张英伯. 世界数学家大会和新世纪的数学问题. 数学通报，2001，(10)：1-3.

[28] Zeng Xiangyong，Zhang Yingbo. A correspondence of almost split sequences between some categories. Communications in Algebra，2001，29(2)：557-582.

[29] Zhang Pu，Zhang Yingbo. Minimal generators of Ringel-Hall algebras of Affine quivers. Journal of Algebra，2001，239：675-704.

[30] 张英伯. 图书馆和我们. 见：北京师范大学图书馆编. 百年情结. 北京：北京师范大学出版社，2002：153-155.

[31] 张英伯. 代数学家刘绍学. 见：王淑芳，邵红英，主编. 师范之光. 北京：北京师范大学出版社，2002：620-624.

[32] Bautista R，Zhang Yingbo，Representations of a k-algebra over the rational functions over k，Journal of Algebra，2003，267：342-358.

[33] Li Longcai，Zhang Yingbo. Representation theory of the system quiver. Science in China，2003，46A(6)：769-803.

[34] 徐运阁，张英伯. 可训表示型与野表示型 Bocs. 中国科学，2004，34A(6)：687-700.（Zhang Yingbo，Xu Yunge. On tame and wild bocses. Science in China，2005，48A(4)：456-468.）

[35] 张英伯. 改版之际话通报. 数学通报，2005，44(1)：1-2.

[36] 张英伯. 数学家关注数学教育. 中国数学会通讯，2005，(2)：20-22.

[37] 张英伯. 编者的话. 数学通报，2005，44(特刊)：1.

[38] 潘俊，张英伯. 一类 domestic bocses. 北京师范大学学报(自然科学版)，2006，42(5)：467-472.

[39] 张英伯. 欧氏几何的公理体系和我国平面几何课本的历史演变. 数学通报，2006，45(1)：4-9；见：大学数学课程报告论坛组委会：大学数学课程报告论坛论文集(2005)，北京：高等教育出版社，2006：80-87.

[40] 王昆扬，张英伯. 您在我们的心中永生. 数学通报，2006，45(4)：2-3.

[41] 张英伯,整理.与伍鸿熙教授座谈摘要.数学通报,2006,45(7):1-3.

[42] 张英伯.庆祝《数学通报》创刊 70 周年开幕词,数学通报,2006,45(11):14.

[43] 徐运阁,张英伯.△-tame 拟遗传代数.中国科学,2006,36A(11):1 254-1 266.(Xu Yunge,Zhang Yingbo. △-tame quasi-hereditary algebras. Science in China,2007,50A(2):240-252.)

[44] Bautista R,Drozd Y A,Zeng Xiangyong,Zhang Yingbo,On Hom-spaces of tame algebras,Central European Journal of Mathematics,2007,5(2):215-263.

[45] 张英伯.傅种孙:中国现代数学教育的先驱.中国数学会通讯,2007,(1):23-28;数学通报,2008,47(1):8-10;数学教育学报,2008,17(1):1-3;数学教学,2007,(9):封 2-5.

[46] 张英伯,叶彩娟.五点共圆问题与 Clifford's 链定理.数学通报,2007,46(9):1-5;数学教学,2007(9):封 2-5;中国数学会通讯,2008,(2):17-28;见:杨学枝,主编.中国初等数学研究,2010,第 2 辑.哈尔滨:哈尔滨工业大学出版社,2010:13-20.

[47] 张英伯.中国的数学课程标准:在第四届世界华人数学家大会上的讲话.数学通报,2008,47(1):2.

[48] 张英伯,李建华.英才教育之忧:英才教育的国际比较与数学课程.数学教育学报,2008,17(6):1-5;张英伯.谈谈英才教育.中国数学会通讯,2008,(4):13-23;李建华,张英伯.英才教育之忧.数学通报,2009,48(1):1-6.

[49] 张英伯,徐运阁.统一化 Tame 定理.中国科学,2008,38A(12):1 372-1 402.(Zhang Yingbo,Xu Yunge. Unified Tame theorem. Science in China,2009,52A(9):2 036-2 068.)

[50] 张学颖,张英伯.两类 domestic bocses.北京师范大学学报(自然科学版),2009,45(1):22-25.

[51] 张英伯,张学颖,赵双美.局部和两点 Bocs 的表示型.中国科学,2009,39A(3):257-266.(Zhang Xueying,Zhang Yingbo,Zhao Shuangmei. Representation type of local and two-vertices bocses. Science in China,2009,52A(5):949-958.)

[52] 赵德科，张英伯. Band-模的典范型. 北京师范大学学报(自然科学版)，2009，45(1)：5-13.

[53] Zhang Yingbo. Professor Liu Shaoxue：his live and work. Ring Theory 2007，15-18；World Scientific Publishing，Hackensack，NJ，2009，16-03.

[54] 张英伯. 发达国家数学英才教育的启示. 中学数学月刊，2010，(2)：1-2，13；数学文化，2010，1(1)：60-64.

[55] 张英伯. 半个世纪前的数学竞赛. 见：丘成桐，杨乐，季理真，主编. 数学与人文，第 2 辑. 北京：高等教育出版社，2010：175-180.

[56] 张英伯. 访日随感. 中国数学会通讯，2010，(4)：32-34.

[57] 张英伯. 以色列的英才教育项目. 见：丘成桐，杨乐，季理真，主编. 数学与教育，数学与人文，第 5 辑，北京：高等教育出版社，2011：123-126.

[58] 李亚玲，张英伯，数学英才教育的国际比较，数学教育学报，2011，20(2)：102；中国数学会通讯，2011，(1)：37-38.

[59] Liu Genqiang，Zhang Yingbo. Canonical forms of indecomposable modules over $K[x,y]/(x^p,y^q,xy)$. Algebra Colloquium，2011，18(3)：373-384.

[60] 张英伯. 卓有成效的民办英才教育：以色列访问纪实. 中国数学会通讯，2012，(4)：41-51；数学通报，2012，51(11)：5-10.

[61] 张英伯，文志英. 法兰西英才教育掠影. 数学文化，2012，3(4)：41-51；数学通报，2013，52(1)：1-15；见：杨学枝，主编：中国初等数学研究，第 7 辑. 哈尔滨：哈尔滨工业大学出版社，2016：70-80.

[62] 张英伯，章璞，肖杰.《刘绍学》. 见：钱伟长，总主编，王元，数学分卷主编. 20 世纪中国知名科学家学术成就概览，数学卷，第 3 分册. 北京：科学出版社，2012：109-117.

[63] 张英伯. 美国英才教育对中国的启示. 见：戴耘，蔡金法，主编. 英才教育在美国(第 15 章). 杭州：浙江教育出版社，2013：203-207.

[64] 张英伯，刘建亚. 渊沉而静，流深而远：纪念中国解析数论先驱闵嗣鹤先生. 数学文化，2013，4(4)：3-15；2014，5(1)：3-21.

[65] 张英伯. 天道维艰，我心毅然：记数学家王梓坤. 数学文化，2015，6(2)：3-51.

[66] 张英伯. 我们 1978 级研究生. 见:李仲来,主编. 北京师范大学数学学科创建百年纪念文集. 北京师范大学出版社,2015:169-172.

[67] 张英伯. Claus 和我们. 见:丘成桐,刘克峰,杨乐,季理真,主编;张英伯,副主编. 数学的教与学,数学与人文,第 20 辑,2016:99-115.

[68] 张英伯. 从颁奖典礼说起. 见:丘成桐,刘克峰,杨乐,季理真,主编;张英伯,副主编. 数学的教与学,数学与人文,第 20 辑. 北京:高等教育出版社,2016:149-153.

[69] 张英伯. 题词. 见:伍鸿熙. 数学家讲解小学数学. 北京:北京大学出版社,2016:封底.

[70] 张英伯.《中国大学先修课程》初探. 中国数学会通讯,2016,(3):17-32;数学文化,2016,7(3):29-37.

[71] 张英伯,王昆扬. 序言. 见:蒋迅,王淑红. 数学都知道. 北京:北京师范大学出版社,2016;数学文化,2017,8(1):119-120.

[72] 张英伯. 女校名师. 见:李红云,主编. 远去的女附中. 北京师范大学附属实验中学,2016.

[73] 张英伯. 元老一席谈. 数学文化,2017,8(1):32-41.

[74] 张英伯,别荣芳,罗里波. 善良为性,原理为心:记数理逻辑专家王世强. 数学文化,2018,9(3):3-23;9(4):3-28.

[75] 赵德科,张英伯. 从正三角形的旋转与反射谈起. 数学通报,2018,57(5):12-15.

[76] 张英伯. AR-箭图的结构及其矩阵双模问题. 中国科学,2018,48A(11):1 651-1 664.

[77] 张英伯,李尚志,翟起滨. 数奇何叹,赤心天然:记数学家、密码学家曾肯成. 数学文化,2019,10(2):3-26;10(3):3-23.

[78] 张英伯. 编委会的故事. 数学文化,2019,10(4):22-28.

[79] Zhang Yingbo,Xu Yunge. Algebras with homogeneous module category are tame. Mathematics. http://arxiv. org/abs/1407. 7576.

著作

[序号] 著者,译者. 书名. 出版地:出版社,出版年份.

[1] 德洛兹德・Ю А,基里钦柯・В В,著. 刘绍学,张英伯,译. 有限维代数. 北京:北京师范大学出版社,1984.

[2] Hattel D, Zhang Yingbo eds. Repersentations of algebras, Vol. 1～2. Beijing: Beijing Normal University Press, 2002.

[3] 人民教育出版社,课程教育研究所,中学数学教材实验研究组编著;高存明,主编. 普通高中课程标准实验教科书,数学(张英伯,主编. 选修 3-4,对称与群). 北京:人民教育出版社,2004.

[4] Кострикин А. И. 著. 张英伯,译. 代数学引论,第 1 卷,基础代数(原书第 2 版). 北京:高等教育出版社,2006.

[5] 张英伯. 代数分册主编(执笔线性代数:90-103,模论:117-123). 数学大辞典. 北京:科学出版社,2010.

[6] 张英伯. 美妙数学花园主编(中学生数学小丛书). 对称中的数学. 北京:科学出版社,2011.

[7] 丘成桐,杨乐,季理真,主编,张英伯,副主编. 数学与教育. 数学与人文,第 5 辑. 北京:高等教育出版社,2011.

[8] 丘成桐,杨乐,季理真,主编,张英伯,副主编. 数学与求学. 数学与人文,第 8 辑. 北京:高等教育出版社,2012.

[9] 张英伯,王恺顺. 代数学基础(上册). 北京:北京师范大学出版社,2012.

[10] 张英伯,王恺顺. 代数学基础(下册). 北京:北京师范大学出版社,2013.

[11] 丘成桐,刘克峰,杨乐,季理真,主编,张英伯,副主编. 数学的教与学. 数学与人文,第 20 辑. 北京:高等教育出版社,2016.

[12] 张英伯. 天道维艰,我心毅然:记数学家、教育家、科普作家王梓坤. 哈尔滨:哈尔滨工业大学出版社,2017.

[13] 数学名词审定委员会,编,张英伯,代数分册主编. 数学名词. 北京:科学出版社,2017.

后 记

Postscript by the Chief Editor

2003年年底，北京师范大学出版社的新一届领导将5位先生（王世强、孙永生、严士健、王梓坤、刘绍学）的文集列入“北京师范大学数学家文库”出版计划，在2005年8～10月陆续出版. 2007年4月至2008年9月陆续出版了3位已经逝世的先生（汤璪真、白尚恕、范会国）的文集. 2011年11月启动出版20世纪30年代出生的先生文集. 2018年3月启动出版20世纪40年代出生的先生文集.

北京师范大学数学科学学院20世纪40年代出生的博士生导师共5人，以年龄大小为序：王昆扬、郑学安、陈木法、张英伯、房艮孙，至少有以下特点. 陈木法教授是中国科学院院士和发展中国家科学院院士，其余4人均任二级教授岗位：前3人是“老五届”，即1966～1970届本科毕业生，后两人是“老三届”，即1966～1968届的高中毕业生（不是“老三届”的初中毕业生），高中毕业后到黑龙江生产建设兵团劳动锻炼.

“老五届”最初留北京师范大学数学系工作的13位毕业生均已先后升迁，或调到校部机关，或调到外校. 其中有4位教师在北京师范大学数学系（以下简称数学系）任教，有两位教师是教育部寄存在数学系的教师，1971～1975年无毕业的学生留数学系，这是一种不合理的师资结构且明显断档.

1978年3月，我国恢复招收研究生. 10月，数学系共招收硕士研究生11人，其中概率方向3人：陈木法、郑小谷和唐守正；分析方向2人：王昆扬和罗俊波；代数与数理逻辑方向6人：罗里波、沈复兴、张英伯、王成德、孙晓岚和程汉生.毕业后先后有6人留数学系任教,张英伯是其中之一.

在我退休之前,与北京师范大学出版社签署出版张英伯文集合同.文集交稿,是在我退休之后的事情.张英伯除了与我是同事外,还有三层关系.

第一,张英伯比我年长6岁,是大姐.在高校,如果将年龄差距定在10岁之内算同一代人,我们应属于同一代人.

第二,我们同属于"老三届",即1966～1968届中学毕业生.张英伯1947年5月生于上海,籍贯湖南,是"老三届"的头,即1966届高中毕业生.张英伯在北京师范大学附属第二小学(1955年改称北京第二实验小学)上小学,在北京师范大学附属实验中学上初中和高中,受的是高标准的正规的基础教育.1966年,她已经有一只脚踏进了大学,就差另一只脚,还没有进去.更重要的一点是,她在1964年读高一时参加北京市中学生数学竞赛获高二组的一等奖,可以免试高考,直接进入北京大学数学力学系.遗憾的是,"文化大革命"开始后的1968年,她去黑龙江生产建设兵团3师20团6营3连"修理地球"了.我是1953年8月生,1959年9月上小学,是"老三届"的尾,即1968届初中毕业生.初中毕业后回乡务农.

第三,我们同属于工农兵大学生,属于"同类"同学.张英伯是北京师范学院(现首都师范大学)数学系1972级工农兵大学生,1975年毕业后到北京通县第二中学(现通州区第二中学)工作三年,于1978年考入北京师范大学数学系攻读硕士学位,毕业后留数学系任教.1986年在北京师范大学数学系攻读博士学位,之后与当时的联邦德国联合培养,在比勒费尔德(Bielefeld)大学获博士学位,在墨西哥做博士后,于1991年回北京师范大学工作,同时被聘为副教授.1992年被聘为教授.2007年年底聘为二级教授.2008年10月退休.她属于"摘帽"的工农兵大学生教授.我是北京师范大学数学系1974级工农兵大学生,1977年毕业后留数学系任教.1984年考入北京师范大学数学系基础数学助教班学习一年半,修完硕士生基础课和专业课.1993年被聘为副教授,1998年被聘为教授.2014年年底聘为二级教授.2018年10月退休.我属于"未摘帽"的工农兵大学生教授.

在北京师范大学工作的所有的工农兵大学生,按晋职教授年份排序,1992年:张英伯,1993年:高琼、罗钢(调走),1994年:李洪兴(调走)、刘永平，1995 年：倪晓健（调走）、史静寰（调走），1996 年：郭小凌（调走）、俞启定，1997 年：包华影、武尊民，1998 年：李守福、李仲来、刘北成（调走）、赵小冬，1999 年：樊善国（2004 年逝世）、刘大禾、赵新华，2000 年：黄海洋，2001 年：刘淑兰（2020 年逝世）、刘小林，2002 年：张德福，2004 年：李奇. 不含调入教授（白暴力、田桂森、张曙光、钟秉林（曾任北京师范大学校长）、周明全），在北京师范大学工作的工农兵大学生中，张英伯晋职教授最早.

严士健教授曾对我说过，数学科学学院 20 世纪 40 年代出生的教授水平，已经超过他们. 张英伯在我们工农兵大学生中，就做学问来讲，尤其是对女性而言，属于出类拔萃者，2012 年被评为第 5 届全国优秀科技工作者. 女同志能做到这样，实属不易.

张英伯文集的书名原计划为《矩阵双模问题与数学英才教育》，后来应她本人要求，“数学英才教育”内容不再收录，改由其他出版社出版，书名改为现名.

张英伯文集的出版，得到了数学科学学院党委书记唐仲伟教授，前院长李增沪教授和现任院长王恺顺教授的大力支持，同时得到了北京师范大学出版社的大力支持，在此表示衷心的感谢.

华罗庚教授说：“一个人最后余下的就是一本选集.”（龚昇论文选集，中国科学技术大学出版社，2008）这些选集的质量反映了数学科学学院某一学科，或几个学科，或学科群的整体学术水平. 而将北京师范大学数学科学学院著名数学家、数学教育家和科学史家论文进行整理和选编出版，是学院学科建设的一项重要的和基础性的工作，是学院的基本建设之一. 它对提高学院的知名度和凝聚力、激励后人，有着重要的示范作用. 当然，这项工作还在继续做下去，收集和积累数学科学学院各种资料的工作还在继续进行.

主编　李仲来

2019-01-08